2nd Conference on Innovation in Technology (CITES 2020)

IOP Conference Series: Materials Science and Engineering
Volume 1041

Padang, Indonesia
4-5 November 2020

Editors:

E. Gunawarman
E. Zulkarnaini
Nurhamidah
B. Pranggono
M. Sasaki
M. A. Yunus
M. Furuuchi
E. Amrina

ISBN: 978-1-7138-4564-5
ISSN: 1757-8981

Printed from e-media with permission by:

Curran Associates, Inc.
57 Morehouse Lane
Red Hook, NY 12571

proceedings.com

Some format issues inherent in the e-media version may also appear in this print version.

Institute of Physics
Dirac House, Temple Back
Bristol BS1 6BE UK

Phone: 44 1 17 929 7481
Fax: 44 1 17 920 0979

techtracking@iop.org

Additional copies of this publication are available from:

Curran Associates, Inc.
57 Morehouse Lane
Red Hook, NY 12571 USA
Phone: 845-758-0400
Fax: 845-758-2633
Email: curran@proceedings.com
Web: www.proceedings.com

TABLE OF CONTENTS

Preface

We are delighted to welcome all of the participants to the 2nd Conference on Innovation in Technology and Engineering Science (CITES). 2nd CITES aims to provide a platform for discussing current development and essential issues on the innovation of technology and engineering science.

Faculty of Engineering, University of Andalas held the 2nd CITES on November 4th -5th, 2020 in Padang, West Sumatera, Indonesia. This annual conference is organized by the Faculty of Engineering Universitas Andalas and supported by Universitas Andalas.

The responses to the call-for-papers had been overwhelming – both from Indonesia and from overseas. Unfortunately, many manuscripts from prestigious institutions could not be accepted due to the reviewing outcomes and our capacity constraints. We would like to express our gratitude and appreciation for all of the reviewers who helped us maintain the high quality of manuscripts included in the proceedings published by IOP. We believe that this event and proceeding will never come to happen without your participation and contribution.

We would also like to extend our thanks to the members of the organizing team for their hard work. We also would like to acknowledge sponsors contribution. Therefore, on the day of completion of this journey, we are delighted with a high level of satisfaction and aspiration.

We are now optimistic and full of hope about getting the proceedings of 2nd CITES covered by IOP Proceeding in due course. We appreciate that the authors of 2nd CITES may want to maximize the popularity of their papers. And we will try our best to support them in their endeavours.

Let us wish that all the participants of 2nd CITES will have a wonderful and fruitful time at the virtual conference and that our oversea guests will enjoy a warm greeting from the University of Andalas.

Conference Chair

IOP Conf. Series: Materials Science and Engineering 1041 (2021) 011001 doi:10.1088/1757-899X/1041/1/011001

CITES 2020 Committee

Peer review declaration

All papers published in this volume of *IOP Conference Series: Materials Science and Engineering* have been peer reviewed through processes administered by the Editors. Reviews were conducted by expert referees to the professional and scientific standards expected of a proceedings journal published by IOP Publishing.

- **Type of peer review: Single-blind, where the reviewers know the identity of each author. However, authors don't know the identity of reviewers**
- **Conference submission management system: Open Source**
- **Number of submissions received: 87**
- **Number of submissions sent for review: 80**
- **Number of submissions accepted: 76**
- **Acceptance Rate (Number of Submissions Accepted / Number of Submissions Received X 100): 87.4**
- **Average number of reviews per paper: 3 times**
- **Total number of reviewers involved: 17**
- **Contact person for queries: Dr. Oknovia Susanti**

Numerical Simulation to Determine the Effectiveness of Groynes and Breakwaters as Protective Structures for Gandoriah Beach, Pariaman City

M Mera[1] and D Chrisnatilova[1]
[1]School of Civil Engineering, Engineering Faculty, Universitas Andalas, Indonesia

Corresponding author: mas_mera@eng.unand.ac.id

Abstract. This research is about numerical simulation to determine the effectiveness of groynes and breakwaters as protective structures for Gandoriah Beach in Pariaman City (West Sumatra - Indonesia) using an existing program. The effectiveness of the protective structure can be seen based on the bathymetry that is formed as a result of this structure performance. In the simulation of the present research, the direction of the incoming waves at the offshore boundary condition is the dominant direction, namely the west direction, with two wave conditions, namely during normal and stormy times. Simulations of morphological changes are carried out in six scenarios. The first three scenarios are for wave simulations under normal conditions with no shore protective structures (scenario 1), with groynes (scenario 2) and with breakwaters (scenario 3). Meanwhile, the second three scenarios are for wave simulation in storm conditions with no shore protective structures (scenario 4), with groynes (scenario 5) and with breakwaters (scenario 6). The simulation results for the 20 days state show that in normal wave conditions, the breakwaters have a fairly good performance for the formation of sedimentation behind the structures so that it is estimated that the formation formed is tembolo. Meanwhile, groyne is only effective to restrain the sedimentation rate along the structures. In the wave simulation during stormy time, there is a dominant current along the nearshore. In this case the performance of the groynes is better than the breakwaters because the groynes can inhibit the rate of erosion along the shore. Due to the use of breakwaters there is erosion behind the structures. Based on the simulation results without shore protective structures, both in normal and storm waves, abrasion and erosion processes occur, so that Gandoriah Beach needs handling.

1. Introduction

Gandoriah Beach (figure 1) is a tourist beach located in Pariaman City, Central Pariaman Sub-District. Almost all sub-districts in Pariaman City (except for East Pariaman Sub-District) have a beach area that is directly opposite the Indonesian Ocean. Pariaman City has a shoreline length of ± 12 km while its land area is only ± 73.36 km^2 [1]. This means that Pariaman City only has a land area of ± 6 km wide. The shore protection that have been built along Gandoriah Beach are a series of groynes. Within a few years these buildings have changed the hydrodynamic pattern of the waters, but at present conditions appear to be ineffective in reducing the rate of abrasion [2]. These ineffective shore protective structures cause the condition of Gandoriah Beach along ± 1.5 km (from the Navy Monument to the ASEAN Monument) as if it does not get a shore protective structure. Meanwhile, there are many landmarks along Gandoriah Beach that must be protected from the threat of abrasion.

The present study objectives are to determine the effectiveness of groynes as commonly used shore protective structures and breakwaters as alternative structures to protect Gandoriah Beach from abrasion due to the ferocious waves of the Indonesian Ocean by numerical simulations. The more effective shore structures will be identified by comparing the formed bathymetry. Numerical simulations are carried out with the CMS Wave and CMS Flow modules. The CMS Wave is a steady module, while the CMS Flow is an unsteady module. Even though these two modules are different, they are integrated with each other as a single unit in the Coastal Modeling System, which is included in the Surface Water Modeling System program package.

The (steady) CMS Wave module is used for wave modeling, while the (unsteady) CMS Flow module is used for modeling currents and elevations, sedimentation and morphological changes. The CMS Wave is based on averaged phase waves and its governing equation is a wave action balance equation which was modified by Mase in 2001 by adding the energy dissipation factor at the time of the wave breaking and the wave diffraction factor [4].As a boundary condition for offshore, the spectral wave density E (σ, θ) is calculated using the parametric spectrum model. The governing equation in the CMS Wave is solved by using the finite difference method, by dividing the domains into rectangular grids with variable sizes. The output of the CMS Wave module is the height, period and direction of the waves on each grid which has taken into account the wave transformation processes such as refraction, diffraction, reflection and breaking waves.

The CMS Flow module is a 2D hydrodynamic and sedimentation model, based on depth averaged. This means that the flow velocity and sediment transport in the vertical direction are not taken into account [5]. The numerical method used in CMS Flow is the finite volume method, and in SMS version 10.1 only an explicit scheme is available. The governing equation for hydrodynamics is derived from the depth averaged shallow water equations in Cartesian coordinates [5]. The boundary in the present study is the water level.

The principle of sediment transport in CMS Flow is the concept of bed shear stress that occurs due to currents, waves and wind. So that the coefficient of friction is very influential in determining sediment transport. The present study considers sediment transport to be an equilibrium bed load and a non-equilibrium suspended load, where only the bed load is in balance, while the floating sediment is not. The elevated sediment transport is calculated using the advection-diffusion (AD) equation, while the basic sediment transport is calculated using the Lund-CIRP equation. The sediment continuity equation for morphological change is AD equation [6]. Nam et al. simulated the sedimentation rates of five series of specimens with different sediment transport formulas [7]. The results of their study indicate that the AD formula for elevated sediments and Lund CIRP for bottom sediments provides the best predictions for longshore sediment transport, compared to the use of other formulas.

2. Materials and methods

2.1. Data collection

Secondary data used in the present study are Batimetri Nasional (Batnas) data, daily maximum wind data for 10 years of observation (2010 - 2019) by BMKG Minangkabau Airport station and tide elevation data for a period of 20 days (from 15 May to 3 June 2020) obtained from Badan Informasi dan Geospasial (BIG). Due to limited measurement data, the influence of the river mouth of Batang Pariaman which is located in the northern Gandoriah Beach and the river mouth of Batang Jirak which is in the southern part is not taken into account in the present study.

2.2. Data processing

The Batnas data processing is carried out using ArcGis to obtain contours with an elevation difference of 1 m. Wind data are processed in two stages. The first stage is to classify daily wind data for 10 years of observation in the form of wind rose diagram to determine the percentage of wind events with a certain speed. Furthermore, the second stage is to predict wave height in deep water based on the dominant direction using the procedure of the Sverdrup MunkBretschneider (SMB) [8, 9].

Figure 1. Gandoriah Beach [3].

2.3. Determination of the simulation domain for the Coastal Modeling System

Numerical set-up for CMS Wave and CMS Flow using bathymetry as the basic data, followed by making a grid, inputting boundary conditions and other parameter data. Basically the simulation is carried out in two stages. The first is to simulate waves without interaction with currents (CMS Wave only) on a parent domain (parent grid), and the second is to simulate wave-current interactions on a child domain (child grid) by coupling the two modules (CMS Wave and CMS Flow) (figure 2).

*The first stage simulation with CMS Wave on parent domain.*The bathymetry at the research site is relatively complex because in front of Gandoriah Beach there are three small islands which are quite influential on the wave diffraction process. Meanwhile, CMS Flow v.3.75 is only intended for small domains and cannot cover domains that are too large for grid cells with relatively small dimensions. In order for the wave transformation process on the bathymetry to be simulated well, in this study, the simulation was carried out in two stages. In the first stage, the waves are simulated in the parent domain, without current interaction. The simulation is carried out in two conditions, namely normal waves and storm waves. The purpose of the first stage simulation is to get a wave output from the parent domain to be forwarded as a wave input in the child domain. The focus of our research is on the child domain, where groynes and breakwaters are located. The parent domain measures 8080 m x 7680 m with a grid cell size of 40 m x 40 m. The input to the boundary conditions in the deep sea is the height and period of the waves generated based on wind data. The spectrum used is the JONSWAP spectrum for fetch limited conditions.

*The second stage simulation with coupling CMS Wave and CMS Flow on the child domain.*Simulations on the child domain are carried out in six scenarios: (1) simulation of normal waves in existing conditions (without shore protective structures); (2) simulation of normal waves with the addition of groynes; (3) simulation of normal waves with the addition of breakwaters; (4) simulation of storm waves in existing conditions; (5) simulated storm waves with the addition of groynes; and (6) simulation of storm waves with the addition of breakwaters. Simulations of morphological changes due to the interaction of waves and currents were carried out for 20 days (480 hours) for each scenario with a state time interval of 6 hours for wave simulation. This 6 hour duration was chosen because of the consideration that the type of tide at Gandoriah Beach is semi-diurnal. The second consideration is the incapable of our computer memory capacity with a duration smaller than that.

Figure2. Parent domain(parent grid) andchilddomain (child grid)

In a child domain, the coupling process is carried out between CMS Wave and CMS Flow. The data will be forwarded in two directions according to the time increment increase of simulation (Δt). The CMS Wave will be processed first, after that the wave data from the CMS Wave will be forwarded to the CMS Flow as input. CMS Flow output in the form of water level elevation, current, sediment transport and morphological changes that occur have taken into account the influence of the waves. The CMS Flow output is forwarded back to the CMS Wave as input in the next increment of time (Δt). In such a way, changes in water level and current are taken into account in the wave simulation performed by CMS Wave, including bathymetry updates due to changes in morphology, and so on until the end of the specified simulation time.

The addition of groynes in the simulation is in the second and fifth scenarios, and the addition of breakwater structures in the simulation is in the third and sixth scenarios. The groyne used is type I. The length of the groyne is determined based on the theory proposed by Horikawa [9], which is between 40% and 60% of the average width of the surfzone with the distance between groynes being one to three times the length of the groyne. The surfzone width is known from the simulation output of the parent domain. The breakwater is expected to form a salient formation so that the sedimentation formed behind the breakwater does not obstruct sediment transport along the nearshore. The comparison between the length of the breakwater and the distance of the breakwater to the shoreline between 0.5 and 0.67 as suggested by Dally and Pope in [10], the gap width between the breakwaters is determined based on the empirical diagram of several breakwaters in the USA put forward by Pope and Dean in [10].

IOP Conf. Series: Materials Science and Engineering 1041 (2021) 012001 doi:10.1088/1757-899X/1041/1/012001

The size of the child domain is 3000 m x 3600 m with a grid cell size of 20 m x 20 m. As a boundary condition for CMS Wave is the simulated output wave from the parent domain. The spectrum used is TMA (Texel, Marsen and Arsloe) for shallow water. While the boundary requirement for CMS Flow is the tide elevation during the period 15 May 2020 to 3 June 2020 (20 days) (see figure 3).

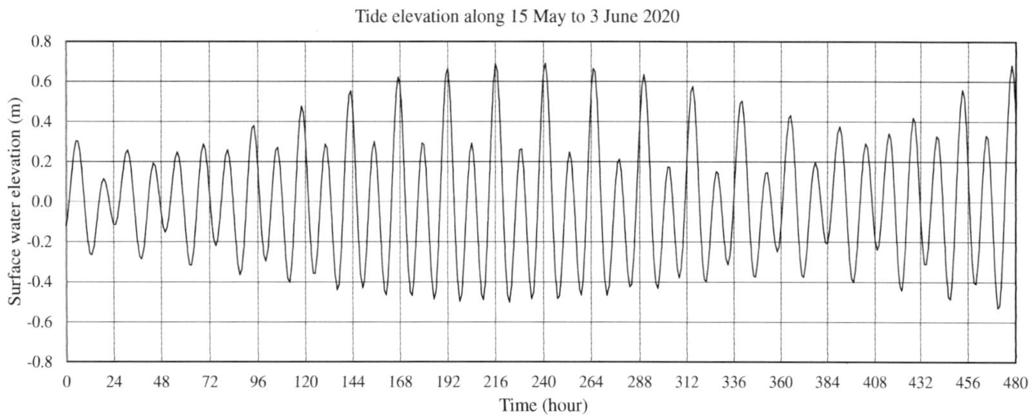

Figure3. Tide elevation from 15 May to 3 June 2020

Input data for simulations in the child domain can be seen in tables 1 and 2.

Table1. Input data for the CMS Wave in the child domain

Wave	Normal			Storm			Notes:
Bathimetry	Existing	With Breakwater	With Groin	Existing	With Breakwater	With Groin	
Coefficient of Refraction	0.025	0.3	0.3	0.025	0.3	0.3	[8], [9], [11]
Coefficient of Diffraction	1	4	2	1	4	2	[4]
Manning resistance-coefficient	0.025	0.025	0.025	0.025	0.025	0.025	[12]

The reflection coefficient for the structureless condition is determined based on the slope angle of the coast ($\theta = 5^{o}$) [11] using the Battjes diagram (1974) in [8]. Meanwhile, the reflection coefficient for the rubble mound structure is determined based on [9]. The diffraction coefficient in CMS wave is determined based on [4].

Table2. Input data for the CMS Flow

Parameters	Input Data	Notes
D50	0.2	Fine sand [12]
Manning resistance-coefficient	0.025	[12]
Water density	1025 kg/m^3	
Soil density	2650 kg/m^3	
Porosity	0.4	
Hydrodynamic time step (Δt)	1 s	

3. Results and Discussion

3.1. Wind wave generation

IOP Conf. Series: Materials Science and Engineering 1041 (2021) 012001 doi:10.1088/1757-899X/1041/1/012001

Based on the wind rose in figure 4, the dominant wind originates from the west with a percentage of 38.72%.

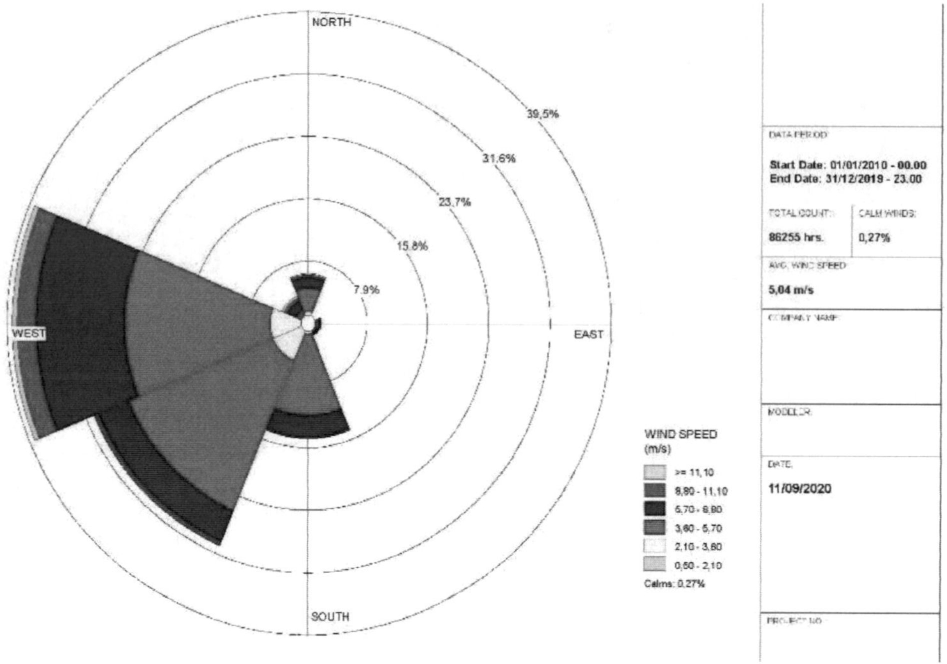

Figure 4. Wind rose from 2010 to 2019

The final result of wave generation by wind is shown in table 3. Wave height *H* in normal conditions is *H* significant or the average height of 33.33% of the highest value of wave data. Wave height *H* during the storm is the maximum wind occurrence from the west on June 12, 2014.The wave data in table 3 is then used as boundary conditions for the parent domain CMS Wave.

Table3. Significant (normal) waves and storm surges from the west.

Incident Wave	Normal	Storm
Angle (°)	0	0
Height, *H* (m)	3.02	5.73
Periode, *T* (s)	8.5	10.57

3.2. First stage simulation

The first stage of the simulation is carried out on the parent domain only with the CMS Wave module only. Simulations are carried out twice. The first is in normal wave conditions, where the offshore boundary condition is normal waves. The second simulation is in a storm surge condition, where the offshore boundary condition is a storm wave. Both simulations use existing bathymetry. So, in these simulations there is no shore protection structures in the parent domain.The wavefront profiles for normal wave and storm wave conditions are shown in figure 5.

The output of the simulation in the parent domain such as height and wave period at the location shown by the red line in figure 5 (left and right) will be passed as boundary conditions for the

IOP Conf. Series: Materials Science and Engineering 1041 (2021) 012001 doi:10.1088/1757-899X/1041/1/012001

simulation input in the child domain. The output of the parent domain used as input for the child domain is shown in table 4.

The dimensions of groyne and breakwater for simulations in the child domain are as shown in table 5.

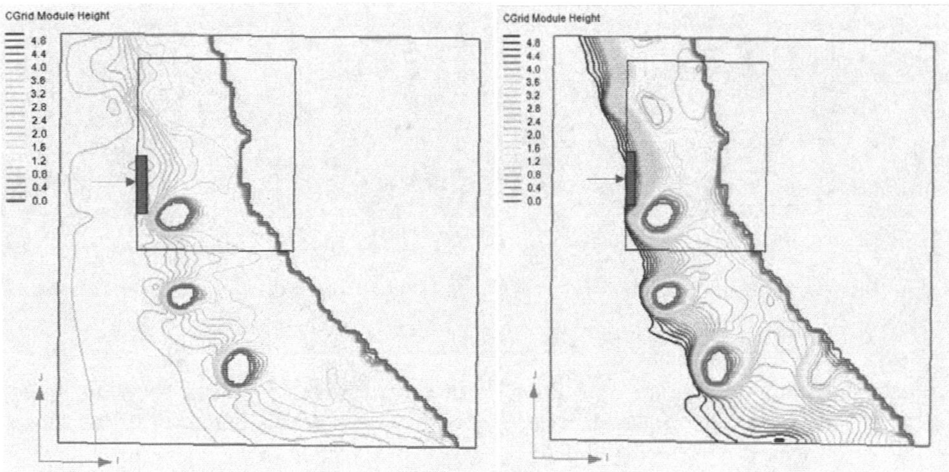

Figure5. Wavefront profile in normal wave conditions (left) and in stormy wave conditions (right).At this stage, no shore protection structures have been added to the parent domain.

Table4. Wave output from the CMS waves in the parent domain.

Incident Wave	Normal	Storm
Angle (°)	4.53	6.26
Height, H (m)	2.9	4.93
Periode, T (s)	8.3	10

Table5. Groyne and breakwater dimensions for simulations in the child domain.

Structures	Groins	Breakwaters
Length (m)	60	80
Distance between them (m)	100 - 120	40 (gap width)
Width (m)	20	20
Top elevation	0.2 m above MSL	0.5 m below MSL (Submerged)
Quantities	8	8

3.3. Second stage simulation

The second stage simulation is only carried out on the child domain by combining the CMS Wave module with the CMS Flow module. The existing morphological conditions with no structure, with groynes and with breakwaters are shown in figure 6.

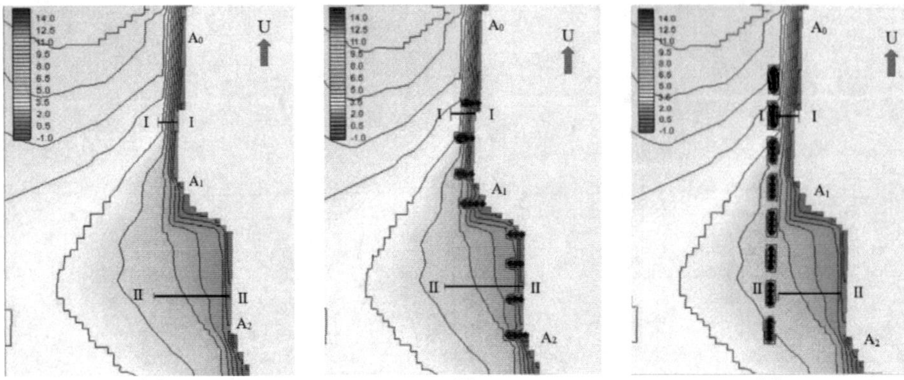

(a) (b) (c)

Figure6. Existing morphology: (a) without structures, (b) with groynes and (c) with breakwaters

Normal wave conditions (scenarios 1, 2 and 3). In normal wave conditions, sediment transport is influenced by waves and currents along the nearshore which predominantly move from north to south. Morphological changes that occur due to normal waves can be seen in figure 7.

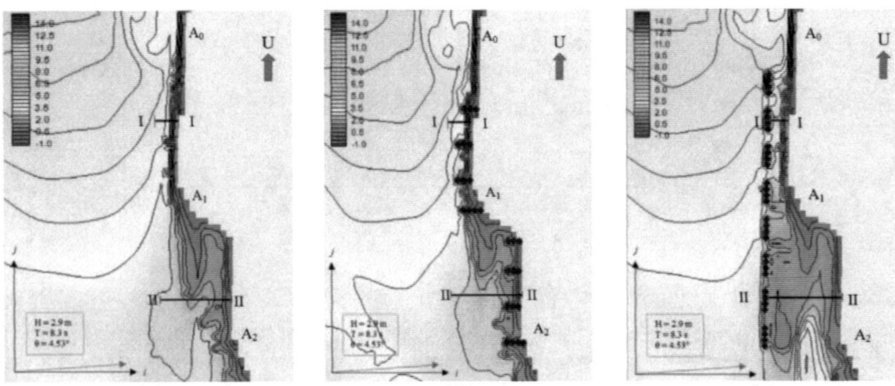

(a) (b) (c)

Figure7. Morphological changes after 480 hours simulation with normal wave: (a) without structures, (b) with groynes and (c) with breakwaters.

Morphological changes due to normal waves in detail are shown in the cross sections as shown in figure 8.

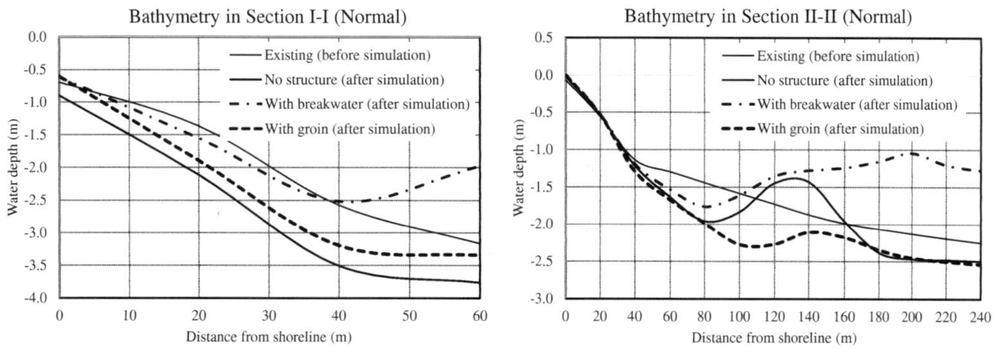

Figure8. Morphological changes due to the normal waves at sections I-I (left) and II-II (right).

The results of the first scenario (namely the simulation results of normal waves propagating on the child domain without shore protective structures) can be seen in figure 7 (*a*), where there has been a significant increase in depth along A0 to A1. Erosion that occurs in detail can be seen in sections I-I (figure 8 left) and II-II (figure 8 right). In the I-I section (figure 8 left), additional depth begins to occur at a distance of 10 m from the shoreline, with a depth difference of 0.5 m from the initial bathymetry. The more it goes offshore, the additional depth that occurs reaches 1 m. From A0 to A1, along ± 500 m, it is a fairly prone location along Gandoriah Beach because of the deep depth, for normal waves the area of breaking waves occurs almost near the shoreline, so the wave conditions near the shore are quite high. In section II-II (figure 8 right), the condition without structure does not really affect the erosion process. This is estimated because at the location of the II-II (A1-A2) section there is silting bathymetry towards the sea as shown in figure 6 (*a*), namely the existing bathymetry conditions, so that at this location the high and the wave energy reaching the shore is weak.

The results of the second scenario (namely the simulation results of normal waves propagating on the child domain with groynes), the effect of groynes can be seen in figure 7 (*b*). Erosion occurs between the first, second and third groynes along A0 to A1. From the I-I section (figure 8 left) it can be seen that despite erosion, groynes can provide protection compared to the condition without structure. The increase in depth that occurs from the shoreline to 60 m (along the groyne structures) to the sea is a maximum of 0.5 m.

The results of third scenario (namely the simulation results of normal waves propagating in the child domain with breakwater protection), the effect of adding breakwaters (substitute for groynes) on morphological changes can be seen in figure 7 (*c*). Sedimentation occurs behind the breakwater which is connected to the shoreline, so it is estimated that the formation formed is tembolo, even though the initial planning used the requirements for salient formation. Details can be seen in cross sections of I-I and II-II. Sedimentation formed by the breakwater can increase the elevation of the existing conditions.

*Storm surge conditions (scenarios 4, 5 and 6).*The simulation results with storm wave conditions show an anomaly. During a storm, the wave height and its energy towards the shore decrease, because some of the waves have broken further from the shoreline. Mera [13] defined that the occurrence of breaking waves caused by depth, starting when the ratio of wave height H and water depth h becomes greater than a certain size. The CMS Wave generally gives the value ratio (H/h) for random waves is 0.64 and uses a more detailed equation with the extended method by Goda [4]. The storm surge 4.93 m high reached the breaking point at a considerable distance from the shore. When a wave breaks, a current occurs which causes the transport of sediment from the location of the breaking wave. Simulated storm waves every 6 hours continuously cause sedimentation to form in the breaking wave area at the offshore boundary, which begins to form on the 5th day of the simulation. Sedimentation that forms blocks the flow of currents from offshore to the shore, so what happens is current forces due to the depth flowing from north to south at very high speed (3 to 5 m / s). These currents are the dominant cause of erosion processes along the nearshore. The dominant morphological changes influenced by currents in storm wave conditions can be seen in figure 9.

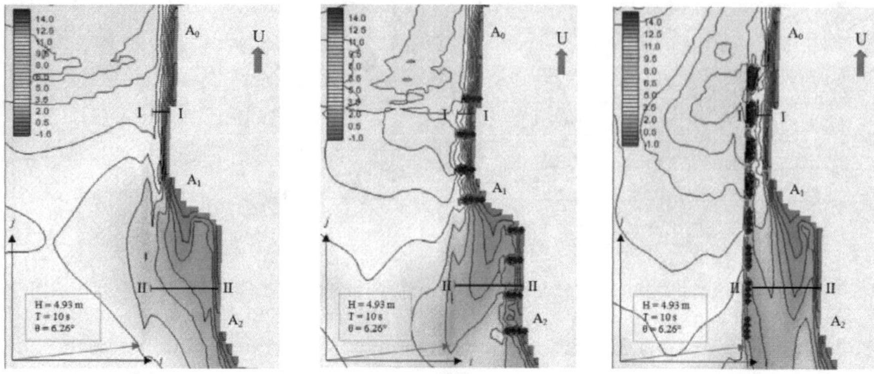

(*a*) (*b*) (*c*)

Figure9. Morphological changes after 480 hours simulation with storm waves: (*a*) without structure, (*b*) with groynes and (*c*) with breakwaters.

Morphological changes due to the storm surges can be seen in more detail in the cross sections shown in figure 10.

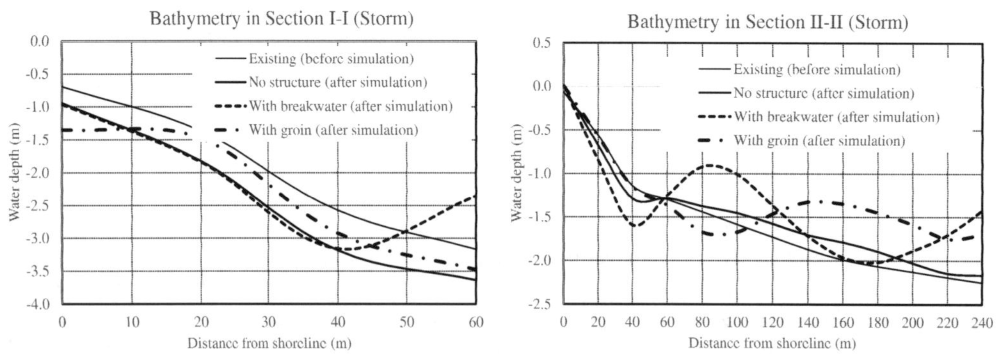

Figure10. Morphological changes due to the storm surges at sections I-I (left) and II-II (right).

The results of the fourth scenario (namely the simulation results of storm waves propagating on the child domain without the shore protective structures) show that erosion has occurred along the nearshore due to currents, especially at locations A0-A1 as shown in figure 9 (*a*). The I-I section (figure 10 left) shows that the results of the fifth scenario (namely the simulation results of storm waves that propagate on the child's domain with groynes), the performance of the groynes look better than the conditions without structures (scenario 4) and with the addition of breakwaters (scenario 6). Likewise, in section II-II (figure 10 right) starting from the shoreline up to 60 m towards the offshore (along the structures), the groynes are able to withstand sediment transport along the nearshore so that there is no significant increase in depth from the existing conditions. In the sixth scenario, the I-I section, starting from the shoreline up to 40 m towards the offshore, shows that the morphological changes due to the breakwaters are the same as those without structure, so it is estimated that the breakwater is unable to withstand the rate of erosion along the nearshore. The sedimentation behind the breakwaters only formed until the 5th day of the simulation because the waves had weakly reached the shore. But after a dominant current occurs from north to south, the process that occurs is erosion along the nearshore, thus forming an irregular erosion pattern behind the breakwater as seen in section II-II.

4. Conclusions

Normal wave simulation results indicate that the sediment transport process is influenced by waves and currents. Breakwater is more effective in shore protection against abrasion. Sedimentation behind the breakwaters leads to the tombolo formation, although the initial setting uses the equation for salient formation. Groin is only effective in holding sediment transport along the nearshore as wide as the groyne length is ± 60 m from the shoreline. In addition, groyne also does not function to reduce incoming wave energy. There is a significant erosion process on Gandoriah Beach if it is not protected by shore protective structures.

The results of the storm wave simulation show that the dominant process that occurs is erosion along the nearshore due to currents, so that the role of groyne is more dominant than the breakwater. This is because the main function of the breakwater is to weaken the incoming wave energy perpendicular to the shore, while the groyne functions well to hold sediment transport along the nearshore.

Acknowledgments

The author would like to express his deepest gratitude to the Balai Wilayah Sungai (BWS) Sumatra V for providing supporting data in the present research. The publication of this article is supported by a grant from the Faculty of Engineering - University of Andalasfor the 2021 fiscal year.

References

[1] BPS 2020 *Kota Pariaman dalam Angka 2020* (Pariaman: Badan Pusat Statistik Kota Pariaman)

[2] Ondara K,Rahmawan G A,Pitri Y A 2018 Kerentanan Pesisir Kota Pariaman Sumatera Barat Menggunakan Pemodelan Numerik Hidrooseanografi *Proc. Seminar Nasional Kelautandan Perikanan IV pp 13-24*

[3] Google Earth 2020 *Google Earth Images*https://earth.google.com/ (accesed on September 11, 2020)

[4] Lin L,Demirbilek Z,Mase H, Zheng J, Yamada F 2008 *CMS Wave: A Nearshore Spectral Wave Process Model for Coastal Inlet and Navigation Project* (Vicksburg, Mississippi: US Army Corps of Engineers - Engineers Research and Development Center)

[5] Sánchez A, Beck T, Lin L, Demirbilek Z, Brown M, Li H 2012 *Coastal Modeling System Draft User Manual* (Vicksburg, Mississippi: US Army Corps of Engineers - Engineers Research and Development Center)

[6] Buttolph A M, Reed C W, Kraus N C,Wamsley T V, Ono N, Larson M, Zundel A K 2006 *Two Dimensional Depth Averaged Circulation Model CMS-M2D: Version 3.0, Report 2, Sediment Transport and Morphology Change* (Vicksburg, Mississippi: US Army Corps of Engineers - Engineers Research and Development Center)

[7] Nam P T, Larson M, Hanson H, Hoan L X 2009A Numerical Model of Nearshore Waves, Currents and Sediment Transport *J. Coastal Engineering,* **56**1-12, pp 1084-1096. http://dx.doi:10.1016/j.coastaleng.2009.06.2007

[8] CERC1984 *Shore Protection Manual* (Vicksburg, Mississippi: Department of the Army, Waterways Experiment Station - Coastal Engineering Research Center)

[9] Triatmodjo B 1999 *Teknik Pantai* (Yogyakarta: Beta Offset)

[10] USACE 2002 *Coastal Engineering Manual* (Washington: Department of the Army - US. Army Corps of Engineer)

[11] Yudhicara 2008 Kaitan Antara Karakteristik Pantai Provinsi Sumatera Barat dengan Potensi Kerawanan Tsunami *J. Geologi Indonesia*, **3**2, pp 95-106

[12] Barua D K 2017 Seabed Roughness of Coastal Water *Encyc. of Coastal Science* DOI 10.1007/978-3-319-48657-4_360-1
[13] Mera M 2020 *Proses Pantai (Coastal Processes)* (Padang: Andalas University Press)

IOP Conf. Series: Materials Science and Engineering 1041 (2021) 012002 doi:10.1088/1757-899X/1041/1/012002

Study of retention and infiltration wells as a result of resettlement of the Batang Kuranji upper watershed

T Ophiyandri[1], B Istijono[1], A Gromiko[2], B Hidayat[1], Junaidi[1] and Nulrajabmil[2]

[1]Civil Engineering Department, Universitas Andalas

[2]Magister Program Postgraduate Civil Engineering, Universitas Andalas

Corresponding author: bistijono@eng.unand.ac.id

Abstract. Padang City is one of the metropolitan cities in West Sumatra Province and has six (6) watersheds, one of which is the Batang Kuranji watershed. Over the last few years Batang Kuranji watershed has experienced significant infrastructure and urban development which began with the massive earthquake of 30 September, 2009 with coastal communities, afraid of being affected by larger aftershocks having the potential of a tsunami, moving to the upper watershed. As a result, the type of land cover that was originally an infiltration area has been changing into a runoff area. The Batang Kuranji channel with a discharge (Q25) 870 m^3/s may not be able to accommodate the change in discharge which can lead to flooding. This paper aims to determine changes in discharge that occur due to changes in land cover and to analyse whether the use infiltration wells and retention pond can reduce the flooding volume. Two methods to calculate runoff discharge were adopted, HEC-HMS 4.1 Modelling and Nakayasu. It was found that the peak discharged increases to (Q25) 1,037.60 m^3/s. Thus, creating retention pond with a dimension of 35 hectare with a depth of 5 m and 34,246 infiltration wells can reduce the flood risk. The percentage of runoff discharge they can accommodate is 90.98% and 9.02% respectively.

1. Introduction

1.1. Background and Objectives

Padang is a metropolitan city and capital of the Province of West Sumatra which have six watersheds, namely Batang Air Dingin, Batang Kandih, Batang Kuranji, Batang Arau, Air Timbalun and Sungai Pisang. In recent years, Batang Kuranji's watershed has experienced significant developments in the field of infrastructure and urban development. One factor in the development of infrastructure in the Batang Kuranji watershed area began with the massive earthquake with tsunami potential on 30 September 2009. As a result, the community in the coastal area, which is the downstream area of the Batang Kuranji watershed, decided to move to the upstream area. Figure 1 shows the development of the area between 2005 and 2019.

Figure 1. Land cover change sample on Batang Kuranji watershed

Not only did the population move to the upstream area of the Batang Kuranji, the city infrastructure also moved, for example the Padang Mayor's office moved to the upper reaches of the Batang Maransi Stream (order 2 of Batang Kuranji), changing land cover that was initially a catchment area to a runoff area.

Peak discharge is important data in a watershed, and with climate change constantly changing, changes in peak discharge also increase. Batang Kuranji watershed also experienced a runoff discharge change, with a downstream capacity of (Q25) 870 m^3/s, which might not be able to accommodate this change in discharge.

The objective of the paper is to examine the changes in discharge that occur to the capacity of the discharge basin in the Batang Kuranji river with a design plan (Q25) of 870m ^{3}/s. To analyze the change in discharge that occurs, it is necessary to describe a hydrograph that is able to analyze runoff conditions, such as field conditions, using the Nakayasu Synthetic Unit Hydrograph method and modeling applications of HEC-HMS 4.1 with the SCS-CN modeling method. Furthermore, this paper analyzes the use of 34,246 infiltration wells and retention ponds with a total area of 35 hectares depth of 5.00 m, with a reduction in the volume of flooding for infiltration wells 9.02% and retention ponds 90.98%.

1.2. Hydrological Analysis

Hydrological analysis is carried out to obtain the calculation of the annual rain return period, the rainfall intensity and rainfall forecast hyetograph using the Alternative Block Method (ABM).

According to Soemarto [1], to calculate flood discharge for a certain period, maximum rainfall data with a certain return period is needed. This maximum rain is often referred to as a rainfall forecast. There are several methods that can be used to estimate the amount of rainfall forecast by frequency testing, including the Gumbel method, the Normal method, the Normal Log method, and the Log Pearson Type III method. Triatmodjo [2] adds that rainfall intensity is one of the data needed to calculate the rainfall intensity forecast and can be expressed in units of mm/hour. The intensity of rainfall is taken from the point depicted on the IDF (Intensity-Duration-Frequency) curve, which illustrates the relationship between rainfall intensity, duration of rainfall and the time of rainfall return. Calculation of rainfall intensity forecast uses the Mononobe method with the following formula.

$$I = \frac{X24}{24} \times \left(\frac{24}{t}\right)^{2/3} \tag{1}$$

where:

I : the intensity of average rainfall in t hours

$X24$: maximum daily rainfall or rainfall forecast (mm)

t : rainfall duration or concentration time (hours)

According to Chow et al. [3], to obtain the flood plan, rainfall forecast data is needed in the form of an hourly distribution model which has been developed to change daily rainfall into hourly rainfall using Alternative Block Method (ABM). Kamiana [4] stated that rainfall forecast hyetograph generated by this model is in the form of a distribution of rainfall forecast heights in n series time intervals with duration Δt during time t (t = n x Δt).

The Nakayasu method was developed in several rivers in Japan, and, in accordance with Soemarto [1], the following formula is used:

$$Q_p = \frac{CxAxR}{3.6(0.3T_p + T0.3))} \tag{2}$$

where:

Qp : peak flood discharge (m^3/s)
Ro : rainfall unit (mm)
T_p : grace period from the beginning of the rainfall to the peak of the flood (hour)
*T*0.3 : the time required by a decrease in discharge, from the peak to 30% of the peak discharge
A : catchment area to outlets (m^2)
C : runoff coefficient

Figure 2 shows the general form of the Nakayasu Synthetic Hydrograph Unit.

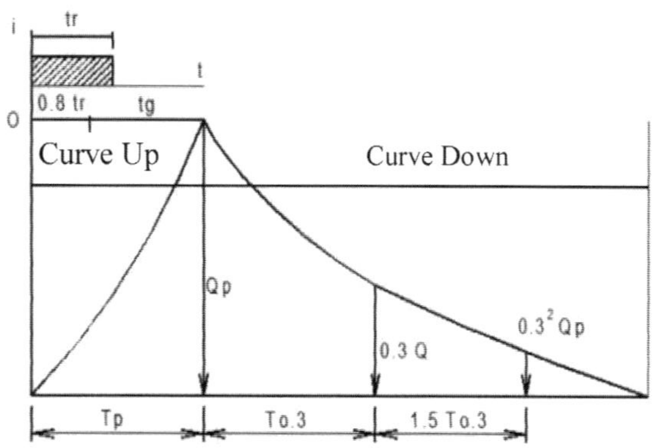

Figure 2. General Form of Nakayasu Synthetic Hydrograph Unit

HEC-HMS stands for Hydrologic Engineering Centre-Hydrologic Modelling System, software developed by the US Army Corps of Engineers Hydrologic Engineering Centre. This software can be used for simulating surface runoff calculations, calculating rain transformations and tracking floods on a watershed, both in existing conditions and under controlled or planned conditions, calculation of base flow, evaluation of water control buildings and water precipitation.

In the HEC-HMS program there are three main components, which are as follows:

1. Basin model is the elements contained in a sub-watershed and parameters in runoff.
2. Meteorological model which contains rain data and evapotranspiration data.
3. Control specifications, which are simulation time intervals to start or end data calculations.

Nugroho [5] conducted research on the use of retention ponds in reducing flooding in the Balakan river, Jepara Regency. This study aims to determine the maximum flood discharge at 5, 20, and 50 years and two annual maximum daily returns in the Balakan river, to determine the volume of retention pond storage and the potential for flood reduction after retention ponds are created.

Kustamar [6] also conducted study on the use of temporary storage capacity and improving of flood control facility. Budianto [7] examined the condition of river normalization and construction of retention pond. As described by Kusnaedi [8], infiltration wells function to store rainwater into a hole that is deliberately made temporarily. Other researchers have also examined the combination of retention ponds and infiltration wells and, according to Wigati [9], making infiltration well planning in the Cisimeut sub-watershed area is less effective in reducing runoff flow loads. Therefore, another alternative to reduce the runoff load will be planned for a retention pool.

2. Materials and methods

The first step of the research is by taking the rainfall data. It was gathered from four rainfall stations in the Batang Kuranji watershed, Khatib Sulaiman station, Batu Busuak station, Bendung Koto Tuo station and Gunung Nago station. The method used in calculating rainfall distribution is Polygon Thiessen. According to the Water Resources Management Agency of West Sumatera Province [10], the design of the cross-section of the channel in the downstream watershed is 870 m^3/s, but, with the occurrence of flooding in the last few years, it needs to be assessed in terms of runoff discharge that occurs. Identification of rainfall use data across 20 years from 1999 - 2018 and using analysis and probability for the rain return period 2, 5, 10, 25, 50, and 100 years to determine the flood Q using two methods, the Nakayasu HSS Method and HEC-HMS application modelling using the SCS-CN method. After the runoff discharged was established, we calculate the size of the retention pond needed in order to reduce the flood risk.

3. Results and discussion

According to Mera [11], the Thiessen polygon method allows the rainfall is ally weighting factor for each measuring tool. Between two stations connected by a line, in the middle of the line a perpendicular line is made which intersects the connecting lines of the station (bisectors) so that they form a polygon around each measuring instrument.

The rainfall distribution that occurs is affected by the rain station and uses the Thiessen polygon method with data such as the following (table 1):

Table 1. Extent of Effect of Rainfall Station.

Station	Area Affected (Km^2)	%
Sta. Khatib Sulaiman	16.97	8.07
Sta. Batu Busuak	129.33	61.48
Sta. Bendung Koto Tuo	51.75	24.60
Sta. Gunung Nago	12.30	5.85
Sum of	210.35	100

Using the rain station above, rainfall is obtained as follows (table 2):

Table 2. Years Annual Rainfall Data.

Year	Rainfall (mm/day)	Year	Rainfall (mm/day)
1999	62.01	2009	60.25
2000	139.26	2010	52.89
2001	189.15	2011	132.45
2002	117.72	2012	97.34
2003	98.39	2013	103.91
2004	96.18	2014	90.02
2005	146.76	2015	131.55
2006	123.01	2016	144.50
2007	121.62	2017	102.52
2008	104.40	2018	94.84

Using the data above, the rainfall calculation plan uses four methods, namely the Gumbel, Normal, Log Normal, and Pearson Type III methods, and, after a match test using the Chi-Square method, the results are as shown in the table 3 below:

Table 3. Chi-Square Test Results.

Probability Distribution	X^2 Counted	X^2cr	Result
Gumbel	7.60	7.815	Accepted
Normal	2.80	7.815	Accepted
Log Normal	6.40	7.815	Accepted
Log Pearson III	6.40	7.815	Accepted

From the above results, all methods can be accepted, but the smallest value of X2 < X2cr is the normal method which will be used as a rainfall forecast with the following data (table 4):

Table 4. Rainfall Forecast with Normal Method.

Return Period	XT
2	110.40
5	137.90
10	152.20
25	166.20
50	177.40
100	186.50

For calculating hourly rainfall distribution patterns by the Alternating Block Method (ABM) method, the results show the calculation of rainfall intensity in the table with the duration of rainfall events taken for six hours. Retrieval of rain duration is in accordance with the results of the flash flood incidents in 2012 with rain that occurred for six hours starting at 05.00 pm until 11.00 pm with peak discharge at 07.00 pm.

Furthermore, the results of the rainfall forecast of the hourly will be used for the calculation of Nakayasu HSU and modelling using the HEC-HMC 4.1 SCS-CN method (figure 3).

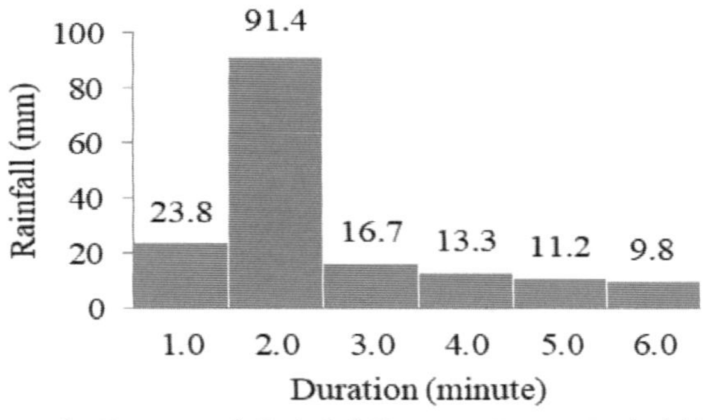

Figure 3. Hyetograph Rainfall Forecast Return Period 25 Year.

Changes in land cover that occurred in 2005 - 2019 are very significant. Many houses have moved or even residents have built houses in the central watershed area of Batang Kuranji. This is inseparable from the change in population thinking due to the potential for a tsunami caused by a 7.6 magnitude earthquake that occurred in the city of Padang.

IOP Conf. Series: Materials Science and Engineering 1041 (2021) 012002 doi:10.1088/1757-899X/1041/1/012002

As a result of this population movement, the area that was previously water infiltration was transformed into concrete land occupied by residents, thereby increasing the value of the runoff coefficient in the area. This is one of the factors increasing runoff discharge that occurs in a watershed.

Runoff discharge calculation is carried out in order to get the discharge flowing in the river, using the following method:

3.1. Nakayasu Method

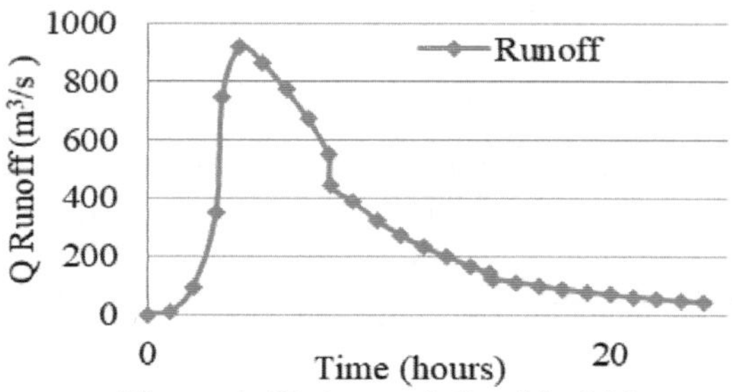

Figure 4. Hydrograph Graphic Q25.

The use of the Nakayasu HSU method aims to obtain a chart of rainfall hydrograph with a rainfall forecast for six hours. This calculation is done using one area of Batang Kuranji watershed. In this calculation, the peak discharge Q25 is 919.07 m^3/s at the 4th hour (figure 4).

Flood hydrograph modelling using HEC-HMS 4.1 consists of seven sub-watersheds arranged in sub-basin elements that are combined in several junctions and terminated at one sink (outlet). For this modelling using the SCS-CN (Cruve Number) method, it applies data such as the value of land cover (CN), imperviousness, lag time, and river slope, following the flow scheme of Batang Kuranji sub-watershed as shown in figure 5 below.

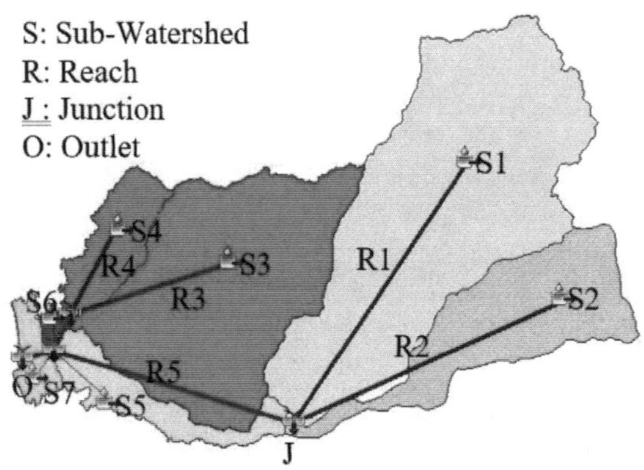

Figure 5. Schematic Sub-Watershed of Batang Kuranji in HEC-HMS Modelling.

The results of the calculation of flood discharge with the 25-year return at the Batang Kuranji watershed outlet point is 1,037.6 m^3/s with the peak flood time at the 3rd hour.

As for the return times, the peak runoff discharge can be seen in the table 5.

Table 5. Peak Discharge Recapitulation.

Discharge Forecast	Peak Discharge Recapitulation (m^3/s)		
	Nakayasu	SCS-CN	Design
Q 2	610.85	544.70	
Q 5	762.61	782.60	
Q 10	842.10	911.70	870
Q 25	919.07	1,037.60	
Q 50	981.22	1,162.80	
Q 100	1,031.80	1,269.50	

3.2. Analysis of Design Discharge with Runoff Discharge

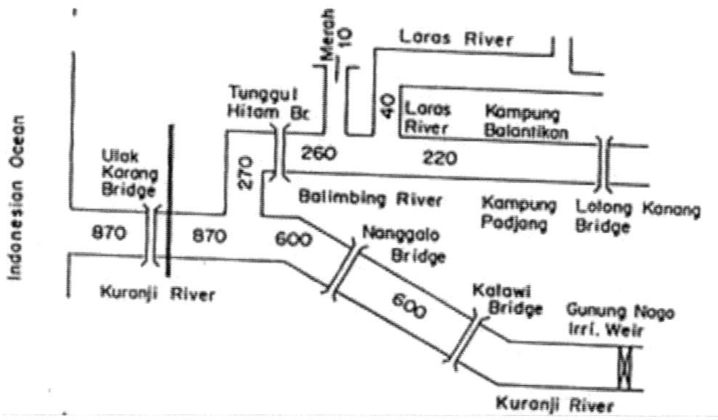

Figure 6. Schematic Distribution of Batang Kuranji Flood Design [9].

Figure 6 shows the schematic distribution of Batang Kuranji flood design. This based on the Q25 design of the Batang Kuranji watershed outlet at 870 m^3/s, Batang Kuranji River at 600 m^3/s, and from the Batang Belimbing at 270 m^3/s, while the HEC-HMS modelling results obtained flood discharges at the Batang Kuranji watershed outlet at 1,037.60 m^3/s, Batang Kuranji River at 494.60 m^3/s, and Batang Belimbing at 562.90 m^3/s.

Based on the data above, the Batang Belimbing sub-watershed experienced an increase in discharge of 292.90 m^3/s, especially at peak hours, at the 3rd hour (table 6). If seen from the hourly discharge on this river, the increase in discharge occurs starting at 2-5 hours as in the table 7.

Table 6. Difference from Channel Plan Discharge.

Rainfall Re-Period	Discharge Design Q25	Discharge SCS CN (m^3/s)	Difference (m^3/s)
Outlet	870	1037.60	167.60
Batang Kuranji River	600	494.60	-105.40
Batang Belimbing River	270	562.90	292.90

Table 7. Hydrograph Data Q25 Batang Belimbing Sub-Watershed.

Time	Batang Belimbing Sub-Watershed	Cross-section Discharge Plan	Excess Discharge
1	0	270	-270
2	174.90	270	-95.10
3	562.90	270	292.90
4	469.90	270	199.90
5	304	270	34
6	219.10	270	-50.90
7	178.80	270	-91.20

Based on the modelling results, a volume of 1,896,480 m^3/s was obtained with multiplication of time and excess runoff discharge.

3.3. Infiltration Wells
In this study, infiltration wells are the first option in flood mitigation, in addition to reducing runoff, infiltration well discharge is also useful for maintaining groundwater levels. The use of recharge wells will be made one for each house, by counting the number of houses in the Batang Belimbing sub-watershed as shown in table 8 and figure 7.

Table 8. Building Calculation.

Sub-Watershed	Building
Belimbing (Sub 3)	24,971
Maransi (Sub 4)	7,987
Belimbing downstream (Sub 5)	1,288
Total	34,246

Using the map above, we calculate the number of houses in the Batang Belimbing sub- watershed, with the following results of processing. If one house uses one infiltration well with the absorbed and floating volume totalled 5.00 m^3, the manageable runoff volume is 171,230 m^3, as seen from the retention pool runoff volume it is able to reduce by 9.02% from 1,896,480 m^3/s.

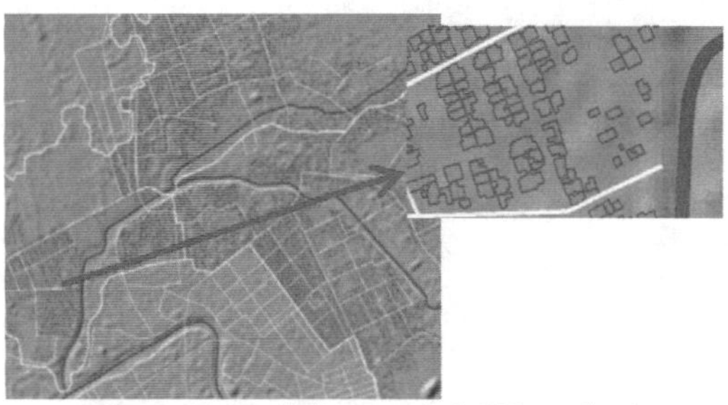

Figure 7. Building on Sub-Watershed.

3.4. Retention Ponds
In this study, retention ponds are used to accommodate excess discharge that is not able to be accommodated by river cross-sections. If the estimated discharge that can be reduced by making infiltration wells in each house is a total of 171,230 m^3 and the total volume of modelling discharge obtains a volume of 1,896,480 m^3, there will be 1,725,250 m^3 remaining.

To accommodate the remaining estimated discharge, we need the dimensions of the retention pond as shown below:

Retention Ponds Dimension = a x t
= 35 hectares x 10,000 x 5 m
= 1,750,000 m^3

To find the location of 35 hectare at the same time it is indeed difficult for the city of Padang, for it is likely to be made per spot or planning system for retention ponds. From several observations through Google Earth, there are several potential locations to be used as retention ponds as shown on figure 8.

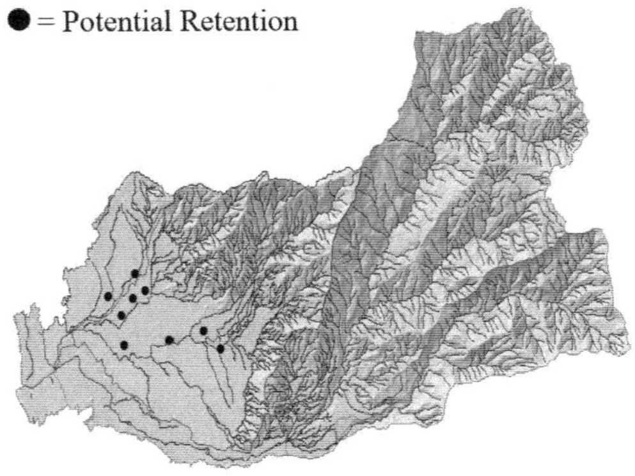

Figure 8. Potential Retention Map.

One example of a good area for retention ponds is near the Padang Mayor's office, because the area is close to a tributary with an area of ± 10 hectare and is government-owned land, as shown in figure 9.

Figure 9. An example of one of the Potential Retention Pools is near the City Mayor's Office.

4. Conclusions

The conclusion of the research is:

- Based on the research plan discharge (Q25), watershed Batang Kuranji in downstream area is not able to accommodate runoff discharge.
- Based on HEC-HMS 4.1 modelling, an increase in peak discharge occurred at the 3rd hour of 1,037.60 m^3/s.

- The increase in discharge occurred at Belimbing sub-watershed (Sub 3, 4, 5) with Q cross-section of Q25 plan of 270 m^3/s to 526.90 m^3/s. It occurs at 3-5 hours with a volume of 1,896,480 m^3/s.
- To overcome the increased discharge, combination of runoff discharge management with recharge wells capable of reducing by 9.02% and retention ponds by 90.98% are proposed.
- There are nine potential locations can be allocated for the planned retention pond area of 35 hectare with a depth of 5.00 m.

Acknowledgment
We would like to express our sincere gratitude to the Universitas Andalas for the Research Grant 2021.

References

[1] Soemarto C D 1987 *Hydrological Engineering (in Bahasa : Hidrologi Teknik)* (Surabaya: Usaha Nasional)
[2] Triatmodjo B 2008 *Applied Hydrology (in Bahasa : Hidrologi Terapan)* (Yogyakarta: Beta Offset)
[3] Te C V, Maidment D R and Mays L W 1988 Applied hydrology *Water Resour. Handbook; McGraw-Hill New York, NY, USA*
[4] Kamiana I M 2011 *Water Building Discharge Forecast (in Bahasa : Teknik Perhitungan Debit Rencana Bangunan Air)* (Yogyakarta: Graha Ilmu,)
[5] Nugroho S A, Hadiani R and Muttaqien A Y 2017 Flood Reduction using a Retention Pool in the Bakalan River, Jepara Regency. (in Bahasa : Reduksi Banjir Menggunakan Kolam Retensi di Sungai Bakalan, Kabupaten Jepara) *Civ. Eng. J.* **14** 195–202
[6] Kustamar K, Handoko F and Soetedjo A 2018 Flood Control Strategy In Sampang City, East Java, Indonesia *Int. J. GEOMATE* **15** 62–7
[7] Budianto B S, Astor Y, Karnisah I, Pramesthy B V and Alfiansyah I 2020 Evaluation Of Flood Mitigation Plan For Cibaligo River, Cimahi City, West Java Province *Int. J. GEOMATE* **19** 188–93
[8] Kusnaedi I 1995 *Infiltration Wells for Urban and Rural Settlements (in Bahasa : Sumur Resapan untuk Pemukiman Perkotaan dan Pedesaan).* (Jakarta: Penebar Swadaya)
[9] Wigati R and Setiawan A C 2017 Effectiveness of the Application of Infiltration Wells in Reducing the Load Flow Overflow of the Sub Cisimeut Surface as an Effort to Manage Floods (in Bahasa : Efektifitas Penerapan Sumur Resapan Dalam Mereduksi Beban Aliran Limpasan Permukaan Sub DAS) *J. FONDASI* **3**
[10] Water Resources Management Agency of West Sumatera Province 1983 Feasibility Study Padang City Flood Control
[11] Mera M 2011 *Engineering Hydrology (in Bahasa : Hidrologi Rekayasa)* (Padand: Cv. Ferila)

IOP Conf. Series: Materials Science and Engineering 1041 (2021) 012003 doi:10.1088/1757-899X/1041/1/012003

Pushover Analysis of 10-Floors Reinforced Concrete Building (Case study: Mahkota Majolelo Sati Bautique Hotel)

Masrilayanti[1], R Kurniawan[1], A L Budi[1]and S H Sourkan[2]

[1]Civil Engineering Department, UniversitasAndalas, Padang, Indonesia
[2]Civil Engineering Department, University of Raparin, Iraq

Corresponding author: masrilayanti@eng.unand.ac.id

Abstract. Analysis of nonlinear pushover (push load) is one component of performance based design that can be used as a means of determining the capacity of a structure. The analysis is carried out by giving a static lateral load pattern to the structure whose value is increased gradually until a displacement target is reached or until the structure collapses. The purpose of this study is to determine the performance level of the building structure of the MahkotaMajolelo Sati Bautique Hotel according to the ATC-40 document. The capacity curve can estimate the basic shear force and maximum deformation that occurs in the structure. The results of the pushover analysis can show the pattern of collapse and the scheme of the occurrence of plastic hinges in the structure when given loading. By looking at the scheme of the occurrence of plastic hinges, then it can be seen the critical parts of the structure of the Majolelo Sati Bautique Hotel Mahkota Building. Pushover analysis is carried out by referring to the new regulations in effect in Indonesia, namely SNI 1726-2019, SNI 1727-2013 and SNI 2847-2019. The performance point is obtained by using the FEMA 440 Equivalent Linearization method, which is a simplification of the capacity spectrum method. The determination of the performance level of the structure follows the ATC-40 document.

1. Introduction

Indonesia is located in the Pacific Earthquake Belt and the Asian earthquake route (Trans Asiatic Earthquake Belt), so the chance of an earthquake occurring is very high. Given the high potential for earthquakes, a strong but also economical structure is required. Currently, a new concept has been developed in the analysis and planning of reinforced concrete building structures, namely the concept of performance-based design. This concept can help designers and practitioners alike in estimating the level of damage to building structures when subjected to earthquake loads. By knowing in advance the effect of earthquake loads on the structure, it is hoped that a strong but also economical building is created.

In this study, a nonlinear analysis of the push load was carried out on the structure of the MahkotaMajolelo Sati Bautique Hotel Building. The nonlinear pushover analysis is one component from the performance based design which can show the effect of the earthquake on the building structure and then can be determined the performance level of the structure. In the analysis process pushover, the building structure is pushed with a static lateral loading pattern whose value is gradually increased until the base shear and deformation are maximum structure is achieved. The results of this study can be used as a reference in planning the repair, strengthening and rehabilitation of structures.

In addition, this research is expected to provide new insights in assessing the performance of structures in resisting lateral loads, especially earthquake loads.

The concept of performance-based earthquake resistant building planning known as Performance Based Earthquake Engineering (PBEE) is a combination of durability and service aspects. This concept can be used to design new buildings (Performance Based Seismic Design) or evaluation of existing buildings (Performance Based Seismic Evaluation). The methodology which is based on performance based seismic design is a design and evaluation criterion which is expressed as the objectivity of the structure's usability. It can be used to define different levels of structural performance. The main objective of performance based seismic design is to create earthquake resistant buildings whose structural capacity can be estimated. Performance based seismic design has two main elements in its planning, namely structural capacity (capacity) and load (demand). The load in question is a representation of the ground motion due to an earthquake, which will be described as a spectrum response curve. The response spectrum curve will be converted into a demand spectrum curve. The capacity of the structure will be described as a capacity curve that shows the structure's ability to withstand earthquake forces without experiencing damage. The process of planning for a performance-based earthquake resistant building begins by modeling the structure in accordance with the prevailing regulations. Furthermore, the performance simulation is carried out on various earthquake events. Each stage of the simulation provides information on the level of damage and the resistance of the structure so that it can be estimated how much safety (life), readiness to use (occupancy) and loss of property (economic loss) that will occur[1,2]. The planner can then rearrange the acceptable risk of damage accordingly with the risk of costs incurred.

2. Methodology of Research

2.1. Description of building

Building structure modeling is done in 2D and 3D using structural analysis software. The structural data in the form of dimensions and quality of materials used for this modeling are in accordance with the building plan drawings.

The type of loading given to the structure of the MahkotaMajolelo Sati Bautique Hotel is in the form of gravity loads and lateral loads. The gravity load used is dead load, namely the load from the structure's own weight and also live loads. While the lateral loads used are static and dynamic earthquake loads.

Dead loads are loads that are caused by the structure's own weight and other loads that are permanently attached to the structure. The dead load used when modeling is the structure's own weight and the additional dead load. The structure's weight has been calculated automatically by software so that manual calculations are not performed. Additional dead load is calculated manually. The calculation of additional dead load can be seen in Table1 below.

Table 1. Material specification.

Material	Density	Width	Total
Ceramics	22 kN/m^3	0,01	0,22 kN/m^2
Spesi	22 kN/m^3	0,03	0,66 kN/m^2
Sand	16 kN/m^3	0,01	0,16 kN/m^2
Plafond and hanger	0,20kN/m^2	-	0,20kN/m^2
Utility	0,25 kN/m^2	-	0,25 kN/m^2
Masonry Wall	250 kg/m	-	250 kg/m

Modeling is done by describing all structural elements, namely columns, beams, floor plates and shear walls. The building has ten floor and is aimed to be a hotel. The model of the building can be seen from the Figure 1.

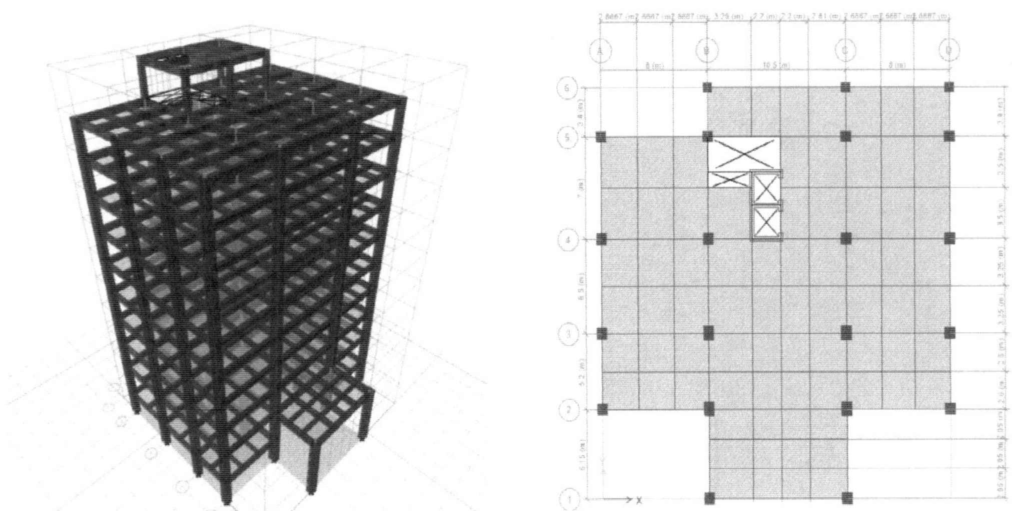

(a) 3D modelling (b) First floor elevation

Figure 1. Building modelling

2.2. Dynamic Analysis

Based on SNI 1726-2019 (seismic resistance planning procedures for building and non-building structures), the calculation of dynamic earthquake loads with a spectrum response is carried out in the following stages[3]:

a. Determining Risk Categories and Priority Factors of the Earthquake (I_e)
Based on Table 3 (Risk categories for buildings and non-buildings for earthquake loads) and Table 4 (earthquake priority factors) from SNI 1726-2019, the analyzed building structure is a hotel, so it is included in the building with risk category II and earthquake priority factor (I_e) of 1,0.

b. Determining the S_s and S_1 Mapped Acceleration Parameters
The S_s parameter (bedrock acceleration in a short period of 0.2 seconds) and the S_1 parameter (bedrock acceleration in a 1 second period) are determined according to the Indonesian earthquake zoning map presented in SNI 1726-2019.

c. Determining Site Class and Site Coefficient F_a and F_v
Soil type is assumed to be a medium soil type so based on Table 5 (Site Classification) SNI 1726-2019, the analyzed buildings are included in the SD site class. Based on Table 6 (site coefficient, F_a) and Table 7 (site coefficient, F_v) from SNI 1726-2019, the values of the F_a and F_v site coefficients are as follows:
F_a = 1,000
F_v = 1,700

d. Determining the S_{MS} Value (Acceleration spectral response parameter in a short period and S_{M1} (Acceleration spectral response parameter in a long period)
Based on Article 6.2 (Site coefficients and spectral response parameters of maximum earthquake acceleration which are considered as targeted risks) from SNI 1726-2019, the parameter values of S_{MS} and S_{M1} are calculated according to the following equation:
$S_{MS} = F_a \times S_s$ = 1,000 x 1,500 = 1,500 g

$S_{M1} = F_v \times S_1 = 1{,}700 \times 0{,}600 = 1{,}020$ g

e. Determining the Design Spectral Acceleration Parameters of S_{DS} (Short-period Design Spectral Acceleration Parameters) and S_{D1} (Long-Period Design Spectral Acceleration Parameters)
f. Defining the Design Response Spectrum S_a
After doing those procedures, the response spectrum of Padang City is obtained as can be seen in Figure 2.

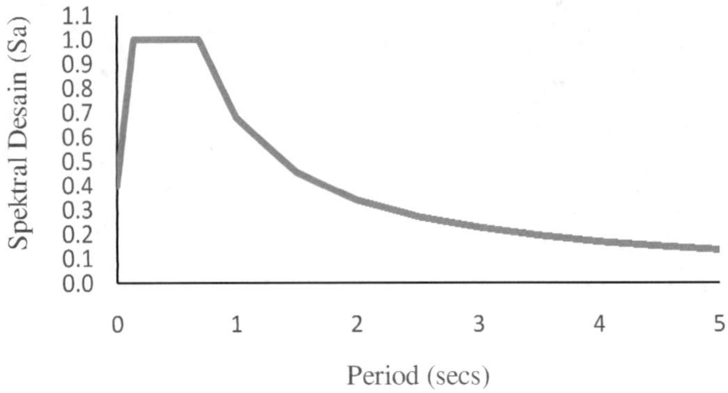

Figure 2. Response spectrum Padang City.

2.3. *Pushover Analysis*
Pushover analysis is a static procedure that uses a simplified nonlinear technique to estimate seismic structural deformations[4,5]. Structures redesign themselves during earthquakes. As individual components of a structure yield or fail, the dynamic forces on the building are shifted to other components[6,7].The initial stage in conducting a nonlinear analysis of thrust loads is to determine the amount of load applied to the structure. There are two types of loading given, namely gravity loading and lateral loading.

3. Results and discussion

3.1. *Capacity Curve*
After modeling and checking the capacity of the structure is done, it is followed by running static nonlinear pushover for the x and y directions. From the results of the running, a capacity curve will be obtained and then the performance level of the structure being analyzed can be determined.

The capacity curve is a curve that shows the relationship between the displacement of the control point and the base shear force of the structure. The control point in question is a point on the top floor of the building which is used as a reference to see the behavior of the structure when given a thrust load. In this study, the control point is set on the top floor of the lift. The capacity curves for both directions can be seen in the graphic image below:

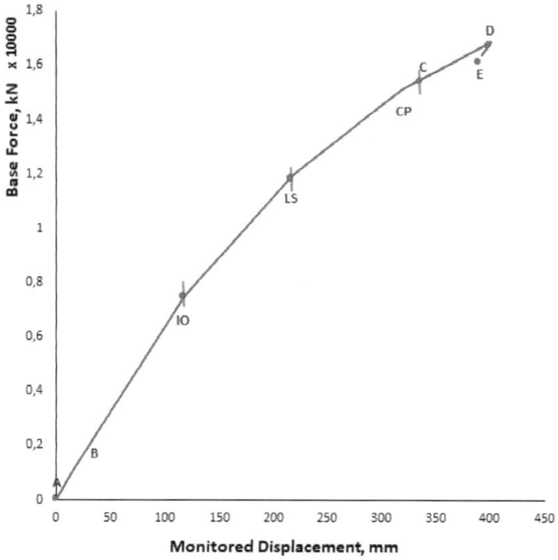

Figure 3. Capacity curve – X.

It can be described in Figure 3 that the curve is nonlinear, this is due to the addition of the load given to the structure so that the strength and stiffness of the structural elements decrease and there is a change in the properties of the structural elements from elastic to plastic.

The displacement values and basic shear forces for pushover direction x can be seen in table 2 below:

Table 2. Displacement and shear forced for pushover – x.

Step	Monitored Dispacement (mm)	Base Force (kN)	A-B	B-C	C-D	D-E	>E	A-IO	IO-LS	LS-CP	>CP	Total
0	0	0	1634	0	0	0	0	1634	0	0	0	1634
1	16,206	1144,656	1634	0	0	0	0	1634	0	0	0	1634
2	116,262	7428,614	1606	28	0	0	0	1634	0	0	0	1634
3	216,646	11937,49	1485	149	0	0	0	1633	0	0	1	1634
4	318,638	15147,667	1360	274	0	0	0	1618	15	0	1	1634
5	330,518	15420,453	1349	285	0	0	0	1614	19	0	1	1634
6	330,528	15411,087	1349	285	0	0	0	1614	19	0	1	1634
7	400,01	16892,698	1305	327	1	0	1	1603	25	4	2	1634
8	391,821	16397,743	1305	326	2	0	1	1603	25	3	3	1634

It can be seen in table 2 above that the structure collapses at step 8 which is marked by a reversal of the capacity curve. In the 8th step, the displacement value at the control point is 391,821 mm with the base shear applied to the structure of 16397,743 kN. The maximum displacement at the control point occurs at step 7 which is 400,01 mm with a basic shear force of 16892,698 kN.

On the other hand, the capacity curve for Y direction is can be observed from the Figure 4 below:

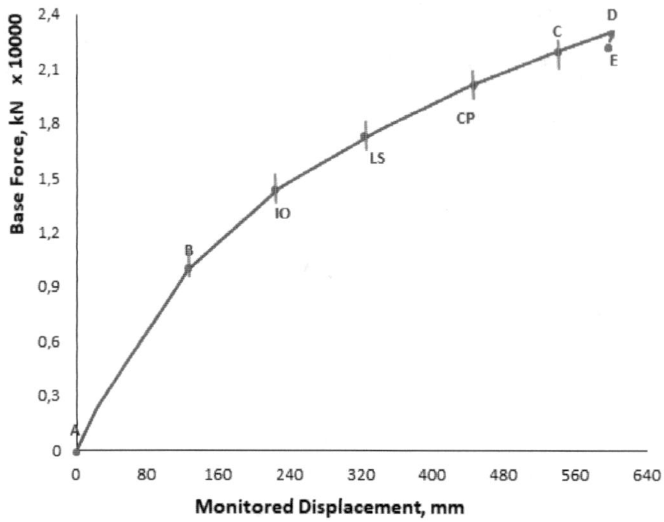

Figure 4. Capacity curve - Y

The displacement values and basic shear forces for pushover direction Y can be seen in Table 3.

Table 3. Displacement and shear forced for pushover – y.

Step	Monitored Displacement (mm)	Base Force (kN)	A-B	B-C	C-D	D-E	>E	A-IO	IO-LS	LS-CP	>CP	Total
0	0	0	1634	0	0	0	0	1634	0	0	0	1634
1	24,478	2365,685	1634	0	0	0	0	1634	0	0	0	1634
2	124,746	9865,413	1567	67	0	0	0	1634	0	0	0	1634
3	227,887	14432,265	1376	256	0	0	0	1634	0	0	0	1634
4	335,61	17559,202	1281	353	0	0	0	1618	15	0	1	1634
5	439,407	20074,28	1206	428	0	0	0	1589	44	0	1	1634
6	540,479	22083,753	1124	510	0	0	0	1509	123	1	1	1634
7	540,489	22061,712	1124	510	0	0	0	1509	123	1	1	1634
8	600,769	23114,937	1073	556	5	0	0	1460	159	10	5	1634
9	600,769	23114,948	1072	557	5	0	0	1460	159	10	5	1634
10	600,779	23071,694	983	626	17	0	8	1406	187	22	19	1634
11	596,781	22564,94	983	625	15	3	8	1406	187	22	19	1634

Based on Table 3, it can be concluded that the structure collapsed at the 11th step with a displacement value at the control point of 596,781 mm with a basic shear force of 22564,940 kN. The maximum displacement occurs in the 10th step, which is 600,779 mm with a base shear force of 23071,694 kN.

3.2. Perfomance Point

The value of performance points in x and y direction can be seen from the Figure 5.The orange line is the capacity spectrum curve, while the blue line is the demand spectrum curve. The point of intersection between the capacity spectrum curve and the demand spectrum is the performance point of the structure for the given direction.

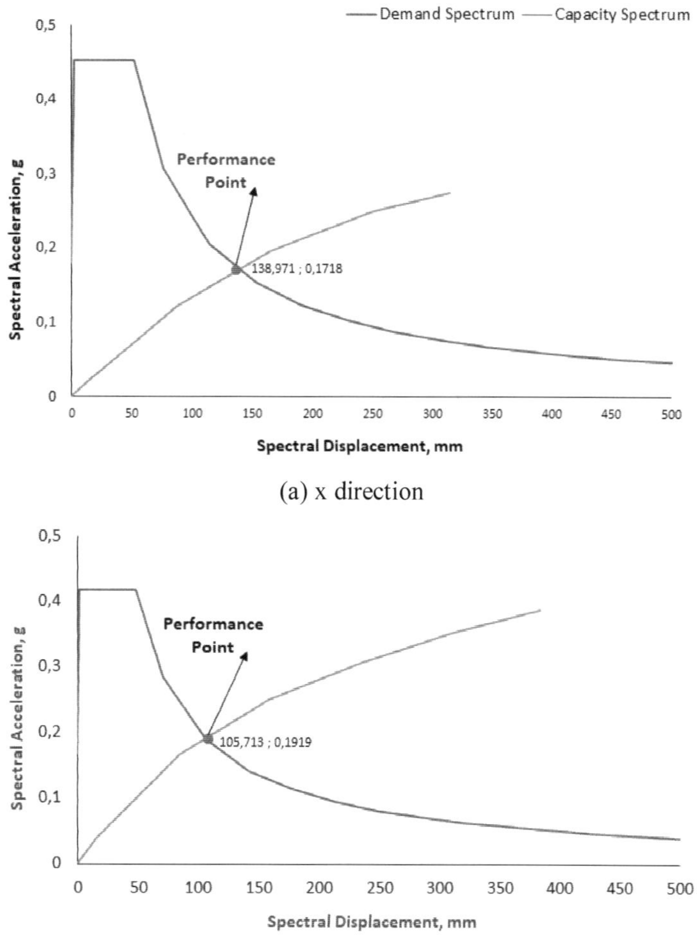

(a) x direction

(b) y direction

Figure 5. Performance Point of the building

The performance points of the structure are calculated through an iteration process which is carried out automatically by the program. It can be concluded from Figure 5 that for both the x and y directions the structure has a sufficiently good capacity to bear the defined earthquake load plan.The method used to get the performance point is the FEMA 440 Equivalent Linearization method. This method is an improvement over the capacity spectrum method (ATC-40) [8]. The way this method works is by converting the capacity curve into a capacity spectrum curve to form a relationship curve between spectral acceleration (Sa) and spectral demand (Sd). Then the spectrum response curve is converted into MADRS (Modified Acceleration Displacement Response Spectrum) format so that a demand spectrum curve is formed.

3.3. Performance Level

The determination of the performance level of the structure follows the ATC-40. The calculation and determination of the structural performance levels for the x and y directions are shown in Table 4 below.It can be seen in Table 4 that for the x and y direction pushover, the structure performance level is obtained including the Immediate Occupancy (IO) performance level. This means that when an earthquake with the strength of the planned earthquake occurs, the structure does not experience structural or non-structural damage so that it can be immediately reused.

Table 4. Performance level of pushover X and Y.

Case	Dt (mm)	D1 (mm)	Ht (mm)	Maximum Total Drift	Maximum Inelastic Drift
Pushover X	183,254	116,262	45000	Immediate Occupancy	Immediate Occupancy
Pushover Y	155,43	124,746	45000	Immediate Occupancy	Immediate Occupancy

4. Conclusions

After analysing the push load on the structure of the Majolelo Sati Bautique Hotel Mahkota Building, the following conclusions can be drawn:

1. Based on the ATC-40 document, the performance level of the structure for pushover in the x direction and y direction is included in the Immediate Occupancy (IO) performance level, which means the Majolelo Sati Bautique Hotel Mahkota Building does not experience structural and non-structural damage when an earthquake occurs as large as the planned earthquake so that immediately reusable.
2. The performance point of the structure based on the FEMA 440 Equivalent Linearization method is obtained for pushover direction x point (V; D) is (10437,643; 183,254) and direction y is at point (11223,943; 155,428).
3. Based on the capacity curve, the x-direction pushover results in a maximum displacement of 400.01 mm with a total base shear of 16892,698 kN while the y-direction pushover produces a maximum displacement of the structure of 600,779 mm with a total basic shear force of 23071,694 kN.

Acknowledgments

This publication of this research is supported by Faculty of Engineering, UniversitasAndalas.

References

[1] Sahoo D R and Chao S H 2010 Performance-based plastic design method for buckling-restrained braced frames *Eng. Struct.* doi: 10.1016/j.engstruct.2010.05.014

[2] Priestley M J N 2000 Performance based seismic design *Bull. New Zeal. Soc. Earthq. Eng.* doi: 10.5459/bnzsee.33.3.325-346

[3] SNI 1726-2019 2019 Tata cara perencanaan ketahanan gempa untuk struktur bangunan gedung dan non gedung *Bandung Badan Stand. Indones.*

[4] Hasan R, Xu L and Grierson D E 2002 Push-over analysis for performance-based seismic design *Comput. Struct.* doi: 10.1016/S0045-7949(02)00212-2

[5] Rofooei F R and Mirjalili M R 2018Dynamic-based pushover analysis for one-way plan-asymmetric buildings *Eng. Struct.* doi: 10.1016/j.engstruct.2018.02.052

[6] Moghaddam H and Hajirasouliha I 2006 An investigation on the accuracy of pushover analysis for estimating the seismic deformation of braced steel frames *J. Constr. Steel Res.* doi: 10.1016/j.jcsr.2005.07.009

[7] Li S, Zuo Z, Zhai C and Xie L 2017 Comparison of static pushover and dynamic analyses using RC building shaking table experiment *Eng. Struct.* doi: 10.1016/j.engstruct.2017.01.033

[8] (Applied Technology Council) ATC-40 1995 ATC-40 (Applied Technology Council) *ATC 40, Appl. Technol. Counc.*

IOP Conf. Series: Materials Science and Engineering 1041 (2021) 012004 doi:10.1088/1757-899X/1041/1/012004

Effect of lumped mass on the structural dynamic modifications of a structure

A R Bahari[1], M A Yunus[1,2], M N A Rani[1,2] and N Jasmin[1]

[1]Structural Dynamics Analysis & Validation (SDAV),Faculty of Mechanical Engineering,UniversitiTeknologi MARA (UiTM),40450 Shah Alam, Selangor, Malaysia
[2]Institute for Infrastructure Engineering and Sustainable Management (IIESM),UniversitiTeknologi MARA (UiTM), 40450 Shah Alam, Selangor, Malaysia

Corresponding author: mayunus@uitm.edu.my

Abstract. The identification of the effectdue to structural changes using modal data is very challenging in the structural dynamic modifications (SDM) study. In this paper,the aim of this studyis to identifythe effect of physical changes of a structural system based on the natural frequencies and mode shapes.A mild steel thin plate structure was used in the study. A mass was attached to the thin plate to alter the dynamic characteristics of the thin platestructure. Experimental modal analysis (EMA) and the normal modes analysis wereperformed in two stages to the thin plate structure, without mass and the mass is attached to the thin plate structure.The normal modes of the thin plate structure were obtained using finite element (FE) analysis.The results of the natural frequencies and mode shapes between the thin plate and thin plate with SDM were compared.It is concluded that the impact of structural changes due to SDM will change the modal parameters of the structure.

1. Introduction

Beamsare widely used in engineering field especially in mechanical and civil engineering applications. Beams come in a broad variety of forms for various uses and capabilities. Beams are historically examples of construction or civil engineering structural features, such as vehicle frames, aircraft parts, computer frames and other mechanical or structural devices, which include beam assemblies intended to support lateral loads. Beams are distinguished by their form of reinforcement, profile (cross-sectional shape), equilibrium requirements, length and content.

The cumulative result of all forces operating on the beam is to create shear forces and bending moments within the beam, which, in essence, cause internal pressures, strains and deflections of the beam. The loads added to the beam result in response forces at the support points of the beam. Throughout architecture, beams are categorized in many forms on the basis of supports, comprising of purely fixed beam, cantilever beam and continuous beam.

Structural dynamic modifications strategies can be characterized as methods by which the dynamic behavior of the system is enhanced by anticipating the changed behavior induced by the introduction of modifications, such as lumped mass, rigid linkages, dampers, supports, or by changes in the design parameters of the structures themselves [1], [2]. Modification of any structure to enhance its natural frequencies is necessary in many places as the structural response is highly affected by its natural frequencies[3]. Structuraldynamic modifications are strategies for achieving the functional properties of the changed structure owing to structural improvements without the requirement for a repetitive theoretical approach or experiments [4], [5].

This paper focuses on the SDMstudy in the experimental and numerical frameworks. Modal testing and normal modes analysis are used to obtain natural frequencies and mode shapes. The comparison of natural frequencies and mode shapes between the thin plate and thin plate with SDM are conducted to investigate the effect of SDM.

2. Materials and methods

2.1. Finite element modelling

A structure used in the SDMwas made from a thin steel plate with the length of 520 mm,60 mm width and1.5 mmof thickness. The finite element (FE) modelling was carried out using MSC PATRAN software. The FE model of the thin plate was constructed using CQUAD4 shell elements with a size of 5 mm as shown in Figure 1a.CQUAD4 element is significantly used for meshing the FE model of thin structure [6],[7], [8].The material properties of mild steel were assigned for the plate using nominal parameter values such as Young's modulus, E = 200 GPa, Poisson's ratio, υ = 0.30 and density, ρ = 7900 kg/m^3.

SDM is implemented in this study, by introducing a block with a solid mass of 100 gram which was glued to the thin plate.Figure 1b shows the FE model of the thin plate with SDM. The rigid body element of RBE2 was used for modelling a rigid glued type connection between the block mass and the thin plate. The weight of the block mass was modelled byconcentrated mass element connection using CONM2.

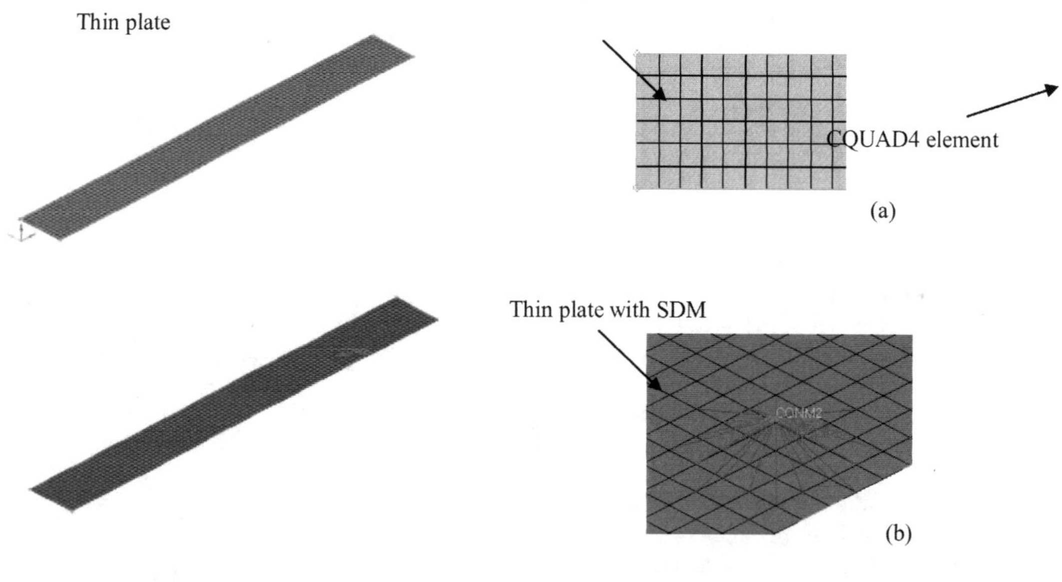

Figure 1. FE model (a) thin plate without mass (b) thin plate with SDM

2.2. Normal modes

The equation of motion for free vibration of undamped system [9], [10] is described by equation (1).

$$M\ddot{x}(t) + Kx(t) = 0 \tag{1}$$

where$K$ and M are all symmetric matrices of stiffness and mass. Meanwhile, x and $\ddot{x}$ are the vector of displacements andaccelerations. For the analysis of eigenvalue problem [11], [12],[13] equation (2) was used to obtain the natural frequencies and mode shapes of the thin plate.

$$K\emptyset_i = \lambda_i M\emptyset_i \tag{2}$$

where λ_i are eigenvalues (natural frequencies) and ϕ_i are the corresponding eigenvectors (mode shapes). MSC Nastran Solution 103 Normal Modes was used for performing normal modes analysis. The common algorithm which is known as Lanczos method was utilised for computing the free vibration modes of the thin plate. The eigenvalues can be calculated using equation (3). The eigenvalues are related to natural frequencies f_i that can be expressed by equation (4).

$$\lambda_{n,i} = \sqrt{\frac{K}{M}} \tag{3}$$

$$f_i = \frac{\lambda_i}{2\pi} \tag{4}$$

2.3. FE natural frequencies and mode shapes

In this study,the natural frequenciesand mode shapes of the first five elastic modes of the thin plate structure iscalculatedby using normal modes analysis as shown in equation 3.Table 1,lists the FE natural frequencies of the thin plate(column II) and thin plate with SDM (column III). The first five FE natural frequencies of thin plate without mass and with SDM are compared for the validation purposes. The results show that the FE natural frequencies of the thin plate without mass were calculated 28.53 Hz, 78.79 Hz, 147.69 Hz, 155.85 Hz and 256.70 Hz. Meanwhile, for the thin plate with SDM, the FE natural frequencies were 28.88 Hz, 76.60 Hz, 136.74 Hz, 154.89 Hz and254.15 Hz.

The natural frequencies of the thin plate with SDM were particularly observed lower than the natural frequencies of the thin plate without mass. The reduction of natural frequencies was influencedby the weight introduced by the block mass which affectedand changed the total mass of the thin plate structure.Mode 1 showed less significant effect of the SDM. Asmall change of natural frequencyis based on the location of the block mass which is at the nodal point of the structure for mode 1 configuration.

Based on equation 3, the eigenvalues are depending on the stiffness and mass properties. The increment of the parameter mass with respect to the initial mass will decrease the eigenvalue. In this case, the additional mass introduced by the block affect the eigenvalues where the natural frequencies decreased to lower values.

Table 1. Results of FE natural frequencies between the thin plate without mass and thin plate with SDM

I Mode	II FE Natural Frequencies Thin Plate without Mass (Hz)	III FE Natural Frequencies Thin Plate with SDM (Hz)
1	28.53	28.88
2	78.79	76.60
3	147.69	136.74
4	155.85	154.89
5	256.70	254.15

Figure 2shows the FE mode shapes of the thin plate without mass and thin plate with SDM. Two types of elastic modes were obtained for the dynamic behaviour. Four elastic modes which were mode 1, mode 2, mode 4 and mode 5 were bending modes while mode 3 was a torsion mode.For the thin plate without mass, the bending and torsion mode shapes were in symmetrical characteristic. Meanwhile, for the thin plate with SDM, the bending and torsion mode shapes were asymmetry. This is due to the effect of mass introduced on the thin plate and asymmetry in mass placement on the thin plate.

Mode 1: Thin Platewithout Mass

Mode 1: Thin Plate with SDM

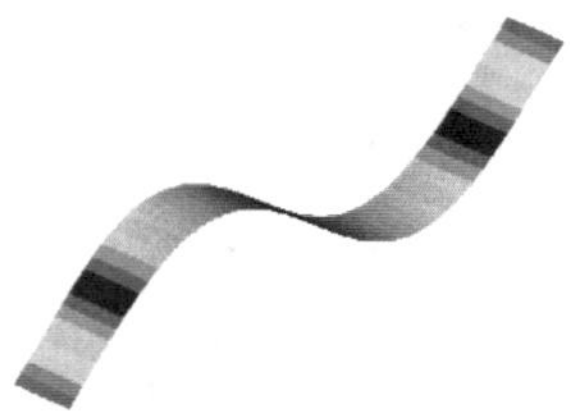

Mode 2: Thin Plate without Mass

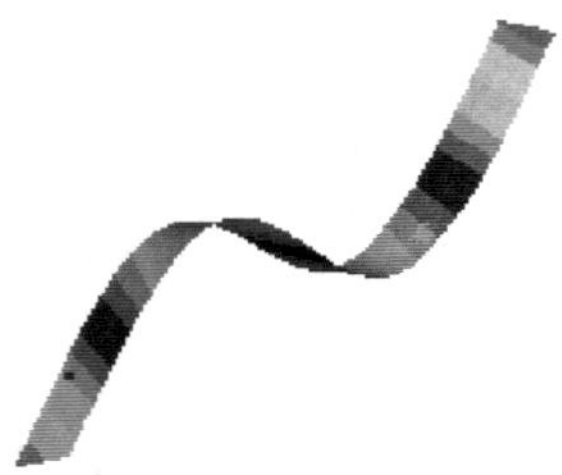

Mode 2: Thin Plate with SDM

Mode 3: Thin Plate without Mass

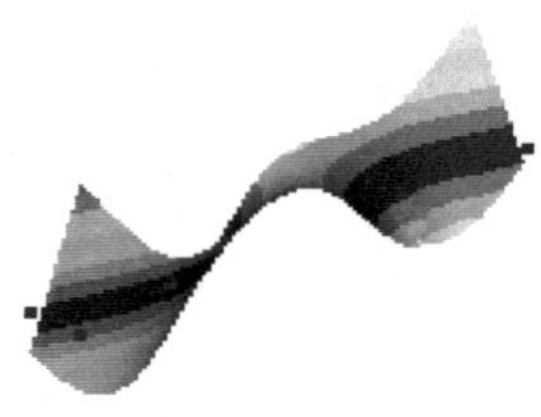

Mode 3: Thin Plate with SDM

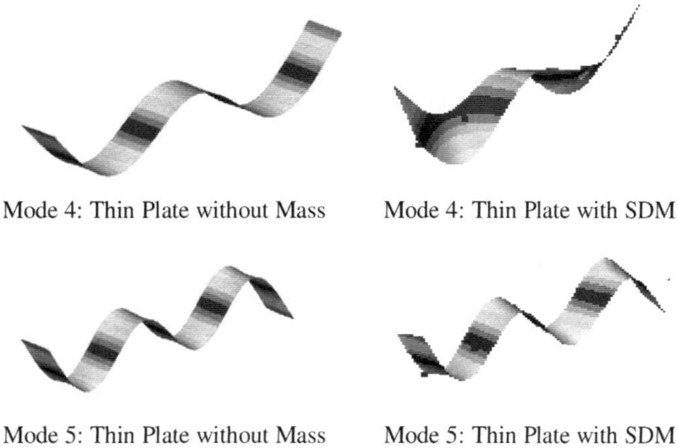

Figure 2. Comparison of FE mode shapes between the thin platewithout mass and thin plate with SDM.

3. Results an discussion

3.1. Test set up

EMA was performed to validate the results obtained from the numerical analysis. Figure 3, shows the experimental setup for impact testing.EMAwas performed on the two case studies: namely, 1) thin plate without mass, and 2) thin plate with SDM.The measurement of natural frequencies and mode shapes wasconductedwith free-free boundary conditionsin order to minimise any constraint. The free-free boundary conditions were achieved by using very soft springs for thesuspension. The impact hammer was used forapplying light impulse excitation and the dynamic response in the acceleration unit was measured using three 10 mV/g sensitivity of miniature uniaxial accelerometers.These light accelerometers contributed to no mass loading issue during the measurement. The data was stored and analysed using LMS Test.Lab software.

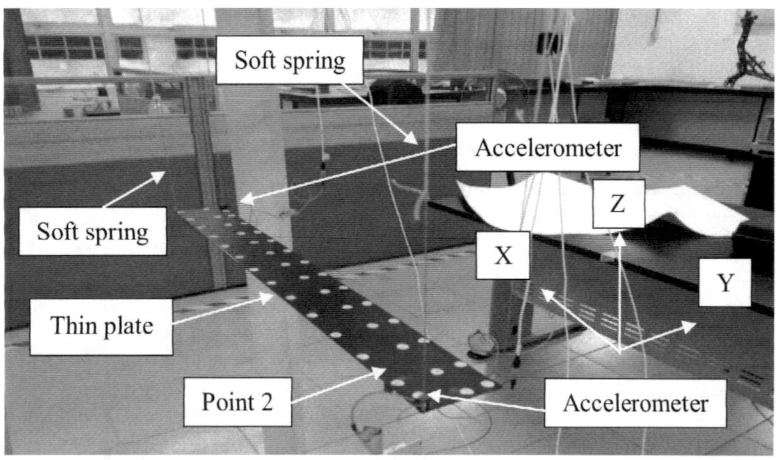

(a)

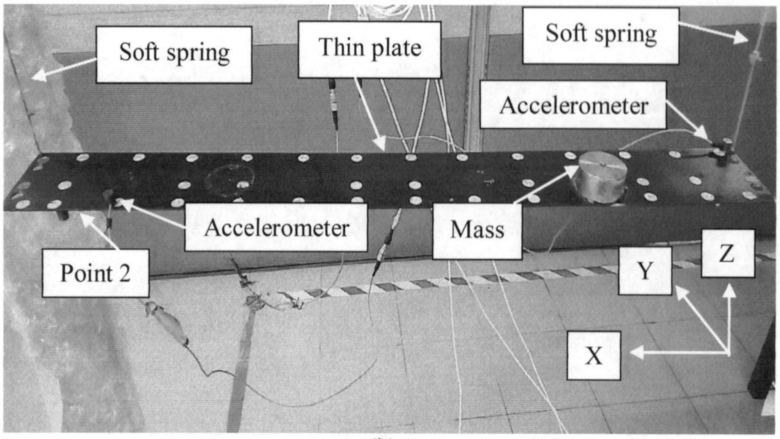

(b)

Figure 3. Experimental set up for modal analysis (a) thin platewithout mass (b) thin plate with SDM.

3.2. Test procedures

The mode of interest are the first five elastic modes.Point 2 which is located at the edge of the thin plate was assigned as a reference point since this point covered the dynamic response for all five elastic modes. During the impact testing, the thin plate was impacted at the same degree of freedom (DOF) in the *Z*-axis. Thedynamic responsecovered 42 DOFsand the measurement of all the measuring points was in the *Z*-axis.

Roving accelerometer was used in EMA. Reference point2 was used as a fixed location for accelerometer no. 1 while accelerometerno. 2 and no. 3 wereroved around the measuring points. The frequency resolution was 0.5 Hz and therigid bodymodes were neglected.For each measurement point, the dynamic response in term of frequency response function (FRF) was recorded by ten measurements. The average FRF was considered for the determination of natural frequencies of the thin plate and thin plate with SDM.

3.3. EMA natural frequencies and mode shapes

The EMA natural frequencies were determinedfrom a set of42 measured frequency response functions in Z-axiswhich were acquired during an impact test. Rigid body modes were also excluded in the analysis. Table 2contains the EMA natural frequencies of the thin plate without mass (column II) and thin plate with SDM (column III). The EMA natural frequencies of the thin plate without mass were 26.69 Hz, 73.26 Hz, 136.17 Hz, 144.36 Hz and 239.54 Hz. Meanwhile, for the thin plate with SDM, the EMA natural frequencies were 27.13 Hz, 71.61 Hz, 123.54 Hz, 142.89 Hz and 235.64 Hz. The first five EMA natural frequencies were less than 240 Hz.

The EMA natural frequencies were compared between the thin plate without mass and thin plate with SDM.A clear indication that the EMA natural frequencies measured from the thin plate with SDM were lower than EMA natural frequencies measured from the thin plate without mass.The reduction of EMA natural frequencies for most of the vibration modes of the thin plate with SDM was occurred because of the introduction ofadditional mass which affected the natural frequencies. Mode 1 showed less significant effect of the SDMbecause the location of the block mass was at the nodal point of the thin plate structure.

IOP Conf. Series: Materials Science and Engineering 1041 (2021) 012004 doi:10.1088/1757-899X/1041/1/012004

Table 2. Results of EMA natural frequencies between the thin platewithout mass and thin plate with SDM

I Mode	II EMA Natural Frequencies Thin Plate without Mass (Hz)	III EMA Natural Frequencies Thin Plate with SDM (Hz)
1	26.69	27.13
2	73.26	71.61
3	136.17	123.54
4	144.36	142.89
5	239.54	235.64

Figure 4 shows the EMA mode shapes of the thin plate without mass and thin plate with SDM. The bending modes were mode 1, mode 2, mode 4 and mode 5; and thetorsion mode was mode 3. For the thin plate without mass, the bending and torsion mode shapes were in symmetrical shape. Meanwhile, for the thin plate with SDM, the bending and torsion mode shapes were asymmetry due to the effect of mass placement on the thin plate. These EMA findings validate the comparison results of natural frequencies and mode shapes obtained from the numerical analysisof the thin plate without mass and thin plate with SDM.

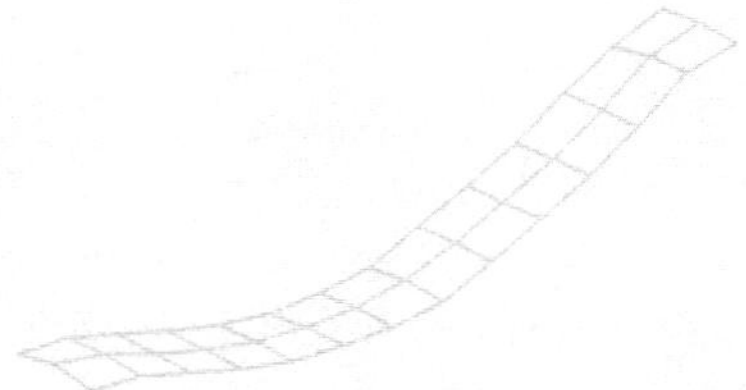

Mode 1: Thin Platewithout Mass

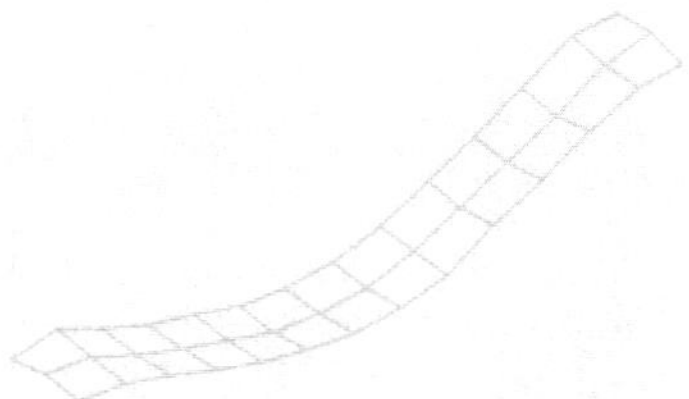

Mode 1: Thin Plate with SDM

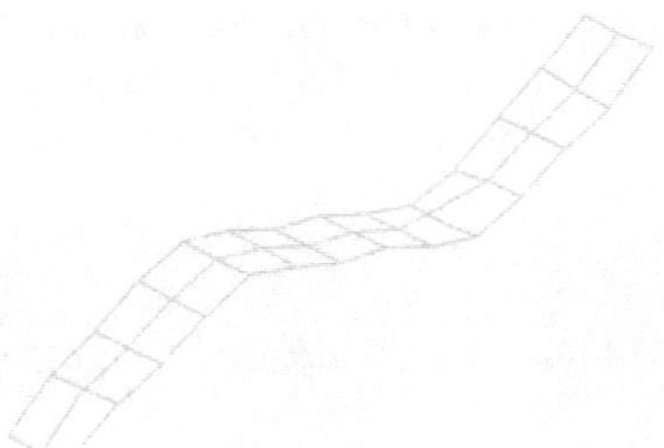
Mode 2: Thin Platewithout Mass

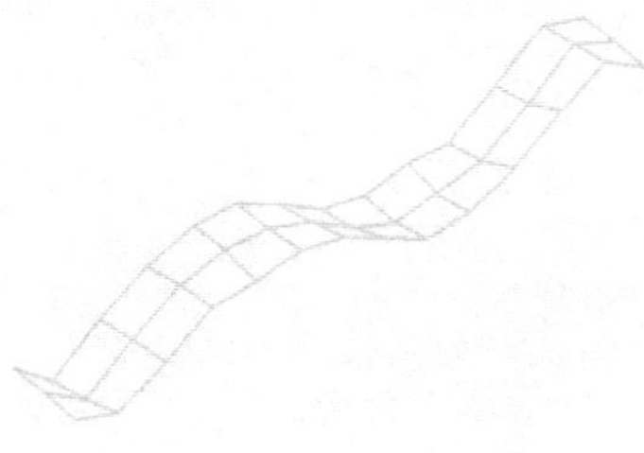

Mode 2: Thin Plate with SDM

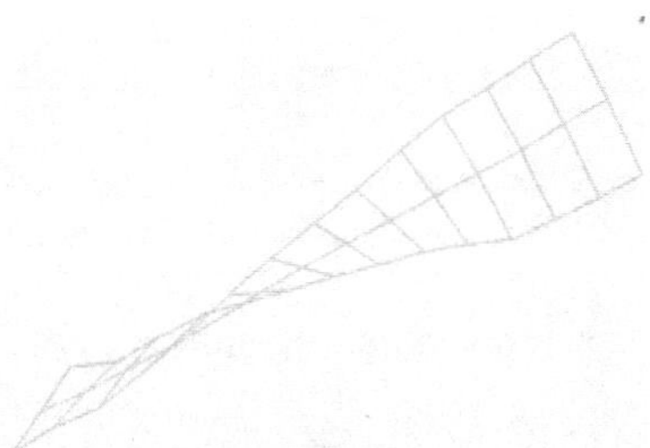
Mode 3: Thin Platewithout Mass

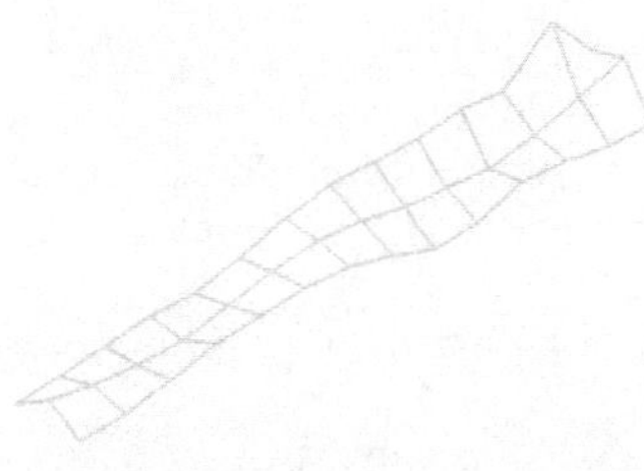
Mode 3: Thin Plate with SDM

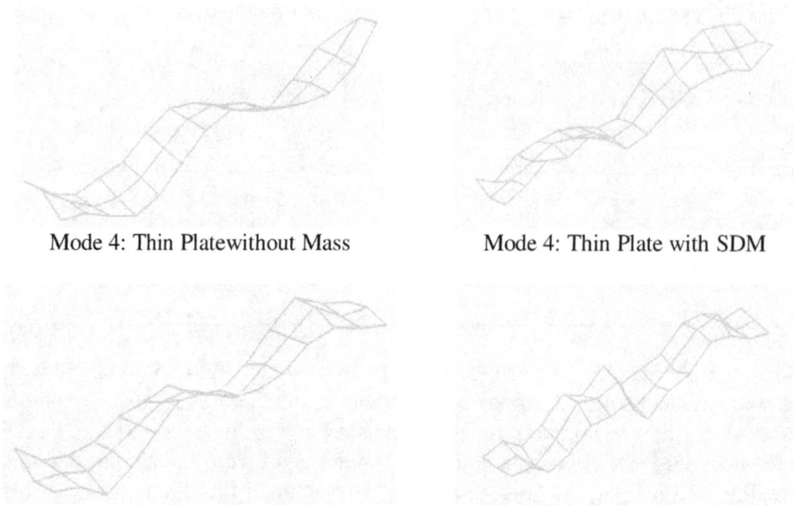

Mode 4: Thin Platewithout Mass — Mode 4: Thin Plate with SDM

Mode 5: Thin Platewithout Mass — Mode 5: Thin Plate with SDM

Figure 4. Comparison of EMA mode shapes between the thin platewithout mass and thin plate with SDM

4. Conclusion

Structural dynamic modifications based on modal approach using experimental and numerical analysis werepresented in this paper.Utilising free-free boundary conditionsin the test set up and finite element modelling of the thin plate, the natural frequencies for the first five elastic modes were less than 260 Hz. From the comparison of results between the EMA and FE analysis, the natural frequencies between the thin plate without mass and thin plate with SDM, the natural frequencies of the thin plate with SDM showed some reductions. Significantchanges of the mode shapes between the thin plate without mass and thin plate with SDM was obtained from the mode shape comparison. This finding showsthe importance of the additional mass and location of the lumped mass to the eigenvalues and eigenvectors of the structure for the structural dynamic modifications.

Acknowledgment

The research work was supported by the Malaysian Ministry of Higher Education (MOHE) under Fundamental Research Grant Scheme, Grant No. 600-IRMI/FRGS 5/3/ (465/2019). This support is gratefully acknowledged.

References

[1] Kumar R J V A, Reddy Y V M and Rao K P 2013 Critical review on structural dynamic modification on beam structures *Crit. Rev.***3** 1453–9

[2] Sestieri A 2000 Structural dynamic modification *Sadhana - Acad. Proc. Eng. Sci.***25** 247–59

[3] Schwarz B, Richardson M and Formenti D L 2007 FEA model updating using SDM *Sound Vib.***41** 18–23

[4] Hutton S G 1994 Analytical modification of structural natural frequencies *Finite Elem. Anal. Des.***18** 75–81

[5] Arora V, Singh S P and Kundra T K 2009 On the use of damped updated FE model for dynamic design *Mech. Syst. Signal Process.***23** 580–587

[6] Bahari A R, Yunus M A, Abdul Rani M N, Ayub M A and Nalisa A 2018 Numerical and experimental investigations of nonlinearity behaviour in a slender cantilever beam *MATEC Web of Conferences* vol 217 p 02008

[7] Omar R, Abdul Rani M N, Yunus M A, Mat Isa A A, Wan Iskandar Mirza W I I, Mohd Zin M S and Roslan L 2018 Investigation of mesh size effect on dynamic behaviour of an assembled structure with bolted joints using finite element method *Int. J. Automot. Mech. Eng.* **15** 5695–708

[8] Yunus M A, Ouyang H, Abdul Rani M N and Mat Isa A A 2013 Finite element modelling and updating of bolted joints in a thin sheet metal structure *20th International Congress on Sound and Vibration* pp 1146–52

[9] Bahari A R, Yunus M A, Abdul Rani M N, Wan Iskandar Mirza W I I and Haizuan A R 2019 A comparative study of the vibrational effects on the boundary conditions of a slender beam *AIP Conference Proceedings* vol 2059 pp 1–9

[10] Bograd S, Reuss P, Schmidt A, Gaul L and Mayer M 2011 Modeling the dynamics of mechanical joints *Mech. Syst. Signal Process.* **25** 2801–26

[11] Asadi E and Qatu M S 2013 Free vibration of thick laminated cylindrical shells with different boundary conditions using general differential quadrature *J. Vib. Control* **19** 356–66

[12] Corradi R, Miccoli S, Squicciarini G and Fazioli P 2017 Modal analysis of a grand piano soundboard at successive manufacturing stages *Appl. Acoust.* **125** 113–27

[13] Yunus M A, Aziz Shah M A S, Abdul Rani M N, Wan Iskandar Mirza W I I and Kushairi S 2017 Experimental and finite element analysis of laser stitch welded structure *J. Mech. Eng.* **4** 155–62

IOP Conf. Series: Materials Science and Engineering 1041 (2021) 012005 doi:10.1088/1757-899X/1041/1/012005

The dynamic load test of two seismic resistance of steel structures EBF and RBS

J I Rastandi[1], M Orientilize[1], C Ashilah[2], D Apritasari[2]

[1] Civil Engineering Department, Universitas Indonesia, Kampus UI Depok, 16424 Indonesia

[2] Postgraduate Students of Civil Engineering Department, Universitas Indonesia

Corresponding author: mulia@eng.ui.ac.id

Abstract. The non-destructive dynamic loading test has been conducted on two seismic resistance of steel structures, reduced beam section (RBS) and eccentric braced frame (EBF). Eccentric mass shaker was used to excite the 1-storey 1-bay steel structure. Two dynamics parameters were obtained which were the natural frequency and the damping ratio. A finite element analysis was performed using OpenSees software to get a comparison result. Similar natural frequency and displacement response were obtained from the test and the FE Analysis. The Natural frequency of RBS and EBF was around 3.614 Hz which is similar to the FE analysis as 3,623 Hz. Meanwhile the damping ratio is 3.149%. The natural frequency of the EBF structure is around 6.0 to 6.13 Hz, while the FE analysis result is 6.13 Hz. The damping ratio from test results is between 2.04 – 2.06%.

1. Introduction

Seismic resistance structural system has been developed for decades and steel structure have several varieties. The concept of the system is by designing the weaker part as a place of energy dissipation during earthquake which is named as plastic hinge. Special Moment resisting frames (SMRF) and Eccentrically braced frame (EBF) are two of seismic resistance system for steel structures.

SMRF which is known as reduced beam section (RBS) was proposed to enforce the form of plastic hinge away from the panel zone by designing RBS as the weaker section. The beam flange is cut in circular radius on the top and the bottom of flange to reduce the area at the beam end. Hence, the plastic deformation form there resulting the energy dissipation during earthquake. Several research about RBS has been carried out since 1996 and during the last ten years the research still have been done to improve its performance. [1],[2],[3],[4],[5],[6].

The EBF combines the high elastic stiffness characteristic of a concentrically braced frame (CBF) and the energy dissipation stability of a moment resisting frame (MRF). The system consists of a link which is designed as a weaker section to allow the plastic deformation. The longer link allows energy dissipation through flexural yielding while a shorter link dissipates energy through shear yielding. Shear yielding allows for the development of plastic deformation without the formation of excessive local strain, which is prevalent in flexural yielding [7]. Shear yielding provides a larger energy dissipation capacity than flexural yielding [8]. For that reason, there are three parameters that must be considered when designing an EBF system: bracing configuration, link length, and link section properties. In long links—often referred to as flexural links—yielding occurs only in the ends of the link where plastic hinges form. It is for this reason that EBF systems with shear links are more stable and more ductile than flexural links. However, for architecture reason, the flexural link is preferable since this system allows for a larger clearance than that of shear link EBFs.

As part of seismic resistance structure, several research about the two systems have been conducted for years but limited research have been reported about the dynamic test of those system. For an economic reason, FEM modelling has been used to obtain the dynamic characteristics of several structures. An effort has been done to get an accurate identification with development of experimental tools to make health assessment of existing structures is more economics and the damage can be easily detected. Eccentric mass shaker (EMS) is one of the innovative tools that can generate excitement for a large building, which is commonly called forced-vibration tests (FVTs). In the last few decades, dynamic tests have been used to obtain the dynamic parameters of structures [9],[10],[11,12]. The parameters are natural frequency, mode shape, and damping ratio. The natural frequency can be used as an indicator for detecting structural damage by examining the decrease in natural frequency [13].

This research aims to investigate the dynamic parameters of the one storey of the RBS and EBF system through FVTs. Eccentric mass shaker was used to induce the sinusoidal dynamic force. Two modal parameters, natural frequency and equivalent damping ratio were obtained.

2. Materials and methods

The test was conducted on a knock down frame that was purposely designed to support the dynamic test of two systems, the RBS and the EBF. The main frame consists of a 3m-length column, beam and composite slab. A WF300x150x6.5x9 was used for the column which is connected to a WF150x75x5x7 for beam on column weak axis. A composite slab with 100mm-thick was constructed on top of metal deck at an elevation of +2682mm. RBS was connected to flange of the column by three end plate system to ensure a rigid connection. Detail of RBS is shown in figure 2. A gap as 200mm was provided between RBS and the composite slab to prevent the composite action. The slab was meant to provide a mass source and to locate the EMS.

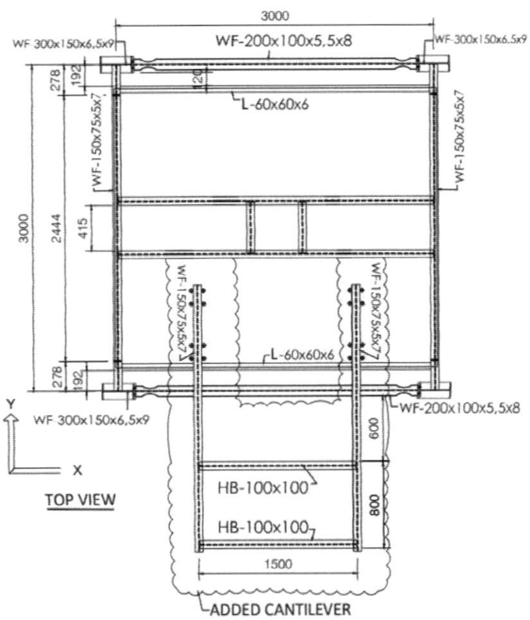

Figure 1. Top View of the Frame

The test was conducted twice, and it was started with the RBS followed by the EBF. The RBS was uninstalled when the second test was conducted. The EBF system was located on the elevation of +1341mm. Detail of EBF system is shown in figure 3. The beam of EBF was constructed by WF200x100x5.5x8 whereas the links use WF70x35x3x5. Double angle profile 2L50x50x5 was chosen for bracing. All the steel material was used grade BJ37 where the yield strength was 235MPa. The steel modulus of elasticity was 206MPa.

IOP Conf. Series: Materials Science and Engineering 1041 (2021) 012005 doi:10.1088/1757-899X/1041/1/012005

The load was induced by Eccentric Mass Vibrator System (EMS) MK-139-10. It was a sinusoidal load with a frequency identical to the structure's natural frequency. The EMS uses a two-armed drive system that produces a single-directional force. The force exerted by the vibrator can be calculated using the following equation:

$$F = MR(2f\pi)^2 \tag{1}$$

where F is the exerted force in Newton, MR was the eccentricity in kgm unit and f is the vibrator rotational velocity (Hz). In this test, the structure was induced by a load of 0.11625 kgm at a frequency of 1-15 Hz. A lighter load was used to identify the elastic response of the structure. The structural response obtained was the mode shape and the damping ratio.

The structural response was recorded using vibrometer and accelerometer. The accelerator can capture response in all directions, longitudinal (X), transversal (Y) and vertical (Z) meanwhile the vibrometer only measured the longitudinal direction according to the laser shooting and at this test the sensor was fired into the centre of the beam as shown in figure 3.

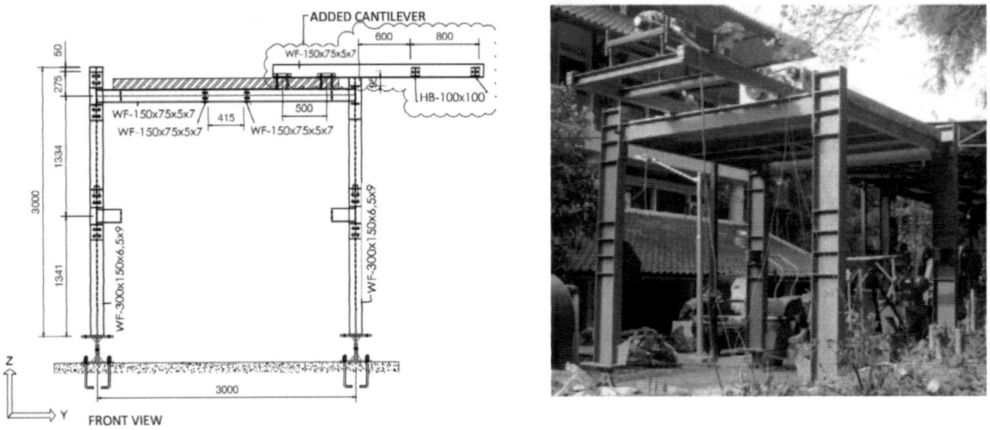

Figure 2. Side View of the Frame and frame the picture.

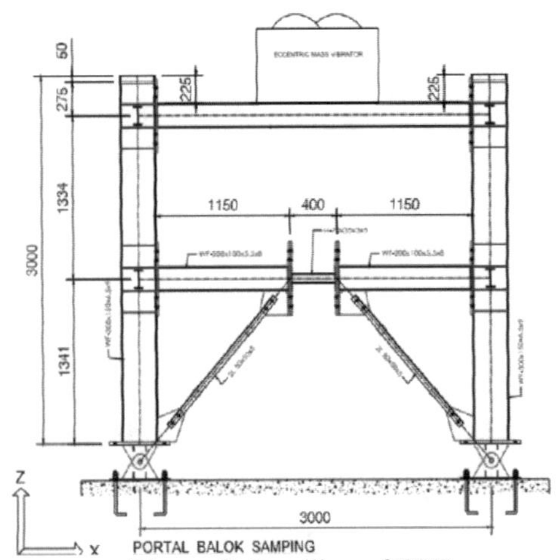

Figure 3. Detailing of EBF.

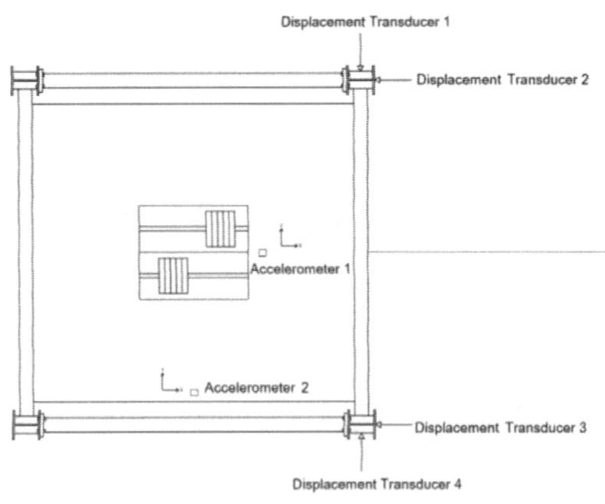

Figure 4. The sensors.

3. Results and discussion

IOP Conf. Series: Materials Science and Engineering 1041 (2021) 012005 doi:10.1088/1757-899X/1041/1/012005

3.1. Test Results of RBS

The structure was induced by sinusoidal forces twice. Firstly, the force frequency was increase in large increment and secondly, the frequency was increased smoothly near to the natural frequency of the structure. All recorded responses were the acceleration in time-domain which then were converted into a frequency domain based on the Fast Fourier Transformation (FFT) by SIGVIEW software. The natural frequency of the structure was obtained by looking at the peak frequency of the Frequency Response Function (FRF). Figure 5 show on example of recording from the first test in time domain and the results after converting to frequency domain is presented in figure. The measurement from two test were compared and presented on Table 1. As can be seen the measurement of accelerometer and vibrometer was very close and the average natural frequency from the test was 3.5933Hz.

There are two methods to determine the damping ratio, using the half-power bandwidth method and logarithmic decrement method. The results from the two methods are slightly different. The average damping ratio from two measurements using the logarithmic and the half-power bandwidth is 3.149% and 3.219% respectively.

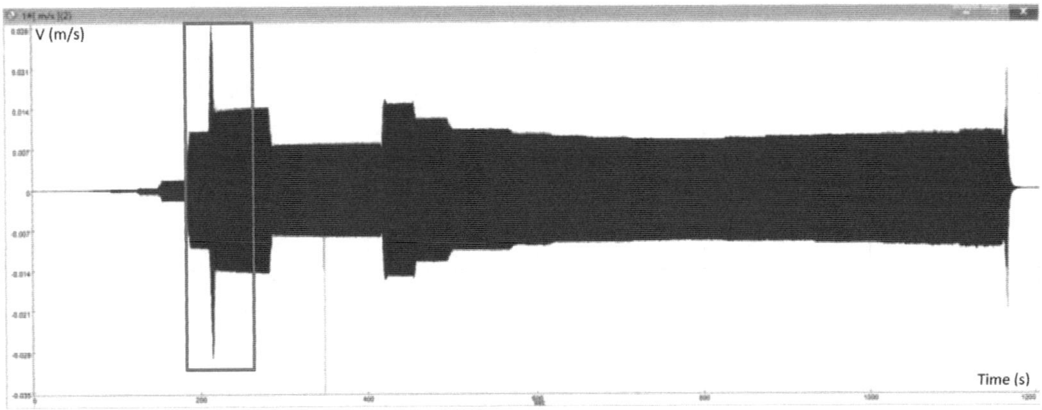

Figure 5. Velocity vs time response obtained from the vibrometer in the first test

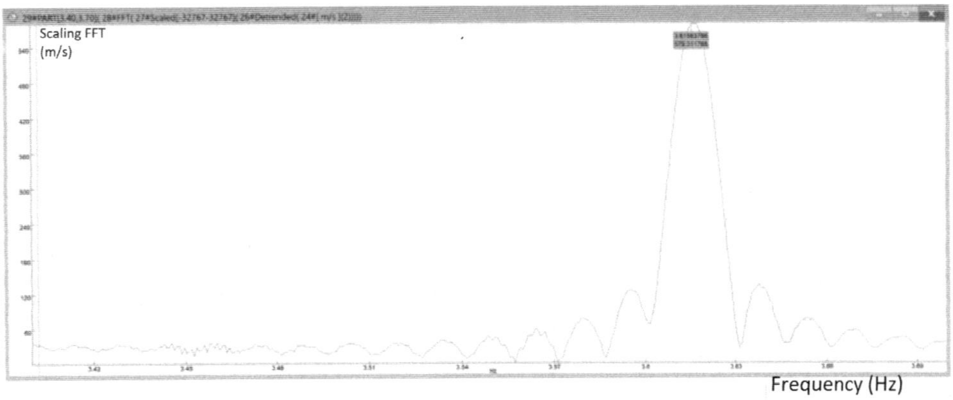

Figure 6. FFT results in structure responses in a function of the frequency of the first test

Table 1. The test results of RBS.

Test	Frequency (Hz)		Damping Ratio (%)	
	Vibrometer	Accelerometer	Vibrometer	Accelerometer
1st Test	3.5625	3.6168	3.553%	3.0810%
2nd Test	3.6000	3.5939	3.553%	2.9645%
The average	3.5812	3.6053	3.4195%	3.0138%

IOP Conf. Series: Materials Science and Engineering 1041 (2021) 012005 doi:10.1088/1757-899X/1041/1/012005

The dynamic FE analysis using OpenSees software was conducted for further analysis. Non-linier beam column element was used to model the column and the beam. Concentrated plastic hinge was assigned to represent the RBS. Properties of plastic hinge were defined by rotational spring based on Ibarra Modified Krawinkler (IMK) [2] to describe the nonlinear behaviour of the frame.

The dynamic load applied in the FE analysis was selected from the test which produced the maximum displacement respond as 0.618mm. The part is shown in figure 5 which is in the red box. The time domain forces were then converted to the frequency domain to get the frequency of the force. The value was 3.6111Hz based on the first test, then the sinusoidal forces with similar frequency was applied to the FE model. The displacement responds based on the FE analysis was 0.619mm which was similar to the test results as 0.618mm. It can be concluded that the FE analysis with OpenSees can imitate the dynamic test very well.

Further analysis was carried out to get an inelastic respond of the RBS. The load was applied with a frequency close to RBS natural frequency which was 3.62 Hz. This value was equal to an eccentricity of EMS (MR) as 40 kg-m or equivalent to lateral load as 20.71kN. The result was presented on figure 8. As can be seen, the RBS reached the plastic zone where the ultimate load was much higher than the force applied (20.17kN). It can be explained that the RBS experienced resonance since its natural frequency was equal to the frequency of the external forces.

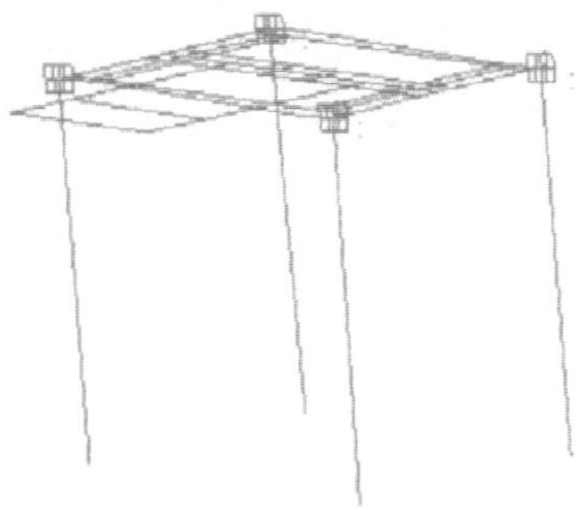

Figure 7. The FE model of RBS

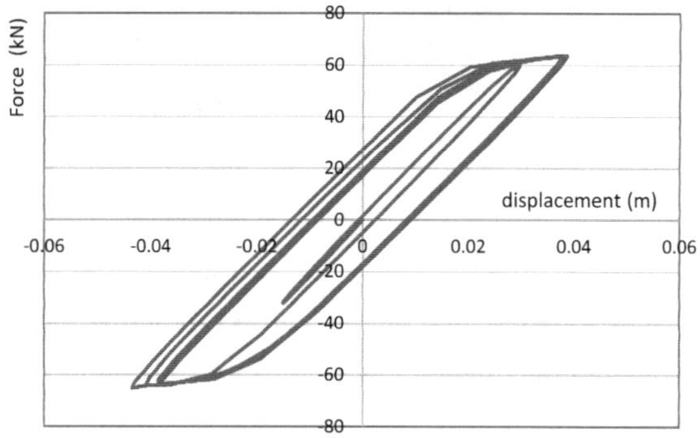

Figure 8. Non- linier response of RBS based on FEA

3.2. Test Results of EBF

Similar with the first test, the EBF was test twice, under free vibration and force vibration. Figure 9 and 10 show the reading of accelerometer #1 under both conditions. The signal obtained in time domain was then converted to frequency domain. The natural frequency and the period of the EBF based on accelerometer #1 was 6.049 Hz and 0.165s, respectively. Using the half-power bandwidth method, the calculation of damping ratio is presented in figure 11 which is equal to 2.06%. The summary of natural frequency and damping ratio based on measurement from accelerometer #1, #2 and #5 and the vibrometer is listed on Table 2. The average natural frequency is around 6.09Hz whereas the damping ratio is about 2.04%.

IOP Conf. Series: Materials Science and Engineering 1041 (2021) 012005 doi:10.1088/1757-899X/1041/1/012005

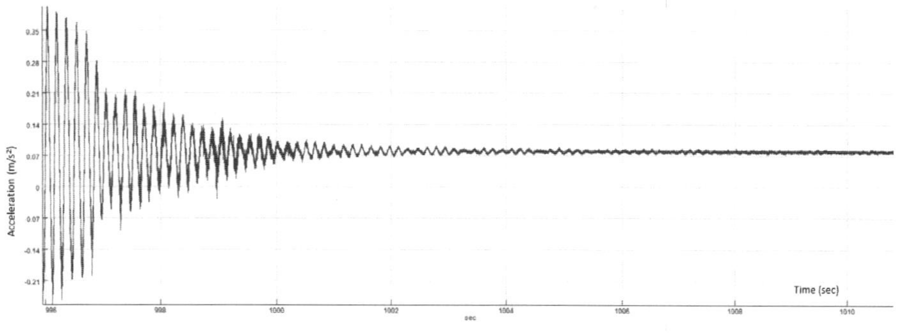

Figure 9. Accelerationvs time from accelerometer #1, under free vibration

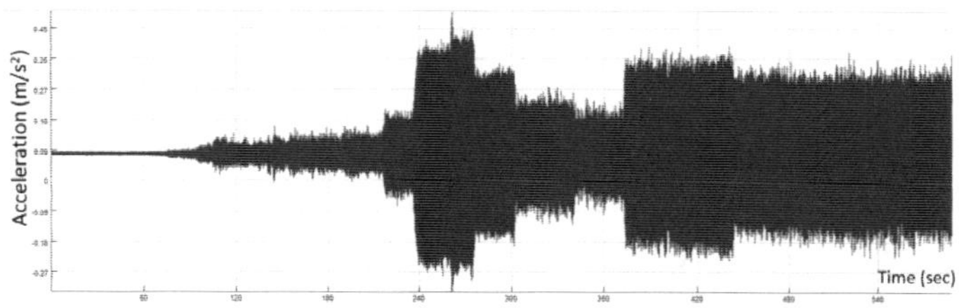

Figure 10. Accelerationvs time from accelerometer #1, under forced vibration

The FE analysis using OpenSees was carried out as a comparison. According to the experiment, the EBF was categorized as a flexural link. Steel02 was selected for steel material which is a uniaxial Giuffre-Menegotto-Pinto model with isotropic strain hardening. Truss element was assigned for bracing. Beam and column used the NonlinearBeamColumn element that assumes plastic deformation to be distributed along the element. Link was defined with BeamWithHinges command, where the plasticity is concentrated at the end of link.

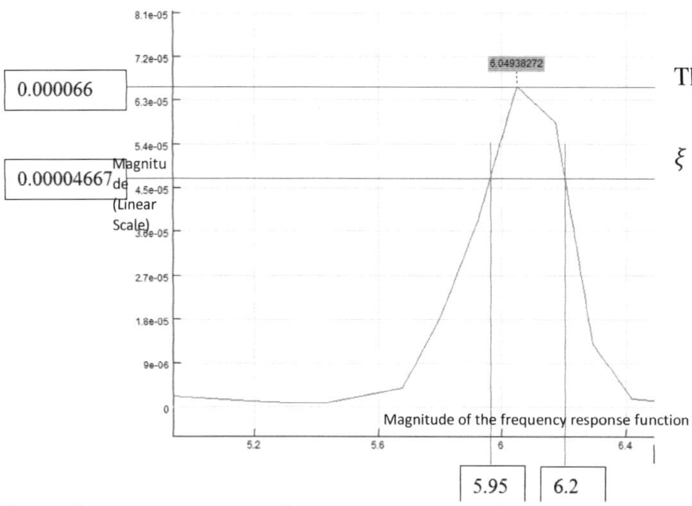

The damping ratio:

$$\xi = \frac{f_B - f_A}{2f_r} = \frac{6.2 - 5.95}{2x6.04938} = 0.0206 = 2.06\%$$

Figure 11. The calculation of damping ratio based on accelerometer #1 under free vibration conditions using the Half-Power Bandwidth Method.

Table 2. The test results of EBF.

	Frequency (Hz)	Damping Ratio (%)
Accelerometer #1	6.049	2.06%
Accelerometer #2	6.111	2.04%
Vibrometer	6.111	2.04%

Similar loading condition as applied on the experiment was used. The excitation with a light load of 0.11625 kgm was induced with gradually increasing frequency. The structural response was in the elastic stage. The displacement reading from the vibrometer was used to compare the FEA and the experiment. Figure 12 shows the vibrometer response where the minimum and the maximum displacement was 0.73 mm and 0.98 mm, respectively thus having a displacement of 0.855 mm. The FE results is shown in figure 13. The displacement obtained is almost similar to the experiment which is 0.853 mm.

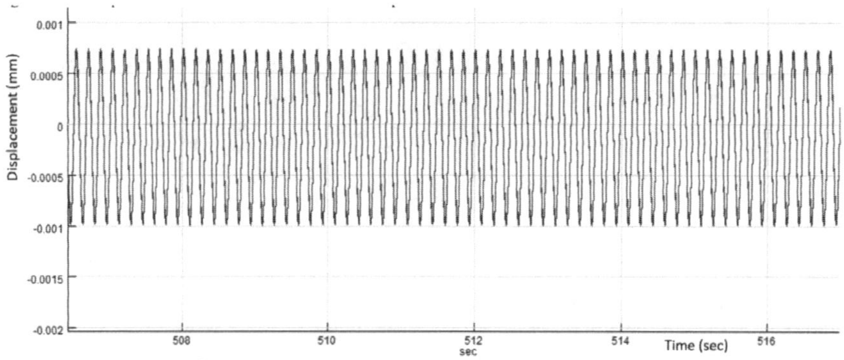

Figure 12. Displacement vs Time from vibrometer response

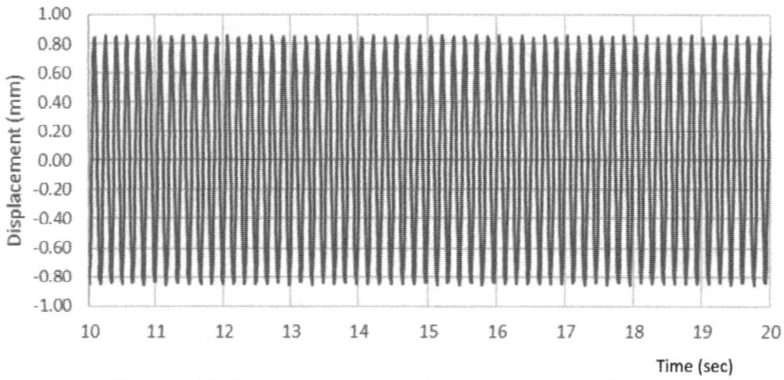

Figure 13. Displacement vs Time from FE Analysis

4. Conclusion

The force vibration test can be conducted by using the EMS. The linier dynamic response obtained from the test are the natural frequency and the damping ratio. The natural frequency of the RBS steel frame is about 3.614 Hz which is similar to the FE analysis as 3,623 Hz. Meanwhile the damping ratio is 3.149%. Similar displacement response was attained from the test and the FE analysis.

The natural frequency of the structure gain from the test is around 6 – 6.133 Hz, while the FE analysis result is 6.13 Hz. Moreover, the damping ratio based on vibrometer and accelerometers measurements is between 2.04 – 2.06%. The elastic displacement obtained from the experiment is 0.855 mm which is almost identical to the displacement from the FE analysis as 0.853 mm. The dynamic response conducted using FE software OpenSEES is potent in modeling a steel RBS and EBF structure with harmonic excitation.

Acknowledgement

Universitas Indonesia supports this work through Hibah PUTI, contract number: NKB-1186/UN2.RST/HKP.05.00/2020

References

[1] Sophianopoulos D S and Der A E 2011 Parameters affecting response and design of steel moment frame reduced beam section connections: An overview*Int. J. Steel Struct.***11** 133–44

[2] Lignos D G and Krawinkler H 2011 Deterioration Modeling of Steel Components in Support of Collapse Prediction of Steel Moment Frames under Earthquake Loading *J. Struct. Eng.***137** 1291–302

[3] Kildashti K, Mirghaderi R and Kani I M 2012 The Efficiency of Reduced Beam Section Connections for Reducing Residual Drifts in Moment Resisting Frames *Open J. Civ. Eng.***02** 68–76

[4] Deri A E and Sophianopoulos D S 2015 Parametric Analysis and Optimization of Reduced Beam Section Steel Frame Connections *Conference: 7th European Conference on Steel and Composite Structures at: Naples, Italy*

[5] Sofias C E, Kalfas C N and Pachoumis D T 2014 Experimental and FEM analysis of reduced beam section moment endplate connections under cyclic loading *Eng. Struct.***59** 320–9

[6] Sophianopoulos D S and Deri A E 2017 Steel beam-to-column RBS connections with European profiles: I. Static optimization *J. Constr. Steel Res.***139** 101–9

[7] Berman J W and Bruneau M 2007 Experimental and analytical investigation of tubular links for eccentrically braced frames *Eng. Struct.***29** 1929–38

[8] Kasai K and Popov E P 1986 Cyclic Web Buckling Control for Shear Link Beams *J. Struct. Eng.***112** 505–23

[9] Morassi A and Tonon S 2008 Dynamic Testing for Structural Identification of a Bridge *J. Bridg. Eng.***13** 573–85

[10] Conte J P, He X, Moaveni B, Masri S F, Caffrey J P, Wahbeh M, Tasbihgoo F, Whang D H and Elgamal A 2008 Dynamic Testing of Alfred Zampa Memorial Bridge *J. Struct. Eng.***134** 1006–15

[11] Hogan L S, Wotherspoon L, Beskhyroun S and Ingham J 2016 Dynamic Field Testing of a Three-Span Precast-Concrete Bridge *J. Bridg. Eng.***21** 06016007

[12] Setareh M and Gan S 2018 Vibration Testing, Analysis, and Human-Structure Interaction Studies of a Slender Footbridge *J. Perform. Constr. Facil.***32** 04018068

[13] Halling M W, Muhammad I and Womack K C 2001 Dynamic Field Testing for Condition Assessment of Bridge Bents *Struct. Eng.***127** 161–7

IOP Conf. Series: Materials Science and Engineering 1041 (2021) 012006 doi:10.1088/1757-899X/1041/1/012006

Behaviour of Cold-formed Steel Back-to-Back Channel Built-Up Sections under Bending

R A Fitrah[1], A P Melinda[2]

[1]Civil Engineering Department, Universitas Dharma Andalas
[2]Civil Engineering Department-Engineering Faculty, Universitas Negeri Padang

Corresponding author: ridho.af@unidha.ac.id

Abstract. Cold-formed steel (CFS) structures have been widely constructed as main elements in light structures. In terms of increasing capacity, the built-up sections of cold-formed steel are used regarding larger span structure.In this study, nine cold-formed steel with back-to-back built-up section specimens with various thicknesses were experimentally investigated under three points bending scheme to obtain their capacity and failure type. Each specimen was assembled by using self-drilling screws in web of two built-up channels.Experimental results show that three various web thicknesses (1.3 mm, 1.5 mm, and 2.00 mm) in cold-formed steel of built-up section give the different moment capacity up to 22%. Then, failure types in all specimens are similar where local buckling is generated in compression flange around loading area and lateral-torsional buckling occurred along span. It is found that web thicknesses in back-to-back section have contribution to increase the stiffness and torsional rigidity. However, there is phenomenon when the load capacity decreased, the back-to-back sections act non-composite. Then, the failure apparently observed on that section is one-side twisting and web crippling in mid-span.

1. Introduction

The application of cold-formed steel (CFS) structures for residential and commercial buildings has been widely developed as main structures that accommodate the internal forces. Furthermore, built-up sections of cold-formed steelare used to uplift its capacity in torsional, compression, and tension. Besides that, connecting two sections is easily made by using self-tapping screws. Built-up sections are always used as compression members. Compared to a single section of cold-formed steel, the rigidity and capacity of built-up sections have been proven increased about 20%-50%[1,2,6], one is back-to-back channel sections. These sections were categorized as open sections which are assembled by connecting both webs. There were studies that revealed back-to-back sections are used as the members for long-span roof trussand many parameters govern their axial capacities and rigidity, including fasteners [2].

However, there is an issue concerning the built-up back to back channelsections when they are designed as flexural member. Most of cold-formed steel sections, which are subjected to bending, experience local and lateral-torsional buckling. These buckling dominantly happened due to beam span and web depth to flange width (d/b) ratio [3,4,5]. In Indonesia, the provision (SNI 7971:2013) consist of moment capacity calculation for double sections but it is not giving the specific equation relate to the failure type parameters for back-to-back instead of providing correction by case. It can be concluded that there is still little comprehension in terms of determining flexural capacity of back-to-back sections. On the other hand, back-to-back sections are futuristic to be developed as the main structure specifically as flexural member [7].

In order to understand the structural behavior of cold-formed steel with back-to-back channel built-up sections under bending, this paper was presented. Furthermore, this study is focused to investigate the flexural capacity and the failure that shown along the beam specimens by doing experimental in three point bending scheme. The results of specimen with several thicknesses which are commercially used as common structures in residential buildings are shown and discussed.

2. Materials and methods

2.1 Test Specimens

The experiment is conducted by testing 9 specimens of back-to-back channel sections which have various thicknesses, each section has 3 specimens. Generally, the cold-formed steel profiles used in this test are single-channel section C 75.75, C 75.80, C 75.100 which has thickness respectively 0.75 mm, 0.80 mm, and 1.00 mm; the sections are shown in figure.1. Then, these sections were assembled by using screws in web; the details of specimen are shown in table 1.The abbreviation of BJR mentions about the specimen which has the same properties of material. Each specimen was tested by simply-supported beam with nominal length 65 cm and load scheme is three points bending.

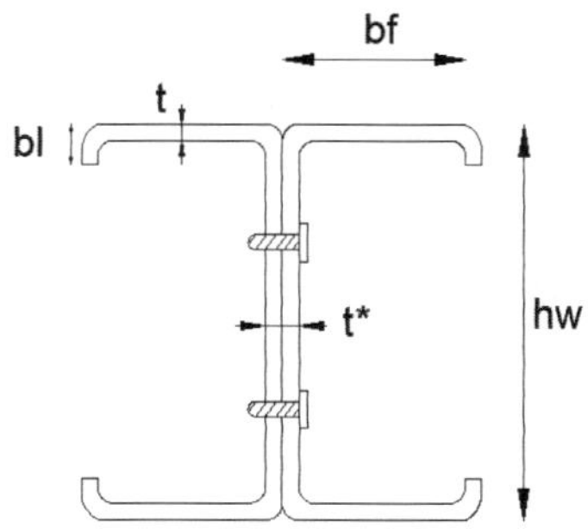

Figure.1. Section of Specimen

Table.1. Measured Dimensions of Specimens.

Specimen	Lip, b_l (mm)	Flange,b_f (mm)	Web Depth, h_w (mm)	Thickness, t (mm)	Web Thickness, t* (mm)
BJR-1-65	3.0	32	75	0.65	1.3
BJR-2-65	3.0	32	75	0.65	1.3
BJR-3-65	3.0	32	75	0.65	1.3
BJR-1-75	3.70	35	75	0.75	1.5
BJR-2-75	3.70	35	75	0.75	1.5
BJR-3-75	3.70	35	75	0.75	1.5
BJR-1-100	3.50	36	75	1.00	2.00
BJR-2-100	3.50	36	75	1.00	2.00
BJR-3-100	3.50	36	75	1.00	2.00

2.2 Experimental Set Up

The bending tests were conducted by universal testing machine (UTM) to investigate the moment capacity and the failure modes of cold-formed steel back-to-back channel built-up section beams. The three-point bending tests were loaded at the mid-span to the load transfer plate through round bar, as shown in figure.2 and figure 3. The result would be generated automatically through computer so-as the load-vertical displacement curve will be obtained for all specimens.

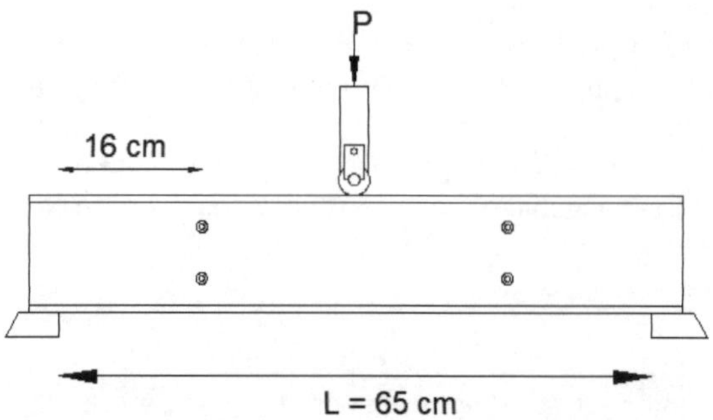

Figure.2. Loading of Specimen in Three-Point Bending

Figure.3. Typical Experimental Setup for Specimens

3. Results and discussion

3.1 Experimental Results

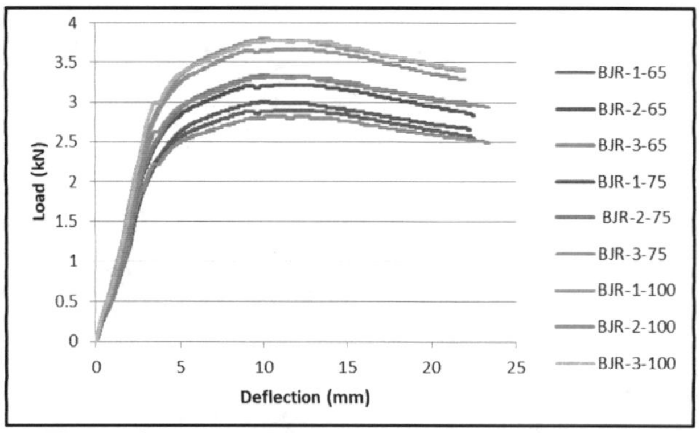

Figure.4. Typical Experimental Setup for Specimens

The graph of load-deflection which is generated from UTM was obtained and shown in figure 4. Generally, the graph shows that each specimen has slightly different stiffness for every thickness. The cold-formed steel built-up section with web thickness (t*) 2.00 mm (BJR-100) has the highest stiffness value compared to others. Moreover, these specimens have the highest load capacity. Moment capacity of these built-up sections is calculated by the simple formula for moment calculation in simple beam subjected one middle-point load and they are shown in table 2. It can be seen that the experimental moment capacity for specimen BJR-100 is higher than BJR-65 and BJR-75 about 21.9 % and 10.7 % respectively. Furthermore, the failure mode is observed having similar type for these specimens,local buckling and lateral torsional buckling.

Table.2. Experimental Moment Capacity and Failure Mode.

Specimen	M_{Exp} (kN.m)	Avg M_{Exp}	Failure Mode
BJR-1-65	0.468		L+ LTB
BJR-2-65	0.488	0,470	L+ LTB
BJR-3-65	0.455		L+ LTB
BJR-1-75	0.539		L+ LTB
BJR-2-75	0.536	0.537	L+ LTB
BJR-3-75	0.536		L+ LTB
BJR-1-100	0,598		L+ LTB
BJR-2-100	0,604	0.602	L+ LTB
BJR-3-100	0,604		L+ LTB

L : Local Buckling LTB: Lateral Torsional Buckling

3.2 Local Buckling

Local buckling failure in built-up specimen BJR-75 can be seen in figure 5(a). Apparently, the local buckling generated in compression flange due to load concentration around that area. However, the local buckling exactly happened when inelastic zone is reached,it is shown in figure 5(b). The other specimens have the similar phenomenon when local buckling is generating in compression flange. It can be stated that thickness (0.65 mm, 0.75 mm, and 1.00 mm) in each specimen contribute to increase the rigidity in compression flange before local buckling is generated.

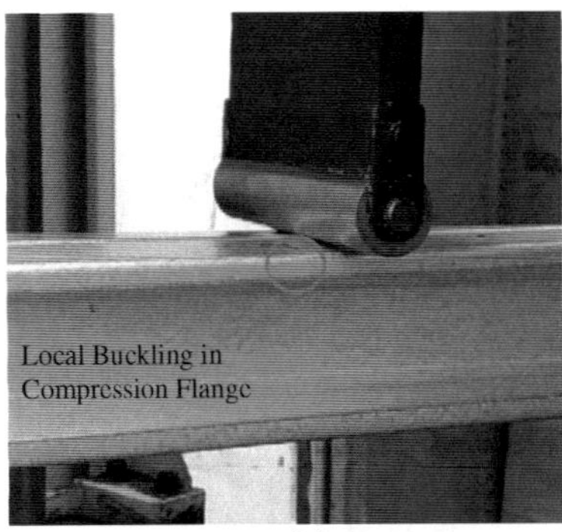

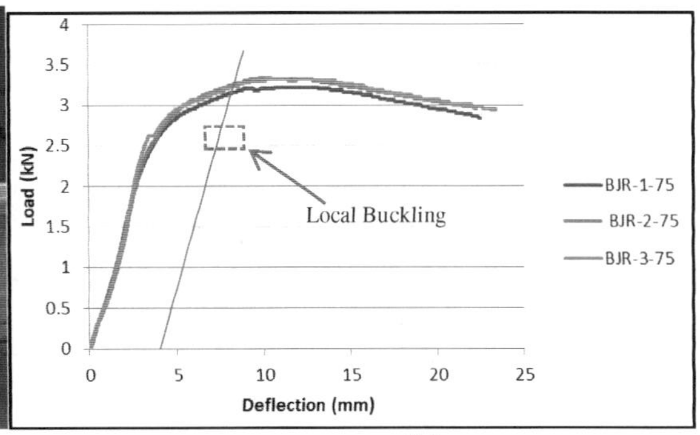

(a) (b)

Figure.5. (a) Local Buckling in Compression Flange of BJR-75 (b) Local Buckling Point

3.3 Lateral-Torsional Buckling

Lateral-torsional buckling is the common phenomenon that happens in cold-formed steel due to thin web. Figure 6(a) shows that the lateral-torsional buckling failure generated along the span. It also confirms that all specimens have similar pattern in terms of lateral deflection and twisting in anti-clockwise. However, the load capacity of these built-up sections still gradually increases until it reach the ultimate load capacity. Increasing thickness whether at web or at compression flange apparently leads higher torsional rigidity and it may cause reduce the angle of twist before it continues increasing the deflection after ultimate load capacity is reached.

(a)

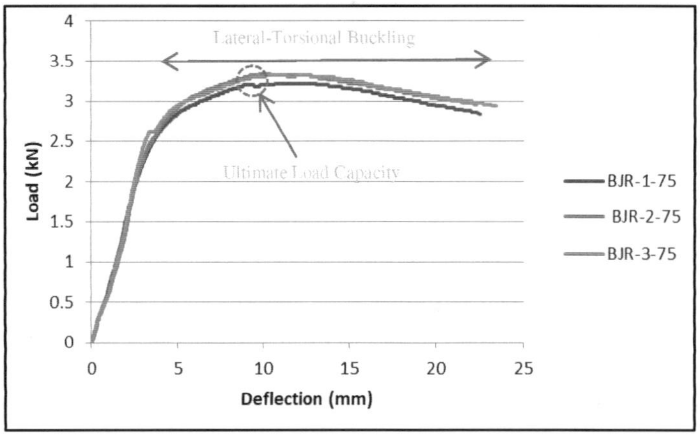

(b)

Figure.6. (a) Lateral-Torsional Buckling along Span (b)Ultimate Load Capacity

Figure.7. Failure after Ultimate Load Capacity

In figure 6 (b) shows that the load starts to decrease while the back-to-back sections start to act non-composite. It is observed in figure 7 that the continuous failure due to twisting happens in just only one side section. Then, when the deflection continues to increase, the web in mid-span generate the crippling failure.

4. Conclusions

This research aimed to comprehend the bending behavior of cold-formed steel of back-to-back built-up section with various web thicknesses. Based on theexperimental results, there are conclusions can be listed as follows if there is the development in terms of bending capacity:

1. Increasing web thickness in cold-formed steel of back-to-back built up sections will increase the moment capacity about 10%-20%.
2. The observed failures for all specimens are local buckling and lateral torsional buckling.
3. Effects of web thickness in all specimens are increasing the stiffness, compression flange and torsional rigidity as long as local buckling and lateral torsional buckling is generated along section and span.

References

[1] R.A. Fitrah and H. Herman, "Studi Eksperimental Perilaku Tekan Baja Ringan Dengan Variasi Profil Penampang" Rang Teknik Journal UMSB., vol. 2, no. 1, pp. 127–131, 2019.

[2] R.A. Fitrah and R. Nofriyandi "The Performance between Two Long-span Roof Trusses of Cold Formed Steel with Back to Back Channel Section " *IOP Conf Ser.: Mater.Sci.Eng* 846 012060, 2020.

[3] L.Wang and B. Young, "Behaviour of Cold-Formed Steel Built-Up Sections with Intermediate Stiffners under Bending. I : Test and Numerical Validation" *J. Struct. Eng.,*pp 04015150-104015150-9, 2016.

[4] L. Wang and B. Young, "Behaviour of Cold-Formed Steel Built-Up Sections with Intermediate Stiffners under Bending. II : Parametric Study and Design" *J. Struct. Eng.,*pp 04015151-1 04015151-11, 2016.

[5] N.DKankanamge and M. Mahendran, "Behaviour and design of cold-formed steel beams subject to lateral torsional buckling" *Thin-walled Structures.*, vol. 51, pp. 25–38, 2012.

[6] F.JMeza, J. Becque, and I. Hajirasouliha, "Experimental study of cross-sectional capacity of cold-formed steel built-up columns" *Thin-walled Structures.*, vol. 155, pp. 1–18, 2020.

[7] K.J.R.Rasmussen, M.Khezri, B.W. Schafer, and H. Zhang, "The mechanics of built-up section cold-formed steel members"*Thin-walled Structure.*,vol.154,pp 1-42, 2020.

Effect of Missing Rainfall Data on the Uncertainty of Design Floods

N Nurhamidah[1], S Hanwar[2], A Junaidi[1], Februarman[1], Sunaryo[1]
[1]Department of Civil Engineering, Faculty of Engineering, University of Andalas, Padang, Indonesia
[2]Department of Civil Engineering, Padang State Polytechnics, Padang, West Sumatra, Indonesia

Corresponding author: nurhamidah@eng.unand.ac.id

Abstract. Hydrological data is a collection of information or facts about the hydrological phenomenon. Mistakes in the hydrological analysis have a fatal effect on the accuracy or accuracy of hydraulic building planning, whether in the form of planning for irrigation buildings, flood control structures, drainage systems, weirs. And others hydraulics structure design. As we already know, rainfall data is the most important data in the hydrological analysis. The minimum amount of rainfall data required in the hydrological statistical analysis is 10 years. To obtain this 10-year data, often encountered incompleteness in recording the rainfall data. Meanwhile, hydrological analysis always required for any design of hydraulic structures. Based on the above statement, an analysis has been carried out on how much influence the completeness of data and availability of rainfall data in an area of accuracy in predicting the discharge plan in different times and make a Rating Curve. In determining the unit hydrograph, the Nakayasu method is used. The results of this study indicate that the completeness of the data is not very influential. This is evidenced by obtaining the discharge value at a certain return period which is not much different. From the unit hydrograph curve, it can be seen that in the first condition with a return period of 100 years, a discharge of 801.775 m3 / s is obtained, for the second condition consisting of 11 rainfall data, a discharge of 833.848 m3 / s is obtained, and for the third condition, ignoring the rainfall data. The missing of rainfall data obtained a discharge of 816,489 m3 / s. Where the difference inaccuracy in the first condition is 3.846% against the second condition, and the difference inaccuracy in the third condition is 2.082% for the second condition. In the rating curve section of Batang Kuranji, the normal water level discharge is 169.948 m3 / s with an elevation of 2.89 m. At the same time, the maximum discharge at the cross-section of Batang Kuranji is 748,260 m3 / s with an elevation of 6.99 m.

1. Introduction

Mistakes in the hydrological analysis have a fatal effect on the accuracy or accuracy of hydraulic building planning, in the form of planning for irrigation buildings, flood control structures, drainage systems, weirs, others design of hydraulics structures. As we already know, rainfall data is the most important data in the hydrological analysis. The minimum amount of rainfall data needed in hydrological statistical analysis is 10 years [1]. To obtain this 10-year data, incompleteness is often encountered in recording the rainfall data. Meanwhile, hydrological analysis is always required in any water construction planning.

Due to data limitations and incomplete data availability, it is necessary to research to analyze how much influence the completeness of the data and the availability of rainfall data in an area on the accuracy of the data to predict the planned discharge in different return periods and to make a Rating Curve of the Kuranji Batang cross-section.

Rainfall data recording at rain stations are not always complete. Then it is necessary to estimate the lost rainfall data. This estimate can be calculated using the inverse square distance method. The following is the formula for the inverse fair distance method.

$$R_A = \frac{\frac{1}{(d_{AB})^2}R_B + \frac{1}{(d_{AC})^2}R_C}{\frac{1}{(d_{AB})^2} + \frac{1}{(d_{AC})^2}} \quad (1)$$

Which is:

R_A	= Rainfall data at station A (mm)	R_B	= Rainfall data at station B (mm)
R_C	= Rainfall data at station C (mm)	d_{AB}	= Distance from A to B (km)
d_{AC}	= Distance from A to C (km)		

The average rainfall can be calculated using several methods:

1.1. Algebra Average Method

According to [2,3], this method is the simplest one and is obtained by calculating the arithmetic average of all total rain gauges in all areas. This method can be used in areas that are relatively flat and have a large number of rain stations, assuming that the rainfall in the watershed is even. The following equation can formulate the algebraic mean method:

$$R_{average} = \frac{R_1+R_2+R_3+\cdots+R_n}{n} \quad (2)$$

$R_{average}$ = Average rainfall (mm)
n = number of data
R_1, R_2, R_3, R_n = Rainfall data at point 1, 2, 3, n (mm)

1.2. Polygon Thiessen Method

The Thiessen Polygon method is suitable for areas with uneven precipitation measurements requiring observation stations in and near the site, and this method does not take topography into account. The following equation can formulate the Thiessen Polygon method:

$$R_{average} = \frac{A_1xR_1+A_2xR_2+\cdots+A_nxR_n}{A_1+A_2+\cdots+A_n} \quad (3)$$

After calculating the average rainfall with both methods. Furthermore, error checking is carried out on each process to determine which way is the best. Error checking can be calculated using the following equation:

$$Error = \frac{\sum_{i=1}^{n}\frac{|\bar{R}-R_i|}{R_i}}{n} \quad (4)$$

R_i	= Rainfall data at i[th] station	$\bar{R}$	= Average rainfall (mm)
n	= Number of station		

1.3. Design Rainfall

To determine the designed rainfall can be calculated by several methods, i.e. the Normal distribution method, Log-Normal distribution, Gumbel distribution, and Log Pearson Type III distribution. In this study, the calculated rainfall design was carried out using four methods. After that, a dispersion test is carried out to select an approach that meets the requirements. Where the calculated values that meet the criteria can be seen in table 1 below:

Table 1. Distribution Limit Value Requirements.

Distribution Methods	Requirement	Distribution Methods	Requirement
Normal	Cs = 0 Ck = 3	Log Pearson Type III	Cs ≅ 0
Gumbel	Cs ≤ 1,1396 Ck ≤ 5,4002	Log-Normal	$Cs = 3Cv + Cv^2$ Ck = 5,383

In this research, the value obtained meets the boundary conditions is the Gumbel distribution method. Then the results of the calculated rainfall plan used for further calculations are the results of the computed rainfall plan using the Gumbel distribution method.

Furthermore, the distribution harmony test was carried out using the Chi-square method, where the Chi-Square test is used to test the vertical deviation [4]. In the calculation of the unit hydrograph, the method used is the Nakayasu method. The following are the steps for calculating the unit hydrograph using the Nakayasu method [5].

2. Materials and methods

Rainfall data collection was obtained by accessing data from BMKG and river sections taken from the journal Flood Control Analysis in the Batang Kuranji River Basin [6]. The location and the cross-section of the Kuranji river can be seen in figure 1.

The rainfall data was taken from the recording of three rain stations, which is located at Gunung Nago station, Batu Busuk station, and Ladang Padi station. The three rain stations are the closest stations to the observed river basin. The maximum daily rainfall data used is the maximum daily rainfall data per month.

In this project, the rainfall data is divided into three conditions. The first condition is the rainfall data for 20 years from 1994-2013, which is equipped with an estimated rainfall loss. The second condition is the rainfall data which is also complemented with the estimated loss of rain, namely from 1994-2004, which consists of 11 data. The third condition, rainfall data regardless of the lost rainfall, namely from 1996-2004, 2006-2007, and 2009-2013 which consisted of 16 data.

According to the literature review that has been carried out, the hydrological analysis that needs to be done in relation to estimating lost rainfall data is as follows:

1. Determine the average rainfall using the Average Algebra method and the Thiessen Polygon method.
2. Rainfall analysis plan using four methods, namely the Normal distribution method, Log-Normal distribution, Log Pearson Type III distribution, and the Gumbell distribution method.
3. Calculation of the debit plan using the Rational Method.
4. Determine the unit hydrograph using the Nakayasu method.
5. Determine the percentage difference in discharge accuracy.
6. Creating a Rating Curve.

IOP Conf. Series: Materials Science and Engineering 1041 (2021) 012007 doi:10.1088/1757-899X/1041/1/012007

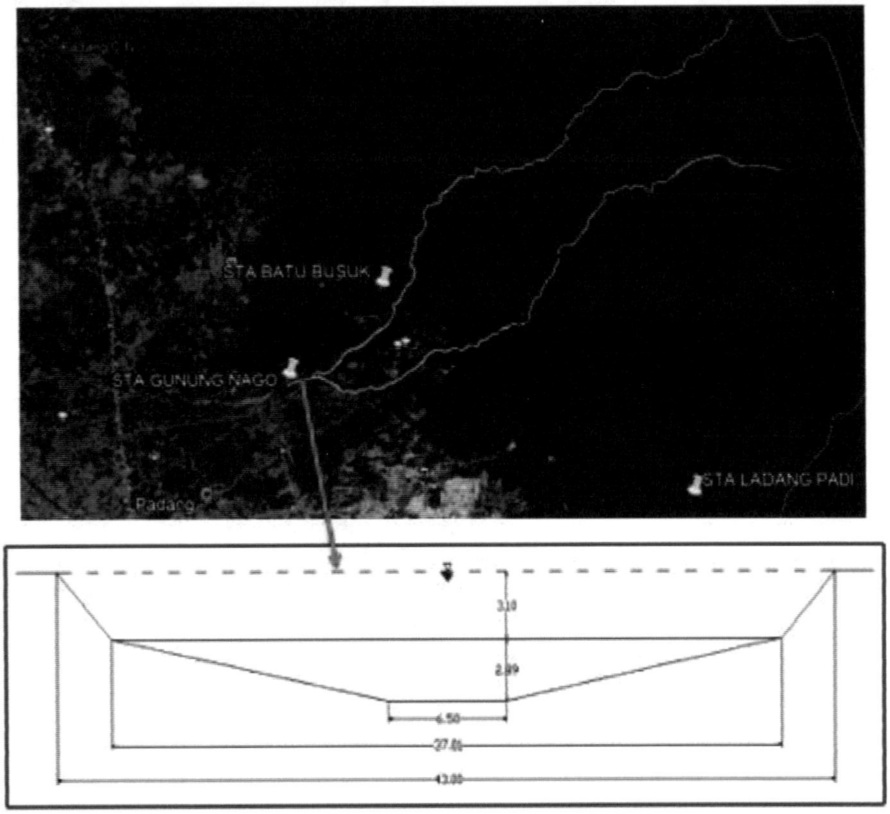

Figure 1. The study area and river cross-section.

3. Results and discussion

3.1 Rainfall Frequency

The calculation of the frequency of rainfall is used to determine the factors required in determining the type of rainfall distribution at the study location. In this calculation, the value of Standard Deviation (S), Skewness Coefficient (Cs), Kurtosis Coefficient (Ck), and Variation Coefficient (Cv) will be obtained that meet the requirements of each distribution.

In selecting the distribution for the first, second, and third conditions, the value that meets the requirements is the Gumbel distribution. Then the planned rainfall data used to determine the planned discharge is the result of calculations using the Gumbel distribution method. Next, the distribution harmony test was carried out using the Chi-square method for the first condition. The results of the conformity test calculation can be seen in table 2.

Table 2. Chi-square.

Interval P	Oi	Ei	(Oi-Ei)	$(Oi-Ei)^2$	$(Oi-Ei)^2/Ei$
P<119.613	5	4	1	1	0,25
119.613<P<139.627	7	4	3	9	2,25
139.627<P<159.640	3	4	-1	1	0,25
159.640<P<179.653	1	4	-3	9	2,25
179.653<P<199.667	4	4	0	0	0
Total	20	20	0	20	5

The conformity test is acceptable if the value (Oi-Ei) 2 / Ei ≤X2. From the above calculation, it is obtained (5≤7,815), then the Gumbel distribution used is acceptable. For the second condition, the suitability test is obtained (4,64≤5,991), then the Gumbel distribution used is acceptable. For the third condition, the conformity test is obtained (4≤7,815), then the Gumbel distribution used is acceptable.

In other calculations using the maximum daily volume per year, a more safety figure is obtained. Because the results obtained, that is a greater value than the calculation using the maximum daily rainfall per month. However, in this final project, the discharge value is obtained from a focused rainfall calculation using the maximum monthly rainfall data. In the discharge calculation with the two calculations, the percentage difference for each condition is obtained, namely 5% -19% for the first condition, 4% -9% for the second condition, and 11% -14% for the third condition in the return period of 2 years to 200 years.

3.2 Rating Curve

Rating Curve is a graph that shows the relationship between discharge data in the cross-section of the Kuranji river and water level.

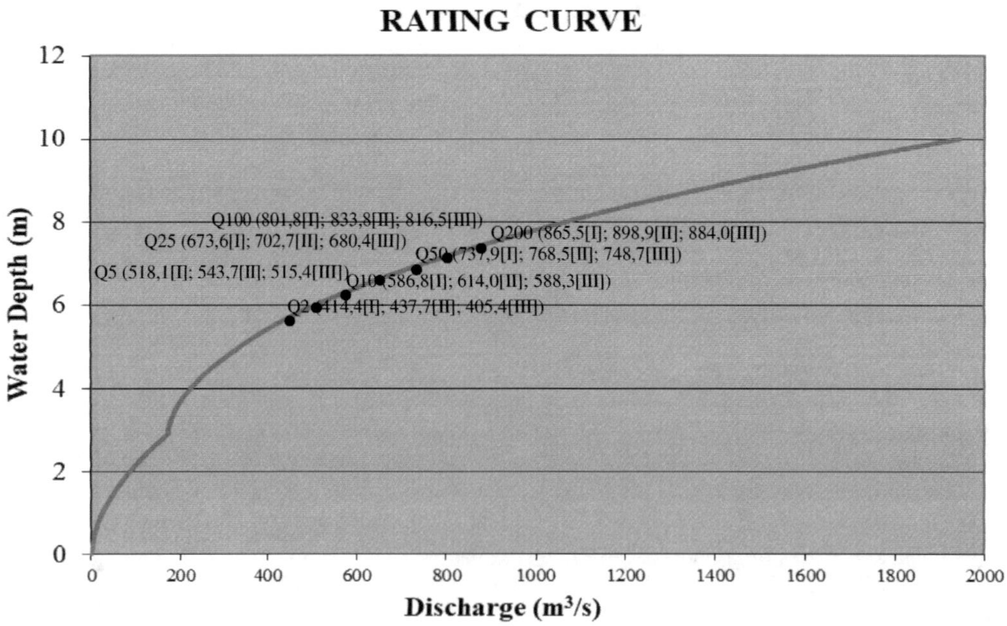

Figure 2. Rating Curve.

Which can be describe for 3 conditons, such as:

I = 1st condition, data was filled by 20 years (1994-2013)

II = 2nd condition, data was filled by 11 years. (1994-2004)

III = 3rd condition, missing rainfall data was neglected and only 16 data available from 1994-2013

The rating curve above is obtained from the calculation of the cross-sectional discharge of Batang Kuranji with a height value per 0.5 meter

Table 3. Discharge and water level relationships measurement.

h (m)	Q (m^3/s)	h (m)	Q (m^3/s)
0	0	5	334.121
0.5	7.143	5.5	413.190
1	24.676	6	508.444
1.5	51.483	6.5	620.605
2	86.945	7	750.426
2.5	130.555	7.5	898.673
2.89	169.948	8	1066.116
3	170.981	8.5	1253.522
3.5	188.491	9	1461.653
4	222.083	9.5	1691.262
4.5	270.585	10	1943.097

From the rating curve, (figure 2), So from the return discharge time of 2, 5, 10, 25, 50, 100, and 200, the water level can be determined.

Table 4. Discharge and water level relationships calculation for 3 conditions.

Return period (T)	Condition 1		Condition 2		Condition 3	
	Q (m^3/s)	h (m)	Q (m^3/s)	h (m)	Q (m^3/s)	h (m)
T2	414.404	4.806	437.669	4.980	405.350	4.737
T5	518.111	5.566	543.734	5.748	515.420	5.547
T10	586.774	6.046	613.958	6.231	588.296	6.057
T25	673.530	6.626	702.686	6.814	680.375	6.670
T50	737.890	7.036	768.510	7.226	748.684	7.103
T100	801.775	7.428	833.848	7.617	816.489	7.515
T200	865.427	7.801	898.947	7.991	884.046	7.907

According to the calculation mentioned above, it shows that the discharge for the first condition is 801,775 m3 / s, for the second condition it is 833,848 m3 / s, and for the third condition it is 816,489 m3 / s. In determining the difference in discharge accuracy obtained, it is reviewed against the second condition because the second condition has more prevalent rainfall data than the first and third conditions. Where the difference inaccuracy in the first condition is 3.846% against the second condition, and the difference inaccuracy in the third condition is 2.082% for the second condition.

4. Conclusions

Based on the description in the analysis and discussion, it can be concluded that the effect of the availability of rainfall data on the planned discharge is not very influential. From the unit hydrograph curve, it can be seen that in the first condition with a return period of 100 years, a discharge of 801.775 m^3/s is obtained, for the second condition consisting of 11 rainfall data, a discharge of 833.848 m^3/s is obtained, and for the third condition by ignoring the rain data that is lost, it is obtained a debit of 816.489 m^3/s. Where the difference inaccuracy in the first condition is 3.846% against the second condition, and the difference inaccuracy in the third condition is 2.082% for the second condition. In the rating curve section of Batang Kuranji, the normal water level discharge is 169.948

m^3/s with an elevation of 2.89 m. In comparison, the maximum discharge at the cross-section of Batang Kuranji is 748,260 m^3/s with an elevation of 6.99 m.

Acknowledgement
Publication of this article is supported by the Faculty of Engineering, University of Andalas, and we would like to thank all those who have involved in this research.

References

[1]. Canchala-Nastar, T., Carvajal-Escobar, Y., Alfonso-Morales, W., Cerón, W. L., & Caicedo, E. 2019 Estimation of missing data of monthly rainfall in southwestern Colombia using artificial neural networks Data in brief 26 104517.

[2]. Hasyimzoem, E.F. 2019 Perbandingan analisis data curah hujan yang hilang menggunakan metode Reciprocal, Normal Ratio, dan Rata-rata Aljabar (Studi kasus: Curah Hujan Beberapa Stasiun Hujan Daerah Pringsewu).[In Indonesian]

[3]. Chowdhury, M. A. I., Kabir, M. M., Sayed, A. F., & Hossain, S. 2016 Estimation of rainfall patterns in Bangladesh using different computational methods (arithmetic average, Thiessen polygon and isohyet) J. Biodivers. Environ. Sci 8 43-51.

[4]. Arvind, G., Kumar, P. A., Karthi, S. G., & Suribabu, C. R. 2017 Statistical Analysis of 30 Years Rainfall Data: A Case Study In IOP Conference Series: Earth and Environmental Science (Vol. 80, No. 1, p. 012067).

[5]. Idfi, G., Yulistyorini, A., & Suryoputro, N. 2020 The comparative study of peak discharge at Ngotok watershed by using the method of SCS, Snyder and Nakayasu for flood control needs In IOP Conference Series: Materials Science and Engineering (Vol. 930, No. 1, p. 012075). IOP Publishing.

[6]. Syofyan, Z. 2017 Analisa pengendalian banjir pada Daerah Aliran Sungai Batu Busuk (Batang Kuranji) Kota Padang Jurnal Teknik Sipil ITP, ISSN 2354-8452 E-ISSN 2614-414X, 3(2).[In Indonesian]

IOP Conf. Series: Materials Science and Engineering 1041 (2021) 012008 doi:10.1088/1757-899X/1041/1/012008

Steady-state analysis of hybrid solar-wind power integration in 20kV distribution system

Syafii[1], H D Laksono[1], A Alfac[1]
[1]Electrical Engineering Department, Engineering Faculty, Universitas Andalas Padang, Indonesia

Corresponding author:syafii@eng.unand.ac.id

Abstract. This paper reports the steady-state performance of a grid-connected photovoltaic-wind power system in a 20 kV active distribution system. The system was initially designed to meet a typical of State Electricity Company (Perusahaan Listrik Negara, PLN) primary feeder in Indonesia. In this study, 200 kWp solar power capacity and 200 kW wind power capacity were selected in the simulation of power flow analysis to evaluate the better system states. The results of the study using four cases show that the better voltage profile is obtained in conditions where the hybrid system is placed on the bus, which has the lowest voltage and on the load bus, which farthest from the source. The presence of power injection from solar and wind power plants on low voltage (LV) buses has reduced the active power flow from the grid utility. All load bus voltages have above 0.95 p.u in these cases, already within PLN standard. The lowest level of losses of 11 kW and 50 kVAr are obtained in the case where hybrid Photovoltaic (PV) and Wind Turbine (WT) with capacitor integration in the distribution network. Power losses are secured far below 10% as the losses permitted according to the PLN standard. Therefore, the integration of PV and wind turbine into a 20 kV distribution network can improve the voltage profile and reduce total system losses.

1. Introduction

The resource of electrical energy in Indonesia is still dominated by fossil energy, especially petroleum. Considering the decreasing availability of fossil energy resources and increasing awareness of environmental preservation causes research related to new and renewable energy generation continues to increase. Among renewable energy sources, solar energy sources, and wind are the most widely developed and reliable alternatives [1]. The condition of Indonesia in the tropics has a vast solar energy potential of around 4.8 kWh/m^2/day or equal to 112,000 GWp [2] and offshore wind energy with an average wind speed intensity of 4.16 m/s [3], so it is feasible to be utilized as an alternative source of electrical power.

Solar and wind energy sources have become priority energy sources that will be developed to achieve the renewable energy target of 23% in 2025 and 31% in 2050 as Government Regulation no. 79 of 2014, concerning the national energy policy [4]. Also, the integration of renewable energy-based power plants in load centers can reduce power losses and increase the electricity network voltage [5]. However, the utilization of solar power plants and wind power is still technically unacceptable to become a more supplier to the electricity grid because its profile varies according to the intermittent or seasonal nature of these renewable energy sources. The speed of the wind decreases when the sun's resources are at its maximum, whereas the wind will usually be stronger when the sun's resources decrease as it is cloudy or cloudy. On the same day, there are different and opposite patterns in terms of potential wind and solar resources in many parts of Indonesia. Based on these conditions, the combined generating system or solar and wind hybrid systems can be the solution to these problems. The reliability of the system will increase when solar and wind power sources are shared or synergized [6].

The power flow is a basic steady-state analysis for the following investigation:

1. Obtain the information of voltage and its phase angles on each bus as well as active power flow and reactive power for each line.
2. Knowing the appropriate generation power in serving the load of an electric power system.
3. Electricity planning and development, where power flow studies provide information about the consequences of adding new loads, adding new generation, adding new substations, adding new transmission lines, interconnecting with other systems, and so on.
4. Determination of the imposition of electrical system equipment such as transmission lines and transformers in the present and future conditions.
5. Provide preliminary data for the calculation of short circuit faults, protection system coordination, harmonics studies, and power system stability studies.

The power flow calculation for transmission systems is generally balanced, while in distribution systems the calculation is unbalanced. Unbalanced distribution system power flow has to consider the presence of the single, and two phases lateral as well as unbalanced system loading. Some methods have been used to improve the algorithm of distribution power flow calculation [7] and continue so now. Some methods have been implemented in unbalanced load flow computation using symmetrical component-based methods and the forward and backward based methods. The forward and backward methods in solving distribution power flows such as OpenDSS [8] and RDAP are public domain software now days. Meanwhile, the symmetric component-based method has been developed by other researchers as published in the paper [9]. The Newton Raphson based methods in [10] are used in this study with additional PV and WT models.

Solar power generation is a technology that converts photon energy from the sun into electricity. This conversion is carried out on solar panels consisting of photovoltaic cells. These cells are layers of pure silicon (Si) or other semiconductors that are processed in such away. If the material is energized, the photons will excite the electrons from their atomic bonds to be free-moving electrons and produce a direct electrical voltage. The advantages of solar power generators are: available free, easy to maintain and simple, no moving equipment, no need to replace spare parts and no lubrication, works silently, and no negative impact on the environment.

Wind power is renewable energy produced by the wind speed that rotates the blades. This rotation is converted into mechanical energy, then mechanical energy is converted into electrical energy by a generator. The working principle of the wind turbine is by utilizing the kinetic energy of the wind that enters the active area of the turbine to rotate the propeller/windmill. Then this rotary energy is passed on to the generator to generate electrical power. Wind and solar energy are intermittent resources and are dependent on geographical and weather conditions [11]. The combination of these energies will improve the reliability of the electrical distribution supply.

Solar power generation can be model as a voltage controller bus or negative load bus, which active power obtains from the multiplication of PV voltage and current. The number of solar panels connected serially and in parallel will determine the amount of electrical power produced by a solar generation. The solar panel connected in parallel will increase the current, and serial connection provide higher output voltages. If the array is composed of Np parallel connections of photovoltaic cells, saturation currents may be expressed as $I_{pv} = I_{PV,cell} * N_p$, and $I_0 = I_{0,cell} * N_p$ while the wind turbine can be modeled with a load bus or PQ bus with a negative Q value. Important parameters of a wind turbine are the power rating, voltage, percent power factor, and generator impedance. Other variations can also be made to the type of turbine and wind curves consisting of ramp wind, gust wind, and noise wind.

IOP Conf. Series: Materials Science and Engineering 1041 (2021) 012008 doi:10.1088/1757-899X/1041/1/012008

The addition of distributed generators from the types of solar and wind power will cause changes in the performance of the 20 kV PLN distribution network [12]. The steady-state performance of 20 kV active distribution feeder is determined by the voltage profile of the main buses and distribution system losses. The voltage profile of the main buses and distribution system losses can be obtained through steady-state analysis. According to standard PLN standard number. 1 of 1995 [13], the distribution system voltage variation permitted is +5% and -10%. The allowed distribution network's power loss should not be higher than 10% of the total system load as mention in PLN standard number 72 of 1987 [14][9]. This paper discusses steady-state analysis related to voltage profile and system losses of the PV and wind power generation integration in 20 kV PLN grid, simulated in ETAP 19.0.1E.

2. Materials and methods

The single line diagram of the 20 kV distribution system used for steady-state analysis of hybrid PV and wind turbine integration is shown in Figure 1. This 20 kV distribution system is one of the PLN distribution systems taken from previous publications [15]. The modification only made for 20 kV line lengths by multiplication with ten.

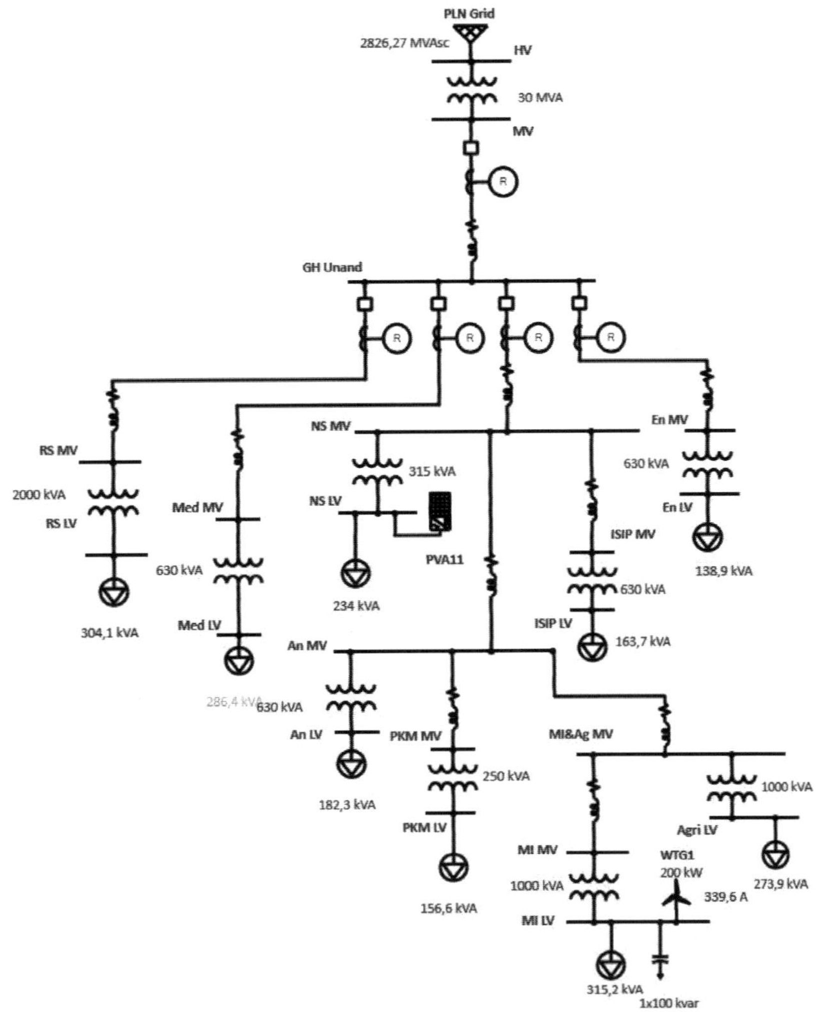

Figure 1. Single line diagramof 20 kV Distribution System.

The input data needed in power flow calculation are line data in the form of, resistance (R), reactance (X) and susceptance (B) and bus data in the form of power injection of generator scheduled and bus voltage generation, load power and shunt capacitor. The 20 kV distribution system delivers electricity from the 150 kV/20kV substations to the load area. The requirements that must be met in the calculation of power flow are:

1. The power generation must be equal to the demand for the load plus losses.
2. The bus voltage is at the allowable voltage rating, usually ± 5% or ± 10%
3. The power plant operates at certain active and reactive power limits.
4. The distribution lines and transformers must not be overloaded.

The calculation of power flow using is the Newton Raphson method and simulated using ETAP 19.0.1E Software.

3. Results and Discussion

The electrical system performance is determined by the main bus voltage profile and losses of entire system. The value of the voltage profile and system losses can be obtained from the load flow computation.

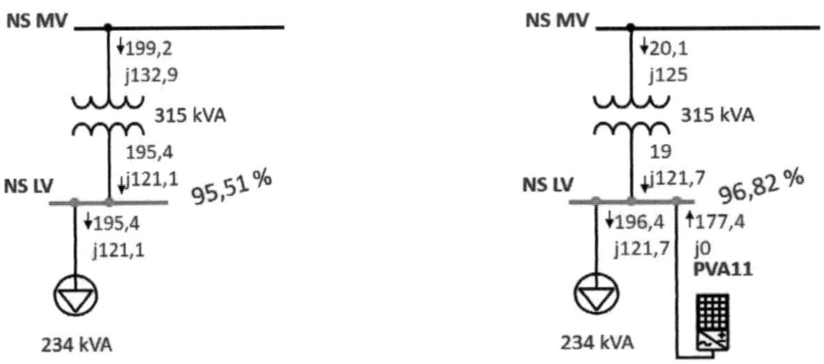

a. Without PV b. With PV connected

Figure 2. Part of ETAP output for NS MV and NS LV buses

Figures 2a and 2b show that the results of power flow simulations on NS LV and NS MV buses with and without PV connections. The existence of power injection from solar power by 177.4 kW has increased the NS LV bus voltage from 95.51% to 96.82%. The active power flow from the grid utility has been reduced from 195.4 kW to only 19 kW with the presence of a solar generator connection on the NS LV bus. While the reactive power flow almost has no change because the PV model used has not produced reactive power.

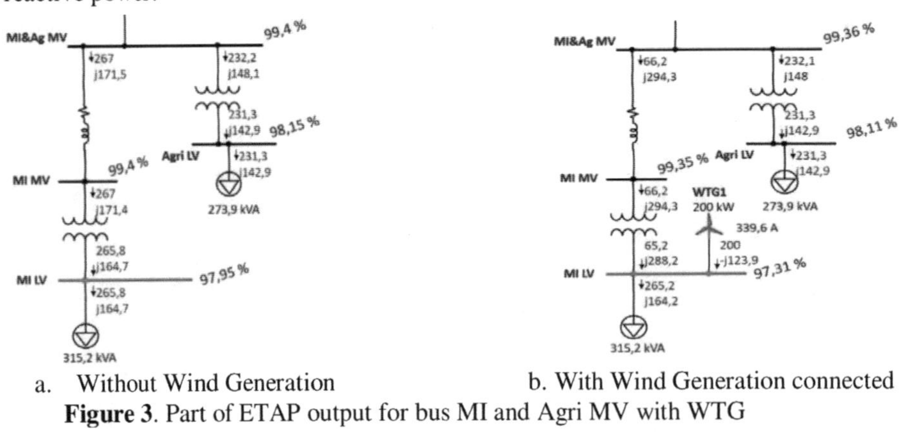

a. Without Wind Generation b. With Wind Generation connected

Figure 3. Part of ETAP output for bus MI and Agri MV with WTG

IOP Conf. Series: Materials Science and Engineering 1041 (2021) 012008 doi:10.1088/1757-899X/1041/1/012008

The results of power flow simulations on MI and Agri buses for the MV and LV sides with and without wind power connection are shown in figures 3a and 3b. The presence of 200 kW power injection from wind power plants on MI LV buses has drastically reduced the active power flow from the grid utility from 265.8 kW to only 65.2 kW. However, the bus voltage almost did not increase because the wind power connection absorbs reactive power from the utility network at 123.9 kVar. Thus, the addition of wind power generation has caused a significant change in reactive power flow from 164.7 kVar to 288.2 kVar because the wind power assemblies model used absorbs reactive power. Therefore, the addition of wind power generation needs to be accompanied by the addition of compensator equipment such as capacitor banks.

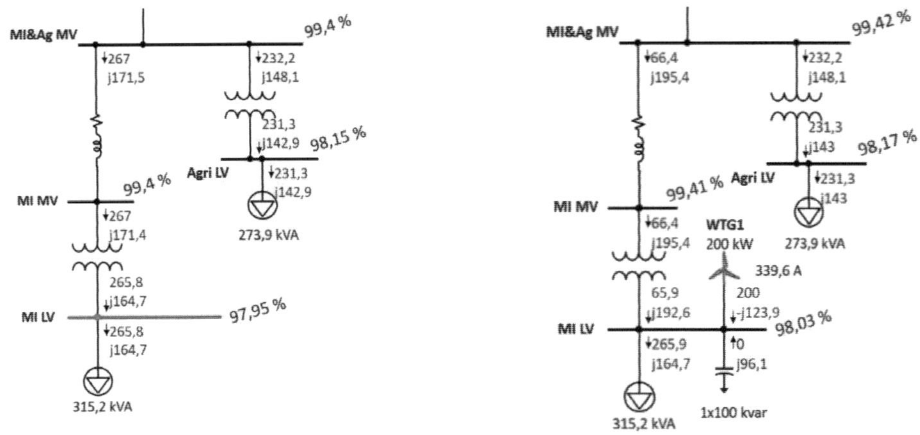

a. Without Wind Generation b. With Wind Generation and Capacitor

Figure 4. Part of ETAP output for bus MI and Agri with WTG and Capacitor

The addition of a bank capacitor of 100 kVar has improved the voltage profile on the MV LV bus to be 98.03%. Reactive power flow from the grid utility has also decreased to 192.6 kVar because the reactive power needed by the wind power plant has been compensated by the reactive power injection from the capacitor bank. Furthermore, the improvement of the voltage profile occurs on other buses adjacent to the bus where wind power plants are added as in the Agri LV bus figure 4a and 4b.

The simulation conducted using four cases: base case, with PV, with WT and Capacitor (WTC) and with hybrid PV + WTC. The voltage profile result as shown in table 1 for all buses. The voltage magnitude of all buses have been higher than 95%.

Table 1. The Result of 20 kV Distribution System Voltage Profile

Bus	Base case	With PV	With WTC	With PV+ WTC
Agri LV	98,15	98,17	98,17	98,18
An LV	98,03	98,04	98,04	98,05
An MV	99,45	99,46	99,46	99,47
En LV	98,52	98,53	98,52	98,52
En MV	99,60	99,61	99,60	99,61
GH Unand	99,61	99,62	99,61	99,62
ISIP LV	98,25	98,26	98,25	98,26
ISIP MV	99,53	99,54	99,53	99,54
Med LV	97,61	97,62	97,61	97,62

Med MV	99,60	99,61	99,60	99,61
MI LV	97,95	97,97	98,03	98,04
MI MV	99,40	99,41	99,41	99,43
MI&Ag MV	99,40	99,42	99,42	99,43
MV	99,66	99,67	99,66	99,66
NS LV	95,55	96,86	95,55	96,86
NS MV	99,54	99,55	99,54	99,55
Average	98,74	98,83	98,75	98,84

The results in table 1 show that the best voltage profile is obtained in conditions where the hybrid solar and wind power generation are installed with the average bus voltage was 98,84 %. The PV integration can increase a little bit of system voltage compared to wind turbine integration. This increase in voltage occurs due to PV does not consume reactive power, while wind turbines absorb reactive power. Therefore, the addition of a capacitor bank is needed when wind turbine integration. The lowest losses level of 11 kW and 50 kVar are also obtained in case four, as shown in figure 5. Thus the addition of distributed power plants to the distribution feeder has cut network losses. However, the further study of distributed generation placement and optimal sizing needs to be done before integrated into a 20 kV electricity network to obtain a better system state.

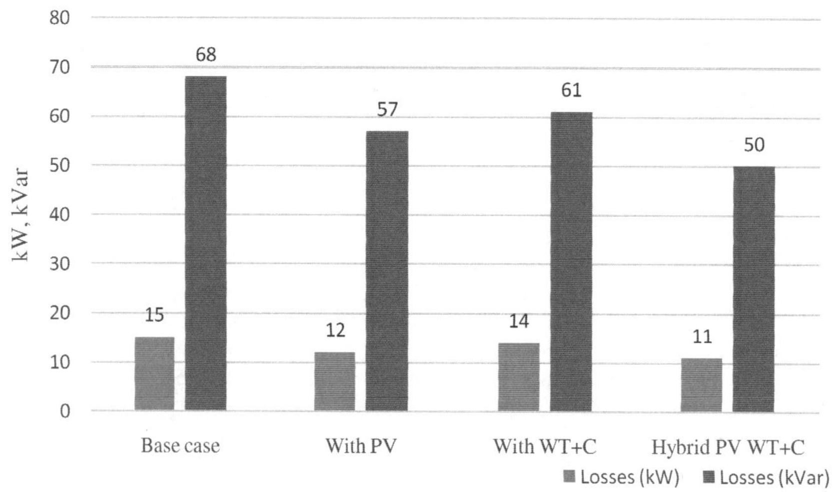

Figure 5. kW and kVar of 20 kV Distribution System Losses

The results of the study using four cases show that the lowest system losses are obtained in conditions where the hybrid system is placed on the bus, which has the lowest voltage and on the load bus, which farthest from the source. The lowest voltage in the four cases has been within the SPLN standard. The lowest level of losses of 11 kW is obtained in case four, where PV and WTC are placed on buses with the lowest voltage, i.e., bus NS LV and farthest bus MI LV. Power losses are obtained far below 10% as the losses permitted according to the PLN standard No. 72 of 1987 [9]. Therefore, the integration of PV and wind power into a 20 kV distribution network have improved the voltage profile and reduce total system losses.

4. Conclusion

The power flow study of 20 kV distribution system with solar and wind power generation integration has been presented in this paper. Power flow analysis is required in a feasibility study for the construction of a hybrid PV and wind power plant on a 20 kV distribution network to obtain a better bus voltage profile and minimum losses. The results of the study using four cases show that the presence of power injection from solar and wind power plants on LV buses has reduced the active power flow from the grid utility. The active power flow from the grid utility has been reduced with the presence of a solar generator connection on the LV bus. While the reactive power flow almost has no change because the PV model used has not produced reactive, as shown in the second case. However, the third case with only the wind turbine connected the bus voltage did not change because the wind power connection absorbs reactive power from the utility network. Therefore, the addition of wind power generation needs to be accompanied by the addition of compensator equipment such as capacitor banks. The fourth case using solar and wind hybrid generators show that the lowest system losses have been reached, and also have an improvement in the voltage on the bus where the solar and wind turbine generators are installed

Acknowledgments

The authors would like to thank for the publication of this article supported by Engineering Faculty of Universitas Andalas under grant 2020.

References

[1] Nayak C K, Kasturi K and Nayak M R 2019 Economical management of microgrid for optimal participation in electricity market *J. Energy Storage*21 657–64
[2] Rahardjo, Irawan ; fitriana ira 2005 Analisis Potensi Pembangkit Listrik Tenaga Surya 43–52
[3] Dida H P, Suparman S and Widhiyanuriyawan D 2016 Pemetaan Potensi Energi Angin di Perairan Indonesia Berdasarkan Data Satelit QuikScat dan WindSat *Rekayasa Mesin*7 95–101
[4] ESDM 2014 Peraturan Pemerintah Republik Indonesia No. 79 Tahun 2014, Tentang: Kebijakan Energi Nasional
[5] Syafii and Nor K M 2015 Unbalanced active distribution analysis with renewable distributed energy resources *Telkomnika (Telecommunication Comput. Electron. Control.***13**
[6] Hadidian Moghaddam M J, Kalam A, Nowdeh S A, Ahmadi A, Babanezhad M and Saha S 2019 Optimal sizing and energy management of stand-alone hybrid photovoltaic/wind system based on hydrogen storage considering LOEE and LOLE reliability indices using flower pollination algorithm *Renew. Energy***135** 1412–34
[7] Khushalani S, Solanki J M and Schulz N N 2007 Development of three-phase unbalanced power flow using PV and PQ models for distributed generation and study of the impact of DG models *IEEE Trans. Power Syst.***22** 1019–25
[8] Liang C, Jun H, Yiran W, Jian T, Hong L, Sanrong G and Yin W 2018 Analysis of access location and capacity of distributed generation based on OpenDSS *2018 China International Conference on Electricity Distribution (CICED)* (IEEE) pp 2264–8
[9] Prabhu J A X, Sharma S, Nataraj M and Tripathi D P 2016 Design of electrical system based on load flow analysis using ETAP for IEC projects *2016 IEEE 6th International Conference on Power Systems (ICPS)* (IEEE) pp 1–6
[10] Abdel-Akher M, Nor K M and Abdul-Rashid A H 2005 Improved three-phase power-flow methods using sequence components. *IEEE Trans. power Syst.***20** 1389-1397.
[11] Ramli M A M, Hiendro A and Al-Turki Y A 2016 Techno-economic energy analysis of wind/solar hybrid system: Case study for western coastal area of Saudi Arabia *Renew. energy***91** 374–85
[12] Syafii and Nor K M 2013 Renewable distributed generation models in three-phase load flow analysis for smart grid *Telkomnika***11**

[13] PT. PLN 1995 Standar-Standar Tegangan *Standar Perusah. List. Negara* 5

[14] PT. PLN (Persero) 1987 *SPLN (1987) No. 72: Design Specification For Medium Voltage Network (JTM) And Low Voltage Network* (Jakarta)

[15] Syafii, Gazaly A El and Abadi A 2019 Load Flow Analysis of PV System Integration in Universitas Andalas Distribution System *2019 3rd Int. Conf. Electr. Telecommun. Comput. Eng. ELTICOM 2019 - Proc.* 17–20

Performance Analysis of Diversity Selection Combining Technique in Rayleigh Fading Channel

Baharuddin[1], A Luthfi[1], H Andre[1], R Angraini[2]

[1] Department of Electrical Engineering, Universitas Andalas, Kampus Unand Limau Manis, Padang, 25163. Indonesia
[2] Politeknik Negeri Padang, Kampus Politeknik Negeri Padang Limau Manis, Padang, 25163. Indonesia

Correspondingauthor: baharuddin@eng.unand.ac.id

Abstract. This paper discussed a performance analysis of diversity selection combining technique in Rayleigh fading channel, in wavelet domain. It proposed diversity selection combining (DSC) method to combat errors during image transmission on wireless channels. The aim of using DSC in the system is to improve the performance of the system. The testing of the system was done by conducting experiment. The experiment was done by testing BER and PSNR. The result showed that BER was decrease to 13.4% in average compared to the system without DSC. PSNR was increased to 2.2 dB in average.The result of this simulation showed that the system with DSC in the image transmission system improvedthe system performance significantly when comparing to the system without DSC.

1. Introduction

In wireless communication, multiple propagation paths exist from transmitter to receiver due to reflection, diffraction, and scattering of wave by different objects. Signal copies received through different path can undergo rapid attenuation, distortion, delays and phase shift over a short period of time due to constructive or destructive interference occurring at receiver [1-2]. In destructive interference signal power decreases significantly and this phenomenon is called as fading. The multiple propagation path are time-varying. This results in multipath time-varying fading channel. The time-varying nature of channel is the major constraint for reliable wireless transmission. Multipath fading results in severe performance degradation in terms of error probability and poor reliability of wireless communication system. [2]

Diversity technique is one of the techniques that is often used to reduce signal quality degradation due to this phenomenon. Diversity is used to improve the performance of radio channels without increasing transmission power [3-9]. The Transmission in telecommunications system, the information in form of data, sound, image and video will be transmitted from the transmitter to the receiver via communication channel. During transmission, the degradation of the quality information might be happened due to the interference, delay spread, attenuation, noise and fading [10-12].

In order to improve the system performance, in the case of significant fading, different methods are used, but diversity techniques are most frequently applied [3-9]. Diversity techniques can be used to improve system performance in fading channels. There are several different types of diversity techniques that are commonly used in wireless communication systems. Diversity transmission is

IOP Conf. Series: Materials Science and Engineering 1041 (2021) 012009 doi:10.1088/1757-899X/1041/1/012009

based on the principle of using multiple copies of the transmitted signals that are statistically independent. The most commonly used techniques of combining signals after application of diversity techniques are: diversity selection combining (DSC), maximum ratio combining (MRC), and Equal Gain Combining(EGC) [5-7].

The idea of diversity is to combine several copies of the transferred signal passing independent fading in order to increase total receiving power [13-14]. Different types of diversity provide different methods of combining. In this paper, DSC is used for improve system performance in fading channels on wavelet domain. Selection combining is based on the principle of selecting the best signal among all the signals received from different branches at the receiver [15]. In this method, the receiver monitors the SNR of the incoming signal using switch logic. In figure 1, there are M independent Rayleigh fading channels providing M diversity branches, the weighted of each branch is adjusted to provide equal average SNR for each branch, co-phased is not required. The branch with the highest instantaneous SNR is connected to the demodulator. Instantaneous selection in not possible, therefore practical diversity selection system should be designed with small time constant of the selection circuit than the fading rate.

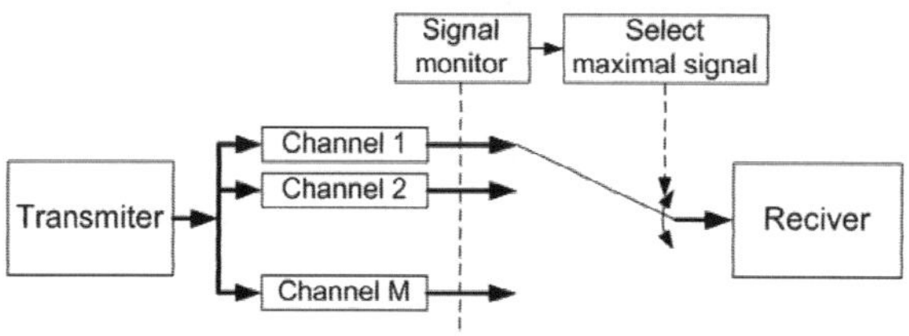

Figure 1. Diversity Selection Combining [2]

Several methods for transmitting images over wireless channels have been proposed in the literature. Most of these are compressed image transmission based on discrete cosine transform (DCT) or wavelet transform[15]. This image transmission method can be divided based on error concealing techniques, error resilient image coding, error correction / detection methods, and hybrid methods. For the error concealing techniques method [16-18] is attempt to hide errors that degrade the perceptual quality of the image after the image has been received. While for the error resilient method [16-18], image makes the transmitted data stream less sensitive to wireless channel errors, so when the data arrive at the receiver, they are easier to recover from errors in the bit stream. For the error correction / detection methods [16-18], it uses different types of error coding to combat the errors which has been introduced during wireless transmission, whereas the hybrid method [16-18] uses a combination of the three methods described above.

This paper proposes SPIHT [19] compressed image transmission using the DSC technique in the wavelet domain. This technique can improve the effect of fading and noise with DSC. For image transmission over wireless channels, two or more diversity channels can be utilized to obtain multiple bit streams at the receiver, where each bit stream independently representing the image data. Then these bit streams can be combined in the wavelet domain to improve the perceptual quality of the received image.

2. Material and methods

In this research, DSC is used to obtain a better quality reconstruction image, when compared to reconstruction images without using DSC. DSC which used in this image transmission system works on wavelet domain which aims to improve the image quality.

2.1. The design system

The design system for this research is given in figure 2.

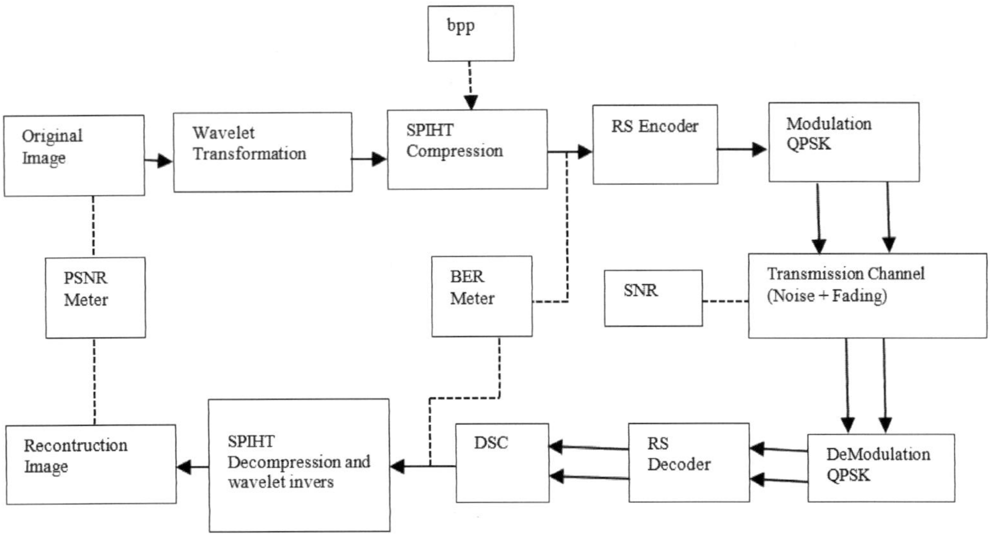

Figure 2. The system design use in the performance analysis of diversity selection combining technique in Rayleigh fading channel.

In the system, the combination of blocks is used by selecting the best blocks from the 2 transmission channels. By using DSC, the bit streams received from the uncorrelated diversity channels are divided into blocks and compared on a block-by-block basis. For each new block, the coefficients of the composite wavelet transform are updated based on the received bits [15].If the coefficients reconstructed from one block are judged more likely, then all of the coefficients from the more likely block are placed in the composite wavelet transform [15]. If all of the chosen coefficient blocks are error free, then the reconstructed image will be error free. Two block-based diversity rules were developed. The first block-based diversity rule only selects blocks of bits, from one bit stream based on a measure that is dependent on wavelet transform characteristics. Figure 3 is the DSC based on block rules.

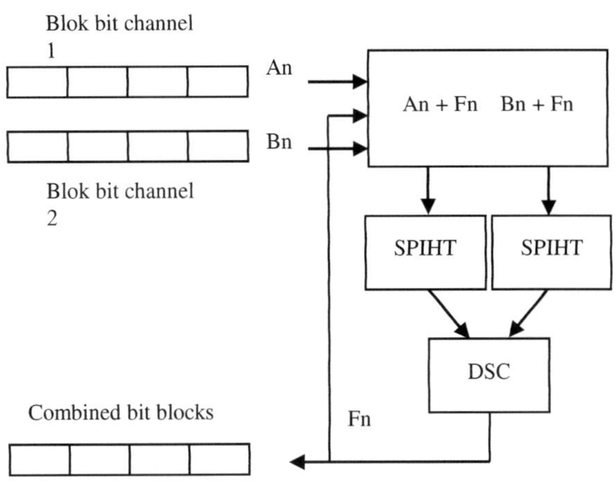

Figure 3. Diversity selection combining techniques based on block rules.

For each new block, the coefficient of the combined wavelet transform is updated based on the bits received. For more details, can be seen in the block diagram in Figure 3. So, if the reconstruction coefficient of a block is considered better, then the coefficient will be placed on the combined wavelet transform. If all selected block coefficients are error free, the reconstruction image will also be error free. This diversity selection combining rule is based on the selection of bit blocks *b(l)* from a bit stream by measuring *w(l)* which depends on the wavelet transformation characteristics. The rules for selecting blocks can be written with the following equation [15]:

$$b(l)=\begin{cases} b_1(l) & if \quad w(l)\geq 0 \\ b_2(l) & if \quad w(l)<0 \end{cases} \qquad for \quad l=1,2,\ldots\ldots,L \tag{1}$$

where:

$$w(l)=\sum_{(i,j)\in lowressubband} h_L^l(i,j)+\sum_{(i,j)\in desubband} h_H^l(i,j) \tag{2}$$

To determine the value of *w(l)*, use the rules for determining the value of $h_L^l(i,j)$ for the approximation coefficient and $h_H^l(i,j)$ for detail coefficient. Here are the rules:

2.1.1. Rule of approximation coefficient

In this rule, if the absolute value of the difference in approximation coefficient at point *(i,j)* with the approximation coefficient at point *(i,j+*1*)* on channel 1 is smaller than the absolute value of difference of approximation coefficient at point*)* with approximation coefficient at point *(i,j)* on channel 2, then *(i,j+*1*)* will be worth 1. However, if the value at channel 1 is bigger than the value of channel 2, $h_L^l(i,j)$ will be worth -1. Whereas, if the value of the approximation coefficient on channel 1 is the same as the approximation coefficient on channel 2, $h_L^l(i,j)$ will be worth 0. The rules for determining the value of $h_L^l(i,j)$ can be given in the equation [15]:

$$h_L^l(i,j)=\begin{cases} 1 & d_1(i,j)<d_2(i,j) \\ -1 & d_1(i,j)>d_2(i,j) \\ 0 & c_{L1}(i,j)=c_{L2}(i,j) \end{cases} \tag{3}$$

where:

$$d_k(i,j)=|c_{Lk}(i,j)-c_{Lk}(i,j+1)| \tag{4}$$

for k=1,2 (number of diversity antenna). The absolute value of the difference between the approximation coefficient at a point (i,j) with the approximation coefficient at a point $(i,j+1)$ or in the next column.

where: $c_{Lk}(i,j)$ = approximation coefficient

2.2. Rules for detailed coefficients

The rules for the detailed coefficients are not much different from the rules for the approximation coefficients. The difference only lies in the coefficients used. The rules for determining the value of $h_H^l(i,j)$ for the detailed coefficients can be given in the equation [15]:

$$h_H^l(i,j)=\begin{cases}1 & t_1(i,j)<t_2(i,j)\\ -1 & t_1(i,j)>t_2(i,j)\\ 0 & c_{H1}(i,j)=c_{H2}(i,j)\end{cases} \tag{5}$$

where:

$$t_k(i,j)=\left|c_{Hk}(i,j)-\left(\sum_{m=0}^{1}\sum_{n=0}^{1}c_{Hk}(2i+m,2j+n)\Big/4\right)\right| \tag{6}$$

for k = 1, 2 (number of diversity antenna)

where: $c_{Lk}(i,j)$ = detail coefficient

2.3. Simulation methods

This research is done by doing a simulation method with the Matlab program to get the optimum performance and system reliability, as well as design accuracy before it is implemented. The transmission method used in this research is RS Code (31,15) in the coder and the decoder [15]. The diversity technique used is DSC [2] on the wavelet domain and the channel model used is additive white Gaussian noise (AWGN) and Rayleigh Fading [2].

2.3.1. System performance test

The system performance test in this study uses two parameters, namely BER and PSNR. BER is used to measure how many errors occurred at the receiver. PSNR is used to determine the quality of the reconstructed image at the receiver. BER and PSNR parameters can be explained as follows.

2.3.1.1. The Calculation of BER

The bit error ratio (BER) is measured by dividing the number of bit errors by the total number of transferred bits.The measurement of BER can be given by the following equation [2]:

$$BER=\frac{n}{N} \tag{7}$$

where: n = number of incorrect bits of information at the receiver
N = number of bits of information transmitted.

2.3.1.2. The Calculation of PSNR

PSNR is a useful parameter to determine the quality of reconstruction image. PSNR calculation is done by comparing the pixel value of the original image with the pixel value of the reconstructed image. PSNR shows the quality of the reconstructed image. The higher the PSNR value obtained, the better the quality of the image. The PSNR can be formulated as follows [15]:

$$PSNR=10\log_{10}\frac{255^2}{\frac{1}{N}\sum_i\sum_j\left(p(i,j)-\hat{p}(i,j)\right)^2} \tag{8}$$

where: $p(i,j)$ is the original image pixel value,
$\hat{p}(i,j)$ is Pixel value of reconstructed image,
N is the number of pixels in image, while the 255 value is the highest value for grey scale image.

2.3.2. Research Variable

The variables used in this study are as follows:

Independent variables are the variables that can be manipulated. The independent variables used in the study are:

1. Bit per pixel (bpp). This variable represents the average number of bits for each pixel of the compressed image.
2. SNR is the variable that compares signal power and noise power in the transmission channels.

Dependent variables are the variables that change based on the changes in independent variables. The dependent variables used in the study are:

1. BER is the variable which compares the numbers of bits error with the total number of bits received. This comparison is done between the bit stream SPIHT compression results in the tranciever with the bit stream of the decoding result of Reed Solomon in the receiver.
2. PSNR is the variable which represents the degree of integrity of the reconstruction image.

3. Results and discussion

To evaluate the system performance of DSC in Rayleigh fading channel, several experiments were carried out to test the BER and PSNR valueresults on Boat image with a compression ratio of 0.4 bpp. BER test results are given in table 1, figure 4 and figure 5.

Table 1. The result of performance analysis of DSC in Rayleigh fading channel.

SNR (dB)	Without DSC		With DSC	
	BER	PSNR (dB)	BER	PSNR (dB)
10	0.0416	5.1555	0.0396	6.6709
12	0.0149	12.6969	0.0138	12.7135
14	0.0031	18.5878	7.6212×10^{-4}	20.6474
16	7.4769×10^{-4}	18.6310	4.1911×10^{-4}	22.4545
18	1.0005×10^{-4}	27.4847	4.9053×10^{-5}	28.9533
20	0	30.5331	0	30.5331

Table 1 shows the simulationsresult of performance analysis of DSC technique in Rayleigh fading channelcomparing tothe system without DSC. These simulation tests are ranged from 10 dB SNR to 20 dB SNR with a compression ratio of 0.4 bpp. The results obtained from table 1 are in the form of BER and PSNR values. The analysis of BER and PSNR values is given in figure 4 and figure 5.

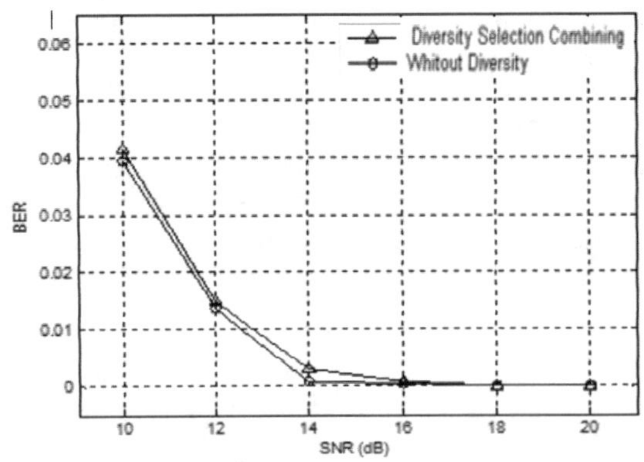

Figure 4. Comparison graph of BER to SNR of Boat image with 0.4 bpp with and without DSC.

Figure 4 shows a comparison graph of BER values with and without using DSC diversity. For the system with DSC, the SNR value at 10dB decrease to 20%, at 12 dB decreased by 11%, at 14 dB decreased by 23%, at 16 dB by 32%, at 18 dB decrease to 3.9% respectively, while the SNR value at 20 dB there are no errors occurred both with and without using DSC. Thus, in the system with DSC, the average error reduction obtained was 13%. At SNR 18 dB, the error reduction is only 3.9%. This value is very low for the system, because at SNR 18 dB, the signal power is greater than the noise power while at 20 dB SNR,there is no errors occurred both with and without using DSC. It happened because the signal power is biggercompared to noise.

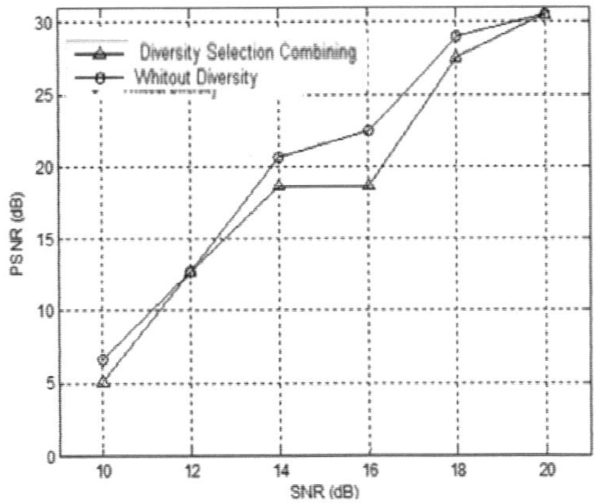

Figure 5. Comparison graph of PSNR to SNR of Boat images with 0.4 bpp with and without DSC techniques.

Figure 5 shows the comparison of PSNR to SNR with and without using DSC diversity. For the system with DSC, the SNR at 10 dB SNR, the quality of the reconstruction image is increased by 1.5 dB, at 12 dB SNR there is a very small increase by 0.0166 dB, at 14 dB SNR increase by 2 dB, at 16 dB SNR increase by 3.8 dB, at 18 dB SNR increase by 1.46 dB respectively. Whereas, SNR at 20 dB there is no increment occurred, both with and without using DSC, because it has reached the maximum PSNR at 30.5331 dB of ratio compression 0.4 bpp.

4. Conclusion

The result of the experiments which were carried out to test BER and PSNR on the Boat image with a compression ratio of 0.4 bpp with DSC in Rayleigh fading channel showed that there was 13.4% in average the reduction of BER values compared to the system without DSC. The lowest decreasing value of BER was at 18 dB SNR. It happened because the power of SNR is large enough to be able to combat the disturbance of fading and noise. The PSNR value also had an average increase to 2.2 dB. However, at SNR 12 dB there was slightly improvement to 0.0166 dB on the quality of the reconstruction image. It was caused by the random nature of the fading and noise signals which was very dominant to the information signal. While, the SNR at 20 dB, there was no improvement for both BER and PSNR values because the signal power to noise ratio (SNR) had reached a maximum value. The results of this study showed that the use of DSC techniques in the wavelet domain can reduce errors and improve the quality of image reconstruction. In order to make this research be more optimal, it is recommended to develop several systems such as modulation technique improvements, error control coding techniques, and standard source coding techniques.

Acknowledgements

Publication of this article is supported by Engineering Faculty of Andalas University grant for publication in fiscal year 2020.Number contract: 012/UN.16.09.D/PL/2020

References

[1]. Bernard S 2001 *Digital Communications: Fundamentals and Applications* 2nd edn Englewood Cliffs New Jersey: Prentice Hall

[2]. Rappaport T S 2002 *Wireless Communications: Principles and Practice* 2nd edn Pearson

[3]. Tse C H, Yip K W and Ng T S 2000 Performance tradeoffs between maximum ratio transmission and switched-transmit diversity *11th IEEE International Symposium on Personal Indoor and Mobile Radio Communications* IEEE pp 1485–1489 doi: 10.1109/PIMRC.2000.881665

[4]. Radaydeh R M and Matalgah M M 2005 Performance analysis of multiple carrier M-ary FSK system with diversity combining over generalized fading channels *IEEE International Conference on Communications* IEEE pp 2357–2361 doi: 10.1109/ICC.2005.1494757

[5]. Najib M A and Prabhu V K 2000 Analysis of equal-gain diversity with partially coherent fading signals *IEEE Transactions on Vehicular Technology* IEEE **49** 3 pp 783–791 doi: 10.1109/25.845098

[6]. Annamalai A, Tellambura C and Bhargava V K 2000 Equal-gain diversity receiver performance in wireless channels *IEEE Transactions on Communications* IEEE **48** 10 pp 1732–1745 doi: 10.1109/26.871398

[7]. Zhang Q T 1997 Probability of error for equal-gain combiners over Rayleigh channels: some closed-form solutions *IEEE Transactions on Communications* IEEE **45** 3 pp 270–273 doi: 10.1109/26.558680

[8]. Zhang Q T 1999 A simple approach to probability of error for equal gain combiners over Rayleigh channels *IEEE Transactions on Communications* IEEE **48** 4 pp 1151–1154 doi: 10.1109/25.775364

[9]. Alamouti S M 1998 A simple Transmit Diversity Technique for Wireless Communications for Wireless Communications *IEEE Journal on Selected Areas in Communications* IEEE **16** 8 pp 1451–1458 doi: 10.1109/49.730453

[10]. Sherwood P G and Zeger K 1998 Error Protection for Progressive Image Transmission over Memoryless and Fading Channels *Proceedings 1998 International Conference on Image Processing* IEEE pp 324–328 doi: 10.1109/ICIP.1998.723484

[11]. Chande V and Farvardin N 2000 Progressive transmission of images over memoryless noisy channels *IEEE Journal on Selected Areas in Communications* IEEE **18** 6 pp 850–860 doi: 10.1109/49.848239

[12]. Sherwood P G and Zeger K 1997 Progressive image coding on noisy channels *IEEE Signal Processing Letters* IEEE **4** 7 pp 189–191 doi: 10.1109/97.596882

[13]. Thomos N, Boulgouris N V and Strintzis M G 2005 Wireless image transmission using turbo codes and optimal unequal error protection *IEEE Signal Processing Letters* IEEE **14** 11 pp 1890–1901 doi: 10.1109/TIP.2005.854482

[14]. Shapiro J M 1993 Embedded Image Coding Using Zerotrees of Wavelet Coefficients *IEEE Transactions on Signal Processing* IEEE **41** 12 pp 3445–3462 doi: 10.1109/78.258085

[15]. Ramac L C and Varshney P K 2000 A Wavelet Domain Diversity Method for Transmission of Images over Wireless Channels *IEEE Transactions on Signal Processing* IEEE **18** 6 pp 891–898 doi: 10.1109/49.848244

[16]. Boulgouris N V, Thomos N and Strintzis M G 2003 Transmission of images over noisy channels using error-resilient wavelet coding and forward error correction *IEEE Transactions on Circuits and Systems for Video Technology* IEEE **13** 12 pp 1170–1181 doi: 10.1109/TCSVT.2003.819187

[17]. Boulgouris N V, Thomos N and Strintzis M G 2012 Unequal error protection random linear coding strategies for erasure channels *IEEE Transactions on Communications* IEEE **60** 5 pp 1243–1252 doi: 10.1109/TCOMM.2012.030712.100454

[18]. Gils W van 1983 Two topics on linear unequal error protection codes: Bounds on their length and cyclic code classes *IEEE Transactions on Information Theory* IEEE **29** 6 pp 866–876 doi: 110.1109/TIT.1983.1056753

[19]. Said A and Pearlman W A 1996 A New, Fast, and Efficient Image Codec Based on Set Partitioning In Hierarchical Trees *IEEE Transactions on Information Theory* IEEE **6** 3 pp 243–250 doi: 10.1109/76.499834

Optimization signal quality on cellular network based on Tilting-Antenna

H Andre[1], M Ghozal[1], Fitrilina[2], Baharuddin[1], R Fernandez[1], and Asriyadi[3]

[1]Electrical Engineering, Andalas University, Padang, Indonesia
[2]Electrical Engineering, Bengkulu University, Bengkulu, Indonesia
[3]Electrical Engineering PoliteknikNegeriSriwijaya,Palembang, Indonesia

Corresponding author:hanalde.andre@eng.unand.ac.id

Abstract. The Difference of Strength and Quality Signal on cellular networks is affected by the traffic on cell and the shape of the region topology. One of the problems encountered in-service performance is overshoot caused by poor signal quality. Overshoot is a condition where a range of power exceeds the specified limits, till the MS (Mobile Station) is served by a distant cell and not from the closest cell. Overshoot occurs due to boosting power which aims to improve service quality when there is an increase in customer activity in the network cell. The antenna tilting method is an effective and responsive method to resolve overshoot problem since it can form cell coverage, cells identification, measure Rx and C/I (Carrier to Interference) levels.). The results showed that after tilting there was a change in cell identification from 5 cells to 3 cells. The value of Rx level increased from -92.8 dBm to -82.03 dBm for the 726-sector and -72.13 dBm to -70.27 dBm for the 727-sector, while the 725-sector remained unchanged. C/I value for <13 dB decreased from 47.20% to 13.68%, for a range of 13 dB values $\leq$ C/I $\leq$ 16 dB decreased from 17.63% to 12.73%, and for> 16 dB increased from 35.17% to 73.59%.

1. Introduction

The Difference of Strength and Quality Signal on cellular networks is affected by traffic on cell and the shape of the region topology. One of the problems encountered in service performance is overshoot caused by poor signal quality such dropping a call, failing to handover, overshoot and missing a neighbor [1–4]. As the number of consumers increases, Maintaining and Improving of Network Services periodically become a necessity. one of network improvement is to control power level increasing which aim to provide good service in extremely dense users region[5,6].

Enhancement of power level needs to be performed carefully otherwise will emerge new problem such overshoot. Overshoot is a condition where a range of power exceeds the specified limits, till the MS (Mobile Station) is served by a distant cell and not from the closest cell. Overshoot can compose the handover range wider, so induce cell resources to run out quickly, handover process fails, blank spot, ping-pong, and high interference to the BTS (Base Transceiver Station) cell[7–9].

Cellular system services can be optimized in various ways to minimize disruption i.e. cell sectoring, setting power levels (adjusting power level), setting base station antenna configuration, and antenna tilt[10–13]. From the above method, the antenna tilt method is an effective and responsive method for building cell coverage, reducing service degradation and interference, and achieving better service quality[14–16].

2. Materials and methods

2.1. Tilting Antenna

The physical tilting value of the antenna in the sector can be modified mechanically, and the effect of mechanical tilt is a change in the overall coverage area. There are two types of tilt in mechanical: such as Down-tilt and Up-tilt as shown in figure 1.

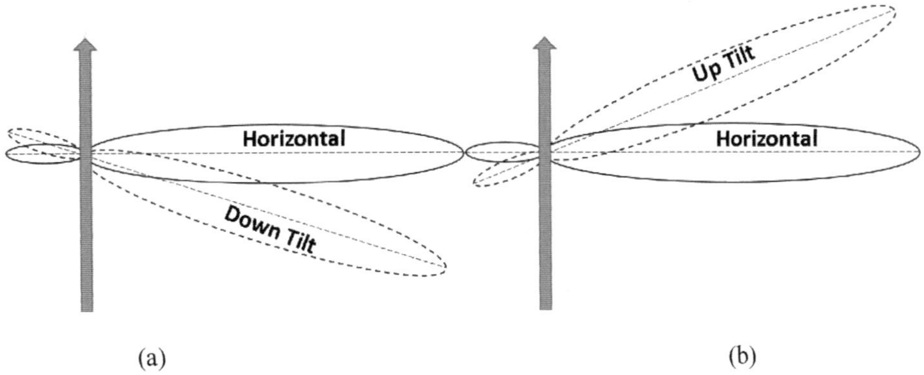

Figure 1. Mechanical Tilting Antenna (a) Down Tilt (b) Uptilt

Furthermore, Electrical tilting antenna is employed to alter the transmit power of the antenna by adjusting the electrical parameters of the antenna sector. In contrast to a mechanical tilt, the electric tilting only transforms the size of the main lobe and the phase emitted by the sector antenna. By using RET device, remote electrical tilting can be adjusted easily and quickly if the value changes.

2.2. Power Signal

Power signal in users is mentioned as Rx-level in which Rx-level is the level of signal strength received by MS in units (dBm). The greater its value indicates better quality. Following is the range of Rx-level quality in table 1.

Table 1. Rx Level Grading

Quality	Rx Level (dBm)
Very Good	$< 0 \; and \geq -85$
Good	$< -85 \; and \geq -95$
Bad	$< -95 \; and \geq -105$
Very Bad	< -105

Stage of the research is carried out by looking at two aspects, i.e. Pre-Optimization and Post-Optimization. In Pre-optimization, events data capture in Base Station is retrieved when overshoot happens by applying drive test. Drive test is an activity to collect signal quality measurement data for a network that aims to improve the quality of a system and develop network capacity. While in post-optimization, data retrieval is gained after antenna tilting adjustment and direction changing of transmit antenna, which aims to see results. Initial condition base station can be seen in table 2. The base station has three sector antenna.

Table 2. Initial Condition Base Station

Sektor	725	726	727
Azimuth	60	130	245
Mechanical Tilting	0	0	2
Electrical Tilting	6	5	5
Height (m)		40	
Antenna		Tongyu& S-Wave	
Bandwidth		1800 Mhz	

3. Result and Discussion

Figure 2 shows the results of the driving test of the Rx-level value before optimization. The Rxlevel range in figure 2 is following table 1. I.e. blue is for excellent Rx-level quality, and orange is medium Rx-level quality. Percentage of Rx-level for a range of 0 dBm> Rx-level 85 -85 dBm is 78.87% blue and for a range of -85 dBm> Rx-level ≥ -95 dBm orange. In this case, the Rx-level value does not occur due to overshoot problems, but the C/I value that is generated very bad even though Rx-level is of good value. It has an impact on signal quality even though the receiving power from MS has a good range of overshoot areas.

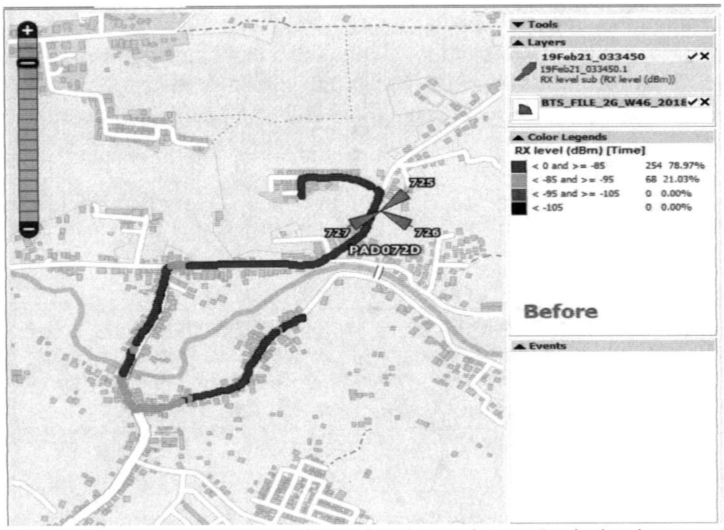

Figure 2. Drive Test Cell Rx-level Result Pre-Optimization

After implementing antenna tilting method, the value of Rx-level experiences improvement as shown in figure 3. It can be seen on figure 4 that the blue colour which indicates the range of 0 dBm ≥ Rx-level ≥ -85 dBm obtains excellent percentage namely 95.53%, while Rx-level value in field -85 dBm> Rx-level ≥ -95 dBm or in Orange criteria is 4,47 %.

IOP Conf. Series: Materials Science and Engineering 1041 (2021) 012010 doi:10.1088/1757-899X/1041/1/012010

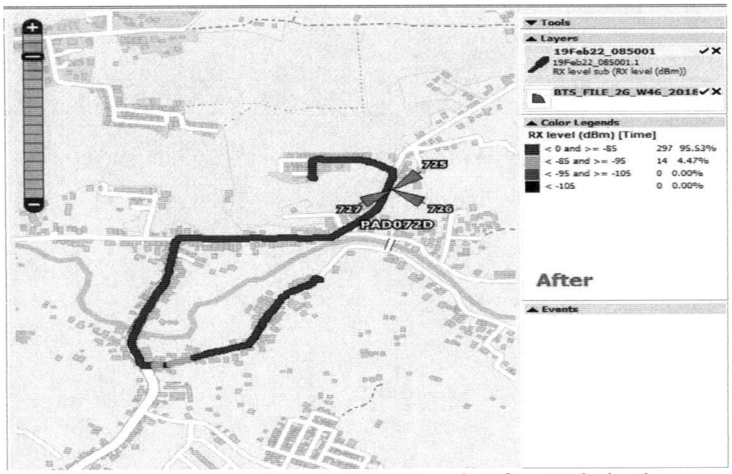

Figure 3. Drive Test Cell Rxlevel Result After Optimization

Drive test record shows that the effect of overshoot experience decreasing, can figure out from Average Value of C/I on the BAD PAD072D> 16 dB. However, some areas still gain C/I values below 13 dB and in yellow color criteria.

Previously, C/I generated on nearly 50% of the PAD072D BTS area is in red colour criteria (<13 dB). When post-optimization, 73.59% of the traversed area during the driving test are green colour criteria (> 16 dB), meanwhile the red colour (<13 dB) turn into 13.68%. It can be concluded from figure 4.8 that that antenna tilting method can increase the value of C/I. Drive test record for C/I measurement before and after optimization can be seen in figure 4.

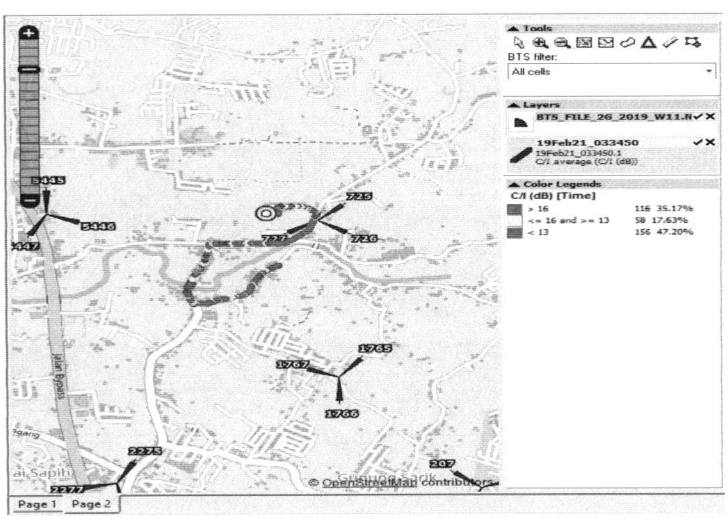

(a)

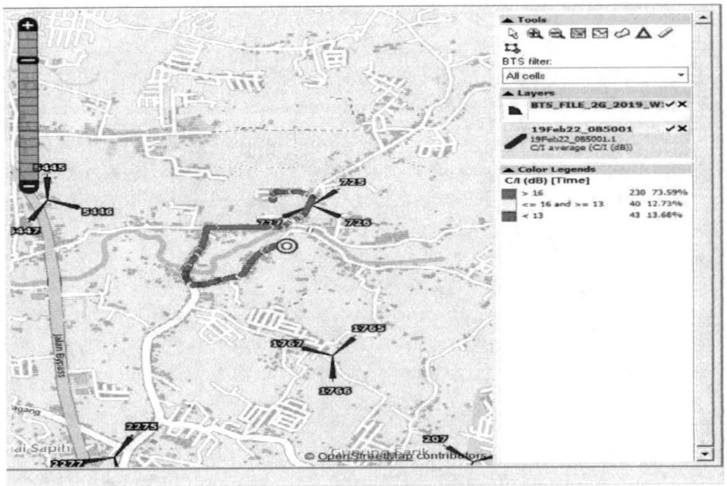

(b)

Figure 4. (a) Drive Test Record C/I Value Pre optimization (b) Post optimization

Mechanical and electrical tilting has been applied to the base station. Antenna sector 726 optimization shifted electrical tilting value from 5 to 3. Antenna sector 727 turned mechanical tilting value from 2 to 1 and electrical tilting value from 5 to 3. Optimization condition base station is shown in table 3.

Table 3. Optimization Condition Base Station

Sector	725	726	727
Azimuth	60	130	245
Mechanical Tilting	0	0	1
Electrical Tilting	6	3	3
Height (m)		40	
Antenna		Tongyu& S-Wave	
Bandwidth		1800 Mhz	

4. Conclusion

Based on the result above, the recommended antenna tilting values on the BTS PAD072D are in sector 725 is 6 °, in sector 726 is 3 °, and in sector 727 is 4 °. By applying the recommended antenna tilting level, C/I parameter on BTS PAD072D, which value about <13 dB decreases from 47,20 % to 13,68 %. The second parameter of C/I,13 dB $\leq$ C/I $\leq$ 16 dB, is also noted reducing from 17.63% to 12.73% and the last parameter of C/I > 16 dB enhances from 35.17% to 73.59%. In addition, there is an increase in Rx-level value in two sectors i.e. from -92.8 dBm to -82.03 dBm in sector 726, from -72.13 dBm to -70.27 dBm in sector 727, while sector 725 stay steady on -84,872 dBm.

Acknowledgement

Publication of this article is supported by the Engineering Faculty of Andalas University publication grant contract No. 034/UN.16.09. D/PL/2020

References

[1] Panjaitan M V, Sukiswo S and Zahra A A 2018 Analisis Quality of Service (Qos) Jaringan 4G Dengan Metode Drive Test Pada Kondisi Outdoor Menggunakan Aplikasi G-Nettrack Pro *Transient J. Ilm. Tek. Elektro*

[2] Putri H and Damayanti T N 2016 Penanganan Block Call Dan Drop Call pada Jari UMTS Berdasarkan Pengukuran Parameter Accessibility, Coverage And Quality *J. Elektro dan Telekomun. Terap.*

[3] Bikos A N and Sklavos N 2013 LTE/SAE security issues on 4G wireless networks *IEEE Secur. Priv.*

[4] Baharuddin, Muharam M, Andre H and Angraini R 2019 Performance analysis of error control coding and diversity in image transmission on wireless channels *IOP Conference Series: Materials Science and Engineering* vol **602**

[5] Hamdana E, Pramono S and Dahlan E 2012 Optimasi Perencanaan Jaringan UMTS Pada Node B Menggunakan Probabilistik Monte Carlo *J. EECCIS*

[6] Partov B, Leith D J and Razavi R 2015 Utility fair optimization of antenna tilt angles in LTE networks *IEEE/ACM Trans. Netw.*

[7] Mondal R U, Ristaniemi T and Turkka J 2017 Cluster-Based RF Fingerprint Positioning Using LTE and WLAN Signal Strengths *Int. J. Wirel. Inf. Networks*

[8] Fernandez R, Firdaus and Andre H 2018 Design of wideband quadrature hybrid coupler with tapered lines on the arms *Proceedings - 2018 International Conference on Applied Science and Technology, iCAST 2018*

[9] Yumin S El and Risviana R 2020 Analisis Mekanisme Rehoming pada BTS Jaringan GSM *Sainstech J. Penelit. dan Pengkaj. Sains dan Teknol.*

[10] Makkatang A and Nugroho R 2019 Analisa Pengaruh Perubahan Tilt Antena Sektoral BTS Secara Electrical Dan Mechanical Site XL 3G *J. Ilm. Giga*

[11] Andre H, Fernandez R and Baharuddin 2019 Dipole planar bowtie printed antenna for ism application *IOP Conference Series: Materials Science and Engineering* vol **602**

[12] Surya Winata I G N, Sukadarmika G and Sudiarta P K 2019 Analisis Drop Call Pada Jaringan Wideband Code-Division Multiple Access (WCDMA) Di Cluster Renon *J. Spektrum*

[13] Qadarfi M 2014 Analisis Pengaruh Perubahan Kemiringan Sudut Pancar Antena Sektoral Terhadap Kualitas Layanan Jaringan Sistem Komunikasi Bergerak Seluler *Tek. Elektro Univ. Tanjungpura Pontianak*

[14] Eckhardt H, Klein S and Gruber M 2011 Vertical antenna tilt optimization for LTE base stations *IEEE Vehicular Technology Conference*

[15] Fernandez R, Ilham M A, Andre H and Firdaus 2020 A Wideband Rectangular Patch Microstrip Antenna using Quad-Slotted Ground Plane *IOP Conference Series: Materials Science and Engineering* vol **846**

[16] Niemela J, Isotalo T, Borkowski J and Lempiainen J 2005 Sensitivity of optimum downtilt angle for geographical traffic load distribution in WCDMA *IEEE Vehicular Technology Conference*

IOP Conf. Series: Materials Science and Engineering 1041 (2021) 012011 doi:10.1088/1757-899X/1041/1/012011

The effect of the core material of the Rogowski coil sensor on the sensitivity of the magnitude of partial discharge

E P Waldi[1], F P Winarta[1], A I Lestari[1], R Fernandez[1], H D Laksono[1], Aulia[1], A Rajab[1], Z Nawawi[4], M H Ahmad[5], N Hozumi[6].

[1]Electrical Department, Andalas University, UNAND, Indonesia
[4]Electrical Department, Sriwijaya University, UNSRI, Indonesia Institute of High
[5]Voltage and High Current, Faculty of Electrical Engineering, UTM, Johor Bahru, Johor, Malaysia.
[6]Electrical and Electronic Information Engineering, Toyohashi University of Technology, TUT, 1-1 Hibarioka, Tempaku-Cho, Toyohashi, Japan

Corresponding author: ekawaldi@eng.unand.ac.id

Abstract. The Rogowski Coil sensor is one of the sensors used to detect aging insulation, which is marked by the occurrence of Partial Discharge. Aging is damaged, which fails power tools when operated. If the insulation is aging and even causing damage, the distribution of electrical energy will be disrupted. This research presents two types of Rogowski Coil sensor core materials to detect Partial Discharge with the same number of turns. The number of turns is 5. The first sensor is a sensor with a ferromagnetic core and the second sensor is a sensor with a non ferromagnetic core. The test of the sensor is carried out by using a magnitude test using a charge calibrator as an imitation of Partial Discharge. The magnitude of the Partial Discharge read by the sensor uses a digital oscilloscope with an impedance of 50 Ohm. Measurement results compared to commercial sensors. The measurement results read by the oscilloscope found that the type of sensor core influences the magnitude value of Partial Discharge. The magnitude of the Partial Discharge sensor with a ferromagnetic core is greater than that of a sensor with a non ferromagnetic core. However, both results are much smaller than the commercial sensor even though the digital oscilloscope can read them.

1. Introduction

Partial discharge (PD) is a local or dielectric discharge that experiences high local voltage caused by an inhomogeneous electric field. The cause of the occurrence of an electric field that is not homogeneous is due to the presence of bubbles or air cavities in the dielectric. If PD occurs continuously and is left for a long time, it will result in a decrease in the quality of the isolation, and the formation of electric tree growth (treeing), so that the complete f ailure of isolation will occur [1–4].

PD has a high frequency so that before the occurrence of complete failure in isolation, detection and identification can be done to find the cause of PD. PD detection can be done using 4 methods, namely, electrical methods, optical methods, acoustical methods, and chemical methods. The method used to detect PD in this study is electrical methods with a frequency range of 3-30 MHz [5],[6]. PD detection and identification are generally conventional, where this method requires a large enough space because the sensor has a large size and is relatively uneconomical, therefore in this study PD detection and identification was carried out by induction and economical [7],[8]. To test the sensor and compare the results from Hafely to the PD signal using an oscilloscope, which functions as the sensor's output, as a control for the amount of charge that is the input of the charge calibrator, a function generator used. The Labview program controls this measurement by connecting to a local area network (LAN) with a HUB, and the results of storage are in the form of digital data. In this study, the induction sensor used is the Rogowski Coil sensor. This Rogowski Coil sensor consists of a different toroid core, namely ferromagnetic and unferromagnetic with a number of turns. The resulting characteristic results are compared with commercial sensors.

2. Methodology

Rogowski Coil (RC) sensors generally work based on Faraday's Law or what is often called the Law of Electromagnetic Induction, where the instantaneous current through the primary winding will produce an immediate magnetic field in the form of flux. This flux field induces a secondary winding. The induced secondary winding will produce an opposing flux, which is indicated by the occurrence of a potential difference between the ends of the windings. This voltage is used as a result of the instantaneous current response to be detected. The resulting voltage is proportional to the change in the resulting current over time. The resulting induced voltage can be calculated by:

$$E = M \frac{di_0}{dt} \tag{1}$$

i_0 is the instantaneous current of the PD pulse passing through the primary side, which changes with the time function, and M is the mutual inductance between the secondary side of the loop sensor and the primary side. The M can calculate with:

$$M = \mu \frac{NW}{2\pi} \ln \frac{b}{a} \tag{2}$$

The value of M depends on the type of material (μ). In this research using ferromagnetic and non ferromagnetic materials, the non ferromagnetic used in this research is a solid polymer. The same sensor size is shown in figure 1, where a is the inside diameter of the sensor, b is the outside diameter of the sensor, and W is the height of the sensor [9]. At the same time, the sensor names are shown in table 1. All researchers agree that the equivalent circuit of the Rogowski Coil sensor is the same [9–11], the equivalent circuit can be seen in figure 2. The view of the Rogowski Coil sensor that has been made can be seen in figure 3.

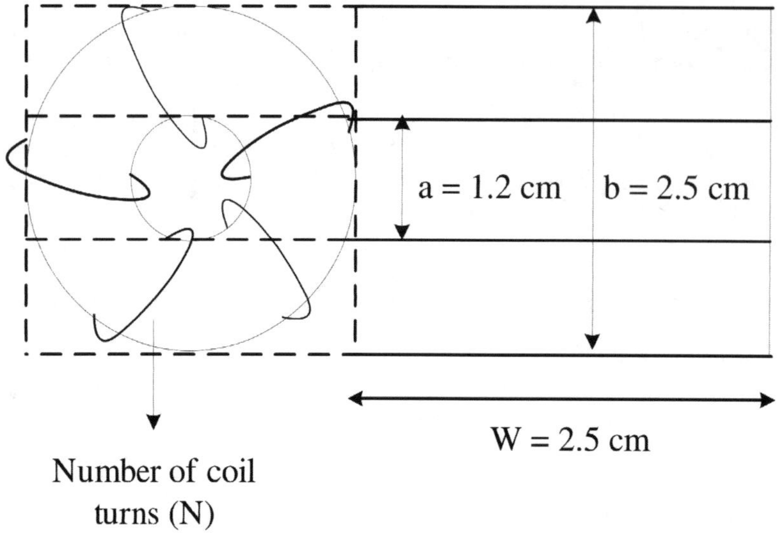

Figure 1. Sensor dimension

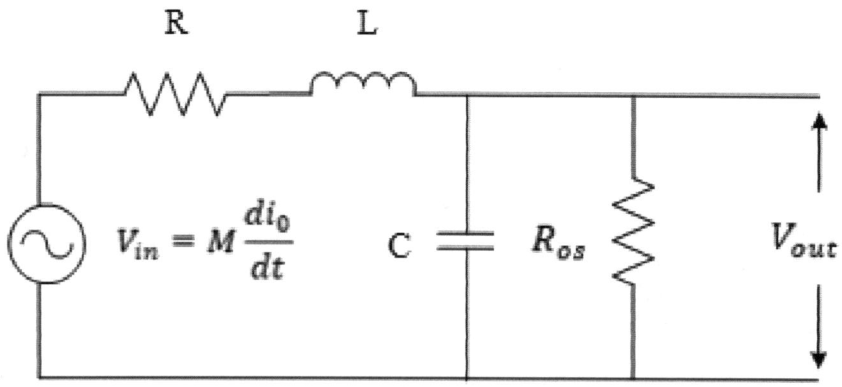

Figure 2. Equivalent circuit of RC sensor with oscilloscope

Table 1. Sensor conditions

No	Name of sensor	Toroid of sensor	Number of turns
1	F	Ferromagnetic	5
2	UF	Non ferromagnetic	5

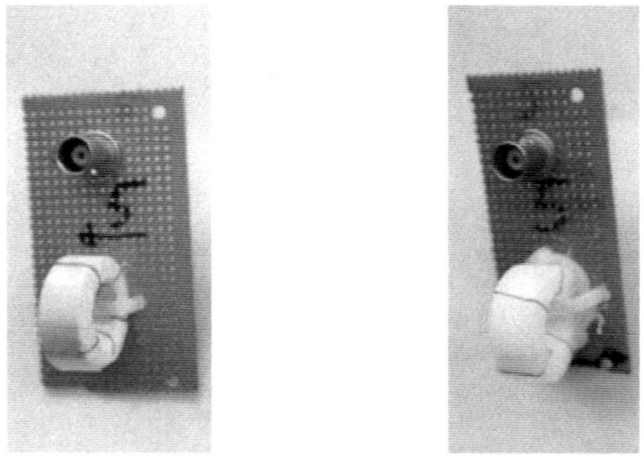

Figure 3. RC sensors with the number of turns of 5 and outside diameter of 2,5 cm (a) in ferromagnetic (b) in non ferromagnetic

Figure 4 is testing circuit for RC sensor. Rogowski Coil sensor testing is carried out using a function generator at 50 Ohm impedance. In addition to using a function generator, this measurement also uses a charge calibrator, oscilloscope, coupling capacitor, PD detector, and haefely sensor (commercial sensor) as a comparison. In this circuit the function generator functions as a source and regulator of the load which is the input for the charge calibrator.

The upper part of the coupling capacitor is served to one of the charge calibrator inputs, the coupling capacitor functions as a voltage divider and is also a sensor of low frequencies, which means that this sensor only accepts high frequencies and does not receive low frequencies. Whereas the lower part of the coupling capacitor is served with a proper PD detector and as output from the sensor is actually served to Chanel 2 on the oscilloscope. Then the ground part of the sensor haefely serves to one of the RC sensor inputs and the other side to the charge calibrator that is not yet connected. The ability to read this sensor can be seen by connecting the output from the RC sensor to Chanel 3 on the oscilloscope, after completion, the measurement results can be saved.

The amount of charge can be calculated by:

$$Q = C.V \tag{3}$$

Q is the amount of charge in pC, C is the capacitance in F, and V is the voltage in V.
The upper part of the coupling capacitor is connected to one of the charge calibrators inputs, the capacitor value used is 141 pF, while the lower part of the coupling capacitor is connected to the sensor Haefely. The output of the sensor must be connected to the channel 2 oscilloscope. Then the ground part of the sensor must be connected to one of the Rogowski Coil sensor inputs and the other side is connected to a charge calibrator that is not yet connected. The ability to read this sensor can be seen by connecting the output from the Rogowski Coil sensor to the channel 3 oscilloscope. All these measuring activities are controlled by the LabVIEW program by connecting a local area network (LAN) and all storage in digital data.

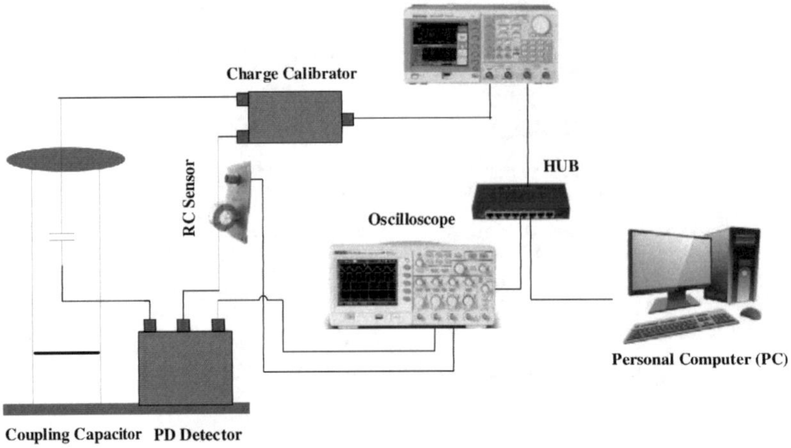

Figure 4. Testing circuit for RC sensor

3. Result and Discussion

Figure 5 is the measurement results of Haefely (commercial) sensor waves and RC sensors with ferromagnetic and non ferromagnetic cores with a total of 5 turns. This wave is a function of time. The left Y-axis is the magnitude reading for the Haefely sensor while the right side is for the two RC sensors. RC sensor waveforms and commercial sensors are very different. Commercial sensors are shaped like an impulse wave which is not oscillating or not underdamped with a duration of 0.5 microsecond. While the RC sensors both oscillate (underdamped) with a duration of about 0.15 microseconds. Consequently, RC sensor waves dissipate faster than Commercial sensors. This is also caused by the small oscilloscope impedance value which makes the number of oscillations small.

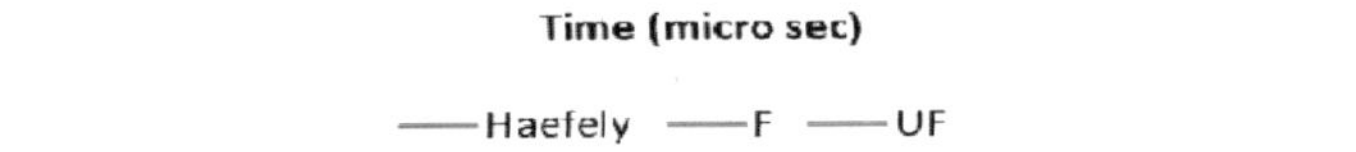

Figure 5. The response of RC sensors in ferromagnetic and non ferromagnetic with 5 turns

Figure 6 is a measurement of the peak value for each sensor. The Haefely sensor has the greatest significance of about 0.71 volts. RC with ferromagnetic cores is greater than non ferromagnetic cores. In accordance with the formula, magnetic ferromagnetic will increase the value of M, which also increases the induced voltage according to formula 4. Where the magnitude of the RC sensor with a ferromagnetic core is 0.109 Volt and a non ferromagnetic core is 0.013 Volt. With the sophistication of digitization technology, small values can be read perfectly (millivolt scale).

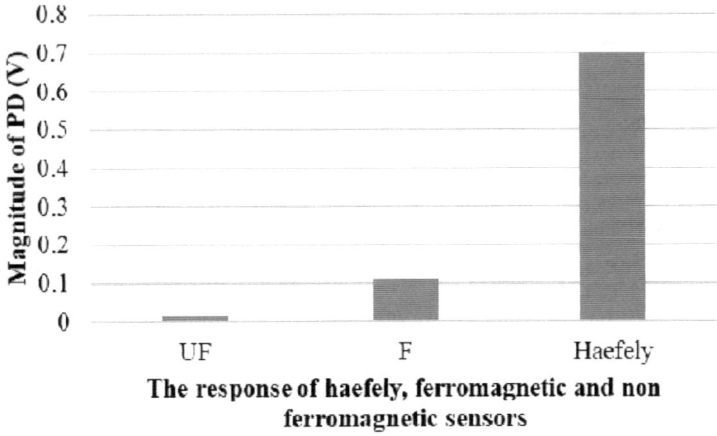

Figure 6. The response of Haefely, ferromagnetic and non ferromagnetic sensors

From figure 5, wave normalization is carried out for commercial sensors, RC ferromagnetic, and non ferromagnetic cores so that the results in figure 7. This normalization has a peak magnitude value of 1 Volt.

The measurement results from figure 5 are normalized for each wave, in order to see the effect of the ferromagnetic and non ferromagnetic cores on the response time. Accordingly that the results in figure 6 are obtained and, as a reference, are waves from commercial sensors. The normalize data purpose to make a sensor response that produces the same waveform even though the amplitude changes [12]. The two RC sensors respond earlier than commercial sensors. However, the peak width of 90% of commercial peak sensors is wider than that of RC sensors. The effect of the core Ferro magnetic peak 90% positive polarity is wider than unferomagnetic, but in contrast to the negative polarity, unferromagnetic is wider [13].

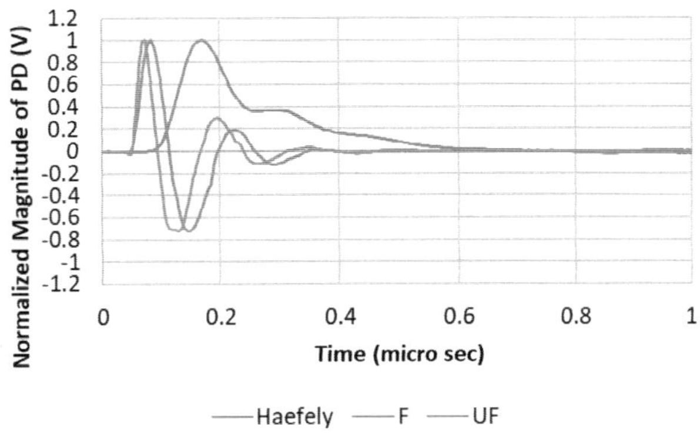

Figure 7. Normalized wave response in ferromagnetic and non ferromagnetic for 5 turns

4. Conclusions

The analysis described above, it can be concluded that the PD magnitude value of the wave response with a ferromagnetic core is greater than that of an non ferromagnetic core. However, even though the non ferromagnetic value is small with the sophistication of digital technology, the value is still legible.

Acknowledgments
This research was supported by "Direktorat Riset dan Pengabdian Masyarakat Deputi Bidang Penguatan Riset dan Pengembangan Kementrian Riset dan Teknologi/Badan Riset dan Inovasi Nasional." Under the research contract number. Number: 034/SP2H/LT/DRPM/ 2020. The fiscal year 2020.

References

[1] Waldi E P, Murakami Y, Hozumi N and Nagao M 2012 Breakdown of air-polymer composite insulation due to partial discharge and influence of thermal insulation *IEEJ Trans. Fundam. Mater.* **132** 1039–44
[2] Awang N A, Ahmad M H, Malek Z A, Sidik M A B, Nawawi Z, Jambak M I, Waldi E P and Aulia A 2017 AC breakdown strength enhancement of LDPE nanocomposites using atmospheric pressure plasma *ICECOS 2017 - Proceeding 2017 Int. Conf. Electr. Eng. Comput. Sci. Sustain. Cult. Herit. Towar. Smart Environ. Better Futur.* 290–4
[3] Waldi E P, Murakami Y and Nagao M 2007 Effect of humidity on breakdown of low density polyethylene film due to partial discharge *Proc. 2008 Int. Conf. Cond. Monit. Diagnosis, C. 2008* 655–8
[4] Nagao M, Jayaram S, Sugio M and Waldi E P 1999 Studies on the dielectric strength of oil-polymer composite insulation under variable frequency AC voltages *Conf. Electr. Insul. Dielectr. Phenom. (CEIDP), Annu. Rep.* **1** 309–12
[5] Siddiqui B A, Pakonen P and Verho P 2017 Novel inductive sensor solutions for on-line partial discharge and power quality monitoring *IEEE Trans. Dielectr. Electr. Insul.* **24** 209–16
[6] Hao L, Contin A, Hunter J A, Lewin P L, Swaffield D J, Walton C and Michel M 2011 a New Method for Automatic Multiple Partial Discharge Classification *XVII Int. Symp. High Volt. Eng. Hann.*
[7] Details C 2011 *on-Line Measurement of Partial Discharges in High Voltage Rotating Machines*
[8] Ahmed N H and Srinivas N N 1998 On-line partial discharge detection in cables *IEEE Trans. Dielectr. Electr. Insul.* **5** 181–8
[9] Waldi E P, Lestari A I, Fernandez R, Mulyadi S, Murakami Y and Hozumi N 2020 Rogowski coil sensor in the digitization process to detect partial discharge *Telkomnika (Telecommunication Comput. Electron. Control.* **18** 1062–71
[10] Shafiq M, Hussain G A, Kütt L and Lehtonen M 2014 Effect of geometrical parameters on high frequency performance of Rogowski coil for partial discharge measurements *Meas. J. Int. Meas. Confed.* **49** 126–37
[11] Shafiq M, Lehtonen M, Kutt L and Isa M 2014 Design, Implementation and Simulation of Non-Intrusive Sensor for On-Line Condition Monitoring of MV Electrical Components *Engineering* **6** 680–91
[12] Jr W H H, Kemmerly J E and Durbin S M 1396 ENGINEERING CIRCUIT ANALYSIS *New York McGraw-Hill Companies, Inc., 2012.*
[13] Waldi E P, Aulia, Hazmi A, Abral H, Arief S and Ahmad M H 2016 An Optimized method of partial discharge data retrieval technique for phase resolved pattern *Telkomnika (Telecommunication Comput. Electron. Control.* **14** 21–8

IOP Conf. Series: Materials Science and Engineering 1041 (2021) 012012 doi:10.1088/1757-899X/1041/1/012012

Preliminary Study on the Basic Characteristics of Isopropyl Esters Used As Insulating Liquid

A Rajab[1], J Ramadhani[1], R Kurniawan[1], K Qibran[1], M Latief[1], MI Hamid[1].

Electrical Engineering Dept., UniversitasAndalas, KampusLimau Manis, Padang, Indonesia.

Corresponding author: a.rajab@eng.unand.ac.id

Abstract. The paper presents the basic properties evaluation of isopropyl esters for an insulating liquid in oil-filled transformer. Two kinds of isopropyl esters, namely, isopropyl palmitate and isopropyl stearate, were synthesized from isopropyl alcohol and palmitate and stearate acids. The resulted isopropyl esters were then evaluated for their basic properties such as breakdown voltage, viscosity, density, acidity, and water-based on the ASTM D 6871 standard specification. The investigation results show that the viscosity and the Density of both isopropyl esters meet the standard's values well. The breakdown voltage of isopropyl stearate is higher than the standard specification, whereas that of isopropyl palmitate is less than the standard value. The water content of both liquids is still much higher than the value specified by the standard.

1. Introduction

One of the essential components of an oil-filled transformer is the insulating liquid. This liquid functioned as insulation in conjunction with paper or pressboard and as a coolant. Besides, the liquid can also be a tool for diagnosing the health condition of the transformer. Mineral oil has been used worldwide for more than a century and still the primary choice nowadays. The oil is widely implemented due to some advantages like high breakdown voltage, less vulnerable to oxidative degradation, low viscosity, and low pour point[1,2]. However, the oil is extracted from petroleum oil, which is considered to be non-renewable. The oil is also known as environmentally unfriendly due to the low biodegradability level [3]. Some oils like silicone oils, halogenated oils, and askarel have been implemented as insulating also the mineral oils. Most of them are relatively more expensive, less biodegradable, and non-renewable. These restrict their use to some specific applications which require a high fire point liquid [4].

The increase of concern on the environmental issue motivates many researchers to reconsider natural esters from vegetable oil as alternative mineral oil substitutes. The natural ester fulfills the criteria which the mineral oil does not. The oil is fully biodegradable. Thus the oil is friendly to the environment. The oil is derived from the seed; hence, the availability can be sustained. The oil is readily produced from edible oils, which means that the oil is non-toxic[5].Vegetable oils are also known to have a higher fire point in comparison with mineral oil, which means less risk to the fire. The oil also possesses a low thermal expansion coefficient. Together with the higher relative permittivity, which is recognized later, the low thermal expansion allows a more compact design of vegetable oil-filled transformer than that of mineral oil[6–9].

Currently, the transformer contained vegetable oil-based insulating liquid has been in service. Most of these vegetable insulating liquids are in triglyceride or tri-ester structure, leading to a high viscosity insulating liquid. The high viscosity liquid results in a less efficient heat transfer, which is less favourable as a cooling medium [10]. There is an urgency in developing an insulating liquid from the vegetable oil or natural ester resource but possesses a low viscosity and low pour point. This paper presents the preliminary investigation results on monoester use, especially isopropyl palmitate and isopropyl stearate, as the low viscosity insulating oil.The properties like breakdown voltage, viscosity, relative Density, acidity, and water content are measured. The measurement results are compared with the ASTM D6871 standard's values, the specification standard for new natural ester used for transformer insulation.

2. Material and methods

2.1. Sample

The samples used in the investigation were isopropyl palmitate and isopropyl stearate. The samples were synthesized from palmitate and stearate acids and isopropyl alcohol, based on the equation (1)[11]. The idea was to prepare samples in the monoester form in order to develop low viscosity insulating liquid. In addition, the choice of saturated fatty acids, namely, palmitate and stearate acids, is intended to develop a relatively stable monoester to the oxidative degradation. It is well-known that saturated fatty acids have higher oxidation stability than unsaturated ones.

(1)

2.2. Procedure

According to the ASTM D 6871 standard, natural esters or liquids derived from natural esters intended to be used as insulating liquid must fulfill some requirements, as listed in table 1. The measurement procedures to evaluate the required properties are also specified by the standard [12]. For instance, the procedure to measure the water content of an insulating liquid should follow ASTM D 1533 standard, and the specified value of the water content should not more than 200 ppm.

Table 1. As-Received New natural ester Property Requirements for transformer use.

Property	Limit	Test Method
Fire Point, min, °C	300	D 92
Pour Point, max, °C	-10	D 97
Relative Density, 15 °C/15 °C, max	0.96	D 1298
Viscosity (40 °C), max, cSt	50	D 445
Breakdown voltage, min, kV	35	D 1816
Dissipation factor at 60 Hz, 25 °C, max, %	0,2	D 924
Acid number, max, mg KOH/g	0,06	974
Water content, max, mg/kg	200	D 1533

3. Results and discussions

3.1. Electrical Property

The most critical parameter in characterizing a liquid intended for insulation purposes is the breakdown voltage. The measurement results of the breakdown voltage of two isopropyl esters samples, namely isopropyl palmitate and isopropyl stearate, are shown in table 2. The mentioned results are the averages of six individual measurements, as suggested by the D1816 standard procedure. It is found that the breakdown voltage of isopropyl stearate complies well with the value specified by the ASTM D6871 standard, whereas that of isopropyl palmitate does not.

It is also seen that the breakdown voltage of the isopropyl palmitate is significantly higher than that of the isopropyl stearate. This result differs from that found in the literature[13]. The change in the number of carbon atoms present in monoester esters does not cause a significant difference in their breakdown voltage. A higher amount of impurities might remain in the isopropyl palmitate, causing a substantial drop in its breakdown voltage. If this is the case, then filtration is required to improve the oils' breakdown voltage performance.

Table 2. The electrical property of tested samples and the comparison with the ASTM D6871 specification standard.

Electrical Properties	Values		
	ASTM	Isopropyl-	
	D6871	Palmitate	Stearate
Breakdown voltage, kV	≥ 35	24.5	38.7

3.2. Physical Properties

The next important parameter is the kinematic viscosity. It should be as low as possible to ensure the efficient cooling property of the liquid. For these reasons, the monoester form was chosen in this investigation instead of the original natural ester of triglyceride form. The measurement results of the kinematic viscosity of both tested samples are listed in table 2. It is clear from the table 3 that the kinematic viscosity of isopropyl palmitate and isopropyl stearate is in proper compliance with the ASTM D6871 specification standard.

Another physical property evaluated in this investigation is the relative Density. A liquid should have a relative density of equal or less than 0.96 g/cm^3 to be qualified as the insulation based on the ASTM D6871 specification standard. Table 3 shows that both tested samples fulfill the qualification as insulation from the relative density perspective. It can also be perceived from table3 that the relative Density of the isopropyl palmitate is slightly lower than that of isopropyl stearate. It might be due to the shorter hydrocarbon chain in isopropyl palmitate molecules compared to isopropyl stearate. The palmitate acid contains 16 Carbon atom numbers, whereas that of isopropyl stearate has 18 Carbon atoms. The shorter hydrocarbon chains owned by two adjacent straight molecules, the weaker Van Der Wall forces they have, which leads to the lower Density. The increase in the monoesters' viscosity with the increase of Carbon atom number owned by the oil molecules is also reported in [13].

Table 3. Physical properties of tested samples and the comparison with the ASTM D6871 specification standard.

Physical Properties	Values		
	ASTM	Isopropyl-	
	D6871	Palmitate	Stearate
Kinematic viscosity, cSt	≤ 50	5.5	2.4
Relative density, g/cm^3	≤ 0.96	0.83	0.86

3.3. Chemical Properties

The evaluation of a liquid's water content intended to be used as insulation in electrical apparatus like a transformer is essential since the presence of water content in a relatively high amount can significantly drop the breakdown voltage of the liquid. Table 4 shows the measurement results on both isopropyl esters' water content and the corresponding value from the ASTM D 6871 specification standard. It is clear from the table 4 that both oils' water content does not meet with the value specified by the standard. The higher water content of both oils could be the main reason for the relatively low breakdown voltage of both oils. It should be noticed that the water is a by-product, which is resulted from the esterification process (See equation (1)). Hence, treatment needs to be conducted to reduce the water content if the oils would be applied in the transformer.

Table 4 also provides information on the acid number of the tested liquids. As we can see, the acid numbers of isopropyl palmitate and stearate are still higher than the value specified by the standard. This indicates that palmitate and stearate acids in the form of free fatty acids remain in both oils in a remarkable amount. The higher amount of free fatty acids could also be another factor causing the low breakdown voltage of tested oils. This phenomenon is reported in[14]. Like the water content parameter, the free fatty acids contents of both oils need further reduction before being used as insulation.

Table 4. Chemical properties of tested samples and the comparison with the ASTM D6871 specification standard.

Chemical Properties	Values		
	ASTM D6871	Isopropyl- Palmitate	Isopropyl- Stearate
Water content, ppm	≤ 200	1900	3071.8
Acidity number, mg KOH /Kg	≤ 0.06	0.86	1.04

4. Conclusion

The basic properties evaluation of isopropyl esters has been performed. It is found that the viscosity and the relative Density of both liquids meet well the specification of ASTM D 6871 standard, whereas their acidity and water content do not. The breakdown voltage of isopropyl stearate has fulfilled the standard's value, whereas that of isopropyl palmitate has not. Reduction of water content, acidity, and impurities are required to improve the breakdown voltage to bring these oils to their implementation as insulating liquids.

References

[1] Wilhelm H M, Tulio L, Jasinski R and Almeida G 2011 Aging markers for in-service natural ester-based insulating fluids IEEE Trans. Dielectr. Electr. Insul. **18** 714–9

[2] Mc Shane Patrick 2002 Vegetable Oil Based Dielectric Coolants IEEE Ind. Appl. Mag. **8** 34–41

[3] Oommen T V 2002 Vegetable oils for liquid-filled transformers IEEE Electr. Insul. Mag. **18** 6–11

[4] McShane C P 2001 Relative properties of the new combustion resistant vegetable oil based dielectric coolants IEEE Cem. Ind. Tech. Conf. **37** 31–40

[5] McShane C P, Corkran J, Rapp K and Luksich J 2009 Natural ester dielectric fluid development update 2009 IEEE Power Energy Soc. Gen. Meet. PES '09 6–10

[6] Rafiq M, Lv Y Z, Zhou Y, Ma K B, Wang W, Li C R and Wang Q 2015 Use of vegetable oils as transformer oils-A review Renew. Sustain. Energy Rev. **52** 308–24

[7] Trnka P, Mentlik V and Svoboda M 2014 Ecologically acceptable insulating liquids for electrical appliances Proc. 2014 IEEE 18th Int. Conf. Dielectr. Liq. ICDL 2014 1–4

[8] Kasahara Y, Kato M, Watanabe S, Iwahashi M, Wakamatsu R, Suzuki T, Kanetani A, Kano T

and Tamura T 2012 Consideration on the relationship between dielectric breakdown voltage and water content in fatty acid esters JAOCS, J. Am. Oil Chem. Soc. **89** 1223–9

[9] Oommen T V., Claiborne C C, Walsh E J and Baker J P 2000 New vegetable oil based transformer fluid: Development and verification Conf. Electr. Insul. Dielectr. Phenom. (CEIDP), Annu. Rep. **1** 308–12

[10] Viertel J, Ohlsson K and Singha S 2014 Studies of the improvement of the viscosity of natural ester liquids Proceedings of the 2014 IEEE 18th International Conference on Dielectric Liquids, ICDL 2014 pp 4–7

[11] Bertrand Y and Lauzevis P 2013 Development of a low viscosity insulating liquid based on natural esters for distribution transformers IET Conference Publications **2013** pp 10–3

[12] ASTM_D6871–03 2003 Standard Specification for Natural (Vegetable Oil) Ester Fluids Used in Electrical Apparatus. Annual Book of ASTM Standard **10** 10–2

[13] Kanoh T, Iwabuchi H, Hoshida Y, Yamada J, Hikosaka T, Yamazaki A, Hatta Y and Koide H 2008 Analyses of electro-chemical characteristics of palm fatty acid esters as insulating oil 2008 IEEE Int. Conf. Dielectr. Liq. ICDL 2008 2–5

[14] Norhafiz A 2012 Ageing Assessment of Insulation Paper With Consideration of In-service Ageing and *Natural Ester Application*

IOP Conf. Series: Materials Science and Engineering 1041 (2021) 012013 doi:10.1088/1757-899X/1041/1/012013

Techno-economic analysis in application of glass solar cell on storey building: A case study of office building

H Fadli[1], R Dalimi[1]

[1]Department of Electrical Engineering, University of Indonesia, Depok, 16424, Indonesia.

Corresponding author:hasbi.fadli@ui.ac.id

Abstract. Urban area populated by edifice and storey building can complicate process of electric energy distribution to the middle of city. Citizen reluctance if their settlement passed by transmission network or being built substation, and bigger cost of underground cable than overhead line increases such difficulty level. If there is problem in delivering electric energy, so the city will face electric energy crisis and disrupt society life. One of solution in overcoming that issue is by installation the glass solar cell (transparent solar cell) with thin film material (CdTe) on storey building, which located in the middle of city and its electric energy supplied by PLN. In this study, glass solar cell will be installed in existing building "W" with solar cell area about 6,916 m^2. Glass transparency level of study object building is 30%, so by installing same transparency level of solar cell can supply maximum electric energy 897,184 kWh/year or equal to 35% electric energy demand of building. Economic feasibility analysis applied in this study, where NPV is IDR 398,961,738, IRR is 7.47% and the payback period is 11 years, so this technology is feasible to be implemented economically. If considering the tilt and azimuth angle of solar cell installation, instalment in B and C side of building generates smaller electric energy than A and D side. This condition caused by bigger reduction on solar irradiation energy that accepted by B and C site, namely 66.35% for B side and 67.22% for C side. The usage of glass solar cell contributes to deduct greenhouse gas emission, maximum 732,999 kg/year.

1. Introduction

Urban area develops rapidly and followed by increase of high-rise building construction. Escalation of building quantity intensifies level of electric energy demand. In 2017, it was founded that a hotel building in South Jakarta consumed electric energy about 7.8 million kWh each year and expensed around 9.4 billion Rupiah per year[1]. Since city become more developed, there is difficulty in delivering electric energy to the midst of it. Free space reduction, citizen reluctance in permitting their settlement passed by high voltage overhead line (HV) and extra high voltage overhead line (EHV) or being built substation enhance that adversity. Reluctance of some people towards HV or EHV is due to assume that health hazards exist if they live around it[2].

Metropolitan area tends to expend energy in big portion almost 60 percent of world energy consumption[3]. Dependence of fossil fuel encourages renewable energy implementation, one of them is utilization of solar energy that has abundant potency. There are various kinds of solar panel installation such as rooftop, façade, field, halt, bridge, and so on. In 2016, potential energy of rooftop solar cell in US was estimated 40% of national electric energy sale[4].

Location of solar cell assembly which enable to be connected directly to serviced load is very compatible with city condition that has obstacle in distributing electric energy to the middle of city. Geographically, Indonesia lies in equator line that owns huge solar energy. Average solar irradiation energy in Indonesia achieves 4.8 kWh/m^2 per day and approximately the value fluctuation is 9 percent in every month[5].

Several types of material are developed for maximizing energy that produced by solar cell. There are five groups of solar cell material, namely Crystalline Silicon, Thin Film, Organic/Polymer, Hybrid PV Cell and Dye-Sensitized[6]. Thin film solar cell is more preferable due to consume fewer production material. Thin film consists of three material categories, namely Amorphous Silicon (a-Si), Copper Indium Gallium Selenide (CIGS), and Cadmium Telluride (CdTe)[7]. CdTe possesses better efficiency rate rather than CIGS, in case it is assembled in high temperature and humidity[8].

Renewable energy applications, specifically for solar cell in building or Building Integration Photovoltaic (BIPV) start to grow increasingly in metropolitan city. This technology tries to perform green energy building by utilization of renewable energy so that it contributes to lessen GHG emission. Research regarding application of vertical solar PV on building façade 120 m and floor area 1296 m^2 had been undertaken as a part of BIPV concept in Malaysia. Solar PV assembly in west and east side generated larger electric energy, namely 800 MWh/year rather than north and south side, around 400 - 390 MWh/year. Economically, payback period for PV system was 12 years[9].

In 2018, R.K. Putri et al., got investigation result that BIPV on high rise building in Jakarta region was capable of lowering energy consumption of building about 5.97% of electric energy demand. Assembly method of solar cell was on rooftop and façade. The building height was 214.6 m and total area of installation 111,960 m^2[10]. Y.U. Putra et al., obtained study result and analysis that by using the same rooftop area and saddle roof, the east and west side of roof acquired larger solar energy than north and south[11].

There is glass solar cell technology (transparent solar cell) that can be performed in BIPV system. This material still enables sunlight to come into building in accordance with its transparency level. It also maintains outside architecture of the building totally. The transparent solar cell is made of nine materials such as Thin Film Photovoltaic (TPV), Near-infrared transparent solar cell, Polymer Solar Cell (PSC), Transparent Luminescent Solar Concentrator (TLSC), Perovskite Solar Cell, Electrophoretic Deposition (EPD) and others. From all those materials, the most developed material is solar cell with thin film type[12]. Solar cell system on BIPV is able to apply on-grid system, where electric energy from network keeps to supply the building constantly. Conversely, it is also known as off-grid system. Centainly, each system has its self excellences.

In this study, it is analysed the integrated solar energy on a building by using on-grid system. Object study is an existing building, where the electric energy supplied by State-Owned Electric Company or PLN's network normally. Technical analysis is performed, such as the calculation of maximum energy quantity that produced by solar cell. The applied material corresponds to transparency level of installed building glass. Economic analysis such as the net present value (NPV), internal rate of return (IRR) and the payback period are calculated for considering the economic feasibility. Besides, the reduction of greenhouse gas (GHG) emission is estimated to assess the contribution of renewable energy system in green building concept as well.

2. Materials and methods

Irradiation data of solar energy is obtained from official website of SOLARGIS. Based on the data, the lowest rate of solar irradiation energy or Global Horizontal Irradiation (GHI) is in February that is 3.816 kWh/m^2 per day. The highest rate is in September 5.530 kWh/m^2 per day as displayed in table 1. The location of study object lies in East Jakarta city and the coordinate is -6.242278, 106.876639. The average of GHI in a range of one year at this coordinate is approximately 4.563 kWh/m^2 per day. In the table 1, the rate of GHI begins to increase gradually from January and achieves the peak rate in September. After that, it declines moderately from October to January. Geographically, the position of study object is in south hemisphere.

Table 1. Solar energy irradiation

Month	Global Horizontal Irradiation (GHI) kWh/m^2 per Day
January	3.931
February	3.816
March	4.553
April	4.557
May	4.466
June	4.301
July	4.606
Augustus	5.234
September	5.530
October	5.180
November	4.480
December	4.101

In figure 1 shown GHI rate hourly in September, where in this month is the time that generates maximum solar irradiation energy in one year. The maximum rate is at 10 to 11 AM, namely 752 Wh/m^2. Meanwhile, the minimum rate of solar irradiation energy is at 5 PM, as much as 23 Wh/m^2. Solar irradiation energy starts from 6 AM, and it rises significantly to peak rate at 10 AM. Then, the energy decreases slowly until 1 PM and it drops sharply to 5 PM.

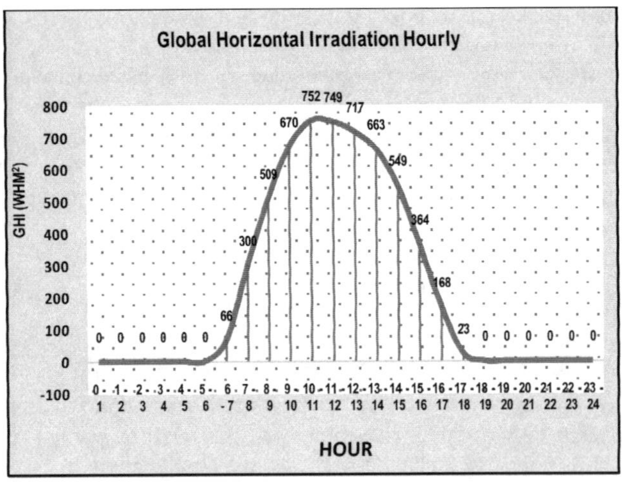

Figure 1. Global horizontal irradiation

Solar cell is installed on the building "W" which is an existing building and it located in the middle of the big city. Installation on A side of building is at azimuth angle (orientation) 129°, and area of installation is 1271 m^2. On B side of the building is at azimuth angle 39°, area of installation is 2045 m^2. Then, on C side of building is at azimuth angle 309°, where installation area is 1584 m^2. The last, on D side which has azimuth angle 219° and installation area is 2016 m^2. The 0° of azimuth angle faces the North Pole orientation. The installation is adjusted to building side that already installed by glass.

Electricity system configuration of glass solar cell (transparent solar cell) as displayed in figure 2. The system is on-grid type. It applies hybrid inverter that equipped with power backup facility such as battery (optional), if it is needed in the future. However, in this study, the glass solar cell system does not include power backup battery on technical design or economic calculation.

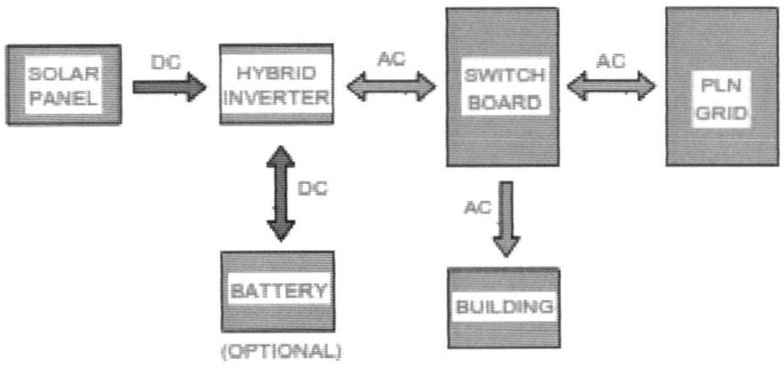

Figure 2. Electricity system configuration of glass solar cell on building

There are some transparency levels of glass solar cell that available in the market. The range is 10% with power output 72W till 80% with power output 16W. Nevertheless, in this study, the applied transparency level is 30% with power output 56W per module. Energy produced by the glass solar cell can be calculated when the value of solar irradiation energy, the efficiency of solar cell module and the area of solar cell installment are provided. The value of solar cell energy (Eth) is linear towards array PV efficiency (η), global irradiation energy (E_{glob}) and area of PV array (A_{array})[13].

$$E_{th} = \eta \, x \, E_{glob} x \, A_{array} \quad (1)$$

E_{th} is in kWh, η is in decimal, E_{glob} is in kWh/m^2 and A_{array} isin m^2.

If taking into consideration of the azimuth angle (γ) orientation as well as tilt (β) of solar cell installation, so solar irradiation energy that generated by glass solar cell will change. The total value of irradiation energy on the tilt surface H_T calculated by using the formula 2.

$$H_T = H_B + H_D + H_R \quad (2)$$

H_B is direct irradiation, H_D is diffusion irradiation, whereas H_R is reflected irradiation or albedo[14]. Electric energy yielded by glass solar cell is linear to solar irradiation accepted by solar cell module.

Economic analysis is undertaken to ensure the investment feasibility of design from economic point of view. The analysis parameters are the net present value (NPV), internal rate of return (IRR) and payback period. NPV is method for counting net cash flow in the present time by implementing money equivalence concept. Investment is categorized to be feasible when NPV value will be greater than zero or positive. Otherwise, in case NPV is lower than zero or negative, then it is assumed to be unfeasible[15]. NPV value calculated by the formula 3. NPV is supposed as present value of income cash flow subtracted by present value of outcome cash flow or present value of all expected cash flow[16].

$$NPV = \sum_{t=1}^{N} \frac{CF_t}{(1+r)^t} \quad (3)$$

IOP Conf. Series: Materials Science and Engineering 1041 (2021) 012013 doi:10.1088/1757-899X/1041/1/012013

Where CF_t is cash flow, 't' is time period, N is time period number and 'r' is interest.

The internal rate of return (IRR) is method in evaluating the investment feasibility pertinent to cash flow capability level on return investment in percent per time period[15]. It can be obtained by equation 4 as following.

$$IRR = iNPV_+ + \frac{NPV_+}{|NPV_+ + NPV_-|}(iNPV_- + iNPV_+)(4)$$

Where IRR is internal rate of return (%). $iNPV_+$ and $iNPV_-$ are interest rate for positive NPV and negative NPV respectively. Payback period is the analyzation method in acquiring the return period of investment when turnover condition is satisfied. Payback period is obtained by equation 5.

$$Payback\ period = n + \frac{a-b}{c-b} x\ 1\ year \qquad (5)$$

The *'n'* is the last year when amount of cash flow does not satisfy initial investment yet. The *'a'* is amount of initial investment. The *'b'* is amount of investment cash flow in year-n and the *'c'* is amount of investment cash flow in year n+1[17].In the table2, investment cost of glass solar cell system is presented.

Table 2. Investment cost of glass solar cell

No	Description	Cost
A.	Material	
1.	Glass solar cell (transparent solar cell)	IDR 5,258,737,500
2.	Hybrid solar inverter	IDR 583,941,600
3.	Distribution panel	IDR 513,024,000
4.	Cable	IDR 337,915,200
5.	PV combiner box	IDR 25,550,000
6.	Material support & accessories	IDR 591,166,629
B.	Installation of glass solar cell	IDR 1,344,700,000
C.	Dismantling of existing glass material	IDR 1,037,400.000
	Total	IDR 9,692,434,929

Table 2 shows that the biggest cost of glass solar cell system constitutes the material cost of glass solar cell. It is 54% of total investment cost. Afterwards, it also represents that investment costs consist of material cost, installation cost and dismantling of existing building glass.

The applied interest rate is 7%. It is based on the corporate interest rate of City Bank, issued in April 2020. The data is registered in official website of Financial Service Authority or "Otoritas Jasa Keuangan"[18].

Energy of glass solar cell also contributes to reduce greenhouse gas emission in urban area. Emission factor of GHG for electricity system of Java-Bali region is 0.817 kg/kWh. The usage of this value is due to the building electric energy is normally supplied by electric network of Java-Bali[19].

Reduction of GHG stipulated by comparing the amount of emission that possibly generated from conventional generator which supply electric energy to the building. Emission total of GHG calculated as formula 6[20].

Emission total of GHG = Electricity of Emission Factor (EF) x Total amount of electricity generation. (6)

3. Results and discussion

The calculation result of energyproduced by glass solar cell based on formula 1 is as indicated in table 3.

Table 3. Energy of glass solar cell with transparency level 30%.

Month	Energy Produced by Solar Cell (kWh)
January	65,621
February	57,553
March	75,841
April	73,690
May	74,227
June	69,387
July	76,917
Augustus	87,137
September	89,288
October	86,599
November	72,076
December	68,311
Yearly	897,184

Table 3 displays the electric energy value that generated by glass solar cell in every month, namely from January till December. The electric energy value that obtained in one year duration is also represented. Electric energy is acquired by installing glass solar cell with transparency level 30% where each modul is able to produce power as much as 56W.

Based on the table3, the lowest electric energy is yielded in February around 57,553 kWh, meanwhile the biggest value is in September, namely about 89,288 kWh. The glass solar cell can produce electric energy as much as 897,184 kWh in one year. In March until May, the generated electric energy does not fluctuate significantly.

If considering the azimuth angle (γ) orientation as well as tilt (β) of solar cell installation there is a degradation of total value of solar irradiation energy that received on each building side. It is namely 55.76% on A side, 66.35% on B side, whereas degradation on C side is 67.22%. The last, degradation on D side is 57.74%. Accordingly, the biggest degradation of received solar irradiation energy is when the installation of glass solar cell performed in B side and C side of building. In other words, the lowest electric energy will be produced if installation is on B side and C side of the building.

From economic analysis, the calculation results of NPV, IRR and payback period for glass solar cell is shown in table 4.

Table 4. Calculation result of NPV, IRR and payback period.

Description	Value
Design life (n)	25 years

Interest rate (i)	7%
Initial investment (II)	IDR 9,692,434,929
NPV	IDR 398,961,738
IRR	7.47%
Payback period	11 years

In the table 4, the design life is as long as 25 years. The calculation results of NPV is positive value, namely IDR 398,961,738. Based on the analysis with NPV method, so implementation of the glass solar cell is feasible.

From computation with IRR method, it is obtained that the IRR value is about 7.47 %. It is higher than applied interest rate 7%. Economically, by using this method, solar cell investment satisfies the feasibility. From payback period view, payback period is 11 years where it is shorter than investment period or design life (25 years), so that investment is also feasible.

Energy produced by glass solar cell is able to reduce the electric energy usage of building "W" that supplied by electric network of PLN. For transparency level of glass solar cell 30% (corresponding to transparency level of building glass), it generates the electric energy as much as 35% per year from the total of electric energy demand of building. In the Table 5 provided the electric energy cost in every month for building "W" as well as the saving that acquired by installing glass solar cell.

In table 5 presented Electric energy cost of building "W" and cost saving by using glass solar cell.

Table 5. Electric energy cost of building "W" and cost saving by using glass solar cell

Month	Total Cost of Building Electric Energy "W"	Transparency 30%	
		Cost Saving	Saving Percentage
January	IDR 245,703,432	IDR 73,241,543	30%
February	IDR 213,166,040	IDR 64,279,468	30%
March	IDR 231,093,105	IDR 84,619,306	37%
April	IDR 236,124,542	IDR 82,152,567	35%
May	IDR 245,912,032	IDR 82,445,441	34%
June	IDR 205,728,529	IDR 77,069,359	37%
July	IDR 238,053,746	IDR 85,746,661	36%
Augustus	IDR 236,530,990	IDR 97,200,985	41%
September	IDR 239,710,287	IDR 99,227,034	41%
October	IDR 251,881,790	IDR 96,328,522	38%
November	IDR 258,062,547	IDR 80,145,189	31%
December	IDR 235,369,130	IDR 75,955,424	32%
Total	IDR 2,837,336,171	IDR 998,411,499	35%

Based on table 5, the total cost of electric energy for building "W" is 2.4 billion Rupiah per year. By using glass solar cell, the cost saving is almost 1 billion Rupiah for one year. The biggest consumption of electric energy is in October and the smallest consumption is in June. The highest saving contribution of electric energy from glass solar cell is in September around 99.2 million Rupiah. It equals to 41% of electric energy cost in that month. Meanwhile, the lowest contribution is in February, approximately 64.3 million Rupiah. It is equivalent to 30% of electric energy in that month.

Energy of glass solar cellparticipates in reducing greenhouse gas emission in urban area. Based on formula 6,the GHG value enables to be lowered as much as 732,999 kg/year or almost 733 tons/year.

4. Conclusion

Application of glass solar cell (transparent solar cell) on the building can become one of solution alternative in overcoming the difficulty of distributing electric energy to the middle of the city. Technically, energy of glass solar cell supplies as much as 897,184 kWh/year or 35% of the electric energy demand of building that normally powered by electric network of PLN. If taking into consideration of the tilt angle and azimuth angle (orientation) in installing the glass solar cell, so electric energy generated in B and C building side is lower than A and D building side. It is caused by B and C side get higher degradation toward solar irradiation energy rather than received by A and D side. Economically, the glass solar cell with 30% transparency is feasible to be implemented. The NPV value is IDR 398,961,738, IRR is 7.47% and payback period is 11 years. The glass solar cell deducts GHG emission, maximum as much as 732,999 kg/year or almost 733 tons/year.

Acknowledgments

This research was supported by University of Indonesia (UI) through PUTI Proceeding grant 2020 launched by DRPM UI. Author gratefully thanks to all the parties that involved in providing the data of study but it is not possible to be mentioned one by one. Then, author deeply indebted to the supervisor who always guides patiently and give many advices in writing the study.

References

[1] Fernando A 2019 *Analisis pengaruh urban heat island terhadap konsumsi energi gedung di sheraton jakarta gandaria city hotel* (Universitas Mercu Buana)

[2] Nawawi A 2018 Dampak Radiasi Listrik Tegangan Tinggi Terhadap Kesehatan Manusia *Swara Patra* **8** 93–106

[3] Baek S, Park E, Kim M G, Kwon S J, Kim K J, Ohm J Y and del Pobil A P 2016 Optimal renewable power generation systems for Busan metropolitan city in South Korea *Renew. Energy* **88** 517–25

[4] Castellanos S, Sunter D A and Kammen D M 2017 Rooftop solar photovoltaic potential in cities: How scalable are assessment approaches? *Environ. Res. Lett.* **12**

[5] Hardianto 2019 Utilization of Solar Power Plant in Indonesia: A Review *Int. J. Environ. Eng. Educ.* **1** 1–8

[6] Pandey A K, Tyagi V V., Selvaraj J A, Rahim N A and Tyagi S K 2016 Recent advances in solar photovoltaic systems for emerging trends and advanced applications *Renew. Sustain. Energy Rev.* **53** 859–84

[7] Lee T D and Ebong A U 2017 A review of thin film solar cell technologies and challenges *Renew. Sustain. Energy Rev.* **70** 1286–97

[8] Siddiqui R, Kumar R, Jha G K, Gowri G, Morampudi M, Rajput P, Lata S, Agariya S, Dubey B, Nanda G and Raghava S S 2016 Comparison of different technologies for solar PV (Photovoltaic) outdoor performance using indoor accelerated aging tests for long term reliability *Energy* **107** 550–61

[9] Ghazali A, Salleh E, Chin Haw L, Mat S and Sopian K 2017 Feasibility of a vertical photovoltaic system on a high-rise building in Malaysia: economic evaluation *Int. J. Low-Carbon Technol.* **12** 349–57

[10] Putri R K, Widhi Astuti Y D R, Wardhany A K and Hudaya C 2018 Building Integrated

Photovoltaic for Rooftop and Facade Application in Indonesia *Proc. - 2018 2nd Int. Conf. Green Energy Appl. ICGEA 2018* 171–5

[11] Putra Y U, Dalimi R, Indonesia U, Cina P, Depok K, Indonesia U, Cina P, Depok K and Pendahuluan I 2018 Desain atap rumah untuk mendapatkan secara maksimum energi matahari dengan panel surya 538–44

[12] Husain A A F, Hasan W Z W, Shafie S, Hamidon M N and Pandey S S 2018 A review of transparent solar photovoltaic technologies *Renew. Sustain. Energy Rev.* **94** 779–91

[13] Skunpong R and Plangklang B 2011 A practical method for quickly PV sizing *Procedia Eng.* **8** 120–7

[14] Darussalam R, Rajani A, Kusnadi K and Atmaja T D 2016 Pengaturan Arah Azimuth Dan Sudut Tilt Panel Photovoltaic Untuk Optimalisasi Radiasi Matahari, Studi Kasus: Bandung - Jawa Barat **V** SNF2016-ERE-31-SNF2016-ERE-36

[15] Giatman M 2011 *Ekonomi Teknik* ed A Aliludin (Jakarta: PT. Raja Grafindo Persada)

[16] Fabozzi F J and Peterson Drake P 2009 *Finance, Capital markets, Financial management, and Investment management* (New Jersey: John Wiley & Sons, Inc.)

[17] Jefri N, Fama A, Setiabudy R, Indonesia U, Cina P, Depok K, Indonesia U, Cina P, Depok K, Cost L C, Pendahuluan I, Sistem A K and Biak K 2018 Surya on Grid Pada Sistem Kelistrikan Biak Papua *Semin. Nas. Teknol. 2018* 501–10

[18] Otoritas Jasa Keuangan 2020 Suku Bunga Dasar Kredit *Otoritas Jasa Keuang.* **2019** 1500855

[19] PT. Perusahaan Listrik Negara 2019 Rencana Usaha Penyediaan Tenaga Listrik 2019-2028 *PT. Perusah. List. Negara* 2019–28

[20] Chuang J, Lien H L, Den W, Iskandar L and Liao P H 2018 The relationship between electricity emission factor and renewable energy certificate: The free rider and outsider effect *Sustain. Environ. Res.* **28** 422–9

IOP Conf. Series: Materials Science and Engineering 1041 (2021) 012014 doi:10.1088/1757-899X/1041/1/012014

An Estimation of Electrical Energy Saving for Lecture Building at Universitas Andalas Using Daily Electric Need Pattern

F Akbar[1], V Pujani[2], E Juliani[3] and R Nazir[3]

[1]Information System Department, Faculty of Information Technology, Universitas Andalas, Indonesia
[2]Management Department, Faculty of Economy, Universitas Andalas, Indonesia
[3] Electrical Engineering Department, Faculty of Engineering, Universitas Andalas, Indonesia

Corresponding author: ijab@it.unand.ac.id

Abstract. Electrical energy is a mandatory requirement for carrying out various kinds of human activities. Nowadays, electrical energy must be used efficiently, that is, as needed. However, in use, it is often done excessively and inefficiently, which causes energy loss. It has an impact on the high spent on electricity costs. This research was conducted using an analytical method to obtain a pattern of ideal daily energy consumption needs for various kinds of academic activities in the Universitas Andalas lecture building, namely during the lecture period, exam period, and lecture holidays. The ideal pattern of needs is obtained by considering electrical equipment data, lecture schedules, the questionnaire results on electricity users' behavior, and the value of lighting intensity. The pattern is plotted on a graph every 30 minutes for 24 hours. The pattern results are then compared with current electrical energy usage to evaluate everyday electrical energy use to improve future use. Based on testing the pattern of ideal electrical energy needs with current, it is found that electrical energy is the least efficient during the lecture holidays with a percentage value of 55% to 75%.

1. Introduction

Nowadays, electrical energy is a mandatory requirement for every human activity, both in the industrial sector, households, to buildings. The building sector currently plays an essential role as the highest energy consumer and the main contributor to greenhouse gas emissions. The building sector (residential, commercial, office) represents about 40% of total world energy consumption and offers significant potential for cost-effective energy savings. Therefore, it is mandatory to understand the factors contributing most to energy consumption in a building [1].

The use of electrical energy in buildings is closely related to user behavior in the building itself. The proper use of electrical energy is used based on their respective needs. The needs are different for each building, often causing energy-losses. Theeffective and efficientelectrical energy used can be determined by knowing in advance the proper electrical energy needs. It is necessary to look at the electricity needs used effectively and efficiently in each building.

Energy management is a policy aboutusing energy to obtain maximum benefits (minimum costs) and enhance a competitive position [2]. In general, things that must be considered in preparing an

IOP Conf. Series: Materials Science and Engineering 1041 (2021) 012014 doi:10.1088/1757-899X/1041/1/012014

energy management program are to consider several aspects, namely: the behavior of energy users, the equipment technology used, installation, or installation of equipment and equipment maintenance management. The energy management system of ENMs is a process of continuous improvement by the PDCA loop (Plan Do Check Act cycle)[3]. Since this topic is related to evaluate the use of electrical energy in existing buildings, it focuses more on the Check and Act section.

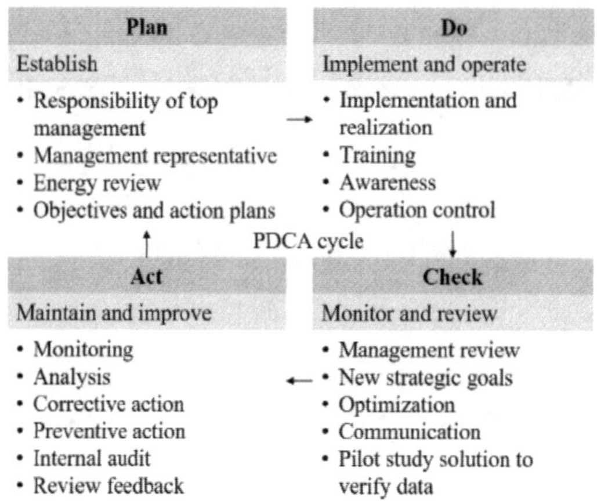

Figure1. PDCA cycle in the energy management system[3]

Energy conservation is an effort to make energy use more efficient for a need to avoid energy loss. Energy conservation is the efficient and rational use of energy without reducing essential energy use. Energy management is an effort to regulate and manage energy use as efficiently as possible in buildings without reducing the comfort level in the residential environment and productivity in the work environment [4]. Awareness and concern for the use of energy and the indoor environment of all people using the building are very important. Education and training can improve building occupants' understanding of building management efforts to maintain and increase the building's energy efficiency and contribution. Zero Energy Building (ZEB) is defined as a building used as a residential or commercial facility that can reduce energy needs drastically so that efficiency is achieved. In this state, a balance of energy needs supplied with renewable energy is achieved. The steps to attain Zero Energy Building are as follows [5]:

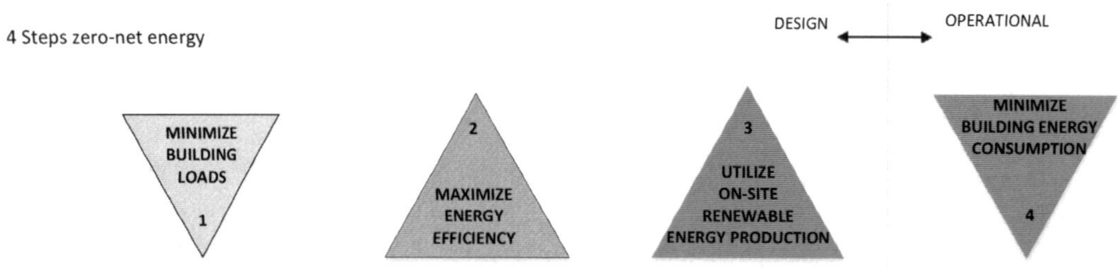

Figure2. Chart of Design and Implementation of Zero Energy Building (ZEB)*[5]*

Where :

1. Minimize Building Loads (Minimizing Building Loads)
2. Maximize Energy Efficiency (Maximizing Energy Efficiency)
3. Utilize On-Site Renewable Energy Production (Utilizing the Production of Renewable Energy On-Site)

4. Minimize Building Energy Consumption (Minimizes Building Energy Consumption.

Universitas Andalas is one of the building sectors in the education sector that consumes electrical energy. As mentioned in the announcement in the entire building at Universitas Andalas by Vice-Chancellor, electricity costs reach 720 million per month. Six hundred million must be paid to fulfill electricity needs only for campus activity and 120 million for electricity payments for the campus teaching hospital[1]. The preliminary research had been conducted to address this issue[4,6,7]. However, the need for electrical energy in the buildings on Universitas Andalas campus varies according to its use and function. One of them is the lecture building, which is always used for the teaching and learning process by Universitas Andalas academic community. Teaching and learning activities have been arranged in the campus academics' schedule, which also regulates campus activities each month, such as any month for regular lectures or exams, be it midterm exams or final semester exams to class holiday schedules. Preliminary research had been conducted to address this problem.

2. Materials an methods

This research is calculates each class's power and energy in the lecture building, defined in figure 3.

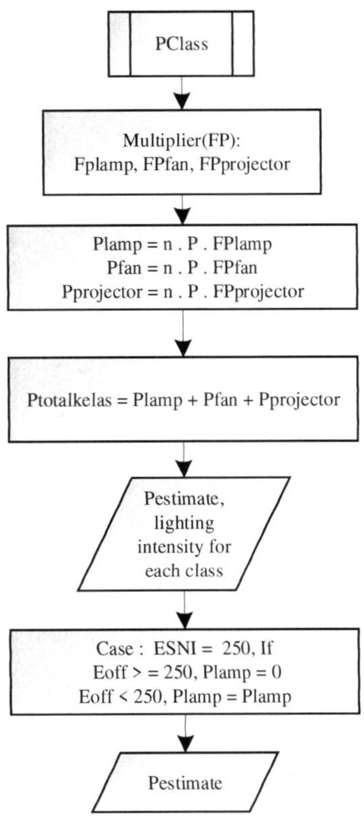

Figure3. The flowchart in calculating electric needed for each classroom

In getting current energy usage, we develop a dashboard for the energy information system. It is based on a client-server architecture and web-based application that collect data from each measurement point. Each measurement point is recorded energy data for each building, including the lecture building. The data is transferred using TCP/IP protocol into the server to visualize as an energy dashboard. The system's architecture can be seen in figure4, and figure 5 shows a dashboard of the energy information system.

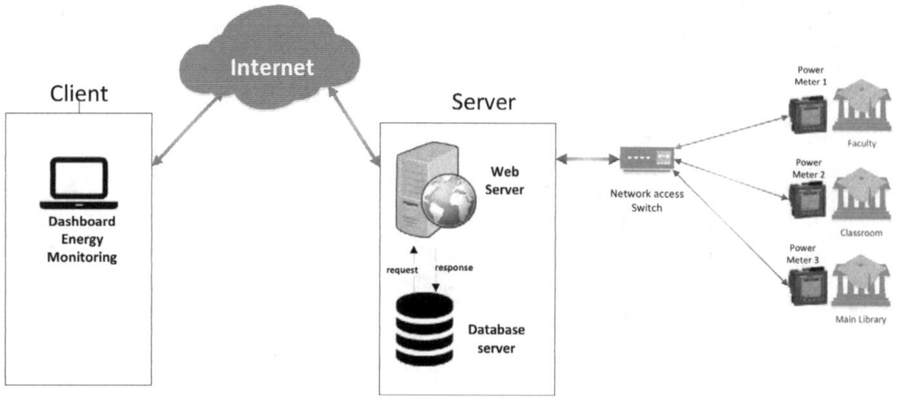

Figure 4. Energy information system architecture

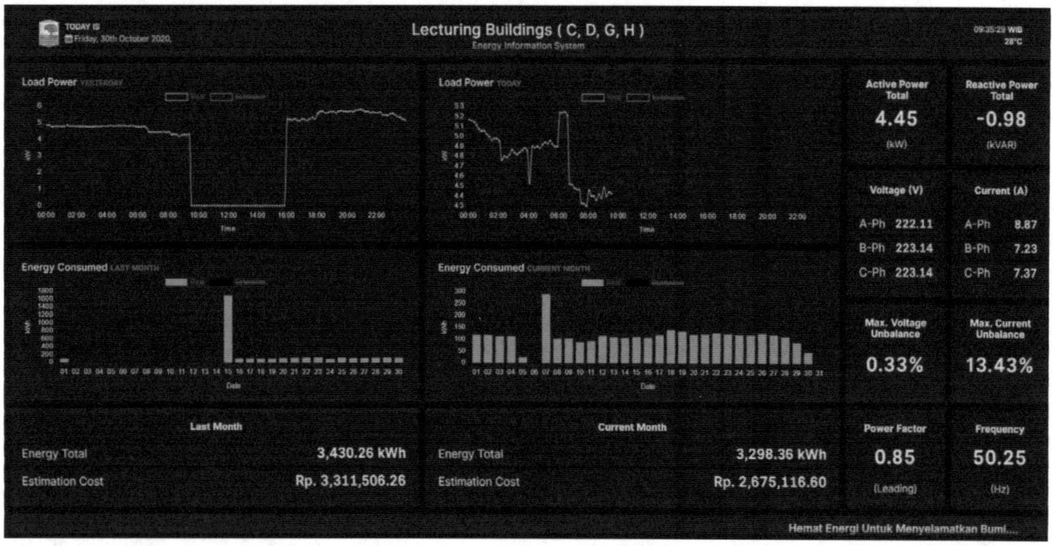

Figure 5. The dashboard of Energy Information System

3. Results and discussion

An estimation data was carried out every 30 minutes for 24 hours for one week. Energy consumption patterns are grouped into three based on the lecture schedule period in the lecture building, namely during the lecture period, the exam period, and the lecture holiday period. The pattern of electrical energy consumption in the three periods varies for each day in all lectures. The pattern of electrical energy needs per day in a week between each lecture activity period is also different.

In modeling the ideal pattern of electrical energy consumption needs in the C, D, G, and H lecture buildings of Universitas Andalasis obtained by calculating and processing data by entering data in the form of existing electrical equipment data, class schedules, multiplier factor (FP) values and intensity values lighting. Based on these data, the pattern of electricity consumption needs can be seen in figure 6-11:

IOP Conf. Series: Materials Science and Engineering 1041 (2021) 012014 doi:10.1088/1757-899X/1041/1/012014

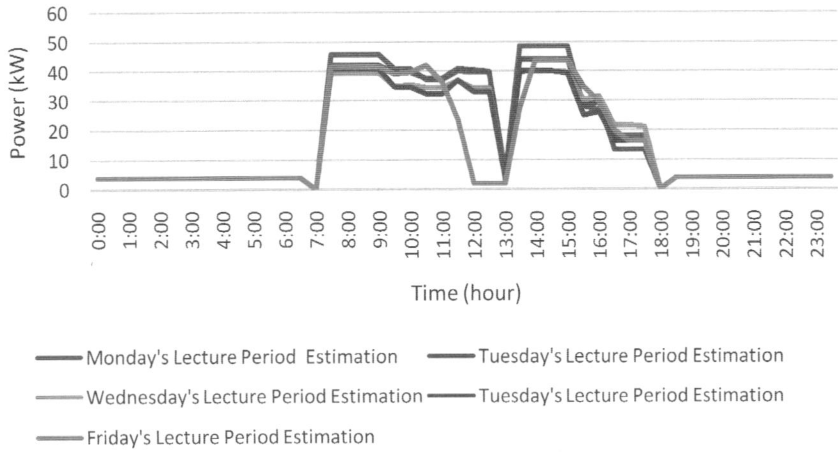

Figure6. Estimation ofpower usage on a workday during the lecture period

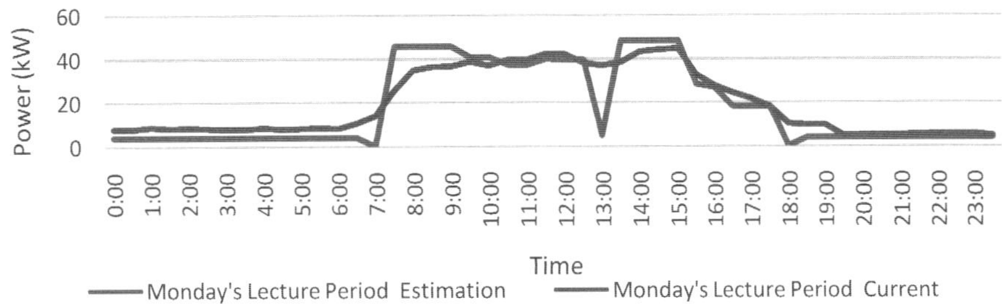

Figure7. Comparing an estimation and current usage on Monday during the lecture period

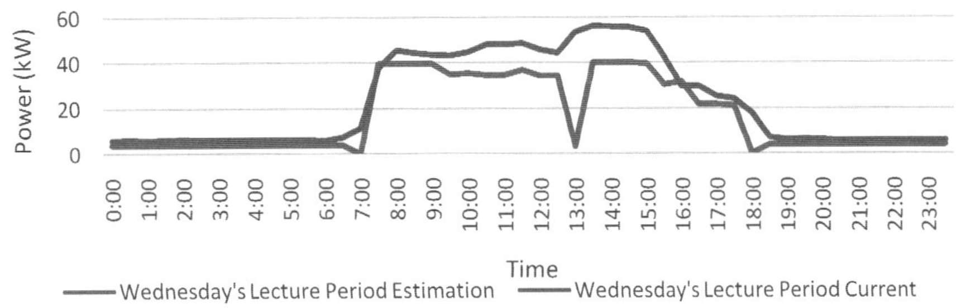

Figure 8. Comparing an estimation and current usage on Wednesday during the lecture period

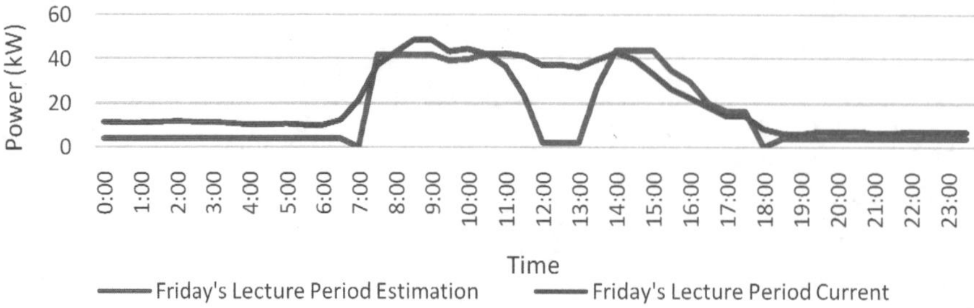

Figure 9. Comparing an estimation and current usage on Friday during the lecture period

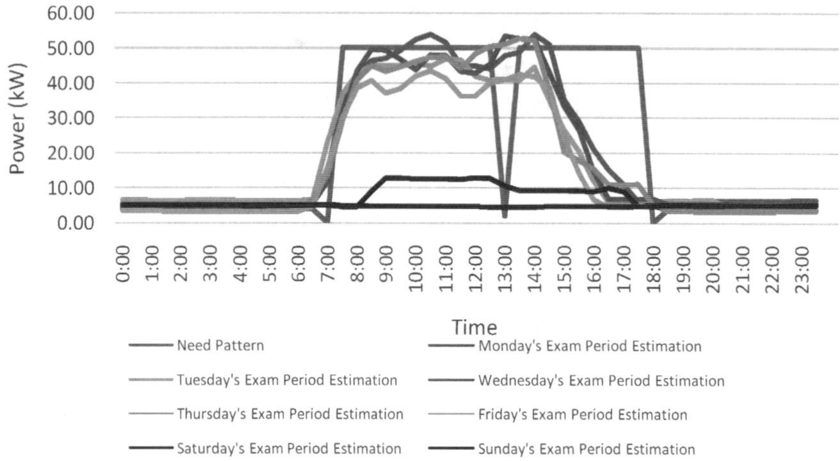

Figure 10. Estimation power usage on a workday during the exam period

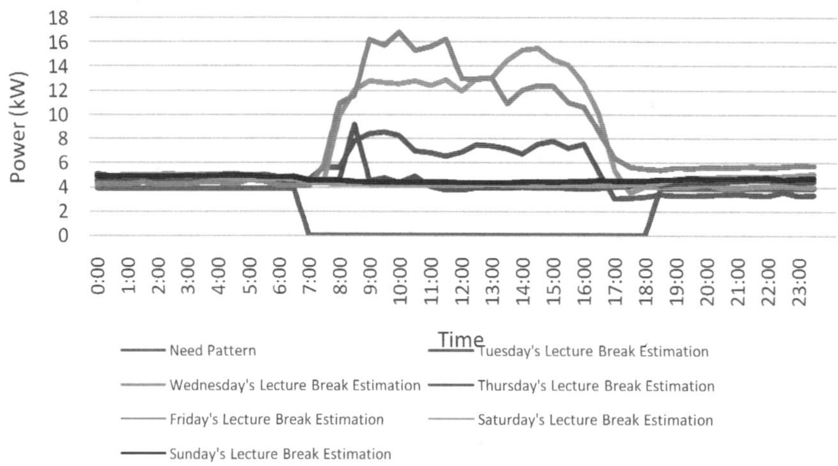

Figure 11. Estimation power usage on a workday during the lecture break

Based on figure 7-12, it can be seen that the pattern of electricity consumption needs is different for each day. However, from the graph, it can be seen that for current use, as shown in the real-time graph, the everyday use of electricity consumption is still used outside of necessity. For example, the most obvious thing is for the lecture holidays, in which during the day, there should be no use of electrical energy. However, in real-time conditions, it can be seen that there is still electrical energy, even having a value of more than 15 kW likewise on other days. During the lecture period, outside needs can be seen at the afternoon break, which should have decreased the power value, but the decrease in power that occurs is minimal from the previous power value.

The proper use of electrical energy is to use electrical energy effectively and efficiently. The practical and efficient use of electricity is meant to use electrical energy as needed. With the previously obtained power need patterns and real-time measuring power, it can be seen whether the current use of electrical energy has been effectively used by using the following equations:

$$W = P_{tot} \cdot t \tag{1}$$

With equation 1, each type of day's energy value is obtained in the need pattern and real-time measurement.

$$Percentage\ of\ loss\ (\%) = \frac{E_{\ pola} - E_{realtime}}{E_{realtime}} \times 100\% \tag{2}$$

By using equation 2, the percentage value is obtained to determine the percentage of loss by looking for differences in energy needs and energy patterns in real-time measurements. The following is the energy and rate of loss requirement patterns and real-time measurements for each type of day.

Based on table 1, it can be seen that the energy difference in need patterns and real-time measurements. Almost all day in the current usage measurement of energy value is more significant than the energy in the need pattern, except during the test period for the type of workday. It is also clarified by the positive percentage value, which indicates that the real-time energy value is greater than the energy need pattern. Meanwhile, the negative percentage indicates that the real-time energy value is smaller than the energy need pattern value.

Overall, the largest percentage value of electrical energy between need patterns and real-time measurements is during the lecture holidays and weekends (Saturday and Sunday) for all activities, with percentage values ranging from 55% to 75%. Meanwhile, on other days the percentage value feels below 30%.

Table 1. Daily energy usage dan rate in comparing estimation based on the design of energy used pattern and existing condition.

Season	Day	Estimation usage based on need pattern	Current usage	Percentage of loss (%)
Lecture Period	Monday	432.76	477.76	9.4%
	Tuesday	408.202	422.335	3.3%
	Wednesday	393.961	551.025	28.5%
	Thursday	380.663	454.86	16.3%
	Friday	363.901	507.42	28.3%
	Saturday	48.3375	178.615	72.9%
	Sunday	48.3375	187.655	74.2%
	Total	2076.162	2779.67	33.3%
Exam Period	Monday	551.789	495.425	-11.4%

	Tuesday	551.789	448.385	-23.1%
	Wednesday	551.789	463.26	-19.1%
	Thursday	551.789	451.875	-22.1%
	Friday	551.789	387.98	-42.2%
	Saturday	48.3375	172.245	71.9%
	Sunday	48.3375	119.155	59.4%
	Total	2855.62	2538.325	1.9%
Lecture break	Monday	48.3375	157.685	69.3%
	Tuesday	48.3375	127.68	62.1%
	Wednesday	48.3375	181.435	73.4%
	Thursday	48.3375	107.12	54.9%
	Friday	48.3375	192.29	74.9%
	Saturday	48.3375	108.78	55.6%
	Sunday	48.3375	111.015	56.5%
	Total	338.3625	986.005	63.8%

4. Conclusion

Based on the test and analyzes that have been carried out, it can be drawn that the pattern of daily electricity consumption needs is influenced by electrical equipment, lecture schedules, user behavior, and the value of lighting intensity in the lecture building. As long as outside working hours, the real-time power value is greater than the required pattern power value, and it can be caused by negligence in energy management. It also found that the pattern of electrical energy consumption needs on lecture days is different for each working day, where the real-time power value is greater than the pattern power requirement.In the need pattern for electrical energy consumption during the exam period, the real-time power value is close to the required pattern power during the morning before noon. During the daytime, the relative power value is relatively low compared to the power need pattern. It also can be concluded that in the pattern of energy usageduringlecture break and holidays, it has a power value that is quite large compared to the value of the power need pattern, due to illegal or unregistered activities and negligence in energy management. The use of electrical energy is currently quite efficient during working hours. However, there is still energy-loss outside working hours and during class holidays.The most significant loss of electrical energy occurs during the lecture holidays and Saturday, Sundays, based on the test results. It found that savings opportunities can be made, especially outside working hours and during lecture holidays.

Acknowledgments

Authors would like thanks to Andalas University for the financial support this work, through PDU KRP1GB KRP1GB Funding No. T/8/UN.16.17/PP.Energi-KRP1GB/LPPM/2020.

References

[1] R. D. Y 2017 Konsumsi Listrik Unand Capai 720 Juta Perbulan *Genta Andalas*

[2] M I, Hartati R and Kumara I N S 2012 Pengelolaan Energi Listrik Pada Gedung Fakultas Kedokteran Universitas Udayana Kampus Sudirman Denpasar *Majalah Ilmiah Teknologi Elektro* **11** 25–32

[3] Mulyani F, Suyono H and Hassanah R N 2018 Audit dan Rancangan Implementasi Sistem Manajemen Energi berbasis ISO 50001 di Universitas Brawijaya Malang *Jurnal EECCIS* **12** 78–84

[4] Pujani V, Akbar F and Nazir R 2019 Management Review of Energy Consumption: The Energy Saving Opportunity in University Buildings Proceedings of the 2019 5th International Conference on Industrial and Business Engineering pp 110–6

[5] Magdalena E D and Tondobala L 2016 Implementasi Konsep Zero Energy Building (Zeb) dari Pendekatan Eco-Friendly pada Rancangan Arsitektur *Media Matrasain* **13** 1–15

[6] Pawawoi A, Zuheldi F A and Nazir R 2020 Analysis of energy losses reduction potential on the distribution line of campus building through electric power quality improvement *Indonesian Journal of Electrical Engineering and Computer Science* **17** 868–76

[7] Pujani V, Pawawoi A, Akbar F, Zuhaidi Z and Nazir R 2020 An electric energy reduction model for campus using the method of controlling energy consumptions *International Journal of Smart Grid and Clean Energy* **9** 411–9

Design a Stethoscope Digital to Keep the Medical Practitioner at an Adequate Distance While Examining the Patient

M I Rusydi[1], E M Ningsih[1], A Pawawoi[1], R Nofendra[1], N Windasari[2]

[1]Department of Electrical Engineering, Faculty of Engineering, UniversitasAndalas, Padang City, West Sumatera, Indonesia
[2]Faculty of Medicine,UniversitasAndalas, Padang City, West Sumatera, Indonesia

Corresponding author: rusydi@eng.unand.ac.id

Abstract. A stethoscope is an instrument in medicine used to listen to the sounds produced in the body or auscultation, primarily to listen to the sound from the lungs and heartbeat. Medical practitioners usually use the sound from a stethoscope as an indicator to diagnose an illness. Regular stethoscope with earpiece is less effective to use since the novel coronavirus quickly spreading from saliva droplets. Medical practitioners have difficulty attaching the conventional stethoscope to the ear because they have to use personal protective equipment, including head covering, to deal with coronavirus patients and patients with other medical complaints. This research designs a digital stethoscope that can be used even the head of medical practitioners covered by safety clothes. This stethoscope has a sound sensor, OLED, a microprocessor, and a speaker. The sound of lungs and heartbeat from 30 volunteers was recorded to investigate the frequency of noise, lung, and heartbeat. A bandpass filter was designed to get the signal with an upper cut-off frequency of 50 Hz to 80 Hz and a lower cut-off frequency of 5 Hz. The result shows that this stethoscope can detect the number of heartbeats and give a clear lung sound. This device can help medical practitioners examine patients while keeping their distance from patients.

1. Introduction

A stethoscope is a prime medical instrument usually used by the medical practitioner to diagnose and treat the patients[1]. Currently, there are two types of stethoscopes; those are acoustic stethoscopes and electronic stethoscopes. The technique of using a stethoscope to distinguish between normal and abnormal sounds is called auscultation[2]. Auscultation can provide accurate and diagnostic information regarding the heart, lung, and tracheal[3]. This diagnostic is usually precisely done in the cardiovascular, respiratory, and gastrointestinal systems. However, because of the Coronavirus pandemic, the regular stethoscope that uses earpiece is less significant, considering that the medical practitioners have to wear clothes that cover all of their bodies, including the head[4,5]. It is officially named Personal Protective Equipment (PPE). For this reason, a stethoscope is needed that can be connected using speakers and a headset so that medical personnel can treat patients while still wearing complete Personal Protective Equipment.

IOP Conf. Series: Materials Science and Engineering 1041 (2021) 012015 doi:10.1088/1757-899X/1041/1/012015

There are several journals and research related to stethoscopes, both acoustic stethoscopes and digital stethoscopes[6–8]. All of this research used some kind of signal filters to optimizing the performance of the stethoscope. The use of the band pass filter to filter the sound from the mic condenser is more frequently implemented. Some of them use medical analysis, such as systolic and diastolic murmur[4]. Data processing using the Internet of Things has been carried out related to this topic[8,9]. The purpose of this final project is to design a digital stethoscope with a sound output system that can be connected to speakers and headsets so that medical practitioners have no difficulty examining patients using a stethoscope even with a complete PPE. This research designs a digital stethoscope that can be used even the head of medical practitioners covered by safety clothes.

2. Materials and methods

The digital stethoscope captured sound from the lungs and heartbeats as the input. The output is the heart or lung sound that can be heard from the speaker and headset. Besides that, the rate of the heartbeat has been shown in OLED (organic light-emitting diode) as the display.

2.1. Data Collection

Data collection is carried out to determine the cut-off frequency of filter in order to get clear sound from the digitalstethoscope. The data were collected by recording the sound of heartbeats and lungs (breath) of 30 respondents. Lungs and heartbeats sounds were recorded using three stethoscopes with different brands. The output from the stethoscope goes directly to the laptop using a modified audio jack cable. The chestpiece of stethoscope is attached to several points on respondent's body or what is called auscultation. Heart and lung auscultations are in the area shown in figure 1.

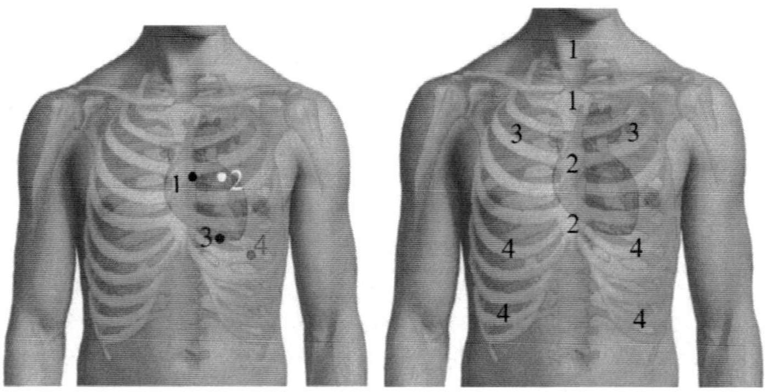

(a) (b)

Figure 1. (a) Heart auscultation's points, and (b) Lung auscultation's points

Figure 1a showsfour locations of heart auscultations. The heart auscultation positions from 1 to 4, respectively, are aortic valve, pulmonary valve, tricuspid valve, and mitral valve. Lung auscultation is carried out by taking data from four breath sounds (figure 1b). Those lung auscultation positionsfrom 1 to 4 are tracheal breath sounds, bronchial breath sounds, bronchovesicular breath sounds and vesicular breath sounds.

2.2. Fast Fourier Transform

The obtained data was in time domain signal form. The timedomain signal was converted into the frequency domain signal by using Fast Fourier Transform (FFT)[10]as formulated in equation (1). It is necessary to distinguish the noise, lungs and heartbeats sound frequency.

$$X[k]=\sum_{n=0}^{N=1} x[n],e^{-\frac{j2\pi nk}{N}} \qquad (1)$$

Where

k = 0, 1, 2, …, N-1
x[n] = input signal
N = length of input signal

MATLAB provides the FFT function to assist mathematical computations with a specific syntax. The FFT result in MATLAB will obtain the frequency domain signal and a graph of the frequency distribution from the sound that has been recorded. This resultis analyzed to differentiate the frequency of the heart, breath, and noise. The analysis results obtain both upper cut-off frequency and lower cut-off frequency to design the bandpass Filter.

2.3. Stethoscope Digital System

The hardware of the digital stethoscope required consists of a stethoscope chest piece, rubber tube, mic condenser, a preamp circuit, a bandpass filter circuit, a microprocessor, a speaker, and a display output. The digital stethoscope receives sound as input captured from the bell and diaphragm where the stethoscope chest piece is located. The system's output is sound and the display of the number of heart rate in a minute. The digital stethoscope hardware design is shown in figure 2.

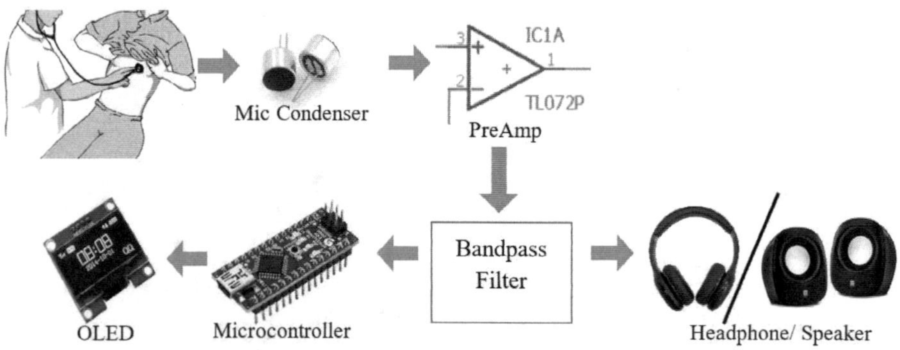

Figure 2. Hardware design.

The stethoscope or tubing will be connected to the condenser mic then proceed to a small box. This box contains the amplifier circuit, filters circuit, and a microcontroller. It also has a button to adjust the volume, an audio jack, and a headphone port. The battery to be used is a rechargeable 3.7 V battery. In addition, the stethoscope will be equipped with an OLED so that it can display the heart rate in a minute. The next component is the speaker or headset. The speaker of the headset and the box are connected by wire using an audio jack cable.The designed digital stethoscope work as follows:

1. The heartbeat and lung sound is captured by the surface of the diaphragm and bell. The sound wave spreads inside the tube, creating the vibration. The condenser mic gets the vibrations as the input and converts to the voltage value.
2. Furthermore, The amplifier circuit using gain the signal using the dual op-amp IC TDA2822.
3. A High Pass Filter (HPF) and a Low Pass Filter (LPF) process the signal.
4. The microcontroller displays the heart rate frequency per minute and forwardsthe signal to the speakers or headphones to produces sound.

2.4. Amplifier Circuit

Since the mic condenses only features a peak to peak voltages of around 200 millivolts, it needs a decent amplifier to increase the voltages. The amplifier circuit consists of two inverting circuits so

IOP Conf. Series: Materials Science and Engineering 1041 (2021) 012015 doi:10.1088/1757-899X/1041/1/012015

that the results in the last output will have the same phase value as the input value. The schematic of the amplifier circuit is shown in figure 3. This circuit also has additional circuit that is intended to activate the condenser mic.

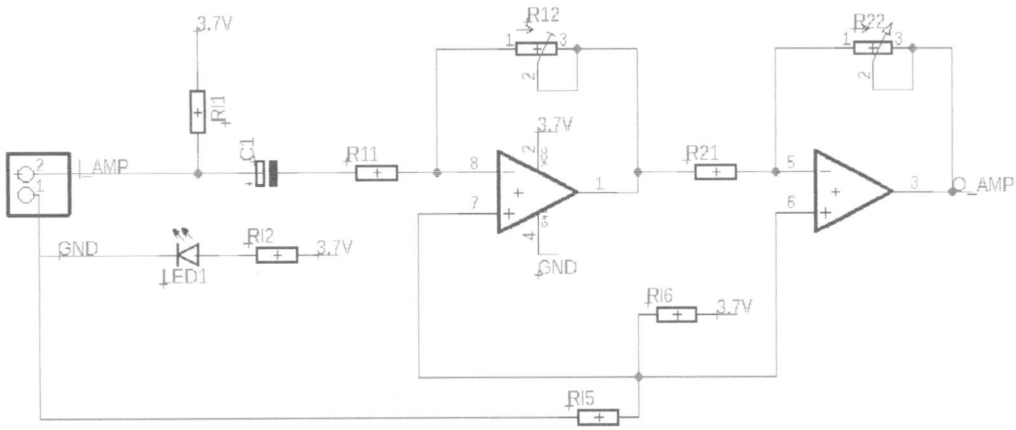

Figure 3. Schematic of an amplifier circuit.

In the circuit, R1 and C1 is the component to activate the mic condenser. The gain value in the amplifier circuit is obtained by the inverting gain formula in equation (2).V_{out} indicates the voltage output of the circuit, and V_{in} indicates the voltage input of the circuit. Gain can also be obtained from the resistors R11, R12, R21, and R22 based on equation (2). The resistors R11 and R12 are the resistors to determine the gain from the first amplifier, while the resistors R21 and R22 are the resistors to determine the gain from the second amplifier. Both resistor RI5 and RI6 are used as the voltage divider on the noninverting input. It is utilized to create an offset voltage at which the output voltage will oscillate around.

$$Gain = \frac{V_{out}}{V_{in}} = \left(\frac{R12}{R11} + 1\right)\left(\frac{R22}{R21} + 1\right) \quad (2)$$

2.5. Band Pass Filter Circuit

In the bandpass filter, the cut-off frequency will be selected according to the analysis of the FFT chart of heartbeat signals and breath sounds. The amplifier bandpass filter circuit is made with the circuit shown in figure 4.

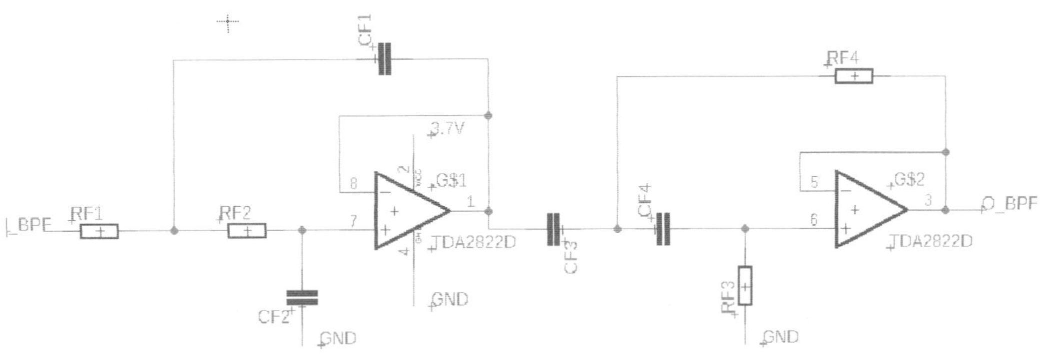

Figure 4. Schematic of band pass filter circuit.

IOP Conf. Series: Materials Science and Engineering 1041 (2021) 012015 doi:10.1088/1757-899X/1041/1/012015

A bandpass filter is a filter that combinesa low pass filter and a high pass filter in one circuit. In figure 4, both low pass and high pass filter that used are second order. It contains two resistors and two capacitors in each circuit. The values of the capacitors and resistors are determine based on the cut off frequency, which is obtained using equation (3)

$$f_{cutoff} = \frac{1}{2\pi\sqrt{C_1C_2R_1R_2}} \quad (3)$$

Equation (3) can be used in both low pass and high pass filters. Low pass filter suppressed the frequency that more than the cut-off frequency value, while high pass filter stopped the frequency that less than the cut-off frequency value.

3. Results and discussion

This research includes the results of the analysis of the FFT graph on the sound recording, the gain for the amplifier circuit, the determination of the capacitor and resistor values used in the filter circuit the heart rate counter features.

3.1 Gain of Amplifier

The amplifier circuit usesa dual operational amplifier in this research. The final output of the circuit has the same phase as the input because both are made in the form of an inverting circuit. In the first op-amp, the value of resistor R1 and R2 are 1K Ω and 10K Ω. In the second op-amp, the value of resistor R1 is 2K Ω and R2 is a potentiometer with a maximum value of 33K Ω. By using equation (2), the maximum gain of the first op-amp is 11×, and the maximum gain of the second op-amp is 17.5×. Therefore, the maximum gain that can be achieved in this circuit is 192.50 multiplies. In the noninverting input, the voltage divider is set to oscillate the output around 2.5 V, so both resistor RI5 and RI6 have a value of 10K Ω.

3.2 Band Pass Filter

The Fast Fourier Transform is used to analyze the distribution of the frequency of the noise, heartbeat, and lung sound. The researcher used noise reduction tools in software Audacity to separate the noise from the desired sound. The frequency distribution of heartbeat and lung signal before and after filtered is shown in figures 5 and 6.

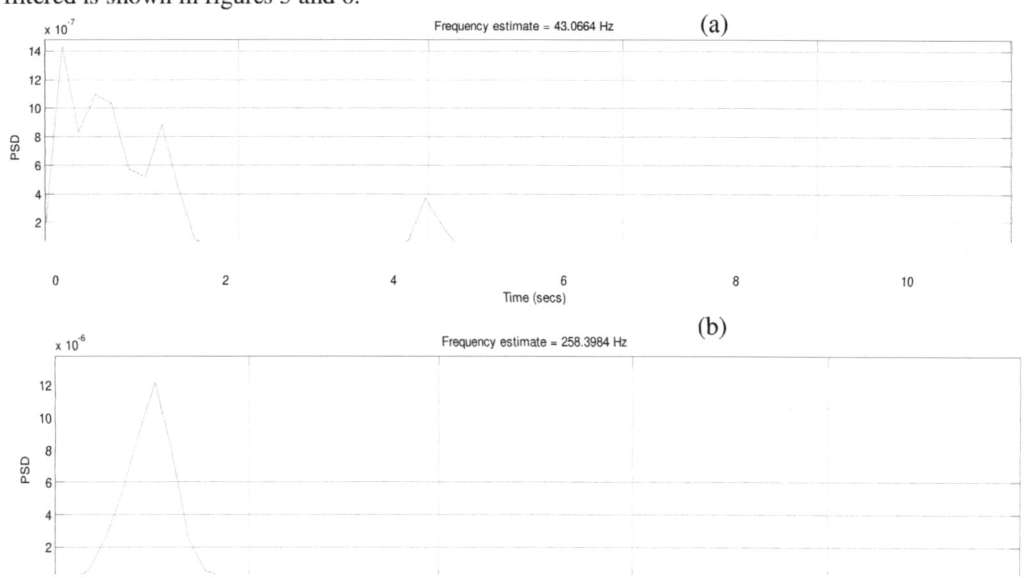

Figure 5. The heartbeat FFT signal (a) before filtered and, (b) after filtered.

From the comparison before and after noise reduction in the heartbeat data sound in figure 5, the noise spread from the frequency 0 - 80 Hz and above 334 Hz. The frequency estimate of the heartbeat sound is 258.4 Hz.

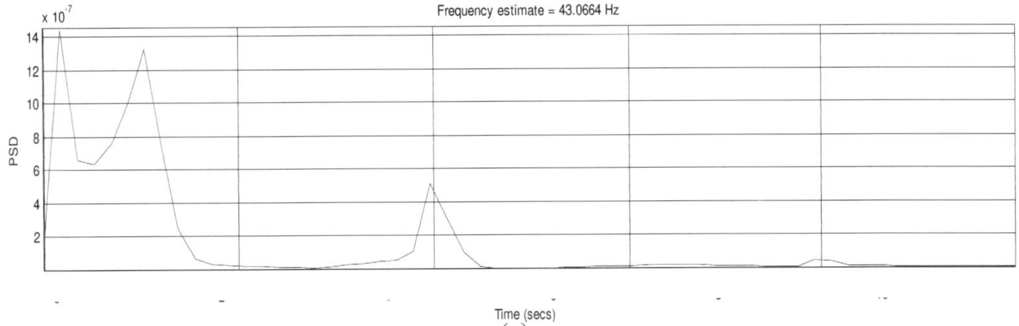

(a)

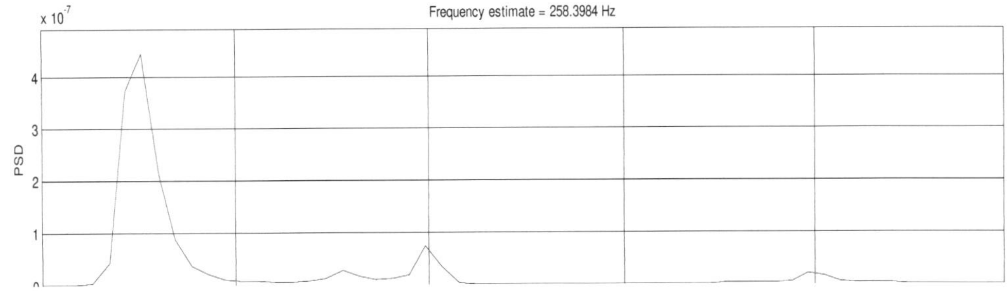

(b)

Figure 6. The lungs FFT signal (a) before filtered and, (b) after filtered.

The lung sound from the FFT plot in figure 6 has the frequency estimate at 258.39 Hz with a frequency distribution between 100 Hz until 370 Hz. From the FFT plot analysis, the lower frequency of the band pass filter is around 70 Hz, and the upper cut-off frequency is around 370 Hz.

Equation (3) is used with the approach value of the resistors and capacitors that sold in the market. The value of all capacitors in this circuit is 100nF to make it easier to find resistor values. For the low pass filter, the value of resistors are 5K Ω and 3.6K Ω. The upper cut-off frequency in the low pass filter circuit is 375.13 Hz. The resistors for high pass filter are 100K Ω and 1M Ω so the lower cut-off frequency is 5 Hz.

3.3 Heart rate counter

The digital stethoscope provides the heart rate per minute counter feature. The heart signal that enters the microcontroller is then processed using the threshold method. Then the heart rate frequency obtained is then obtained from the interface in the form of an OLED. The display on the OLED is shown in figure 7.

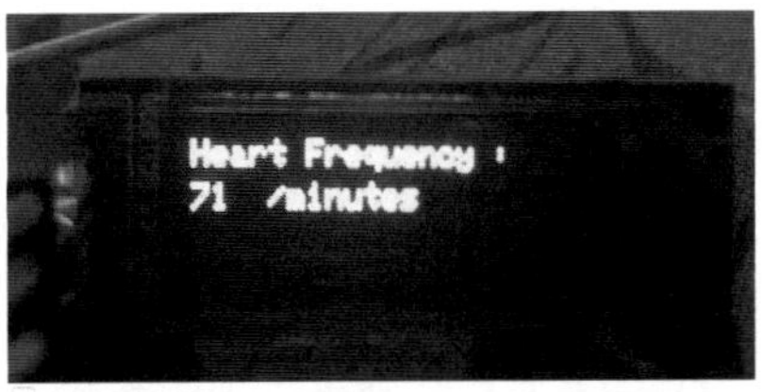

Figure 7. OLED Display

4. Conclusion

The research aims to design a digital stethoscope that can be used even the head of medical practitioners covered by safety clothes. Based on the analysis of FFT signal of heartbeat and lung sound, the band pass filter si designed to reduce the lower cut-off frequency is 5 Hz and the upper cut-off frequency is 376.13 Hz. The amplifier circuit has a maximum gain of 192.50 × with the output voltages oscillate around 2.5 V. The result shows that this stethoscope can detect the number of heartbeats and give a clear sound of lungs. This device can help medical practitioners examine patients while keeping their distance from patients.

Acknowledment

This research is funded by Universitas Andalas contract number: T/4/UN.16.17/PP.RPP/LPPM/2020

References

[1] Kalra S and Reddy S 2020 New Scope for the Stethoscope *Mayo Clin. Proc. Innov. Qual. Outcomes***4** 1–2

[2] Wilson M A, Fouts B L and Harding C J 2019 Investigation of Stethoscope Technology for En Route Care *Air Med. J.***38** 30–5

[3] Sánchez I and Vizcaya C 2003 Tracheal and lung sounds repeatability in normal adults *Respir. Med.***97** 1257–60

[4] Buonsenso D, Pata D and Chiaretti A 2020 COVID-19 outbreak: less stethoscope, more ultrasound *Lancet Respir. Med.***8** e27

[5] Vasudevan R S, Horiuchi Y, Torriani F J, Cotter B, Maisel S M, Dadwal S S, Gaynes R and Maisel A S 2020 Persistent Value of the Stethoscope in the Age of COVID-19 *Am. J. Med.***133** 1143–50

[6] Makaryus A N, Makaryus J N, Figgatt A, Mulholland D, Kushner H, Semmlow J L, Mieres J and Taylor A J 2013 Utility of an advanced digital electronic stethoscope in the diagnosis of coronary artery disease compared with coronary computed tomographic angiography *Am. J. Cardiol.***111** 786–92

[7] Silverman B and Balk M 2019 Digital Stethoscope—Improved Auscultation at the Bedside *Am. J. Cardiol.***123** 984–5

[8] Yun H, Kim H, Kim E and Jun M B G 2020 Development of internal sound sensor using stethoscope and its applications for machine monitoring *Procedia Manuf.***48** 1072–8

[9] Andrès E, Reichert S, Brandt C, Hill N and Gass R 2016 Développement et expérimentation d'un nouveau stéthoscope numérique, communiquant et intelligent *Eur. Res. Telemed.***5** 145–55

[10] Lyons R G 2004 *Understanding digital signal processing* (Prentice Hall)

IOP Conf. Series: Materials Science and Engineering 1041 (2021) 012016 doi:10.1088/1757-899X/1041/1/012016

Design of Polarimetric-Based Optical Current Sensor for Electric Power System Application

M Nasir[1] and Adrianti[1]

[1]Electrical Engineering Department, Universitas Andalas, Padang, 25163, Indonesia

Corresponding author: m.nasir.sonni@eng.unand.ac.id

Abstract. Many problems arise in the electric power protection system due to applying a ferrous core current transformer (CT). One of the most crucial problems is a magnetic saturation issue, which tends to cause a measurement error on the secondary side of the CT and could reduce the performance of the protection system. An alternative solution to this issue is the needs to develop an electric current sensor utilizing optical materials. A polarimetric-based optical current sensor (OCS) is a technique considered to be compatible with the future smart grid environment. The OCS principles are based on the magneto-optic Faraday effect that could produce optical rotation in the optical media when a polarized light propagates in those materials. Due to the linear and circular birefringences effect in optical fiber medium are exist, therefore, in this paper the OCS and its associated elements are modelled by considering the birefringence effect in the optical sensor system including optical fiber links. Then, the OCS model is translated to a computer simulation program using a computer application software. Orientation of the second polarizer (analyzer) to the first polarizer in the simulation were conducted using two different angles which are angle α of 45° and 90°. Performance assessment of the OCS model is presented using two various electric current of the power system test to investigate the presence or absence of the saturation effect that may arise in the polarimetric-based optical current sensor model. The simulation results have provided that the saturation effect disappear in the simulation and the analyzer orientation angle α of 45° gives better performances of optical power modulation than the other angle.

1. Introduction

The conventional voltage and current transformers are intended to provide a scaled down replica of primary voltage and current, respectively, from transmission levels down to a suitable level for protective relays. As a consequence of reducing the electrical quantities in magnitude the protection relays can be designed as relatively small and inexpensive devices. Both instrument transformers also provide insulation of the low voltage relay equipment from the high voltage power system.

Magnetic saturation phenomenon in CTs due to the use of ferrous materials as CT's core is one of the main limitations of the conventional CTs. The saturation process is initiated by the creation of high flux density in the iron core that can be caused by several factors such as high fault currents with or without a dc offset, residual flux, high secondary burden, or a combination of these factors. The saturation occurs when the flux density exceeds the design limits of the CT core. As a result, accuracy of the CT becomes poor because the secondary current is lower in magnitude than the value expected from the nominal ratio of the CT. Another undesirable effect is the introduction of harmonic components into the output current that may cause a distortion [1,2]. This, in turn, can potentially affect the protection relay performance.

Other potential problems of the conventional instrument transformers may result from size, weight, and insulation requirements. Increasing the power system generating and transfer capacity requires a bigger CT size CT while increasing system voltage requires a better insulation and suitable dimensional design. It will result the massive CT's structures that ultimately consume high cost.

An alternative solution for avoiding saturation problems is the utilization of a different CT construction without ferrous a core. Some CT manufacturers have developed alternative designs such as air-gap CT, linear coupler, Rogowski coil, hybrid optical and pure optical current transducer techniques [3]. The optical current sensing techniques attract researcher to develop a new design of optical current transducer for electric power system applications.

Optical systems for high voltage current measurements that is known as optical current transducer have been developed in early 1970s when several different approaches were well documented [4-6],[7]. The optical current transducers have several advantages over the conventional CT due to their small size, light weight, low cost, absence of ferromagnetic resonance problems, large linear measurement range, broad response frequency bandwidth and electromagnetic interference immunity [8,9],[10]. Therefore, this paper introduced a design of polarimetric based optical current sensing technique.

2. Materials and methods

A formal system for analyzing both the light and optical material interaction is known as Jones Formalization. shown in figure 1. It provides an efficient method for describing the polarized light and propagation of the electric field through several polarizing components at the amplitude level [11]. When the polarized wave incident $\boldsymbol{E_i}$ propagates in a single polarizing optical system J, one or more modified TE exiting waves $\boldsymbol{E_o}$ are emerged from the optical system. This Jones formalization is constrained in equation (1)

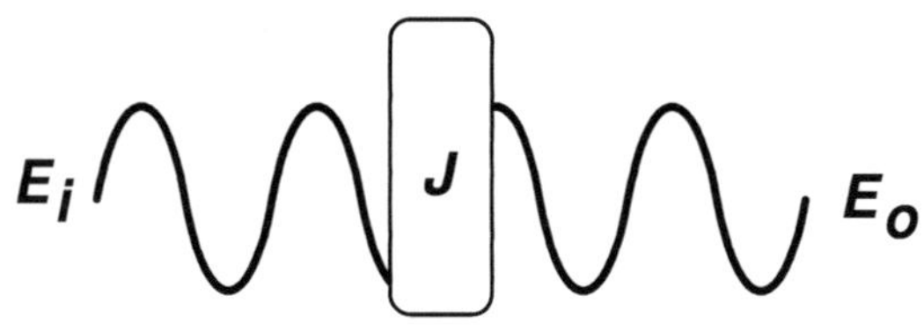

Figure 1. Interaction of polarized light and optical devices J

$$\boldsymbol{E_o} = J\,\boldsymbol{E_i} \tag{1}$$

A fundamental difference between optical current measurements and conventional CTs is in the signal power involved. The secondary signal of a conventional CT has a power level of several watts, whereas the power in an optical current measurement is only a few micro watts (μW). In general, the current being measured in an optical current measurement is represented as modulated light.

The most common optical technique for current sensing is polarimetric method which is based on the Faraday rotation effect that is exhibited by magneto-optic crystals. This method has been reported in literatures [3,12-16].

Crystal materials for the optical current sensor application could be diamagnetic, paramagnetic or ferromagnetic crystals [8] such as optical glass MOC-series, SF-6 glass, or Terbium Gallium Garnet (TGG) crystal. These magneto optic crystals become an actively optical medium when it is under magnetic field effect. If a linearly polarized light incident $\mathbf{E_i}$ passes through the actively optical medium, simultaneously the direction of the polarized light state emerging $\mathbf{E_o}$ is rotated parallel to the light propagation direction in proportion to the magnetic field **H** as displayed in figure 2 [17].

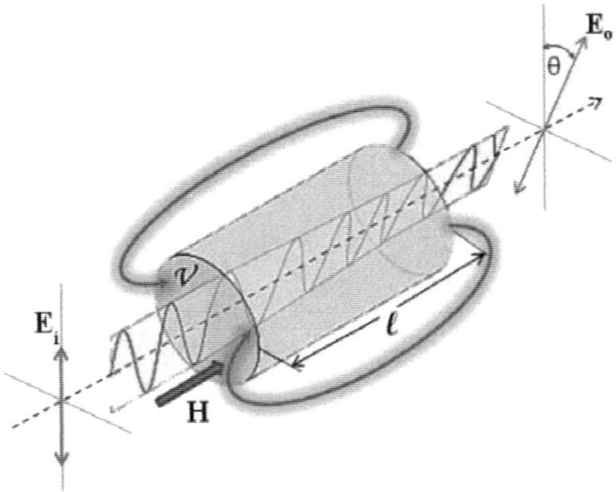

Figure 2. Faraday magneto-optic effect

In term of a quantitative relationship, the rotation angle of the polarized light θ is proportional to the magnetic field $\boldsymbol{H}$ and the cosine of angle between the direction of the magnetic field and the direction of the light propagation. This statement is described by Becquerel's formula which is well known as the Faraday effect [18,19]:

$$\theta = V\,\mu \int_0^L \boldsymbol{H}\, dl \tag{2}$$

where, V is the material Verdet constant related to material characteristics, wavelength, and temperature; μ is the permeability of the magneto-optic material; $\boldsymbol{H}$ is magnetic field; and dl is the path element of Faraday magneto-optic material.

A typical arrangement of optical components in a Faraday sensor is shown in figure 3. A light beam is generated from the light source then passing to a Faraday magneto-optic crystal via a section of fibre optic, a GRIN lens collimator and a first polariser (a polariser). When a magnetic field is applied as the effect of the current flowing, the linearly polarized light is rotated by an angle θ proportional to the applied field. The light output is passed to a photo detector via a GRIN lens collimator and output fibre. The photo detector is utilized to translate modulated light into optical power modulation.

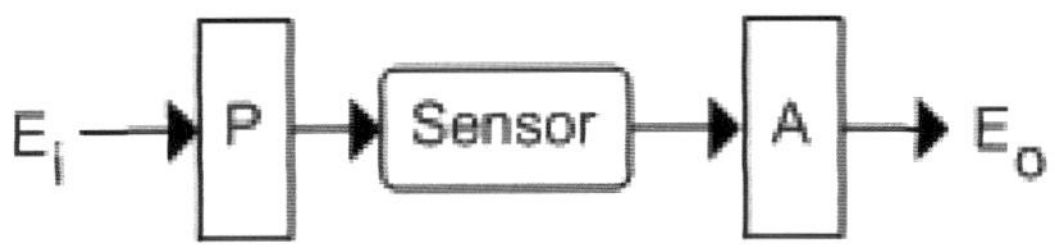

Figure 3. An arrangement of optical component in a Faraday sensor technique

IOP Conf. Series: Materials Science and Engineering 1041 (2021) 012016 doi:10.1088/1757-899X/1041/1/012016

3. Results and discussion

The proposed method has been implemented to a case study which is a power system network as shown in figure 4.

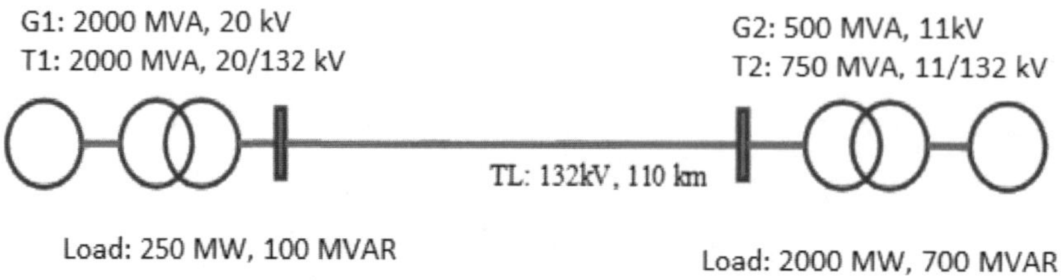

Figure 4. Single line diagram of a power system for case study

The optical sensor model has to be developed using Jones Formalization and applied into Matlab application software. Several optical device models and the power system model have to be combined and defined in the Matlab environment in order to facilitate for computer simulation.

Simulations of the test power system were conducted to determine two kinds of current i.e. normal operation and fault currents, that flow in the line where the optical sensors were equipped. Orientation of the second polarizer (termed as analyzer) is arranged at two different angles which are at angle α of 45° and 90°. These two angles of analyzer orientation have been applied in the simulation. For the simulation purposes, light source has center wavelength of 1550 nm.

The optical power modulation output for these two angles of the analyzer orientation are displayed in figure 4 and figure 5 as response to the normal operation current. These results could be used to compare the analyzer performance in the optical sensor application.

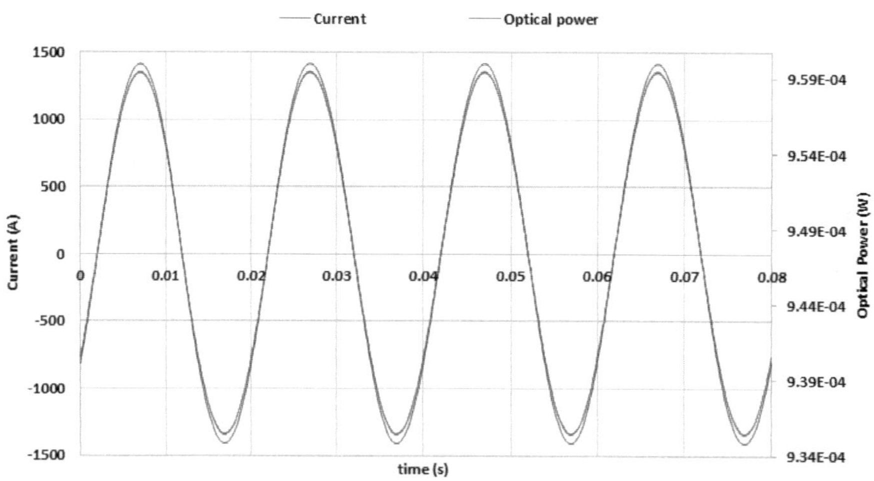

Figure 5. Optical power modulation with analyzer angle α of 45° for normal operation current.

Comparing both figure 4 and figure 5 show that the analyzer arrangement at angle α of 45° with respect to the polarizer provides better output in terms of optical modulation shape than angle α of 90°, where the first analyzer arrangement produce a similar pattern of the optical power modulation with normal operation current pattern.

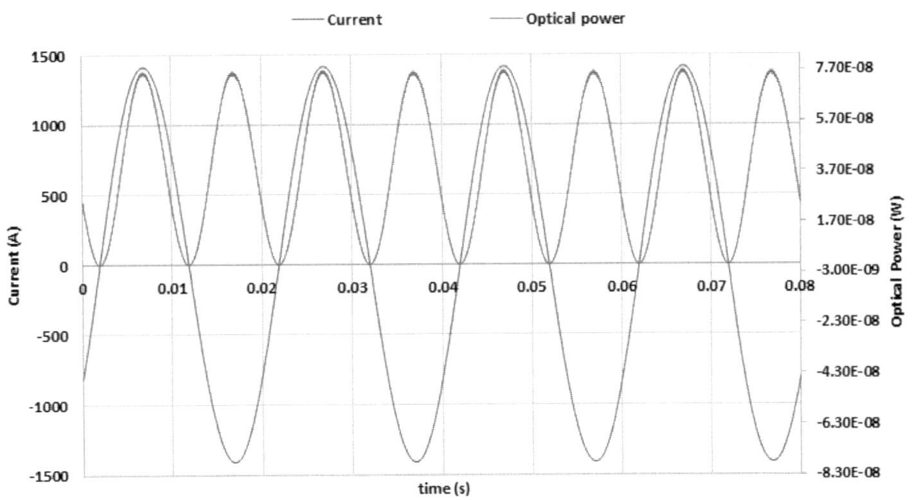

Figure 6. Optical power modulation with analyzer angle α of 90° for normal operation current.

Another result could be derived from both figures above at the same current magnitude of 1410 A that the first angle of analyzer arrangement provided higher optical power modulation output which is the peak value of 9.60E-04 W, whereas the second arrangement could produce the peak value of 7.7E-08 W.

The same results of optical power modulation shapes were also obtained for the fault current simulations as shown in figure 6 and figure 7.

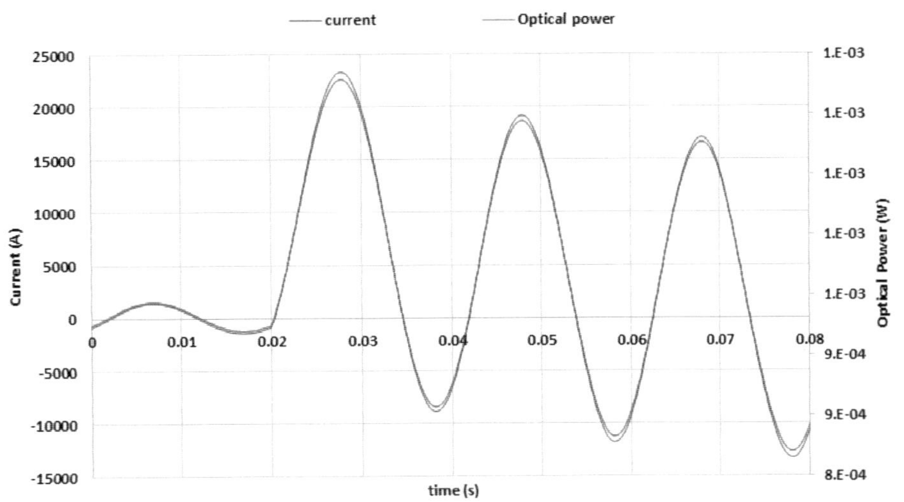

Figure 7. Optical power modulation with analyzer angle α of 45° for the fault current.

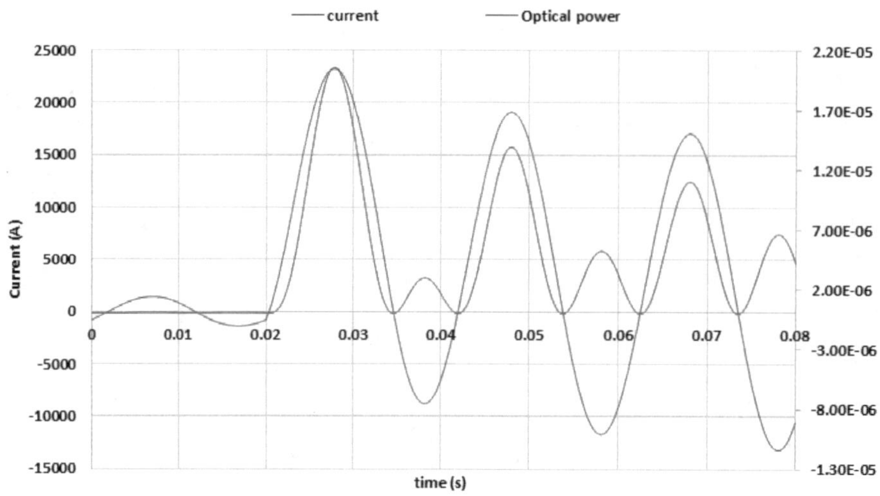

Figure 78. Optical power modulation with analyzer angle α of 90° for the fault current.

The optical power modulation shape has the same form when the analyzer was set to the angle α of 45°. For the same fault current with peak value of 23.3E+04, the optical power modulations for the first and the second arrangement respectively were obtained around 1.14E-03 W and 2.05E-05 W. Therefore, the analyzer arrangement at angle α of 45° with respect to the polarizer could be selected in order to maximize the sensitivity of the optical sensor response.

4. Conclusion

The polarimetric-based optical current sensor have been successfully modelled and simulated. Based on the simulation result evaluation, it can be concluded that the magnetic saturation effect disappears on optical current sensor for this case study. Other result could be provided that the analyzer angle α of 45° with respect to the polarizer position could produce the optical modulation shapes similar to the electric current shapes in both normal and fault condition of power system. Moreover, this analyzer orientation gives also higher optical modulation outputs compare to other angles. This angle generates the sensitivity of the optical sensor response at a maximum value.

The future work will continue to investigate several different optical media that can be utilize in the optical current sensor system. Some assumption in the previous research that neglected the fiber losses and laser beam orientation to optical media will be explored to obtain the general considerations and requirements in the optical current sensor design.

Acknowledgments

This research was supported by the Electrical Engineering Department, Universitas Andalas using research scheme with contract number 086/UN.16.09.D/PL/2020. Therefore, authors would like to thank to the Electrical Engineering Department for providing the financial support.

References

[1] J. L. Blackburn, *Protective relaying : principles and applications*, 3rd ed.. ed. Boca Raton, FL: Boca Raton, FL : CRC Press, 2007.
[2] IEEE, "Transient response of current transformers," *IEEE Transactions on Power Apparatus and Systems,* vol. 96, pp. 1809-1814, 1977.

[3] IEEE, "Optical current transducers for power systems: a review," *Power Delivery, IEEE Transactions on,* vol. 9, pp. 1778-1788, 1994.
[4] S. Takeshita and T. Sasano, "Measurement of impulse current by laser current transformer," *Proceedings of the IEEE,* vol. 56, pp. 1404-1405, 1968.
[5] A. Braun and J. Zinkernagel, "Optoelectronic Electricity Meter for High-Voltage Lines," *IEEE Transactions on Instrumentation and Measurement,* vol. 22, pp. 394-399, 1973.
[6] A. J. Rogers, "Optical technique for measurement of current at high voltage," *Electrical Engineers, Proceedings of the Institution of,* vol. 120, pp. 261-267, 1973.
[7] T. W. MacDougall, D. R. Lutz, and R. A. Wandmacher, "Development of a fiber optic current sensor for power systems," *Power Delivery, IEEE Transactions on,* vol. 7, pp. 848-852, 1992.
[8] R. M. Silva, H. Martins, I. Nascimento, J. M. Baptista, A. L. Ribeiro, J. L. Santos, P. Jorge, and O. Frazão, "Optical Current Sensors for High Power Systems: A Review," *Applied Sciences,* vol. 2, p. 602, 2012.
[9] M. H. Samimi, A. A. S. Akmal, and H. Mohseni, "Optical Current Transducers and Error Sources in Them: A Review," *IEEE Sensors Journal,* vol. 15, pp. 4721-4728, 2015.
[10] Y. Shen, Y. Lu, Z. Liu, X. Yu, G. Zhang, and W. Yu, "Performance of magneto-optical glass in optical current transducer application," *Journal of Magnetism and Magnetic Materials,* vol. 389, pp. 180-185, 2015.
[11] R. C. Jones, "A New Calculus for the Treatment of Optical Systems: I. Description and Discussion of the Calculus," *Journal of the Optical Society of America,* vol. 31, pp. 488-493, 1941/07/01 1941.
[12] A. J. Rogers, "Method for simultaneous measurement of current and voltage on high-voltage lines using optical techniques," *Electrical Engineers, Proceedings of the Institution of,* vol. 123, pp. 957-960, 1976.
[13] A. J. Rogers, "Optical Measurement of Current and Voltage on Power Systems," *Electric Power Applications, IEE Journal on,* vol. 2, pp. 120-124, 1979.
[14] K. Kyuma, S. Tai, M. Nunoshita, T. Takioka, and Y. Ida, "Fiber optic measuring system for electric current by using a magnetooptic sensor," *Quantum Electronics, IEEE Journal of,* vol. 18, pp. 1619-1623, 1982.
[15] A. Cruden, J. R. McDonald, I. Andonovic, D. Uttamchandani, R. Porrelli, and K. Allan, "Current measurement device based on the Faraday effect," in *1993 Fifth International Conference on Developments in Power System Protection*, 1993, pp. 69-72.
[16] T. W. Cease and P. Johnston, "A magneto-optic current transducer," *Power Delivery, IEEE Transactions on,* vol. 5, pp. 548-555, 1990.
[17] M. Aerssens, A. Gusarov, B. Brichard, V. Massaut, Me, x, P. gret, and M. Wuilpart, "Faraday effect based optical fiber current sensor for tokamaks," in *Advancements in Nuclear Instrumentation Measurement Methods and their Applications (ANIMMA), 2011 2nd International Conference on*, 2011, pp. 1-6.
[18] E. Hecht, *Physics*: Brooks/Cole Pub., 1994.
[19] R. D. Guenther, *Modern optics*: New York : Wiley, 1990.

The use of edge cut on microstrip antenna patch with the modified partial ground plane for bandwidth enhancement

R Fernandez[1], F Ra'id[1], H Andre[1], Baharuddin[1], Firdaus[2].
[1]Electrical Department, UniversitasAndalas, UNAND, Indonesia
[2]Electrical Department, PoliteknikNegeri Padang, PNP, Indonesia

Corresponding email: rfernandez@eng.unand.ac.id

Abstract. This research proposes the use of edge cut on the antenna patch with a modified partial ground plane to overcome the inherent characteristic of the microstrip antenna, which is narrowband. The antenna has to cover all of the LTE frequency bands, which is from 0.824 – 2.4 GHz. FR4 epoxy with a dielectric constant of 4.4 and a thickness of 1.67 mm is used as a substrate for the antenna. The proposed antenna is simulated by an EM Simulator. The simulated results show that the return loss of -18.82 dB and the VSWR of 1.25. These parameters confirms that the antenna is well-matched along with the LTE bands. This wideband antenna has a directional radiation pattern with a maximum gain of 4.65 dB.

1. Introduction

Long Term Evolution (LTE) technology, also known as 4G technology, has been launched in Indonesia since 2013. This technology works in 5 different band frequencies, namely LTE band 1, 3, 5, 8, and 40 as regulated by PeraturanMenteriKomunikasidanInformatikaRepublik Indonesia nomor 27 Tahun 2015[1]. Each band has different uplink and downlink frequency ranges.

Mobile communication systems that work on LTE frequencies use antennas to transmit and receive information in the form of radio waves. The antenna used is small, thin, and light. A microstrip antenna is an antenna that meets these criteria. The manufacture of this antenna uses printed-circuit technology so that it can be made in a variety of forms. In contrast, this antenna has a narrow bandwidth [2]. Therefore, it is impossible to cover all of the LTE band frequencies.

Various methods have been proposed to increase the bandwidth. The use of antenna array [3] achieves the bandwidth but the antenna dimension getting relatively larger. The air gaps [4]between the substrate and ground plane acts as an additional layer that presents a challenge in fabrication as well as thickness increased. Then the use of Electromagnetic Bandgap (EBG) structure[5]-[6]. The use of metamaterials with negative refractive index is also an option but provide challenges in analysing [7]. In [8] slit and slot are applied in the patch to get wider bandwidth. Finally, a slot in the form of X is applied in the patch with proximity coupled used as a feeding line to increase the bandwidth [9].

In this study, a rectangular microstrip antenna with the modified partial ground plane and the edge of the patch cut is proposed to have an antenna with a wideband characteristic to meet the required bandwidth for LTE band frequencies. Furthermore, a feeder line that directly connected to the edge of the patch is used as a feeding technique in the antenna.

IOP Conf. Series: Materials Science and Engineering 1041 (2021) 012017 doi:10.1088/1757-899X/1041/1/012017

2. Material and methods

The proposed antenna is based on a rectangular patch microstrip antenna. The antenna will be operated in all of LTE frequency bands (0.824 – 2.4 GHz) which is about 1576 MHz.FR4 epoxy with a dielectric constant of 4.4 and thickness of 1.67 mm is used as a substrate for the antenna. The substrate is relatively cheap and easy to find in the market.

The procedure in [2]is followed to have a preliminary design with a resonant frequency of 1612 MHz, as seen in figure 1. The designed antenna is simulated using an EM simulator software. The simulator evaluates the performance of the antenna whether it meets a required parameter for a good antenna which is return loss (RL) ≤ -10 dB. The simulated result shows that the bandwidth of the antenna is narrowband, as seen in figure 2.

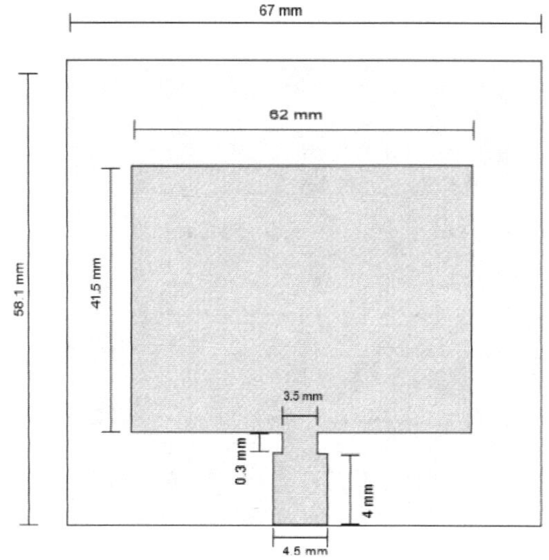

Figure 1. The preliminary design

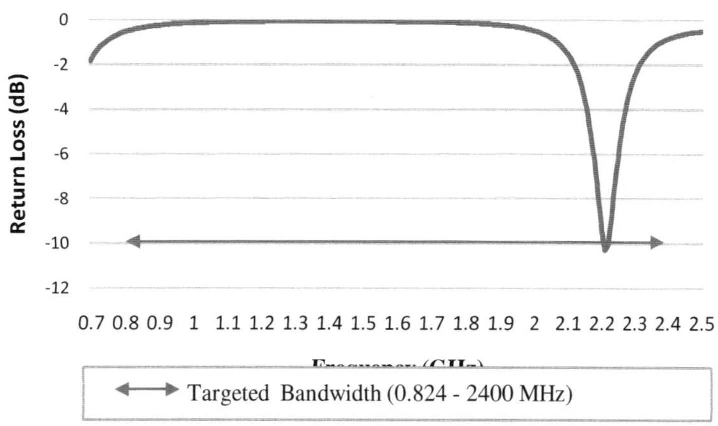

Figure 2. The simulated result of the antenna for Return Loss

The proposed antenna should be able to cover all of the LTE frequency bands from 824 MHz to 2400 MHz. The minimum bandwidth target is 1576 MHz. Therefore, the preliminary antenna has to be optimised by modification on the patch and the ground plane to achieve the required bandwidth. The patch located on the upper side of the antenna is cut on both sides. The cut and its dimension are seen in figure 3(a). Then the ground plane section that is on the lower side where the copper part is reduced in such a way that it stays partially. This partial ground plane is applied two slots with the same dimensions and three slits with size and position as in figure 3(b).

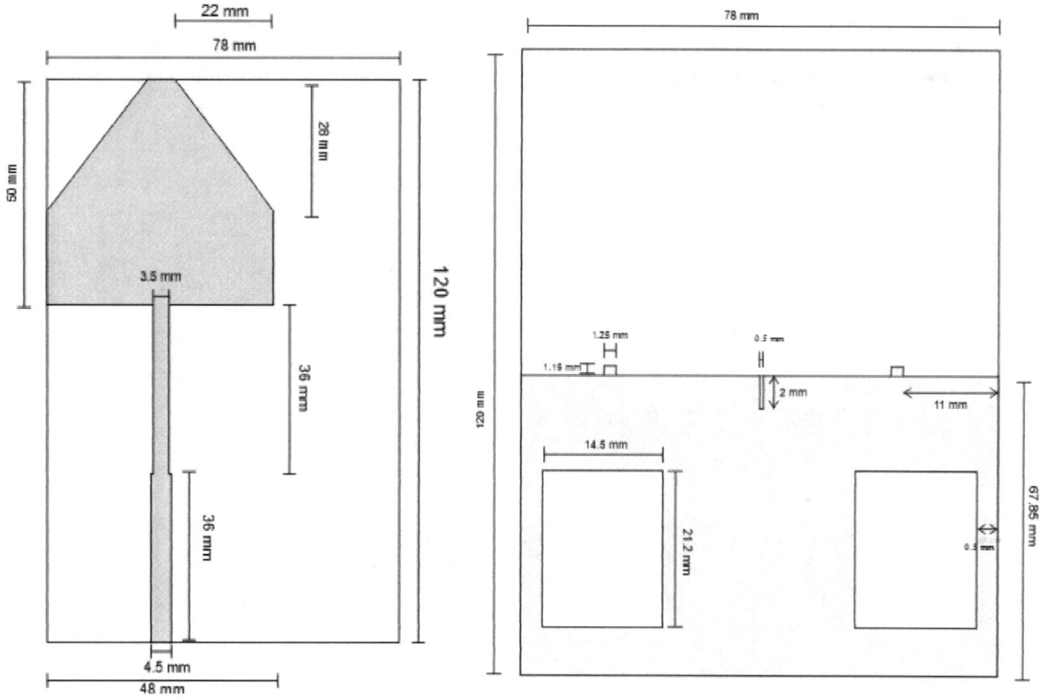

(a) The patch (b) The ground plane

Figure 3. The proposed antenna design

3. Results and discussion

The EM simulator is used to evaluate the antenna in figure 3. The evaluation is measured based on RL (figure 4), Voltage Standing Wave Ratio (VSWR) (figure 5), gain (figure6) and radiation pattern of the antenna (figure7).

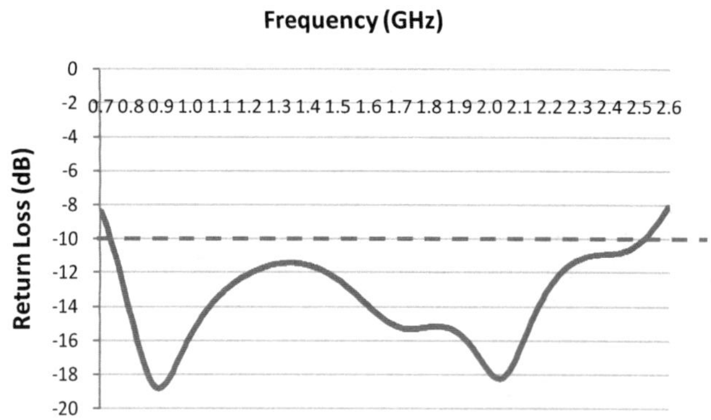

Figure 4. The RL of the proposed antenna

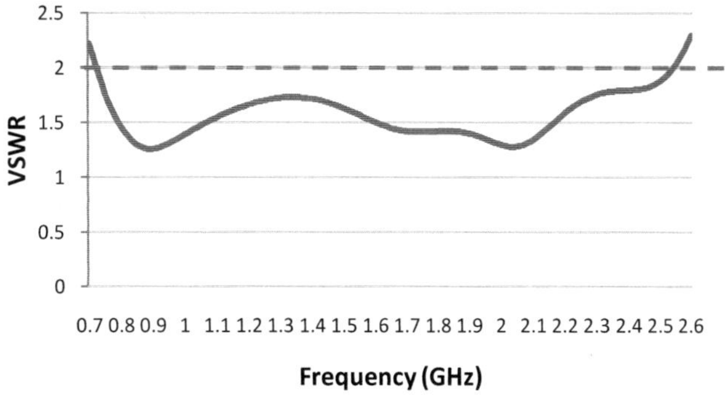

Figure 5. The VSWR of The proposed Antenna

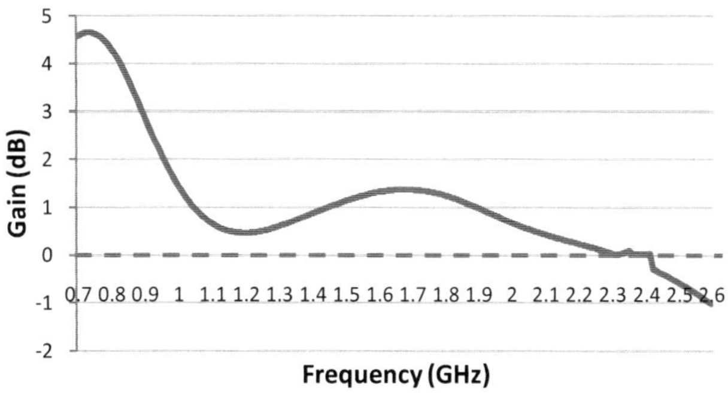

Figure 6. The gain of the proposed antenna

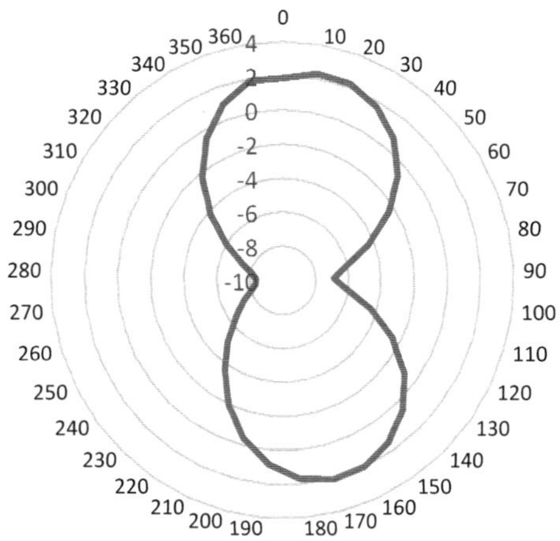

Figure 7. The radiation pattern of the proposed antenna

As stated before, an antenna has to meet the required parameters of good antenna performance [2]. Figure 4 shows the return loss of the antenna is less than -10 dB along with the targeted bandwidth.

The minimum and maximum frequency are determined based on the required RL at these frequencies. They are 731.3 and 2517.5 MHz respectively. The bandwidth of the proposed antenna is 1786.3 MHz which is wider than the targeted bandwidth. The performance of the antenna also is confirmed by the result of simulated VSWR in Fig.5. It shows that the value of VSWR is below 2 as needed, along with the bandwidth. These parameters, RL and VSWR, show that the antenna is well-matched.
The antenna is also required to have better gain, which indicates how strong a signal an antenna can send or receive in a specified direction. The maximum gain of the antenna from figure 6 is about 4.65 dB. Also, the radiation pattern of the antenna is directional, as shown in figure 7.
All of the parameters of the proposed antenna is presented in table 1. The table shows the comparison of the parameters between the preliminary antenna and the proposed antenna. The value of RL and VSWR in the table is the minimum value that each antenna can get.

Table 1. The comparison between the preliminary and the proposed antenna.

Antenna Parameter	Preliminary	Proposed
Return Loss (dB)	-10.310	-18.82
VSWR	1.8	1.25
Bandwidth (MHz)	30	1786.3
(Frequency Range)	(2200 – 2230)	(731.3 – 2517.5)

These parameters show that the proposed antenna has a wideband characteristic to operate in the desired bandwidth.

4. Conclusions

The proposed antenna that can work in all of the LTE frequency bands has been successfully designed and simulated. The simulation results show that the antenna has RL of -18.82 dB, VSWR of 1.25 with the bandwidth of 1786.3 MHz. The result confirms that the use of edge cut on the patch with the modified partial ground plane can enhance the bandwidth of the antenna.

Acknowledgements

This research is supported by a research grant for dosenjurusanteknikelektro from Dana DIPA UniversitasAndalasTahunAnggaran 2020 No. 083/UN.16.09.D/PL/2020

References

[1] Menkominfo 2015 Peraturan Menteri Komunikasi dan Informatika Republik Indonesia Nomor 27 Tahun 2015 tentang Persyaratan Teknis Alat dan atau Perangkat Perangkat Telekomunikasi Berbasis Standar Teknologi Long Term Evolution *Kementeri. Komun. dan Inform. Republik Indones.* 1–11
[2] Balanis C A 2005 *ANTENNA THEORY* (New Jersey: John Wiley and Sons, Inc., Publication)
[3] Gupta S, Singh M, Yadav R, Patel S K, Singh S and Email W S 2013 Bandwidth Enhancement in Multipatch Microstrip Antenna Array **2** 353–8
[4] Fakharian M M, Rezaei P, Azadi A and Dadras M R 2014 A Capacitive Fed Microstrip Patch Antenna with Air Gap for Wideband Applications *Int. J. Eng. Trans. B Appl.* **27** 715–22
[5] Gupta R and Kumar M 2012 Bandwidth enhancement of microstrip patch antennas by implementing electromagnetic bandgap (EBG) structures *Proceedings - 4th International Conference on Computational Intelligence and Communication Networks, CICN 2012* pp 15–8
[6] Goswami D and Chaudhari A 2015 High Bandwidth Microstrip Antenna Using Electromagnetic Band Gap Structures for Cellular System and Satellite Communication *Proceedings - 2015 2nd IEEE International Conference on Advances in Computing and Communication Engineering, ICACCE 2015* pp 417–20
[7] Chen P Y and Alù A 2010 Sub-wavelength elliptical patch antenna loaded with μ-negative

metamaterials *IEEE Trans. Antennas Propag.* **58** 2909–19
[8] Kumar Mishra V, Kumar Singh S and Kumar S 2014 Design of Slit Slotted Rectangular Microstrip Patch Antenna for L-band Communication *Int. J. Eng. Trends Technol.* **17** 348–52
[9] Rama Sastry I V S and Jaya Sankar K 2014 Proximity Coupled Rectangular Microstrip Antenna withX-slot for WLAN Application *Glob. J. Res. Eng. F Electr. Electron. Eng.* **14**

IOP Conf. Series: Materials Science and Engineering 1041 (2021) 012018 doi:10.1088/1757-899X/1041/1/012018

A Circular Patch Microstrip Antenna with Partial Ground Plane for WiMAX and WLAN Applications

R Fernandez[1], T Putra[1], H Andre[1], Firdaus[2].
[1]Electrical Department, UniversitasAndalas, UNAND, Indonesia
[2]Electrical Department, PoliteknikNegeri Padang,PNP, Indonesia

Corresponding author: rfernandez@eng.unand.ac.id

Abstract. Narrow bandwidth is the main problem of a microstrip antenna. Enhanced bandwidth is done by applying a partial ground plane. The antenna patch layer is a circular formed based on the formula to obtain its initial dimension. The dimension is used as a base for the antenna design with ground plane reduction. The antenna is designed using FR4, which has a thickness of 1.6 mm with a dielectric constant of 4.3. Simulation results from the design of partial ground plane antennas on WiMAX and WLAN band frequencies showed an increase in bandwidth of 3.14 GHz with minimum return loss values of -17.01 dB, gain antenna of 1.32 dB and a directional radiation pattern.

1. Introduction

Worldwide Interoperability for Microwave Acces (WiMAX) is a broadband wireless technology that has high access speeds with a wide range. WiMAX frequency allocation is 2.3 GHz, 2.5 GHz, 3.3 GHz, 3.5 GHz and 5.8 GHz according to IEEE 802.16[1]. Wireless Local Area Network (WLAN) is a network that uses radio waves as its data transmission media. WLAN using IEEE standard 802.11 has frequency allocation in 5 GHz area with a range of 5.12 GHz-5.25 GHz, 5.25 GHz-5.35 GHz and 5,725 GHz-5825 GHz. For 2 GHz regions work in the frequency range of 2.4 GHz-2.4835 GHz.

This application at WLAN and WIMAX frequencies requires an antenna as its transmission medium. The antenna is an essential component in wireless telecommunications. The antenna can function as a receiver and a transmitter simultaneously. Without the antenna, wireless communication will not run.

A microstrip antenna is one of antenna type that can be used for WLAN and WIMAX applications. The microstrip antenna has the advantage of small, light, low cost, easy inthe manufacturing process, and allows to be made dual or triple frequency. But this antenna also has some drawbacks which are small gain and narrow bandwidth[2].

Various methods have been proposed to increase bandwidth. The selection of the feeding method is important in the antenna design, such as coupled stripline fed, resulting in an impedance bandwidth of about 13%[3]. However, it is not so easy to create because it requires proper alignment at the time of fabrication. Another method is the use of electromagnetic band-gap (EBG) structure [4]-[5]. The use of metamaterials with negative refractive index is also an option but takes complexity in analyzing [6] The use ofadditional substrate leads to more thickness to getan increment in bandwidth[7].

In this study proposes a designed antenna that has a wide bandwidth by applying a partial ground plane (PGP). The patch used on this microstrip antenna is a circular patch that has the advantage of being relatively easy to optimize compared to other forms. The optimized is reached by adjusting the size of the patch radius. The Antenna has a single band to cover WIMAX and WLAN frequencies.

2. Materials and methods

The patch of the microstrip antenna is circular to be designed according to the shape shown in Figure 1. In this figure shows the circular patch has the radius,which is a, t is the thickness of the patch, h is the thickness of the substrate, and ε_r is the substrate dielectric constant.

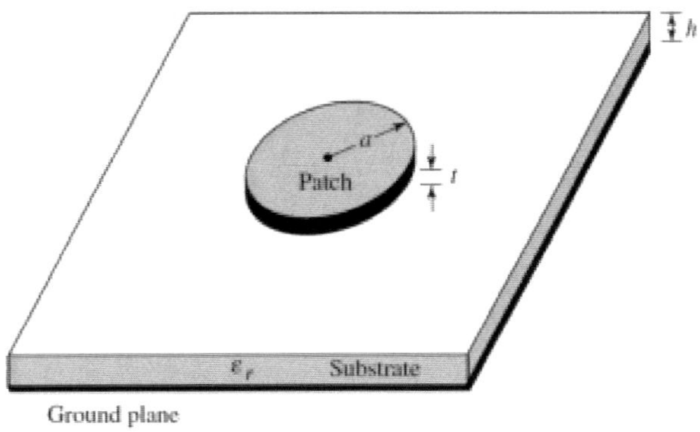

Figure 1. A circular patch [8].

The antennamicrostrip is designed using these following formulas as in equation (1) and (2).

$$a = \frac{F}{\left\{ 1 + \frac{2h}{\pi \varepsilon_r F} \left[\ln\left(\frac{\pi F}{2h}\right) + 1{,}7726 \right]^{1/2} \right\}} \quad \textbf{(1)}$$

F is the logarithmic function of the radiating element and is indicated by :

$$F = \frac{8{,}791 x 10^9}{f_r \sqrt{\varepsilon_r}} \quad \textbf{(2)}$$

The substrate used for the antennais FR4 Epoxy with a thickness (h) of 1.6 mm, a dielectric constant (ε_r) of 4.3 and a dielectric tangent loss of 0.02. Before applying the data from substrate, the value of F is needed to get the radius as in the equation 1. F is determined by the resonance frequency of the designed antenna. The designed antenna operates from 2.3 GHz to 5.35 GHz. Therefore, the resonance frequency (f_r) is 3,825 GHz.

The calculationresults show that the radius is about 18.24 mm with F = 1,906. The feed point is set 18.24 mm from the midpoint of the patch. The antenna is fed by a line feed. The width of the line feed is 3 mm, and the length is 21.35 mm. The total dimension of the antenna is 40.48 x 59.73 mm^2 after taking into account the dimension of the ground planeas in figure 2(a).

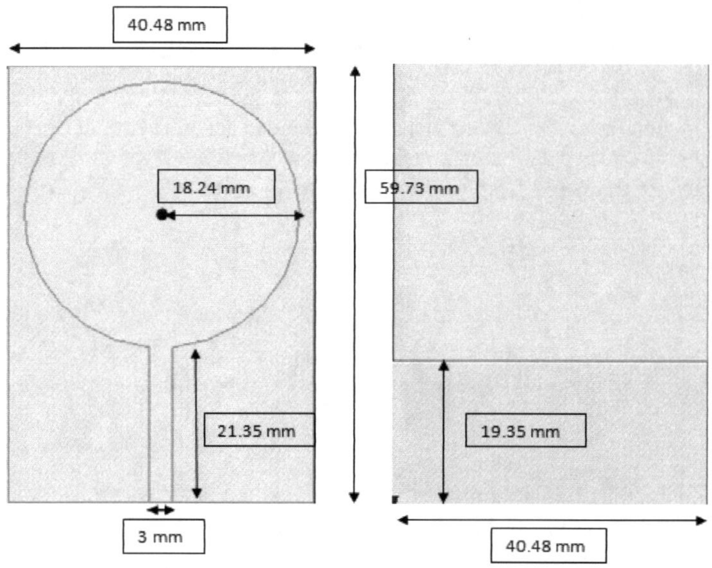

(a)The patch (b) The reduced ground plane

Figure 2. The proposed antenna design.

On the ground plane in figure 2(b), the cooper part is etched. This part is reduced until giving a better impact on the bandwidth of the antenna. The area with the cooper part is becoming decreased as small as 19.35 x 40.48 mm2.

The performance of a microstrip antenna is based onthe value of return loss and Voltage Standing Wave Ratio (VSWR) parameters. The antenna has a good performance when the return loss is < -10 dB and VSWR ≤ 2 at the expected working frequency, and the radiation pattern is set adirectional [8].

3. Results and discussion

The proposed antenna is simulated using an EM Simulator to evaluate the antenna in figure 1. The evaluation is measured based on RL (figure3), Voltage Standing Wave Ratio (VSWR) (figure4), and radiation pattern of the antenna (figure 5).

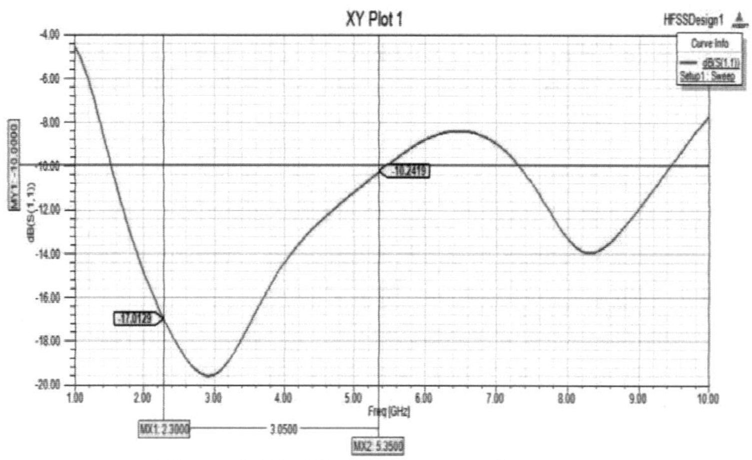

Figure 3. The RL of the proposed antenna

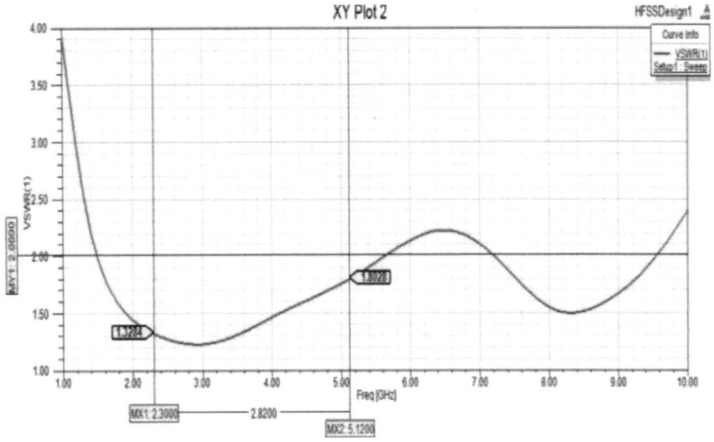

Figure 4. The VSWR of The proposed Antenna

After applying the partial ground plane, parasitic capacitance is introduced by increment of fringing field due to defected ground structure. The coupling between the ground plane and the patch is increased by this parasitic capacitance which leads to enhancement of the bandwidth.

In figure 3above, the bandwidth has achieved the desired result with a return loss below -10 dB from WiMAX frequency (2.3 GHz) to WLAN (5.35 GHz). The achieved bandwidth is measured from the minimum frequency of 1.53 GHz to the maximum frequency of 5.57 GHz (4.04 GHz). The bandwidth is wider than the bandwidth target. The bandwidth for covering WiMAX and WLAN frequencies is about 3.05 GHz.

Similarly, with figure 4, VSWR has reached the target, which is below 2 from WiMAX frequency (2.3 GHz) to WLAN (5.35 GHz). So, The result of simulated VSWR also confirms the achieved bandwidth. It shows that the value of VSWR is below 2 as needed, along with the range of the achieved bandwidth. The value of simulated RL and VSWR parameters affirm that the antenna has a good impedance matching in target bandwidth.

The radiation pattern of the proposed antenna is also simulated to see its directivity. It is a directional, as shown in figure 5. The gain of the simulated antenna is 1.32 dB.

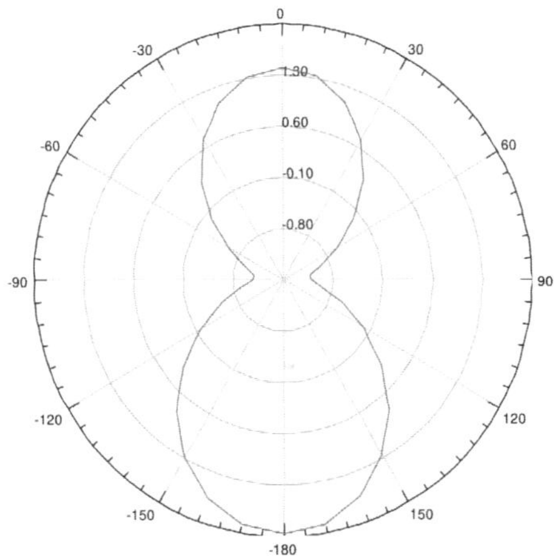

Figure 5. The radiation pattern of the proposed antenna

All of the parameters of the proposed antenna is presented in table 1. The table shows the comparison of the parameters between the antenna without PGP and with PGP. The value of RL and VSWR in the table is the minimum value that each antenna can get.

Table 1. The comparison between antenna without and with PGP

Antenna Parameter	Without PGP	With PGP
Return Loss (dB)	-11.16	-17.01
Gain (dB)	0.39	1.32
Bandwidth (GHz)	0.9	4.04
(Frequency Range)	(3.78 – 3.87)	(1.53 – 5.57)

The parameters in the table show that the partial ground plane has an impact on broadening the bandwidth of the antenna. Therefore, the antenna can operate in WiMAX and WLAN frequencies.

4. Conclusions

The partial ground plane on circular microstrip antenna that works on WiMAX and WLAN frequencies has managed to increase bandwidth by 3.14 GHz from its original bandwidth. The results of the simulation show the following parameters, minimumreturn loss -17.01 dBwith the antennagain of 1.32 dB.The antenna has a directional radiation pattern.

Acknowledgements

This research is supported by a research grant for DosenJurusanTeknikElektro from Dana DIPA UniversitasAndalasTahunAnggaran 2018No. 086/UN.16.09.D/PL/2018

References

[1] S. Y. Tang, *WiMAX SECURITY AND QUALITY OF SERVICE: An End-to-End Perspective*, vol. 66. West Susex, United Kingdom: John Wiley and Sons, Inc., Publication, 2010.

[2] R. Garg, *MICROSTRIP ANTENNA DESIGN HANDBOOK.pdf*. Norwood, United Kingdom: Artech House, Inc., 2001.

[3] S. He, P. Tang, B. Chen, and P. Wang, "Design of an X-band Stripline-fed Co-aperture Microstrip Multilayer Transceiver Array Antenna," *2016 Prog. Electromagn. Res. Symp. PIERS 2016 - Proc.*, pp. 4676–4681, 2016.

[4] N. Llombart, A. Neto, G. Gerini, and P. de Maagt, "Planar Circularly Symmetric EBG Structures for Reducing Surface Waves in Printed Antennas," *IEEE Trans. Antennas Propag.*, vol. 53, no. 10, pp. 3210–3218, 2005.

[5] N. Rao and K. V. Dinesh, "Performance enhancement of a microstrip antenna by suppression of surface waves using EBG structure in multiple layer substrate," in *Proceedings - 2011 IEEE-APS Topical Conference on Antennas and Propagation in Wireless Communications, APWC'11*, 2011, pp. 935–939.

[6] P. Y. Chen and A. Alù, "Sub-wavelength elliptical patch antenna loaded with μ-negative metamaterials," *IEEE Trans. Antennas Propag.*, vol. 58, no. 9, pp. 2909–2919, 2010.

[7] R. Fernandez and F. A. Pratama, "Penggunaan Dual-Layer Substrate untuk Meningkatkan Bandwidth Antena Mikrostrip pada Frekuensi LTE," *J. Nas. Tek. Elektro*, vol. 7, no. 2, pp. 117–121, 2018.

[8] C. A. Balanis, *ANTENNA THEORY*, 3rd ed. New Jersey: John Wiley and Sons, Inc., Publication, 2005.

IOP Conf. Series: Materials Science and Engineering 1041 (2021) 012019 doi:10.1088/1757-899X/1041/1/012019

The economics comparison of power plants fuelled by wellhead gas and liquefied natural gas in Aceh Province

D Budiana[1], R Dalimi[1]

[1]Departement of Electrical Engineering, University of Indonesia, Depok, 16424, Indonesia

Corresponding author: dodi.budiana@ui.ac.id

Abstract. The Aceh Province Electricity System is supplied by Gas Engine Power Plant (GEPP) fuelled by Liquefied Natural Gas (LNG), Coal-fired power plant and Diesel Engine with Diesel fuel and transfers from the North Sumatra system during peak loads. LNG fuel is sent by ship from the Tangguh Papua Refinery which is regasification in Perta Arun Gas, Lhokseumawe, Aceh Province, so the production costs are expensive due to the distance from the gas supply sources. To get cheaper electricity production costs, it can be done by using the gas potential in Aceh province. In this study, the economic value will be calculated by comparing the use of two types of gas fuel, namely wellhead gas fuel and LNG gas for Gas Turbine and Gas Engine power plants, to determine the cost of electricity production for each power plant with the same fuel. Based on the calculation results, gas engine power plant with fuel from the Wellhead Gas has the best production costs with value of IDR 1,262.21 / kWh lower than the electricity tariff of IDR 1,467 /kWh, and has an IRR of 24.78% and a payback period of 4,88 years. By knowing the most economic value for the construction of power plant with Wellhead Gas, it can be used as a basis for policy making in choosing an economical power plant development for areas with gas potential, such as in Aceh province.

1. Introduction

Energy is a primary need in modern human life, almost all aspects of life require energy as the prime mover. Global energy consumption is predicted to increase by approximately a third by 2040, with fossil fuels being the predominant energy sources [1]. The vast territory of Indonesia contains energy potential for fossil and renewable energy that has the opportunity to be used as a source of electrical energy. Natural gas is a Non renewable energy source, it is identified as low carbon emitter for power generation and transport globally [2]. Aceh Province is a province located north of the island of Sumatra, which has potential primary energy sources consisting of potential coal, water, geothermal, petroleum and gas which can be used for electricity generation. Natural gas from Aceh and Northern Sumatra has long been produced. The Arun field in Aceh has been producing since the 1970s to meet the needs of fertilizer factories, power plants and LNG exports[3]. Based on data from the National Energy General Plan for 2017, total natural gas reserves in Aceh amounted to 7,516.3 BCF in the form of proven reserves of 1,420.8 BCF, probable reserves of 4,720.4 BCF and possible reserves of 1,375 BCF. Apart from natural gas supplies from fields in Aceh and Northern Sumatra, currently natural gas is also imported from Tangguh Papua through to the Arun Regasification Terminal.

The electric power system in Aceh consists of a 150kV North Sumatra-Aceh interconnection system and an isolated system with a distribution voltage of 20 kV. Most of Aceh's electricity power system is supplied by the 150 kV North Sumatera interconnection system and a small part is still in isolated areas. The Aceh Province Electricity System is supplied by the Coal Power Plant located in Meulaboh, the Gas Engine Power Plant (GEPP) fuelled by Liquefied Natural Gas (LNG) located in Lhokseumawe and the Diesel Engine Power Plant is spread using diesel fuel as well as transfers from the North Sumatra system during load peak. The highest peak load of the Aceh system in 2018 was 424 MW in July[4]. The average growth of electrical energy in Aceh province for 10 years is 8.11% with the consideration of economic growth and in Aceh province, the Arun Lhokseumawe Special Economic Zone will be built, several cement factories and industrial areas [4].

Liquefied Natural Gas (LNG) fuel for Gas Engine Power Plant is shipment from the Tangguh Papua Refinery which is regasification at Perta Arun Gas, Lhokseumawe, Aceh Province and then flowed through gas pipes to the power plant location, so the production costs are expensive because of the distance the source of gas supply as well as the regasification process. To get cheaper electricity production costs, it can be done by using the gas potential in Aceh province so that electricity production costs can be more economical. Aceh Province is an area that has gas potential, currently this potential has not been optimally utilized for the electricity sector. Electricity generation cost assessment are typically use to rank different power generation technologies based on to expected generation cost and to estimate economic subsidies or penalty charges needed. As such, generation cost assessments are important in energy decision and policy making [5].

In this study The economic calculation of gas-fired power plants is used to determine the economic value of the production costs of electricity generated by the power plant for the use of LNG gas fuel and the well head gas fuel located in the province of Aceh, Economic analysis such as Net Present Value (NPV), Internal Rate of Return (IRR) and Payback Period are calculated for considering economic feasibility. Knowing the economic production costs will be the basis consideration of the policy for selecting the type of fuel and type of power plant to be built in Aceh province.

2. Materials and methods

Natural Gas is one of the world's abundant sources of energy. Natural gas is the fastest growing fossil fuel with its global consumption expected to increase by 1.9%/year. Natural Gas is usually transported in either pipeline as gas or special marine tanker as cryogenic liquid after exploration and treatment [6]. Natural gas can be distributed in two forms, namely the liquid phase and the gas phase. In the gas phase, natural gas from gas wells can be distributed in piped gas or Compressed Natural Gas (CNG). Well Head gas is natural gas fuel located near the upstream oil and gas production facilities or gas that is flowed through pipes that have never undergone a regasification process[4].

In the liquid phase, natural gas is distributed in the form of Liquefied Natural Gas (LNG). LNG is natural gas that converted to a liquid state by cooling it down to approximately -162 °C, In this liquid state, it takes up much less volume compared to a (compressed) gaseous state, which make LNG particularly suitable as a fuel for long-haul transportation[7]. It is more efficient for transportation. The LNG supply chain can be described as a network of natural gas from the gas field/gas well to the liquefaction plant to turn the gas phase into liquid[8]. The purpose of changing the gas phase to the liquid phase is to facilitate the transportation and storage processes. For the shipping process, usually using the shipping process before being used for power plant needs, LNG will go through a regasification process, where the LNG phase changes from the liquid phase to the gas phase again and after that it will be channel through the pipeline. Natural gas market consists of producers, pipeline operators and local distribution companies on the supply-side and electrical, industrial, commercial and residential customers on demand-side Natural gas is first purchased at the wellhead by independent brokers and pipeline companies for direct delivery to electrical and industrial customers[9].

To meet the increasing demand for electricity supply, PLN needs gas on a very large scale for electricity generation[10]. Gas Fired Unit has become the most potential fossil generation, for its

IOP Conf. Series: Materials Science and Engineering 1041 (2021) 012019 doi:10.1088/1757-899X/1041/1/012019

cleanliness and high efficiency [11]. Utilization of the potential for natural gas for power generation will encourage increased achievement of natural gas potential utilization. Unlike other usual gas well, natural gas supply of LNG terminal is influenced by the risk of LNG supply, which caused by shipping or bad weather on sea [11]. Gas fuel supply constraints will have an impact on the operational side of the plant, with the construction of a generator in accordance with the gas potential that is owned, it will reduce the potential for this problem. Aceh Province with the availability of gas energy potential that is owned and has not been maximized for power generation can be a priority for planning the development of power plants by considering its economy compared to natural gas which is currently used for power generation in Aceh province.

Comparison of the economics of the use of LNG gas fuel and Well Head gas fuel used for gas-fired power plants. In order to produce affordable electric power, in choosing the energy resources to be used to carry the electrical load, an economic analysis of the cost of generation is necessary.

Comparison of economic calculations by comparing the use of two types of fuel gas, namely well head gas and LNG gas for Gas Turbine Power Plant (GTPP) and Gas Engine Power Plant (GEPP), to determine the electricity production costs of each power plant with burn the same.

3. Results and discussion

Economic analysis is undertaken for making sure the feasibility investment of design from economic review. Electricity production cost prices are calculated by applying electricity calculations where transmission costs are not included in the calculation. The price of electricity is the average cost which is the cost required per kWh which takes into account all cost components including capital costs, operating costs and fuel costs[12]. The calculation of the cost of generation by taking into account the electricity generation Et (kWh) in year t, the average cost in year t, denoted B in mills / kWh or IDR /kWh is

$$\boldsymbol{B} = \frac{\mathbf{1000}\ \boldsymbol{x}\ \boldsymbol{Ct}}{\boldsymbol{Et}} \quad (1)$$

Where Ct ($) is the costs incurred for operating costs, fuel and capital return costs for year t. With the price of electricity for each year, during the life of the generator (N years), the present value of revenue from electricity sales Rv' is

$$Rv' = \sum_{t=1}^{N} \frac{B\ x\ Et}{1000\,(1+di)^t} \quad (2)$$

Where di is the amount of the discount rate. Rv 'will be equal to the present value of costs incurred for the Ct' operation.

The formula for the levelized cost or tariff price B '(mills / kWh or IDR/ kWh) is

$$B' = \left[\frac{\sum_{t=1}^{N} \frac{Ct}{(1+di)^t}}{\sum_{t=1}^{N} \frac{Et}{(1+di)^t}}\right] \quad (3)$$

Levelized cost of electricity generation is essentially the total cost of producing power over the lifetime of the unit divided by the total electricity generated in the same time period[13]. The levelized tariff or cost can be calculated by the equation.

$$\textbf{Levelized tariff} = \frac{\textbf{Total Cost in net present value}}{\textbf{Total Energy in NPV salable energy}} \quad (4)$$

Economic analysis is carried out to determine the economic value of an investment, for investment in power plants. Some of the values analysed are Net Present Value (NPV), Internal Rate of Return (IRR) and Payback Period.

Net Present Value (NPV) is a method for calculating the net present value of investment cash flows by applying the money equivalent concept. Investments will be considered feasible if the NPV value is greater than zero and investment is considered unfeasible if the NPV is less than zero[14]. The formula for determining the NPV value is like formula 5 below. The NPV value is equal to the present value of cash inflows minus the present value of cash outflows or the Present Value (PV) of all expected cash flows[15].

$$NPV = \sum_{t=1}^{N} \frac{CFt}{(1+r)^t} \tag{5}$$

Where *CFt* is cash flow t is time period, N is number of time periods and r is interest (interest rate).

Internal Rate of Return (IRR) is a method used to evaluate the feasibility of investing in terms of the level of cash flow ability to return investment, which is described in terms of a percentage per time period [14]. The formula for obtaining the IRR value is as in formula 6 below.

$$\boldsymbol{IRR = iNPV_+ + \frac{NPV_+}{|NPV_+ + NPV_-|}(iNPV_- + iNPV_+)} \tag{6}$$

Where IRR is the Internal Rate of Return in percent (%), iNPV + and iNPV- are the interest rates for positive and negative NPV.

Payback period is an investment feasibility analysis method to determine the investment return period when the return on investment occurs [15]. The payback period value can be obtained by using an equation such as formula 7 below.

$$Payback\ Period = n + \frac{a-b}{c-b} x\ 1\ year \tag{7}$$

Where n is the last year where the total cash flow has not covered the original investment, a is the initial investment amount, b is the total cash flow investment in year n and c is the total cash flow investment in year n + 1.

The Power plant used for the calculation of investment costs are gas-based power plants with the type of Gas Turbine and Gas Engine Power Plants with a total load approximately 100 MW and capacity factor 81%. The gas turbine used is Gas Turbine Type SGT A-65 DLE with a load of 55 MW per unit, while the gas engine used is the type 18V51 / 60 with a load of 17 MW per unit. The total investment cost is the total cost of Engineering, Procurement and Construction (EPC) and Non EPC costs. EPC costs consist of civil work costs, mechanical work, electrical and instrument work and other costs while Non EPC costs consist of owner costs, value added taxes, interest during construction and contingency fees. In the Table 1, investment cost of Gas Turbine and Gas Engine Power Plant is presented.

Table 1. Investment Cost.

No	Description	Cost (IDR)	
		Gas Turbine	Gas Engine
A EPC Costs			
1	Civil Works	146.859.330.695	223.234.538.181
2	Mechanical Work	1.103.449.225.571	849.656.650.821
3	Electrical and Instrument Work	273.807.103.444	190.533.729.876

4 Others	5.478.923.095	3.799.939.910
Total Cost A	1.529.594.582.805	1.267.224.858.788
B. Non EPC Cost		
1 Owner Cost	152.959.458.281	126.722.485.879
2 VAT	37.828.666.530	58.808.124.239
3 IDC	106.226.409.262	89.010.851.072
4 Financing Fee	2.713.748.458	2.224.291.618
5 Contingency	168.255.404.109	139.394.734.467
Total B	467.983.686.640	416.160.487.275
Total Costs	1.997.578.269.445	1.683.385.346.275

Based on the table above, the largest cost is for mechanical work, which is the purchase of a power plant machine with a percentage of 67% of the EPC cost for gas engines power plant and 72% of the EPC costs for gas turbines power plant. The total cost for the construction of a Gas Turbine Power plant is IDR 1,997,578,269,445,-. Meanwhile, the total cost for the construction of a Gas Engine Power plant is IDR 1,683,385,346,063,-.

Operation and Maintenance (O&M) costs are divided into two, namely fixed costs (Component B) and variable costs (Component D), where O&M fixed costs for Gas Engine power plant are IDR 90.77 / kWh and O&M variable costs are 45.53 IDR / kWh. As for the Gas Turbine power plant, the O&M fixed cost is IDR 27.20 / kWh and the O&M variable cost is IDR 84.69 / kWh. Use of the operating and maintenance cost assumptions based on the operation and maintenance costs that have been implemented in the PLN project.

The price of Liquefied Natural Gas fuel used in this study is Liquefied Natural Gas in Arun, Lhokseumawe which is shipment from the Tangguh Papua Gas Refinery which is sent via LNG ship and regasification at the Arun regasification plant. The gas price at the plant gate based on Decree of the Minister of Energy and Mineral Resources No.91K / 12 / MEM / 2020 concerning the price of natural gas at the Power Plant for Northern Sumatra is $ 7.5 / MMBTU.

The price of gas fuel comes from gas well head in the Aceh region, with the assumption that the gas price for power plant with the well head gas according to Decree of the Minister of Energy and Mineral Resources ESDM No.10 of 2020 is a maximum of $ 6 / MMBTU.

To make an economic comparison of the use of gas fuel between Liquefied Natural Gas and well head gas for gas engine and gas turbine power plants, an economic calculation is carried out for power plants fuelled by Liquefied Natural Gas and wellhead gas. Calculating the NPV, IRR and payback period for investment in gas turbine power plants and gas engine power plants with Liquefied Natural Gas fuel with a design life of a power plant of 25 years and a discount rate of 9.8%, the comparison of the economic value for a power plant fuelled by Liquefied Natural Gas is obtained as in table 2.

Table 2. Results of the Calculation of LNG Gas-fired Power Plants.

No	Description	Value	
		Gas Turbine	Gas Engine
1	Design Life	25 Years	25 Years
2	Discount Rate	9,8%	9,8%
3	NPV Total Cost	IDR. 9.545.944.449.000	IDR. 8.544.473.071.000
4	NPV Saleable energy	6.479.307.304 kWh	6.178.872.117 kWh
5	Production cost	IDR 1.473,30 /kWh	IDR 1.382,85 /kWh
6	IRR	16,53%	20,61%
7	Payback Period	7,23 Years	5,87 Years

NPV, IRR and payback period calculations are also carried out for investment in gas turbine power plants and gas engine power plants using wellhead gas with a design life of a power plant of 25 years and a discount rate of 9.8%, a comparison of the economic value for wellhead gas-fired power plants is obtained the results are as in table 3.

Table 3. Results of the Calculation of a Well Hell Gas fired Power Plant.

No	Description	Value	
		Gas Turbine	Gas Engine
1	Design Life	25 Years	25 Years
2	Discount Rate	9,8%	9,8%
3	NPV Total Cost	IDR. 8.650.463.154.000	IDR. 7.799.055.516.000
4	NPV Saleable energy	6.479.307.304 kWh	6.178.872.117 kWh
5	Production cost	IDR 1.335,09 /kWh	IDR 1.262.21 /kWh
6	IRR	20.94%	24.78%
7	Payback Period	5.75 Years	4.88 Years

From the above results, it is obtained a positive NPV value, which means that the construction of gas-fired power plants is feasible and has an IRR value that is greater than the interest rate for all types of power plants with each fuel used so that economically the development of a gas fire power plant with well head gas and LNG fulfil the element of feasibility.

Based on the calculation results, the production cost of the construction of a Gas Turbine power plant using LNG gas is IDR 1,473.30 / kWh while for a Gas Engine power plant using LNG gas fuel is IDR 1,382.85 / kWh. If the power plant uses gas from the well head gas, then the production cost for a Gas Turbine power plant is IDR 1,335.09 / kWh while for Gas Engine power plant is IDR 1,262.21 / kWh. Where the production costs are below the cost of providing the power generation PT. PLN for the Aceh region, namely IDR 1,673 / kWh and could potentially reduce the cost of providing generators in Aceh province.

When compared with the PLN electricity selling rate of IDR 1,467.28 / kWh, only gas turbine power plants with LNG fuel have a value greater than the selling rate, while Gas engine power plants using LNG and well head gas fuels and gas turbine power plant using well head gas have lower production costs than the selling rate of electricity. This will provide benefits due to lower production costs compared to PLN's electricity sales tariff.

Utilizing natural gas for electricity generation in Aceh province will make a positive contribution to the electricity system economically.

4. Conclusion

Based on the above economic analysis, gas engine power plant with gas from the well head gas has an electricity tariff of IDR 1,262.21 / KWh cheaper than the selling value of PLN electricity of IDR 1,467.28 / KWh, for a gas engine power plant with gas fuel from the well head gas has a better economic value with an IRR of 24.78% and payback period of 4.88 years so that the gas engine power plant with gas from the Well Head gas can be a consideration for the policy of selecting the construction of gas fired power plants in Aceh province.

Acknowledgments

This research was supported by University of Indonesia (UI) through PUTÌ Proceeding grant 2020 launched by DRPM UI.

References

[1] Katebah M A, Hussein M M, Shazed A R, Bouabidi Z and Al-musleh E I 2020 Rigorous simulation, energy and environmental analysis of an actual baseload LNG supply chain *Comput. Chem. Eng.* **141** 106993

[2] Becerra-Fernandez M, Cosenz F and Dyner I 2020 Modeling the natural gas supply chain for

sustainable growth policy *Energy***205** 118018
[3] Purwanto W W, Muharam Y, Pratama Y W, Hartono D, Soedirman H and Anindhito R 2016 Status and outlook of natural gas industry development in Indonesia *J. Nat. Gas Sci. Eng.***29** 55–65
[4] PT. PLN (Persero) 2019 Electric Power Supply Business Plan (2019-2028) 2019–28
[5] Larsson S, Fantazzini D, Davidsson S, Kullander S and Höök M 2014 Reviewing electricity production cost assessments *Renew. Sustain. Energy Rev.***30** 170–83
[6] Nagesh Rao H and Karimi I A 2018 Optimal design of boil-off gas reliquefaction process in LNG regasification terminals *Comput. Chem. Eng.***117** 171–90
[7] Alvarez J A L, Buijs P, Deluster R, Coelho L C and Ursavas E 2020 Strategic and operational decision-making in expanding supply chains for LNG as a fuel *Omega***97** 102093
[8] Budiyanto M A, Riadi A, Buana I G N S and Kurnia G 2020 Study on the LNG distribution to mobile power plants utilizing small-scale LNG carriers *Heliyon***6** e04538
[9] Mohammadi H 2011 Market integration and price transmission in the U.S. natural gas market: From the wellhead to end use markets *Energy Econ.***33** 227–35
[10] Kementrian Energi dan Sumber Daya Mineral 2014 *Peta Jalan Kebijakan Gas Bumi Nasional 2014-2030* ed H S . Ilwan (Jakarta)
[11] Chen G, Liu Q, Lu E and Chen Q 2017 Optimization of LNG Terminal Reserve Planning for Combined Gas and Electricity System *Energy Procedia***105** 3332–8
[12] Nasrullah M and Sriyana 2010 Harga dan Tarif Listrik PLTN di Dunia *J. Pengemb. Energi Nukl.***12** 20–30
[13] Rashid K, Safdarnejad S M, Ellingwood K and Powell K M 2019 Techno-economic evaluation of different hybridization schemes for a solar thermal/gas power plant *Energy***181** 91–106
[14] Giatman M 2011 *Ekonomi Teknik* ed A Aliludin (Jakarta: PT. Raja Grafindo Persada)
[15] F. Fabozzi. P. Peterson Drake 209AD *Finance, Capital market, Financial management, and Investment management* (New Jersey: John Wiley & Sons, Inc.)

IOP Conf. Series: Materials Science and Engineering 1041 (2021) 012020 doi:10.1088/1757-899X/1041/1/012020

Design analysis of solar panel structure LAPAN-Constellation Satellite using finite element method

S Ramayanti[1], **P Budiantoro**[1]

[1] Satellite Technology Center, LAPAN, Indonesia

Corresponding author: sri.ramayanti18@gmail.com

Abstract. Satellite Technology Center (Pusteksat) of LAPAN currently develops a satellite constellation with communication mission. This satellite will implement a deployable solar array to perform its missions. The objective of this research is to find solar panel structure with natural frequency in accordance with the requirements of the launcher and have the minimum mass. This method starts by determining the requirement of solar panel structure and then designing the solar panel into three models as follow: solar panel structure with solid aluminium sheet model, isogrid and orthogrid model, and honeycomb model. The material used for solar panel structure for aluminium sheet and isogrid and orthogrid model is Al 7075-T6 because the material is light in weight and has a high stiffness to withstand loading conditions. Meanwhile, for solar panel structure honeycomb use aluminium alloy 5052 for both face and core follow the previous satellite. The next step is analyzing the natural frequency of three models solar panel structure with modal analysis. This study shows that solar panel with honeycomb model give the best option as a solar panel structure on a constellation satellite. From the result, it can also predict the modal shapes that can be used as a place to lay the accelerometer when testing and measuring natural frequency on vibration test machines.

1. Introduction

Satellite Technology Center (Pusteksat) of LAPAN that has heritage and experience to develop small satellite [1][2][3] currently proposed design of satellite constellation with communication mission in equatorial countries including Indonesia. This satellite will orbit in equatorial Low Earth Orbit at an altitude of ± 600 km and a 0° inclination angle. This satellite will carries several communication payloads as follow: VHF Data Exchange System (VDES), Voice Repeater (VR), Automatic Packet Reporting System (APRS), Automatic Identification System (AIS) and Automatic Dependent Surveillance-Broadcast (ADSB).

The power requirement for a satellite depending on the mission. The more missions the satellite carriers, the more power it will need, thus requiring a larger solar array but there will be increased in dimension. However, due to the limitation of the dimensions of the launcher rocket, the solar array must be folded or stowed during launch, which will then be deployed when the satellite is separated from the launch vehicles and has reached the predetermined orbit. Moreover, based on research on satellites in equatorial orbit, solar panels with simple deployments produce double percentage of illumination compared to solar panels with body mounted[4]. Therefore, this satellite constellation will implement the deployable solar array to perform its missions as illustrated in figure 1.

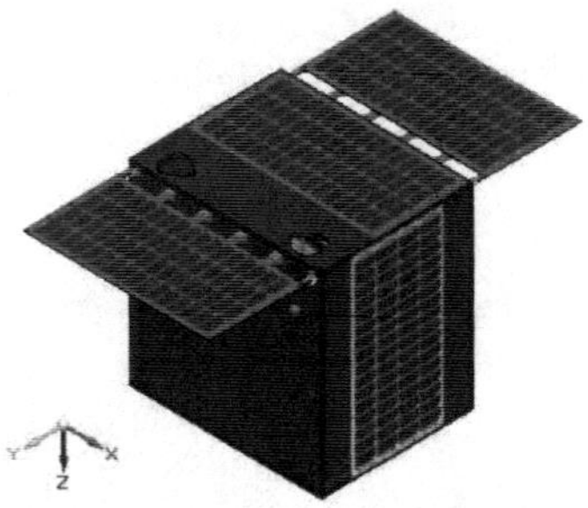

Figure 1. Constellation Satellite with deployable solar panel

The main structure of deployable solar panel usually made by aluminium alloy sheet [5] and aluminium honeycomb plate [6][7]. Aluminium alloy used in spacecraft construction because of durability, resistance to various types of corrosion and light in weight and have good thermal stability. Meanwhile, the aluminium honeycomb plate is a sandwich structure that consists of upper face and lower face of thin sheets and aluminium honeycomb structure in the middle as a core [8]. Some of the unique characteristic which make honeycomb structure a choice for this application are due to low surface density, high specific stiffness and good fatigue resistance [9]. Therefore, in this study, there will be a comparison of solar panel structure made of aluminium sheet with aluminium honeycomb in solar panel deployable structure.

In satellite launches, the point to consider is the natural frequency of the system or structure. The value of the natural frequency structure should not be the same or adjacent to the natural frequency of the launcher. Therefore, the structure including solar panel structure will not resonate and will not have deformation or failure. Modal analysis by using finite element method has become a widely useful to get this natural frequency and can be used as tools to select the solar panel structure. Moreover, another point to consider is the mass of the solar panel structure. Total mass of spacecraft structure is one of critical design factors due to the limitations inherent in the payload capabilities of launch and orbital transfer vehicles as well as for reduced launch costs [10][11]. Therefore the objective of this research is to find solar panel structure with natural frequency in accordance with the requirements of the launcher and have the minimum mass.

2. Materialas and methods

2.1. Determining the requirement of solar panel structure

One of the launchers that meets the mission requirements of a communications satellite is LauncherOne. This launcher requires a minimum natural frequency of 35 Hz and maximum payload mass of 500kg [12]. Thus the solar panel structure must satisfy these requirements. Mass of 500kg is for the total mass of the satellite including payload, structure and all of supporting component. Therefore, to fulfil the requirement, the selected solar panel must have the lightest mass.

2.2. Designing the solar panel structure

Solar panel structure is designed based on the size of the solar panel used where the size is 700x350mm as illustrated in figure 2. In this study, solar panel structure is designed in three models as follow: solar panel structure with solid aluminium sheet model, isogrid and orthogrid model, and honeycomb model. Solar panel structure with solid aluminium sheet model was designed with a thickness of 6mm and 8 mm to determine the effect of thickness of solid aluminium on natural frequency.As for reducing mass, solar panels are designed with isogrid and orthogrid models at 10mm thickness. Orthogrid model is rectangular cut-out where the pattern is square which aims to maximize strength to weight ratio. This model is the best option for fabrication on milling machine [13]. Meanwhile isogrid model is a pattern of equilateral triangles that machined from flat surface. This model has advantages such as high stiffness, light in weight and thermal stability [14]. Orthogrid and Isogrid model can be seen in figure 3. The next design is the design of solar panel structure using aluminium honeycomb. The thickness of the honeycomb used is 8mm for the core and 1 mm for each face. So the total thickness of this solar panel structure is 10mm.

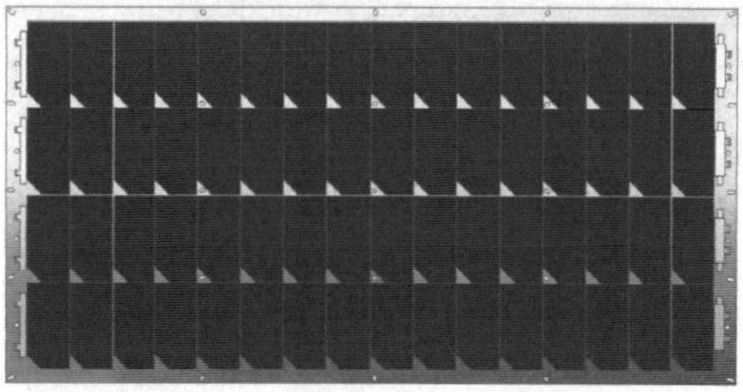

Figure 2. Dimension of solar panel

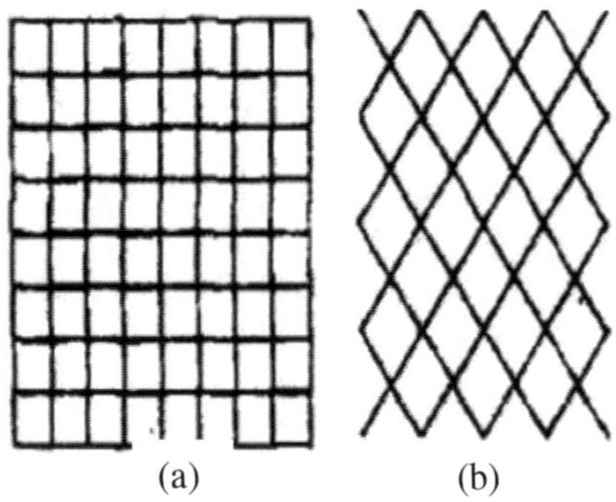

Figure 3. (a) Orthogrid and (b) Isogrid Models [15]

2.3. Collecting/establishing the material of solar panel structure

The spacecraft industries have long relied upon aluminium alloys. The material used for solar panel structure for aluminium sheet and isogrid and orthogrid model is Al 7075-T6 because the material is light in weight and has high stiffness to withstand loading conditions [16]. The properties of this material can be seen in table 1. Meanwhile, for solar panel structure honeycomb use aluminium alloy 5052 for both face and core follow the previous satellite [17]. This property can be seen in table 2 for the face and table 3 for core properties.

Table 1. Properties of Aluminium Alloy Al 7075

Item		Unit
Density	2810	Kg/m3
Poisson Ratio	0.33	
Modulus Elasticy	71.7	GPa
Shear Modulus	26.9	GPa

IOP Conf. Series: Materials Science and Engineering 1041 (2021) 012020 doi:10.1088/1757-899X/1041/1/012020

Table 2. Properties of Aluminium Alloy Al 5052

Item		Unit
Density	2680	Kg/m3
Poisson Ratio	0.33	
Modulus Elasticy	70.3	GPa
Shear Modulus	25.9	GPa

Table 3. Properties of Aluminium Alloy Al 5052

Item		Unit
Density	97.7	Kg/m3
Poisson Ratio 12	0.30	
Poisson Ratio 12	0.30	
Poisson Ratio 12	0.30	
Modulus Elasticy	0.67	GPa
Modulus Elasticy	0.283	GPa
Modulus Elasticy	0.1655	GPa
Cell Size	1/8	Inch

2.4. Analyzing the natural frequency of solar panel structure

Finite Element analysis of the design has been established using MSC. Patran and the element employed in all of the models (design) are quad-4 element topology (four corner nodes). In modal analysis, the detail like holes and pores can be ignored, because the natural frequency of a structure is mainly decided by stiffness and structural mass. The boundary condition in the modal analysis concern in 3 place, two place is the location for hinge to structure satellite which is constrained in displacement of x, y and z are zero, and rotation of x, y, and z are zero and the remaining place is the location for hold mechanism which is constrained in displacement of x, y and z are zero.

In this research, honeycomb model use equivalent plate methods which replace the honeycomb model into an equivalent single plate [17]. This method has the objective of creating simplified finite element model with less numbers of elements, less number of material types and therefore more efficient. To estimate the equivalent parameter of honeycomb equivalent models can be seen in equation (1), (2), and (3) [18].

$$t_{eq} = \sqrt{3h_c^{\;2} + 6h_c + 4t_f^{\;2}} \tag{1}$$

$$E_{eq} = \frac{2t_f}{t_{eq}} . E_f \tag{2}$$

$$G_{eq} = \frac{2t_f}{t_{eq}} . G_f \tag{3}$$

Where,

t_{eq} = equivalent plate thickness

h_c = core thickness

t_f = face thickness

E_{eq} = equivalent plate Young Modulus

G_{eq} = equivalent plate Shear Modulus

E_f = Modulus young face

G_f = Modulus shear face

From the equation is derived the value of parameter of honeycomb equivalent models as seen in table 4.

Table 4. Properties of Aluminium Alloy Al 7075

Item		Unit
Equivalent plate density	393.19	Kg/m3
Poisson Ratio	0.33	
Equivalent plate Young modulus	9	GPa
Equivalent plate thickness	0.01562	mm

3. Results and discussion

The coloured fringe gives the amplitude of the displacement vector describing the shape each mode. The blue colour corresponds for null displacement and the red one presents the maximum amplitude.

3.1. Effect of thickness on aluminium sheet model

The influence of the sheet thickness on the natural frequency is considered in this sub-section. The thickness of the skins is used as follows: 6mm and 8mm. Figure 4. shows the mode shape of the modal analysis and table 5 is the natural frequency. According to the results obtained, we note that sheet thickness of the aluminium sheet affects the natural frequency but not the mode shape. The thicker the aluminium sheet, the resulting of natural frequency also increase, but with consequences on the increased weight as can be seen in table 6.

Table 5. Frequency Natural of Solar Panel Structure

	Solar Panel Solid 6mm (Hz)	Solar Panel Solid 8mm (Hz)	Solar Panel Isogrid 10 mm (Hz)	Solar Panel Orthogrid 10 mm (Hz)	Solar Panel Honeycomb (Equivalent Method) (Hz)
1st Natural Freq.	79.442	104.58	98.632	105.52	166.1
2nd Natural	79.911	106.22	100.76	106.94	178.38

Freq.					
3rd Natural Freq.	218.84	277.92	276.49	272.58	583.66

Table 6. Mass of Solar Panel Structure

	Massa (kg)
Solar Panel Solid 6 mm	4,246
Solar Panel Solid 8 mm	5,661
Solar Panel Isogrid 10 mm	4.862
Solar Panel Orthogrid 10 mm	4,020
Solar Panel Hoenycomb	1.504

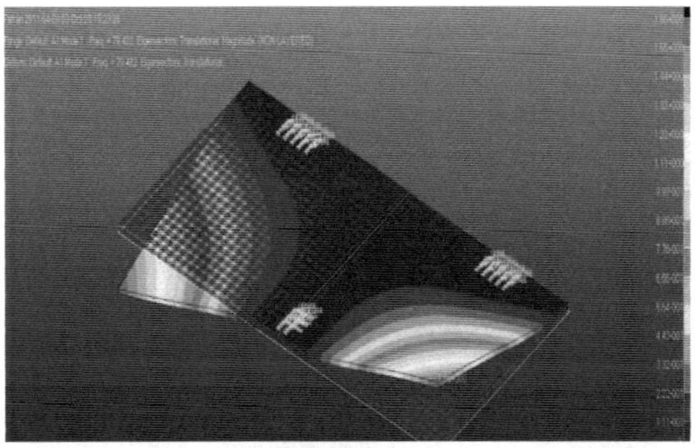

Model 1

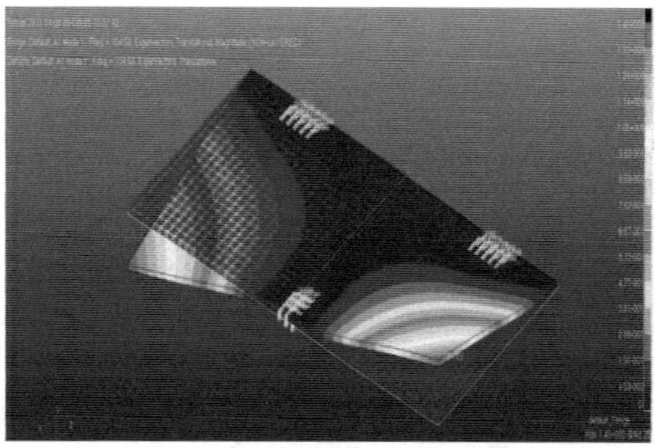

Model 1

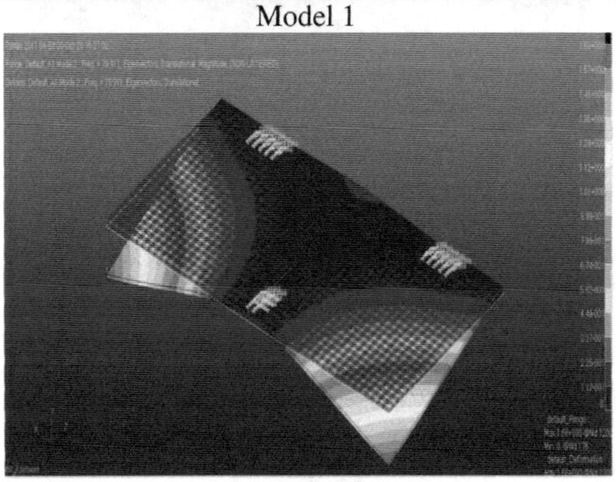

Model 2

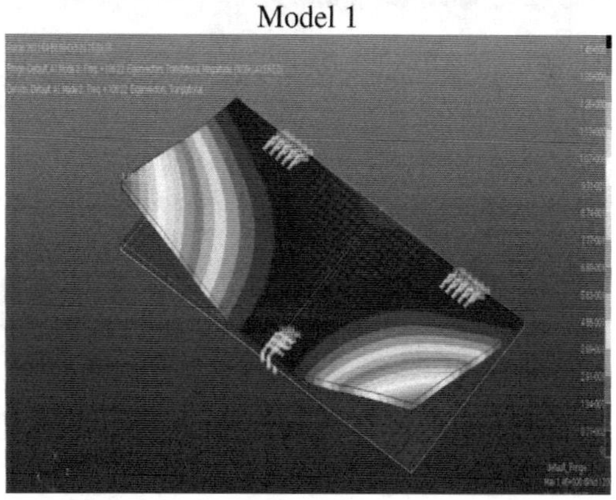

Model 2

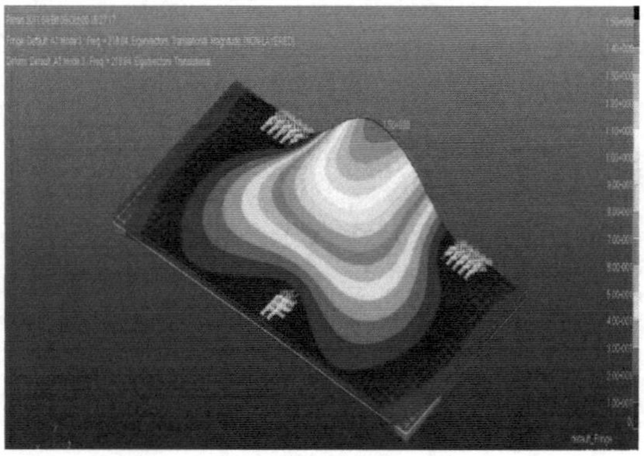

Model 3
(a)

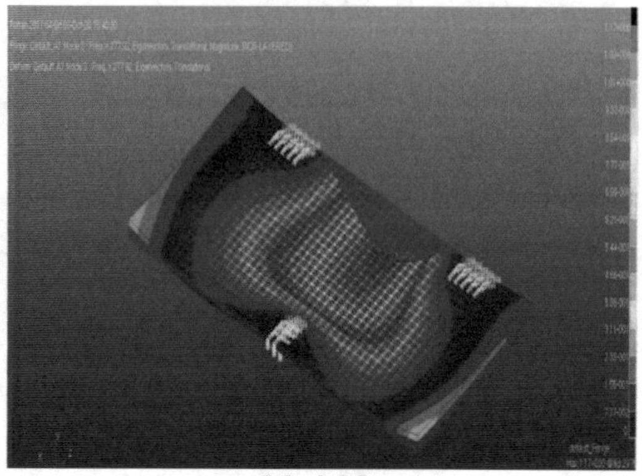

Model 3
(b)

Figure 4. Mode shape of the aluminium sheet model at thickness (a) 6mm, (b) 8mm

3.2. Effect orthogrid and isogrid model of solar panel structure

This section performs the influence of geometry of solar panel structure which is orthogrid and isogrid models. The thickness of both these structures is 10mm.

From Figure 5. it can be seen that the shape modes of both the orthogrid and isogrid models are the same, but the solar panel structure with orthogrid model has a higher frequency value as can be seen in table 5. By changing the geometry from solid sheet into isogrid and orthogrid and keeping the same material properties, the natural frequency is not much different from orthogrid model, but higher than isogrid model as can be seen in table 5. However, the mass of the solar panel structure with isogrid and orthogrid model is lower compared to solid aluminium sheet even with thicker thickness.

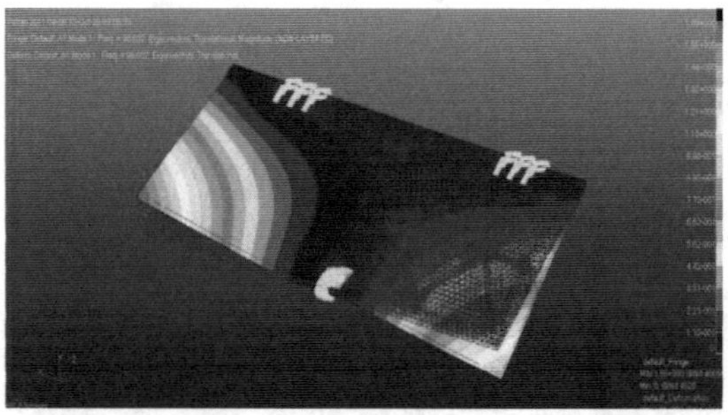

Model 1

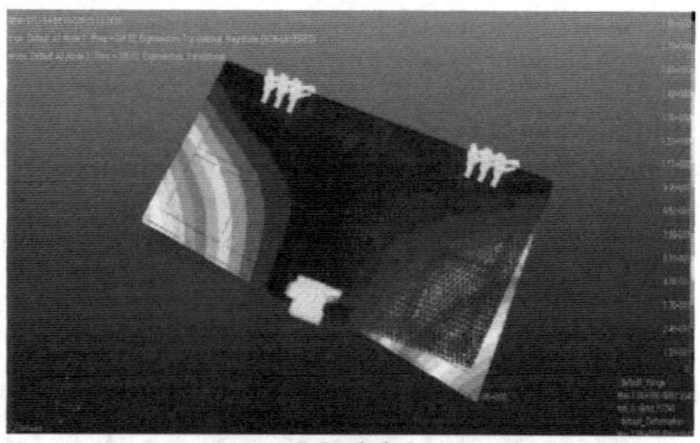

Model 1

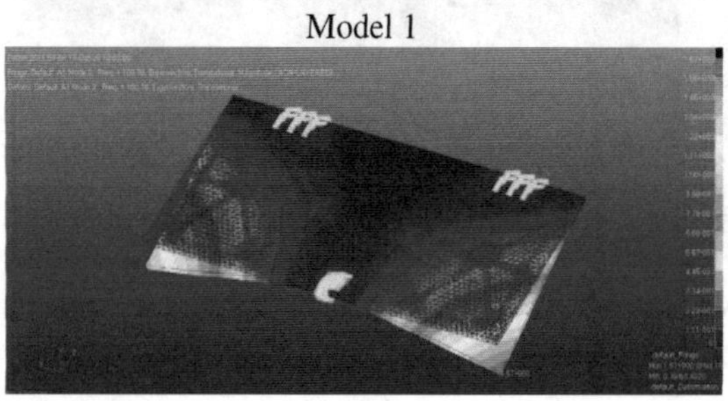

Model 2

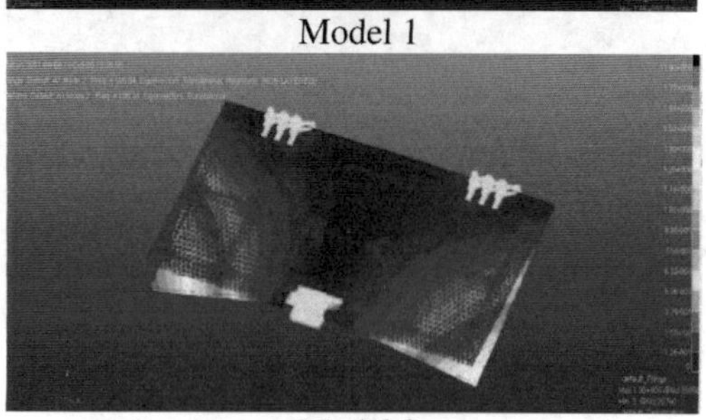

Model 2

Model 3
(a)

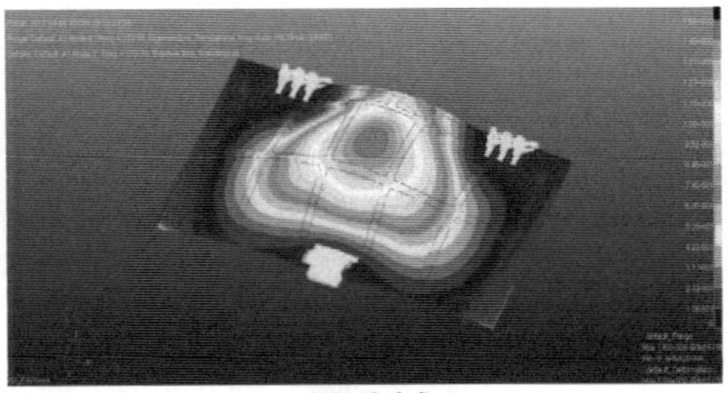

Model 3
(b)

Figure 5. Mode shape of the a. isogrid model b. orthogrid model

3.3. Effect of changing aluminium solid sheet ke aluminium honeycomb model

In this subsection, considering changing aluminium solid sheet into aluminium honeycomb model bring massive impact on not only in natural frequency but also on mass. The natural frequency is increasing yet the mass is decreased. The model shape can be seen in figure 6.

Model 1

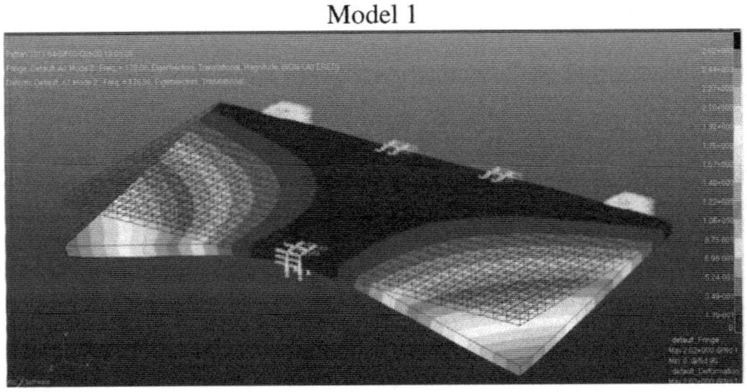

Model 2

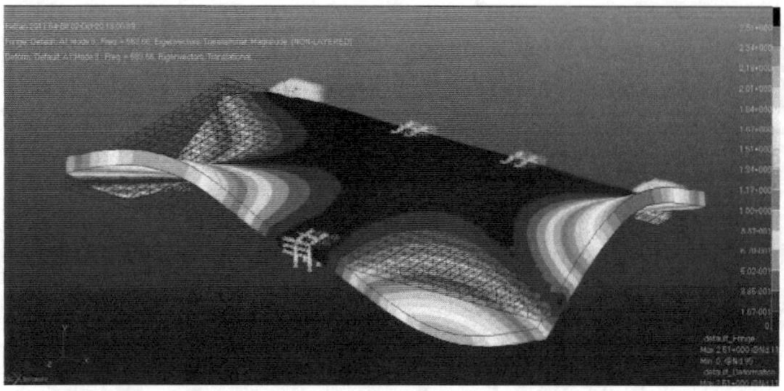

Model 3

Figure 6. Mode shape of honeycomb solar panel structure

4. Conclusion

Modal analysis is the basic requirement for the solar panel structure design which is can calculate the natural frequency of solar panel structure with three different models as follow: aluminium solid sheet, aluminium isogrid and orthogrid model and aluminium honeycomb model. The different results obtained in this paper show that all of solar panel structure models have fulfilled the natural frequency of LauncherOne requirement, however, the honeycomb model gives the best option as a solar panel structure on a constellation satellite due to it has the lowest mass. The mode shapes from the modal analysis can be used as a reference to lay the accelerometer when testing and measuring natural frequency of the structure on vibration test machines.

References

[1] Triharjanto R H, Hasbi W, Widipaminto A, Mukhayadi M and Renner U 2004 LAPAN-TUBSAT: Micro-satellite platform for surveillance & remote sensing *Eur. Sp. Agency, (Special Publ. ESA SP* 277–83

[2] Hardhienata S, Triharjanto R H and Mukhayadi M 2011 LAPAN-A2 : Indonesian Near-Equatorial Surveilance Satellite *18th Asia-Pacific Regional Space Agency Forum (APRSAF)* (Singapore) pp 1–10

[3] Hasbi W and Suhermanto S 2013 Development of LAPAN-A3/IPB Satellite an Experimental Remote Sensing Microsatellite *34th Asian Conf. Remote Sens.*

[4] Suryanti D I, Ramayanti S and Mukhayadi M 2019 Iluminasi Panel Surya pada Satelit Orbit Rendah Ekuatorial *Elkomika***7** 480–92

[5] Santoni F, Piergentili F, Donati S, Perelli M, Negri A and Marino M 2014 An Innovative Deployable Solar Panel System for Cubesats *Acta Astronaut.***95** 210–7

[6] Tanzman J R 2008 Material considerations in the STEREO solar array design *Acta Astronaut.***63** 1239–45

[7] Brough Z and Paoloni C 2015 Advanced Deployable/Retractable Solar Panel System for Satellite Applications *Int. J. Mech. Mechatronics Eng.***9** 157–62

[8] Teng L, Zheng X D and Jin Z H 2019 Performance Optimization and Verification of a New Type of Solar Panel for Microsatellites *Int. J. Aerosp. Eng.* 1–14

[9] Yuan J, Zhang L and Huo Z 2020 An Equivalent Modeling Method for Honeycomb Sandwich Structure Based on Orthogonal Anisotropic Solid Element *Int. J. Aeronaut. Sp. Sci.*

[10] Chen S and Huang H 2010 Optimum Design of a Space Frame and its Application in Satellite Structure *J. Spacecr. Rockets***47** 1063–6

[11] Aborehab A, Abbas M K, Nemnem A F and Kamel H 2019 Configuration Design and Modeling of an Efficient Small Satellite Structure *Eng. Solid Mech.***8** 1–14

[12] Orbit V 2020 LauncherOne Service Guide

[13] Harrison S, Scott P and Zapien V 2012 *Nanosatellite Fabrication and Analysis* (Santa Clara University)

[14] Tsinas A and Welch C 2001 Efficient Satellite Structural Design Optimised for Volume Production *15th Annual USU Conference on Small Satellites* pp 1–20

[15] Ghadi N, Mattikalli P A and Ghadi P A 2017 Design and Fe Analysis of Composite Grid Structure for Skin Stiffening Applications *Int. Res. J. Eng. Technol.* **4** 792–802

[16] Israr A 2014 Vibration and Modal Analysis of Low Earth Orbit Satellite *Hindawi Publ. Corp.* **2014** 1–8

[17] Triharjjanto R H and Budiantoro P A 2019 Stiffness Evaluation of Lapan-A5/Chibasat Deployable Solar Panel Composite Plate Using Simplified Finite Element Model 169–75

[18] Paik J K, THayamballi A K and Kim G S 1999 The Strength Characteristics of Aluminum Honeycomb Sandwich Panels *Thin-Walled Struct.* **35** 205–31

IOP Conf. Series: Materials Science and Engineering 1041 (2021) 012021 doi:10.1088/1757-899X/1041/1/012021

Design the motor acceleration for the desired frequency air volume inspiration of asynchronous mechanical ventilator based on a pressure sensor

M I Rusydi[1], A Jordan[1], H D Laksono[1], N Windasari[2], H T Ngetha[3]

[1]Electrical Engineering Department, Faculty of Engineering, Universitas Andalas, Padang City, West Sumatera, Indonesia
[2]Faculty of Medicine, Universitas Andalas, Padang City, West Sumatera, Indonesia
[3]Department of Electrical and Electronic Engineering, Dedan Kimathi University of Technology, Dedan Kimathi, Kenya

Corresponding author: rusydi@eng.unand.ac.id

Abstract. The SARS-CoV-2,a virus that attacks the respiratory system, has some difficulty breathing and chest pain. A mechanical ventilator is a simple version of a ventilator and most used to support patient breathing because it iseasy to use.This research design a mechanical ventilator operate based on the frequency and air volume of inspiration.The operator can set the ventilator using a potentiometer and choose the menu on an LCD. A couple of press wall push a bag valve mask in the desired frequency. The motor speed has a vital role to perform the number pushing based on the desired setting. The amount of airflow out from this bag depends on the duration and the depth of pushing. Some standard frequencies of human breathing were investigated in the range of 12 to 20 times of inspiration in a minute. The acceleration of the ventilator based on calculation was implemented to the machine, and it reduced the number frequency to 64% of the desired frequency. An error compensator is used to improve the accuracy of the ventilator. It compensates by multiply calculated acceleration by 2.31. It is successfulin improving the performance and resulted in the ventilator work in the desired frequency and volume.

1. Introduction

The number of novel coronavirus cases is increasing since December 2019, firstly identified in Wuhan, China[1]. The saliva droplet from the infected person is easy to spread to other people and infect the community. WHO reported that there are 1.3 million death cases from 50 million patients in 220 countries only in 11 months.This virus damage level varies from just a simple cold to severe respiratory problems such as pneumonia. In any case, infected people have no symptoms [2].

Neither proven treatment nor vaccine applicable to solve the problem caused by this virus. But in common, the patients with a bad condition need medical assistance for respiratory since the virus harms the lungs caused breathing obstacles. Ventilator plays a vital role in these situations. It helps patients to keep breathing in a critical condition[3].

A mechanical ventilator is one of the many interventions for the patient with lung damage. This ventilator can maintain the amount of oxygen supplied to the patient in the Intensive Care Unit (ICU)[2]. This pandemic raises the need for this machine. The advanced production of the existing ventilator or the new one is being pushed to fulfill the gap[3].

The ventilator has three modes pressure, volume, or hybrids[4]. Many researchers and the community tried to develop a safe and low-cost mechanical ventilator to fulfill the pandemic's mechanical ventilator requirement[5]. The non-invasive ventilators are more likely to be developed quickly according to patient needs. Some standard variables used in a ventilator to examine patients are airflow, air pressure [6–8], cycle variable,and the inspiratory phase of breath[9].

In this research, an asynchronous non-invasive mechanical ventilator was designed. This article reports primarily findinghow to set control the motor to get the desired number of inspiration. This control important to get the expected time, volume, and pressure.

2. Materials and methods

This research designed a motor acceleration considering the inspiration rate and air volume gave to patients. This is a portable ventilator with a graphic menu for the operator. A bag valve mask is compressed from two sides. A pressure sensor measures the inspiration airflow.

2.1. Electrical Component

A ventilator is designed to work in the desired frequency and volume of inhalation. The components used for this ventilator are shown in figure 1. An LCD shows the menu for the operator. The operator can rotate the potentiometer to set the frequencyand volume of inspiration. Two motors with the driver move the wall to press the bag valve mask (BVM). A differential pressure sensor measures air flowing through the BVM's neck. As the center of the device, a microcontroller receives the potentiometer input and givessignal to the LCD and the two motors as output. This ventilator requires a 24 V power supply and one voltage regulator, 24 V to 5 V. Figure 1 shows the designed ventilator scheme.

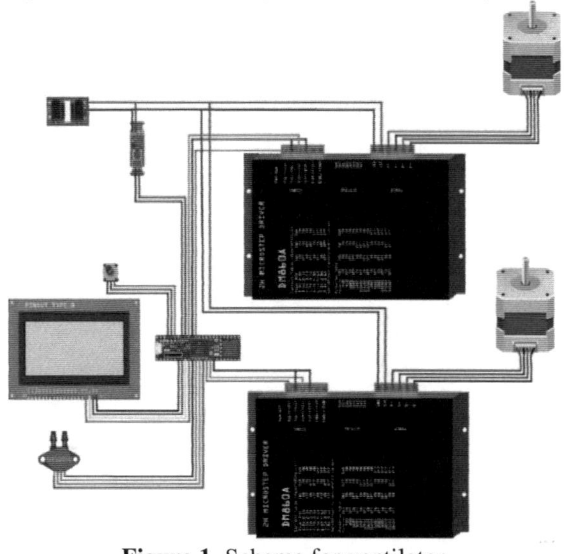

Figure 1. Scheme for ventilator.

2.2. Ventilator Box

The designed ventilator has a box packaging. Figure 2(a) is the front view,and figure 2(b) is the top view of the ventilator. The operator rotates the potentiometer (1) and choosessome menu on LCD (2).A differential pressure sensor measures the gap air pressure inside respirator tube 3(a) and outside 3(b).A microprocessor (4) is inside the box. This microprocessorreceives an input signal from the sensor and potentiometer. This device processes the signal and generatesa command fortwo motor drivers. These drivers operate the left motor (5a) and right motor (5b). Both motors shift the left wall (6a) and right (6b). The shifting of the wall press will push the BVM (7). Air flows to the patient through the tube (9) when the BVM is pushed.

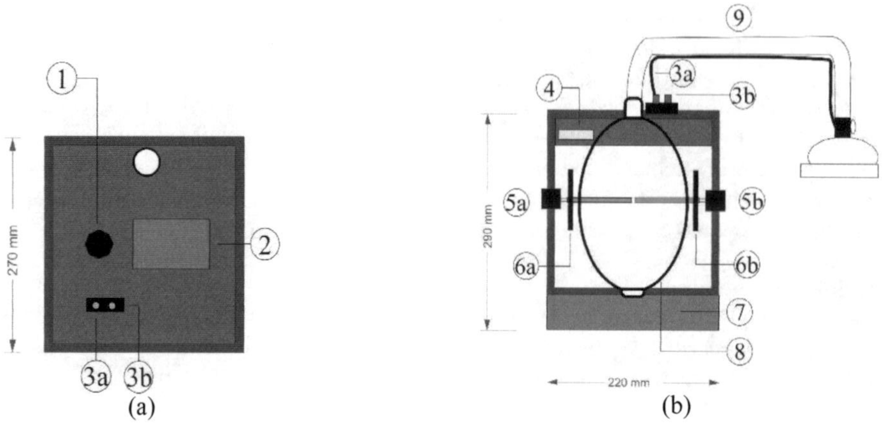

Figure 2. (a) Front view and (b) top view.

2.3. Air Volume

Air will flow to the patient from BVM if air pressure is generated [10]. This pressure is generated if the wall press pushes the BVM. It flows through the tube to the patient. This flow is irreversible because of the BVM design. It just only can go in one direction. The illustration is shown in figure 3.

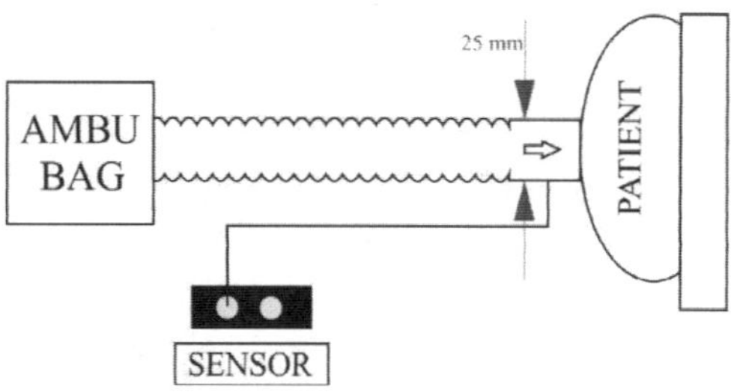

Figure 3. Airflow from Ambu-bag to patient.

A differential pressure sensor[11]measures the gauge pressure inside the tube. The air velocity inside the tube is calculated based on the assumption that the atmosphere's air velocity is equal to zero. Based on the Bernoulli equation, the air velocity inside the tube is determined using equation (1).

$$v = \sqrt{\frac{2\Delta P}{\rho}} \tag{1}$$

Where

v = air velocity (m/s),

ΔP = air pressure (Pa)

ρ = mass density (kg/m^3)

The flow rate of air transfer to the patient for each push is calculated by (2).

$$Q = v.A \tag{2}$$

Where

Q = flow rate (m^3/s)

v = air velocity (m/s),

A = area inside the tube (m^2)

The volume of airflow to the tube is equal to (3).

$$V = Q.t \tag{3}$$

Where

V = volume (m^3)

Q = flow rate (m^3/s)

t = time (s),

2.4. Motor acceleration

The inspiration rate affects the frequency of pushing BVM. Figure 4 illustrates the ventilator while pushing the BVM.The two press walls push the BVM with distance (d) from the left and right sides. The walls move from the normal position to the center. After pushing the BVM at the desired distance, the two walls areback to the normal position until the BVM is released. The total distance pass by a wall press is twice of d, as shown in equation (4).

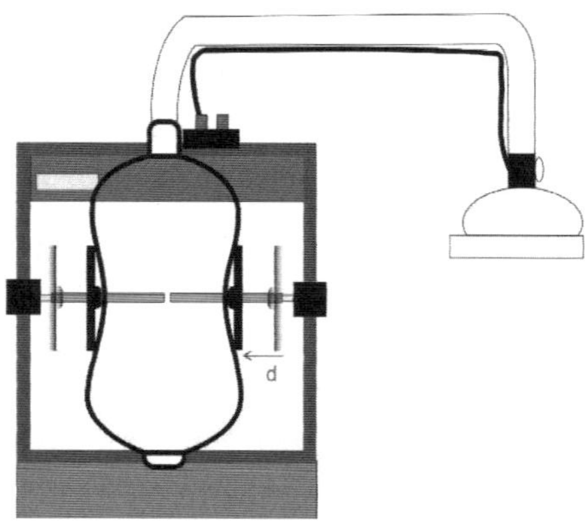

Figure 4. Airflow from Ambu-bag to patient.

$$d_{total} = 2.d \tag{4}$$

Where

d =the desired pushing distance (cm)

d_{total} = the total length passed by a suppressor after back to the normal position (cm)

One cycle for inspiration and expiration is when the initial position wall pushes the BVM to the desired distance and back to the normal position. The time for one process is called a period. The number of inspirations in a minute is the inspiration rate. It is equal to equation (5).

$$f = \frac{60}{T} \tag{5}$$

Where

f = inspiration rate (inspiration/minute)
T = Period of one cycle (second)

The period of one cycle affects the acceleration motor to push BVM. The movement of the press wall indicates motor acceleration. The acceleration of the press wall to pass the desired distance is (6).

$$a = \frac{2d_{total}}{T^2} \tag{6}$$

Where

a = motor acceleration (cm/s^2)
d_{total} = the total distance passed by a suppressor after back to the normal position (cm)
T = Period of one cycle (second)

3. Results and Discussion

An experiment to determine the acceleration of the press wall to receive the desired inspiration rate was performed. The total distance (d_{total}) was set constant at 4 cm, and the desired frequencies were varied from 12 to 18 inspirations in a minute. The inspiration rates were varied to know the error of frequency. The result shows in table 1.

Table 1. Formatting sections, subsections and subsubsections.

f_d	f_a	f_e	T_d	T_a	T_e	a_d	a_a	a_e
12	7	5	5.00	7.50	2.50	0.32	0.14	0.18
13	8	5	4.62	6.59	1.97	0.38	0.18	0.19
14	10	4	4.29	6.06	1.77	0.44	0.22	0.22
15	10	5	4.00	5.70	1.70	0.50	0.25	0.25
16	10	6	3.75	5.34	1.59	0.57	0.28	0.29
17	11	6	3.53	5.44	1.91	0.64	0.27	0.37
18	11	7	3.33	5.25	1.91	0.72	0.29	0.43

The ventilator that was designed before being implemented the error compensator turned out not to work as desired. There is a difference in the amount of inspiration rate generated by the ventilator, and the desired frequencyis r_f. The average system only reaches 64% of the desired frequency, with 153%of the desired time. This means the system is running slower than expected. Based on the calculation, the acceleration obtained was only 43% of the set point. The ratio between actual acceleration and desired acceleration is r_a. Therefore, it is necessary to compensate for the error with 2.31 of the desired acceleration. This is the final acceleration value of the press wall after getting compensation. The complete ratio between actual to the desired frequency, time cycling, and acceleration is shown in table 2.

Table 2. The ratio of actual to the desired variable.

f_d	r_f	r_t	r_a
12	0.58	1.71	0.34
13	0.62	1.61	0.39
14	0.71	1.41	0.50
15	0.67	1.43	0.49

16	0.63	1.42	0.49
17	0.65	1.54	0.42
18	0.61	1.57	0.40

Experiments to determine the amount of air pumped by the ventilator in each inspiration are carried out by varying distances. The resulting volume increases from the smallest to the largest distance. It is proof the Bernoulli's equation work in this ventilator. The system works more stable, the further the pressing distance. The need for human inspiration, which is 5 to 10 L for each inspiration, can be fulfilled with encouragement from this wall press, especially at a distance of 4 cm. Table 3 shows the volume of air transfer to the patient from BVM at various lengths.

Table 3. The volume of air from BVM.

d	$\bar{V}$ (L)	SD
2.5	2.49	0.86
3.0	2.96	0.31
3.5	4.16	0.70
4.0	5.51	0.16

4. Conclusion

An asynchronous mechanical ventilator is designed in this research. This article mainly focuses on the acceleration performance to get the desired inspiration rate. An error compensator for the acceleration is implemented to the system 2.31 of the desired acceleration. This compensator fulfilled the required frequency of inspiration. The ventilator also produces enough air volume for the patient.For future development, the measurementof the expiration air volume and pressure are urgently required. The implementation of an early warning system if a system works improperly is also essential.

Acknowledgment

This research is funded by the Faculty of Engineering, Universitas Andalas contract number: 001/UN.16.09.D/PL/2020

References

[1] Li Q, Guan X, Wu P, Wang X, Zhou L, Tong Y, Ren R, Leung K S M, Lau E H Y, Wong J Y, Xing X, Xiang N, Wu Y, Li C, Chen Q, Li D, Liu T, Zhao J, Liu M, Tu W, Chen C, Jin L, Yang R, Wang Q, Zhou S, Wang R, Liu H, Luo Y, Liu Y, Shao G, Li H, Tao Z, Yang Y, Deng Z, Liu B, Ma Z, Zhang Y, Shi G, Lam T T Y, Wu J T, Gao G F, Cowling B J, Yang B, Leung G M and Feng Z 2020 Early Transmission Dynamics in Wuhan, China, of Novel Coronavirus–Infected Pneumonia *N. Engl. J. Med.***382** 1199–207

[2] Branson R, Dichter J R, Feldman H, Devereaux A, Dries D, Perry J F, Benditt J, Hossain T, Ghazipura M, King M, Baldisseri M, Christian M D, Domingiuez-Cherit G, Henry K, Martland A M O, Huffines M, Ornoff D, Persoff J, Rodriquez D, Maves R C, Kissoon N "Tex" and Rubinson L 2020 The US Strategic National Stockpile (SNS) Ventilators in COVID-19: A Comparison of Functionality and Analysis regarding the Emergency Purchase of 200,000 devices. *Chest*

[3] Iyengar K, Bahl S, Raju Vaishya and Vaish A 2020 Challenges and solutions in meeting up the urgent requirement of ventilators for COVID-19 patients *Diabetes Metab. Syndr. Clin. Res. Rev.***14** 499–501

[4] Singh G and Cao M 2020 Noninvasive Ventilator Devices and Modes *Sleep Med. Clin.***15** 545–55

[5] Carraway J S, Carraway M W and Truelove C A 2020 Nursing implementation of a validated agitation and sedation scale: An evaluation of its outcomes on ventilator days and ICU length of stay *Appl. Nurs. Res.* 151372

[6] Szlosarek R, Teichert R, Wetzel A, Fichtner A, Reuter F and Kröger M 2020 Design and construction of a simplified, gas-driven, pressure-controlled emergency ventilator *African J. Emerg. Med.* 1–7

[7] Zuckerberg J, Shaik M, Widmeier K, Kilbaugh T and Nelin T D 2020 A lung for all: Novel mechanical ventilator for emergency and low-resource settings *Life Sci.***257** 118113

[8] Zhang L, Mao K, Duan K, Fang S, Lu Y, Gong Q, Lu F, Jiang Y, Jiang L, Fang W, Zhou X, Wang J, Fang L, Ge H and Pan Q 2020 Detection of patient-ventilator asynchrony from mechanical ventilation waveforms using a two-layer long short-term memory neural network *Comput. Biol. Med.***120** 103721

[9] Fierro J L and Panitch H B 2019 Transitioning from an ICU ventilator to a portable home ventilator *Semin. Fetal Neonatal Med.***24** 101041

[10] Mora S, Duarte F and Ratti C 2020 Can Open Source Hardware Mechanical Ventilator (OSH-MVs) initiatives help cope with the COVID-19 health crisis? Taxonomy and state of the art *HardwareX***8** e00150

[11] Choi Y and Song D 2020 How to quantify natural ventilation rate of single-sided ventilation with trickle ventilator? *Build. Environ.***181** 107119

IOP Conf. Series: Materials Science and Engineering 1041 (2021) 012022 doi:10.1088/1757-899X/1041/1/012022

Performance Analysis of Diversity Selection Combining on Radio Frequency for Image Transmission

Baharuddin[1], A Luthfi[1], H Andre[1], M Muharam[1], R Angraini[2]

[1] Department of Electrical Engineering, Universitas Andalas, Kampus Unand Limau Manis, Padang, 25163. Indonesia

[2] Politeknik Negeri Padang, Kampus Politeknik Negeri Padang Limau Manis, Padang, 25163. Indonesia

Corresponding author: baharuddin@eng.unand.ac.id

Abstract. This paper describes a simulation of the image transmission system with diversity selection combining (DSC) method on radio frequency. DSC is used to obtain multiple data streams corresponding to the transmitted image at the receiver in order to combat errors during image transmission on wireless channels. These individual image data streams are combined to form a composite image with higher perceptual quality. When DSC is used, it is expected that it can affect the number of errors, so that the signal transmitted to the wireless channels becomes more robust against interference caused by transmission channels. By using DSC technique, it can increase the quality of image transmission at the receiver. The result of the experiment which was carried out to test bit error rate (BER) and peak signal to noise ratio (PSNR) on Lena image with rate 0.8 bpp by using DSC on radio frequency in Rayleigh fading channel showed that there was a decrease in BER at the average of 0.0152 or 3.39%, while for PSNR there was an increase at the average of 1.6737 dB or 6.73% when comparing to the system without DSC.

1. Introduction

The advancement of digital image technology has triggered a rapid development in the imaging business, which leads to high demand for effective digital image storage and transmission media [1]. It happens because digital images is a type of information that has a relatively larger size than other types of information, such as text and sound [2]. In this case, compression or coding of sources becomes a solution to reduce the number of information symbols, so that, it can streamline storage media and increase the transmission rate for certain channel capacities [3]. On the other hand, with a certain transmission rate, compression allows the use of channels with a smaller bandwidth.

Set Partitioning in Hierarchical Trees (SPIHT) wavelet-based image compression method [4] is the development of the embedded zerotree wavelet (EZW) method [5]. SPIHT has better performance and simpler algorithm than EZW, which using partitioned grouping for significance tests in accordance with the decreasing threshold order. With the reduced number of symbols to represent an image, the compressed image will experience the significant impact on information damage due to noise than uncompressed images. Therefore, coding controls and error correction is absolutely necessary for transmission of compressed images [6-10]. Reed Solomon (RS) codes is one of the alternative codings that has comparatively capabilities on correcting additional redundant data. This encoding is based on bit groups [11-12], so that it has ability in error correction, especially burst errors. In wireless communication, the signal can propagate from the transmitter to the receiver through various reflection paths, which is called multipath propagation [13-16]. This phenomenon causes fluctuations in signal quality in the receiver which is known as fading. Fading is a noise that contaminates the signal by multiplying the signal, thus it gives a greater impact than additive noise. Diversity technique is one of the techniques that is often used to reduce signal quality degradation due to this phenomenon. Diversity is used to improve the performance of radio channels without increasing transmission power [17].

Several techniques have been used to improve the performance of systems such as the use of diversity technique equal gain combining on radio frequency [15-17]. However, it is not a good technique because the increment of signal to noise ratio (SNR) will not automatically increase PSNR, but it only implies to degrade BER [18]. The use of diversity maximum ratio combining on wavelet domain improves the performance of image reconstruction [18]. Diversity techniques use on wavelet domain is to get better reconstruction image that used a two-state Gilbert-Elliott channel [18].

For this research purpose, DSC on radio frequency [2] uses fading distribution reyleigh and noise which has normal distribution gaussian [2]. To combat the errors that occurs on wireless channel, the use of coding channel is advantageous. In this research, RS codes and DSC are chosen due to its high ability in detecting and correcting errors on the image that caused by fading and noise.

The experiment is done by comparing the system with and without using DSC. The evaluation of the improvement quality system on the image transmission is calculated by the use of BER and PSNR. BER indicates how many bits with errors or symbols occurred at the receiver and compares to the number of bits or symbols received in a transmission, while PSNR is used to evaluate the quality of reconstruct image [4-7,18].

In the simulation, image information with size 8 bpp is compressed by SPIHT with bit rate 0.8 bpp. The purpose of using SPIHT compression is to minimize signal bandwith so that the use of transmission channel become more efficient. Diversity selection combining technique is used to improve image quality on the receiver because this technique has the ability to choose the largest SNR. RS codes (31.15) [18-19] is used to protect data during transmission process.

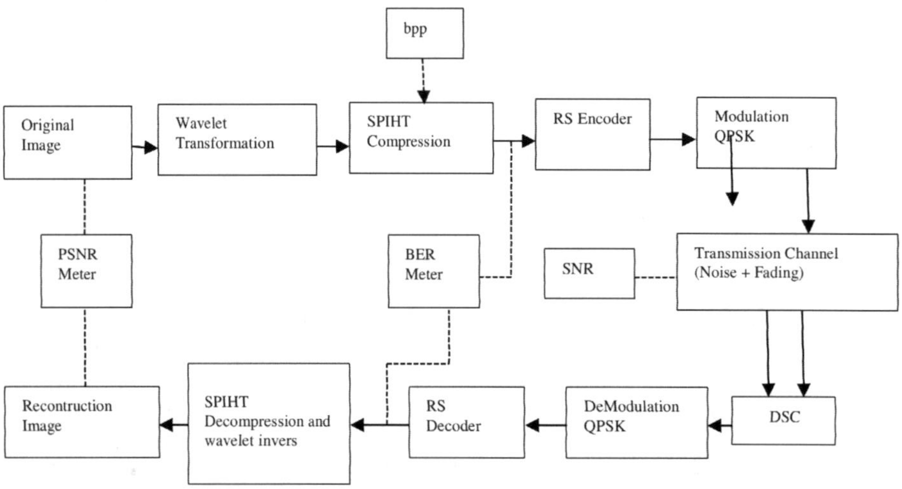

Figure 1. The design system of SPIHT compressed image transmission with DSC technique in Radio Frequency Domain.

The design system of SPIHT compressed image transmission with DSC technique on Radio Frequency Domain can be seen in figure 1. The image to be transmitted is a grayscale image of 512 × 512 with a depth of 8 bits. Before compressing the image, it is decomposed by wavelet transformation [18]. The wavelet transformation image is compressed with the SPIHT algorithm with bit rate 0.8 bpp [4], so that it produces the series of bit streams with the number corresponding to the bits per pixel (bpp) which is independent variable of the system. To reduce errors that occur during transmission, the RS coding is used. In this system, the output of RS Codes is then modulated QPSK before it is transmitted to the wirelesss channel.

The process of DSC uses two diversity channels by implementing error correction and grouping the bit stream into blocks can be seen on Figure 2. The algorithm of this diversity is based on the block chosen of bit *b(l)* from one bit stream which based on the characteristic of wavelet transformation. The algorithm of DSC captures two signals that are independent on fading and noise from the transmission channel. Then, this DSC choose signal that has largest SNR. The simulation model for the simulation system is as follows [2]:

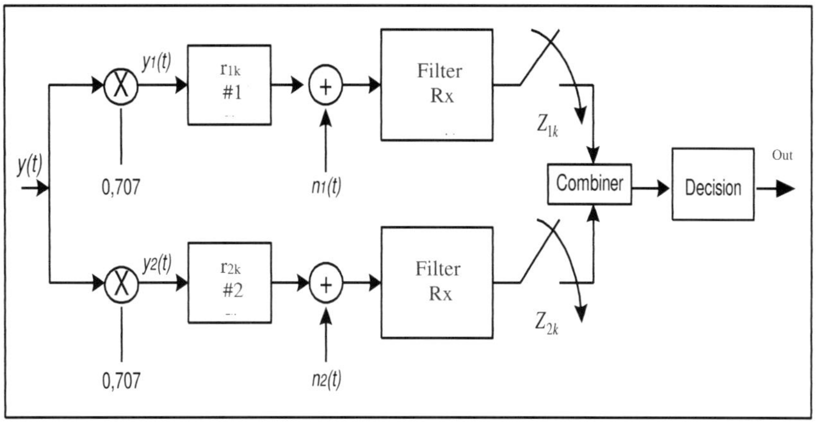

Figure 2. The model system with Selection Diversity Combining (DSC) [2].

On decision circuit, decoding process is used to detect bit on the reciever. This process is done by applying a rule. The rule gives the optimum performance, so it could reduce the error [2]. The simulation model for DSC is as follows:

$$Z_k = \begin{Bmatrix} Z_{1k}, & if & |Z_{1k}| & \geq & |Z_{2k}| \\ Z_{2k}, & if & |Z_{2k}| & \geq & |Z_{1k}| \end{Bmatrix} \quad (1)$$

where :

Z_k = detected Bit on reciver

Z_{1k} and Z_{2k} are decision variables at the first and second diversity paths and Z_k is the decision variable at the output of the diversity combiner.

The diagram block for this method is shown on figure 2. If there are some diversity branches for signal that gets into chosen circuit, so the output of this circuit Z_k is the chosen of the largest SNR. On fading channel, some diversity branches of Rayleigh channel at the receiver and assumed on every branch has the same average value of SNR, it can be formulated as follows [2]:

$$SNR = \Gamma = \frac{E_b}{N_0}\overline{\alpha^2} \quad (2)$$

where $\overline{\alpha^2}$ is mean from the random variabel. In this case, the random variable is the envelope from fading Rayleigh. If each brances has instant SNR γ_i, the power of spectral density (PDF) from γ_i is [2]:

$$p(\gamma_i) = \frac{1}{\Gamma} e^{\frac{-\gamma_i}{\Gamma}} \qquad \gamma_i \geq 0 \quad (3)$$

where Γ is average SNR from every branches. Probability that a branch has SNR less than the value of threshold γ is [2]:

$$P_r[\gamma_i \leq \gamma] = \int_0^{\infty} p(\gamma_i)d\gamma_i = \int_0^{\infty} \frac{1}{\Gamma} e^{\frac{-\gamma_i}{\Gamma}} d\gamma_i \quad (4)$$

Probability that every branches recieved simultanous signal less than SNR threshod γ is [2]:

$$P_r[\gamma_1, \ldots \gamma_M \leq \gamma] = (1 - e^{\frac{-\gamma}{\Gamma}})^M = P_M(\gamma) \quad (5)$$

$P_M(\gamma)$ is probability for all branches that fail to get the value of SNR = γ. If the only branch reached SNR > γ, the probability become [2]:

$$P_r[\gamma_i > \gamma] = 1 - P_M(\gamma) = 1 - (1 - e^{\frac{-\gamma}{\Gamma}})^M \quad (6)$$

From the equation above, for the probability of SNR more than threshold γ when selection diversity is used. To indentify average of signal to noise ratio from received signal when using diversity, it needs pdf from signal fading. For selection diversity, the average SNR is gained from the first $P_M(\gamma)$ [2].

$$P_M(\gamma) = \frac{d}{d\gamma} P_M(\gamma) = \frac{M}{\Gamma}(1 - e^{\frac{-\gamma}{\Gamma}})^{M-1} e^{\frac{-\gamma}{\Gamma}} \quad (7)$$

So, the average SNR ($\bar{\gamma}$) is formulated as follows [2]:

$$\bar{\gamma} = \int_0^{\infty} \gamma P_M(\gamma) d\gamma = \Gamma \int_0^{\infty} Mx(1 - e^{-x})^{M-1} e^{-x} dx \quad (8)$$

where x is the comparison between γ and Γ.

The above equations are used to evaluate the increase of the average SNR given by selection diversity. DSC offers the improvement on link margin without needing the added power on transmiter. This methods is easy to be implemented, but it did not result the optimum result since it did not use all possible branches.

2. Materials and methods

This research is done by doing a simulation method with the Matlab program to get the optimum performance and system reliability, as well as design accuracy before it is implemented. The transmission method used in this research is RS Code (31, 15) in the coder and the decoder [18-19]. The diversity technique used is DSC [2] on the Radio Frequency and the channel model used is additive white Gaussian noise (AWGN) and Rayleigh Fading [2].

2.1. System performance test

The system performance test in this study uses two parameters, namely BER and PSNR. BER is used to measure how many errors occurred at the receiver. PSNR is used to determine the quality of the reconstructed image at the receiver. BER and PSNR parameters can be explained as follows.

2.1.1. The Calculation of Bit Error Rate (BER)

The bit error ratio (BER) is measured by dividing the number of bit errors by the total number of transferred bits. The measurement of BER can be given by the following equation [2]:

$$BER = \frac{n}{N} \quad (9)$$

where: n = number of incorrect bits of information at the receiver
N = number of bits of information transmitted.

2.1.2. Peak-to-peak Signal to Noise Ratio (PSNR).

Peak Signal to Noise Ratio (PSNR) is a useful parameter to determine the quality of reconstruction image. PSNR calculation is done by comparing the pixel value of the original image with the pixel value of the reconstructed image. PSNR shows the quality of the reconstructed image. The higher the PSNR value obtained, the better the quality of the image. The PSNR can be formulated as follows [4, 18].

$$PSNR = 10\log_{10}\frac{255^2}{\frac{1}{N}\sum_i\sum_j\left(p(i,j)-\hat{p}(i,j)\right)^2} \quad (10)$$

where: $p(i,j)$ is the original image pixel value,
$\hat{p}(i,j)$ is Pixel value of reconstructed image,
N is the number of pixels in image, while the 255 value is the highest value for grey scale image.

2.2. Research Variable

The variables used in this study are as follows:

1. Independent variables are the variables that can be manipulated. The independent variables used in the study are:
   a. Bit per pixel (bpp) where this variable represents the average number of bits for each pixel of the compressed image.
   b. SNR where this variable is a comparison of signal power and noise power in the transmission channels.
2. Dependent variables are the variable that changes based on the changes in independent variables. The dependent variables used in the study are:
   a. BER where this variable is a comparison between the numbers of bits error with the total number of bits received. This comparison is done between the bit stream SPIHT compression results in the tranciever with the bit stream of the decoding result of Reed Solomon in the receiver.
   b. PSNR where this variable represents the degree of integrity of the reconstruction image

3. Results and discussion

To evaluate the system performance analysis of DSC on radio frequency in Rayleigh fading channel, several experiments were carried out to test the BER and PSNR value results for Lena image with rate 0.8 bpp. BER test results are given in table 1 and Figure 3 and Figure 4.

Table 1. The performance analysis of DSC on radio frequency in Rayleigh fading channel with rate 0.8 bpp.

SNR (dB)	Without DSC		With DSC	
	BER	PSNR (dB)	BER	PSNR (dB)
10	0.0216	14.6946	0.0204	14.7464
12	0.0175	19.3743	0.0170	19.7135
14	0.0153	21.2786	0.0148	24.5918
16	0.0141	24.6029	0.0139	26.2011
18	0.0132	27.1660	0.0129	30.2538
20	0.0126	32.0900	0.0123	33.7379

Table 1 represents the comparison of the DSC performance on radio frequency in Rayleigh fading channel with rate 0.8 bpp, with SNR 10dB, 12dB, 14dB, 16dB, 18 dB and 20dB with and without DSC. From the table, the system with DSC shows the improvement on the system performance. It can be seen on the error that decreased to the average of 0.0152 or 3.39 %, while for the quality of image reconstruction is increased to the average of 1.6737 dB or 6.73%.

Without Diversity SNR 14 dB	Diversity selection combining SNR 14 dB
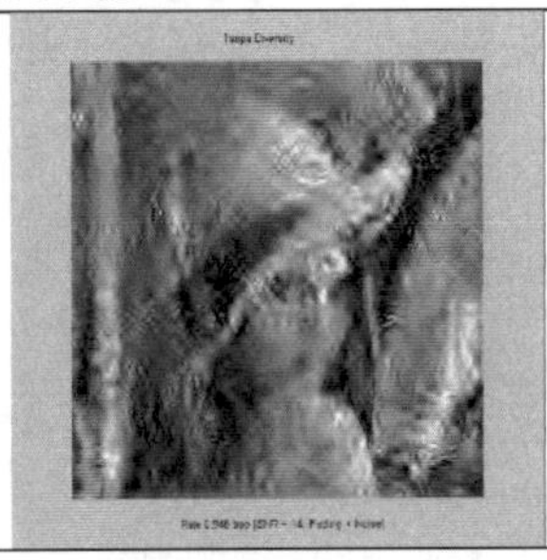	
BER = 0.0153 PSNR = 21.2786 dB	BER = 0,0148 PSNR = 24.5918 dB

Figure 3. The comparison result of image reconstruction with and without DSC.

Figure 3 shows the comparison of BER and PSNR with and without DSC. The image with SNR 14 dB is the representation of one of the image reconstructions. For the system with DSC, there is a significant improvement at 14dB SNR in image reconstruction quality compared to other SNRs. This significant improvement is seen in the decrease of error by 0.0005 or 3.4%, and an increase in the quality of image reconstruction by 3.3132 or 13.47%.

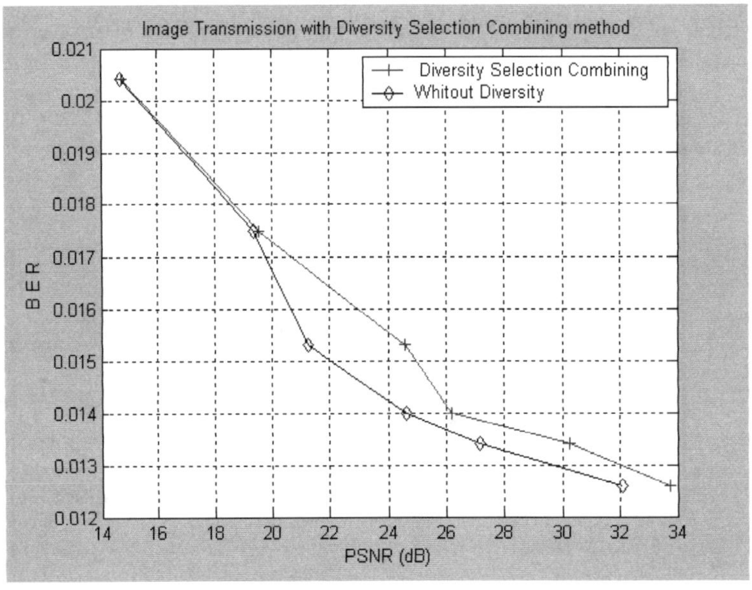

Graph 1. The comparison of graphs of the system performance with and without DSC.

Graph 1 represents the comparison of the DSC performance on radio frequency in Rayleigh fading channel with rate 0.8 bpp and with SNR 10dB, 12dB, 14dB, 16dB, 18 dB and 20dB with and without DSC. Comparing to system without DSC, the system with DSC mengalami perbaikan system yang dapat dilihat dari penurunan error, sedangkan kualitas image reconstruction terjadi peningkatan yang signifikan kecuali pada SNR 10 dB dan 12 dB. Comparing to the system without DSC, the system with DSC has an improvement in the system performance which could be seen from the decrease in error, while the quality of image reconstruction is improve significantly except for SNR 10 dB and 12 dB.

4. Conclusion

The result of the experiments which were carried out to test BER and PSNR on Lena image with rate 0.8 bpp by using DSC on radio frequency in Rayleigh fading channel showed that there was a decrease in BER at the average of 0.0152 or 3.39%, while for PSNR there was an increase at the average of 1.6737 dB or 6.73%. However, the SNR does not always make the PSNR increase significantly, this can be seen from SNR 10 dB and 12 dB. However, BER decreases in every SNR tested. The lowest decreasing of BER was at SNR 10 dB. It happened because the SNR is able to combat the disturbance of fading and noise. However, at SNR 10dB and 12 dB, the slightly improvement was caused by the random nature of the fading and noise which was very dominant to the information signal. While at SNR 20 dB, the improvement for both BER and PSNR were not significant because they had reached a maximum value. The results of this study show that the use of DSC in the Radio Frequency can reduce errors and improve the quality of image reconstruction. In order to make this research be more optimal, it is recommended to develop several systems such as modulation technique improvements, error control coding techniques, and standard source coding techniques.

Acknowledgements

The authors would like to thank the Directorate General of Higher Education, Ministry of National Education, LP2M Universitas Andalas Under contract number T/58/UN.16.17/PT.01.03/PRK-RD/2020.

References

[1]. C –H Tse, K -W Yip and T -S Ng 2000 *Performance tradeoffs between maximum ratio transmission and switched-transmit diversity* Vol 2 (in *Proc. 11th IEEE Int. Symp. on Personal Indoor and Mobile Radio Comm)* pp 1485–1489

[2]. Theodore S Rappaport 2011 Wireless Communications Principles and Practice *Equalization Diversity and Channel Coding* (Pearson Prentice Hall Second Edition) chapter 7 pp 355-412

[3]. R M Radaydeh and M M Matalgah 2005 *Performance analysis of multiple carrier M-ary FSK system with diversity combining over generalized fading channels* Vol 4 (in Communications, 2005. ICC 2005. 2005 IEEE International Conference on Communication) pp 2357–2361

[4]. A Said and W A Pearlman 996 *A New, Fast, and Efficient Image Codec Based on Set Partitioning In Hierarchical Trees* Vol 6 (IEEE Trans. Circuits Syst Video Technol) pp 243-250

[5]. Jerome M Shapiro 1993 *Embedded Image Coding Using Zerotrees of Wavelet Coefficients* Vol 41 (IEEE Trans on Signal Processing) pp 3445-3462

[6]. P G Sherwood and K Zeger 1998 *Error Protection for Progressive Image Transmission over Memoryless and Fading Channels* Vol 46 (IEEE Trans: Commun) pp 1555-1559

[7]. V Chande and N Farvardin 2000 *Progressive transmission of images over memoryless noisy channels* Vol 6 (IEEE Journal on Selected Areas in Communications) pp 850-860

[8]. G Sherwood and K Zeger 1997 *Progressive image coding on noisy channels* Vol 4 (IEEE Signal Process: Lett) pp 189–191

[9]. N V Boulgouris, N Thomos and M G Strintzis 2003 *Transmission of images over noisy channels using error-resilient wavelet coding and forward error correction* Vol 13 (IEEE Trans Circuits Syst: Video Technol) pp 1170–1181

[10]. N Thomos, N V Boulgouris and M G Strintzis 2005 *Wireless image transmission using turbo codes and optimal unequal error protection* Vol 14 (IEEE Trans. Image Process) pp 1890–1901

[11]. V Stankovic and D Vukobratovic 2012 *Unequal error protection random linear coding strategies for erasure channels* Vol 60 (IEEE Transactions on Communications) pp 1243-1252
[12]. W J V Gils 1983 *Two topics on linear unequal error protection codes: Bounds on their length and cyclic code classes* Vol IT-29 (IEEE Transactions on Information Theory) pp 866-876
[13]. M A Najib and V K Prabhu 2000 *Analysis of equal-gain diversity with partially coherent fading signals* (*IEEE Trans. Veh. Tech*) pp 783–791
[14]. A Annamalai, C Tellambura and V K Bhargava 2000 *Equal-gain diversity receiver performance in wireless channels* Vol 48 (*IEEE Trans. Comm*) pp 1732–1745
[15]. Q T Zhang 1997 *Probability of error for equal-gain combiners over Rayleigh channels: some closed-form solutions* Vol 45 (*IEEE Trans. Comm*) pp 270–273
[16]. Q T Zhang 1999 *A simple approach to probability of error for equal gain combiners over Rayleigh channels* Vol 48 (*IEEE Trans. Veh. Tech*) pp 1151–1154
[17]. Siavash M Alamouti 1998 *A simple Transmit Diversity Technique for Wireless Communications* Vol 16 (IEEE Journal on selected areas in Communications) pp 1451-1458
[18]. Liane C Ramac and Pramod K Varshney 2000 *A Wavelet Domain Diversity Method for Transmission of Images over Wireless Channels* Vol 18 (IEEE Journal on Selected Areas in Communication) pp 891–898
[19]. Sklar B 2001 Digital Communications: Fundamentals and Applications. *Channel Coding* 2nd (Upper Saddle River: Prentice-Hall) chapter 5 pp 314-376

IOP Conf. Series: Materials Science and Engineering 1041 (2021) 012023 doi:10.1088/1757-899X/1041/1/012023

Performance Analysis of Trellis Coded Modulation and Diversity Combining on Wireless Channel

Baharuddin[1], H Andre[1], M Muharam[1], A Luthfi[1], R Angraini[2]

[1] Department of Electrical Engineering, Universitas Andalas, Kampus Unand Limau Manis, Padang, 25163. Indonesia
[2] Politeknik Negeri Padang, Kampus Politeknik Negeri Padang Limau Manis, Padang, 25163. Indonesia

Corresponding author: baharuddin@eng.unand.ac.id

Abstract. The use of trellis coded modulation (TCM) and diversity maximal ratio combining (MRC) are applied in this research in order to overcome the impairment of information signals caused by fading and noise. TCM is chosen due to its ability to correct error without increasing the transmit power and channel bandwidth. MRC is used to obtain multiple data streams corresponding to the transmitted image at the receiver. The testing of the system is carried out by calculating peak signal to noise ratio (PSNR) and bit error rate (BER). The result of this simulation shows that the use of TCM and MRC in the image transmission system improves significantly compare to the system which only use TCM without MRC.

1. Introduction

Currently, people tend to use multimedia communication over wireless channels and becomes the important technologies in the telecommunications market.Wireless communication uses information such as data, voice, images, and video [1]. However, wireless communication systems are prone to fading and noise phenomena which cause burst errors in the transmission channel [2]. This fading phenomenon is random and causes problem to image transmission in wireless channels when compared to image transmission in cable environment channels [3].

One of the simple and the efficient techniques to overcome the impairment of information signals caused by fading is to use diversity maximal ratio combining (MRC) [4-9] and Trellis Code Modulation (TCM) [10-11]. This technique will take advantage of the random propagation of radio waves that is by finding a method to generate and extract independent signal paths [11-13]. The concept of diversity is very simple, that is, if one path of radio waves experiences interference caused by fading, it is possible that there are other trajectories that have better signals. With the number of signal paths to choose, the average signal to noise ratio(SNR) can be improved at the receiver [1-13].

TCM is used for data communication which aims to obtain noise immunity through uncoded transmission without changing data rates. The use of TCM is also to improve the system performance without increasing the transmit power and channel bandwidth. Modulation Quaternary Phase Shift Keying (QPSK) is used in TCM to increase data transmission rate [11].

One of the ways to determine the quality of a digital transmission system is to measure the peak signal to noise ratio (PSNR) and bit error rate (BER) [10-11]. PSNR is a useful parameter to determine the quality of reconstruction image. PSNR calculation is done by comparing the pixel value of the original image with the pixel value of the reconstructed image. PSNR shows the quality of the reconstructed image. The higher the PSNR value is obtained, the better the quality of the image [14-15]. BER is calculated by comparing the transmitted bit sequence and the received bit, in which the number of errors is calculated [14-15].

Several techniques have been used to improve the performance of systems such as the use of diversity technique equal gain combining on radio frequency [7-9, 15]. This technique is good apply because the increment of signal to noise ratio (SNR) will increase peak signal to noise ratio (PSNR) and alsowill degradebit error rate (BER) [15]. The use of diversity maximum ratio combining on wavelet domain improves the performance of image reconstruction[15]. Diversity techniques use on wavelet domain is to get better reconstruction image that used a two-state Gilbert-Elliott channel [14].

In this research, TCM and MRC are chosen due to its high ability in detecting and correcting errors on the image that caused by fading and noise.The experiment is done by comparing the TCM with and without using MRC on radio frequency. The evaluation of the improvement quality system on the image transmission is calculated by the use of BER and PSNR. In the simulation, image information with size 8 bpp is compressed by SPIHT with bit rate 0.8 bpp. The purpose of using SPIHT [16] compression is to minimize signal bandwith so that the use of transmission channel become more efficient.

The design system of SPIHT compressed image transmission use TCM with and without MRCon radio frequency can be seen in figure 1. The transmission image is a 8-bit 512 x 512 Grayscale Lena Image.Before compressing the image, the image is decomposed with a wavelet transform. This transformation uses Haar wavelet and to optimize the system performance, a transformation with the maximum decomposition level is used. This maximum level will divide the image into 28 subbands which consist of 1 low resolution subband and 27 detail subbands, of which 1 low resolution subband consists of 1 pixel. The output of wavelet transform on image is compressed with the SPIHT [16] algorithm, which resulting bits stream. These bits stream are the result of bit rate 0.8 bpp. The TCM encoder uses Trellis diagram. This system uses poly2trellis which has constraints length (l) = 7 and the code generator $g_0 = 1 + x^1 + x^2 + x^3 + x^6$ and $g1 = 1 + x^2 + x^3 + x^5 + x^6$ [11].

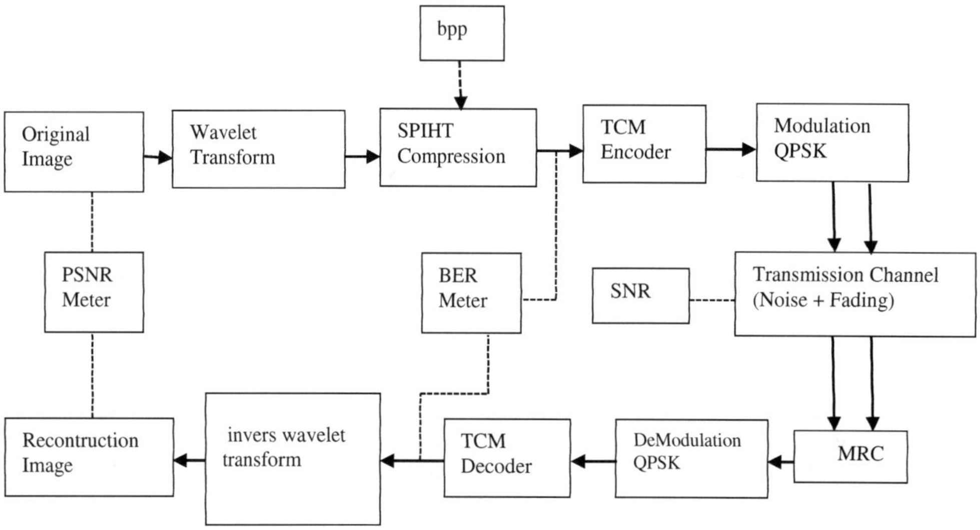

Figure 1. The design system of SPIHT [16] compressed image transmission with TCM and MRC techniquein Radio Frequency Domain.

The output of TCM Code is then modulated QPSK before it is transmitted to the wirelessschannal. This modulation removes the carrier component so that the signal that has been produced is in a complex form [10-11]. The equation for modulation signal is given below:

$$s_{bb}(t) = X_I(t) + jX_Q(t) \tag{1}$$

$$X_I(t) = A \cdot \cos\phi(t) \tag{2}$$

$$X_Q(t) = A \cdot \sin\phi(t) \tag{3}$$

$S_{bb}(t)$ is the equation of the modulation signal, $X_I(t)$ and $X_Q(t)$ are the inphase and quadrature components of the information signal. Fading used in this channel model is a type of slow flat fading with Rayleigh distributed. Generating of signal fading is done by using the summing sine method as shown in the following equations [10-11]:

$$h_I(nT_s) = \frac{1}{\sqrt{M}} \sum_{m=1}^{M} \cos\left\{2\pi f_D \cos\left[\frac{(2m-1)\pi + \theta}{4M}\right] \cdot nT_s + \alpha_m\right\} \tag{4}$$

$$h_I(nT_s) = \frac{1}{\sqrt{M}} \sum_{m=1}^{M} \sin\left\{2\pi f_D \cos\left[\frac{(2m-1)\pi + \theta}{4M}\right] \cdot nT_s + \beta_m\right\} \tag{5}$$

$$h(nTs) = h_I(nTs) + j \cdot h_Q(nTs) \tag{6}$$

where:

$h_I(nTs)$	=*inphase* fading components
$h_Q(nTs)$	=*quadrature* fading components
$h(nTs)$	= Rayleigh fading
M	= number of multipath components
f_D	= maximum Doppler *spread*
T_s	= periodic sampling
θ, α, dan β	= random variable uniform distribution [0,2π]

The receiver captures two signals that are independent on fading and noise from the transmission channel. Then, the signals from each antenna diversity are co-phased and weighted, both of which are carried out by estimation process. The co-phase process and weighting are carried out by the equation [5]:

$$s_{MRC1} = A_1 \cdot e^{-j\phi_1} \cdot s_{kanal1} \tag{7}$$

$$s_{MRC2} = A_2 \cdot e^{-j\phi_2} \cdot s_{kanal2} \tag{8}$$

where A and ϕ are the weights and estimation phases of signals diversity branch. These two results are then summed to produce the signal as the results of diversity MRC. MRC output is demodulated QPSK

$S_{bb}(t)$ isthen multiplied by the fading signal $h(nTs)$, so that amplitude and phase fluctuations occur, as formulated below [11]:

$$s_{fd}(nT_s) = s_{bb}(nT_s) \cdot h(nT_s) \tag{9}$$

After eliminate the bits training sequence from the output of the QPSK demodulator, the bit sequence is converted to decimal form for the input from the TCM decoder. The decoder process of TCM produces bit stream that has minimum error. To test the errors that occur in the system performance of TCM with and without MRC on radio frequency, BER is calculated by comparing the bit streams of TCM encoder input with the TCM decoder output as seen in equation (10).TCM decoder output is then processed in inverse wavelet transform by using partitioned grouping which based on wavelet characteristics. This process produces image reconstruction. PSNR is used to test the image quality of the system performance. PSNR measurement can be seen in equation (11).

2. Materials and methods

This research is carried out by doing a simulation with the Matlab program to get the optimum performance and system reliability, as well as design accuracy before it is implemented. The transmission method used in this research is TCM in the coder and the decoder. The diversity technique used is MRC [2] on the Radio Frequency and the channel models used are additive white Gaussian noise (AWGN) and Rayleigh Fading [11].

2.1. System performance test

The system performance test in this study uses two parameters, namely BER and PSNR. BER is used to measure how many errors occurred at the receiver. PSNR is used to determine the quality of the reconstructed image at the receiver. BER and PSNR parameters can be explained as follows:

1. The Calculation of BER

The BER is measured by dividing the number of bit errors by the total number of transferred bits. The measurement of BER can be given by the following equation [11]:

$$BER = \frac{n}{N} \tag{10}$$

where: n = number of incorrect bits of information at the receiver

N = number of bits of information transmitted.

2. The Calculation of PSNR

PSNR is a useful parameter to determine the quality of reconstruction image. PSNR calculation is done by comparing the pixel value of the original image with the pixel value of the reconstructed image. PSNR shows the quality of the reconstructed image. The higher the PSNR value obtained, the better the quality of the image. The PSNR can be formulated as follows [14]:

$$PSNR = 10\log_{10}\frac{255^2}{\frac{1}{N}\sum_i\sum_j\left(p(i,j)-\hat{p}(i,j)\right)^2} \tag{11}$$

where: $p(i,j)$ is the original image pixel value,

$\hat{p}(i,j)$ is Pixel value of reconstructed image,

N is the number of pixels in image, while the 255 value is the highest value for grey scale image.

2.2 Research Variable

The variables used in this study are as follows:

1. Independent variables are the variables that can be manipulated. Bpp and SNR are the independent variables used in this study, where:
   a. Bpp is the variable which represents the average number of bits for each pixel of the compressed image.
   b. SNR is the variable that compare the signal power and noise power in the transmission channels.
2. Dependent variables are the variable that changes based on the changes in independent variables. BER and PSNR are dependent variables used in the study, where:

a. BER is calculated by comparing the bit streams of TCM encoder input with the TCM decoder output.
b. PSNR is calculated by comparing the pixel value of the original image with the pixel value of the reconstructed image.

3. Results and discussion

To evaluate the system performance of TCM with and without MRC on radio frequency in Rayleigh fading channel, several experiments had been carried out. The experiments that had been done were to test the BER and PSNR for Lena image with bit rate 0.8 bpp. BER and PSNR test results are given in table 1 and figure 2 and graph 1.

Table 1. The simulation results of the system performance of TCM with and without MRC with bit rate 0,8bpp in Rayleigh fading channel on Radio Frequency.

SNR (dB)	BER		PSNR (dB)	
	With Diversity	Without Diversity	With Diversity	Without Diversity
10	0.0193	0.0247	15.5542	12.7464
12	0.0165	0.0233	19.5597	14.3743
14	0.0142	0.0208	24.5915	17.2786
16	0.0122	0.0191	27.9301	20.6029
18	0.0113	0.0172	30.8535	23.166
20	0.0093	0.0166	34.7372	26.090

With diversity SNR (14 dB)	Without diversity SNR (14 dB)

BER =0.0142 PSNR = 24.5915dB	BER =0.0211 PSNR = 17.2786 dB

Figure 2. The simulation results of the system performance of TCM with and without MRC with bit rate 0.8 bpp in Rayleigh fading channel on Radio Frequency.

Table 1 represents the comparison of the TCM with and without MRC on radio frequency in Rayleigh fading channel with rate 0.8 bpp, with SNR 10dB, 12dB, 14dB, 16dB, 18 dB and 20dB. From the table, the system with MRC shows the improvement on the system performance. It can be seen on the error that decreased to the average of 0.006483 or 46.97 %, while for the quality of image reconstruction is increased to the average of 6.4946 dB or 25.43%.

Figure 2 shows the comparison of BER and PSNR of the system performance of TCM with and without MRC. The image with SNR 14 dB is the representation of one of the image reconstructions from table 1. For the TCM with MRC, there is a significant improvement at 14dB SNR in image reconstruction quality compared to other SNRs. This significant improvement is seen in the decrease of error by 0.0069 or 48.59%, and an increase in the quality of image reconstruction by 7.3129 dB or 29.37% compare to the system without MRC.

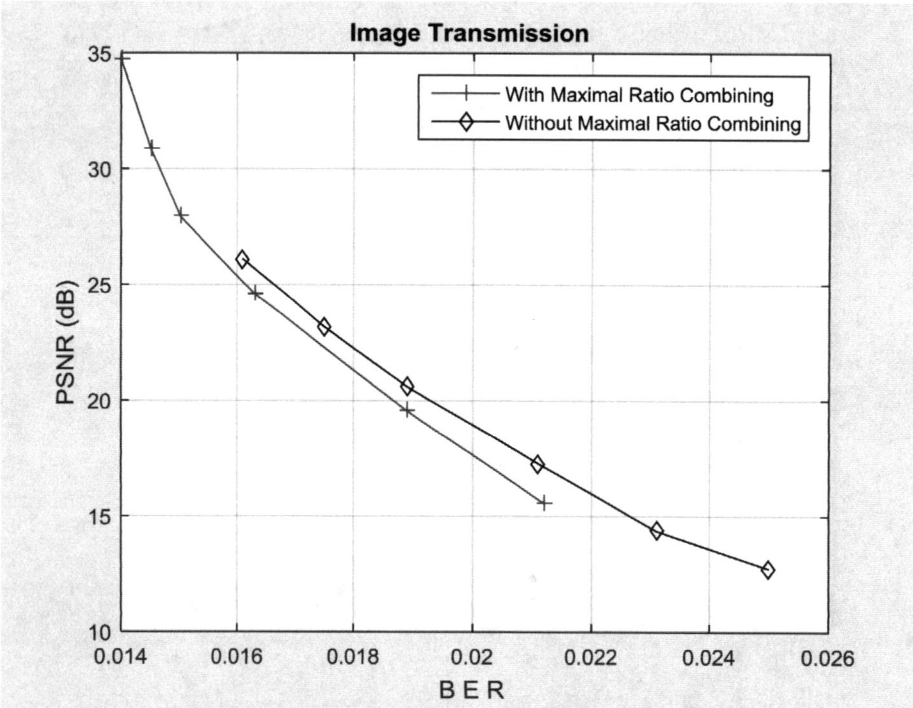

Graph 1. The simulation results of the system performance of TCM with and without MRC with bit rate 0,8 bpp in Rayleigh fading channel on Radio Frequency.

Graph 1 represents the comparison of the system performance of TCM with and without MRC with bit rate 0,8 bpp in Rayleigh fading channel on Radio Frequency, with SNR 10dB, 12dB, 14dB, 16dB, 18 dB and 20dB. Comparing to the system TCM with and without MRC. The system with MRC has an improvement in the system performance which could be seen from the decrease in error, while the quality of image reconstruction is improved significantly.

From the simulation above, it is clear that the system with and without MRC increase the PSNR and decrease the BER. This is because the PSNR or the quality of the reconstruction image is not only affected by the number of bits error, but it also depends on the location of the error bits in the decompressed SPIHT. In accordance with the characteristics of the SPIHT compression algorithm, which is progressive transmission, the important information is located at the beginning, so that the error on these initial bits will greatly affect the quality of the reconstruction image. While BER is decreased because both of these systems use TCM which can combat errors that occur during transmission in the Rayleigh fading channel.

4. Conclusion

The result of the experiments which had been carried out is to test BER and PSNR on Lena image with bit rate 0.8 bpp by using TCM with and without MRC on radio frequency in Rayleigh fading channel. It is showed that there was a decrease in BER at the average of 0.006483 or 46.97 % while for PSNR there was an increase at the average of of 6.4946 dB or 25.43%. While at SNR 20 dB, the improvement for both BER and PSNR were significant because they had reached a maximum value. The results of this study showed that the use of TCM with and without MRC on radio frequency in Rayleigh fading channel can reduce errors and improve the quality of image reconstruction.

Acknowledgements

Publication of this article is supported by Engineering Faculty of Andalas University grant for publication in fiscal year 2020. Number contract: 039/UN.16.09.D/PL/2020

References

[1]. N V Boulgouris, NThomos and M GStrintzis2003*Transmission of images over noisy channels using error-resilient wavelet coding and forward error correction*Vol 13 (IEEE Trans Circuits Syst: Video Technol) pp 1170–1181

[2]. G Sherwood and KZeger1997*Progressive image coding on noisy channels*Vol 4 (IEEE Signal Process: Lett) pp 189–191

[3]. P G Sherwood and K Zeger1998*Error Protection for Progressive Image Transmission over Memoryless and Fading Channels* Vol 46 (IEEE Trans: Commun) pp 1555-1559

[4]. Siavash M Alamouti 1998 *A simple Transmit Diversity Technique for Wireless Communications*Vol 16 (IEEE Journal on selected areas in Communications) pp 1451-1458

[5]. C –H Tse, K -W Yip and T -S Ng 2000 *Performance tradeoffs between maximum ratio transmission and switched-transmit diversity* Vol 2(in *Proc. 11th IEEE Int. Symp. on Personal Indoor and Mobile Radio Comm)* pp 1485–1489

[6]. R M Radaydeh and MMMatalgah 2005 *Performance analysis of multiple carrier M-ary FSK system with diversity combining over generalized fading channels*Vol 4(in Communications, 2005. ICC 2005. 2005 IEEE International Conference on Communication) pp 2357–2361

[7]. A Annamalai, C Tellambura and V K Bhargava 2000 *Equal-gain diversity receiver performance in wireless channels* Vol 48 (*IEEE Trans. Comm*) pp 1732–1745

[8]. Q T Zhang 1997 *Probability of error for equal-gain combiners over Rayleigh channels: some closed-form solutions* Vol 45 (*IEEE Trans. Comm*) pp 270–273

[9]. MANajib and V K Prabhu 2000 *Analysis of equal-gain diversity with partially coherent fading signals* (*IEEE Trans. Veh. Tech*) pp 783–791

[10]. Sklar B 2001Digital Communications: Fundamentals and Applications. *Channel Coding* (Upper Saddle River: Prentice-Halled2nd) chapter 5 pp 314-376

[11]. Theodore S Rappaport 2011 Wireless Communications Principles and Practice *Equalization Diversity and Channel Coding*(Pearson Prentice Halled2nd)chapter 7 pp 355-412

[12]. V Chande and N Farvardin 2000*Progressive transmission of images over memoryless noisy channels* Vol 6(IEEE Journal onSelected Areas in Communications) pp 850-860

[13]. N Thomos, N V Boulgouris and M G Strintzis 2005 *Wireless image transmission using turbo codes and optimal unequal error protection* Vol 14(IEEE Trans. Image Process) pp 1890–1901

[14]. Liane C Ramac and Pramod K Varshney 2000*A Wavelet Domain Diversity Method for Transmission of Images over Wireless Channels* Vol 18(IEEE Journal on Selected Areas in Communication) pp 891–898

[15]. Q T Zhang 1999 *A simple approach to probability of error for equal gain combiners over Rayleigh channels* Vol 48 (*IEEE Trans. Veh. Tech*) pp 1151–1154
[16]. A Said and W A Pearlman 996*A New, Fast, and Efficient Image Codec Based on Set Partitioning In Hierarchical Trees* Vol 6(IEEE Trans. Circuits Syst Video Technol) pp 243-250

Analysis of Particulate Matter 2.5, carbon monoxide, carbon dioxide, CO/CO_2 ratio and rate of fuel consumption from the use of biomass stove fueled by corncob and rice husk

F Goembira[1], R W Putra[1], T Ihsan[1]

[1]Department of Environmental Engineering, Faculty of Engineering, Universitas Andalas

Corresponding author: fgoembira@eng.unand.ac.id

Abstract. This study was conducted to analyze $PM_{2.5}$, CO and CO_2 concentrations, CO/CO_2 ratio, and the rate of fuel consumption due to the use of a biomass stove. Water Boiling Test Method was used to simulate cooking activities, which consisted of three phases, i.e., cold start (CS), hot start (HS), and simmering (SM). Biomass wastes used in this research were corncobs and rice husks. Based on the measurement results, for the corncob, $PM_{2.5}$ concentrations were 239.823±60.83, 262.962±79.17, and 120.75±45.20µg/Nm^3 at CM, HS, and SM phases, respectively. While for CO concentrations during the CM, HS, and SM phases were 36.984 ±3.67, 35.918±1.74, and 37.959±5.56 ppm, correspondingly. Moreover, 595,468±13.25, 611,492±7.73, and 565,205±18.50 ppm CO_2 concentrations were detected during the CM, HS, and SM phases, respectively. Furthermore, for the rice husk biomass waste, $PM_{2.5}$ concentrations of 158.579±14.07, 115.014±16.08, and 69.603±17.11 µg/Nm^3 were observed during the CS, HS, and SM phases. While for CO concentrations during the CS, HS, and SM phases, the measured values were, in turn, 38.152±2.73, 38.152±5.02, and 36.422±4.72 ppm. Lastly, the measured CO_2 concentrations were 593,647±8.28, 589,241±12.45, and 556,886±13.06 ppm for the CS, HS, and SM phases, respectively. $PM_{2.5}$ and CO concentrations did not meet the air quality standards for both biomass wastes, while CO_2 meets the air quality standard according to Indonesian Minister of Health Regulation No. 1077/2011. The CO/CO_2 ratio of the biomass stove is above 0.02, while the specific fuel consumption rate, for corncob were 0.186, 0.135, and 0.238g/g, while for the rice husk were 0.238, 0.305, and 0.335g/g, during the cold start, hot start, and simmering phases.

1. Introduction

Biomass is generated from anthropogenic activities, which results in the form of organic material, for example: plants, livestock manure, and agricultural waste. Energy contained in this biomass can be utilized as alternative energy to replace conventional energy, i.e., fossil-derived fuels, which is non-renewable [1]. Several studies have shown the potential utilization of biomass to replace the use of fossil fuels [2–5].

Indonesia needs an alternative energy fuel to replace fossil-derived fuels. Biomass is an alternative energy that potential availability and amount in Indonesia reaches 98 million tons/year. The amount of biomass is approximately 20 million tons/year solid as the waste of cooking oil production, 20 million tons/year from agriculture sectors, around 12 million tons/year in form of bagasse, about 12 million tons/year as the waste from industrial forests, 16 million tons/year from plantation sectors, and approximately 10 million tons/year from livestock sectors [6,7].

Rice husks and corncobs are potential biomass to be used as alternative fuel. The reason is because their abundant availability and they have not been utilized appropriately [8]. According to data from Indonesian Central Statistics Agency [9], the average corn production is estimated around 19,008,426 tons per year with an estimated amount of waste generated as many as 5,702,527 tons of corncobs per year. While for the rice husk, based on data from the statistics agency [9], paddy harvest that produces rice also generates rice husk, with the ratio of paddy to rice husk of 1: 0.24. This means that each ton of paddy harvested will generate 0.24 tons of rice husk.

An alternative measure to utilize the biomass wastes is by using a biomass stove fueled by the wastes. Hendri Sawir is an innovator who makes biomass stoves in West Sumatra Province. Sawir biomass stoves have been used extensively in Padang Pariaman Regency. The biomass stove use has a positive impact on users, such as they no longer dependence on fossil-derived fuels, due to the use of waste derived charcoal and briquettes as the fuel for the biomass stove. The benefits of using this biomass stove on the environment are improving urban sanitation and reducing organic waste [10].

The use of biomass, in fact, also produces emissions that are potentially harmful to health in the form of carbon monoxide (CO) and particulate matter (PM). According to Afifah [11], the smaller the size of the particulate matter, the higher the potential impact on health. PM_{10} with size less than 10 μm can reach the central respiratory organ, namely bronchioles, while for $PM_{2.5}$ that has size lesser than 2.5 μm can penetrate to the alveoli in the human body. CO gas produced from incomplete combustion is also harmful due to its affinity with hemoglobin that is higher than oxygen. If inhaled, it will disrupt nerve function and make disrupt oxygen-bound hemoglobin in human body [12].

Therefore, this research on the indoor use of biomass stoves was conducted to examine $PM_{2.5}$, CO, CO_2 concentrations, and CO/CO_2 ratios. This research also analyzed the rate of fuel consumption due to the use of biomass stoves fueled by corncob and rice husk wastes, using biomass stoves developed by Sawir [10].

2. Materials and methods

2.1. $PM_{2.5}$ Measurement

The measurement procedure was carried out in 2 steps, namely the first step of taking background concentration data for 1 hour as an initial data, then the second step was measuring the pollutant concentration while operating the biomass stove. Measurements were made three times in repetition to calculate their standard deviation. The instrument used for $PM_{2.5}$ analysis was Low Volume Air Sampler (LVAS) that is an ambient-air particulate matter sampler that can be used for $PM_{2.5}$ sampling gravimetrically by using filters. In sampling ambient-air $PM_{2.5}$, the LVAS setting of air flow rate was 3.5 L/min.

2.2. CO and CO_2 Measurements

Measurement of CO and CO_2 was done by using an Air Quality Hygrometer equipped with both gas detectors. The instrument uses the principle of the electro-thermal sensor to measure CO and the principle of non-dispersive infrared sensors (NDIR) to measure the CO_2. The equipment has a CO measurement range of 0 - 500 ppm and accuracy of ± 50 ppm, and 0 - 5,000 ppm measure range and accuracy of ± 50 ppm for CO_2 measurements.

2.3. Measurement of Fuel Consumption Rate

Analysis of the rate of fuel consumption was conducted by using the Water Boiling Test Method [13] with the Hood Method, which is a measurement made in a closed room with little ventilation. Measurement was made on Sawir's biomass stove fueled by corncobs and rice husks. The measurement consisted of 3 testing phases, namely the cold start, the hot start, and the simmering phases.

2.4. Analysis of $PM_{2.5}$, CO, and CO_2 Concentration Data

$PM_{2.5}$ concentration calculation for 1-hour sampling:

$$C = \frac{(Ws - Wo)}{Vstp} \tag{1}$$

C = concentration of suspended particles (µg/Nm)
Ws = weight of glass fiber filter after sampling (g)
Wo = weight of glass fiber filter before sampling (g)
10^6 = conversion from g to µg

Calculation of particulate concentration for 24 hours.

$$C24hours = Cn \; x \; [\frac{n}{24}]^p \qquad (2)$$

C_n = average concentration measured when sampling n hour (ppm)
n = length of sampling time (hours)
p = Canter conversion factor (0.17 - 0.20)

The measured CO and CO_2 concentrations were then compared with indoor air quality standards, according to the Minister of Health Regulation No.1077/MENKES/PER/2011 [14] concerning guidelines for indoor air sanitation. CO and CO_2 concentrations were further analyzed by calculating CO/CO_2 ratio as a reference for biomass combustion efficiency. High combustion efficiency is a situation when CO/CO_2 ratio smaller than 0.02 [15].

3. Results and discussions

3.1. Concentration before Testing (Background Concentration)

Background concentration measurement was carried out for 1 hour to get the $PM_{2.5}$, CO, and CO_2 concentrations of the testing room before the research was conducted. The measured $PM_{2.5}$ CO, and CO_2 background concentrations were 6.468µg/Nm^3, 0 ppm and 376.925 ppm, respectively.

3.2. Concentration during Testing

$PM_{2.5}$Concentration

From figure 1 it can be seen that corncob biomass fuel utilization shows higher $PM_{2.5}$ concentration as compared to rice husk biomass. While in both biomasses, the lowest particulate concentrations were measured during the simmering stage. Nevertheless, the use of both biomasses did not in compliance with the indoor air quality standard for particulate parameter as regulated by the Minister of Health Regulation[14].

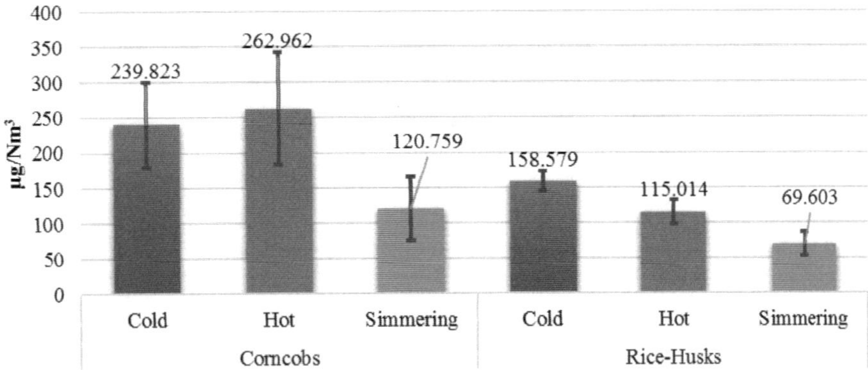

Figure 1. Comparison of $PM_{2.5}$ Concentrations in µg/Nm^3

IOP Conf. Series: Materials Science and Engineering 1041 (2021) 012024 doi:10.1088/1757-899X/1041/1/012024

CO Concentration

CO concentrations were observed with almost the same values for both biomass fuel utilized in this research, as shown in figure 2. During the cold and hot start stages, corncob biomass exhibits lower CO concentration than that of rice husk one. Contrary, during the simmering stage, the rice husk shows lower CO concentration. Both biomasses, at the moment, did not meet the Minister of Health Regulation No. 1077/2011 [14].

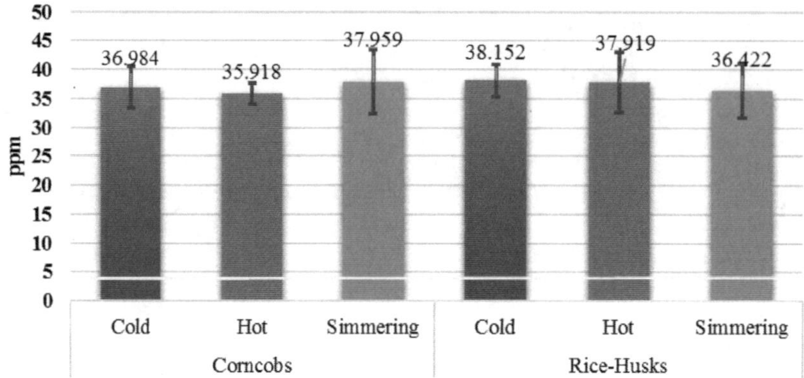

Figure 2. Comparison of CO Concentrations in ppm

CO/CO₂ ratio

CO/CO_2 ratio is determined in research as a reference in the efficiency of combustion on a stove. The ideal condition of the ratio is less than 0.02. The information of CO/CO_2 ratio ca be used to know possible excessive unburnt carbon emission from the fuel that we utilize. If it doesn't emit too much carbon, then CO/CO_2 ratio is considered as excellent [15–18]. figure 3 shows observed CO/CO_2 ratio from biomass fuel utilization in this research. Unfortunately, from this research, it is found out that the CO/CO_2 ratio of the two tested biomasses during all stages in WBT was above 0.02. The highest CO/CO_2 ratio shown by the use of corncobs during the simmering stage, while the lowest ratio shown by the use of the same biomass during the hot start stage.

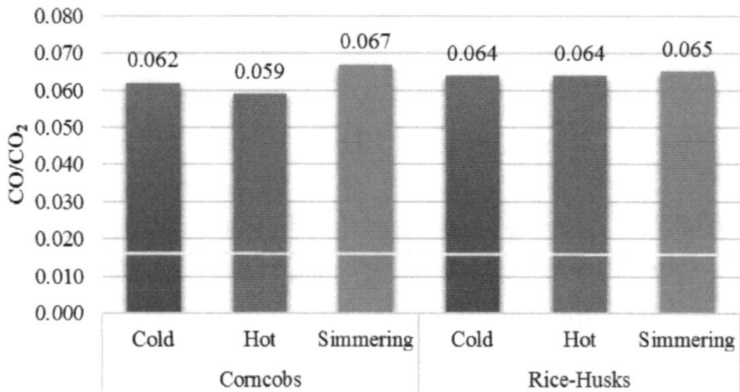

Figure 3. CO/CO_2 Ratio for Biomass

Fuel Consumption Rate

As shown in figure 4, corncob biomass showed lower fuel consumption rate of 0.135 g/g, which was lower than that of rice husk that had 0.335 g/g fuel consumption rate. The amount of rice husk used in cold start, hot start and simmering stages were 962, 934, and 1,023 g, respectively. While for corncobs, only 660, 558, and 845 g were used in each stage, correspondingly. This situation is due to smaller size of rice husk, which caused the fuel burnt quickly [19–21]. The consumption rate is further compared with that of other biomass reported by Goembira et al[22] for the use of firewood, coconut shell, and oil palm kernel shells.

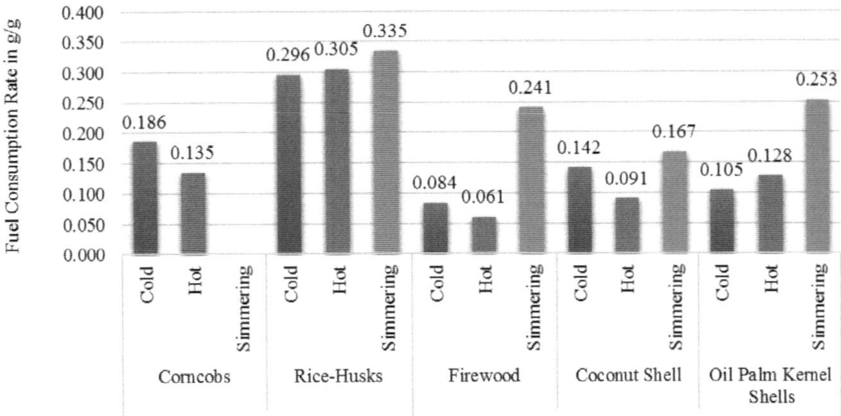

Figure 4. Comparison of Biomass Fuel Consumption Rate in g/g

4. Conclusions

$PM_{2.5}$ concentrations measured from the burning of rice husk biomass was lower as compared to that of corncob biomass for the cold phase, the hot stage, and the simmering stage. The values have also been exceeded the indoor air quality standards based on the Regulation of the Minister of Health of the Republic of Indonesia No. 1077/MENKES/PER/V/2011. However, the CO gas concentration measured from the burning of corncob biomass was lower than that of rice husk biomass for all stages in water boiling test method, and the concentrations have exceeded the indoor air quality standards based on the same referred regulation. As for CO_2 gas concentrations, rice husk biomass showed lower results as compared to corncob biomass in all stages of WBT method, and still in compliance with the regulation. Meanwhile, CO/CO_2 ratio calculated from the burning of corncob biomass was lower than that of rice husk biomass during the cold start and the hot start stages, while for the simmering stage, the rice husk CO/CO_2 ratio was lower than that of the corncob biomass ratio of CO/CO_2. Nevertheless, both of the tested biomass did not meet the ideal CO/CO_2 ratio. In addition, the corncob biomass showed lower fuel consumption rate than that of rice husk biomass.

Acknowledgment

Publication of this article is supported by Engineering Faculty of Universitas Andalas grant 2020.

References

[1] Djafar R, Djamalu Y, Haluti S and Botutihe S 2018 Analisis Performa Kompor Gasifikasi Biomassa Tipe Forced Draft Menggunakan Variasi Jumlah Bahan Bakar Tongkol Jagung *J. Technopreneur* **5** 90–6

[2] Goembira F and Saka S 2010 Pongamia pinnata as Potential Biodiesel Feedstock *Zero-Carbon Energy Kyoto* 111–6

[3] Phoo Z W M M, Razon L F, Knothe G, Ilham Z, Goembira F, Madrazo C F, Roces S A and Saka S 2014 Evaluation of Indian Milkweed (Calotropis gigantea) Seed Oil as Alternative Feedstock for Biodiesel *Ind. Crops Prod.* **54** 226–32

[4] Goembira F and Saka S 2015 Advanced Supercritical Methyl Acetate Method for Biodiesel Production from Pongamia pinnata Oil *Renew. Energy* **83** 1245–9

[5] Goembira F, Mahmud C and Fitriani A 2015 Study on the Potential of Land Utilization for Energy Plantation as Biodiesel Feedstock: Case study of Andalas university campus at Limau Manis *J. Agron.* **14** 146–51

[6] Budiman A 2014 Potensi Pengembangan Bio-Compressed Methane Gases (Bio-CMG) dari Biomassa sebagai Pengganti LPG dan BBG *Seminar Nasional Pemanfaatan Gas Alam menuju Kedaulatan Energi Indonesia* (Yogyakarta)

[7] Putrasari Y, Praptijanto A, Santoso W B and Lim O 2016 Resources, Policy, and Research Activities of Biofuel in Indonesia: A review *Energy Reports* **2** 237–45

[8] Supriyatno and Crishna M 2010 Studi Kasus Energi Alternatif Briket Sampah Lingkungan Kampus Polban Bandung *Pros. Semin. Nas. Tek. Kim. "Kejuangan" Pengemb. Teknol. Kim. untuk Pengolah. Sumber Daya Alam Indones.* 1–9

[9] *Bureau of Statistics of Republic of Indonesia* 2012

[10] Hendri S 2016 Kompor Biomassa (sebagai Upaya Pemanfaatan Limbah Menjadi Energi) [in Bahasa Indonesia] *Formulir Apl. Penghargaan Inov. K3 dan Lingkung. Hidup PT Semen Padang*

[11] Afifah T 2009 *Pencemaran Udara [in Bahasa Indonesia]* (Bandung: Institut Teknologi Bandung)

[12] Sugiarti 2015 Gas Pencemar Udara dan Pengaruhnya Bagi Kesehatan Manusia *J. Chem.* 50–8

[13] Ogle R, Olge D, MacCarty N and Still D 2007 The Water Boiling Test version 3.0.0. The Clean Cook Stove Standard **4.2.2** 86

[14] *Regulation of the Minister of Health of the Republic of Indonesia* No.1077/MENKES/PER/V/2011 *tentang Pedoman Penyehatan Udara dalam Ruang Rumah* [in Bahasa Indonesia]

[15] Kirumbi M . and Ondu C K . 2016 Comparative Analysis of Indoor Air Pollutants Emitted by the Advanced Stove Relative to the Conventional Bioethanol Gel Stoves *Int. J. Adv. Eng. Res. Technol.* **Vol 4**

[16] Jetter J J and Kariher P 2009 Solid-fuel Household Cook Stoves: Characterization of Performance and Emissions *Biomass and Bioenergy* **33** 294–305

[17] Jetter J, Zhao Y, Smith K R, Khan B, Yelverton T, Decarlo P and Hays M D 2012 Pollutant Emissions and Energy Efficiency under Controlled Conditions for Household Biomass Cookstoves and Implications for Metrics *Environ. Sci. Technol.* **46** 10827–34

[18] Chica E and Pérez J F 2019 Development and Performance Evaluation of an Improved Biomass Cookstove for Isolated Communities from Developing Countries *Case Stud. Therm. Eng.* **14** 100435

[19] Purnomo R H, Kuncoro E A and Wahyuni D 2014 Rancang Bangun dan Uji Teknik Kompor Berbahan Bakar Limbah Biomasa Pertanian *Buana Sains* **14** 71–8

[20] Nandi P, Patil A N, Raikar N, Kasturi S and Mahalingesh B 2016 Performance Evaluation of Improved Biomass Cook Stove *Int. J. Sci. Eng. Res.* **4** 137–43

[21] Suresh R, Singh V K, Malik J K, Datta A and Pal R C 2016 Evaluation of the performance of improved biomass cooking stoves with different solid biomass fuel types *Biomass and Bioenergy* **95** 27–34

[22] Goembira F, Oktafianto F and Hakim K 2017 Studi Konsentrasi PM2,5, CO, dan CO2 dari Penggunaan Briket Biomassa sebagai Bahan Bakar Alternatif [in Bahasa Indonesia] *Pros. Semin. Nas. Pengelolaan DAS secara Terpadu 2017. Univ. Riau* 201 – 210

Application of Life Cycle Assessment (LCA) On Green Tea Product (Case Study in the X Company)

G Mila[1], B Primasari[1], and R Aziz[1]

[1]Environmental Engineering Departement, Faculty of Engineering, Universitas Andalas, Kampus Limau Manis, Padang, 25163, Indonesia

Corresponding author: gustimila.98@gmail.com

Abstract. This study analyzes the environmental impact of 1 ton green tea 50 gr canned in x company using the Life Cycle Assessment (LCA) method. A cradle to gate approach was used to scope the study, from plantations to packaging. SimaPro version 9 software is used as a tool for assessing LCA. The evaluation step refers to ISO 14044 of 2006, which determines the objectives and scope, inventory analysis, impact, and interpretation. The method used to assess the environmental impacts is the CML-IA Baseline. The classification results of the dominant and significant impact characterization on the environment are global warming with 475.557,7 kg CO_2 eq. Drying process II, drying process I, PLN electricity, and transportation contribute to the environmental impact. The recommendation for technical improvement based on the sensitivity analysis results is using PLN electricity with renewable energy, namely micro hydro (PLTMH). The combination of wood biomass with other biomass is rice husks, then changing the truck fuel from diesel to fuel that is morm e environmentally friendly Pertamina Dex.

1. Introduction

Tea is a popular drink with health benefits after mineral water. Overall global tea production (black, green, and instant tea) reached 5,5 million tons in 2016, increasing about 4% compared to 2015 [1]. World tea production is estimated to continue to experience an increased rate of around 2,9 % in black tea and 8,2% in green tea with a production projection in 2023. The increase in tea production is also supported by the rise in tea consumption by 5 % in 2013 [2]. The Indonesian tea industry in 1999 was estimated to absorb about 300.000 workers and support around 1,2 million people. The tea industry nationally contributes to a Gross Domestic Product (GDP) of approximately 1,2 trillion (0,3% of total non-oil and gas GDP). It contributes to a net foreign exchange of around USD 110 million per year [3]. Indonesia is the fifth tea producing country after India, China, Sri Lanka, and Kenya. Most of Indonesia's tea production (65%) is for the export market. Most of Indonesia's tea export volume (94%) is still in the form of dry leaves. Apart from being a producer, Indonesia is also the fifth tea exporting country globally in terms of volume after Sri Lanka, Kenya, China, and India [4].

With the significance of tea in the global beverage market, there is an urgent need to put its environmental implications into perspective by adopting suitable analytical tools such as life cycle assessment (LCA). LCA is a quantitative tool for systematically assessing the environmental impact potential associated with a product or a service through its life cycle. This approach has been standardized in ISO 14044 series and has been widely adopted to determine environmental implications in various sectors [5]. LCA can account for impact footprints under different scenarios. It has also been employed to quantify the environmental consequences associated with consumption behaviors. LCA's development is believed to have emanated from the need for identifying alternative packaging designs in the beverage industry around the 1960s [6]. Given the market interest, several studies attempted to quantify black tea's environmental effects by adopting LCA. A recent study examined the environmental impacts of black tea produced in Malawian factories, focusing primarily on the processing stage [7].

Based on the results of research by Farshad et al. in 2018, the results of the calculation impact assessment of the green tea production process for the plantation stage, the most significant contributor was the use of machines amounting to 44,67% for all impact categories. For the production process steps at the green tea factory, the most significant contributor was that the use of electricity was 57% for all impact categories [8]. According to research by Yi When In 2019, the results concluded that the most significant impact of global warming potential (GWP) and eutrophication potential (EP) was due to the use of fertilizers and pesticides during the planting process is 52% and 83% in the oolong tea production process in Taiwan [9].

The impact produced at the green tea production stage comes from several components: fertilizers and pesticides on plantations, the use of electricity, wood, and generators for the production stage, and the organic waste of tea and plastic produced in the packaging process. The three main steps produce CO_2, CH_4, antimony, arsenic, NOx, and SOx, which are produced from the use of PLN electricity, wood, and generators, which can cause global warming, acidification, depletion of the ozone layer, and damage freshwater ecosystems[10].

This study aims to determine the main environmental impacts of the 50-gram canned green tea production process in The XCompany. The use of the life cycle assessment (LCA) application serves to make it easier to determine the hotspots of several techniques that produce the most significant impact, then provide suggestions for improvements to the process. Green tea used is dry tea as much as 1 ton is packaged using 50-gram cans. The improvement scenario can serve as input for The XCompany to improve the future so that the production process becomes environmentally friendly.

2. Materials and methods

2.1. *Green Tea Product of The X Company*

The raw material used in the tea industry is the leaves of the tea plant. The tea plant is a plant originating from Southeast Asia. The tea plant in Latin camellia sinensis L can grow from the coast to the mountains. The higher the tea planting area, the better the quality produced [11]. The tea shoots that process into green tea are usually processed immediately after being picked. After the leaves have been oxidized in a minimal amount, the oxidation process is stopped by heating or drying. Tea that has been dried can be sold in the form of tea leaves or rolled tightly in the shape of small balls (tea called gun powder) [12].

X Company is a company engaged in tea plantations and tea processing at the factory, with a production capacity of 20.000 tons of wet tea shoots or 4.000 tons of finished tea products per year. The consideration of choosing just one tea is that the X company's quality is superior in green tea products. Besides, selecting green tea and selecting canned packaging as the study object because the tin packaging uses the most materials and produces an impact. The data collection process was carried out through interviews and direct observation of green tea processors at x companies. Also, the secondary data collection process is carried out with journals related to the object of research. The X company's green tea production process is divided into three main stages, namely plantation, green tea production process, and packaging. The plantation includes green tea cultivation and maintenance, fertilizers and pesticides, and picking using picking tools.

The green tea production process consists of 7 stages. Weighing and unloading using fresh tea leaves transported from the plantation are brought to the factory by truck, and at this station, the fresh tea leaves are weighed together with the transporter (truck) at the weighbridge. The weighed tea leaves are unloaded from the transport truck and then loaded onto the monorail and taken to the Withering

IOP Conf. Series: Materials Science and Engineering 1041 (2021) 012025 doi:10.1088/1757-899X/1041/1/012025

Trough. In WT the tea shoots are sent to the blower. Withering and cooling is done by inserting the tea leaf shoots that have been turned towards the blower, which will then be raised to the conveyor to enter the rotary planner (RP) machine.

The machine works with a wood stove, electricity, and a generator to rotate the engine cylinder. The cooled withered shoots are put into the Open Top Roller (OTR) machine for the shoot rolling process. Filling the mill is carried out in idle conditions and filled with one-third of the mill's volume, then the engine is turned. The first drying was carried out with a continuous chain pressure (ECP) machine with heating from a heat exchanger (HE) machine.The result of the winding is fed into the ECP machine continuously and regularly until it is full. ECP results will be transported to the Ball tea machine, and then the ball tea machine will rotate to carry out the drying process for 14 hours until the moisture content is 5%. The sorting process is carried out in several stages. The packaging process is done manually, while the last step uses a hairdryer as a heat source to unify the outer plastic from the packaging. Each process requires energy input from fossil fuels, electricity, and natural fuels (wood). The gas emissions produced by each method can be seen in figure 1.

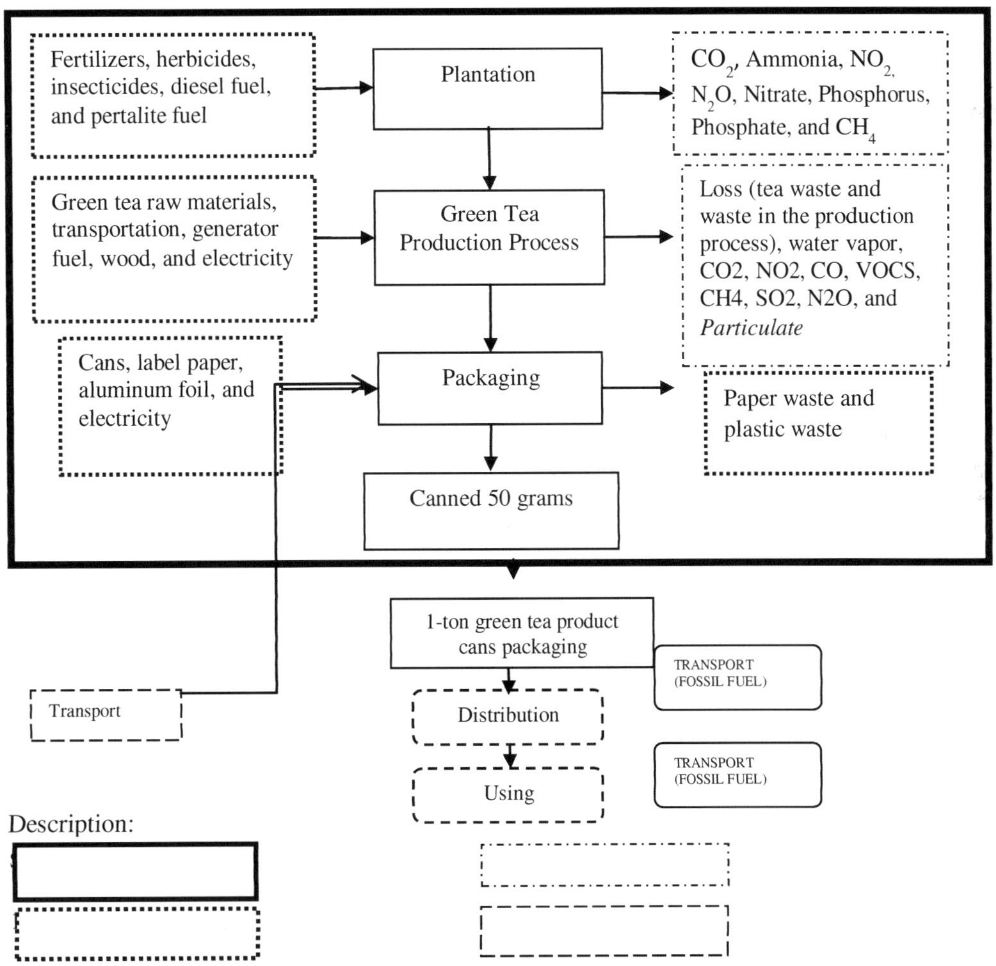

Figure 1. System Boundaries and flowchartof 1 ton of Green Tea Packaged Cans 50 gram

2.2 LCA Methodology.
Life Cycle Assessment (LCA) is a cradle to grave approach covering the entire product life cycle [13]. LCA can be used to study the impact, both from the product and the product's expected function [14]. LCA consists of four main steps, goal and scope definition, inventory analysis, impact assessment, and interpretation [15]. In this study uses the Simapro v 9.0.0 application.

1. Goal and Scope Definition

This research was conducted to analyze the life cycle of tea products in The XCompany from raw materials to produce canned green tea products (50 gram) in functional units of 1 ton of green tea products. The research used a cradle to gate approach, starting from taking raw materials to produce tea products. The study will use a method based on ISO 14044. The system boundaries of the green tea product's life cycle assessment can be seen in figure 2.

2. Inventory Analysis

The need for complete and relevant data in LCA analysis causes data collection to be carried out by several methods, including direct observation, data archived in The XCompany, previous research data, and the simaPro database. These comprehensive data sources mean that some of the data in this study do not have the same quality in representing the actual data. This study uses assumptions of the data used to complete the inventory.

The advantage is that the data obtained from The XCompany is the month's data collected in one year. Green tea production data obtained is data in 2019. This data includes the withering process of packaging. Production data has a consistent and complete input pattern. Assumptions of data derived from related journals are carried out for emissions from electricity use, fuel, emissions from production processes, emissions from plantations, and inputs for fuel in improvement scenarios [7], [8], [16]–[18].

3. Impact Assessment

The impact assessment performs using the Centrum Voor Milieukunde Leiden (CML) IA Baseline method, or CML IA baseline developed by the Centre of Environmental Sciences at Leiden University. This study's impact categories consider related studies by Farshad [8]. Impact category classified into global warming potential (GWP), freshwater aquatic eco-toxicity, photochemical oxidation, acidification, eutrophication, and human toxicity.

4. Interpretation

The interpretation is the conclusions of the inventory analysis and impact assessment stages by looking at the processes that contribute to the impact, comparing the product system's steps against the impact, the impact sensitivity assessment, and the improvement efforts recommended.

3. Result and discussion

3.1. Inventory Analysis
Based on **Table 1**, the inventory process's data input is carried out by dividing the main stages in the 50 gram canned green tea production process into three main steps, namely plantation, green tea production process, and packaging. Each main stage will be further divided into various inputs material, energy, and output in gases and solids (loss). The plantation stage consists of fertilizers, pesticides, fuel for pesticide machines, and fuel for picking appliances. The plantation process also considers the possibility of CO_2 absorption by tea plants to reduce emissions produced by other methods. The resulting input and output must correspond to a predetermined functional unit is 1 ton of dry green tea.

The production process stages consist of the input of tea leaves for each co-process and expressed in tonnes. Furthermore, energy ranging from electricity, wood, and generators, not all processes that use wood are only withering, drying I, and drying II using wood as additional energy. Loss output is obtained from direct observation into the field and asking for data recapitulated by x companies. Other results are obtained from several journals that are following the research. In the packaging stage, this process's input is based on direct observation of the field by weighing the packaging components.

3.2. *Impact Assessment*

The impact characterization results for 1 ton of canned green tea can be seen in table 2.Based on the contents of table 3, it can be seen that the packaging process, plantations do not produce significant environmental impacts.

Table 1. Life Cycle Inventory Inventory of Initial Green Tea Process.

No.	Process	Parameter	Total Green Tea 50 gram Can Packaging	Unit
1.	Plantation			
	Input:	Fertilizer (N, P, K, Mg)	426,38[a]	kg
		Herbicide	0,056352228[a]	kg
		Insecticide	0,008782711[a]	kg
		Diesel fuel for picking	18,4[a]	l
		Pertalite fuel for spraying	12[a]	l
		The absorption of CO_2 by the tea plant	-602,72[b]	kg
	Output:	Ammonia	0,015[b]	kg
		NO_2	0,004[b]	kg
		N_2O	0,004[b]	kg
		Nitrate	0,050[b]	kg
		CO_2	85,816[b]	kg
		Fosfor	0,002141[b]	kg
		Fosfat	0,0071[b]	kg
		CH_4	0,0035712[b]	kg
		N_2O	0,014[b]	kg
2.	Production process			
	Input:	Tea Raw Materials for Weighing	4,178[a]	ton
		Tea Raw Material for Withering	4,164[a]	ton
		Tea Raw Materials for Cooling	2,915[a]	ton
		Tea Raw Material for Rolling	2,915[a]	ton
		Tea Raw Materials for Drying II	2,914[a]	ton
		Tea Raw Materials for Drying II	1,749[a]	ton
		Tea Raw Materials for Sorting	1,084[a]	ton
		Wood	909,8[a]	m^3
		Electricity	219,344[a]	kWh
		Transport of tea raw materials	75,25[a]	tkm
		Diesel fuel for genset	90,08[a]	l
	Output:	Tea Raw Materials from Weighing	4,164[a]	ton
		Loss from weighing	14[a]	kg
		Tea Raw Materials from Withering	2,915[a]	ton
		Loss from withering	1,5[a]	kg
		Water vapor from withering	1247,5[a]	kg
		Tea Raw Material from Cooling	2,915[a]	ton
		Tea Raw Material from Rolling	2,915[a]	ton
		Loss from Rolling	1[a]	kg
		Tea Raw Materials from Drying I	2,914[a]	ton
		Loss from Drying I	4[a]	kg
		Water Vapor from Drying I	1161[a]	kg
		Tea Raw Materials from Drying II	1,749[a]	ton

Table 1. Life Cycle Inventory Inventory of Initial Green Tea Process.

No.	Process	Parameter	Total Green Tea 50 gram Can Packaging	Unit
		Loss from Drying II	3[a]	kg
		Water Vapor from Drying II	662[a]	kg
		Dried Tea from Sorting	1[a]	ton
		Waste tea leaves that do not pass the sorting	84[a]	kg
		CO_2	1.875,203565[b]	kg
		CO	0,90027585[b]	kg
		NO_2	0,061656092[b]	kg
		VOCs	0,70074349[b]	kg
		CH_4	0,463337562[b]	kg
		N_2O	0,54727[b]	kg
		SO_2	0,00000393[b]	kg
		Particulate	54,588[b]	kg
3.	Packaging			
	Input:	Unsorted tea	1[a]	ton
		20,000 Can Packaging	676[a]	kg
		Aluminum Foil	2[a]	kg
		Plastic	2[a]	kg
		Label	20[a]	kg
		Electricity	4,448[a]	kWh
		Green Tea 50 gram Canned	20.000[a]	Tin
	Output:	Paper Waste	0,2[a]	g
		Plastic Waste	0,2[a]	g

Source: field observation, related study, and interview
Note: a = from field observation;
b = from the related study about emission gasses.

Table 2 Characterization of Environmental Impact of 1 ton of Green Tea Products

Impact category	Unit	Total	Plantation	Green Tea Production Process	Packaging
Global warming (GWP100a)	kg CO_2 eq	475.557,7	-256,138	**464.676,6**	11.1137,23
Human toxicity	kg 1,4-DB eq	794,3687	0,536873	**779,0943**	14,73746
Freshwater aquatic ecotox.	kg 1,4-DB eq	422,4157	0,000877	**412,4283**	9,98651
Photochemical oxidation	kg C_2H_4 eq	31,85757	0,010429	**31,1574**	0,689745
Acidification	kg SO_2 eq	75,30837	0,168723	**73,89024**	1,249408
Eutrophication	kg PO_4^{-3} eq	299,6215	0,058696	**292,6177**	6,945077

At the plantation stage, the tea plant's possibility of CO_2 absorption is also calculated to reduce the total CO_2 levels caused by other processes. The use of energy in the production process is carried. The use of a generator is done if the energy in the production process does not meet the electricity usage. The environmental impact that contributes most of the three energy uses is from electricity use because of its extensive use and large emissions.

The process that contributes to environmental impacts is the production process. Details of the production process can be seen in table 3, where the process that has the most impact is drying II. This impact happens because considerable electric power and wood as an added energy are also extensive. The high impact caused by the drying process occurs due to the extended production process at this stage, which is 10-14 hours. The most significant contribution to this process is the use of electricity and the use of wood.The environmental impact characterization result shows that the production process is the major contributor to impacts. The impact includes human toxicity, GWP, eutrophication, freshwater aquatic eco-toxicity, photochemical oxidation, and acidification. The packaging process uses electricity to attaching plastic to the outside of cans in the form of a hairdryer, but it does not have a significant environmental impact.

3.3. *Interpretation*

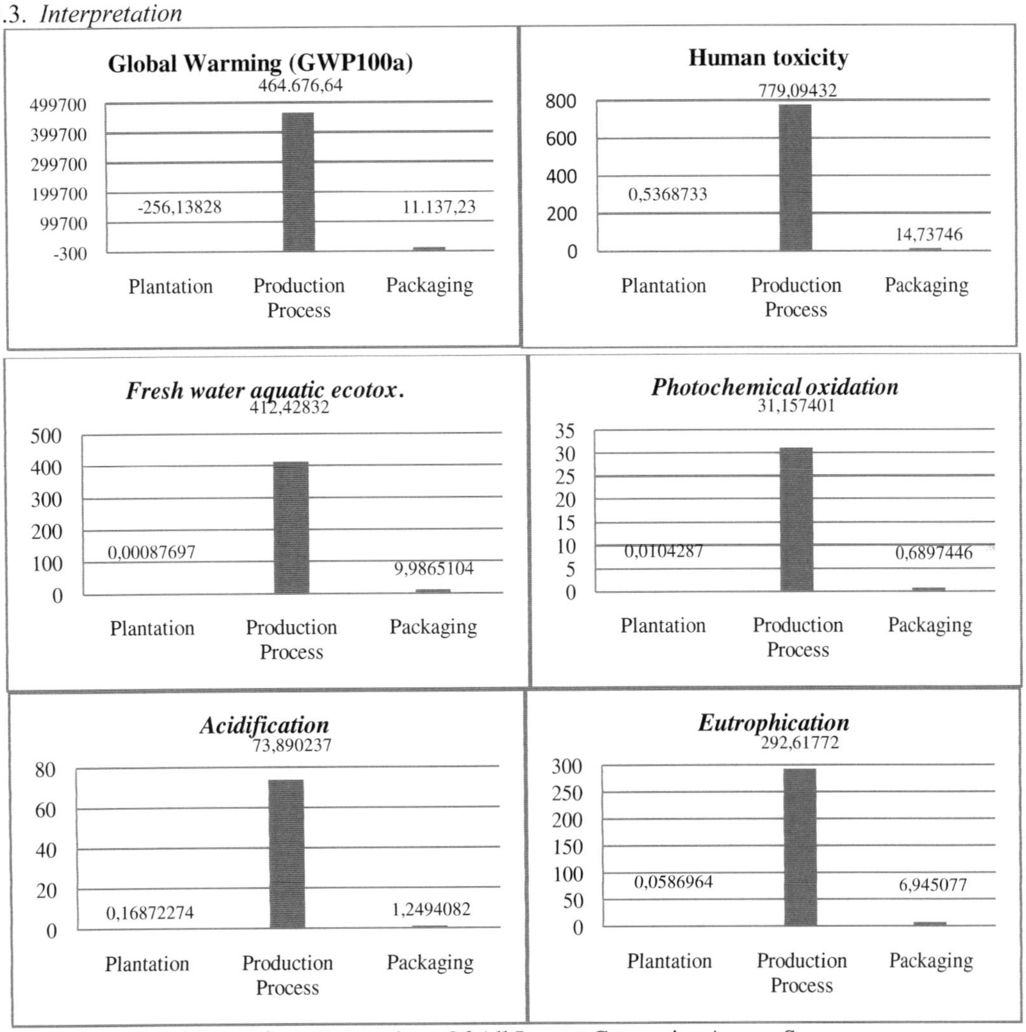

Figure 2. A Comparison Of All Impact Categories Across Stages.

Based on figure 2, it was comparing the characteristics of the global warming environmental impact for each stage of the product system. This comparison shows that the production process produces the highest global warming impact on 464.676,64 kg CO_2 eq. This impact happens because energy comes from generators, electricity, and wood so that the CO_2 is produced in this process, more significant than any other process. The smallest impact is the plantation phase. At this stage, the absorption of CO_2 by the tea plant is calculated to produce an impact of -256,13828 kg CO_2 eq. The negative value indicates that at the plantation stage, the CO_2 had is less than the ability to absorb by the tea plant so that further reduction can be made for another process. Comparing the environmental impact of human toxicity shows that the production process produces the highest impact is 779,09432 kg 1,4-DB eq. In this process, generators, electricity, and wood are relatively high toxic substances such as antimony, arsenic, carbon monoxide, and others. The smallest impact on plantations is 0,14931966 kg 1,4-DB eq because this process does not produce toxic emissions. The environmental impact comparison of freshwater aquatic Ecotox shows that the production process has the highest impact is 412,42832 kg 1,4-DB eq. The smallest impact on plantations is 0,006426377 kg 1,4-DB eq.Comparing the environmental impact of photochemical oxidation shows that the production process produces the highest impact is 31,157401 kg C_2H_4 eq. This impact occurs because the production process produces CH_4 emissions from electricity and wood. In this impact category, the smallest emission is the plantation process, which is 0,002950961 kg C_2H_4 eq because, in this process, the fuel is used a little so that the emissions are also small.

The comparison of acidification environmental impact shows that the production process produces the highest effect on 73,890237 kg SO_2 eq. This condition occurs because the production process has emissions to the air from diesel fuel, wood, and electricity. The smallest emission is from plantations, which are 0,048160035 kg SO_2 eq. Eutrophication environmental impact comparison shows that the production process produces the highest impact as 292,61772 kg PO_4 eq. This impact happens because the production process produces ammonia, carbon monoxide emissions from electricity and wood. In this impact category, the smallest emission is the plantation, which is 1,9690279 kg PO_4 eq.

Based on the comparison results, the stage that has the most contribution for each impact category is the production process. In detail, based on table 3, it can be seen that in the production stage, the most significant environmental impact is caused by the use of PLN electricity. The production process in the drying II and drying process I except for the most significant impact of acidification is diesel fuel use on 6 wheels pick up trucks. Wood using itself has a considerable effect on the resulting emissions. The use of generators for additional energy also contributes to emissions for the environment. And the use of diesel in trucks used to carry raw materials does not have much impact except for the impact of acidification because the truck's capacity is large enough so that transportation can be carried out to a minimum. Drying process II, drying process I, PLN electricity, and 6 wheels pick up trucks (for the impact of acidification) are the impact contribution components. The effect generated in each process occurs due to the high energy use according to table 3used in this analysis, taken from the production of 1 ton of green tea product because it produces a more significant impact on the environment.

Table 3. Contribution of Environmental Impact on Product System 1 ton Green Tea Product System

Impact category	Unit	Total	Weighing	Whitening	Cooling	Rolling	Drying I	Drying II
Global warming (GWP100a)	kg CO2 eq	475.557,7	0	22.553,5	156.113	10.571,24	**23.802,71**	**197.697,7**
Human toxicity	kg 1,4-DB eq	794,3687	0	30,10481	20,65782	13,98851	**31,8012**	**329,6951**
Freshwater aquatic ecotox.	kg 1,4-DB eq	422,4157	0	20,21381	13,99831	9,479003	**21,33238**	**174,8087**
Photochemical oxidation	kg C2H4 eq	31,85757	0	1,396147	0,96683	0,654692	**1,473409**	**12,08081**
Acidification	kg SO2 eq	75,30837	0	2,529133	1,751323	1,185914	**2,669109**	**21,92038**
Eutrophication	kg PO4--- eq	299,6215	0	14,05766	9,735065	6,592133	**14,83557**	**121,583**

Table 3. Continued

Impact category	Unit	Sorting	Genset	Wood	6 Wheel pick-up truck	Plantation	Packaging	Electricity PLN
Global warming (GWP100a)	kg CO2 eq	8.017,813	171,7316	1.716,814	67,6891	- 256,138	11.137,23	**184.466,1**
Human toxicity	kg 1,4-DB eq	10,60966	0,000958	44,83738	53,30208	0,536873	14,73746	**244,0968**
Fresh water aquatic ecotox.	kg 1,4-DB eq	7,189398	0	0	0	0,000877	9,98651	**165,4067**
Photochemical oxidation	kg C2H4 eq	0,496555	8,42E-05	0,004494	2,660129	0,010429	0,689745	**11,42425**
Acidification	kg SO2 eq	0,899463	0,000404	0,031343	**22,2092**	0,168723	1,249408	**20,69397**
Eutrophication	kg PO4--- eq	4,999837	0,000479	0,008149	5,774392	0,058696	6,945077	**115,0314**

Recommendations for improvement can be made by reducing the electricity consumption of PLN in drying process II because the production time in drying II is 10-14 hours of electricity use. It can be minimized by increasing the production capacity of one ball tea to be larger so that it per day. Another way to reduce electricity consumption is to evaluate the use of ball tea technology combined with a new type of machine that is more efficient and environmentally friendly. The recommendation for technical improvements is a combination of PLN electricity use with renewable energy such as micro-hydro (PLTMH). Furthermore, wood biomass as additional fuel can be replaced by combining it with other biomass. One of the common biomass that can be used as fuel is rice husks. Another recommendation that can be applied to reduce environmental impact is the replacement of truck fuel from diesel to fuel that is more environmentally friendly; this is Pertamina Dex.

4. Conclusion

The method used to assess environmental impacts is the CML-IA Baseline**.** Based on the comparison results, the stage that has the most contribution for each impact category is the production process. In detail, in the production stage, the production process in drying II and drying process I, the most significant environmental impact is caused by PLN electricity. The most significant impact of acidification is the use of diesel fuel in the 6 wheels pick up truck.The technical recommendation based on the sensitivity analysis results combines PLN electricity with renewable energy, namely micro-hydro (PLTMH). Furthermore, wood biomass as additional fuel can be replaced by combining it with other biomass. One of the common biomass that can be used as fuel is rice husk. Another recommendation that can be applied to reduce environmental impact is replacing truck fuel from diesel to a more environmentally friendly fuel, namely Pertamina Dex.

References

[1] International Tea Committee 2018 *Annual Bulletin Statistic*
[2] K Chang 2015 Contribution of Tea Production and Exports to Food Security, Rural Development and Smallholder Welfare in Selected Producing Countries
[3] Asosiasi Teh Indonesia (ATI) 2000 Reform of the Tea Marketing System for the Sustainability of The Indonesian Tea Industry
[4] International Tea Committee 2003 World Tea Production & Consumptiom Global Production

Growth 1995-2015

[5] International Organization for Standardization 2006 ISO 14040 *Environmental Management — Life Cycle Assessment — Principles and Framework* **3** 1 Geneva Switzerland

[6] Khasreen M M, Banfill P F G and Menzies G F 2009 Life-Cycle Assessment and the Environmental Impact of Buildings: A Review *J. sustanabaility* i pp 674–701 doi: 10.3390/su1030674.

[7] Taulo J L and Sebitosi A B 2016 Material and Energy Flow Analysis of the Malawian Tea Industry

[8] Soheili-Fard F, Kouchaki-Penchah H, Ghasemi Nejad Raini M and Chen G 2018 Cradle to Grave Environmental-Economic Analysis of Tea Life Cycle in Iran *J. Clean. Prod.* **196** pp 953–960 doi: 10.1016/j.jclepro.2018.06.083

[9] Yi-wen C 2019 Environmental Implications of Taiwanese Oolong Tea and the Opportunities of Impact Reduction *J. sustanabaility* **11** 21 p 6042

[10] The X company From the North Sumatra Student Job Training Report

[11] Murtina L T, Zuhrawati, Susanti F, Rusli, Asmilia N and Muttaqien A 2009 Pengaruh Pemberian Ekstrak Teh Hijau (Camelia sinensis) Terhadap Penurunan Kadar Hemoglobin dan Nilai Hematokrit pada Tikus Wistar (*Rattus norvegicus*) *J. Med. Vet.* pp 141–143

[12] Pringgajaya K A and Ciptomulyono U 2012 Implementasi Life Cycle Assessment (LCA) dan Pendekatan Analytical Network Process (ANP) *J. Tek. ITS* **1** 1 pp 515–520

[13] Tapia M, Siebel M A, Van Der Helm A W C, Baars E T and Gijzen H J 2008 Environmental, Financial and Quality Assessment of Drinking Water Processes at Waternet *J. Clean. Prod.* **16** pp 401–409 doi: 10.1016/j.jclepro.2006.07.053

[14] ISO 14044 2006 *Environmental Management — Life Cycle Assessment — Requirements and Guidelines* **2006** Switzerland: ISO Central Secretariat

[15] Kementrian Lingkungan Hidup 2012 Pedoman Penyelenggaraan Inventarisasi Gas Rumah Kaca Nasional

[16] Witjonarko R D E and Haryono Edi 2017 Kajian Eksperimental Emisi Gas Buang Two Stroke Marine Diesel Engine Berbahan Bakar Campuran Minyak Solar (HSD) dan Biodesel Minyak Jelantah Pada Beban Simulator Full Load *J. Inovtek Polbeng* **7** 2 pp 84–97

[17] Kastianto R E 2017 Pengaruh Penggunaan dan Perhitungan Efisiensi Bahan Bakar Pertamax 92 dan Pertalite 90 Terhadap Kinerja Motor Bakar Honda Beat Injeksi

[18] DEDE 2012 Department of Alternative Energy Development and Efficiency (DEDE) Ministry of Energy

Cleaner production analysis at hospital wastewater treatment plant

K Amala[1], S Budhi[1] and H Kusnoputranto[1]

[1]School of Environment Science, University of Indonesia

Corresponding author: kusumaputriamala@gmail.com

Abstract. The construction of WWTP (Wastewater Treatment Process) hospital should prioritize the environmental aspect and apply the principle of sustainable development so that the environmental impact can be minimized and optimize its management. The wastewater management strategy used in Hospital X uses an end of pipe strategy that focuses only on sewage treatment and disposal, so as not to completely solve existing environmental problems. Cleaner production analysis is applied to improve wastewater management and WWTP performance thus improving WWTP's operational cost-efficiency. The implementation of cleaner production can be one of the alternatives to reduce management costs through the identification of opportunities also inefficiencies found in WWTP management in hospitals. The analysis was carried out using descriptive analysis through the depiction of the water balance in WWTP treatment, the analysis then continued using fishbone diagrams to see the efficiency opportunities that can be applied. The results showed wastewater produced from the hospital met the standard requirements of wastewater quality so that it can be reused as one of the raw water sources for the hospital's operational activities.

1. Introduction

Hospital wastewater contains chemical compounds as well as pathogenic microorganisms, which can cause disease for it to be required Wastewater Treatment Plant (WWTP) at each hospital that has been contained in Ministry of Environment and Forestry of The Republic of Indonesia Regulation No. 5 2014th about wastewater quality standards. Many problems found in WWTP management are the lack of attention from the hospital management weighing that WWTP management will increase the operational costs of the hospital. Along with the development of environmental awareness and regulation, demands to cut costs associated with public spending on health care have forced hospitals and other entities to improve environmental performance[1]. The implementation of clean production can be one alternative to reduce management costs through the identification of opportunities also inefficiencies that exist in WWTP management in hospitals.

X Hospital located in West Jakarta which is the site of the research, conducted efforts to use liquid wastewater in 2008, but utilization efforts are no longer carried out because of the operational costs of the equipment and the hospital has not done the calculation of WWTP operational costs. Research conducted by [2]at Pucang Sewu Surabaya Public health centreconcluded that wastewater produced from public health centre meets the standard requirements of wastewater quality so that it can be reused as one of the raw water sources.Other research that has been done before [3]implementation of clean production through reuse of water resulted in a 50% savings in water usage costs. The implementation of cleaner production can reduce WWTP management costs through the identification of inefficiencies also opportunities contained in WWTP to be further analyzed economically to see the number of expenses and decreases in operating costs that occur when the implementation of cleaner production is carried out. The final result of the analysis is the cost efficiency options/opportunities for hospital WWTP management costs.

Hospital liquid waste refers to infectious, harmless and hazardous liquid waste with free liquids arising from dental, medical, pharmaceutical or similar practices, teaching and or research that require additional packaging to ensure the safety of transportation, care and storage[4]. Hospital liquid wastewater can contain organic and inorganic materials that are generally measured by BOD, COD, TSS, and other parameters [5]. The realization of sustainable waste management requires a holistic and balanced approach in evaluating management policy strategies [6]. Wastewater management carried out in health care facilities is reviewed from three aspects including the waste type and source, waste characteristics and wastewater treatment system [7].

Cleaner production is defined as an ongoing application of an integral and preventive environmental management strategy to processes, products and services to increase ecoefficiencies and reduce human and environmental risks according to[8]. Clean production implementation technology is a combination of waste reduction techniques on sources and recycling techniques. Environmental management strategies through clean production approaches are carried out to prevent or avoid the formation of waste [9]. Cleaner production therefore focuses on before-the-event techniques that can be categorized including source reduction (Good housekeeping), Process changes (that can be done with better process control, equipment modification, technology change also input material change), Recycling (including on-site recycling, and useful byproducts through off-site recycling) also product modification [10].

Reducing waste at its source is an attempt to reduce the volume, concentration, toxicity, and level of waste hazards that will spread in the environment, preventively directly at the source of polluters. After the effort of minimization of waste is carried out to the maximum, then the waste that is formed is then processed about concerning to the standard quality of waste that applies [11]. Efforts to implement waste minimization several strategies that willcontribute to waste minimization including

Avoid waste creation (changing the production process, so that less production waste is created, by using cleantechnologies, methods to improving selectivity, upgrading by-products), Use less material to produce a product (through dematerialization, miniaturization, light-weighting), Create durable/re-usable/repairable products (using measures that postpone the product becoming waste) and use less harmful substances (to an expected hazardous waste flow is replaced with a less dangerous one[12].

2. Materials and methods

Sampling is done by laboratory analysis of wastewater output from WWTP treatment system taken in January 2020. The analysis that was done included components of physics, chemistry and biology parameter which are BOD(Biological Oxygen Demand), COD (Chemical Oxygen Demand), TSS (Total Suspended Solid), and other parameters.Analysis of WWTP processing process conducted through Observing and analyzing the WWTP processing process found in X hospital. WWTP processing process is described into a flow diagram processed using descriptive analysis. The results of the WWTP treatment process are then made a balance of water to see the direction of the inflow and outflow of the water contained in the hospital. The analysis obtained is then processed and analyzed using the fishbone diagram to see the inefficiency process that occurs in the processing process and to know the efforts that can be made so that the WWTP processing process becomes more efficient.

3. Results and discussion

3.1. Wastewater Treatment System Water Balance

The selection of an effective and efficient wastewater treatment technology aims to improve existing environmental sanitation quality systems. Technology that takes precedence can produce a product that can be recycled in this case a waste product that can be reused for various purposes [13].

Wastewater treatment plant (WWTP) in RS X uses a combination of aerobic and anaerobic treatment systems. WWTP is divided into two processing plants namely WWTP I and WWTP II. WWTP I have a capacity of 750 m^3/day while WWTP II has a capacity of 100 m^3/day. Each treatment system has a different source of wastewater. The water balance sheet can be seen in figure 1. The water balance shows that the water managed by WWTP comes from several different sources where the incoming water from each unit is not all directly processed into the WWTP machine. In WWTP I Liquid waste comes from all hospital activities derived from toilets, bathrooms, sinks, urinals, laboratory waste (blood/serum/plasma, urine and reaction/reagent remnants) both from service rooms and offices flowed to the container before being put into the processing machine. WWTP I used aeration tubs and sedimentation ponds as a processing process. at WWTP II Special units such as laundry and nutrition through the primary treatment process first before finally being incorporated into the wastewater treatment machine (WWTP). The source of waste derived from nutrition unit,canteen installed grease trap to filter fat before being put into the processing process. WWTP II also uses aerobic and anaerobic systems in the processing process using aeration pools and sedimentation pools.

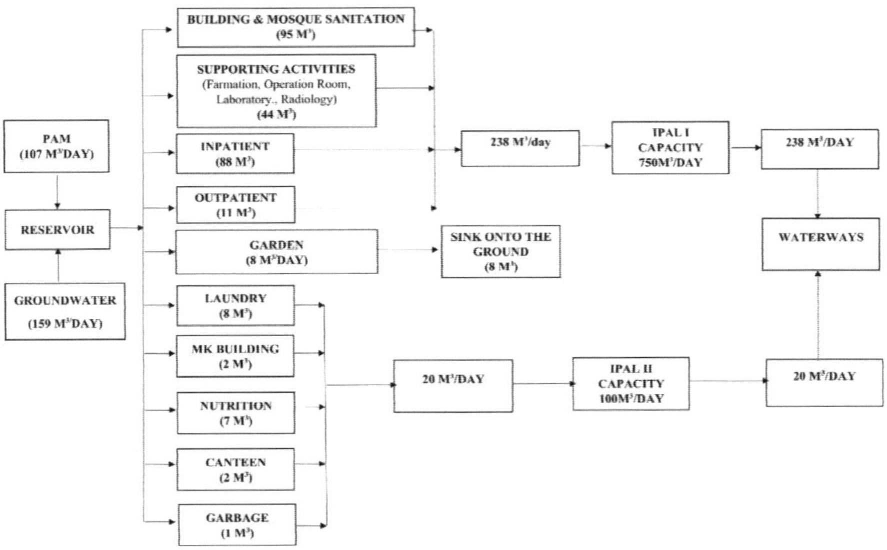

Figure 1. Wastewater Treatment System Water Balance at X Hospital

To ensure that the quality of treated wastewater meets quality standards, some of the wastewater is channelled into an indicator pond in which there are several types of fish that have high sensitivity to wastewater. Wastewater effluent is carried out by self-monitoring test for BOD, COD, pH, ammonia, TSS and tested every month where the results contained in this study are tests carried out in January 2020. The test was carried out in an accredited laboratory to see the quality of wastewater flowed to the waterways

3.2. Result of the wastewater treatment system

The results of outlets contained in each WWTP are then carried out laboratory tests using several parameters namely pH, BOD_5, COD, TSS, Oil & Grease, Ammonia, Coliform, Discharge. The results of these observations are then compared to the prevailing quality standards used in this study arePERMENLHK RI No. P.68/Menlhk/Setjen/Kum.1/8/2016. Laboratory testing was carried out in January 2020 where samples taken came from the results of the WWTP outlet I and II. The results of laboratory analysis are described in table1 and 2 below.

Table 1. WWTP I Test Results.

No	Parameter	Unit	Quality Standards	Results	Method
1	pH (26°)	-		6	SNI 06-6989. 11-2004
2	BOD_5	mg/L	30	9	SNI 6989.72-2009
3	COD	mg/L	100	45	SNI 06-6989. 15-2004
4	TSS	mg/L	30	<2	UP.IK.21.01.07 (Spectrophotometry)
5	Oil & grease	mg/L	5	<1,8	SNI 6989.10-2011
6	Ammonia	mg/L	10	<0,01	SNI 06-6989.30-2005
7	Coliform	MPN/100mL	3.000	<1,8	APHA Ed. 23rd 9221.B-2017
8	Discharge	L/Person/Day	100	-	

Table 2. WWTP II Test Results.

No	Parameter	Unit	Quality Standards	Results	Method
1	pH (26°)	-		7	SNI 06-6989. 11-2004
2	BOD5	mg/L	30	3	SNI 6989.72-2009
3	COD	mg/L	100	12	SNI 06-6989. 15-2004
4	TSS	mg/L	30	<2	UP.IK.21.01.07 (Spectrophotometry)
5	Oil & grease	mg/L	5	<1,8	SNI 6989.10-2011
6	Ammonia	mg/L	10	<0,01	SNI 06-6989.30-2005
7	Coliform	MPN/100mL	3.000	<1,8	APHA Ed. 23rd 9221.B-2017
8	Discharge	L/Person/Day	100	-	

The test results show that the quality of wastewater produced in WWTP I and II are still below the standard stipulated quality. Wastewater quality also meets the requirements for reuse as raw water because it is included in the category ofC where Water that belongs to that category can be used for fishery and livestock purposes following Government Regulation No. 20 of 1990 on Water Pollution Control. Clean production implementation technology is a combination of waste reduction techniques on sources and recycling techniques. Environmental management strategies through clean production approaches are carried out to prevent or avoid the formation of waste Utilization that can be done on

WWTP processed water products such as to water plants, washing cars or other operational equipment, cooling or washing machines, or absorbed back into the soil as a raw water source for soil. Utilization needs to be re-adjusted to the needs and availability of resources both material, physical and human resources.

The depiction of the WWTP processing process shows that there are still several inefficiencies in each aspect involved in the WWTP management process. The aspects are categorized into sections divided into human, methods, materials and machines depicted using fishbone diagrams to find out the root of the problem in each aspect, depicted diagram in figure 2.

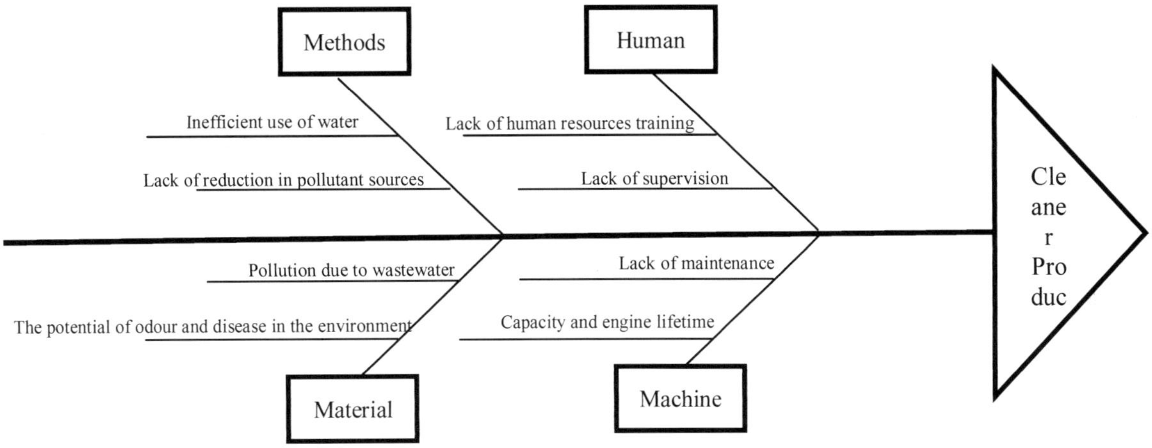

Figure 2. Fish-bone diagram

Fishbone diagram image shows the root of the problem contained in the WWTP treatment process in hospital X, from the root of the problem is then set options – options that can be implemented to optimize WWTP performance. The basis of cleaner production implementation as mentioned in the literature is divided into five parts: good housekeeping, material input changes, technological changes, product changes, on-site reuse. In a human aspect the problem is lack of human resources training and supervision, an option that can be taken is to conduct regular training or outreach to all employees to increase understanding of WWTP to improve environmental performance. This training is also included in one of the cleaner production implementations, namely good housekeeping. In the method aspect, the problems found are an inefficient use of water and lack of reduction in pollutant sources, Laboratory results show that the WWTP process or machine is still functioning properly, but on each waste source there has not been an SOP (Standard Operating Procedure) is making efforts to save water so that the wastewater entering the WWTP is still not be controlled. This can be overcome by installing SOPs at each point of the source of waste. Clean production implementation technology is a combination of waste reduction techniques on sources and recycling techniques. By reducing the source of waste at the source by installing SOPs at each point of the source of waste, it can reduce the quantity of incoming wastewater, so that the process at the WWTP machine can function optimally so that water quality improvement can occur so that water sources can be recycled.

In the material aspect, the problems encountered are pollution due to wastewater and the potential of odour and disease in the environment. Currently, the condition of the engine is still in good condition but if the capacity and hospital patients continue to increase it will decrease engine performance and bring out the possibility of odours and environmental problems. Options that can be taken are the selection of new technology or technology updates contained in the WWTP system. The choice of new technology will increase the opportunities for reuse of water treatment or recycling results

In the engine aspect, it is found that lack of maintenance and capacity and engine lifetime, an option that can be done is conducting regular inspection and maintenance of the machine. Existing field operators also have not yet held special training on efforts to conserve water, and machine checks are only carried out if there is a machine leak and there is no fixed machine control schedule. The realization of sustainable waste management through the application of cleaner production demands a holistic and balanced approach in evaluating management policy strategies. This is reinforced by collaborating problem findings in every aspect as well as in implementing options that can be done to improve WWTP management, which is not only done by one party but requires collaboration from everyone considering all aspects are involved in the effort to implement this cleaner production.

4. Conclusion

The results of WWTP treatment show that the water produced from the treatment system is still well categorized and can be reuse processed against the water treatment results. Aspects in WWTP management that still need to be improved to advance the WWTP including human, method, machine and WWTP material. Cleaner production might be a tool to find the inefficiencies of a process through identification directly from the sourceso that the implications are not only seen from the main process but from the beginning to the end of the whole process. The implementation or improvement of the options that have been obtained through the cleaner production process if the option implemented properly and consistently, not only to improve the performance of the WWTP but also can become a new energy source for the hospital through reuse of wastewater. The water that obtains from the reuse of wastewater can also reduce the operational costs of clean water requirements in hospitals.

References

[1] Karlsson M and Öhman D P 2005 Material consumption in the healthcare sector: Strategies to reduce its impact on climate change - The case of Region Scania in South Sweden *J. Clean. Prod.* **13** 1071–81
[2] Kholif M Al and Asmoro P 2016 Penerapan Produksi Bersih Sebagai Strategi Penghematan Penggunaan Air Bersih di Puskesmas Pucang Sewu Surabaya *J. Tek. WAKTU* **14** 56–62
[3] Imamuddin M 2016 Re-Use Air Pada Instalasi Pengolahan Air Limbah (Ipal) Guna Mengurangi Daya Rusak Air Di Upt Puskesmas Rawat Inap Ajibata Sumatera Utara *J. Konstr.* **7** 83–94
[4] Journal B, Vol E S, Centre E and Uk D 2016 2,3* , **4** 11–34
[5] Rahmat B and Mallongi A 2018 Studi Karakteristik Dan Kualitas BOD Dan COD Limbah Cair Rumah Sakit Umum Daerah Lanto DG. Pasewang Kabupaten Jeneponto *J. Nas. Ilmu Kesehat.* **1** 1–16
[6] Donde O O 2017 Wastewater Management Techniques: A Review of Advancement on the Appropriate Wastewater Treatment Principles for Sustainability *Environ. Manag. Sustain. Dev.* **6** 40
[7] Siregar sakti a 2008 Instalasi Pengolahan Air Limbah *Seri Sanitasi Lingkung. Pedoman Tek. Dengan Sist. Biofilter Anaerob Aerob Pengolah. Air Limbah Instal. Pada Fasilitas Pelayanan Kesehat. Kementeri.* **24** 1–9
[8] Sulaeman 2005 Kajian Produksi Bersih Untuk Industri Batik 23–7
[9] Sumadi - and Hermanuadi D 2017 Penerapan Teknologi Produksi Bersih (Cleaner Production) Untuk Peningkatan Produktivitas Dan Kualitas Kacang Oven Pada Agroindustri "Ud. Rajawali" Kabupaten Jember *J-Dinamika* **2**
[10] El-Haggar S M 2007 Sustainable Development and Environmental Reform *Sustain. Ind. Des. Waste Manag.* 125–48
[11] Cara D and Kembali P 2000 Bahan dan Peralatan yang Digunakan
[12] Pongrácz E, Phillips P S and Keiski R L 2004 From waste minimization to resources use

optimization: Definitions and legislative background *Proc. Waste minimization Resour. use Optim. Conf. June 10th* 942–51

[13] Machdar I, Sekiguchi Y, Sumino H, Ohashi A and Harada H 2000 Combination of a UASB reactor and a curtain type DHS (downflow hanging sponge) reactor as a cost-effective sewage treatment system for developing countries *Water Sci. Technol.* **42** 83–8

Spatial Variation of Electrical Conductivity, Total Suspended Solids, and Total Dissolved Solids in the Batang Arau River, West Sumatera, Indonesia.

D Helard[1], S Indah[1], M Wilandari[1]

Department of Environmental Engineering, Faculty of Engineering, UniversitasAndalas, KampusUnandLimau Manis, Padang 25163

Corrseponding author: dennyhelard@eng.unand.ac.id

Abstract. This study aims are toanalyze the spatial variation of electrical conductivity (EC) values, total suspended solids (TSS), and total dissolved solids (TDS) concentrations along the Batang Arau River.Water samples were collected from eight stations along the river at biweekly intervalsbetween March and May 2014.The results indicate an increasing trend in the value of three parameters from upstream to downstream, reflecting the effect of natural and anthropogenic activities along the river. The EC values were in the range of 173-5,097 µS/cm and had exceeded the water quality standard established by World Health Organisation (WHO) at the downstream. The TSS and TDS concentrations were in the range of 76-2,078 mg/L and 124-3,560 mg/L, respectively. The TSS concentrations at all stations had exceeded the river quality standard of class II established by Governor Regulation of West Sumatera, while the TDS concentrations had exceeded only at the downstream area. Significant spatial variation of the EC and TDS concentrations was observed between the upstream-midstream and the downstream but was not obtained along the river for the TSS concentrations. It indicates that the anthropogenic activities along the river affected the variability of EC and TDS, but not for TSS. Based on these results, reducing the number of sampling points into three stations could be performed for the EC and TDS, while TSS monitoring at only one station could be applied.

1. Introduction

In many parts of the world, the water quality control is a top-priority policy agenda [1], because if the water quality is contaminated, it can affect human health. The potential sources of water contamination are industrial, domestic, and agricultural activities as well as geological conditions, and water treatment plants. These contaminants are further categorized as physical, chemical, and microbial parameters [2].

Physical parameters such as electrical conductivity (EC) and total solids including total suspended solids (TSS) and total dissolved solids (TDS) are important parameters that must be analyzed in monitoring water quality. EC is a measure of water's capability to pass electrical flow that is directly related to the concentration of ions in the water. TSS is the measurement of solids that can be retained on a water filter and can settle out of the water column onto the stream bottom. They include silt, clay, organic wastes, inorganic precipitates, and plankton. TDS are those that pass through a water filter. They include some organic materials, as well as salts, inorganic nutrients, and toxins. The high concentration of solids in water can increase the turbidity and inhibit the penetration of sunlight into the water body which in turn will affect the photosynthetic process in the water. The sources of material in EC, TSS, and TDS can come from nature, i.e. geological condition, and seawater, and human activities, i.e. domestic and industrial waste, and agriculture [3-6]. Many standards govern EC and TDS in water. For health reasons, EC is no more than 1,500 µS/cm and the desirable limit for TDS is between 500 mg/L and 1,000 mg/L [1].

The Batang Arau River is one of the most important rivers in Padang, West Sumatra. There are various activities along the Batang Arau River, including villages, factories, industries, restaurants, pharmacies, agriculture, fisheries, and workshops, which can contribute to various types of pollutants that are carried by the current into the river [7,8]. From the monitoring result of the river in 2013, the TDS concentrations were 190 mg/L and 191 mg/L at the upstream point, 190 mg/L, and 185 mg/L at the midpoint, 171 mg/L and 164 mg L at the downstream point. Meanwhile, for the TSS, the concentrations of 16 mg/L and 26 mg/L at the upstream point, 44 mg/L and 63 mg/L at the midstream point, 97 mg/L and 109 mg/L at the downstream point were obtained [9]. These results indicate that the concentration of the parameters studied in general tends to increase from the upstream to the downstream area of the river.

Efforts to monitor water quality are very necessary to avoid more severe river water pollution. However, so far, monitoring of the water quality of the Batang Arau River has only been limited to presenting concentration and pollution load profile data, without analyzing the correlation between pollutant parameters and variations in sampling locations (spatial).This study aims to analyze spatial variation on physical parameters including EC, TSS, and TDS in the Batang Arau River due to the many activities around the Batang Arau watershed. Analysis of spatial variations will provide information on how significant the differences in sampling locations are, resulting in variations in the concentration of physical parameters. Besides, a spatial analysis should be performed to examine the effects of various anthropogenic activities and or land-use types on water quality and to design strategic sampling locations in the management of water resources for an effective monitoring program [10-12]. This research is expected to provide more comprehensive information on the quality of the Batang Arau River in terms of physical parameters and develop water management of the Batang Arau River basin.

2. Material and methods

2.1. Water sample collection

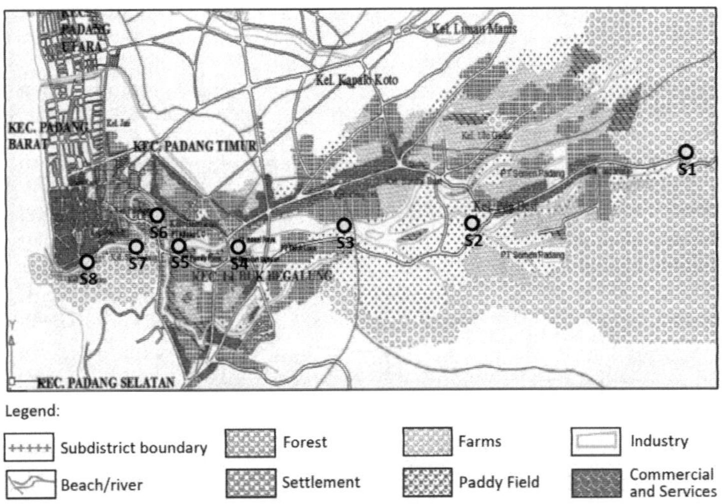

Figure 1. The sampling locations and land use alongBatang Arau River

Water sampling was carried out biweekly from March to May 2014 at eight stations along the Batang Arau River. The stations were divided into one baseline station (S1) and seven impact stations (S2, S3, S4, S5, S6, S7, and S8). The baseline station is located upstream of the river referring to the natural and unpolluted state of the river. The impact stations are located next to the upstream where some anthropogenic activities exist i.e. domestic, agricultural, industrial, and commercial activities along the river. The locations of sampling are presented in figure 1 and a detailed description of sampling points is displayed in table 1. Some environmental parameters including temperature, pH, and dissolved oxygen were measured in the field. Water samples were collected in 1 L glass bottles and were placed in cooler boxes with ice at approximately 4°C before analysis.

Table 1. Description of eight sampling locations.

Stations	Latitude	Longitude	Elevation m.a.s.l.* (m)	Distance (from S1) (km)	Description
S1	0°56'49.9"	100°30'31.5"	229	0	As a reference point for the upstream of the river located in a forested area and no human activity around this location.
S2	0°57'30.4"	100°27'08.0"	124	4.2	This point is 4.2 km from S1. The quality of the river changes marked by the brownish-yellow color of the water as the effect of the limestone crushing and agricultural activities around the location.
S3	0°57'39.7"	100°25'29.7"	72	10.1	This location was representing the water quality of the midstream of the Batang Arau River where has received the domestic and commercial discharge.
S4	0°57'40.8"	100°24'02.3"	18	13.8	Located in residential areas. This location has passed through agricultural and industrial areas.
S5	0°57'43.3"	100°22'54.1"	7	16.7	At this location, the river has received wastewater from the rubber industry.
S6	0°57'26.8"	100°22'41.1"	6	17.6	At this location, the stream joins the flow from a secondary drainage channel and has received wastewater from commercial and domestic activities
S7	0°57'41.4"	100°22'28.4"	3	18.9	This point has received domestic wastewater, restaurants, workshops, hospital, and hotel wastewater.
S8	0°57'44.8"	100°21'51.5"	1	19.9	Downstream of the river.

2.2. Analytical methods

The environmental parameters including, pH, temperature, and DO were measured at the field by using the corresponding portable meters (Hanna, USA) pH and temperature and DO-meter (Lutron, Taiwan) for DO. The ECwas measured by the conductometric method based on Indonesia's standard method (SNI 06-6989.1-2004). The TSS and TDS were measured using the gravimetric method based on the standard method (SNI 06-6989.3-2004 for TSS and SNI 06-6989.27-2005 for TDS). Each sample was measured in triplicate, and average readings were automatically determined.

2.3. Statistical analysis

Data are presented in tables and graphs after analyzing using MS Excel 365. Spatial analysis was analyzed using the one-way analysis of variance (ANOVA) with SPSS version 20.0 at 95% confidence level.

3. Results and discussion

Table 2 summarizes the overall means, standard deviations, and minimum and maximum values of environmental parameters including pH, temperature and DO as well as the physical parameters including EC, TSS and TDS concentration in the Batang Arau River at the eight stations. The decreasing of pH and DO concentration were observed at eight stations from upstream to downstream of the river, while the temperature increased. The presence of several effluents entering the river and the different conditions of land-use from upstream to downstream may result in this finding. However, the values of the environmental parameters had not exceeded the quality standard for river water class II based on the Governor Regulation of West Sumatera No. 5/2008.

Table 2. The means, standard deviations, minimum and maximum values of environmental parameters and concentrations of nitrogen at the eight sampling locations on the Batang Arau River (n = 5).

Parameters		Station								Quality standard
		S1	S2	S3	S4	S5	S6	S7	S8	
pH	mean	8.3	8.7	8.7	7.9	7.7	7.6	7.6	7.5	6-9*
	stdev	0.35	0.35	0.87	0.43	0.55	0.42	0.39	0.58	
	min	8.0	8.1	7.8	7.4	7.1	7.2	7.2	6.9	
	max	8.9	9.1	9.6	8.5	8.2	8.1	8.2	8.1	
DO (mg/L)	mean	8.2	7.2	7.4	6.2	6.9	7.2	7.0	6.7	min. 4*
	stdev	0.39	0.72	0.87	1.82	1.09	0.27	0.54	0.93	
	min	7.9	6.4	6.1	3.3	5.2	6.9	6.6	5.5	
	max	8.9	8.3	8.2	7.9	7.9	7.6	7.9	8.0	
T (^{0}C)	mean	24.9	28.6	30.7	30.6	30.5	30.5	30.4	30.4	± 3*
	stdev	0.78	1.82	2.48	1.67	1.83	1.73	1.07	1.13	
	min	24.2	26.4	28.1	29.3	29.2	29.6	29.7	29.5	
	max	26.2	31.2	34.7	33.4	33.7	33.6	32.3	32.4	
EC (µS/cm)	mean	193	216	216	249	262	316	1,362	4,051	< 1,500**
	stdev	18	60	21	43	49	142	594	1,211	
	min	173	176	182	182	183	183	369	2,018	
	max	206	308	234	298	312	559	1,910	5,097	
TSS (mg/L)	mean	94	392	462	597	622	716	1,077	1,611	50*
	stdev	15	212	250	252	208	213	535	510	
	min	76	224	228	342	326	418	568	1,052	
	max	112	756	840	938	824	966	1,654	2,078	
TDS (mg/L)	mean	140	184	169	220	244	262	1,106	2,884	1,000*

stdev	19	58	20	16	35	105	522	820
min	124	125	146	200	204	204	300	1,518
max	160	246	188	238	288	446	1,658	3,560

* Governor Regulation of West Sumatera for river water quality standard of class II, 2008
** WHO, 2011

The spatial variation of EC values along the Batang Arau River is presented in figure 2. The EC values increased from upstream to downstream. The lowest conductivity value is at S1station with a EC value of 173μS/cm. The natural and rarely polluted by wastewater as well as no human activities around that location, as it is an upstream area, may cause this finding. Next to S1 station, the average EC values at S2 until S6 were in the range of 216-316 μS/cm, not much different from the EC value at S1. At S7 the average EC value increased to 1,362 μS/cm and the highest EC value was at S8 station with a value of 5,097μS/cm. In this location, there are many pollutants or other compounds in the form of ions, that can increase the ability of water to conduct electricity. Besides, S8 station is very close to the sea, so that the possibility of seawater intrusion is quite high. Usually, the higher the conductivity in the water, the water will taste brackish to salty. This value has exceeded the limit of water quality standards by WHO, i.e no more than 1,500 μS/cm. Figure 2 also shows the differences in data distribution, due to the differences in conditions at the sampling time such as weather conditions before and at the sampling time.

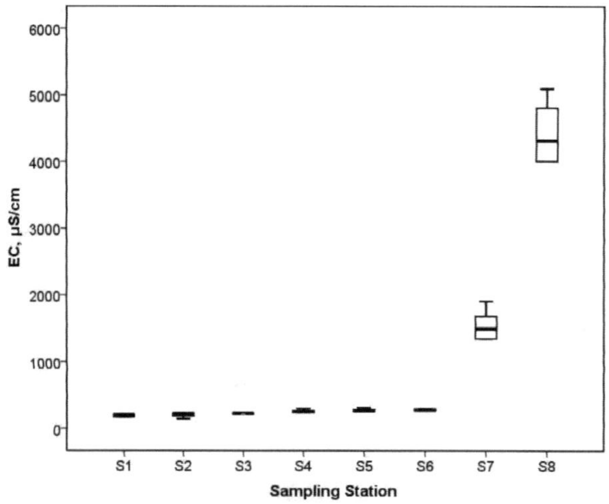

Figure 2. Spatial variability of EC values in Batang Arau River.

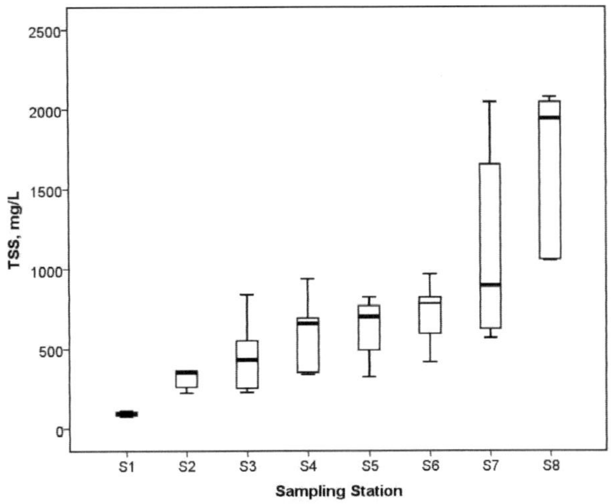

Figure3. Spatial variability of TSS concentrations in Batang Arau River.

Figure 3 shows the spatial distribution of the TSS concentrations at the eight stations of the Batang Arau River. From Tabel 2, it is demonstrated, generally, the TSS concentration contained in the Batang Arau watershed exceeds the quality standard for river water class II based on the Governor Regulation of West Sumatera No. 5/2008 (50 mg/L). An increasing trend from upstream to downstream was also observed for TSS concentrations along the Batang Arau River. The lowest concentration is at S1 stations with a value of 76 mg/L and the highest concentration is at S8 with a concentration of 2,078 mg/L. The S1 is the upstream area, located in a forested area where no human activities existed there. In the upstream, the concentration of the pollutants in rivers is very low and mostly derived from the weathering of rock and soil [7, 8, 11, 12]. The various activities including domestic, industry, commercial, and agriculture along the river increased TSS concentration.

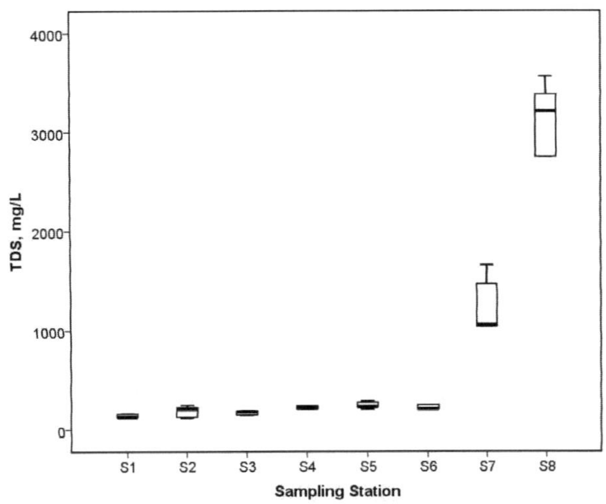

Figure 4.Spatial variability of TDS concentrations in Batang Arau River.

The TDS variability along the Batang Arau River is illustrated in figure 4. The concentration of TDS in the Batang Arau River for the upstream to middle areas is still below the quality standard, as shown in table 2. Meanwhile, in the downstream areas, the TDS concentration exceeded the quality standard. The main source for TDS in waters is runoff from agriculture, domestic and industrial waste. Figure 4 shows that the TDS concentration has increased from upstream to downstream. The results of TDS concentrations ranged from 124-3,560 mg/L with the lowest value at S1 station (124 mg/L) and the highest at S8 with a value of 3,560 mg/L, that not meet the water quality standard for class II based on Governor's West Sumatra Regulation No.5 of 2008 (1,000 mg/L). Figure 4 also demonstrates the distribution of TSS concentration as the result of the different human activities along the river.

For spatial analysis, a one-way analysis of variance (ANOVA) was performed to verify the significant variability in EC and solids concentrations along the Batang Arau River. Tables 3, 4, and 5 show the comparison of *p* values of EC, TSS, and TDS concentrations at eight stations. Table 3 indicates that the EC values at S1 until S6 had no significant difference, while with S7 and S8, the significant differences were obtained ($p<0.05$). As mentioned above, S1 is the upstream area and from S2 to S6 the river receives wastewater from industrial, agricultural, commercial, and domestic sources. However, these conditions did not lead to significant variability in the spatial distribution of EC values. While at S7 and S8 were located near to estuary and the coastal area, the accumulation of the wastewater from the various activities exist, result in the significant differences of the EC value with those at S1 until S6. The EC values between S7 and S8 also have significant differences, indicating the different effects of activities around those locations.

For TSS concentrations, there was no significant difference among the eight stations, as shown in table 4. These results revealed that different anthropogenic activities upstream to downstream of the Batang Arau River did not cause a significant difference in the spatial distribution of TSS concentrations in the river. However, for TDS concentrations, the significant differences were found between S1-S6 and S7-S8 stations, similar to the spatial distribution of the EC values.The value of ECand TDS are correlated [13-15]. EC is the measure of liquid capacity to conduct an electric charge, which depends on temperature, ionic strength, and dissolved ion concentrations (TDS) [16,17].Moreover, the significant differences were also observed between the S7 and S8 stations, as an effect of the different anthropogenic activities at those stations.

Table 3. *p* values of EC concentrations at eight stations.

EC	S1	S2	S3	S4	S5	S6	S7	S8
S1								
S2	1.000							
S3	1.000	1.000						
S4	1.000	1.000	1.000					
S5	1.000	1.000	1.000	1.000				
S6	1.000	1.000	1.000	1.000	1.000			
S7	**0.011***	**0.014***	**0.014***	**0.018***	**0.020***	**0.031***		
S8	**0.000***	**0.000***	**0.000***	**0.000***	**0.000***	**0.000***	**0.000***	

* The mean difference is significant at the 0.05 level.

Table 4. *p* values of TSS concentrations at eight stations.

TSS	S1	S2	S3	S4	S5	S6	S7	S8
S1								
S2	1.000							
S3	1.000	1.000						
S4	1.000	1.000	1.000					
S5	1.000	1.000	1.000	1.000				
S6	1.000	1.000	1.000	1.000	1.000			
S7	0.999	1.000	1.000	1.000	1.000	1.000		
S8	0.994	1.000	1.000	1.000	0.997	1.000	1.000	

* The mean difference is significant at the 0.05 level.

Table 5. *p* values of TDS concentrations at eight stations.

TDS	S1	S2	S3	S4	S5	S6	S7	S8
S1								
S2	1.000							
S3	1.000	1.000						
S4	1.000	1.000	1.000					
S5	1.000	1.000	1.000	1.000				
S6	1.000	1.000	1.000	1.000	1.000			
S7	0.999	**0.004***	**0.004***	**0.007***	**0.009***	**0.011***		
S8	**0.003***	**0.000***	**0.000***	**0.000***	**0.000***	**0.000***	**0.000***	

* The mean difference is significant at the 0.05 level.

From the result of spatial analysis, an effective future spatial monitoring program can be designed for evaluation of the physical parameters including EC and solids along the Batang Arau River. For this, the determination of sampling points can be defined at the location where had a significant difference only from all monitored stations. For the EC values dan TDS concentrations, from eight stations, it can be reduced to three stations only to monitor the concentrations, i.e., S1, S6, and S8. While for TSS concentrations, it can be selected one from eight stations for the sampling point, as there was no significant difference was observed from all monitored stations. Reducing the number of locations for sampling can reduce the time and money and, in turn, can be an effective monitoring program for the management of water resources[8,12].

4. Conclusions

The monitoring result of the EC, TSS, and TSS concentrations along the Batang Arau Riverfor 3 months sampling indicates an increasing trend in concentration from upstream to downstream. It demonstrates that the EC values and solids concentration was affected by the natural and anthropogenic activities along the river. The EC values were in the range of 173-5,097 μS/cm. The TSS and TDSconcentrations were in the range of 76-2,078 mg/L and 124-3,560 mg/L, respectively. The EC values at S1-S7 stations were still below the maximum permissible value of the water quality established by WHO. However, the TSS concentrations at all stations had exceeded the river quality standardof class II established by Governor Regulation of West Sumatera, while the TDS concentrations had exceeded only at the downstream area, S7 and S8 stations.Significant spatial variation of the EC values and TDS concentrations was observed between the upstream-midstream and the downstream of the Batang Arau River but was not obtained along the river for the TSS concentrations. It indicates that the anthropogenic activities along the river did not affect the variability of TSS concentrations. While for the EC and TDS, no significant difference was found in the upstream area as the natural condition of the river and in the midstream where received the wastewater from industrial, agricultural, commercial, and domestic activities. However, the activities near the downstream affected the variability of both parameters may be due to the accumulation of wastewater at the locations. For an effective monitoring program of the EC and TDS parameters, reducing the number of sampling points into three stations could be performed, while for TSS parameters, monitoringat only one station could be applied.These results can be a reference to water management in the Batang Arau River for local authorities.

Acknowledgments

The authors are thankful to Faculty of Engineering, UniversitasAndalas, Indonesia for supporting this research.

References

[1] World Health Organization (WHO).2011.Guidelines for Drinking-Water Quality, WHO Press, Geneva, Switzerland, 4th edition.

[2] NolletL M L. 2013. Handbook of Water Analysis, Marcel Dekker, New York, NY, USA, 2000.

[3] Marandi A, Polikarpus M and Jõeleht A. 2013. A new approach for describing the relationship between electrical conductivity and major anion concentration in natural waters.*Appl. Geochemistry*.**38**. 103–109.

[4] Appelo C A J and Postma D. 2005. Geochemistry, groundwater and pollution (Amsterdam: CRC Oress, Taylor &Francais Group).

[5] Carreira P M, Marques J M and Nunes D. 2014. Source of groundwater salinity in coastline aquifers based on environmental isotopes (Portugal): Natural vs. human interference. A review and reinterpretation.*Appl. Geochemistry*.**41**. 163–175.

[6] Rusydi A F. 2018. Correlation between conductivity and total dissolved solid in various type of water: A review.*IOP Conf. Ser.: Earth Environ. Sci.* **118**. 012019.doi :10.1088/1755-1315/118/1/012019.

[7] Helard D., Indah S., andWilandari M. 2019 Spatial distribution of coliform bacteria in Batang Arau River, Padang,West Sumatera, Indonesia. *IOP Conf. Ser: Mat. Sci. Eng.***602**, 012062.

[8] Indah S., Helard D., Herfi M. A. and Hamid H. 2018. Spatial Variation of Metals in the Batang Arau River, West Sumatera, Indonesia, Wat. Env. Res.**90(3)**, 234–243.

[9] Bapedalda (BadanPengendalianLingkunganHidup Daerah Kota Padang). 2013. PenelitianKualitasPerairan Sungai Batang Arau. Padang: PusatPenelitianKedudukandanLingkunganHidup. (in Bahasa Indonesia).

[10] Delgado J., Nieto J M andBoski T. 2010. Analysis of the spatial variation of heavy metals in the Guadiana Estuary sediments (SW Iberian Peninsula) based on GIS-mapping techniques, *Estuari Coastal Shelf Sci.***88(1)** 71–83.

[11] Varol B., Gökot A., BekleyenandŞen, B. 2012. Spatial and temporal, variations in surface water quality of the dam reservoirs in the Tigris River basin, Turkey, Catena, **92**, 11–21

[12] Helard D., Indah S. and Oktavia. 2020. Spatial variability of ammonium, nitrite and nitrate concentrations in water of Batang Arau River, West Sumatera, Indonesia, *Wat. Supp*., **20**, 1484-1492.

[13] Patil P N, Sawant D V, and Deshmukh R N. 2012.Physico-chemical parameters for testing of water – a review, *Int. J. Environ. Sci.* **3** 1194–1207

[14] Marandi A, Polikarpus M and Jõeleht A. 2013. A new approach for describing the relationship between electrical conductivity and major anion concentration in natural waters,*Appl. Geochemistry*.**38** 103–109

[15] Daniels W L, Zipper C E, Orndorff Z W, Skousen J, Barton C D, McDonald L M and Beck M A 2016. Predicting total dissolved solids release from central Appalachian coal mine spoils *Environ. Pollut.* **216** 371–379

[16] Kumar S K, Logeshkumaran A, Magesh N S, Godson P S and Chandrasekar N. 2015. Hydrogeochemistry and application of water quality index (WQI) for groundwater quality assessment, Anna Nagar, part of Chennai City, Tamil Nadu, India Appl. Water Sci. **5** 335–343

[17] Hem D. Study and Interpretation the Chemical of Natural of Characteristics Natural Water 19853rd edition USGS Water Supply Paper 2254 66-69 US Govt Printing Office Washington DC

IOP Conf. Series: Materials Science and Engineering 1041 (2021) 012028 doi:10.1088/1757-899X/1041/1/012028

Food Waste Composting with The Addition Of Cow Rumen Using The Takakura Method and Identification of Bacteria that Role in Composting

Y Dewilda[1], S Silvia[1], M Riantika[1], Zulkarnaini[1]

1Department of Environmental Engineering, Faculty of Engineering, Andalas University, KampusLimau Manis Padang 25163, Indonesia

Corresponding E-mail: yommidewilda@eng.unand.ac.id

Abstract. The purpose of research to measure and analyse the quality and quantity of food waste composting by adding cow rumen activator and efective microorganisme (EM4).This research do to identify the dominant bacteria that play a role in composting. The composting method applied in this research is Takakura composting. The method for bacterial identification was referred to as the Bergey's manual which consisted of a gram stain test and a biochemical test. Three types of composter variation are being use in this composting, variation 1 consist of (food waste and maturecompost), variation 2 consist (food waste, mature compost and EM4 activator) and variation 3 consist (food waste, mature compost and cow rumen activator). The parameter which include maturity, quality (moisture content, temperature, pH, texture, color, C/N ratio, P and K), and the quantity of compost. The results showed that not all variations met the compost maturity and compost qualitybased on the SNI 19-7030-2004 standart such as organic carbon in variation 1B and C/N ratio in variations 1A and 1B. The affects of addition activators to composter give short time and good quality of compost. Composting with the addition of cow rumen activator is the fastest composting for 7 days with C/N ratiois 19.22%, P is 0.34%, Kis 2.08%. Composting with EM 4 activator occurred for 8 days with C/N ratio is 20.53%, P is 0.30%, K is 1.82%. Composting without the addition of activator is the longest composting which is 11 days with C/N ratio is 23.74%, P is 0.21%, Kis 1.76%. The results show variation 3 with the addition of cow's rumen is the best composting with a score of 15. The results of the identification of dominant bacteria that a role play in composting are some Bacillus sp bacteria which are bacteria that degrade organic matter.

1. Introduction

Being one of the tourist destinations in Indonesia, positively affect the economic sector of Padang City. This is noticeable with the constant emergence of many restaurants, cafes, fast-food restaurants, and various other types of eating places. According to the records of Tourism Destinations in 2017, there are 230 restaurants spread across the city of Padang. This will potentially create a lot of waste from this sector, especially from the food scraps (food waste). Along with the increase in food waste pile, various problems will raise especially environmental pollution. According to the Food and Agriculture Organization (FAO), food waste is a waste produced during the process of making food and after eating activities, which is related to the behavior of sellers and consumers. Globally, the food waste problem is a particular concern to be overcome to improve environmental sustainability.

thecomposition of food waste in Padang City, 70.69% of the total food waste is organic waste (food scraps, vegetables, and fruit peels)[1]. Of all the organic waste, 33.59% is unprocessed food waste and will be disposed into the landfill.

One of the alternative solutions to overcome this waste problem is by processing organic waste into compost and adding the bio-activators to assist the decomposition of food waste. There are many composting methods, one of them is Takakura composting. The takakura composting is a practical method of composting that could be applied on a household scale or by a small restaurant business, because it does not require a wide area of land. The process of compost break down could be accelerated by adding the bio-activators. One of the alternatives that could be used for bio-activator is cow rumen [2].Besides using the cow rumen, materials such as EM4, orgadec, and Stardec and rice mole also can be used as the bio-activators. This study will observe the effect of adding the cow rumen activator and EM4 activator on the quantity of compost result, analyze the level of compost maturity and the compost quality as based on the SNI 19-7030-2004 standard, identify the dominant bacteria that play a role in the composting process and determine the optimum composition of the compost raw material.

2. Materials and methods

This study used a Takakura Composter which was made from a basket measured as 36 cm x 26 cm x 47 cm. The amount of waste placed into the composting basket is 4 kg including the addition of the mature compost. The components of the Takakura composter are cardboard, paddy husk pads, black fabrics, and a basket cover. The raw materials consist of food waste (vegetable waste, fruit peels, and leftover rice), cow rumen, and bio activators. This research was conducted in duplicate. The total raw materials required for each variation can be seen in table 1. The first composting step is chopping the raw materials and preparing the takakura basket. After that, the raw material is added to the Takakura basket according to the variation of the raw material according to Table 2. Then the cow rumen activator and EM-4 are added into the raw material according to the variations. Then, the compost components are well stirred until they became homogeneous, the stirring is conducted once in 24 hours during the composting process.

Table 1 Composition of Raw Materials.

Variation	Composition of rawmaterials (%)		Activator		Weight (kg)	
	FW	MC	EM4	CR	FW	MC
1A 1B	50	50			2 kg	2 kg
2A 2B	50	50	5 ml		2 kg	2 kg
3A 3B	50	50		5 ml	2 kg	2 kg

Note: FW ;Food Waste, MC; Mature Compos ;CR: Cow Rumen

The compost maturity test was carried out daily. As based on SNI 19-7030-2004, the parameters measured are temperature, pH, texture, and color. The composts were stated as mature when the moisture temperature is ≤ 30 ° C, the pH reaches the neutral state (6.8- 7.49), and the texture and color are similar to soil. The standard for compost quality test is based on SNI 19-7030-2004, namely the ratio of C/N, nitrogen, phosphorus, and potassium contains[3]. While the quantity test is carried out by weighing the solid compost has been produced. The objective of the bacterial identification test is to observe the dominant bacteria that play a role in composting. The method for bacterial identification was referred to as the Bergey's manual which consisted of a gram stain test and a biochemical test[4]. Identification of bacteria was conducted in the mid-process of the composting (at the peak temperature), this is the time when microorganisms became more active and playing role in decomposing the organic compounds.

After obtaining the data, it was analyzed for the raw material, compost maturity, quality, and quantity, and the dominant bacteria that play a role in the composting. The data which were analyzed in this study includes the analysis of the effect of adding the cow rumen activator and the EM4 activator towards the raw materials. This analysis is carried out based on the duration of composting time, the amount of material reduction that occurs, and the percentage of material reduction. The level of compost maturity was tested through the compost temperature, color, texture, and pH. Compost quality was tested by examining the moisture content, the parameters of Organic C, Nitrogen, C / N ratio, Phosphorus, and Potassium. The compost quality is compared with the standard of SNI 19-7030-2004 on the Compost Specifications for Household Organic Waste. Furthermore, the analysis and discussion of compost quality are carried out by referring to the standard of SNI 19-7030-2004 regarding the specifications for compost from household. The selection of the best compost variations was based on the highest score. Results of the bacteria identification included the tests of morphological, Gram stain, and biochemistry. Furthermore, the results will be integrated and used to determine the bacteria contained in each existing isolate. Ascertainment of the bacteria is determined by comparing the obtained data of the test results with biochemical test data.

3. Results and discussion

3.1 Effect on Addition of Cow Rumen Bioactivator.

Analysis of the effect of adding the cow rumen bioactivators could be observed from the composting duration and the reduction of food waste during the composting. The fastest composting process occurred in the composter 3A and 3B, which last only for 6 days. Variation number 3 was the cow rumen activator. The C/N ratio obtained for the variation 3 was 19.22% and 19.88% so that the decomposition process runs rapidly. Variation 3 also has adequate moisture content at 29.29%. Humidity plays a very important role in the metabolism processes on the microorganisms and indirectly affects oxygen supply. Microorganisms could utilize organic material if it is dissolved in water. The longest composting process was for the variation 1A that lasts for 11 days, it contains a composition of 100% food waste + finished compost. Variation 2 did not have a too significant difference in composting time as compared to variation 3, which lasts for 7 days. Details should be figure 1 in the text.

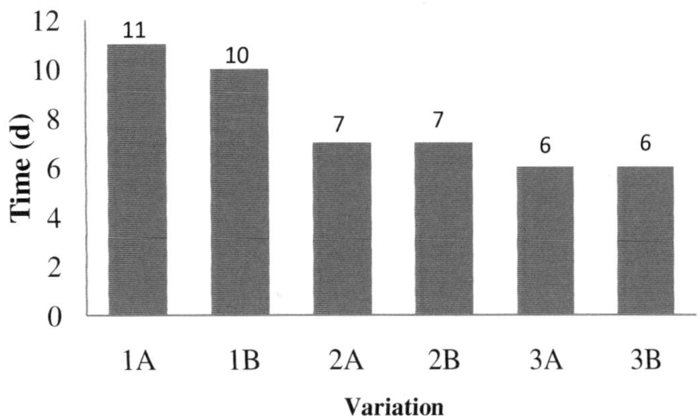

Figure 1. Analysis of Time Duration

Compost maturity occurred simultaneously with the compost shrinkage. Based on figure 2, it can be seen that the material shrinkage on each composter during the composting process. The highest reduction value as 35% was found in variation 3, a variation with the addition of cow rumen activator.

The lowest reduction as 20% occurred in the composter 1, a variation without activator addition. While the variation with EM4 bioactivators addition has experienced a 25% reduction.

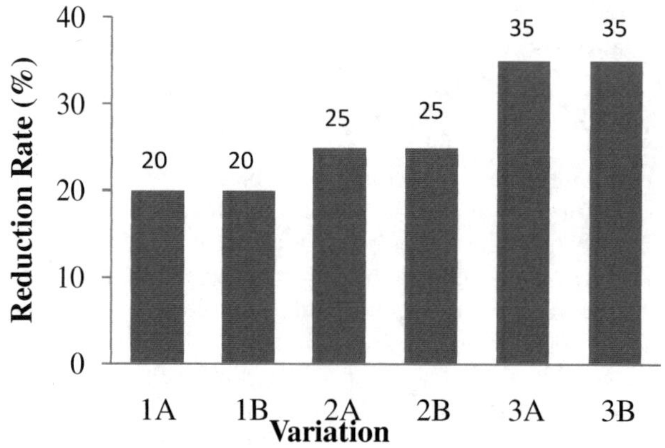

Figure 2. Analysis Material Reduction.

3.2 Compost Maturity Analysis
Analysis of compost maturity was observed from the parameters of moisture content, temperature, color, and pH. The temperature, pH, and color were monitored daily.

3.3 Moisture Content Analysis
Judging from the measurement, the moisture content of all variations has meets the optimum range of moisture content for composting. The highest moisture content was 31.77% for variation with the activator EM4 addition. While the lowest moisture content was 19.98% for variation 1 which is without using any activator.

The decreasing moisture content in each variation is caused by moisture absorption by the rice husks pad so that liquid produced in the composter during the composting process was absorbed, then the moist status inside the composter could be maintained. This condition was in line with the study results of Widarti et al. (2015) which stated that the decrease of moisture content in each composter was due to the release of moisture from organic matter of the wasted and absorbed by the pads. In the early stages of composting, microorganisms are very active in absorbing organic matter, where the results of this degradation process produce liquid (leachate)[5]. The moisture content of composting should be seen in figure 3.

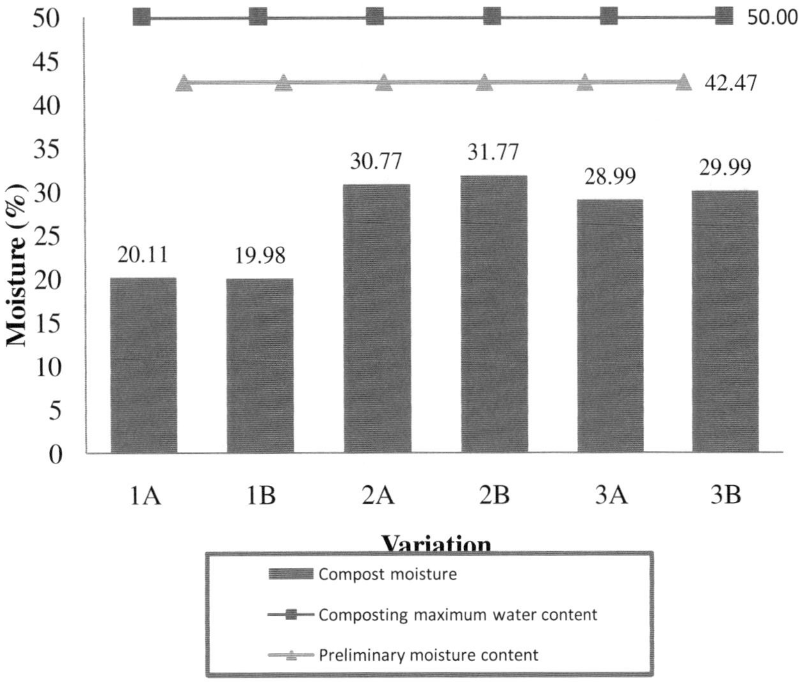

Figure 3. Compost Water Content.

3.4 Temperature Analysis

In the semi-aerobic composting process, the microorganisms involved in the composting process are mesophyll types (composting temperature is below 40° C). In this study, the optimum temperature achieved was 34 ° C for the variations of 3A, 3B, 2A, and 2B on different days during composting. This shows that the compost variation with the addition of cow rumen activator has reached the same optimum temperature compared to the variations 1A and 1B which is without activator addition. The microorganisms that playing role in the composting process are mesophyll bacteria which act at the working temperatures (25°C - 40°C). These bacteria break down the cellulose and hemicellulose into more simple sugars. The composting temperature will increase as the activity of microorganism increases until it reaches the peak temperatures and then gradually decreases because there is no more breakdown of the organic substance by bacteria [6].The thermophilic temperature of 34 °C was reached, this thermophilic temperature will cause the microorganisms to be active to degrade the organic substance and kills the pathogenic bacteria in the compost. In the mesophilic phase, the temperature drops to 33° C, which indicates the compost is entering the maturation stage. In this study, it was found that the temperature on final composting has begun to decline at the end of composting. Each research variation has reached a temperature of 26°C averagely. Based on the standard of SNI 19-7030-2004 about compost quality standards, the maximum temperature value is 30 ° C. In this study, the composting temperature of each composter has met the standard value.

3.5 Compost Color Analysis

The starting color of compost is brownish-white since the decomposition process of the material has not occurred yet, while the mature compost that has been changed into blackish as resembling the soil's color.

3.6 pH analysis

At the beginning of composting, the reaction tends to be acidic because the organic materials were broken down to produce simple organic acids such as ammonia[7]. The pH value may change as the composting process occurs. The increase in the pH level was seen on the 3rd and 4th days. The increase in pH also occurs due to the activity of microorganisms. These microorganisms have converted the organic acids that have been formed in the previous stage. The pH value of composting will increase simultaneously with the composting day and eventually stable at a neutral pH (7). The change in compost's pH begins with an acidic pH due to the formation of simple organic acids, then the pH increases on further incubation due to protein breakdown and the release of ammonia [8].

Based on the SNI-19-7030-2004 regarding the standards for compost quality, the minimum pH value is 5.4 and the maximum value is 7.4, while in this study as shown by Figure 4, the pH value at the end of the composting day is in the range 7-7.5, has met the SNI-19-7030-2004. The increase in pH level occurred due to the activity of the microorganisms that convert the organic acids that have been formed in the previous stage. The pH value will increase with composting day and will eventually stable at neutral pH (7) [9].

3.7 Analysis of the Compost Quality Test

Analysis of the C / N ratio

The C/N ratio is the ratio between carbon and nitrogen content in compost, which is useful for microorganisms for the process of decomposing organic compounds. The microorganisms need carbon and nitrogen as energy sources in the decomposition process. Carbon is a source of energy for microorganisms, while nitrogen is the most important component for building protein.

The C/N ratio in the organic compound is an indicator of nutrient availability. If an organic compound that contains a high C/N ratio is directly applied to the soil (without composted before), the decomposition process will take place in the soil. In the process of decomposing by microorganisms, a combustion reaction occurs between the elements of carbon and oxygen to produce heat and carbon dioxide. This carbon dioxide is then released in the form of gas, while the element of decomposed nitrogen is captured by microorganisms. At the time when microorganisms die, this nitrogen element will remain in the compost (along with the dead bodies), so that the C/N ratio decreases. If the C/N ratio is too low, there will be not sufficient compounds as energy sources for microorganisms to bound the free nitrogen. However, if the ratio is too high, it can be a factor that restricts the development of microorganisms [2]. The ratio can be seen in table 2.

Table 2. Composting C/N Rasio.

Variation	Rasio C/N	SNI 19-7030-2004 standard	Statement
Variation1A	23,74	Range 10-20	Does not Meet standard
Variation 1B	23,46		Does not Meet standard
Variation 2A	20,53		Meet standard
Variation 2B	18,45		Meet standard
Variation 3A	19,22		Meet standard
Variation 3B	19,88		Meet standard

3.8 Potassium Analysis

Based on the SNI 19-7030-2004 standard on compost quality standards, good compost must contain a minimum value of potassium as 0.2%. At the end of the composting process, all compost variations in the study have meet the standard. Good quality compost contains higher levels of potassium. The highest potassium content was in the 3B variation as 2.20%, and the lowest was in the 1B variation as 1.56%. Potassium levels can also be affected by the addition of activators. In the composting process, the raw material will be decomposed by microorganisms.This process requires optimal conditions such as the availability of adequate nutrients, sufficient air, and proper humidity[10]. The potassium element is beneficial for plants because it helps the formation of protein and cellulose in plant stems to make it stronger [11].

3.9 Phosphorus Analysis
Based on the standard of SNI 19-7030-2004 regarding the compost quality, good compost should have a minimum phosphorus content of 0.1%. The test results show that all variations meet the quality standards that have been set.The highest phosphorus content as 0.36% is found in the 3A variation. This result is compatible with the study results of Nurdiansyah's (2015), stated that the P content will be higher in the decaying event of the composed organic substance[12]. In the maturation stage, the microorganisms will die and the P content in the microorganisms will merge in the compost and directly will increase the Phosphorus content in compost. Based on the study results on phosphorus[13], the content (P_2O_5) in compost is related to the N content in the compost material.Recapitulation of the total scoring can be seen intable 3.

Table 3. Recapitulation of total compost scoring.

variation	compost maturity	Compost quality	Compost quantity	Total scoring
1A	5	4	3	12
1B	5	3	3	11
2A	7	5	2	14
2B	7	5	2	14
3A	9	5	1	15
3B	9	5	1	15

Base on 3, the highest score in the total monitoring score for compost maturity was found in the variations 3A and 3B, scored as 9. Compared to variation 1, the score was significantly different. The variation 1 scored as 5 because it has the lowest reduction level which results in a low final score in the compost maturity. The variation 2 scored as 7, it was considered quite good based on the composting duration. Based on the length of composting duration, variation 3 got the highest score, because it has the fastest composting time as 7 days. This also proves that adding the cow rumen activator in the compost variation is very helpful to accelerate the decomposition process, but in terms of compost quantity, the variation with the fastest composting time is not certainly being the best compost. The best compost must be seen in terms of quality over the measured parameters of the C/N, P, and K values. Based on the total score for compost quality, the highest score found in the variation 3A and 3B which are the addition with cow rumen activator scored as 15. This proves that microorganisms play an important role in the composting process. The mixing is considered as reasonable since the organic material can be broken down by microorganisms, then the amount of compost produced at the end of composting was less because the decomposition process runs optimally.

3.10Bacteria Identification
In this study, The biochemical test was conducted to identify a bacterial culture from isolation through their physiological properties and to produce the dominant bacteria that play a role in composting. Based on the biochemical test, the results found that the dominant bacteria for the composting process was Bacillus Sp1.Based on the biochemical test, the results found that the genus of bacteria in composting was Bacillus sp 1, Bacillus sp 2, Bacillus sp 3. This Bacillus sp are bacteria that can break down the organic compounds of food waste in its composting process. Bacillus sp can grow and develop in aerobic conditions [14]. Bacillus bacteria are single-celled

bacteria and are structured as rod-shaped bacilli, spherical cocci, or spiral-shaped spirilla, gram-positive bacterial cells that are facultatively anaerobic, capable of fermenting sucrose. Bacillus bacteria also produce gas in their metabolism. Some bacillus bacteria use citrate as a carbon source and are motile. This classification of bacteria is based on the bacterial identification book *Bergey's Manual of Determinative Bacteriology Seventh Edition* [4]these bacteria are classified in Division Bacteria, Classis Schizomycetes, Order Eubacteriales, Family Bacillaceae, Genus Bacillus. Refers to the results study of Supriyatna's (2012) on isolating and identifying cellulotic bacteria in the waste of organic vegetable and fruit peel that producing Bacillus bacteria[15], and other studies conducted by Lynd et al (2002) which also found that 4 isolates belonged to the genus Bacillus. A large number of Bacillus isolates could be found in the garbage because this genus can grow faster in a high number of species, besides that, the Bacillus is scattered in all types of habitat, especially in the garbage[16]. Bacterial identification was carried out using Bergey's Manual method with 12 tests. The results found that the composting also contained the Bacillus sp bacteria.

The addition of activator EM4 and cow rumen have affected the duration of compost maturity and the quantity of compost produced. The fastest composting time which lasts for 6 days was found in variations 3A and 3B that use additional cow rumen activator, producing the compost quantity that resembling soil and black color with a ratio C/N as 19.22%, P as 0.34%, K as 2.08% for the 3A variation, while the 3B variation has the ratio C/N as 19.88%, P as 0.36%, K as 2.22%. Composting using the EM4 activator was lasts for 8 days, the compost quality was resembles soil and black in color with the variation 2A found C/N ratio as 20.53%, P = 0.30% and K = 1.82% while the variation 2B found C/N ratio as 18.45%, P= 0.32% , K= 1.75. Composting without the addition of EM4 and rumen has made the variations 1A and 1B become the longest of the three variations, its composting duration last for 11 days, the compost quality resembled soil and black in color with a C/N ratio = 23.74, P= 0.21 and K= 1.76. Based on the bacteria identification using the Gram stain test and biochemical test, it was obtained that the dominant bacteria in the composting process is the Bacillus sp.

4. Conclusion

The addition of activator EM4 and cow rumen have affected the duration of compost maturity and the quantity of compost produced. The fastest composting time which lasts for 6 days was found in variations 3A and 3B that use additional cow rumen activator, producing the compost quantity that resembling soil and black color with a ratio C/N as 19.22%, P as 0.34%, K as 2.08% for the 3A variation, while the 3B variation has the ratio C/N as 19.88%, P as 0.36%, K as 2.22%. Composting using the EM4 activator was lasts for 8 days, the compost quality was resembles soil and black in color with the variation 2A found C / N ratio as 20.53%, P = 0.30% and K = 1.82% while the variation 2B found C/N ratio as 18.45%, P= 0.32% , K= 1.75. Composting without the addition of EM4 and rumen has made the variations 1A and 1B become the longest of the three variations, its composting duration last for 11 days, the compost quality resembled soil and black in color with a C / N ratio = 23.74, P= 0.21 and K= 1.76. Based on the bacteria identification using the Gram stain test and biochemical test, it was obtained that the dominant bacteria in the composting process is the Bacillus sp.

Acknowledgment

Thanks addressed to the Faculty of Engineering UniversitasAndalas for financial support of this publication.

References

[1]Dewilda Y, Aziz R and Fauzi M 2019 Kajian Potensi Daur Ulang Sampah Makanan Restoran di Kota Padang, *Jurnal Serambi Engineering*. ***4*** 2 482–487.

[2] Tarigan S Pembuatan Pupuk Organik Cair Dengan Memanfaatkan Limbah Padat Sayur

Kubis (Brasica aleracege.L) dan Isi rumen Sapi 2011 Tesis.Universitas Sumatera Utara

[3]Badan Standarisasi Nasional 2004 Tabel I Standar Kualitas Kompos *Sni 19-7030-2004. Cd*, 7030.

[4]Breed RS, Murray EGD, Smith NR 1957 Bargey'S Manual of Determinative Bacteriology Seven Edition the Williams & Wilkins Company United States of America

[5]Widarti B N, Wardhini W K and SarwonoE2015 Pengaruh Rasio C/N Bahan Baku Pada Pembuatan Kompos Dari Kubis dan Kulit Pisang *Jurnal Integrasi Proses*. ***5*** 2 75–80.

[6]Kusumawati. 2011 Evaluasi Perubahan Temperatur, pH, dan Kelembaban Media Pada Pembuatan Vermikompos dan Campuran Jerami Padi dan Kotoran Sapi Menggunakan Lumbricus Rubellus*Jurnal Inovasi Teknologi Bandung. Bandung. 15*(45–46).

[7]Simamora and Salundik 2006*Meningkatkan Kualitas Kompos*. Agromedia Pustaka.

[8]Supadma AA and ArthagamaDM2008 Uji Formulasi Kualitas Pupuk Kompos Yang Bersumber Dari Sampah Organik Dengan Penambahan Limbah Ternak Ayam, Sapi, Babi Dan Tanaman Pahitan*Jurnal Bumi Lestari*.***8*** 2113–121.

[9]Djurnani N, K Setiawan 2005 *Cara Cepat Membuat Kompos* (cetakan 1), Agromedia Pustaka.

[10]Sugeng Winarso 2005 *Kesuburan Tanah Dasar Kesehatan dan Kualitas Tanah*. Vol. 350. Gava AgroMedia Pustaka.

[11]Ekawandani N 2018 *Pengomposan Sampah Organik (Kubis Dan Kulit Pisang Dengan Menggunakan EM4*. ***12*** 1 38–43.

[12]Nurdiansyah A 2015*Pengaruh Berbagai Tingkat Dosis Effective Microorganism 4 terhadap Rasio C/N, Rasio C/P, pH, dan Fosfor Kompos Pelepah Kelapa Sawit (Elaesiguineensis Jack)*Universitas Lambung Mangkurat. Banjarmasin.

[13]PJ Stoffella, SB Wilson, MOzores-Hampton2001Compost Utilization in Horticultural Cropping System*Lewis Publisher*. 95–108.

[14] Cowan and Steel's 1993 Manual for the Identification of Medical Bacteria Third Edition Cambridge University press

[15] Supriyatna A, Rohimah A, Suryani y, Saadah S 2012 Isolation and Identifikation of Cellulolityc Bacteria From Waste Organic Vegetables and Fruits For role In Making

Material Biogas.volume VI no 1-2 ISSN 1979-8911
[16]Lynd LR, Weimer PJ, H Willem, Zyl van and Pretorius IS 2002 Microbiology and Molecular Biology Reviews P.506-577

LCA Analysis on Improving Scenario of Domestic Solid Waste Management in Padang City based on the case in Kitakyushu City

S Wulandari[1], S Raharjo[1], I Rachman[2]and T Matsumoto[2]

[1]Department of Environmental Engineering, Faculty of Engineering, Universitas Andalas, Kampus Limau Manis Padang, Indonesia
[2]Graduate Programs in Environmental Systems, Graduate School of Environmental Engineering, The University of Kitakyushu, Kitakyushu, Japan

Corresponding authors: suciwdd@gmail.com

Abstract. The implementation of domestic solid waste management in Padang City conducted by the government still uses a conventional method, including acollection-transportation-disposal system with only about 5% of solid waste recycling rate. In Japan, Kitakyushu City manages its domestic solid waste appropriately by recycling the solid waste into beneficial products and applying sea reclamation landfill for solid waste residues from the final treatment process. By conducting Life Cycle Assessment (LCA) analysis, this study examines the appropriate improved scenario of domestic solid waste management in Padang City throughevaluation and improvement of the existing implementation and the shortcomings of Padang City based onthe case study inKitakyushu City, comprisingfour scenarios. Scenario 0 presents the existing condition; scenarios 1, 2, and 3 present the improvement of Scenario 0 in recycling percentage rate and technology implementation in a row by composting, incineration, and anaerobic digestion. CML 2001 is used to assess the environmental impact of Global Warming Potential, Acidification Potential, and Eutrophication Potential. This study implies that Scenario 1 is the most appropriate scenario for improving domestic solid waste management in Padang City because of the low environmental impact as the smallest normalization score of $5.81 x 10^{-10}$.

1. Introduction

Solid waste is anissue for many cities in Indonesia because of the increaseinsolid waste generation and improper management. The increaseinsolid waste every year is related to the increasing number of people and their consumptive habits [1]. The average solid waste production in Indonesia reaches the number of 64 million tons per year. 69% of the solid waste was transported to a landfill. Furthermore, the other solid waste was treated by composting and recycling (7.5%), open burning (5%), open burying (10%), and unhandled (8.5%)[2].

Padang City is one of the big cities in Indonesia that face the above issue. The amount of 60% solid waste generation in Padang City finishesits cycle in the landfill. It isonly 5% of solid waste was composted and recycled.The other 35% was open burned and thrown awayby the communityinto the river [3]. Most domestic and non-domestic solid waste is not in sortingfrom source[4].The big rivers in Padang City were polluted by the solidwaste generation thrown away by the community. It shows the lack of communityawareness to manage domestic solid waste properlyfrom source [5].

Kitakyushu City in Japan is one of the best practices ofdomestic solid waste management. It has the Eco-Town Project as its transition catalyst from an industrial city to a green city[6].The Kitakyushu Eco-Town Project is a comprehensive solid waste management technology, education, and research facility that points out sustainability conceptsthrough environmental conservation to construct a resource-recycling-based society[7].This comprehensive system is completed with the appropriate technology, tight regulation, and involvement of multisector, including privates, government, academics, NGO's, and community.In contrary to Padang City, Kitakyushu City manages their domestic solid wasteappropriately, recyclesinto beneficial products, and usessea reclamation landfillto dispose the final residue of solid waste processing.

This study examinesthe appropriate improving domestic solid waste management scenario in Padang City by Life Cycle Assessment (LCA) analysis. Improving scenario as technical strategies are obtained through developing the existing implementation and the shortcomings of Padang City based onthe case in Kitakyushu City.This study could alsorecommendthe City of Padang related to a domestic solid waste management issue.

2. Materials and methods

2.1. LCAMethodology

LCA is a technique for assessing environmental aspects and potential impacts of a product from the raw material of production, use, and disposal. Some categories being considered in LCA are resource use, human health, and ecological consequences. LCA is used for identifying opportunities to improve their life cycle of environmental aspects, decision-making in governmental, industry, and NGO organizations, selecting the relevant indicators of environmental performance and marketing [8].

The researcher has used LCA for many studies related to municipal solid waste management that eventually assess two or more processing or system [9][10][11][12][13][14]. Specifically, LCA is used to assess the input and output flows of environmental impact and release some suggestions for improving the process [15]. LCA software as Gabi 5 Education was used in this study.Gabi 5 Education is developed by ThinkStep Company, located in Boston, Germany. LCA methodology comprises four stages, as follows:

1. Goal and Scope Definition

This stageclearly describes the goal and scope of the study. The goals should determine why the study was conducted, who the target of this study. The functional unit is needed to be decided because it is used to assess all LCA Analysis processes. The scope defines the limitation that shows how deep the analysis is conducted [16].

This analysis aims to assess and determine the best scenario applied in Padang city using the LCA method.The input of each process and flow that happened to the system was based on solid waste flow from source to landfill and whether the best scenario for Padang City is. The assessment was performed for Scenarios conductedby the government involving the community and scavengers in 2023, although the industry involvement was not be assessed. This study limited the analysis to discuss only the flow and transformation of solid waste. However, the raw material for the facilities and infrastructures would not be discussed. The functional unit used in this study is the 1-ton solid waste generation. The assessment was conducted by comparing emission released for each scenario and the solid waste reduction process by applying processing technology. The smallest solid waste to landfill and the lowest environmental risk would be the best scenario.

2. Inventory Analysis

Inventory analysis is conducted by collecting supporting data for LCA analysis, called inventory data. The inventory data may include input and output flow, such as material needed and released, emission, and energy consumed and produced [16]. The data are collected from several references related to this study, such as existing conditions, journals, manual books, also articles. Scenario schemas are also being considered to collect inventory data. In this stage, the impact assessment method is determined.This study uses CML 2001as the method, developed by the Institute Environmental Science of Leiden University.

Domestic solid waste in Padang City classifies into organic and inorganic waste, also into wet waste and dry waste [17]. Wet waste, known as solid waste for composting in this paper,comprises food waste, wood, and yard waste. However, dry waste, known as solid waste for recycling, includes all solid waste types except wet waste and other solid waste. Both wet waste and dry waste have a recycling potency of 65.16%. Wet waste has a significant recycling potency of 59.86%,while dry waste has a relatively small 5.3% recycling potency[17].

The mileagefor collection and transport was assumed similar for all scenarios. This study adoptedmileage from Komala, Aziz, & Ramadhani [18] using the farthest distance of 54 km. This mileage assumed 27 km for collection and 27 km for transport. The solid waste collection usedgasoline-fueled vehicles with an energy consumption of 433.33 kWh, andsolid waste transport used diesel-fueled trucks with an energy consumption of 171.43 kWh. The other inventory data used is adopted fromGabi 5 Education Software, some researches, and manual books [19][20][21][22][23][24].

3. Impact assessment

Impact assessment is modeled in Gabi 5 Educationthroughthe input of the inventory data. The impact category for this study selecting from CML 2001 comprised Global Warming Potential (GWP), Acidification Potential (AP), and Eutrophication Potential (EP). The three categorieswere chosen because many related studies discussedthese impacts, and they had significant effects on the environment and humans. Global warming potential analyzes contributioncompoundsto global warming, including CO_2 (Carbon dioxide) and CH_4 (Methane). Global warming potential is defined as kg CO_2 unit. Therefore, other compounds would be converted to a kg CO_2 equivalent. However, acidification potential analyzes contribution compounds for acidification, including SO_x, NO_x, HF, and HCl. Acidification potential is defined as kg SO_2 unit so that other potential compounds would be converted to kg SO_2 equivalent. Besides, Eutrophication potential analyzes contribution compounds for eutrophication, including nitrogen and phosphor. Eutrophication potential is defined as kg PO_4^{3-} unit, and other potential compounds would be converted to kg PO_4^{3-} equivalent.

Table 1. Normalization Factors

Impact Categories	Factor
CML2001 - Apr. 2015, *Acidification Potential* (AP)	4,18E-12
CML2001 - Apr. 2015, *Eutrophication Potential* (EP)	6,32E-12
CML2001 - Apr. 2015, *Global Warming Potential* (GWP 100 years)	2,39E-14

Source: Gabi Education Database, 2018

Table 2. Scoring Factors

Impact Categories	Factor
CML2001 - Apr. 2015, Acidification Potential (AP)	6,1
CML2001 - Apr. 2015, Eutrophication Potential (EP)	6,6
CML2001 - Apr. 2015, Global Warming Potential (GWP 100 years)	9,3

Source: Gabi Education Database, 2018

These three impactswere normalized and scored to determine whether the best scenario would be offered. Normalization is a tool to manage the inconsistency of inventory data available. It is used by multiplying the result of impact assessment to the normalization factor in table 1. However, scoring is used to assess affection for each impact on affection for all impacts by multiplying the result of normalization to scoring factors in table 2.

4. Interpretation
Interpretation is used to analyze and recommend improvingthe selected scenarioto minimize the environmental impact[16].

3. Results and discussion

3.1. Case Study
Kitakyushu City conducts comprehensive solid waste management, beginning with solidwaste sortingbased on its treatment. Type of solid waste includes kitchen waste, can and glass bottle, PET plastic bottle, package and plastic, lamp, used vegetable oil, home appliance, and paper.

Kitchen waste would be treated in Kitchen Waste Building by using incineration technology that releases ashes as its residue. Can and glass bottles would be processed in the Can and Glass Bottle Building through compacting (for the can) and crushing (for glass bottle). The product of this process would be processed in another Recycling Building to release beneficial products. PET Plastic bottle would be processed in PET Plastic Bottle Recycling Building into pellets and flakes transferred to another recycling building to produce the new PET Plastic Bottle and another beneficial product (coat, ball, sneaker). Package and plastic would also be treated in its recycling building to compact the waste and be transferred to another recycling building. Others would be treated in their recycling building.

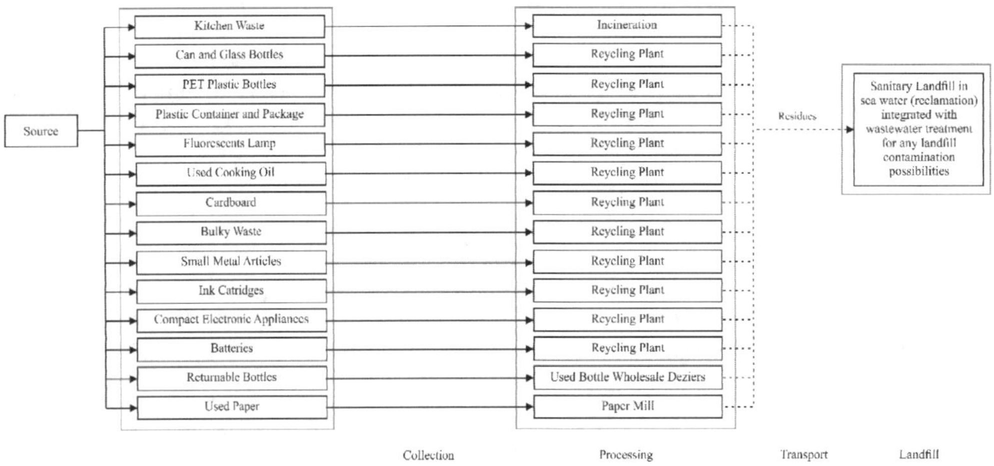

Figure 1. Schema of Technical Aspect of Domestic Solid Waste Management in Kitakyushu City

The residue of all treatment (ashes) would be transported to a landfill. Kitakyushu City has a Sanitary Landfill System in Hibikinada Landfill by adopting the sea reclamation method. However, it does not need more area to disposeof the residue, yet it creates new land. This landfill has seawater treatment for considering the probability of solid waste residue contamination to seawater. Schema of Technical Aspect of Domestic Solid Waste Management that Kitakyushu City used generally can be seen in figure 1.

On the other hand, Padang city made an effort to groupthe solid waste into organic and inorganic waste, but domestic solid waste is not sorted in practice. Solid waste is transferred to a temporary container (TPS) and treated (for small percentages) by composting in Solid Waste Processing Station based on Reduce Reuse Recycle (TPS3R) and recycling in the solid waste bank and by scavengers. Untreated solid waste and treatment residue would be transported to a landfill. Padang City has a

Controlled Landfill System that uses a large area. Schema of Technical Aspect of Domestic Solid Waste Management that Padang City used generally can be seen in figure 2.

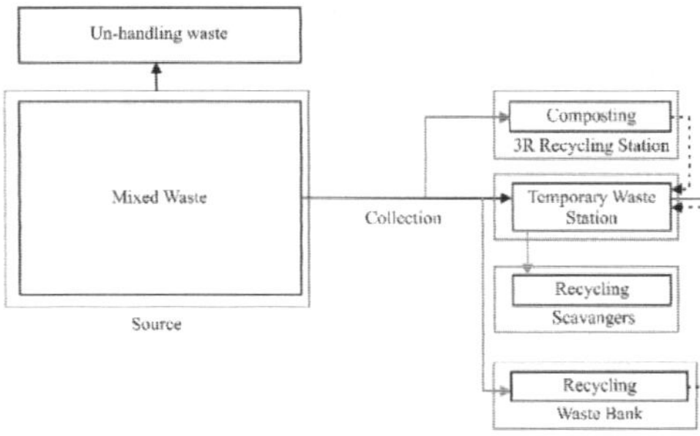

Figure 2. Schema of Technical Aspect of Domestic Solid Waste Management in Padang City

According to appropriate conditions in Kitakyushu City, we adopted some technical strategies to improve the quality of domestic solid waste management in Padang City by arranging four scenarios adjusted to the City of Padang capability.The scenarios would be described for the next five years of projections. The details can be found in the next section.

3.2. Scenarios

There are four different scenarios, including the existingdomestic solid waste management practices in Padang City.

Scenario 0 (ExistingScenario): Solid waste was being collected in mixed conditions. It causedinconveniencefor management and processing that carried out. However,the unmanaged solid waste percentage was still relatively high, namely 35% of the total domesticsolid waste generation. The community treated itthrough burning or throwing away into the river. Besides, the remaining 65% of solid waste was handled by the government. The diagram of Scenario 0is illustratedin **Figure 3**.

The existing condition adopted the collection-transport-disposalsystem, which means that solid waste generationfrom the source would be directly collected and transported to the landfill. Solid waste processing was conducted through a small percentage. Composting was carried out at 2% while recycling by the informal sector (scavengers) was 3%. Meanwhile, the other 60% and the residue from the compostingwere transported to the landfill.

Scenario 1: The solid waste collection was divided into three types: waste for composting (compostable waste), waste for recycling (recycled waste), and other waste. This condition was structured to facilitate the management and processing carried out. The percentage of unmanaged solid waste was reduced by 2% per year [26]. Because the scenario calculation assumption is in the next five years (2023), the total solid waste reduction is 10% of the existing condition. Unmanagedsolid waste decreases to 25% of the total domestic solid waste generation. The government handles the remaining 75% by involving the informal sector (scavengers) and the community. The diagram ofScenario 1 is illustratedin figure 4.

Scenario 1 performs solid waste processing by composting and recycling. Solid waste processing is carried out in two sites, composting at TPS3R and recycling at TPS and Solid Waste Bank. Composting and recycling are assumed to increaseby 0.14% per year [26]. Therefore in 2023, the processing percentage will increase to 1.4% at each solid waste processing site. Composting at TPS3R is 3.4%, while recycling is 4.4%, at the Solid Waste Bank by 1.4% and in the informal sector 3%. However, the untreated solid waste (67.2%) and solid waste processing residue aretransported to the landfill equipped with landfill gas recovery facilities.

Scenario 2: The solid waste management assumption applied is similar to Scenario 1 exclude the increase in composting percentage. This scenario uses incineration to processsolid waste by a higher percentage without involvement from the community. The diagram of Scenario2is illustratedin figure 5.

Scenario 2 performssolid waste processing by incineration, composting, and recycling. Incineration and composting are carried out at TPS3R and recycling at TPS and Solid Waste Bank. Incineration is assumed to accommodate 50% of the total managed compostable solid waste, about 25.8%. This assumption is selected through the consideration for applying new technology in a short period (5 years) has not been able to perform in a high percentage directly.

Composting at TPS3R is 2%, while recycling is 4.4%, at the Solid Waste Bank(1.4%) and in the informal sector (3%). However, the untreated solid waste (42.8%)and solid waste processing residue aretransported to the landfill without landfill gas recovery facilities. The landfill gas facility is not implemented due to the considerable investment of incineration being used.

Scenario 3: The solid waste management assumption applied is similar to Scenario 1.However, Scenario 3 uses anaerobic digestionto process some compostable solidwaste. Scenario 3 diagramis illustrated in figure 6.

Scenario 3 performs solid waste processing by anaerobic digestion, composting, and recycling. Anaerobic digestion and composting are carried out at TPS3R with the percentage in a row by 1.4% and 2%. While recycling (4.4%) is carried out at TPS (by informal sectors 3%) and Solid Waste Bank (1.4%).However, the untreated solid waste (67.2%) and solid waste processing residue aretransported to the landfill equipped with landfill gas recovery facilities.

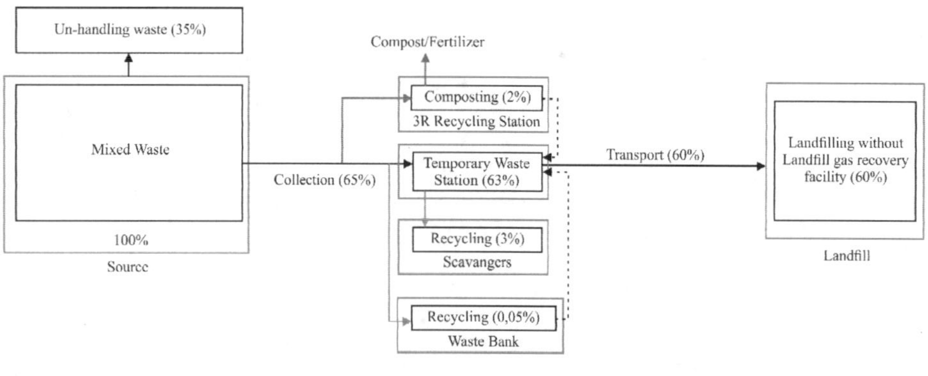

Figure 3. The Diagram of Scenario 0

Annotation:

———————— Mixed Solid Waste

———————— Compostable Solid Waste

———————— Recyclable Solid Waste

- - - - - - - - Residue

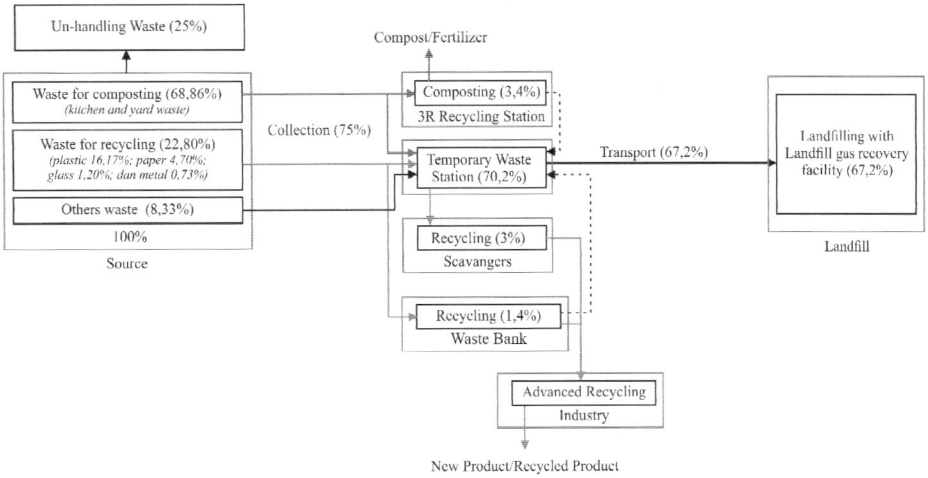

Figure 4. The Diagram of Scenario 1

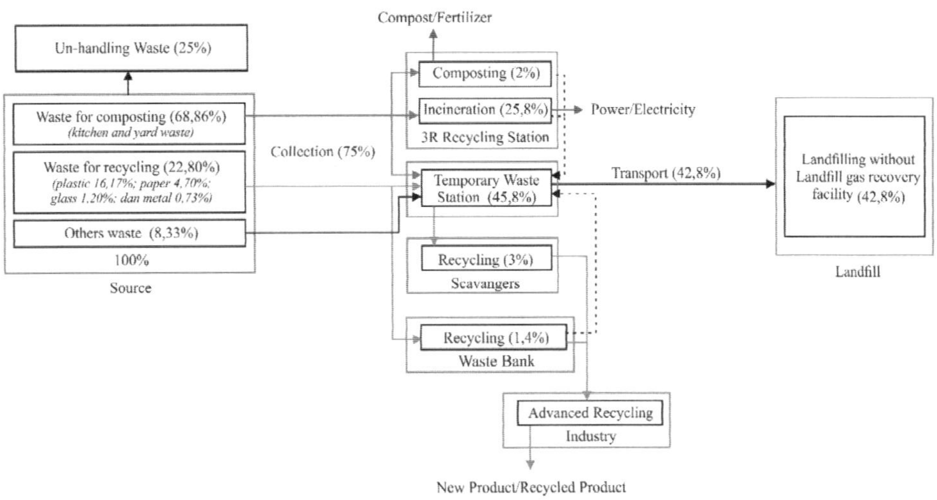

Figure 5. The Diagram of Scenario 2

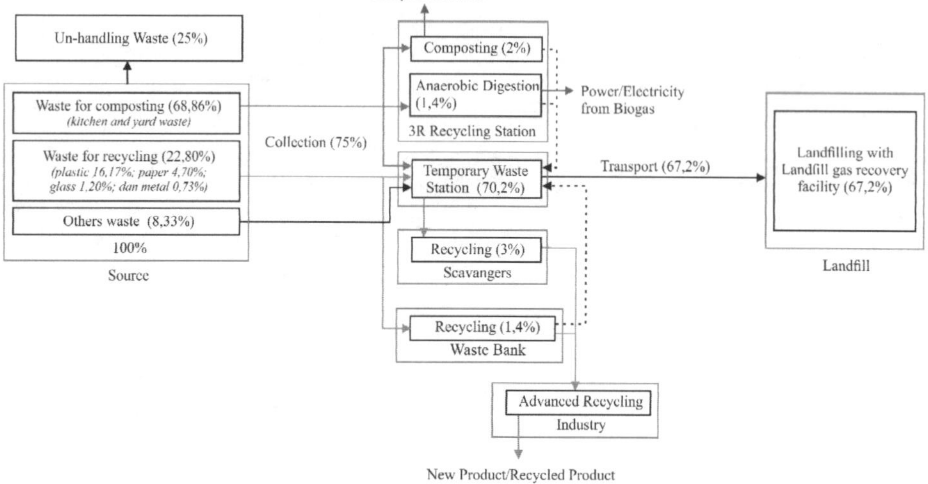

Figure 6. The Diagram of Scenario 3

3.3. Impact Assessment

Environmental impact assessment for each scenario usedthe CML 2001 method bythe Institute Environmental Science of Leiden University. CML 2001 was chosen because it could adequately assess the impact of midpoint indicators by applying ISO 14040. Impact categories assessed were global warming potential (GWP), acidification potential (AP), and eutrophication potential (EP). The resultsofthe impact assessment are describedas follow:

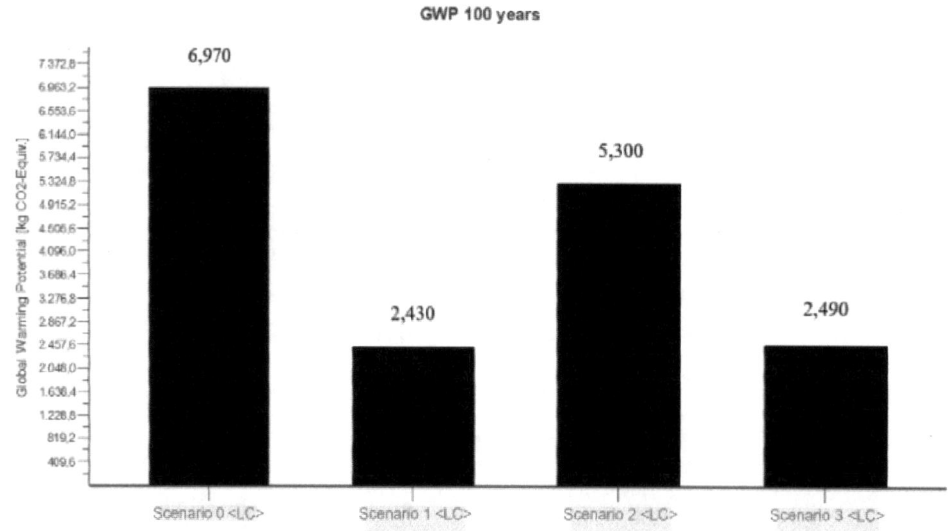

Figure 7. Global Warming Potential for all scenarios

As described infigure 7,Scenario 0 has the highest GWP of 6,970 kg CO_2-eq. It is caused by Scenario 0 has no proper solid waste management system with un-handling solid waste 35%, processing solid waste 35%, and 60% dispose of in a landfill without gas recovery facility.

Unavailability of gas recovery facility directly indicate all the pollutant of the landfill (CO_2 and CH_4) contribute to GWP affects the atmosphere. It is getting deplorable because CH_4 has impact twenty-eighttimes bigger than CO_2. This condition also happens in Scenario 2, and it becomes the second-highest GWP 5,300 kg CO_2-eq.The incineration process also releases CO_2 gases. However, this CO_2 does not affect global warming because CO_2 tends to be biogenic carbon and recycled in the environment. Different from CO_2 released from burning fossilfuel (gasoline and diesel fuel) that tend to non-biogenic carbon. Non-biogenic carbon potentially affects global warming [26].Scenario 1 has the lowest GWP 2,430 kg CO_2-eq and Scenario 3 has GWP 2,490 kg CO_2-eq. Both scenarios can minimize the GWP rate using solid waste handling technology, including anaerobic digestion, composting, and gas recovery facility in the landfill. The gas recovery facility significantly helps the GWP because air pollutants would be converted to electricity.

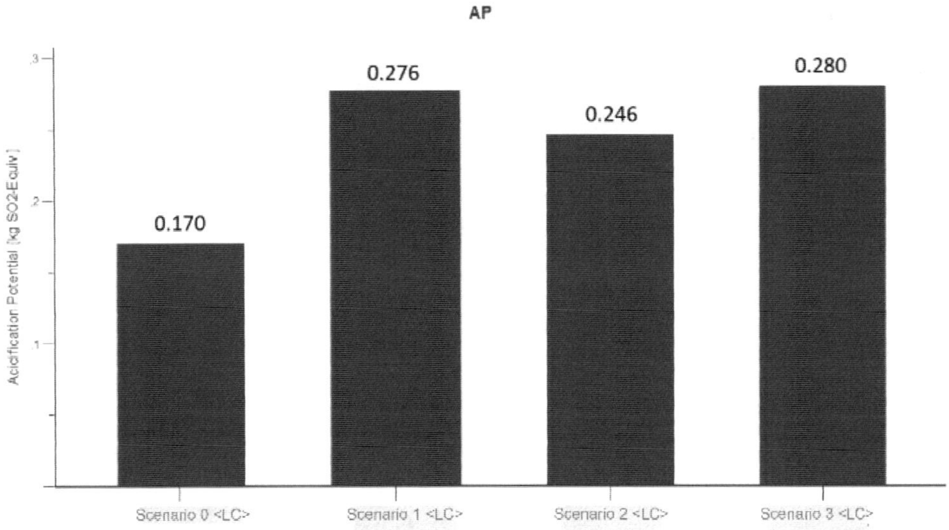

Figure 8. Acidification Potential for all scenarios

Figure 8 shows theScenario 0 has the lowest AP rate of 0.170 kg SO_2-eq. It is caused by minimalization using fueled facilities that released emissions. Scenario 3 has the highest AP rate of 0.280 kg SO_2-eq. This AP rate is dominated by emission released from using fueled facilities. Acidification can impact the environment because of the acid characteristic. Acidify the environment cannot support much life of nature. For instance, acid rain is caused by air acidifying, and it is potentially dangerousto the biotic and the abiotic environment. The leading cause of acidification is SOx, NOx, HF, and HCl contained in solid waste. The sources are also fromthe collection and transport process.

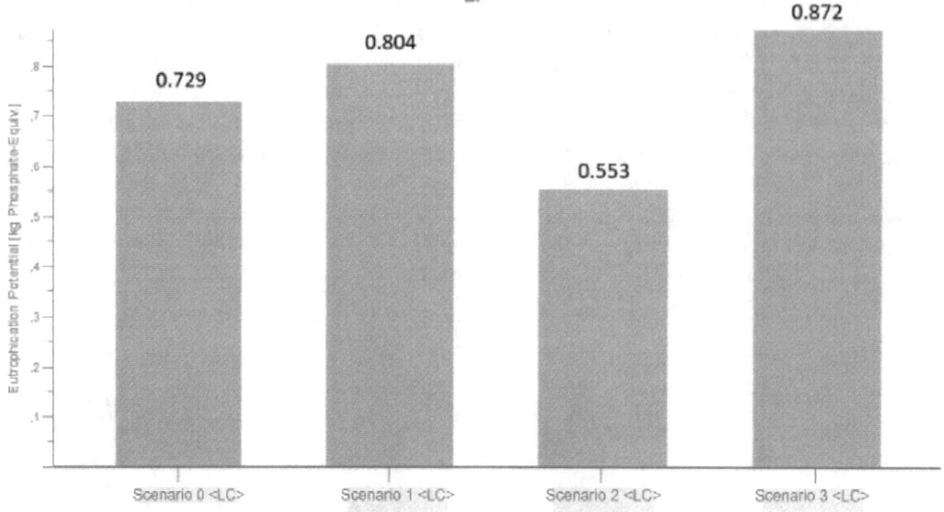

Figure 9. Eutrophication Potential for all scenarios

As described infigure 9, Scenario 3 has the highest EP rate of 0.872 kg PO_4^{3-}eq. Many water emissions are released from un-handling solid waste, anaerobic digestion wastewater, and most solid waste untreated properly. Scenario 2 has the lowest EP rate of 0.553 kg PO_4^{3-}eq caused by an incineration process with low wastewater and minimal solid waste disposal to landfill. The high EP rate can affect the unbalance of the ecosystem. As the high of nutrients in the water can trigger the increase of poisoned phytoplankton, it can decrease the water's oxygen rate and kill aquatic. It is called alga blooming that colors the water green.

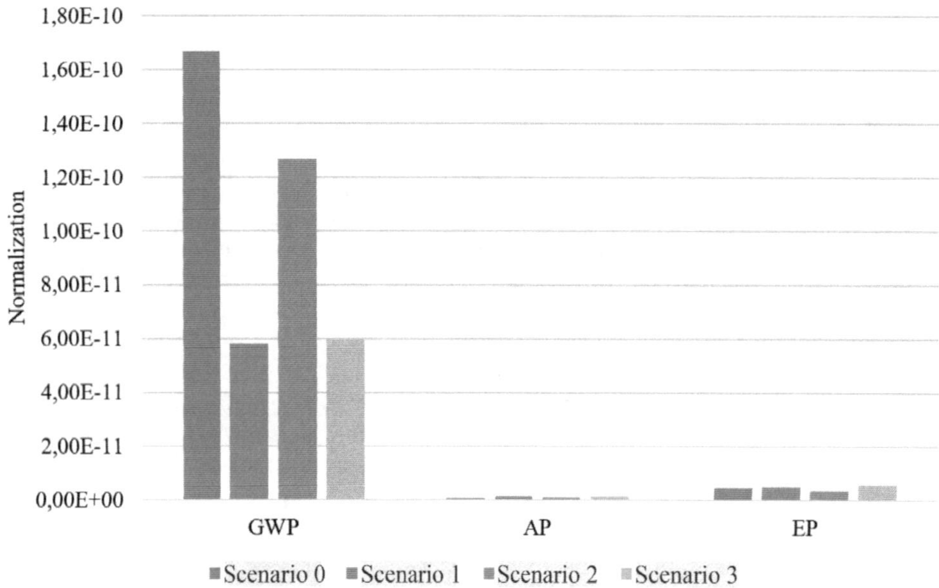

Figure 10. The Result of the Normalization of the four scenarios based on GWP, AP, and EP

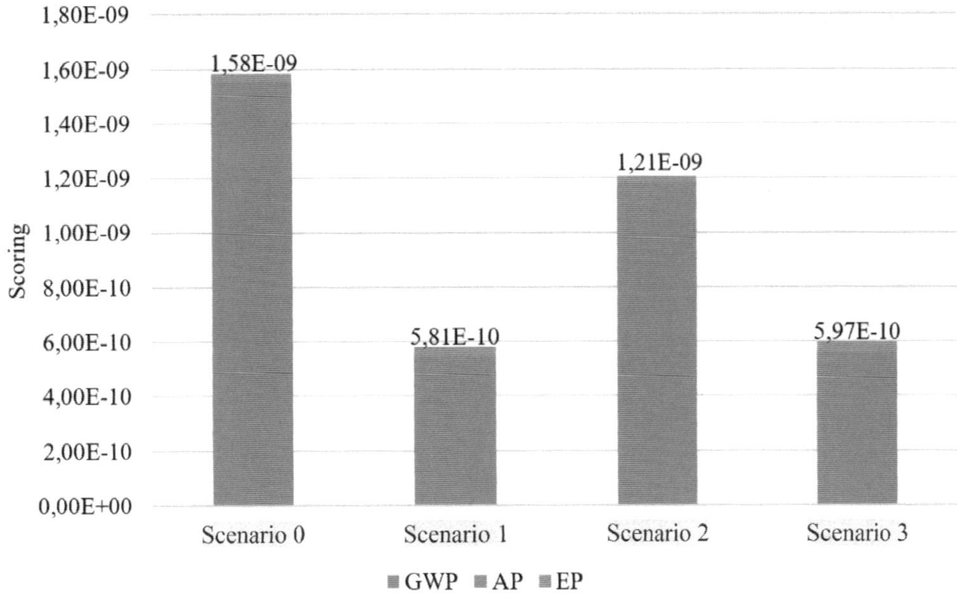

Figure 11. The Result of Scoring for every four scenarios.

After gaining the result of the assessment for each scenario, then normalization is conducted. The result of normalization is in figure 10. Determination for the best scenario is performed by using the scoring method. The smallest score is the best scenario instead. The result of scoring is in figure 11.According to this graphic, scenario 0 has anexcellent score of $1.58x10^{-9}$, scenario 1 has the smallest score $5.81x10^{-10}$, while scenario 2 and 3 have each score of $1.21x10^{-9}$ and $5.97x10^{-9}$. Therefore, scenario 0 becomesanimperfect scenario. As an impact assessment result, it is not recommended because of its high environmental impact. However, scenario 1 becomes the appropriate scenario as its small environmental impact and recommend to be applied in Padang City.

3.4. Interpretation

Since scenario 1 is the appropriate one according to Scoring on LCA Analysis, it has an environmental impact on its every single process instead. The recommendations are offered to improve Scenario 1, including Landfilling, Composting, and Collection and Transport. Landfilling can be improved by maximizing wastewater (leachate) treatment so the impact could be adequately managed and improved gas hopping and treatment efficiency in landfill gas recovery facilities.Composting could minimize the emission by improving composting efficiency and using a better bio-activator to cut the composting times.The recommendation for Collection and Transport is to convert high emission fuel to a low emission fuel (high combustion efficiency). Furthermore, applying pollution control for collection and transport facility become one of recommended suggestion to conduct.

Conclusion

We adopt some technical strategiesfrom Kitakyushu City to improve the domestic solid waste management in Padang City by arranging four scenarios adjusted to the City of Padang's capability. LCA analyzes the four scenarios for the next five years of projections (2023).Based onthe LCA analysis score, Scenario 0 is the poorest due to inappropriate for the environment as the score of $1.58x10^{-9}$. However, Scenario 1 is an appropriate scenario to be applied in Padang City score of

5.81×10^{-10}.The recommendation given for the selected scenario includes landfilling, composting, and collection and transport aspects due to the environmental impact oftheirseverysingle process.

References

[1] Badan Pusat Statistik 2016 Kota Padang dalam Angka (Padang Municipality in Figure) 2016 Badan Pusat Statistik Kota Padang.

[2] Kementerian Lingkungan Hidup dan Kehutanan 2017 Indonesia - Finlandia bahas kerjasama pengelolaan sampah menjadi energi [Online] Available at: http://www.menlhk.go.id/berita-189-indonesia--finlandia-bahas-kerjasama-pengelolaan-sampah-menjadi-energi.html#: [accessed 06/04/2018]

[3] Raharjo S, Matsumoto T, Ihsan T, Rachman I, Gustin L 2017 Community-based solid waste bank program for municipal solid waste management improvement in Indonesia: a case study of Padang city *Journal of Material Cycles and Waste Management*

[4] Aziz R, Febriardy F 2016 Analisis Sistem Pengelolaan Sampah Perkantoran Kota Padang Menggunakan Metode Life Cycle Assessment *Jurnal Dampak*

[5] Candra S A, Aini N 2018 Mengurai Permasalahan Sampah di Kota Padang [Online] Available at: https://www.republika.co.id/berita/nasional/daerah/18/02/04/p3m0jk382-mengurai-permasalahan-sampah-di-kota-padang: [accessed 08/18/2018]

[6] Shiroyama H, Kajiki S 2016 Case Study of Eco-town Project in Kitakyushu: Tension Among Incumbents and the Transition from Industrial City to Green City In.

[7] Environment Bureu 2019 Kitakyushu Eco-town Project *Environment Bureu of City of Kitakyushu* [Online]. Available at: https://www.kitaq-ecotown.com/docs/ecotown-pamphlet-en-2019.pdf

[8] ISO 2009 The ISO 14000 family of International Standards *Environmental Management*

[9] Feo G De, Malvano C 2009 The use of LCA in selecting the best MSW management system *Waste Management*

[10] Hong J, Li X, Zhaojie C 2010 Life cycle assessment of four municipal solid waste management scenarios in China *Waste Management*

[11] Laurent A, Clavreul J, Bernstad A, Bakas I, Niero M, Gentil E, et al. 2014 Review of LCA studies of solid waste management systems - Part II: Methodological guidance for a better practice *Waste Management*

[12] Erses Yay A S 2015 Application of life cycle assessment (LCA) for municipal solid waste management: A case study of Sakarya *Journal of Cleaner Production*

[13] Liu G, Hao Y, Dong L, Yang Z, Zhang Y, Ulgiati S 2017 An emergy-LCA analysis of municipal solid waste management *Resources, Conservation and Recycling*

[14] Bartolozzi I, Baldereschi E, Daddi T, Iraldo F 2018 The application of life cycle assessment (LCA) in municipal solid waste management: A comparative study on street sweeping services *Journal of Cleaner Production*

[15] Wang H, Thakkar C, Chen X, Murrel S 2016 Life-cycle assessment of airport pavement design alternatives for energy and environmental impacts *Journal of Cleaner Production*

[16] Standardization I O for 2006 International Standard ISO 14040 *Environmental management - Life cycle assessment - Principles and framework*

[17] Hafizh M 2017 Potensi Daur Ulang Sampah Domestik Kota Padang Tahun 2016 (Padang: Universitas Andalas)

[18] Komala P S, Aziz R, Ramadhani F 2012 Analisis Produktivitas Sistem Transportasi Sampah Kota Padang Productivity Analysis of Municipal Waste Transportation System in Padang City *Jurnal Teknik Lingkungan UNAND*

[19] Mendes M R, Aramaki T, Hanaki K 2004 Comparison of the environmental impact of incineration and landfilling in São Paulo City as determined by LCA *Resources, Conservation and Recycling*

[20] McDougall F R, White P R, Franke M, Hindle P 2007 Integrated solid waste management: A

life cycle inventory: Second edition Integrated Solid Waste Management: A Life Cycle Inventory: Second Edition.

[21] Sundqvist J O 2006 Assessment of Organic Waste Treatment In: *Renewables-Based Technology: Sustainability Assessment*

[22] Komilis D P, Ham R K 2004 Life-cycle inventory of municipal solid waste and yard waste windrow composting in the United States *Journal of Environmental Engineering-Asce*

[23] Levis J W, Barlaz M A, DeCarolis J F, Ranjithan S R 2013 A generalized multistage optimization modeling framework for life cycle assessment-based integrated solid waste management *Environmental Modelling and Software*

[24] Ahrens M B, Orger M B, Robson D N, Li J M, Keller P J, Atakulreka A, et al. 2015 Modeling of energy consumption and environmental life cycle assessment for incineration and landfill systems of municipal solid waste management-A case study in Tehran Metropolis of Iran *J Physiol*

[25] Pusat Pengendalian Pembangunan EkoregionSumatera 2018 Pendampingan Penyusunan Jakstrada di Wilayah Sumatera (Pekanbaru)

[26] Bacovsky D, Ludwiczek N, Ognissanto M, Wörgetter M 2013 Status of Advanced Biofuels Demonstration Facilities in 2012 IEA Bioenergy Task 39 Report.

Solid Waste Management System Model of Village Tourism Area of Nagari Tuo Pariangan, Indonesia

R Aziz[1], Y Ruslinda[1], Y S Yeni[1], Y Dewilda[1]

[1]Environmental Engineering Departement, Faculty of Engineering, UniversitasAndalas, KampusLimau Manis, Padang, 25163, Indonesia

Corresponding author: rizkiaziz@eng.unand.ac.id

Abstract. NagariTuoParianganis one of the three leading tourist attractions in Tanah Datar Regency. The increase of tourists visits NagariTuoPariangan Area has resulted in increased waste generation, improperly managed existing solid waste management endanger the environment, and sustainable development. This study aims to develop a solid waste management system model of NagariTuoPariangan as an effort to preserve the environment and support tourism development. The generation of solid waste of this area was 786,803 kg/day, with the largest were food waste as 35% and plastic waste of 32% of total waste. The solid waste management system model considers three categories of sorting types (compostable, recyclable, and residue), while the storage use of individual type (trash bags and bin) and communal (bin 40 L, 6 m^3 containers). The collection system uses a 1.5 m^3 motorized pedicab. The processing method applies composting, enumerating plastic waste, selling paper, and metal waste to the collectors. Moreover, the residue transport using a 6 m^3 arm roll truck from the recycling center (TPS 3R) to Bukit Sangkiang Landfill. Meanwhile, the non-technical aspects planned for regulation on waste management by WaliNagari, the formation of Non-Governmental Organizations, adjusting funding according to applicable regulations, and increasing the participation of traders, communities, and visitors.

1. Introduction

NagariTuoPariangan is one of the nagari/villages located on Mount Merapi, Pariangan District, TanahDatar Regency, which covers1,792 hectares. It is one of the favorite tourist attractions in Tanah Datar Regency. In 2012, it was chosen as one of the five most beautiful villages in the world because of its natural beauty, according to the United States' influential tourism media, namely the Travel Budget. The tourism objects in NagariTuoPariangan consist of historical/ cultural tourism and nature. It has a strong potential to meet the demand for the tourism market. The number of tourists visits to this tourism area in 2014 was 6,162 visitors and increased in 2016 to 8,915 visitors[1]. The increasing number of tourists visit has increased tourists' waste. The waste produced from this area comes from the tourist attraction and residential areas.

The NagariTuoPariangan area has not been included in the waste service area ofTanah Datar Regency and does not have a waste management system. The waste generated by the surrounding community and visitors is managed independently by collecting garbage, then throwing the garbage into the river/ditch, and some of the garbage is burned. As a waste producer, the community does not concern with reducing and utilizing waste that can still be used. People tend to use product packaging that cannot be recycled to increase the amount of waste generated, and the community has not carried out sorting the waste produced. If waste cannot be appropriately managed, it will become a burden and cause significant problems. However, if managed properly, it will become an asset and benefit the government and society. The problem of solid waste must be anticipated to prevent the danger of environmental pollution to get worse in the future. It is necessary to develop a waste management system with the 3R (reduce, reuse, and recycle) concept[2]. The improperly managed waste in the tourism area will also affect tourists' attractiveness to NagariTuoPariangan. Therefore, it is necessary to plan a waste management system.

Tanah Datar Government issued a policy that explains the solid waste network system, which consists of increasing the community's participation and the business / private sector to implement a solid waste management system[3]. Solid waste processing carried out with environmentally friendly technology following technical principles, increasing service coverage, and quality of service systems and reducing waste input to the final waste processing site with the 3R (reduce-reuse-recycle) concept around the waste source area. While according to the district/city tourism regional development plan for tourism, it must pay attention to the carrying capacity of the physical, socio-cultural, and economic environment of tourism.

Considering the solid waste management problems, the regulations, and the district/city tourism development plan above, it is necessary to develop a solid waste management system model for the NagariTuoPariangan area to increase tourist attraction and maintain the environmental sustainability. Study on tourism waste has developed since last fifteen years located in the various region around the world[4,5,6,7,8], while in Indonesia it is located in Kalimantan[9], Bali[10], Java[11,12], and Sumatra, especially in West Sumatra[13,14,15,16]. The study reported the quantity of solid waste generation. Some of the studies reported the composition and recycling potential and solid waste management systems[17].

2. Materials and methods

The research stages include a literature study sourced from textbooks, journals, and previous research on waste management system planning. Furthermore, data collection is carried out in the form of secondary data and primary data. Secondary data is a general description of the location, number of tourist facilities, area, number of visitors, map of tourist areas. Primary data are solid waste generation, composition, and recycling potential, and existing practice of solid waste management. Primary data collected by interviews and field observations. Interviews were conducted at the Tourism Office and the Environmental Affairs Agency of Tanah Datar Regency.

Identification of problems and development needs is carried out by comparing the existing conditions of waste management in the TuoPariangan Nagari Area with the criteria for Indonesia's existing waste management system. The criteria for the waste management system refer National Standard of Indonesia[18], National Constitution[2], and Ministry Regulation[19].

The model of the waste management system for NagariTuoPariangan area is developed for seven years of the design period, namely 2020-2026. The solid waste management system model includes technical aspects and non-technical aspects. Technical aspects include sorting and packaging, collection, transportation, and processing. Simultaneously, Non-technical aspects include regulatory aspects, institutional aspects, financing aspects, and community participation aspects.

3. Results and discussion

3.1. The existing practice of solid waste management

The waste management system of NagariTuoPariangan is carried out independently by the community by collecting garbage in an open area. The garbage is burned, and some of it is dumped into a ditch/river. The source of waste varied from tourist objects, tourist visitors, and residential waste. The waste produced is in food scraps, beverage bottles, plastic, and food wrapping paper. The existing condition of waste in the NagariTuoPariangan area can be seen in Figure 1.

Figure 1. The practice of waste dispose and burning on drainage, river, and tourism spot

3.2. Solid waste generation, composition, and recycling potential

The waste source can be grouped into residential areas and tourist areas. The average waste generation for the residential area is 0.928 Liter/cap ./day. The tourism area is 0.697 Liter/m^2/day or 2.611 Liter/cap./day. Waste generation in NagariTuoPariangan area can be seen in table 1.

Table 1. Solid waste generation of NagariTuoPariangan Area.

Sources	Unit Generation	Unit	Generation rate	Unit
Settlement	0.928	L/cap./d	5,428.743	Lit./day
	0.105	kg/cap./day	615.051	Kg/day
Tourism Area	2.611	L/cap./d	386.395	Lit./day
	1.160	kg/cap./day	171.752	Kg/day

Waste generated from the NagariTuoPariangan area has a different composition. The separation of waste according to the type of waste consists of food waste, yard waste, plastic, hazardous waste, and other waste. The highest average waste composition is food waste, with 35% and plastic waste at 32%. Meanwhile, for recycling potential for each waste type, yard waste, tin can, and paper & waste can fully be recycled, while food waste has recycling potential as 88,59% while plastic only 32,23%.

3.3. Problem Identification

The following is an explanation of the identification of waste management problems in the Tuo Pariangan Nagari Area:

1. A highpercentage of plastic waste is generated in the Nagari Tuo Pariangan area.
2. Nocommunity concern as a waste producer to reduce and utilize waste that can still be used.
3. People tend to use product packaging that cannot be recycled, thereby increasing the amount of waste generated.
4. Some waste generated is burned, and some of it is thrown into the river.
5. Nagari Tuo Pariangan area is not included in the service area ofTanah Datar Regency.
6. Wastehas not been separated. The waste is still mixed.

7. There are no waste containers provided in the Tuo Pariangan area.
8. Absenceof a collection system and no means of collecting garbage
9. There is no vehicle for transporting waste from the Nagari Tuo Pariangan area to the final waste processing site.
10. There is noregulation on area scale waste management.
11. There is no financialbudget for regional waste management.
12. No organization manages regional level waste.

3.4. Waste Management Development Needs

The development needs include technical aspects and non-technical aspects. The development of waste management in the Nagari Tuo Pariangan area needs to create a waste management system, which applies for a regional scale. The waste management system consists of sorting and storage, collection, processing, transfer, and transportation systems. Meanwhile, for non-technical aspects, making rules regarding waste management in the Tuo Pariangan Area, forming a waste management organization, making a budget for regional waste management, and increasing community participation in waste management.

3.5. Solid waste management system model

This system model aims to reduce waste from the source and increase community participation in 3R-based waste management. The waste management system model considers the Spatial Planning and Territory of Tanah Datar Regency for 2011 to 2031. The solid waste management system model for the NagariTuoPariangan area can be seen in table 2.

Table 2. The solid waste management system model of NagariTuoPariangan Area.

Solid wastes management system	Model
Segregation and Storage	It is separated into compostable waste, recyclable waste, and residue. Communal storage uses plastic bins 40 Lit capacityat the tourism spot.
Collection	Indirect individual pole. Collection vehicle motorized cart with a partition on the box capacity 1,5 m^3. Service areas: settlement, shops. Collection frequency: daily for all waste types. Indirect communal pole. Collection vehicle motorized cart with a partition on the box capacity 1,5 m^3. Service areas: settlement, tourism spots. Collection frequency: daily for all waste types.
Treatment and Processing	The treatment takes place in the Recycling Center (TPS 3R). Type of treatment: composting, chopping, processing of recyclable waste, and trading.
Transfer and Transport	Transfer uses 6 m^3containers at TPS 3R. Transport uses the HCS system with *an arm roll truck*
Landfilling	Located in TPA Bukik Sangkiang
Regulation	Provide regulation and sanction for solid waste management system offenders
Financial	Sources could be from Corporate Social Responsibility, traders' retribution, the community served, visitors, and waste processing product sales.
Institution	Develop community groups to manage the recycling center (TPS 3R)
Community participation	Provide training and education to elevated community awareness on solid waste management

3.6. Infrastructures and Facility Needs

The planned storage of waste is an area scale by applying individual and communal container patterns. Containerization is carried out in a segregated manner in order to facilitate the further processing of waste. The number of containers is adjusted to the type of waste that is sorted. The number of communal storage located on the tourism area fade of fiberglass bin with 40 L volume is 41. This storage has three types of purposes, green-colored containers for compostable waste, yellow for recyclable waste, and gray for other waste. While for settlement, it needs only one 6 m^3 container.

The waste collection system uses indirect communal and individual indirect collection patterns. The indirect communal collection pattern is the collection of waste from communal containers to 3R TPS. The number of means of collecting is one motorized pedicab with a capacity of 1.5 m^3 with four times a day rite, and the age of using a motorized pedicab is 5-8 years.

Waste processing activities at TPS 3R include composting, plastic waste for sale, other recyclable material, and other waste. Compostedwaste is food waste, yard waste. The processing of biodegradable waste is carried out using the Takakura composting method. The Takakura composting method at TPS 3R uses a basket of 14 units with a composting time of 14 days. Plastic waste for sale is the plasticwaste that will be sold is in the form of used plastic drink bottles. Plastic waste will be cleaned, chopped, and packaged for third parties (dealers).

Other recyclable material consists of paper, metal, and aluminum cans, which can be recycled will be cleaned and packed to be sold to third parties (dealers). In contrast, other waste will be combined with residual waste and disposed of in a 6 m^3 container located in the TPS 3R area.

The TPS 3R to be built is in the Nagari Tuo Pariangan area. TPS 3R has 200 m^2, including the parking area and the road inside TPS 3R. The supporting facilities available at TPS 3R are chopping machine, manual sieving, shovel, broomstick, water hose, scales, and cart. The number of requirements for waste processing facilities at TPS 3R can be seen in table 3.

Table 3. Equipment Needs for Recycling Center (TPS 3R).

No	Items	Number	Unit
1.	Total area	200	m^2
2.	Takakura basket	17	unit
3.	Chopping machine for compost material	1	unit
4.	Chopping machine for plastic	1	unit
6.	Manual siever	1	unit
7.	Cart	3	unit
8.	Shovel	2	unit
9.	*Thermometer*	2	unit
10.	Scale	2	unit
11.	Sewing machine	1	unit
12.	Sack	10	unit

The transportation of residual waste and other waste from TPS 3R uses the Hauled Container System (HCS) transportation pattern to the TPA to Bukik Sangkiang with an arm roll truck with a capacity of 6 m^3.

4. Conclusion

Based on the waste management planning for Nagari Tuo Pariangan Area that has been carried out, it can be concluded that the container pattern used is a communal container with three types of waste. Waste collection is carried out every day uses an indirect communal pattern and an indirect individual using a motorized pedicab with a capacity of 1.5 m^3. The transportation system uses an arm roll truck with a capacity of 6 m^3 to TPA Bukik Sangkiang. A recycling center building (TPS 3R) can be in Jorong Guguak and apply composting using the stacking takakura method.Meanwhile, non-technical aspects consist of developing the institutional structure at TPS 3R, the financialsource planning, and the participation of visitors in the waste management system.

It will be interesting to assess the solid waste management system correlation with the environmental impact that contributed to the study further. It will help find the sustainable system one, as the requirement stated in the national law of solid waste management, one of the methods that can be applied is Life Cycle Assessment.

5. Acknowledgment

Thanks addressed to the Institute for Research and Community Service of UniversitasAndalas for this study's financial support by Grant No. T/10/UN.16.17/PT.01.03/PRK-RD/2020.

References

[1] Badan Pusat Statistik. Statistik Kepariwisataan Kabupaten Tanah Datar , 2018. Published online 2018.

[2.] Undang-Undang Republik Indonesia No. 18 Tahun 2008 tentang Pengelolaan Sampah. Vol 76.; 2008:61-64.

[3] Peraturan Daerah Kabupaten Tanah Datar No. 2 Tahun 2012 tentang Rencana Tata Ruang Wilayah Kabupaten Tanah Datar Tahun 2011 - 2031. Published online 2012.

[4] Bhat RA, Nazir R, Ashraf S, Ali M, Bandh SA, Kamili AN. Municipal solid waste generation rates and its management at Yusmarg forest ecosystem, a tourist resort in Kashmir. *Waste Manag Res*. 2014;32(2):165-169. doi:10.1177/0734242X13518089

[5] Zorpas AA, Voukkali I, Loizia P. The impact of tourist sector in the waste management plans. *Desalin Water Treat.* 2015;56(5):1141-1149. doi:10.1080/19443994.2014.934721

[6] Arbulú I, Lozano J, Rey-Maquieira J. The challenges of municipal solid waste management systems provided by public-private partnerships in mature tourist destinations: The case of Mallorca. *Waste Manag.* 2016;51(2016):252-258. doi:10.1016/j.wasman.2016.03.007

[7] Hoang, Fujiwara PP. Municipal Waste Generation and Composition in a Tourist City. *J JSCE,*. 2017;5:123-132.

[8] Giurea R, Precazzini I, Ragazzi M, et al. Good practices and actions for sustainable municipal solid waste management in the tourist sector. *Resources.* 2018;7(3):1-12. doi:10.3390/resources7030051

[9] Irsan R, Studi P, Lingkungan T, Tanjungpura U. Kabupaten Sintang. Published online 2014:1-10.

[10] Wijaya IMW, Trihadiningrum Y. Strategi Penanganan Sampah di Obyek Wisata. *Acad Edu.* Published online 2014:1-6. https://www.academia.edu/21986270/Strategi_Penanganan_Sampah_di_Obyek_Wisata_Eks_Pelabuhan_Buleleng_Bali

[11] Akbari R. Analisis Timbulan Dan Komposisi Sampah Di Kawasan Bhumi Merapi Dan Stonehenge Kaliurang, Sleman, Di Yogyakarta. Published online 2018:1-2. https://dspace.uii.ac.id/handle/123456789/10357

[12] Thumbnail N. Evaluasi Pengelolaan Sampah Wisata Kabupaten Magelang Jawa Tengah. Published online 2020:1-2.

[13] R JK dan A. Studi Potensi Daur Ulang Sampah Kawasan Wisata Taman Marga Satwa dan Budaya Kinantan (TMSBK) dan Kawasan Wisata Taman Panorama dan Lobang Jepang (TPLJ) di Kota Bukittinggi. Published online 2018.

[14] Aziz R, Mira. Study of recycling potential of solid waste of tourist area in Pariaman City. *IOP Conf Ser Mater Sci Eng.* 2019;602(1). doi:10.1088/1757-899X/602/1/012059

[15] Aziz R, Ruslinda Y, Oktavianus N. Development of Sustainable Solid Waste Management

System of Air Manis Tourism Area for Supporting Sustainable Tourism in Padang City, West Sumatra, Indonesia. *ASEAN/Asian Acad Soc Int Conf Proceeding Ser*. 2019;1:611-617.

[16] Aziz R, Dewilda Y, Teknik F, Andalas U, Carocok P. Kajian Awal Pengolahan Sampah. 2020;20(1):77-85.

[17] Aziz R, Dewilda Y, Khair H, Faklin M. Pengembangan Sistem Pengelolaan Sampah Kawasan Wisata Pantai Kota Pariaman dengan Pendekatan Reduce-Reuse-Recycle. *J Serambi Eng*. 2020;5(3):1188-1194. doi:10.32672/jse.v5i3.2141

[18] Badan Standarisasi Nasional. Tata Cara Teknik Operasional Pengelolaan Sampah Perkotaan. *ACM SIGGRAPH 2010 Pap - SIGGRAPH '10*. 2002;(ICS 27.180):1. http://portal.acm.org/citation.cfm?doid=1833349.1778770

[19] Menteri Pekerjaan Umum. Peraturan Mentri Pekerjaan Umum Republik Indonesia Nomor/3/PRT/M/2013. *Tentang Penyelenggaraan Prasarana dan Sarana persampahan dalam Penanganan Sampah Rumah Tangga Dan Sampah Sejenis Sampah Rumah Tangga*. Published online 2013:1-374.

IOP Conf. Series: Materials Science and Engineering 1041 (2021) 012031 doi:10.1088/1757-899X/1041/1/012031

An Assessment of Heavy Metals Pollution in the Waters and Sediments of Lake Maninjau, Indonesia

P S Komala[1], A Nur[1], N L Badriah[1], M Harefa[1], S Silvia[1], Ridwan[1], Zulkarnaini[1]

[1]Environmental Engineering Department of Universitas Andalas, Limau Manis Kec. Pauh, Kota Padang, Sumatera Barat, Indonesia, 25163

Corresponding author: putisrikomala@eng.unand.ac.id

Abstract. This research performs an analysis of heavy metals in the waters and sediments of Lake Maninjau and the resultant pollution index value. The research was carried out in 11 locations, e.g., floating net cages, endemic fisheries, near settlements, hydropower plants, and seven rivers at the lake's inlet and outlet at a depth of 0-1.5 metres. Determining the pollution index was conducted based on heavy metals and environmental parameters. Aside from Zn, the concentration of Cd, Hg, Pb, Cu in the waters in all locations exceeded the quality standard. The Hg metal in sediments in all areas exceeded the quality standard of contaminated soil. Concentrations of heavy metals in sediment are higher than that of heavy metals in water. 8 of 11 sampling locations were in the medium polluted category, with the fish cage location having the highest pollution index (PI) value. Anthropogenic activities such as aquaculture, water transportation, and settlements around the lake have an effect on the waters and sediments of the lake, characterised by heavy metal contamination. The lake's quality must be continuously monitored, and wastewater management improved from activities around the lake to control heavy metal contamination. Further evaluation is required of the heavy metals contamination originating from anthropogenic activities and natural sources.

1. Introduction

One lake which has a crucial role in supporting the surrounding community's development and economy is Lake Maninjau in West Sumatra. Various activities have recently severely polluted the lake and have degraded water quality and sediment. Three main activities are performed in Lake Maninjau: hydropower plants, freshwater aquaculture centers, and tourism destinations [1]. Other activities on the lake include agriculture, human settlements, and water transportation; these are thought to have contributed to the deterioration of water quality and sediment due to the presence of heavy metals. Both human and natural activities from urban and industrial activities are the primary sources of heavy metals entering the aquatic environment, these include forest fires, continental weathering, and volcanic activity [2].

At a high level, heavy metal content can cause harm to organisms which live in lakes; as a result there is a high likelihood of them entering the food chain and being consumed by humans [3]. This material is distributed partly in aqueous phases, is partially adsorbed on suspended materials, and accumulates in the lake sediment [4]. Sediments are considered as the sink and source for heavy metals discharged into the environment [5]. How much these metals are released from the sediments into the water body depends on microbial activity, hardness, alkalinity, ionic strength, pH, and Eh [6]. Therefore, estimating the level of heavy metal pollution in the aquatic environment is determined by analyzing water, sediment, and marine organisms [7]. Water pollution levels can be determined through an environmental quality index. This study aims to obtain an overview of heavy metal concentrations of Cd, Hg, Pb, Cu, Zn in the waters and sediments of Lake Maninjau. The results from the heavy metal analysis were used to determine the pollution index in the waters of Lake Maninjau and assess the pollution in the waters and sediments due to heavy metals.

2. Materials and Method

Study Area

Lake Maninjau is situated in West Sumatra Province, has a surface area of 99.5 km^2 with a catchment area of 24,800 ha, and a maximum depth of ± 165 m. Lake Maninjau is surrounded by nine villages, with Maninjau being the capital city. Lake Maninjau primarily serves as a hydroelectric power plant (PLTA), with power generated by 64 MW to supply electricity to West Sumatra *[8]*. The lake has also been developed as a center of aquaculture and tourism. This area is a volcano-tectonic depression area, with similarities to Lake Toba but smaller in size. The height of the Maninjau caldera wall is 1200-1400 meters above sea level, or 459 meters from the lake's surface, with a maximum depth of 157 meters. The average monthly rainfall is 325.8 mm, and the average maximum temperature of Lake Maninjau is 31.27 °C, while the average minimum is 22.66 °C. 95.20% is the average level of humidity, with wind speeds around Lake Maninjau averaging 23.5 km/hr. 88 rivers, large and small, flow into the lake. The lake outlet flows into Batang Antokan and empties into the Indian Ocean. Various resident activities occur on the edge of the lake, for example, settlements, hotels, agriculture, and livestock, which are all sources of pollutants entering lake waters.

Sampling and Parameters Analysis

Sampling was conducted 3 times from February-May 2018 over a period of 2 weeks. The sampling site was decided on the basis of the Indonesian National Standard [9], as shown in figure 1. Eleven locations were determined by considering how the lakes are used and the site closest to the source of the pollutants, i.e., dense settlements, endemic fisheries, fish cages, hydropower, inlets (Batang Kularian, Batang Maransi, Jembatan Ampang, Batang Kurambit, Batang Asam, Bandar Ligin River) and the outlet (Batang Antokan River) of the lake. A vertical water sampler on the boat was used to obtain water samples.The water sampling procedure adhered to government regulation standards [9]. The water samples were directly measured for temperature and DO using Lutron DO-5510 and pH with HI 9813-5 pH-meter. 1.5 ml of concentrated HNO_3 was added to the water samples for metal analysis and then placed into the cool box. Sediment samples were obtained from the same location as the water samples. The sampling procedure used the sediment sampling guide and methodologies [10]. Analysis of the water samples was performed and then compared to the "Government Regulations of the Republic of Indonesia Number 82 (2001) Class 2" [11]. In comparison, heavy metals in sediment analysis referred to the "Government Regulation about Total Concentration for Determination of Management of Soil Contaminated by Hazardous and Toxic Waste" [12]. Identification of heavy metals in water and sediment was measured using "Inductively Coupled Plasma Shimadzu, ICPE-9000", which is referred to as "Standard Methods for the Examination of Water and Wastewater" [13].

Pollution Index

The pollution indices were used for determining the water quality of Lake Maninjau based on environmental parameters (temperature, DO, pH) and heavy metal. These indices were calculated by using equation 1 based on the "Decree of the State Minister for the Environment Regarding Guidelines for Determining Water Quality Status" [14]. This value shows the water contamination level as a result of heavy metal and environmental parameters based on the "Government Regulation for Surface-Water Class 2" standard.

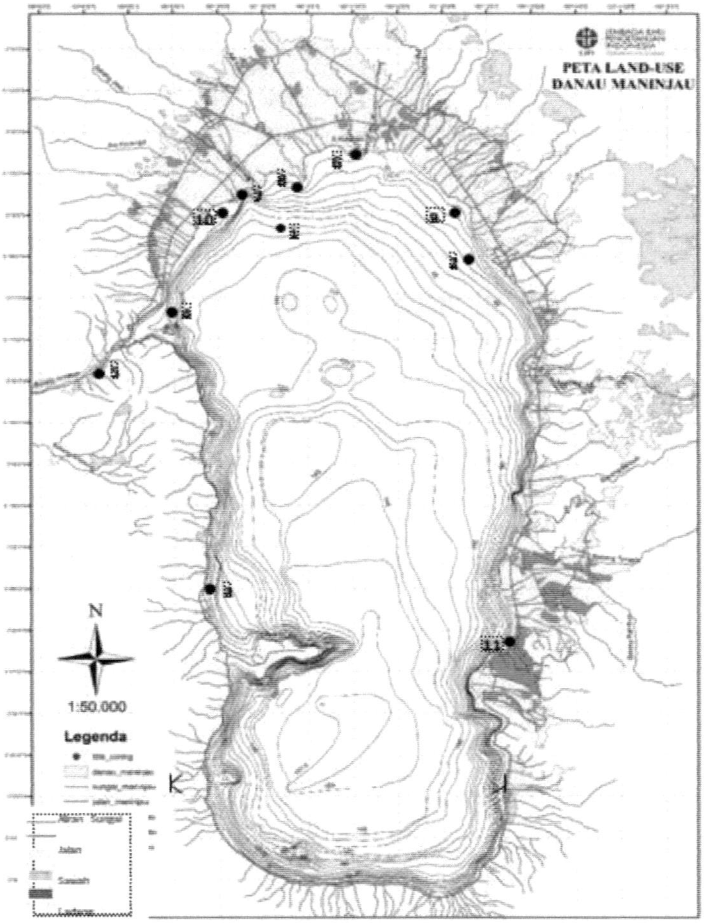

Figure 1. Map of Sampling Location. 1. Fish Cage, 2. Endemic Fisheries, 3. Dense Settlement, 4. Hydropower, 5. Batang Antokan River, 6. Batang Kularian River, 7. Jembatan Ampang River,8. Batang Asam River, 9. Batang Maransi River, 10. Batang Kurambit River, 11. Bandar Ligin River

$$PI_j = m\sqrt{\frac{\left(\frac{C_i}{L_{ij}}\right)_M^2 + \left(\frac{C_i}{L_{ij}}\right)_R^2}{2}} \qquad (1)$$

Where:

Lij = water quality concentration parameters listed in the water designation standard (j). Ci denotes water quality concentration parameters (i) of the water sample at the sampling locations. The calculated pollution index value is then categorised: $0 \leq PIj \leq 1,0$ is baseline level, $1,0 < PIj \leq 5,0$ lightly polluted, $5,0 < PIj \leq 10$ moderately polluted, $PIj > 10$ heavily polluted.

3. Results and discussion

3.1. Environmental Parameters

The pH of Lake Maninjau and the rivers entering and leaving the lake has a pH value of 7.17-9.4. The pH originating from the rivers is lower than the lake waters and ranges between 7.24 and 8.37. Lake and river water temperatures range between 27.17°C and 31.47°C. DO ranges from 5.23 to 9.9. The pH value of DO originating from rivers is lower, i.e., 5.23 - 7.7, while in lake waters, DO is higher, i.e., 8.8 to 9.9. This lower DO concentration exhibited incoming contaminants that originate in anthropogenic sources transported via river flow. High pH and DO levels also indicate a high level of primary production of algae blooms from the surroundings [15]. The green color evidences this in the lake waters.

3.2. Heavy Metal Concentrations in water and sediments

Figure 2a and figure 2b show heavy metal concentrations in waters and sediments. The Cd, Hg, Pb, and Cu metals concentrations have exceeded the quality standard in almost all sampling locations in the waters. In contrast, the metal concentrations in sediments have not exceeded the quality standard except for Hg.

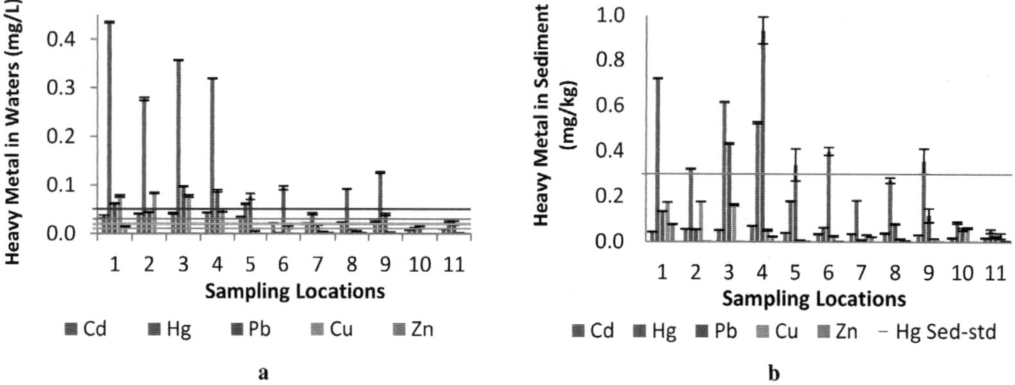

Figure 2. Heavy Metal Concentrations a. in water b. in sediment, sampling locations: 1. Fish Cage, 2. Endemic Fisheries, 3. Dense Settlement, 4. Hydropower, 5. Batang Antokan River, 6. Batang Kularian River, 7. Jembatan Ampang River,8. Batang Asam River, 9. Batang Maransi River, 10. Batang Kurambit River, 11. Bandar Ligin River.

Aside from Zn, the Hg, Pb, Cd, and Cu lake water concentrations exceeded the quality standards in every sampling location, in decreasing order of Hg>Pb>Cd>Cu. Concentrations of Hg metal ranged between 0.002 and 0.435 mg/L, with the highest concentrations at the location of the fish cage. The cage area is the centre of dense fish farming activity, where there are also medium density settlements and traditional markets. The primary sources of Hg contamination are mining activities, industrial activities, wastewater, atmospheric deposits, etc. [16]. Aside from anthropogenic sources, weathering of Hg-carrying minerals in the soil also adds to natural Hg input [16]. The metal concentration in the lakeshore area is higher than the middle of the lake. Various activities around the lake and the vicinity of the contaminants increase the high concentrations of heavy metals. Apart from Hg; Pb, Cd, and Cu metals were not detected in the middle of the lake, exceeding the quality standard. Hg metal in the middle of the lake indicates that the Hg metal concentration is evenly distributed in the waters of Lake Maninjau. Pb concentrations in lake waters range between 0.0005-0.0973 mg/L, with the lowest concentration at the Bandar Ligin river and the highest concentration near residential areas. The high lake water concentration of Pb has resulted from oil and speedboat fuel spills. The primary source of Pb metal pollution stems from coal, fuels, and leaded gasoline being released into the water [17]. Metal concentrations entering the lake from the river flow are also dominated by Hg and Pb metals. Consequently, metal concentration in lake waters is higher than the metal concentration in river waters. The highest metal concentration comes from the Batang Maransi River, which is located in Nagari Bayur. There are hotels, markets, agriculture, and settlements of moderate density along the river, which contribute to heavy metals in the river and water sediments.

Heavy metal concentrations are usually higher in sediments than in water. Based on the contaminated soil standard, Hg metal concentration is excessive in all sampling locations, with concentrations ranging from 0.048-0.720 mg/Kg. This shows that Hg metal contamination is evenly spread in both the waters and sediments throughout the waters of Lake Maninjau. High heavy metal levels in sediment are frequently ascribed to anthropogenic influences instead of sediment being naturally enriched by geological weathering [2]. Hg metal content is higher in comparison to previous studies, namely 0.0005-

0.0591 mg/kg [18]. The high concentration of metals in sediments results from heavy metals combining with organic materials into metal complexes. Most remain relatively stable in the sediment [19].

3.3. Coefficient correlation of heavy metals in waters and sediments

The correlation of heavy metals in sediment and waters is shown in table 1. The association of heavy metals Cd, Hg, Pb, Cu, and Zn in both mediums ranges between 0.8767-0.9761, which indicates that there is a strong relationship between them. However, variation across components depends on an element's chemical form or speciation. These components include mobility, transportation, and partitioning of metals and metalloids in the environment (e.g., sediment, soil, water, air) [6]. Heavy metals change and partition into various physicochemical forms and phases [6]. Metal in the particulate phase can be absorbed into the water column or precipitate into the sediment.

Conversely, metal can also be released from the deposit into the water column by resuspension [17]. Heavy metals are unable to be removed from the water and therefore accumulate in sediments. Heavy metals can be distributed in sediments as the exchange-, organic matter- or residual-bound species [20].

Table 1 Correlation of Heavy Metals in Waters and Sediments.

No.	Heavy metals	Coefficient of Determination (R2)	Correlation Coefficient (r)
1.	Cadmium (Cd)	0,8910	0,9439
2.	Mercury (Hg)	0,9155	0,9568
3.	Lead (Pb)	0,7686	0,8767
4.	Coppper (Cu)	0,9528	0,9761
5.	Zinc (Zn)	0,8991	0,9482

3.4. Pollution Index

The Pollution Index (PI) is used for determining pollution levels relative to the parameters of allowable water quality. Figure 3 shows the PI value distribution in every sampling location. The PI values ranged from 2.66-9.32 and met the moderate and light-polluted categories based on the Regulation of State Minister for the Environment. The Batang Antokan River, the Ampang Bridge River, the Batang Asam River, and the Batang Maransi River are included in the moderate polluted category with a PI 5.0 to 10. Contrastingly, the other three rivers remain in the lightly polluted category with a PI>5.0. A higher PI value is a result of many polluting activities occurring around the Maninjau river and lake. The increase in metal concentration results from anthropogenic activities such as agricultural waste, household waste, and aquaculture activities [21]. Lake Maninjau is almost in the same pollution category as Lake Tempe in South Sulawesi. The Tempe Lake pollution index value is included in the medium polluted category in the range of 5.45-6.65 [22]. Lake Tempe's primary water source is several local rivers that are subject to many human activities, including transportation, bathing, washing, farming, etc. Similar to Lake Maninjau, the highest pollution source comes from aquaculture, water transportation, and settlements.

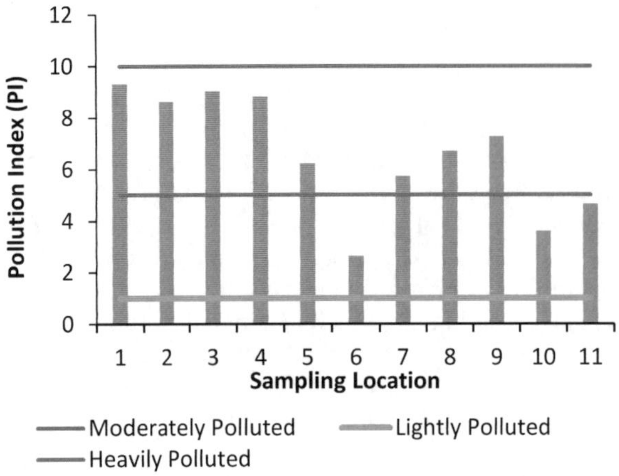

Figure 3. Pollution Index Distribution sampling locations: 1. Fish Cage, 2. Endemic Fisheries, 3. Dense Settlement, 4. Hydropower, 5. Batang Antokan River, 6. Batang Kularian River, 7. Jembatan Ampang River,8. Batang Asam River, 9. Batang Maransi River, 10. Batang Kurambit River, 11. Bandar Ligin River.

4. Conclusion

Cd, Hg, Pb, Cu concentrations in the waters in every sampling location exceeded the "Government Regulation of Class 2" standards, apart from Zn. However, only heavy metal Hg in sediment exceeds the quality standard in all sampling locations. This significant correlation between heavy metals and sediments indicates mutual interaction. The pollution index is based on environmental parameters (temperature, DO, pH) and heavy metals from 8 of 11 sampling locations, which were in the medium polluted category, which indicates that the entire lake area is contaminated, particularly by heavy metals. The lake's activity, primarily from aquaculture, water transportation, and settlements, has affected the water and sediment quality of the lake. Therefore, it is necessary to continuously monitor the quality of the lake and improve wastewater management of the activities occurring around the lake to control heavy metal contamination.

Acknowledgments

This research received support from the "Directorate General of Research and Development Strengthening, at the Ministry of Research and Higher Education of the Republic of Indonesia," through the scheme Cluster of Accelerated to Professor Research-Publication No. T/12/UN. 16.17/PT.01.03/IS-PTU-KRP2GB/2020.

References

[1] Komala P S, Nur A and Nazhifa I 2019 Distribution of organic contamination based on depth stratification in Maninjau Lake, Indonesia *IOP Conf. Ser. Mater. Sci. Eng.* **602** 9–16
[2] Goher M E, Farhat H I, Abdo M H and Salem S G 2014 Metal pollution assessment in the surface sediment of Lake Nasser, Egypt *Egypt. J. Aquat. Res.* **40** 213–24
[3] Monroy M, Maceda-Veiga A and de Sostoa A 2014 Metal concentration in water, sediment and four fish species from Lake Titicaca reveals a large-scale environmental concern *Sci. Total Environ.* **487** 233–44
[4] Pintilie S, Brânză L, Beţianu C, Pavel L V, Ungureanu F and Gavrilescu M 2007 Modelling and simulation of heavy metals transport in water and sediments *Environ. Eng. Manag. J.* **6** 153–61
[5] Yin H, Gao Y and Fan C 2011 Distribution, sources and ecological risk assessment of heavy metals in surface sediments from Lake Taihu, China *Environ. Res. Lett.* **6**

[6] Azeez P A, Anjan Kumar Prusty B and Jagadeesh E P 2006 Chemical speciation of metals in environment, its relevancy to ecotoxicological studies and the need for biosensor development *J. Food, Agric. Environ.* **4** 235–9
[7] El Nemr A, Khaled A, Moneer A A and El Sikaily A 2012 Risk probability due to heavy metals in bivalve from Egyptian Mediterranean coast *Egypt. J. Aquat. Res.* **38** 67–75
[8] Raya Consult 2016 *Detail Desain Pemanfaatan Pengembangan Danau Maninjau Provinsi Sumatera Barat* (Padang)
[9] Badan Standardisasi Nasional 2008 SNI 6989.57:2008 Mengenai Air dan Air Limbah-Bagian 57:Metode Pengambilan Contoh Air Permukaan.
[10] Ohio Environmental Protection Agency 2001 *Sediment Sampling Guide and Methodologies* (Columbus, Ohio)
[11] Peraturan Pemerintah Republik Indonesia Nomor 82 2001 *Tentang Pengelolaan Kualitas Air dan Pengendalian Pencemaran Air*
[12] Peraturan Pemerintah Republik Indonesia 2014 *Pengelolaan Limbah Bahan Berbahaya dan Beracun* (Indonesia)
[13] APHA 2005 *Standard methods for the examination of water and wastewater* (Washington, D. C.: APHA-AWWA-WEF)
[14] Keputusan Menteri Negara Lingkungan Hidup 2003 *Keputusan Menteri Negara Lingkungan Hidup Nomor 115 Tentang Pedoman Penentuan Status Mutu Air*
[15] Komala P S, Primasari B and Ayunin Q 2020 The Influence of the Physicochemical Parameters on the Ortho Phosphate and Total Phosphate Concentrations of Maninjau Lake *J. Phys. Conf. Ser.* **1625**
[16] Driscoll C T, Mason R P, Chan H M, Jacob D J and Pirrone N 2013 Mercury as a global pollutant: Sources, pathways, and effects *Environ. Sci. Technol.* **47** 4967–83
[17] Hu J, Zhou S, Wu P and Qu K 2017 Assessment of the distribution, bioavailability and ecological risks of heavy metals in the lake water and surface sediments of the Caohai plateau wetland, China *PLoS One* **12** 1–15
[18] Syawal M S, Wardiatno Y and Hariyadi S 2016 Pengaruh Aktivitas Antropogenik Terhadap Kualitas Air , Sedimen dan Moluska di Danau Maninjau Sumatera Barat Jurnal Biologi Tropis, Januari 2016: Volume 16 (1):1-14 ISSN: 1411-9587 **16** 1–14
[19] Ying X and Lu J L 2010 Study on form distribution and correlation of heavy metals in the sediment of urban water *2010 4th Int. Conf. Bioinforma. Biomed. Eng. iCBBE 2010* 1–4
[20] Kamaruzzaman Y, M.A. Z and Akbar J 2020 A review on the accumulation of heavy metals in coastal sediment of Peninsular Malaysia *Ecofeminism Clim. Chang.* **1** 21–35
[21] Syandri H 2015 Heavy Metals in Maninjau Lake , Indonesia : water column , sediment and biota **3** 273–8
[22] Yani A, Amin M, Rohman F, Suarsini E and Haerunnisa 2019 Water quality and pollution index of lake tempe in south Sulawesi, Indonesia *Pollut. Res.* **38** 568–74

IOP Conf. Series: Materials Science and Engineering 1041 (2021) 012032 doi:10.1088/1757-899X/1041/1/012032

Study on solid waste generation in Lake Toba area

I Suryati[1], H Khair[1*], R L Siregar[1], H Sibagariang[2] and R Utami[1]

[1]Environmental Engineering, Universitas Sumatera Utara, Medan 20155
[2]Dinas Lingkungan Hidup Provinsi Sumatera Utara

Corresponding author: hafizhul@usu.ac.id

Abstract. Girsang Sipangan Bolon and Merek do not have basic data for solid waste management design. This research aimed to analyze the generation, composition of domestic and non-domestic waste. Sampling methods are determined based on SNI 19-3962-1994. The average amount of waste generation in units of volume for the area divider factor and the number of inhabitants for Girsang Sipangan Bolon and Merek are 92,771 and 87,227 m3/day, respectively. This study found the composition of domestic waste in Girsang Sipangan Bolon consists of 65% organic waste; 7% paper; 17% plastic; textiles 6%; rubber 0%; wood 0%; glass 2%; 0% metal and 3% other waste. The composition of non-domestic waste consists of 61% organic waste; 12% paper; 15% plastic; textile 1%; rubber / leather 0%; wood 1%; glass 0%; metal 1% and other waste 9%. In Merek, the composition of domestic waste consists of 69% organic waste; paper 14%; 8% plastic; textile 1%; rubber / leather 0%; wood 0%; glass 2%; 0% metal and 6% other waste. The composition of non-domestic waste consists of 55% organic waste; paper 14%; plastics 21%; textile 0%; rubber / leather 0%; wood 1%; glass 3%; 0% metal and 6% other waste.

1. Introduction

Lake Toba is the largest lake in Indonesia and the largest volcanic-tectonic caldera (giant volcano crater) worldwide. It was formed about 75,000 years ago due to volcanic eruptions and tectonic soil. Lake Toba has a length of 87 km northwest-southeast with a width of 27 km, an altitude of 904 meters above sea level, and the lake's most profound depth of 505 meters. This area covers part of 8 (eight) districts, namely Samosir Regency, Toba Samosir Regency, Dairi Regency, Karo Regency, Humbang Hasundutan Regency, North Tapanuli Regency, Simalungun Regency, and Pakpak Bharat Regency [1].

Lake Toba area is a strategic national tourism area. A national strategic has a significant national influence on state sovereignty, defense and national security, economic, social, cultural, and environmental, including areas that have been designated as a world heritage [2]. In order to realize Lake Toba as a tourism spot that attracts local, national, and international level tourists, it needs to be designed a sustainable layout. Lake Toba Spatial Plan, set up by Presidential Regulation No. 81 of 2014, mentions the purpose of the Lake toba national strategic areas spatial arrangement is to realize a safe, comfortable, productive, and sustainable lake area community. One of the policies formulated to achieve that goal is to construct waste management systems (articles 24, paragraphs 1, and 2) [3]. The existence of tourism areas has a positive and negative impact. The positive impact is in the form of the addition of local native income or economic improvement. Meanwhile, the negative impact on tourist areas is the decrease in environmental quality, such as water pollution, air pollution, and increasing amounts of waste.

A study about waste management has been done in a country like Tunisia [4], Ukraine [5], China [6], Pahalgam[7], and Vietnam [8]. In contrast, a study about waste management has been done only a few in Indonesia, especially in North Sumatera province. The research about waste management in Indonesia has been performed by some researchers like Chrisdianti [9]. It showed that the most dominated waste in Bromo-Tengger National Park is plastic, and travelers and vendor owners throw organic waste. This study parallels, according to Megi [10]. In Singkarak Lake, the most extensive waste composition is dominated by food waste (25.488 %) and plastic waste (24,834 %). The waste composition in Pariaman city is dominated by food waste amount 30,86% and plastic, 26,31%[11].

The environmental problem in the Lake toba tourism area is unwell managed waste. It could be seen through existing conditions around Lake toba, such as in sub-districts of Girsang Sipangan Bolon and Merek. Both districts still used the old waste management system, collecting, transporting, and dumping the waste. Besides temporary waste disposal, the limited and final waste process cannot be used in the long term. Related study of waste density, composition, a waste management plan in the Merek sub-district in Karo regency, and the Girsang Sipangan Bolon sub-district in Simalungun regency has not been done comprehensively. It is because there has not been any waste management following standard in the location. Data on generation and composition waste will affect and change the strategy and waste management system planned in the Lake Toba area. It can be found from the waste collection, transport system setting, waste management system design, landfill design, recycling program, and equipment that needed. This study is intended to figure out the generation and composition of waste in the Toba area.

2. Materials and methods

2.1. Study area

The determination of the location of the study under the directive in Presidential Regulation of the Republic of Indonesia No. 81 of 2014 on the Spatial Plan of Lake Toba area and its surroundings states that the body of Lake Toba is surrounded by 7 (seven) districts and 28 (twenty-eight) sub-districts. The Merek sub-district selection is representative of the Karo Regency, and the Girsang Sipangan Bolon sub-district representing the Simalungun Regency.

2.2. Data collection

The primary data are embossing data, composition, garbage characteristics, and questionnaires about household and non-household waste in the Girsang Sipangan Bolon sub-district and Merek Sub-district. Primary data is obtained by taking and measuring directly in the field and conducting surveys and interviews with the people who enter the sample criteria. This secondary data is population data in 2018 in the Girsang Sipangan Bolon sub-district and Merek Sub-district, map of Girsang Sipangan Bolon sub-district and Merek Sub-district, regional development direction and policies, institutional data, and others of general nature (literature study).

2.3. Calculation methods

The measurement and calculation of waste arising samples are conducted following SNI 19-3964-1994[12]. Based on the SNI, there was a sampling number for these areas as many as 35 locations representing housing, shops, schools, markets, offices, roads, hotels, restaurants, and other public facilities. Sampling was carried out for three consecutive days.

Measurement of the waste composition was carried out based on SNI 19-3964-1994. The waste component is directly measured in the field by separating based on its components, namely food waste, paper, wood, fabric-textile, rubber-leather, plastic, non-ferrous metal, glass, and others (e.g., soil, sand, stone, ceramic). Percent composition of waste was obtained from the weight of each component of waste that has been weighed and divided by the total weight of the overall waste. Calculation of specific weight was obtained by comparing the weight of waste by its volume (kg/L). The measurement of generation based on SNI 19-3964-1994 includes measurement of the weight and volume of waste from each sample based on the sources. Unit of volume generation is in L/cap./day and L/m2/d, while weight generation unit is kg/cap./day and kg/m2/d.

$$Generation\,formula\,per\,day = \frac{weight\ or\ volume\ of\ waste\ (kg)}{area\ or\ number\ of\ waste\ producer\ (m^2 or\ person)}$$

Solid waste composition determined by measurement of the weight of each waste component compared with the total weight of the waste.

$$Composition\,formula\ (\%\ weight) = \frac{weight\ or\ waste\ component\ (kg)}{total\ weight\ of\ waste\ (kg)} \times 100\%$$

3. Result and discussions

3.1. Waste generation

In this study, the calculation of the arising and composition of garbage was guided in SNI 19-3964-1994 [12]. Waste collection in the Girsang Sipangan Bolon sub-district and Merek Sub-district can be found in table 1.

Table 1. Waste generation.

No	Source	Girsang Sipangan Bolon		Merek		SNI 19-3983-1995		Unit
		Weight(kg)	Volume (l)	Weight (kg)	Volume (l)	Weight(kg)	Volume (l)	
1	Household	0,414	3,02	0,166	2,93	0,625 - 0,70	2,5 – 2,75	person/day
2	Shops	1,727	24,19	0,150	20,81	0,15 - 0,35	2,5 – 3,00	officer/day
3	Restaurant	1,404	11,39	1,176	16,50	-	-	officer/day
4	Hotel	0,190	3,37	0,460	6,19	-	-	room/day
5	School	0,006	0,05	0,009	1,17	0,010 - 0,020	0,10-0,15	student/day
6	Market	0,106	0,13	0,370	0,20	0,10 - 0,30	0,20 -0,60	m²/day
7	Port	0,011	0,47	-	-	-	-	m²/day
8	Tourism area	-	-	0,004	0,09	-	-	m²/day
9	Bank	0,026	4,44	0,268	2,24	0,025 - 0,10	0,50-0,75	employee/day
10	Road	0,010	0,29	0,016	0,33	0,010 - 0,050	0,10-0,15	m/day
11	Office	0,175	7,20	0,032	2,80	0,025 - 0,10	0,50-0,75	employee/day

Based on table 1, the amount of waste for household scale in Girsang Sipangan Bolon amounted to 0.414 kg/person/day, and Merek Sub-district amounted to 0.166 kg/person a day. Compared to SNI 19-3983-1995[12], the generation of household waste for the Girsang Sipangan Bolon sub-district and Merek Sub-district is still below the standard. Population and community activity levels could affect the result. The number of household waste in the area around the Toba area is higher than the Singkarak Lake tourism area, whose waste is only 0.021 kg/person/day [10].

Meanwhile, for non-household waste generation, the highest amount of waste generated is in Girsang Sipangan Bolon, namely shops and restaurants. The largest non-household waste generation in Merek comes from hotel activities. Merek area is close to tourist spots, so tourist numbers dramatically influence it. This result is in line with research conducted by Kasam [13] in Gunungpring tourist park that is an immense waste pile on Sunday, about 155.1 kg, and the smallest on Wednesday is about 90 kg because the number of visitors on holidays is more than a typical day.

Another study conducted by Subzar in Pahalgam, India, found that non-household waste in the tourist area is more significant than household waste [7]. The results showed that 74% of non-household waste came from hotels and restaurants.

3.2. *Waste composition*

The result of the waste composition from this study can be seen in figure 1 and figure 2.

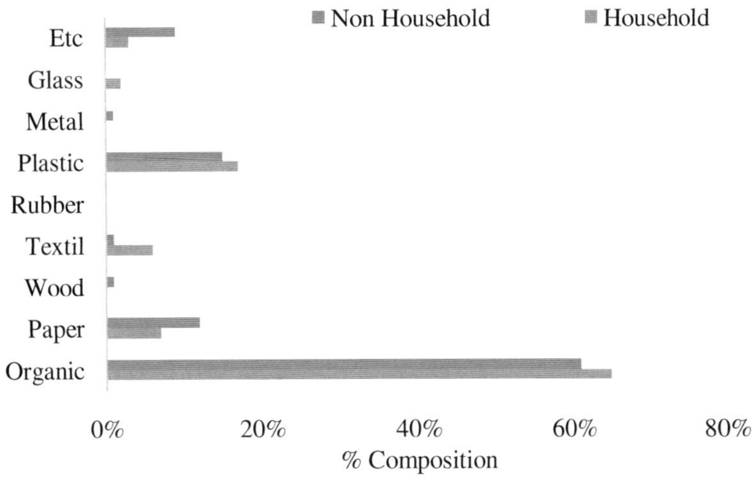

Figure 1. Composition of household and non-household waste in Girsang Sipangan Bolon Subdistrict

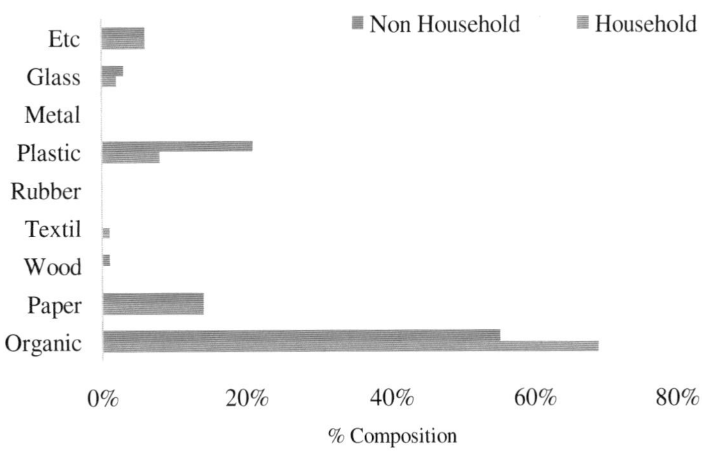

Figure 2. Composition of household and non-household waste in Merek Subdistrict

Based on the composition of waste from both, the dominant waste is organic waste from household and non-household sectors ranging from 55 - 70%. This figure follows data from KLHK [14], where the most extensive waste composition in Indonesia is the organic waste of 60%.A study on waste generation and composition conducted in the provincial capital, Medan City, also shows a similar composition [15]. In line with research conducted by Minh [8], the largest composition in Hoi An area, Vietnam's tourist area, is an organic waste by 42%. Meanwhile, Kasam's research in the Gunungpring park area is plastic waste, a recyclable paper by 49.84 %, organic waste that can be composted by 29.53%, and the rest discarded [13].

When considering the composition of the dominant waste is organic waste, composting is to reduce waste generation. The composting process can be done in an integrated manner by making compost houses in each district. The composting process can be done aerobically with either a windrows or a mechanical system. Windrows composting has the principle of flowing air using an air tunnel at the beginning of composting. Windrows systems need to pay attention to temperature and humidity by manually reversing[16].

Meanwhile, for the mechanical system, the composting process is carried out in a reactor equipped with a mechanical reversal. Usually the amount of compost produced is 30% of the raw material for compostable organic waste. The quality of the compost produced must of course comply with SNI 19-7030-2004 standards regarding the specifications of compost from domestic organic waste with good compost requirements, namely a water content of <50%, Nitrogen content <0.4, C-organic content> 9.8, Calcium content <25.50 and pH ranges from 6.8 to 7.49[17].

4. Conclusions

The results of the study of waste generation and composition around the Lake Toba Tourism Area, namely Girsang Sipangan Bolon District and Merek District, obtained waste generation for household waste ranging from 0.166 to 0.414 kg/person/day. In contrast, the dominant waste generation from non-household waste originated from hotels, restaurants, and shops. The dominant composition of waste in the study area is organic waste ranging from 55 to 69%, plastic waste 8-21%, and paper waste 7-14%. Waste handling based on the composition of waste in the study area that can be done is composting, recycling, and the rest is disposed of in the landfill.

Acknowledgments

The authors gratefully acknowledge that the present research is a collaboration between the Environmental Service of North Sumatra Province and the author as part of the experts of the Integrated Waste Management System Planning in the Lake Toba Area in 2019.

References

[1] Dinas Lingkungan Hidup Provinsi Sumatera Utara 2019 *Perencanaan Sistem Pengelolaan Sampah Terpadu di Kawasan Danau Toba*

[2] Peraturan Presiden (PERPRES) 2014 *Rencana Tata Ruang Kawasan Danau Toba dan Sekitarnya* (Indonesia)

[3] Peraturan Presiden 2014 *Peraturan Presiden No. 81 Tahun 2014 tentang Rencana Tata Ruang Kawasan Danau Toba dan Sekitarnya* (Indonesia)

[4] Chaabane W, Nassour A, Bartnik S, Bünemann A and Nelles M 2019 Shifting towards sustainable tourism: Organizational and financial scenarios for solid waste management in tourism destinations in Tunisia *Sustain.*

[5] Murava I and Korobeinykova Y 2016 The analysis of the waste problem in tourist destinations on the example of Carpathian region in Ukraine *J. Ecol. Eng.*

[6] Dong R, Chen X, Ding Z, Guo H, Li X, Lian X and Zhao X 2019 Study on the Solid Waste Properties and Disposal Methods of Island at Home and Abroad *IOP Conference Series: Earth and Environmental Science*

[7] Bashir S and Goswami S 2016 Tourism Induced Challenges in Municipal Solid Waste Management in Hill Towns: Case of Pahalgam *Procedia Environ. Sci.*

[8] Hoang M G, Fujiwara T and Phu S T P 2017 Municipal Waste Generation and Composition in a Tourist City - Hoi An, Vietnam *J. JSCE* **5** 123–32

[9] Chrisdianti 2017 *Alternatif Pengelolaan Sampah Di Kawasan Wisata Laut Pasir Taman Nasional Bromo Tengger Semeru* (Institut Pertanian Bogor)

[10] Megi P 2019 *Studi Timbulan, Komposisi, Karakteristik dan Potensi Daur Ulang Sampah, Kawasan Wisata Danau Singkarak, Sumatera Barat* (Universitas Andalas)

[11] Aziz R and Mira 2019 Study of recycling potential of solid waste of tourist area in Pariaman City *IOP Conf. Ser. Mater. Sci. Eng.* **602** 012059

[12] Badan Standarisasi Nasional 1994 *Metode pengambilan dan pengukuran contoh timbulan dan*

komposisi sampah perkotaan (Indonesia: Badan Standarisasi Nasional)

[13] Kasam, Iresha F M and Setyoadi N H 2019 Management of Municipal Solid Waste in Religious Tourism Park Based on Reduce, Reuse and Recovery: An Indonesian Attraction Case Study *MATEC Web Conf.*

[14] Kementerian Lingkungan Hidup dan Kehutanan 2017 Komposisi Sampah di Indonesia Didominasi Sampah Organik *Katadata*

[15] Khair H, Rachman I and Matsumoto T 2019 Analyzing household waste generation and its composition to expand the solid waste bank program in Indonesia: a case study of Medan City *J. Mater. Cycles Waste Manag.*

[16] Kustiasih T, Darwati S and Arventi 2018 *Penerapan Teknologi Pengolahan Sampah di Kawasan Wisata Pantai* (Pusat Penelitian Dan Pengembangan Perumahan & Permukiman)

[17] Badan Standarisasi Nasional Indonesia 2004 *Spesifikasi Kompos dari Sampah Organik Domestik SNI 197030-2004*

The effect of chopping raw material on composting result with the biopore infiltration hole method

Y Ruslinda[1], **R Aziz**[1], **N Sari**[1] **and L S Arum**[1]

[1]Environmental Engineering Department, Faculty of Engineering, Universitas Andalas, Limau Manis, Padang City 25163, West Sumatera, Indonesia

Corresponding author: yenni@eng.unand.ac.id

Abstract. This research aims to analyze the effect of chopping raw material on compost quality and quantity that was made with the biopore infiltration hole method. The composting was done in duplo on the backyard with clay soil type, and the water infiltration rate was 0.3 cm/hour. The holes were made with 10 cm of diameter, 100 cm of depth, and the distance between holes were 50 cm. There were four composition variations on this research i.e 50% yard waste and 50% food waste with and without chopping, 70% food waste and 30% yard waste with and without chopping. The chopping was using an organic waste chopping machine so that the waste size became 0.3-1.5 cm, while the size waste without chopping was ≥ 3 cm. The compost analysis was done toward maturity, compost quality (physical elements and macro elements), and compost quantity. The composting process took 65-80 days to get riped. The best variation was chosen by the scoring method. The result showed that compost with 50% yard waste and 50% food waste was the best variation because the compost quality has met the standard for physical elements and macro elements, it produced more compost quantity. The chopping process could speed up the composting time until 15 days.

1. Background

Organic waste consists of food waste, paper, carton, plastic, and yard waste [1]. Organic waste has the most significant percentage of whole waste production in Indonesia. Based on data from the Ministry of Environment, 57% of total waste is organic waste[2]. The amount of organic waste that can be processed is only 14%, and 66% ends up in the landfill, while the rest are in the environment and cannot be handled. The most widely used, biodegradable organic waste processing is composting. Composting is a natural process of decomposition of organic materials or decomposition by microorganisms on biodegradable organic materials. Composting converts biodegradable organic materials into biological materials that contain humus substances [1].

One of the composting methods is the biopore infiltration hole method. Biopore infiltration hole has a vital function in assisting waste processing. It has many advantages to the environment, such as improving the soil ecosystem, absorbing water, preventing flood, increasing groundwater reserves, facilitating waste handling, maintaining cleanliness, and turning waste into compost [3]. The biopore infiltration hole method is different from the other ground composting method. It uses soil as a medium to convert waste into compost. In this method, waste is put into a hole with a diameter 10-30 cm and 50-100 cm depth, made in the house yard or garden. After that, the waste in the hole is eaten by biota in the ground. That biota could process the waste by decreasing its size and mixing it with a microbe to speed up the composting process naturally [4].

From the various benefits of this method above, many efforts have been made by the Indonesian Government to promote biopore infiltration hole method as a filler for groundwater reserves and as a compost maker to reduce the amount of domestic organic waste. However, in its implementation, people commonly put the waste directly into the hole without any treatment or addition like activator or starter addition, so that the result was not optimal and required a longer composting time. Some factors could affect the process, such as the C/N ratio, the size, and composition of waste, humidity, temperature, and microorganisms [3]. In this research, the wastes were chopped first before being composted. So, the waste size became smaller and even. The optimal size waste for composting are 2.5-7.5 cm [1]. This research's expected result is that there is a recommendation for waste treatment before the composting process. The quality and quantity of compost can be improved, and the composting time can be shortened.

2. Material and methods

2.1 Preparation Stage

The preparation for this research were:

1. Choosing the composting location

The composting process was located in the yard/garden, and soil type and water infiltration rate were tested to support the data analysis.

2. Preparation of compost raw materials

The waste used in this research was domestic organic waste, which was food waste, and yard waste. The food waste used was biodegradable food leftovers such as rice, meats, fruits, and vegetables, and the yard waste that was used consists of leaves. This study's percentage of food waste refers to Eryuningsih's research [5] (look table 1). This composition was used for the food waste that was put into the biopore hole. In this research, half of the waste were chopped, and the other half were not chopped.

Table 1. The Composition of Domestic Food Waste.

Waste Component	Composition Percentage (% weight)
Vegetable	54.40
Fruits	28.55
Rice	11.77
Meats	5.27

Source:[5]

3. Determination of research variations

This research was to analyze the effect of chopping raw material on compost that made used biopore infiltration hole method. So, the variations of this research were with and without chopping. The chopping process used an organic waste chopping machine so that the waste is 0.3-1.5 cm in size. Meanwhile, waste without chopping was >3 cm in size. The composting was done in two composition variations: 50% of food waste with 50% yard waste and 70% food waste with 30% yard waste. These two variations were chosen based on research about the composition of domestic waste, and the ministry of environment [2] said that the percentage of food composition was about 50-70%. In contrast, the percentage of yard waste was about 30-50% [6]. Each variation was done in duplo. Variation for this research showed in table 2.

4.The biopore hole digging

Eight biopore holes were made because there were four variations, and all of that was done in duplo. The holes were made using biopore hole drill. It was made with 10 cm of diameter, 100 cm of depth, and the distance between holes are 50cm [7]. The design of the biopore infiltration hole is shown in figure 1.

Table 2. Research Variation.

Variation	Chopping	Raw material
1	Unchopped	50% FW[a]: 50%YW[b]
2	Chopped	50% FW[a]: 50%YW[b]
3	Unchopped	70% FW:[a] 30%YW[b]
4	Chopped	70% FW[a]:30%YW[b]

[a] YW = Yard Waste
[b] FW = Food Waste

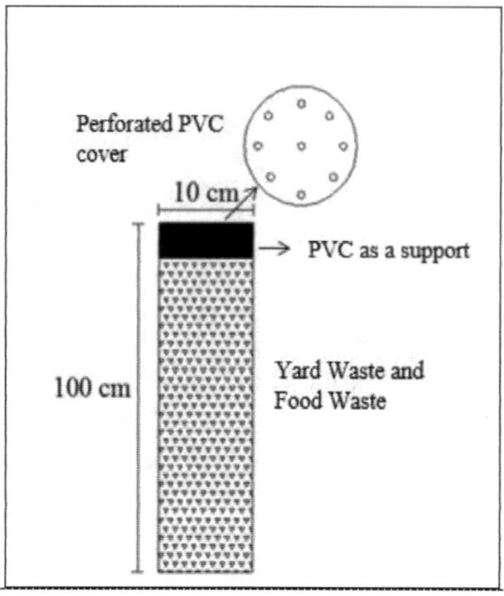

Figure 1. Biopore Infiltration Hole Design

5. Water infiltration rate and soil type test
Testing the water infiltration rate was done by inserting water until it was full into the biopore hole and leaving it for 24 hours, then the measurements were taken with a ruler so that the water infiltration rate was 0.3 cm/hour. The soil types were obtained from research by Liliwarti and Satwarniratwhich conducted soil types testing at the Faculty of Engineering, Andalas University. From the result of the analysis, the type of soil is clayey loam[8].

2.2Research Stage

The steps for the main research were:

1. Waste analysis
Before the composting process, some analysis was done, including temperature analysis, pH, water content, and the C/N ratio. This step was done to determine if the waste has met the standard for composting or not. The water content for waste before composting is between 50-60%, and the optimum C/N ratio is between 25-50 [9]. The other parameters, like temperature and pH based on Sudaryono said that the optimum temperature for waste before composting is between 25-28°C and pH, are between 6-7[10]. If the waste has met the standard so it could be the compost material.

2. Inserting compost into the bipore hole
The next step was putting the waste into the biopore hole according to each variation. Afterwards, the holes were closed with PVC to strengthen the holes so as not to slide and prevent animals or unwanted objects from entering the holes.

3. Compost maturity test
The observation of compost maturity was done every day for temperature, pH, compost volume reduction rate, humidity, color, texture, and odor, according to SNI 19-7030-2004[11]. The observation was also done for composting time.

4. Compost removal

When the compost was ripe, the compost is removed from the hole. Afterwards, compost was dried in the shade or a place that was not exposed to direct sunlight.

5. Compost quality test
Compost quality was tested based on SNI 19-7030-2004[11] about compost and domestic organic waste. Parameters that were tested are physical elements (water content, temperature, pH, color, and odor) and macro elements (carbon (C), nitrogen (N), C/N ratio, phosphor, and kalium).

6. Compost quantity test
The compost quality test was done with weighing the solid compost produced from each variation.

2.3Data Processing and Analysis

1. Analysis of compost maturity, quality, and quantity
Analysis of compost maturity, quality, and quantity was done by comparing the data with the standard. Raw material data were comparing with Tchobanoglous [9] standard, quality, and maturity compost data were comparing with SNI 19-7030-2004[11], and for the quantity were analyzed by the weight percentage of compost.

2. Determining the optimal composting variation
Determining the optimal composting variationwas done by a scoring method. The scoring system includes three criteria. Those criteria are:

a. Criterion 1: Score 1 was given to the variation that satisfies the standard [11]. This score was for quality and maturity parameter;
b. Criterion 2: Score0 was given the variation that does not satisfy the standard [11]. This score was for quality and maturity parameter;
c. Criterion 3: This was for the parameter which has no standard. The maximum score (4) was given the best variation, and the minimum score (1) was given the variation not right. This score was for composting time and compost quantity parameter.

The best variation was the variation that has the highest score.

3. Results and discussion

3.1 Environmental Conditions

The environmental conditions could affect the composting process. Hence, before starting the composting, it was important to check and analyze the composting place condition. The soil analysis result showed that the soil type is clay, and the water infiltration rate was 0.3 cm/hour, which was a slow category. Soil texture has an important role in efforts to infiltrate rainwater. Based on Brata dan Nelistya, a free soil texture has more pore than the clay texture[7]. Therefore, the sandy soil texture will absorb water faster than the clay texture. Based on Environment Ministry Rules No. 12, 2009 about Rainwater Utilization said that the soil texture for biopore infiltration hole should have the water infiltration rate $\geq$ 2.0 cm/hour. For details about the relationship between soil texture with water infiltration rate shown by table 3

Table 3. Relation Between Soil Texture and Water Infiltration Rate.

Soil Texture	Water Infiltration Rate (cm/hour)	Category
Sandy Clay	>3.3	Faster
Clay	2.5-3.3	Fast
Dusty Clay	1.5-2.5	Medium
Clay Loam	0.25-1.5	Slow

Source: [7]

Soil texture and water infiltration rate could affect the composting process in biopore infiltration hole. The rain intensity could also affect biopore infiltration hole because of the infiltration hole made on open land, which was directly exposed to rainwater. High rain intensity will make the holes fulled by water, and the humidity will increase, then it could affect the composting process.This research was done in the transition season from the rainy season to the dry season. So that in this research, all the conditions and rain duration that happened during the composting process were noted. The result was there are 38 rainy days from 60-80 days of composting. Based on data from the Meteorology and Geophysics Agency of Padang City, the rain that happened were classified as light rain until medium rain.

3.2 Raw Material Analysis

The raw material analysis was done to know if the raw material has qualified as compost material depends on the standard. The parameter that was analyzed includes temperature, pH, humidity, and the C/N ratio. The result of the raw material analysis is shown in table 4. The result showed that all parameters has met the standard so that the raw material can use for compost. Food waste has a lower temperature, pH, and C/N ratio than yard waste, but it has higher water content. The more water content in raw material causes the temperature became lower. It caused by food waste has more water content than yard waste, which consists of dry leaves. Based on Tchobanoglous [9], food waste contains 58% water content and water content in yard waste, only 51%. The more water content in raw material caused the initial temperature for raw material close to the water temperature. The pH of raw materials at the beginning of composting was generally acidic to neutral (pH 6-7). This result inline with Kurnia's research, where food waste has a pH of 3 to 6, while yard waste has a pH of 6 to 7[12]. C/N ratio is the critical parameter in the composting process. If the C/N ratio in raw material were high, the microorganisms would lack of N for protein synthesis. The decomposition process became slow, while if the ratio C/N was too low because of too much nitrogen, so the nitrogen would be released to the atmosphere and caused a foul odor [13].

Table 4. Raw Material Analysis

Parameter	Raw Material		Standard (*)	Information
	Yard Waste	Food Waste		
Temperature	28°C	26°C	25-28°C	Qualified
pH	6.6	6.5	6-7	Qualified
Water content	51%	58%	50-60%	Qualified
C/N ratio	40	34.8	25-50	Qualified

Source: *[11]

3.3 Maturity Analysis

During the composting process, the maturity observation was done every day until the compost was ripe. The parameters that measured were temperature, pH, reduction rate, humidity, texture and color, odor, and composting time. Because the research was done in duplo, the analysis results were averaged for each variation in the study. The compost maturity result showed that the compost temperature was in the range 27-28°C and neutral pH (7-7.4). The compost volume reduction was in the range of 32-54%. Texture, color, and odor have resembled soil. Thus, all the composting variation has met the maturity standard based on SNI 19-7030-2004[11]. The temperature less than 30°C or same with soil water temperature, neutral pH, compost reduction in 20-40%, and the texture, color, and odor has resembled soil.

The composting process took time about 65-80 days. The composting time for chopped raw material (variation 2 and 4) faster than unchopped raw material (variation 1 and 3). Composting time for chopped raw material is 65 days, and the unchopped material is 80 days. Chopping the material can speed up the composting process for 15 days. Chopping made the size of raw material smaller and even, making it easier for the decomposition process. Based on Damanhuri and Padmi, the smaller the material's diameter, the larger the surface area, and it made the contact between bacteria and organic matter became better[1]. The smaller the raw material size, so the decay process became faster. The composition of the raw material also affected the maturity of the compost. The results of the analysis of composting maturity is shown in table 5.

Table 5. Maturity Analysis.

Variation	Temperature (°C)	pH	Compost Reduction (%)	Humidity	Texture and Color	Odor	Composting Time (Day)
1	27	7.2	32	Less humid	Soil and black	Soil	80
2	28	7.2	41	Less humid	Soil and black	Soil	65
3	27	7.3	35	Less humid	Soil and black	Soil	80
4	28	7.0	41	Less humid	Soil and black	Soil	65

Composting with biopore infiltration hole method took time a little longer than the other composting method like *Rotary Kiln* and *Takakura* method. This condition was caused by biopore infiltration composting was done on the outdoor, so it was affected by weather conditions. This research was done in the transition season from rainy season to dry season, and 38 rainy days happened during the composting process. It was affected by humidity and composting time. Weather conditions in indoor composting (*Rotary Kiln* and *Takakura*) can be minimized so that the composting time can be faster to 15-30 days. Biopore infiltration composting can be done in the dry season to avoid the holes fulled by water because of rain [14].

3.4 Quality Analysis

The compost quality tested after the compost was ripe and after it was dried and sifted. The parameter for this quality analysis was physical elements and macro elements. The result of physical elements analysis showed that solid compost water content was in the range 39.03 - 45.67%, the temperature in 27-28°C, pH in the neutral range (7.0-7.3) soil texture, the color was black, and the odor had resembled soil. The result for macro elements showed that C-organic was in the range 10-10.02%, N-total in 0.51-0.64%, C/N ratio in 15.59-19.58%, Phosphor in 0.15-0.32%, and Kalium in 6.7-9.6%. All the variation has met the standard according to SNI 19-7030-2004[11]. There was no significant effect between the chopped waste with the unchopped waste in the final result. The chopped waste was more affected by composting time. The chopped waste ripen faster than the unchopped waste. The result of the quality analysis is shown in table 6 and table 7.

Table 6. Physical Element Analysis of Compost Quality.

Variation	Water Content (%)	Temperature (°C)	pH	Color	Odor	Information
1	40.28	27	7.2	Black	Soil	Qualified
2	39.03	28	7.2	Black	Soil	Qualified
3	45.67	27	7.3	Black	Soil	Qualified
4	41.43	28	7.0	Black	Soil	Qualified
Standard[10]	<50%	<30°C	6.8-7.49	Black	Soil	

Table 7. Macro Element Analysis of Compost Quality.

Variation	C-Organic	N Total	C/N Ratio	Phosphor	Kalium
1	10.2	0.55	18.74	0.16	7.3
2	10.0	0.51	19.58	0.15	6.7
3	10.1	0.64	15.59	0.32	9.6
4	10.2	0.59	17.24	0.22	7.1
Standard[10]	9.8-32%	>0.4%	10-20	>0.1%	>0.2%

3.5Quantity Analysis

The quantity compost that produces from biopore infiltration hole composting is shown in table 8. The solid compost produced was about 1.10-1.40 kg, with a percentage between 75% to 81.4%. The compost quantity was affected by the type and composition of raw material. Variation 2 with 50% food waste and 50% yard waste and chopped raw material has the highest quantity than the others.

Table 8. Quantity Analysis.

Variation	Raw Material (Kg)	Solid Compost (Kg)	Quantity Percentage (%)
1	1.4	1.10	78.6
2	1.4	1.14	81.4
3	1.8	1.35	75.0
4	1.8	1.40	77.5

3.6 Selection of the Best Variation

The selection of the best compost variation was done by scoring method. It was included the maturity analysis, quantity, and quality analysis. Based on the scoring criteria that explained in the methodology, the variation with the highest score was variation 2, with 50% chopped food waste and 50% chopped yard waste. Quality and quantity on variation 2 has met the standard [11]. The percentage quantity of compost is 81.4%. It was more than the other variation. Afterwards, the composting time shorter than the other variation with 60 days of composting. The result of the composting score is shown in table 9.

Table 9. Scoring for the Best Variation Selection.

Variation	Compost Maturity	Compost Quality	Compost Quantity	Total Score
1	7	5	3	15
2	8	5	4	17
3	7	5	1	13
4	8	5	2	15

3.7Recommendation

The result of this research shown that the biopore infiltration composting needs 65-80 days of composting. It was caused by the rain that happened during the composting process. The soil type (clay) and the slow water infiltration rate (0.3 cm/hour) were the other reasons the composting process being slow. It took time a little longer to infiltrate the rainwater, increasing the humidity, and the decomposition process became slower.

The purpose of biopore infiltration hole method were to increase groundwater reverse, increase water infiltration rate, increase soil fertility, and produce compost. Therefore, this research was also analyzing the water infiltration rate after composting. The result showed that the water infiltration rate increase to 0.8 cm/hour from 0.3 cm/hour. It showed that the other purpose of this method to increase the water infiltration rate has succeeded. So it could recommend that, if wanted a faster-composting process, so the composting could do in the dry season. In the dry season, rain intensity lower than the rainy season, so the hole was not often waterlogged, and the compost will ripe faster.

4. Conclusion

The conclusion of this research were:

1. The result of compost maturity and quality for all variations has met the standard according to SNI-19-7030-2004 about compost specification from domestic organic waste. The composting time was 65-80 days, and the compost reduction is 32-41%, and the percentage is 75-81.4%. But, the rainy weather that happened for 38 days caused the compost always fulled of water, and the decomposition became slow;
2. Chopping on raw material affected composting time and compost quantity. It can speed up the composting time until 15 days and increase the solid compost quantity about 2.5-2.8%;
3. The scoring result for maturity, quality, and compost quantity, showed that the optimum variation for composting is 50% chopped food waste and 50% chopped yard waste.

Acknowledgments
The authors acknowledge Universitas Andalas for financial support (contract number: T/18/UN.16.17/PT.01.03/Energi-RD/2020)

References

[1] E. Damanhuri T P 2016 *Integrated Waste Management*

[2] Ministry of Environment and Forestry 2018 National Waste Management Information System *http//sipsn.menlhk.go.id/. access date 11 August 2019*

[3] S. Nugroho W H 2019 Application of Bio-pore Infiltration Hole as an Urban Runoff Management *IPTEK J. Proc. Ser.* **0** 324

[4] D. R. Hartono 2012 *Composting of Fruits Leftover in Biopore Infiltration Holes in Various Land Use*

[5] W. Eryuningsih, S. Raharjo Y R 2017 *Design and Performance Testing of Communal Composter for Processing Domestic Waste*

[6] F. Zahra T P D 2011 Study of Composition, Characteristic, and Recycling Potential of Waste at Cipayung Final Disposal Site, Depok *J. Environ. Eng.* **17** 59–69

[7] K. R. Brata 2008 *Biopore Infiltration Hole*

[8] Liliwarti, Silvianengsih S 2015 Characteristics of Mechanical Properties of Clay to Moisture Content (Unand Limau Manis Campus, Padang) *J. Civ. Eng.* **4** 21–6

[9] Tchobanoglous and Keith 2002 *Integrated Solid Waste Management-Engineering Principles and Management Issues*

[10] Sudaryono 2009 Ultisol Soil Fertility Level in Sangatta Coal Mining Land, East Kalimantan *J.*

Environ. Technol. **10** 337–46

[11] National Standardization Agency 2004 *National Standardization Agency 2004 SNI 19-7030-2004 about Compost Spesification from Organic Domestic Waste*

[12] V. Ci. Kurniati, S. Sumiyati G S 2017 The Effect of Moisture Content on the Results of Composting Organic Waste Using the Open Windrow Method *J. Mech. Engine***06** 119–23

[13] A. Ismayana, N. S. Indrasti, Suprihatin, A. Maddu A F 2012 Initial C / N Ratio Factor and Aeration Rate in Bagasse and Blotong Co-Composting Process *J. Agric. Ind. Technol.* **22** 173–9

[14] S. Santosa 2018 Effect of Fruits Waste in Biopore Infiltration Hole Toward the Effectiveness of Water Infiltration Rate on Baraya Campus Land of Hasanuddin University *J. Phys. Conf. Ser.* **979** 1–5

IOP Conf. Series: Materials Science and Engineering 1041 (2021) 012034 doi:10.1088/1757-899X/1041/1/012034

Solid waste management in Lake Toba area

H Khair[1], I Suryati[1*], R L Siregar[1], H D. P. Sibagariang[2] and R Utami[1]

[1]Environmental Engineering Department, Universitas Sumatera Utara, Jl. dr. Mansur Kampus USU Medan 20155

[2]Dinas Lingkungan Hidup Provinsi Sumatera Utara

Corresponding author: isratl@usu.ac.id

Abstract. Lake Toba is a tourist area that continues to be developed, attracting local and foreign tourists. The increase in tourists visiting lake Toba affects the social and economic aspects and impacts the environment, one of which is the increase in waste generation. On the other hand, waste in the Tourist area is not well managed. This study determines the condition of existing waste management in the Lake Toba tourism area. Girsang Sipangan Bolong and Merek sub-districts were chosen as study area. This study found that waste management in both areas is not sustainable. The sources of waste come from residential and non-residential areas. The two regions manage waste independently. Some of the community's waste handling includes burning, composting, separating kitchen waste for animal feed, and recycling. The two regions' government is still trying to provide the best service in solid waste management, such as providing communal waste containers in public facilities and waste transportation to landfills. However, it was found that there are still many deficiencies that need to be fixed. Solid waste management in these two regions is still far from Indonesia's standard.

1. Introduction

Toba is a national strategic area that is described in Presidential Regulation No. 81 of 2014. National Strategic Area is a prioritized area. It has a significant national influence on state sovereignty, national defense, security, economy, social, culture, and environment, including areas designated as world heritage.

Toba is developed as a world tourism destination. Having a positive impact on economic improvement also negatively impacts the environment, such as water pollution, air pollution, and waste. The purpose of tourists coming to tourist attractions, of course, wants to enjoy the beauty and cleanliness of the tourist attractions themselves. The beauty and cleanliness of the tourist area are reflected in the culture of a nation. In order to increase tourist attraction, of course, the Lake Toba area needs good environmental management, including solid waste management. Several researchers have conducted studies related to solid waste management around tourist areas, such as Pariaman [1]; Bromo [2]; Thailand [3]; Romania and Italy [4], and Vietnam [5]. These studies are related to the best and practical management for tourist areas.

Girsang Sipangan Bolon is a sub-district in Simalungun Regency, North Sumatra, Indonesia, with 129.89 km^2. This area is located on the shores of Lake Toba with Parapat as its icon and a tourist destination, which is the capital of Girsang Sipangan Bolon District. Girsang Sipangan Bolon District is geographically located at 920 above sea level, with an area of 123.56 km^2 and a population in 2019 of 15,023 people [6].

Merek is a sub-district in Karo Regency, North Sumatra. This sub-district is the division of the Tiga Panah sub-district into three sub-districts. It consists of Sipituhuta, Pertibi Lama, Aek Popo, Pangambat, Tongging, Situnggaling, Pertibi Tembe, Tambusan-Garingging, plus some villages resulting from the division of Tiga Panah sub-district. Merek has a hilly topography and is a center for producing Arabica coffee, vegetables, pine forest areas. It is an agribusiness terminal station to support the agropolitan national program located in Bandar Tongging. The population of Merek sub-district in 2019 is around 21,944 people with 5,149 household heads. Overall, the population can be associated based on sex, type of work, religion, age, and education level. This study aims to determine the existing conditions of solid waste management in Lake Toba's tourist area, especially in Girsang Sipangan Bolon and Merek sub-districts, North Sumatra.

2. Materials and methods

In this study, data collection was carried out by direct observation in the field, interviews, and analyzing reports related to waste in the study location. The study locations were Girsang Sipangan Bolon District, Simalungun Regency and Merek District, Karo Regency. This study was conducted in mid-2019.

3. Results and discussions

3.1. Girsang Sipangan Bolon Subdistrict

The source of waste in the Girsang Sipangan Bolon Subdistrict area comes from residential and non-residential areas such as commercial areas and public facilities. The Girsang Sipangan Bolon Subdistrict is part of Simalungun Regency. The regency does not yet have a sustainable waste management system from the source to the final processing site. Some of Simalungun Regency people still manage the waste produced independently in various ways that are considered capable of destroying the waste produced, including throwing waste into the trash, burning, composting, separating kitchen waste for animal feed, and recycling.

The household waste container system in Simalungun Regency consists of individual and communal containers. Individual containers are provided independently by the community, while the government provides communal containers at several public facilities. The individual containers provided independently by the community are generally in the form of rattan sacks and baskets. Communal storage can be made of drums and made permanent from concrete provided by the Sanitation and Gardening Service along the main road. Many communal containers are in a damaged condition, so waste is scattered in several public facilities. For market waste in Kecamatan Raya, there is no adequate container. The resulting waste is thrown away and collected just like that around the place where the traders sell, which the next day, the piles of waste will be transported by waste trucks. Meanwhile, markets located in tourism areas such as Parapat, which operate every day, already have their communal complexes.

Meanwhile, the pattern of waste collection in Simalungun Regency is divided into two, namely [7]:

1. Direct individual collection
   Waste from the source is directly transported to the landfill without any transfer process. This pattern is applied around protocol roads where a 6 m^3 dump truck can still pass through.
2. Direct communal collection
   Waste from several points of communal containers is collected to be transported to the landfill. The community brings waste into communal containers independently.

The number of fleets operating every day is 13 units of dump trucks measuring 6 m^3 and 100 units of motorbikes that carry waste with an average operation of 2 trips/day. The distance traveled from the waste centroid to the landfill is about 70 km, with a distance of about 2 hours.

Figure 1. Waste collection and transportation in Girsang Sipangan Bolon

The Sanitation and Gardening Agency manages the waste system of Simalungun Regency under Regional Regulation No. 3 of 2014 concerning the Organization and Administration of the Regional Apparatus of Simalungun Regency. Waste management on the protocol road of Girsang Sipangan Bolon District is carried out by a third party in collaboration with the local government of Girsang Sipangan Bolon. Waste handling on protocol roads is collected and transported using a truck to the temporary disposal site as seen in figure 1. The Girsang Sipangan Bolon sub-district transports the waste collected at the temporary disposal site within two weeks before being transported to the landfill. The local government manages the waste generated by households, offices, restaurants, grocery stores, and hotels. It is subject to a fee of IDR 80,000/month. However, some households do not collect this levy so that waste management is carried out independently by how to dispose of directly to the temporary disposal site.

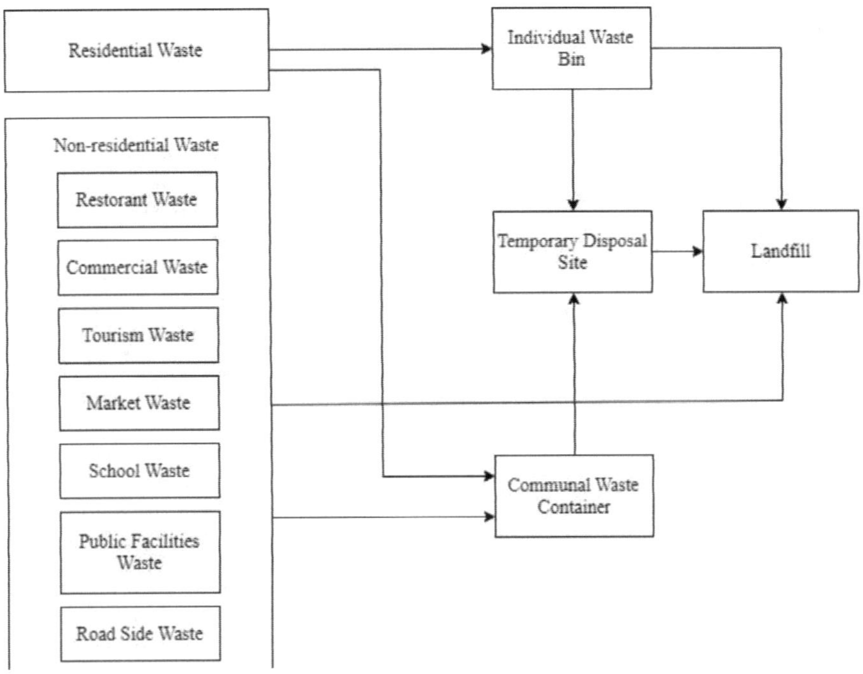

Figure 2. Chart of Waste Management Process in Girsang Sipangan Bolon District

As seen in figure 2, the waste management system in Girsang Sipangan Bolon still uses the old paradigm of collect-transport-dispose.There is a landfill in Girsang Sipangan Bolon District. The landfill is still implementing an open dumping system and there are waste burning activities. Girsang Sipangan Bolon landfill has a land area of about 4 Ha and serves 23 sub-districts in Simalungun Regency. Its location which is more than 25 km from several transportation service areas has resulted in the transportation service to the TPA not running optimally. The available land does not meet long-term needs.

3.2. Merek Subdistrict

Waste management that is carried out in Merek District, which is part of the Karo Regency area, tends to be carried out independently and conventionally. Waste processing is still carried out individually, namely by collecting it in the home yard. Organic waste is brought to the field and then piled up while inorganic waste is burned/dumped in a place without any handling. Besides, waste collection activities are carried out by putting waste into bags/sacks or throwing them directly into the temporary site. The government's lack of attention towards waste management in the Merek District resulted in a low public awareness level.

The household waste container system in Karo Regency consists of individual and communal containers as seen in figure 3. Generally, the container is not equipped with a lid, and no sorting is done, causing unpleasant odors and aesthetic problems. This condition becomes a breeding ground for vectors such as flies, mice, and cockroaches, making it easier for disease transmission to occur. For instance, it can be seen at a pile of waste, pets roaming along the road will scavenge for food scraps to eat, which causes the trash to shatter even more.

The waste collection schedule is carried out once a week using a dump truck. The transported waste is waste that has been collected at the temporary disposal location, not directly from each household/non-household. The conditions for this temporary are also very minimal, where the location is far from residential areas. The infrastructure facilities are also inadequate so that the collection and transportation of waste are delayed to the landfill location. The shape of the land in Merek District, which is physically hilly and the landfill location, is relatively far away, makes it difficult for waste trucks to reach it.

Figure 3. Individual and communal waste container system in Merek

The legal basis for waste management in Karo District still refers to government regulations and regulations issued by the Ministry of Environment. There is no regional regulation that specifically regulates waste management. However, regarding the waste/cleaning service fees, it refers to the Karo Level II Regional Regulation No 14 of 2006. The tariff structure is classified based on the services provided, the type and volume of waste produced, and the community capacity. Meanwhile, under the Karo Regency Regional Regulation Number 6 of 2008 concerning the Organization and Work Procedure of Regional Technical Institutions, solid waste management, especially those dealing with cleaning staff, is carried out by the Karo Regency Sanitation and Gardening Office. The coverage of waste services is currently still limited to the Karo area and Market. Meanwhile, the waste in the sub-district capital and the village market is not well managed and is still dumped into rivers/ravines/hillside [8].

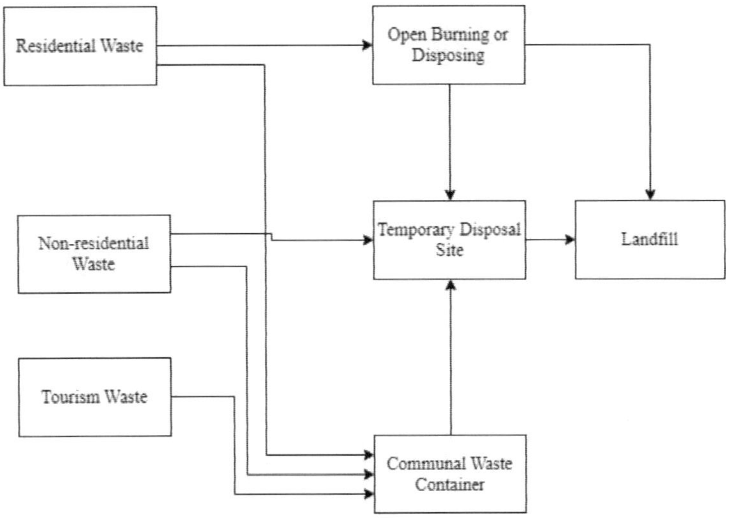

Figure 4. Chart of Waste Management Process in Merek District

Karo Regency has a landfill, and it is located in Nang Belawan, Kabanjahe District, Karo Regency, with ± 4 Ha with a load capacity ± 750,000,000 m^3. Besides that, Karo Regency also has ± 72 places of temporary waste disposal. Each has a ± 1.5 m^3 capacity, located in Kabanjahe City, Berastagi City, and the District Capital. Karo District also has 18 bin container units with a capacity of 6 m3, which are in each of the subdistrict capitals. The intensity of transportation once a day by two dump trucks. In practice, waste management at the Nang Belawan landfill is still using an open dumping system. Nang Belawan, apart from not having professional management staff, also does not have supporting facilities and infrastructure for sanitary landfill patterns such as bulldozers, excavators, scales, workshops, and a management office. Waste management process in Merek Subdistrict described as figure 4.

4. Conclusion

This study found that the two regions are still trying to provide the best service in solid waste management to the community. However, it was found that there are still many deficiencies that need to be fixed. Solid waste management in these two regions is still far from Indonesia's standard. Both of them are where the open dumping landfill is still in operation. Also, both regions still lack professionals in solid waste management.

Acknowledgments

The authors gratefully acknowledge that the present research is a collaboration between the Environmental Service of North Sumatra Province and the author as part of the experts of the Integrated Waste Management System Planning in the Lake Toba Area in 2019.

IOP Conf. Series: Materials Science and Engineering 1041 (2021) 012034 doi:10.1088/1757-899X/1041/1/012034

References

[1] Aziz R and Mira 2019 Study of recycling potential of solid waste of tourist area in Pariaman City *IOP Conf. Ser. Mater. Sci. Eng.* **602** 012059

[2] Chrisdianti 2017 *Alternatif Pengelolaan Sampah Di Kawasan Wisata Laut Pasir Taman Nasional Bromo Tengger Semeru* (Institut Pertanian Bogor)

[3] Manomaivibool P 2018 Tourism and Municipal Solid Waste Management in Developing Economies: Challenges and Opportunities *Open Access J. Waste Manag. Xenobiotics* **1**

[4] Giurea R, Precazzini I, Ragazzi M, Achim M, Cioca L-I, Conti F, Torretta V and Rada E 2018 Good Practices and Actions for Sustainable Municipal Solid Waste Management in the Tourist Sector *Resources* **7** 51

[5] Hoang M G, Fujiwara T and Phu S T P 2017 Municipal Waste Generation and Composition in a Tourist City - Hoi An, Vietnam *J. JSCE* **5** 123–32

[6] Badan Pusat Statistik Kabupaten Simalungun 2018 *Kecamatan Girsang Sipangan Bolon dalam Angka 2018*

[7] Dinas Lingkungan Hidup Kabupaten Simalungun 2018 *Rencana Induk Sistem Pengelolaan Sampah Kabupaten Simalungun*

[8] Pemerintah Kabupaten Karo 2011 *Buku Putih Kabupaten Karo*

IOP Conf. Series: Materials Science and Engineering 1041 (2021) 012035 doi:10.1088/1757-899X/1041/1/012035

Performance of anammox biofilm reactor under tropical temperature

Z Zulkarnaini[a], B Primasari[a], D J Saputra[a]

[a]Department of Environmental Engineering, Universitas Andalas, Padang, Indonesia

Corresponding author: zulkarnaini@eng.unand.ac.id

Abstract. Application of anaerobic ammonium oxidation (anammox) process for mainstream wastewater in the lower temperature is a current challenge where most of the anammox reactors planted in the sub-tropical country. The purpose of this study was to analyze the performance of nitrogen removal using anammox process at room temperature in a tropical country. The experiment was carried out using 300 mL an up-flow anaerobic sludge blanket (UASB) reactor in treating low ammonia concentration artificial wastewater at tropical temperature without any adjustment. The reactor filled with 50% in volume spiral-sheet plastic (polyethylene terephthalate) as the carrier. Anammox bacteria *Candidatus Brocadia sinica* as seeding acclimatized in room temperature for three months before the experiment. The reactor fed with 70 mg-N/L for both ammonium and nitrite with variations in hydraulic retention time (HRT) 24 hours and 12 hours. The maximum nitrogen removal rate (NRR) achieved 0.12 kg-N/(m^3·d) at nitrogen loading rate (NLR) 0.14 kg-N/(m^3·d) and NRR 0.25 kg-N/(m^3·d) at NLR 0.28 kg-N/(m^3·d). The anammox process was stable at room temperatures (15-28°C) in a tropical country for nitrogen removal and anammox bacteria grown on plastic media.

1. Introduction

Anaerobic ammonium oxidation (anammox) is a new effective biological process in nitrogen removal. In the process, nitrite is used as an electron acceptor in the conversion of ammonium to nitrogen gas under anaerobic condition [1] by anammox bacteria belong to phylum *Planctomycetes*. Compared with conventional nitrification-denitrification, anammox process could reduce supplied aeration up to 64%, 100% organic carbon, and 90% sludge production [2]. Anammox is a beneficial process for removing ammonium from wastewater containing low carbon per nitrogen (C/N) levels, such as leachate from landfills, wastewater from the electroplating industry [3]. Based on the first experience anammox reactor was built in the Dokhaven for domestic wastewater treatment plant in Rotterdam, the Netherlands, which is capable of removing nitrogen at nitrogen removal rate (NRR) 9.5 kg-N/m^3.d proved that the anammox process is a powerful method for treatment high-strength ammonium wastewater [4]. In the laboratory scale, Tang et al. achieved a higher NRR of 77 kg-N/m^3.d using UASB reactor Up-flow Anaerobic Sludge Blanket (UASB)[5]. Accumulation of sufficient anammox biomass is a key for high nitrogen removal rate. However, due to the slow-growing rate with doubling time (10–14d) and low cell yield (0.11 gVSS/gNH_4^+-N) of anammox bacteria, achieving fast cultivation and maintenance the stable operation becomes a challenge for the application of anammox process. To maintain sludge retention and prevent the biomass wash-out, sequencing batch reactors (SBR), upflow biofilter (UBF) [6], up-flow anaerobic sludge blanket

IOP Conf. Series: Materials Science and Engineering 1041 (2021) 012035 doi:10.1088/1757-899X/1041/1/012035

(UASB) reactors [7], and membrane bioreactors (MBR)[8] have been conducted for achieving fast start-up and cultivation.

Temperature influence performance and activity of anammox bacteria where the nitrogen removal rate decreased due to decreased temperature, also the specific red carmine of anammox biomass color changed into reddish-brown [9]. Based on the characteristics of anammox bacteria, the removal of nitrogen from wastewater reaches optimal conditions in the temperature range of 30-40 °C [10], although the performance of the anammox process continues at 10 °C [11]. Nowadays, more than 114 anammox reactors installed worldwide, mostly in Europe, China, and North America were located in sub-tropical regions [2]. Therefore, the operational cost will be higher to maintain the temperature, especially in the winter. Indonesia, as a tropical country, has not been focused on eliminating nitrogen levels in wastewater and the environment yet. On the other hand, the impact of higher nitrogen concentration in the environment promote eutrophication that disturbance ecosystem balance and produce high undesirable biomass. Tropical countries have potential opportunity for application of anammox process due to the suitable temperature for anammox bacteria.

The purpose of this study was investigated the nitrogen removal capability of anammox process using an Up-flow Anaerobic Sludge Blanket (UASB) reactor in the tropical temperature at Padang City, Indonesia.

2. Materials and methods

2.1. The configuration of anammox reactor

The UASB reactor had an inner diameter of 4 cm, height of 24 cm and working volume of 0.3 l.The spiral-sheet plastic (polyethylene terephthalate)filled into the reactor (50% in volume) as supporting media for anammox biofilm to start the study. Filters set up in the bottom and upper of the reactor to prevent biomass wash out. Substrates tank connected to a Tedlar bag containing nitrogen gas to maintain anaerobic conditions and regulate the balance of atmospheric pressure in the tank.

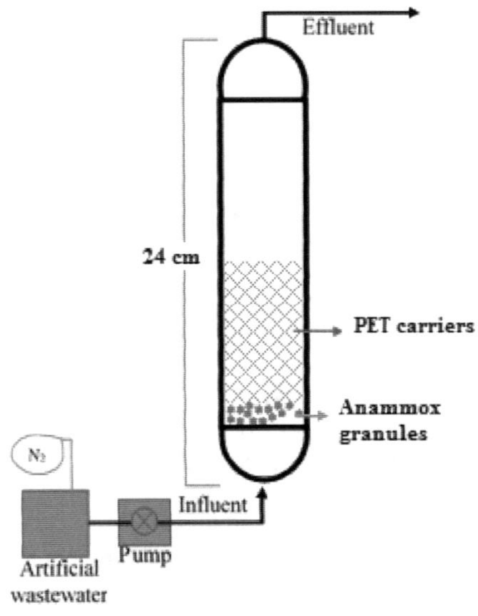

Figure1. Schematic of anammox biofilm reactor.

The reactor covered with aluminum foil to prevent the growth of heterotrophic bacteria. Reactor installation scheme can be seen in figure 1. The experiments were performed with a variation of Hydraulic Retention Time (HRT) 24 hours and 12 hours (table 1).

Table 1. Operation of reactor

Period	Time (d)	HRT (h)	NH_4^+-N (mg/L)	NO_2^--N (mg/L)	NLR kg-N/(m^3·d)
I	1-56	24	70	70	0.04
II	57-123	12	70	70	0.30

2.2. Seeding sludge for inoculation
Species of anammox bacteria (*Candidatus Brocadia sinica*)acclimatized in a 100 mL column reactor for three months before the experiment in room temperature operation. The substrate solution, containing 70 mg-N/L of ammonium and nitrite fed into the reactor as food for anammox bacteria, continuously. Then, the UASB reactor was inoculated with a 20 mL anammox granular sludge taken from the acclimatized column reactor for start-up the anammox reactor.

2.3. Composition of synthetic wastewater
Ammonium and nitrite were supplemented to mineral medium with concentration 70 mg-N/L $(NH)_2SO_4$ and 70 mg-N/L $NaNO_2$. The composition of mineral medium (per l of tap water) was KH_2PO_4, 27.2 mg; $MgSO_4{}_{\cdot 7}H_2O$, 300 mg; $CaCl_2$, 136 mg; $KHCO_3$, 500 mg; and 1 mL trace element solutions I and II [12]. The trace element I contained (per l demineralized water): EDTA 2Na, 6.37 g; FeSO4, 5 g and trace element II contained (per l demineralized water): EDTA 2Na, 19.11; $ZnSO_4$, 0.241 g; $CoCl.6H_2O$, 0.24 g; $MnCl_2.4H_2O$, 0.99; $CuSO_4.5H_2O$, 0.25 g; $NaMoO_4.2H_2O$, 0.22 g; $NiCl_2.6H_2O$, 0.19 g; $NaSeO_4.10H_2O$, 0.24g; H_3BO_4, 0.14 g. The synthetic wastewater was purged with N_2 gas for 30 minutes to remove dissolved oxygen before delivered to the bottom of reactor using a peristaltic pump continuously.

2.4. Chemical analyses
The influent and effluent samples were collected twice a week and were analyzed immediately. Water samples were analyzed according to the standard methods for the examination of water and wastewater [13]. Ammonia, nitrite, and nitrate were analyzed using the ultraviolet spectrophotometric method. The temperature and pH were recorded using a pH meter.

3. Results and discussion
3.1. Stoichiometric
The reactor was operated for 123 days that divided into two periods based on nitrogen loading rate (NLR): first period for start-up at 0.14 kg-N/(m^3·d) NLR and changed to 0.28 kg-N/(m^3·d) at followed period. During the whole operational, temperature and pH were not adjusted. The temperature operation and pH were in the range 15-28°C and 7.1-8.2, respectively. The anammox process can be observed with the parameters of the conversion of ammonium, nitrite, and produced nitrate in the reactor. Identifying the occurrence of the anammox process in the reactor based on the relationship of the stoichiometric reaction proposed by Lotti et. al.[14] as follows:

$$NH_4^+ + 1.146NO_2^- + 0.071HCO_3^- + 0.057H^+ \rightarrow 0.986N_2 + 0.161NO_3^- + 0.071CH_2O_{0,5}N_{0,15} + 2,002H_2O \quad (1)$$

The results in figure 2showed the average ratio of ΔNO_2^--N/ΔNH_4^+-N and ΔNO_3^--N/ΔNH_4^+-N in the stable operation lower than higher than the stoichiometric, they were 1.10 and 0.19 respectively. The higher ratio indicated the possibility of a nitrification process in the operation of the reactor. Where nitrite was oxidized by nitrite oxidation bacteria (NOB) to nitrate, therefore higher nitrate observed in the effluent, this difference in the stoichiometric ratio can be influenced by bacterial physiological conditions (nitrogen loading, temperature, and pH), experimental conditions and microbial populations.

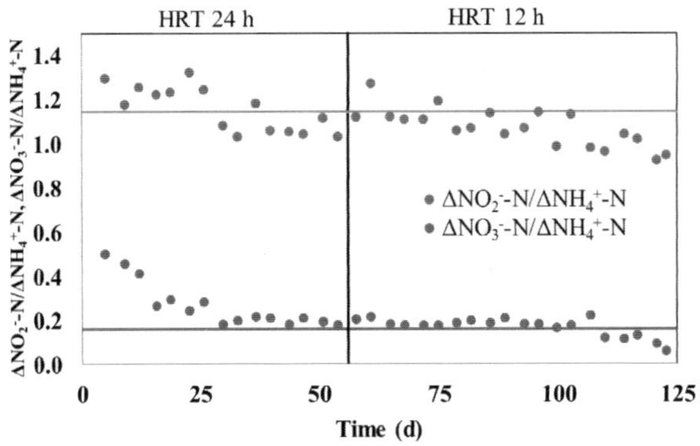

Figure 2. The ratio of experimental nitrogen conversion confirmed to the stoichiometry of anammox process.

3.2. Profile of nitrogen

Since the start of operations using HRT 24 hours, both NH_4^+-N and NO_2^--N in effluent declined gradually. In the first period, with supplied 70 mg/L NH_4^+-N and NO_2^--N, the anammox process capable of removing both ammonium and nitrite and produced nitrate in the anaerobic condition in one-week operation, with the effluent concentration 50.84 mg-N/L, 38.67 mg/L and 12.04 mg/L respectively. Stable produced NO_3^--N in the range 8.79-12.04 mg/L indicated that proses anammox occurred continuously from the beginning until the end of period 1. The acclimatization in the room temperature of anammox seeding before the experiment could enhance the start-up proses where anammox bacteria could adapt with the changed and varied temperature. In the end of the first period on day 54, small amount concentration of NH_4^+-N and NO_2^--N detected in the effluent. Concentration NH_4^+-N and NO_2^--N in the effluent declined rapidly from 50.84 mg/L to 7.47 mg/L and from 38.67 mg-N/L to 9.56 mg/L. In period 2, NLR was decreased from 24 h to 12 h, corresponding to increased NLR from 0.14 kg-N/(m^3·d) to 0.28 kg-N/(m^3·d). As a result, NH_4^+-N and NO_2^--N concentration in effluent increased from 7.47 mg/L to 24.34 mg/L and 9.56 mg/L to 17.88 mg/L. The increased NLR disturbance anammox process, and this condition were stable until three weeks of operation in period 2. Anammox bacteria high sensitivity to changing of environmental operation due to the slow growth rate with doubling time 10-14 d at 30-40°C [15][16]. Then, anammox activity increased where NH_4^+-N and NO_2^--N concentration in the effluent decreased gradually until all of them removed on day 110 and stable until the end of period 2.

3.3. Reactor performance

The removal of nitrogen from the anammox process was expressed by NRR and Nitrogen removal efficiency (NRE). At the beginning of the experiment, NRR reached 0.04 kg-N/(m^3·d) in 0.14 kg-N/(m^3·d) NLR and stable for one-week operation. The low value of NRR at the beginning of the operation due to the anammox bacteria have a slower growth rate compared with other types of bacteria such as AOB, NOB. On day 54, NRR increased gradually from 0.06 kg-N/(m^3·d) to 0.12 kg-N/(m^3·d). Anammox bacteria can remove high ammonium concentration at the condition of their high specific ANAMMOX activity (SAA) and high biomass concentration [5]. Therefore, on day 58, HRT decreased to 12 hours, and consequently, NLR increased to 0.28 kg-N/(m^3·d). As a result, NRR increased to 0.18 kg-N/(m^3·d) and achieved maximum NRR 0.29 kg-N/(m^3·d). As shown in figure 3.

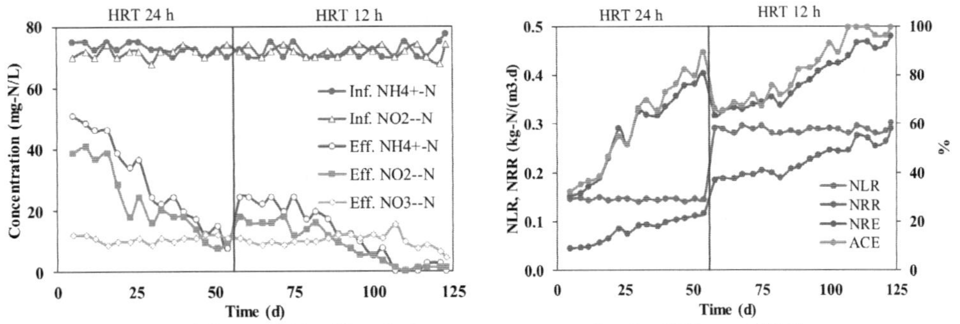

Figure 3. Profile of nitrogen concentration in different HRT

Nitrogen removal efficiency (NRE) demonstrated nitrogen removal that expressed in units of percent. At the beginning of the operation, NRE was 30%, and an increased rapidly in the following days. The maximum efficiency, 81%, was obtained on day 54 with HRT 24 hours. While NRE dropped to 63% after HRT was decreased to 12 hours. The shock loading disturbance the anammox process, and it took one month for recovery of anammox activity. After recovery, NRE achieved 82% on day 96, and removal of nitrogen continues to increase until almost all nitrogen removed. The maximum NRE 96% was reached at the end of period two on day 123.

$$NRR=\frac{[NH_4^+\text{-}N]_{in}+[NO_2^-\text{-}N]_{in}-[NH_4^+\text{-}N]_{out}-[NO_2^-\text{-}N]_{out}-[NO_3^-\text{-}N])_{out}}{HRT} \quad (2)$$

$$NRE=\frac{[NH_4^+\text{-}N]_{in}+[NO_2^-\text{-}N]_{in}-[NH_4^+\text{-}N]_{out}-[NO_2^-\text{-}N]_{out}-[NO_3^-\text{-}N])_{out}}{[NH_4^+\text{-}N]_{in}+[NO_2^-\text{-}N]_{in}}\times 100\% \quad (3)$$

Generally, raw wastewater containing only ammonia, therefore ammonium removal efficiency (ACE) become one of the parameters to measure the performance of anammox process. At the beginning of reactor operation, only 30% ammonium converted to nitrogen gas using nitrite as the electron acceptor. With increasing time, ACE increased gradually with the pattern similar to NRE. At the end of period 2, ACE reached a maximum of 89%. Same as NRE, ACE decreased rapidly after HRT decreased from 24 h to 12 h and increased slowly by time. In this period, all of ammonia could be removed from the reactor. The operation of anammox biofilm reactor for treating low ammonia concentration wastewater in the tropical country become an opportunity for the application where the temperature of tropical country suitable for operation of anammox bacteria without any extreme temperature changed like winter in the sub-tropical country.

$$ACE=\frac{[NH_4^+\text{-}N]_{in}-[NH_4^+\text{-}N]_{out}}{[NH_4^+\text{-}N]_{in}}\times 100\% \quad (4)$$

3.4. Effect of temperature operation

The anammox process is very sensitive to both pH fluctuation and temperature, and it is activity optimized by high temperature [17]. This experiment was conducted in room temperatures in the tropical area, even though generally it was in the range of physiological temperature of anammox bacteria, except in the night where the temperature dropped drastically below 20°C. The fluctuation of temperature influences the specific growth rate and doubling time of anammox bacteria where the anammox specific activities decreased rapidly in the temperature 10-20°C [11].

Bacteria Anammox can adapt to low-temperature conditions with acclimatization. In the natural system, anammox bacteria have the ability to tolerate high temperatures between -2 and 85°C [18]; thus it was a strategy acclimatized anammox bacteria before using it to treat wastewater with a lower temperature. Whereas in this study, the temperature was in the range 21-28°C and anammox process occurred stably for low ammonia containing artificial wastewater.

3.5. Effect of FA, FNA, and PH

Inhibition of anammox process caused by the Free Ammonia (FA) and the Free Nitrous Acid (FNA) [19]. Concentration FA and FNA are influenced by temperature, pH and ammonium, and nitrite concentration, respectively. Waki et al. 2007[20] reported that the FA as one of the factors inhibiting the anammox reaction in levels 13-90 mg/L and toxic to the bacteria anammox. Fernandez et al. 2012 also said the anammox process performance would be unstable in the FA concentrations exceeding 20-25 mg/L. The concentration of FA obtained during the operation was in the range of 0.18 mg/L - 6.17 mg/L. The concentration of FA lower than inhibition concentration of FA, therefore, operation of the reactor free from FA inhibition. While under low pH (below 6.7), FNA affects the equilibrium in the anammox process [21], Previous research also reported that the anammox process was inhibited by FNA at concentrations of 0.0015 to 0.213 mg/L [22]. FNA obtained in the study are relatively low and stable at the range of 0 to 0.011 mg-N/L, as shown in figure 4, lower than inhibition concentration.

One important factor in the operation of the anammox process was pH. Where suitable of pH range for physiological anammox bacteria is 6,7-8.3 [10]. In a continuous flow reactor using granular anammox bacteria, anammox process running stable although higher pH in the effluent of 8.5 to 9.3 [23] where according to Equation 1 anammox increased the pH [10]. From the experimental pH was 7.1 to 8.2, in the range of physiology anammox process. There no inhibition observed because of pH, FA, and FNA during the whole operation.

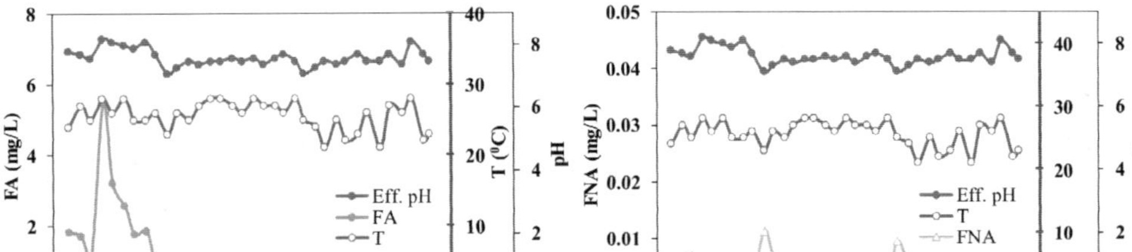

Figure 4. Time courses of free ammonia (FA) and free nitrous acid (FNA). The temperature was recorded at the sampling time

3.6. Reuse of PET bottle for a carrier

The experiments using spiral-sheet plastic made from PET bottle that stacked in three levels in the reactor. Each level was 5 cm in diameter and 4 cm in high, as seen in figure 5. During the operation, the carrier was covered gradually by reddish-brown biofilm, which is the typical color of the anammox bacteria. The color of anammox bacteria as seeding was changed from carmine red to reddish-brown after operated in room temperature. The distinct color variation of anammox biomass from black, brown to carmine-red during varied reactor operation due to heme proteins [24][25].Ma et al. 2019[9] also reported that the changed of anammox granule color due to varied temperature operation. It was proved that the PET bottle could be reused as the carrier for threating low nitrogen-containing wastewater. The used of carrier enhanced performance of nitrogen removal (0.25 kg-N/(m^3·d))compared with granular sludge (0.196 kg-N/(m^3·d)NRR) [26]and palm fiber (0.20 kg-N/(m^3·d)NRR) [27].

Figure 5. The changed of anammox bacteria color, (A) original anammox granule before acclimatization, spiral-sheet plastic as carrier (B), (C) carrier covered by anammox biofilm

4. Conclusions

Anammox process was stably operated in room temperature in a tropical country using anammox biofilm reactor with spiral-sheet plastic as the carrier. The carrier enhance the performance of nitrogen removal at HRT 24 h and 12 h were NRR 0.12 kg-N/(m^3·d) in NLR 0.14 and 0.29 0.12 kg-N/(m^3·d) in NLR 0.30 kg-N/(m^3·d), respectively.

Acknowledgments

This work was supported by Engineering Faculty of Universitas Andalas, Indonesia by grant number 044/UN.16.09.D/PL/2019.

References

[1] Jin R-C, Yang G-F F, Yu J-J J and Zheng P 2012 The inhibition of the Anammox process: A review *Chem. Eng. J.* **197** 67–79

[2] Ali M and Okabe S 2015 Anammox-based technologies for nitrogen removal: Advances in process start-up and remaining issues *Chemosphere* **141** 144–53

[3] Lackner S, Gilbert E M, Vlaeminck S E, Joss A, Horn H and van Loosdrecht M C M 2014 Full-scale partial nitritation/anammox experiences - An application survey *Water Res.* **55** 292–303

[4] Van der Star W R L L, Abma W R, Blommers D, Mulder J W, Tokutomi T, Strous M, Picioreanu C and van Loosdrecht M C M M 2007 Startup of reactors for anoxic ammonium oxidation: Experiences from the first full-scale anammox reactor in Rotterdam *Water Res.* **41** 4149–63

[5] Tang C J, Zheng P, Wang C H, Mahmood Q, Zhang J Q, Chen X G, Zhang L and Chen J W 2011 Performance of high-loaded ANAMMOX UASB reactors containing granular sludge *Water Res.* **45** 135–44

[6] Jin R C, Zheng P, Hu A H, Mahmood Q, Hu B L and Jilani G 2008 Performance comparison of two anammox reactors: SBR and UBF *Chem. Eng. J.* **138** 224–30

[7] Casagrande C G, Kunz A, De Prá M C, Bressan C R and Soares H M 2013 High nitrogen removal rate using ANAMMOX process at short hydraulic retention time *Water Sci. Technol.* **67** 968–75

[8] Van Der Star W R L L, Miclea A I, Van Dongen U G J M J M, Muyzer G, Picioreanu C and

Van Loosdrecht M C M M 2008 The membrane bioreactor: A novel tool to grow anammox bacteria as free cells *Biotechnol. Bioeng.* **101** 286–94
[9] Ma H, Zhang Y, Xue Y, Zhang Y and Li Y Y 2019 Relationship of heme c, nitrogen loading capacity and temperature in anammox reactor *Sci. Total Environ.* **659** 568–77
[10] Strous M, Kuenen J G and Jetten M S M 1999 Key Physiology of Anaerobic Ammonium Oxidation Key Physiology of Anaerobic Ammonium Oxidation *Appl. Environ. Microbiol.* **65** 0–3
[11] Lotti T, Kleerebezem R and van Loosdrecht M C M 2015 Effect of temperature change on anammox activity *Biotechnol. Bioeng.* **112** 98–103
[12] Van De Graaf A A, De Bruijn P, Robertson L A, Jetten M S M M and Kuenen J G 1996 Autotrophic growth of anaerobic ammonium-oxidizing micro-organisms in a fluidized bed reactor *Microbiology* **142** 2187–96
[13] American Public Health Association, American Water Works Association W E F 1999 Standard Methods for the Examination of Water and Wastewater Standard Methods for the Examination of Water and Wastewater *Public Health*
[14] Lotti T, Kleerebezem R, Lubello C and van Loosdrecht M C M M 2014 Physiological and kinetic characterization of a suspended cell anammox culture *Water Res.* **60** 1–14
[15] Ni S Q and Meng J 2011 Performance and inhibition recovery of anammox reactors seeded with different types of sludge *Water Sci. Technol.* **63** 710–8
[16] Strous M, Kuenen J G, Jetten M S M M, Gijs Kuenen J and Jetten M S M M 1999 Key Physiology of Anaerobic Ammonium Oxidation *Appl. Environ. Microbiol.* **65** 3248–50
[17] Tomaszewski M, Cema G and Ziembińska-Buczyńska A 2017 Influence of temperature and pH on the anammox process: A review and meta-analysis *Chemosphere* **182** 203–14
[18] Gao D W and Tao Y 2011 Versatility and application of anaerobic ammonium-oxidizing bacteria *Appl. Microbiol. Biotechnol.* **91** 887–94
[19] Fernández I, Dosta J, Fajardo C, Campos J L, Mosquera-Corral A and Méndez R 2012 Short- and long-term effects of ammonium and nitrite on the Anammox process *J. Environ. Manage.* **95** S170–4
[20] Waki M, Tokutomi T, Yokoyama H and Tanaka Y 2007 Nitrogen removal from animal waste treatment water by anammox enrichment *Bioresour. Technol.* **98** 2775–80
[21] Puyol D, Carvajal-Arroyo J M, Sierra-Alvarez R and Field J A 2014 Nitrite (not free nitrous acid) is the main inhibitor of the anammox process at common pH conditions *Biotechnol. Lett.* **36** 547–51
[22] He S, Niu Q, Ma H, Zhang Y and Li Y Y 2015 The treatment performance and the bacteria preservation of anammox: A review *Water. Air. Soil Pollut.* **226**
[23] Tang C J, Zheng P, Hu B L, Chen J W and Wang C H 2010 Influence of substrates on nitrogen removal performance and microbiology of anaerobic ammonium oxidation by operating two UASB reactors fed with different substrate levels *J. Hazard. Mater.* **181** 19–26
[24] Sabine Marie P, Pümpel T, Markt R, Murthy S, Bott C and Wett B 2015 Comparative evaluation of multiple methods to quantify and characterise granular anammox biomass *Water Res.* **68** 194–205
[25] Zhang Y, Ma H, Chen R, Niu Q and Li Y Y 2018 Stoichiometric variation and loading capacity of a high-loading anammox attached film expanded bed (AAEEB) reactor *Bioresour. Technol.* **253** 130–40
[26] Zulkarnaini, Afrianita R and Putra I H 2020 Application of Anammox Process in Nitrogen Removal Using Up-Flow Anaerobic Sludge Blanket Reactor *J. Teknol. Lingkung.* **21** 31–9
[27] Zulkarnaini, Nur A and Ermaliza W 2019 Nitrogen Removal in the Anammox Biofilm Reactor using Palm Fiber as Carrier in Tropical Temperature Operation *J. Ris. Teknol. Pencegah. Pencemaran Ind.* **10** 7–15

The Sublethal Effects of Tannery Wastewater Exposure on Oxygen Consumption Levels and Operculum Movement of Indonesia Mahseer Fish

T Ihsan[1*], T Edwin[1], J R Zupit[1], S F Ananda[1], Nofrita[2]

[1]Department of Environmental Engineering, Faculty of Engineering, UniversitasAndalas, KampusUnandLimau Manis, Padang, West Sumatra 25163, Indonesia
[2]Departement of Biology, Faculty of Mathematics and Natural Sciences, UniversitasAndalas

Corresponding author:taufiqihsan@eng.unand.ac.id

Abstract. This study aimed to analyze the effect of tannery wastewater exposure of Padang Panjang, Indonesia, on the level of oxygen consumption (OC) and the operculum movement of the Mahseer fish (Tor tambra CV). Mahseer is an endemic fish in the BatangAnai River in West Sumatra Province. The study is started by determining the acute toxicity value of LC_{50}-96h using a static method within 96 hours of observation. Based on the results of the Probit Method, the LC_{50}-96h amount of mahseer was 15.41%. Furthermore, variations in the wastewater concentration were 0%, 10%, 20%, and 30% of the LC_{50}-96h value. Observations were carried out for 30 days in triplicate. Based on OC value, the concentration of 10%, 20% and 30% has decreased by 0.13 mg O_2/g.hours; 0.18 mg O_2/g.hours, and 0.27 mg O_2/g.hours consecutively otherwise in control, treatment increased by 0.05 mg O_2/g. While the operculum movement, the concentration of 10%, 20%, and 30% have increased by 90 times/minute, 108 times/minute, and 116 times/minute, respectively. In control, treatment has risen by 44 times/minute. OC's correlation value and mahseer's operculum movement ranged from 0.94 to 0.97, which means the correlation was reliable. The significance value obtained $p < 0.05$, which means that there were significant differences in OC values and movements of the mahseer's operculum.

1. Introduction

The tannery industry is an industry that treats raw leather into ready-made leather by using chemicals in the tanning process. In the processing, natural leather was carried out by biological, chemical, and physical processes. The by-product of the tannery process were wastewater and solid waste. Usually, wastewater can still be recycled or directly discard into the river or water body[1].

One of the leather tanning industries was XYZ Inc at Padang Panjang City, West Sumatra Province, Indonesia. XYZ Inc used a tanning agent in Chromium (Cr), with the raw materials used are animal skins (cows, buffaloes, goats), mainly from slaughterhouses. The XYZ Inc produced leather processing ± 500 kg in one production and produced ± 12-16 m^3 of wastewater that could pollute the environment. The outlet of Wastewater Treatment Plant heading to BatangAnai River, Padang Panjang City, West Sumatra.

Mahseer fish (Tor Tambra CV) is a type of endemic fish in the BatangAnai River, Padang Panjang City, West Sumatra. Mahseer has the local name "garing" (*garing* means crispy). It raises the potential of heavy metal contamination because of the outlet of XYZ Inc's wastewater plant. Besides, Mahseer fish is a type of fish that becomes a favorite food for West Sumatra[2,3]. Consuming fish contaminated with heavy metals can cause diseases to humans, such as kidney failure, liver damage, and dermatitis. Fish, as an indicator, has a significant role in determining water pollution because it has a high sensitivity to changes in the water environment [4].

Sublethal influences in fish body organs cause interference with these organisms' physiology and histology [5]. One of the physiology of fish includes metabolic processes[6]. The metabolic process of living things could be estimated by measuring the amount of oxygen consumed by living things per unit of time (also known as the level of oxygen consumption). If the fish's metabolic process were disrupted, it would reduce the fish's ability to consume oxygen[7,8].

Fish operculum was the process of fish swallowing water with its mouth and pressing it through the gills and then out through the hole under the operculum. The movement of the operculum was related to the level of oxygen consumption. If the fish is exposed to wastewater, it will damage the gills, so the fish's ability to consume oxygen would be reduced, and the movement of the operculate in the fish will increase[8–10].

This study aimed to analyze the acute toxicity and sublethal toxicity of mahseer fish in tannery wastewater exposure. Fish used are the wild fish and cultured fish in the BatangAnai River, Padang Panjang City, Indonesia.

2. Materilas and methods

This study used 12 glass tanks with 35 x 30 x 30 cm for control and three variations of concentration. The study used mahseer fish (Tor tambra CV.) As a test animal with an average size of 5-7 cm with a weight of 2-4 grams each. The acute toxicity test and the sublethal test used 60 fish for each test. One tank contains 16-18 liters of water for five fish (one liter of water for a fish weight of 0.8 grams) [11–14].

At the time of acclimation, the test animals were given food in the form of fish pellets and given oxygen through an aerator. Mahseer fish were fed pellets every 2-3 times a day, i.e., morning, afternoon, and evening. The amount of feed given is 2-3% of the weight of fish per day[15]. Replacement of test water was also carried out every four days [16]. Acclimatization was carried out for seven days [17]. Environmental factors measured during acclimatization were dissolved oxygen levels, temperatures, and pH levels, measured using Lutron type DO-5510 and pH meter PH-009 (1) A. During the acclimation period, the aquarium's environmental conditions must be adjusted to the ecological needs suitable for maintaining mahseer fish.

Wastewater samples in this study were taken and analyzed to determine the characteristics of liquid waste. Sampling was conducted on weekdays at the outlet point of wastewater treatment. The components of wastewater tested are Biochemical Oxygen Demand (BOD), Chemical Oxygen Demand (COD), Total Suspended Solid (TSS), Chrome (Cr), Oil, and Grease, Nitrogen, Ammonia, Sulfide, and pH. The results of the wastewater characteristics showedthat XYZ Inc's wastewater had exceeded the quality standard according to the Indonesian Ministry of Environment Regulation No. 5 of 2014 concerning Wastewater Quality Standards.

2.1. The acute toxicity test

The preliminary test aimed to determine the limits of the range of crisis concentrations of the test material used for the determination of LC_{50}-96h. This experiment's treatment was carried out with five dilution variations of the tannery industrial liquid waste and control. United States Environmental Protection Agency (USEPA) recommends the concentration of pollutants (wastewater), among others, 100%, 50%, 25%, 12.5%, 6.25%, and one as a control. Experiment with three repetitions or triplicate. During the experiment, the test animals were not fed. Observations were made at 0, 24, 48, 72, 96 hours.

The primary test was carried out on the tannery industrial wastewater with variations in concentration in the range that refers to the preliminary test. The primary analysis was carried out to obtain a more precise LC_{50}-96h value with a concentration range close to 50% of test animals (LC_{50}). The basic test concentration range was obtained from the preliminary test LC_{50} output value on Probit, which was then varied to five concentrations based on the upper threshold concentration. The lower threshold concentration was obtained [14].

2.2. *The sublethal toxicity test*

The implementation of sublethal toxicity tests in the study for 30 days[18,19]. Research by repeating three times to each concentration. During the investigation, Mahseer fish were fed pellets every 2-3 times a day, i.e., in the morning, afternoon, and evening[20,21]. The amount of feed given is 2-3% of the weight of fish per day[15].The parameters observed were the level of oxygen consumption and the operculum movement in mahseer fish. The operculum's move followed the oxygen consumption level in this study using a counter and a camera. During the trial period, the parameter measurement data used in the regression and correlation analysis are the data on day 0, 10th, 20th, and 30th.

2.3. *The level of oxygen consumption*

The level of oxygen consumption was calculated using the formula [22]:

$$OC = \frac{(V) \times (DO_0 - DO_t)}{W \times T} \quad (1)$$

OC = level of oxygen consumption (mgO_2/ g.hour)
V = volume of water in the container (l)
DOo = concentration of dissolved oxygen at the beginning of the observation (mgO_2/l)
DOt = dissolved oxygen concentration at time t (mgO_2l)
W = weight of test fish (g)
T = observation period (hours)

2.4. *The operculum movement*

Calculation of the operculum movement was using a counter and a video camera. During one hour of video capture three times. The video results were applied to the Movie Maker tools; the footage was slowed down so that the operculum movements could be calculated using a counter.

3. Results and discussion

3.1. *The acute toxicity test*

Acclimatization was carried out for seven days. Environmental parameters measured during acclimatization were dissolved oxygen levels and temperatures measured using a DO meter and pH levels measured with a pH meter. During the acclimation period, DO measurements ranged from 7.2 to 8.3 mg/l, pH ranged from 7.5 to 8.15, and temperatures ranged from 25.5 to 26.75 °C. According to USEPA [17], the allowable DO range was > 4 mg O_2/l; the pH was around 6-9, and the temperature was 25-30 °C. So, it could be concluded that all parameters obtained were still within the allowed range at the time of all parameters. This condition was still eligible for the maintenance of mahseer fish.

The preliminary test was carried out for 96 hours of observation time by measuring pH, DO, and temperature as well as the data of dead test animals for each observation. At the preliminary tests of pH values ranged from 7.28 - 8.78, DO levels covered between 6.3 - 7.7 mg/l, and the temperature ranged between 24.6 - 26.6°C. pH values obtained still meet the range of fish rearing, where fish experience optimal growth in the pH range of 6.0 - 9.0 [17]. The USEPA[17]explained that the dissolved oxygen content was suitable for freshwater fish life in waters is> 4 mg/l. This temperature was still within the temperature tolerance for mahseer's life, where fish could live with temperatures ranging from 25-30°C [17].

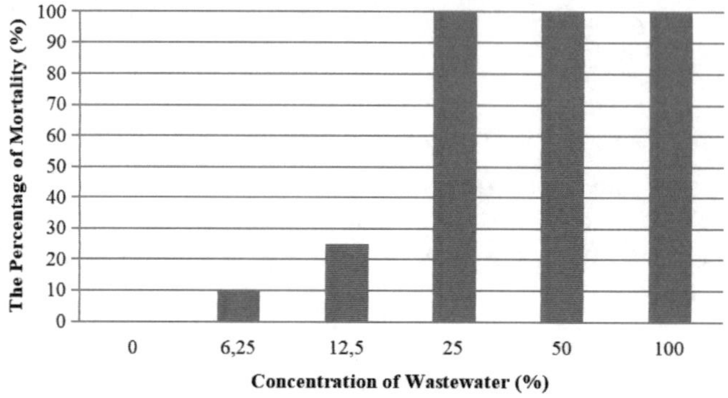

Figure 1.The preliminary test.

Based on figure 1, it could be concluded that in the preliminary test for 96 hours, the mahseer fish did not experience death at a concentration of 0%, but the fish died at a level of 6.25%, 12.5%, 25%, 50%, and 100%. Based on the study results, the tannery wastewater concentration was directly proportional to the percentage of test animal mortality. Based on data on test animals' death, there were two or more concentrations that experience death, so the LC_{50}-96h value can be determined by the probit method[17]. The LC_{50}-96h value was 13.403% (the highest concentration was 16.210%, the lowest was 11.060%). Then, this lowest and highest concentration limit was used for attention in the primary test.

After a preliminary test, the primary test was carried out using a range of waste concentrations that caused 50% fish mortality based on a preliminary test conducted on the sample. The concentration range used were between 11,060% -16,210%. In order to facilitate the dilution of waste, concentrations of 0% (control), 9%, 11.25% were chosen; 13.5%; 15.75%; 16% and 18% as the concentration of wastewater in the primary test. The basic test was carried out for 96 hours. Parameters that affected the life of fish were pH, DO, and temperature.

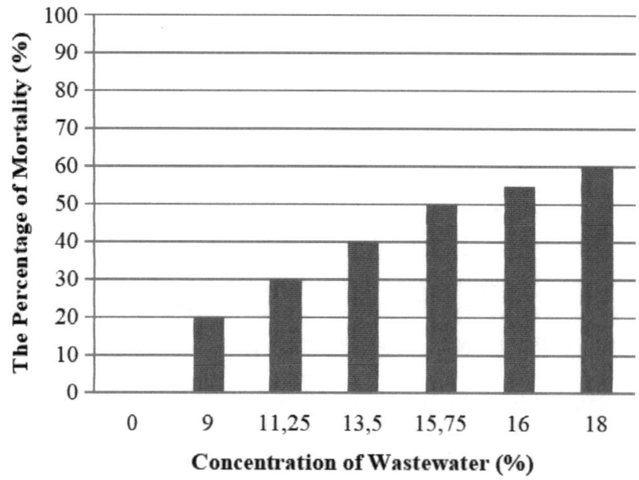

Figure 2. The primary test.

The water temperature ranged between 25 - 26.3°C, DO levels varied from 6.3 mg/L to 7.8 mg/L, and the pH value ranged between 7.33 – 8.77. This condition was still within the tolerance of temperature for the life of fish [17]. The increased pH value indicates the possibility of organic decomposition. This pH value still meets the range of fish maintenance, where the fish experienced optimal growth in the pH range of 6.0 - 9.0 [17].

Figure 2 shows that the higher the concentration of tanning wastewater added to the tank, the more mahseer fish deaths, and vice versa. This observation resulted in the LC_{50}-96 h value was 15.410%.

3.2. *The sublethal test*

During the acclimation period, DO measurements ranged from 8.10-8.60 mg/l, pH ranged from 7.20-8.10, and temperatures ranged from 26.30-27.20°C. According to USEPA [17], the allowable DO range was> 4 mg O_2/l, the pH was around 6-9, and the temperature was 25-30°C. It can be said that at the time of acclimation, DO values, pH, and temperature obtained is still in the range that is allowed so that it still meets the requirements for the maintenance of fish

A sublethal toxicity test was carried out for 30 days. The concentration used was below the LC_{50}-96h value of 15.41%, which varied to 1.5 % (0.1 from 15.41%); 3% (0.2 from 15.41%); and 4.6% (0.3 of 15.41%). During the observations, fish were given the same treatment as the acclimatization treatment. This treatment includes cleaning up impurities, feeding, replacing water, measuring pH, DO, temperature, and aeration.

During the sublethal test, DO measurements ranged from 8.70 to 9.80 mg O_2/l; pH measurements ranged from 7.03 to 8.73; the temperature ranged from 25.93-26.87°C. According to USEPA [17], during the observation of DO values, pH, and temperature obtained was still within the allowed range.

3.3. *Oxygen consumption (OC) level*

The measurement of OC levels was calculated based on the ratio of dissolved oxygen (at the beginning and end of the observation) using iodometric titration. The value of the level of OC in mahseer fish was obtained based on equation 1. The graph of the relationship between the level of oxygen consumption of mahseer fish and the observation time can be seen in figure 3.

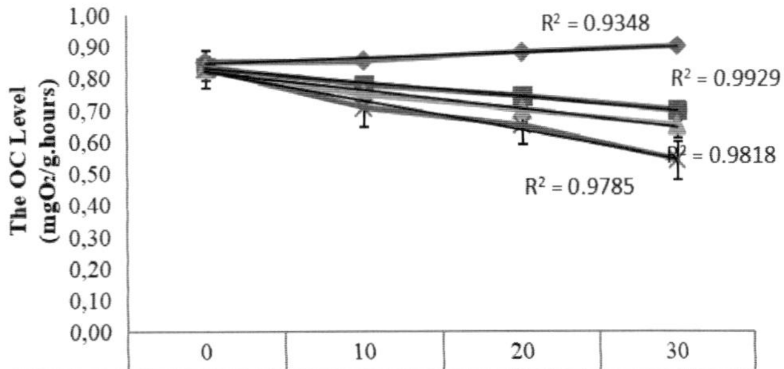

	0	10	20	30
0%	0,85	0,85	0,88	0,90
10%	0,83	0,78	0,74	0,70
20%	0,83	0,74	0,70	0,65
30%	0,83	0,70	0,65	0,54

Figure 3.The relationship between the OC level and the observation time

3.4. *Movement of operculum*

Based on figure 4, it can be seen that there was an increase in operculum movement in mahseer fish caused by exposure to wastewater, from observations from day 0 to day 30. In control, the movement of the operculum increased from 125 to 169 times/min. At a 10% concentration, the fish operculum movement also experienced a significant increase ranging from 130 times/min to 220 times/min. Whereas at 20% also increased from 137 to 245 times/min, 30% increased from 155 to 271 times/min. Between these variables, at the concentration of 0%, 10%, 20%, and 30% has a correlation value of 0.99, 0.98, 0.97 and 0.97 respectively. Based on those correlation values, it means that the duration of exposure and concentration of the wastewater to the operculum movement has a solid relationship

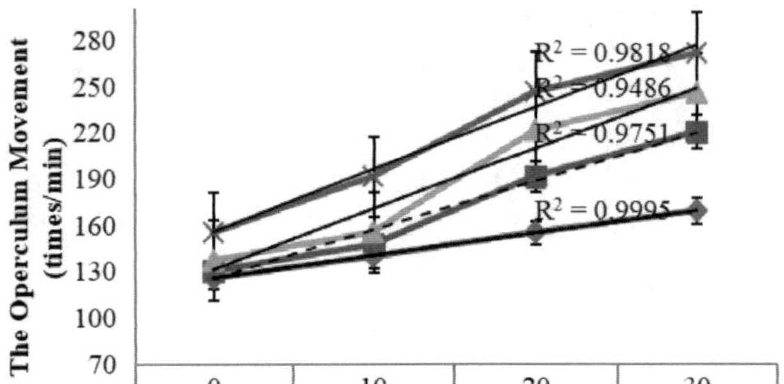

	0	10	20	30
0%	125	139	155	169
10%	130	147	191	220
20%	137	155	221	245
30%	155	191	246	271

Figure 4. The relationship between the operculum movement and the observation time

3.5. The relationship between OC level and operculum movement

The OC level and the movement of the operculum were related to each other. Both are related to breathing in fish. The relationship of OC level with the operculum movement can be seen in figure 5.

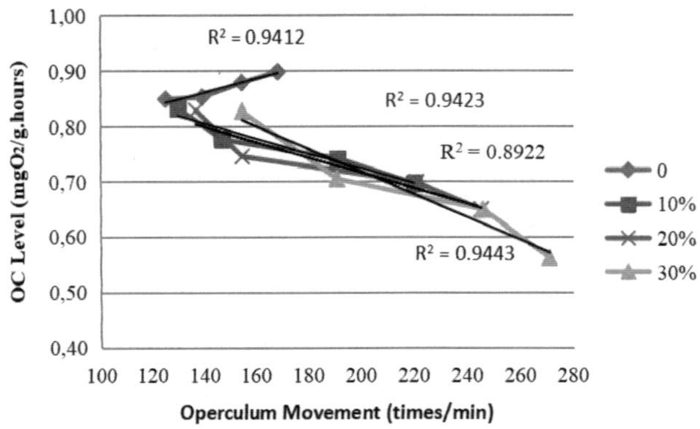

Figure 5. The relationship between OC level and operculum movement

Based on the exposure with a concentration of 0%, 10%, 20% and 30% obtained a correlation value of 0.97, 0.97. 0.94 and 0.97, respectively. The lower level of oxygen consumption causes this; the operculum movement has increased from the beginning of observation (day 0) to the end of observation (30th day).

Based on figure 5, a concentration of 0% the OC level of mahseer fish on day 0 to day 30 increased from 0.85 mg O_2/g.hour to 0.90 mg O_2/g.hour. At a 10% concentration, the OC level of fish decreased by 0.83 mg O_2/g.hour to 0.70 mg O_2/g.hour. Likewise, with a 20% concentration, the OC level also decreased from 0.88 mg O_2/g.hour to 0.65 mg O_2/g.hour. The same happened at a 30% concentration, decreased from 0.83 mg O_2/g.hour to 0.54 mg O_2/g.hour. At the concentration of 0%; 10%, 20%, and 30% wastewater has a correlation value of 0.97; 0.99; 0.99 and 0.98 respectively. So the concentration of XYZ Inc's wastewater exposure affected the OC of mahseer fish with a solid relationship. The decrease in oxygen consumption level occurred after the fish were exposed to XYZ Inc's wastewater exposure. This situation was because the metabolic processes in fish were disrupted, which causes the reduced ability of fish to consume oxygen caused by gills in fish that were damaged due to the influence of wastewater [23–25].

Fish take dissolved oxygen in water by filtering water that enters through the mouth and carries oxygen dissolved in water using gills. Breathing will begin to be disrupted if the water used as a place for living animals contains toxic substances. That would continuously enter the gills because of damage to the gills. The dissolved oxygen absorbed through the gills was reduced and caused the fish's operculum to work so that the respiration process keeps it running [26–28]

4. Conclusions

The LC_{50}-96h value of mahseer fish was 15.410%. The OC level of fish from day 0 to day 30, between concentration 10 - 30% decreased ranged from 0.83 mg O_2/g.hour to 0.56 mg O_2/g.hour. Movement of the operculum for 30 days of observation ranged between 125 to 271 times/minute. Correlation values of OC levels and operculum movement on the exposure time and wastewater concentration based on 0%, 10%, 20%, and 30% obtained correlation values of 0.97, 0.97. 0.94 and 0.97, respectively. The longer the time and the higher the concentration of tannery wastewater exposure, the OC level will decrease, and the operculum movement increases.

Acknowledgments

The Engineering Faculty of Universitas Andalas support the publication of this article.

References

[1] Lofrano G, Meriç S, Emel G and Orhon D 2013 Science of the Total Environment Chemical and biological treatment technologies for leather tannery chemicals and wastewaters : A review **462** 265–81

[2] HARYONO H 2006 Morphological Study for Identification Improvement Tambra Fish (Tor spp.: Cyprinidae) from Indonesia *Biodiversitas, J. Biol. Divers.***7** 59–62

[3] Azhar 2015 *Studi Bio-Ekologi Ikan Garing (Tor tambra CV: Cyprinidae) di Perairan Batang Anai Sumatera Barat* (Andalas University)

[4] Chandra P and Kulshreshtha K 2004 Chromium accumulation and toxicity in aquatic vascular plants *Bot. Rev.***70** 313–27

[5] Mondon J A, Duda S and Nowak B F 2001 Histological, growth and 7-ethoxyresorufin O-deethylase (EROD) activity responses of greenback flounder Rhombosolea tapirina to contaminated marine sediment and diet *Aquat. Toxicol.***54** 231–47

[6] Yushinta F 2004 *Fisiologi Hewan Air* (Jakarta: Rineka Cipta)

[7] Sivakumar P 2016 Phytoremediation of Tannery Waste Polluted Soil using Hyptis suaveolens (Lamiaceae) *Int. J. Pure Appl. Biosci.***4** 265–72

[8] Rand G M 1995 *Fundamentals Of Aquatic Toxicology Effects, Environmental Fate And Risk Assessment* ed G M Rand (CRC Press)

[9] Soemirat J H D A 2015 *Toksikologi Lingkungan* ed J Soemirat (Yogyakarta, Indonesia: UGM Press)

[10] Byung-Mu Lee; Sam Kacew; Hyung Sik Kim 2017 *Lu's Basic Toxicology: Fundamentals,*

Target Organs, and Risk Assessment 7th Edition (Washington DC: CRC Press)

[11] Ihsan T, Edwin T, Husni N and Rukmana W D 2018 Uji Toksisitas Akut Dalam Penentuan LC50-96H Insektisida Klorpirifos Terhadap Dua Jenis Ikan Budidaya Danau Kembar, Sumatera Barat *J. Ilmu Lingkung.* **16** 98

[12] Ihsan T, Edwin T and Yanti R D 2019 The effect of sublethal exposure of chlorpyrifos to nile tilapia (Oreochromis niloticus): Case study of twin lakes of West Sumatra, Indonesia *Nat. Environ. Pollut. Technol.* **18** 1399–403

[13] Ihsan T, Edwin T and Anggraeni W 2018 Behavioral responses of Nile tilapia (Oreochromis niloticus) by sublethal exposure to chlorpyrifos: a case study in Twin Lakes of West Sumatra *Environ. Heal. Eng. Manag.* **5** 205–10

[14] Federation A W W A P W A E 2017 *Standard Methods For The Examination OF Water and Wastewater, 23rd Edition* ed A D E E.W. Rice, R.B. Baird (USA: American Water Works Association/American Public Works Association/Water Environment Federation)

[15] Suyanto S R 2010 *Pembenihan dan Pembesaran Nila* (Jakarta: Penebar Swadaya)

[16] Vinodhini R and Narayanan M 2009 Heavy metal induced histopathological alterations in selected organs of the Cyprinus carpio L. (common carp) *Int. J. Environ. Res.* **3** 95–100

[17] Agency U S E P 2002 Methods for Measuring the Acute Toxicity of Effluents and Receiving Waters to Freshwater and Marine Organisms Fifth Edition October 2002 *Environ. Prot.* 266

[18] Amalia R, Marsi and Taqwa F H 2013 KELANGSUNGAN HIDUP, PERTUMBUHAN DAN TINGKAT KONSUMSI OKSIGEN IKAN PATIN (Pangasius sp.) YANG TERPAPAR LIMBAH CAIR PABRIK KELAPA SAWIT Survival Rate, Growth and Oxygen Consumption Rate of Catfish (Pangasius sp.) Ex- posed Industrial Palm Oil Mill Effluent *J. Akuakultur Rawa Indones.* **1** 203–15

[19] Ganeshwade R M, Rokade P B and Sonwane S R 2006 Behavioral responses of Cyprinus carpio to industrial effluents *J. Environ. Biol.* **27** 159–60

[20] Dedy Heryadi Sutisna R S 1995 *Pembenihan ikan air tawar* (Yogyakarta: Kanisius)

[21] Ng C K 2004 *Kings of the Rivers: Mahseer in Malaysia and the Region* (Malaysia: Inter Sea Fishery)

[22] Liao, I.C., Huang H J 1975 Studies on the respiration of economic prawns in Taiwan: I. Oxygen consumption and lethal dissolved oxygen of egg up to young prawn of Penaeus monodon Fabricius *J. Fish. Soc.* **4** 33–50

[23] Muthukumar G, Anbalagan R and Radha Krishnan K 2009 Adaptive changes in respiratory movements of an air breathing fish Tilapia mossambicus exposed to Endosulfan *J. Ind. Pollut. Control* **25** 67–72

[24] Witeska M, Kondera E, Lipionoga J and Jastrzebska A 2010 Changes in oxygen consumption rate and red blood parameters in common carp Cyprinus carpio L. after acute copper and cadmium exposures *Fresenius Environ. Bull.* **19** 115–22

[25] Julius, N. J.; Arimoro, F.O.; Edegbene, A.O.; Danbaba, N.; Jiya, A.G.; Usman, M.D.; Tsowa E 2014 Electronic Journal of Biosciences EFFECTS OF TANNERY WASTE ON NILE TILAPIA (Oreochromis niloticus) *Electron. J. Biosci.* **02** 63–6

[26] Qadir S, Bukhari R and Iqbal F 2015 Effect of sub lethal concentration of imidacloprid on proximate body composition of Labeo rohita *Iran. J. Fish. Sci.* **14** 937–45

[27] Kim B M, Saravanan M, Lee D H, Kang J H, Kim M, Jung J H and Rhee J S 2018 Exposure to sublethal concentrations of tributyltin reduced survival, growth, and 20-hydroxyecdysone levels in a marine mysid *Mar. Environ. Res.* **140** 96–103

[28] Schulte C M P 2016 *Principles of Animal Physiology 3rd Edition* (San Fransisco: Pearson)

Performance of Nitrogen Removal in Ceramic Anammox Reactor with Two-Inflow

Z Zulkarnaini[1] and S Silvia[1]

[1]Department of Environmental Engineering, Faculty of Engineering, UniversitasAndalas, Limau Manis, Padang, 20362, Indonesia

Corresponding author: zulkarnaini@eng.unand.ac.id

Abstract. Anaerobic ammonium oxidation (anammox) converted ammonium into nitrogen gas using nitrite as an electron donor in anaerobic conditions by anammox bacteria. Granular anammox was used commonly to remove ammonium from wastewater. It easy to be broken and tend to settle in the bottom of the reactor in the application. A pored ceramic carrier was proposed to cultivate anammox bacteria with two-inflow substrate feeding. The carrier with a diameter of 1.0 mm was set up in the center of the 1.0 L conical filtering flask reactor. The substrate with 70 mg-N/L of ammonium-supplemented substate was introduced into the reactor through the center of the ceramic carrier from the bottom and 70 mg-N/L nitrite-supplemented substrates into the outside of ceramic carrier with HRT 3 hours. The influent and effluent concentration measured once a week using ion chromatography. The experiment was conducted in the 35°C controlled temperature room. Anammox bacteria grew by filled and covered the ceramic carrier's pore, and a flock of anammox settled in the reactor. The reactor's performance achieved a maximum of 76.496 % NRE and 0.901 kg-N/m^3·d NRR, respectively.

1. Introduction

Anaerobic ammonium oxidation (anammox) process converted ammonium into nitrogen gas using nitrite as electron acceptor autotrophically under anoxic conditions [1]. The stoichiometry of anammox process is defined by Lotti et al. (2014),as shown in equation 1.

$$NH_4^+ + 1.146NO_2^- + 0.071HCO_3^- + 0.057H^+ \rightarrow 0.986N_2 + 0.161NO_3^- + 0.071CH_2O_{0.5}N_{0.15} + 2.002H_2O \quad (1)$$

Comparing with conventional traditional nitrification-denitrification technology for biological nitrogen removal, Anammox has economic advantages with lower operational costs, a smaller footprint, and useful to remove nitrogen from wastewater. The first experience, high NRR reached 9.5 kg-N/m^3·dat the first full-scale Anammox process in the Dokhaven municipal wastewater treatment plant in Rotterdam, Netherlands [3]. It proved that anammox is a powerful method for nitrogen removal. As new technology in wastewater treatment, more than 100 reactors are currently applied to treat ammonium-rich wastewater, such as landfill leachate, sludge-digestion liquid, coke-oven wastewater, and swine wastewater, monosodium glutamate wastewater, pharmaceutical wastewater, and mainstream wastewater [4].

In the reactor, the anammox process occurs in the form of suspended sludge, granular, and biofilm. Granular is commonly used in the application. However, it could break in the reactor and tend to settle in the bottom part of the reactor, where the performance decreased to the reactor's high. Other carriers used to retain and to distribute biomass, e.g., sponge [5], gel [6], membrane [7], string wound filter [8], palm fiber[9].

Pored-ceramic was studied as biological supporting media to cultivate anammox bacteria with a two-inflow supplied substrate. For the first time, a two-inflow system was introduced by Zulkarnaini et al. (2018) to cultivate anammox bacteria in a string wound filter for start-up one-stage nitritationanammox process. However, the thick anammox biofilm grows only on the surface and decreases substrate penetration to the biofilm's inner part. The performance of the anammox process was evaluated during the whole operation of the reactor.

2. Materials and methods

2.1. Reactor configuration

Ceramic anammox reactor (CAR) was conducted on a laboratory scale with a 1.0 L conical filtering flask. A pored-ceramic (2 cm x 2 cm x 5 cm, pore size 1 mm) was used as microbial supporting media to grow anammox biofilm, with a total volume of 5 ml^3. The ceramic carrier is put in the center of the reactor. The substrates were delivered to the reactor by two-inflow lines, ammonium into the center of the ceramic carrier from the bottom and nitrite into the outside using a peristaltic pump (EYELA, Japan). The reactor was set-up in the 35 ^oC temperature-controlled room. CAR was operated until the carrier was covered with anammox bacteria. Anammox bacteria Genus *Candidatus*Brocadiasinica was inoculated in CAR for start-up.Table 1 showed the operation condition of the CAR.

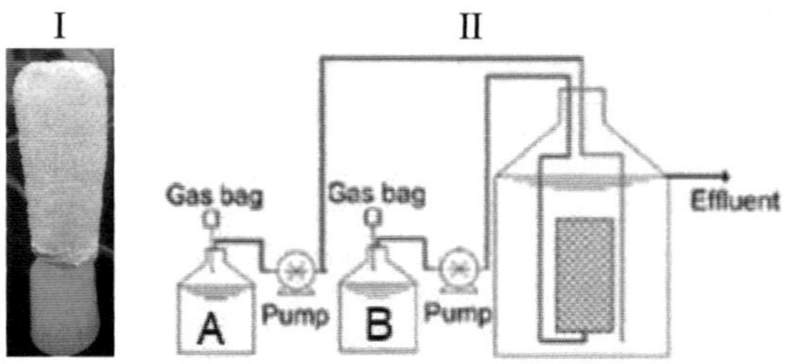

Figure 1. Ceramic carrier (I), Reactor configurations (II), tank A containing NO_2^--N and, tank B containing NH_4^+-N.

2.2. Substrates composition

The substrate solution's composition was (per L of tap water) KH_2PO_4, 27.2 mg; $MgSO_4.7H_2O$, 300 mg; $CaCl_2.2H_2O$, 180 mg; $CaCl_2$, 136 mg; $KHCO_3$, 500 mg; and 1 mL trace element solutions I and II[10]. The tank of the substrate is connected with an N_2-containing gas bag to maintain an anoxic condition. The solution was flushed with N_2 gas for 30 minutes before supplemented with substrate's compositions during preparation media. The substrate was supplied with 70 mg-N/L for both ammonium and nitrite with a Hydraulic retention Time (HRT) of 3 h.

2.3. Analytical methods

Samples for analysis were collected from influent and effluent lines once a week. The samples were filtered before the analysis with a 0.2-µm-pore-size membrane filter (Merc Millipore Ltd., Ireland). The concentration of ammonium, nitrite, and nitrate was analyzed using ion chromatography

(Shimadzu,IC-A3, IC-C3 Type Column, Japan). Parameter pH was measured using a pH meter (F-71, Japan).

The performance of anammox was defined as nitrogen removal rate (NRR, kg-N/m^3·d), nitrogen removal efficiency (NRE, %), and ammonium conversion efficiency (ACE, %) were calculated according to the equations describedby Zulkarnaini et al. (2020)[12]:

Table 1. Operation Condition of CAR.

Process	Time (d)	NH_4^+-N (mg/L)	NO_2^--N (mg/L)	HRT (h)
Anammox	1-67	70	70	3

3. Results and discussion

The reactor was operated for 67 days for anammox process with two-inflow lines with separated supplied ammonium and nitrite. Anammox bacteria were grown in the carrier. Two-inflow supplied ammonium and nitrite caused the movement of enriched anammox biomass (*Candidatus* Brocadia sinica) to the carrier due to the physiological characteristics of anammox bacteria, which was inhibited by excess nitrite while ammonium is not an inhibitor for anammox bacteria. Figure 2 shown the correlation between ammonium conversion, nitrite conversion, and nitrate production during operation CAR. It was evident that the stoichiometric ratio of the anammox process close to the ratios obtained in this study with two-inflow CAR operation. This experiment indicated that anammox bacteria active and growing in the ceramic carrier. The lower conversion ratio of NO_3^--N/NH_4^+-N may be due to the denitrification process in the reactor where denitrification bacteria used decay of anammox biomass as a carbon source [13].

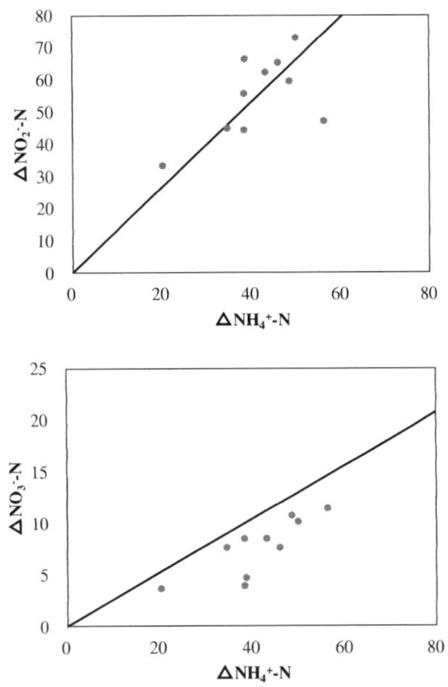

Figure 2. Correlation between ammonium conversion, nitrite conversion, and nitrate production.

Figure 3 shown the performance CAR. During the first week of operation, a small amount of ammonium and nitrite was removed from the reactor. This condition could be an adaptation of inoculant with operation condition where nitrite contacted directly. Nitrite could be an inhibitor for anammox process and decreased the cell metabolism of anammox bacteria. Then ammonium and nitrite converted to nitrogen gas where ACE achieved 85% and almost all supplied nitrite consumed for anammox process achieved 98%. Carmine biomass growth and covered ceramic carrier indicated anammox bacteria moved to and growth in the ceramic carrier. After a 1-month operation, nitrite concentration increased in the effluent while ammonium concentration stable.

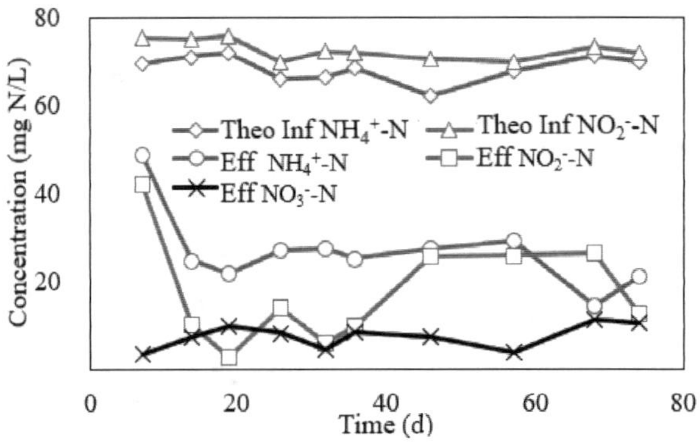

Figure 3. Nitrogen profile on CAR operation.

This problem occurred because a pump failure led to an unbalance of the flow rate two-line feeding substrate. The supplied nitrite was higher than ammonium slightly. This technical problem caused reactor exposure to nitrite concentration for a long time. Strous et al. 1999 reported at a level nitrite concentration of 5 mmol NO_2^-/L anammox activity completely lost in the reactor for more than 12 hours of exposure. This situation continued until the peristaltic pump changed with a stable flow rate. As a result, nitrite concentration decreased in the effluent, and anammox process could be recovered.

Figure 4 showed the performance of CAR. The reactor supplied with 1.120 kg-N/m^3·d during the whole operation of the reactor. Anammox process occurred at the first-week operation where NRE, NRR reached 34.590%, 0.400 kg-N/m^3·d, respectively. The maximum nitrogen removal was achieved at 19 days of operation with 76,496 % NRE and 0.901 kg-N/m^3·d NRR. The nitrogen removal tended to decrease and reached the lowest NRE and NRR of 54.124% and 0.574 kg-N/m^3·d, respectively. The nitrogen removal could be recovered after set-up the new peristaltic pump, and the performance showed NRE and NRR reached 68.667% and 0.779 kg-N/m^3·d, respectively.

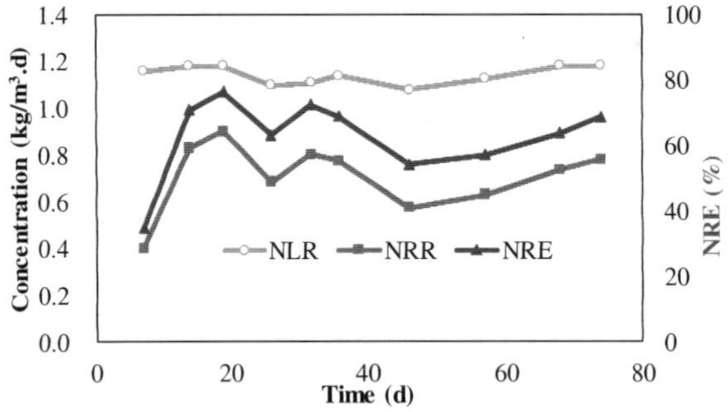

Figure 4. Performance of nitrogen removal in the CAR.

At the end of the operation, the ceramic carrier was taken out from the reactor to observeanammox bacteria's growth. Figure 5 showed the carrier covered by anammox bacteria and filled the ceramic carrier's pore until the inner part of the carrier. The anammox bacteria belong to

*Candidatus*Brocadiasinica that was used as inoculum was sensitive to nitrite concentration. Two-inflow in CAR provided a better environment for anammox bacteria living in the ceramic carrier and pushed-up anammox bacteria to move and grow from seed to the carrier.

Figure 6 showed the proposed process of anammox bacteria moving in the reactor.Anammox bacteria prefer to grow in the filter supplied with ammonium than outside due to exposure to nitrite. It was reported that *Candidatus*Brocadia had lower tolerance on nitrite (70 mg/L) compared to other anammox bacteria such as *Candidatus* Kuenenia (70 mg/ L)[15]. This result better than others reported anammox biofilm using a filter where anammox biofilm grows only in the surface [8].

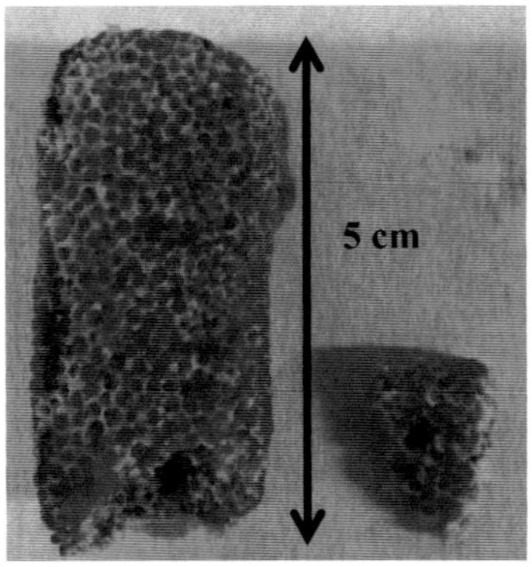

Figure 5. Anammox biofilm growth in the ceramic carrier.

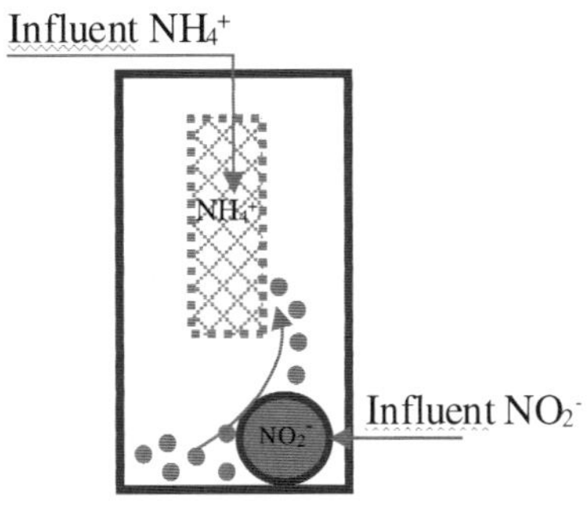

Figure 6. The Proposed moving process of anammox bacteria.

4. Conclusion

Fast start-up of anammox bacteria reached at 19 days operation. The reactor's performance reached maximum NRE, and NRR were 76,496 % and 0.901 kg-N/m^3·d, respectively. Anammox bacteria growth covered the pored-ceramic carrier and filled the pores. Two-inflow lines provide a suitable environment for anammox bacteria growth.

References

[1] Van De Graaf A A, Mulder A, Bruijn P D, Jetten M S M M S M, Robertson L A, Kuenen J G, Jetten M S M M S M and Kuenen J G 1995 Anaerobic Oxidation of Ammonium Is a Biologically Mediated Process *Appl. Environ. Microbiol.***61** 1246–51

[2] Lotti T, Kleerebezem R, Lubello C and van Loosdrecht M C M M 2014 Physiological and kinetic characterization of a suspended cell anammox culture *Water Res.***60** 1–14

[3] Van der Star W R L L, Abma W R, Blommers D, Mulder J W, Tokutomi T, Strous M,

Picioreanu C and van Loosdrecht M C M M 2007 Start-up of reactors for anoxic ammonium oxidation: Experiences from the first full-scale anammox reactor in Rotterdam *Water Res.* **41** 4149–63

[4] Lackner S, Gilbert E M, Vlaeminck S E, Joss A, Horn H and van Loosdrecht M C M 2014 Full-scale partial nitritation/anammox experiences - An application survey *Water Res.* **55** 292–303

[5] Lu Y F, Ma L J, Ma L, Shan B and Chang J J 2018 Improvement of start-up and nitrogen removal of the anammox process in reactors inoculated with conventional activated sludge using biofilm carrier materials *Environ. Technol. (United Kingdom)* **39** 59–67

[6] Isaka K, Kimura Y, Matsuura M, Osaka T and Tsuneda S 2017 First full-scale nitritation-anammox plant using gel entrapment technology for ammonia plant effluent *Biochem. Eng. J.* **122** 115–22

[7] Suneethi S and Joseph K 2011 ANAMMOX process start up and stabilization with an anaerobic seed in anaerobic membrane bioreactor (AnMBR) *Bioresour. Technol.* **102** 8860–7

[8] Zulkarnaini, Yujie Q, Yamamoto-Ikemoto R and Matsuura N 2018 One-stage nitritation/anammox process using a biofilm reactor with two-inflow *J. Water Environ. Technol.* **16** 106–14

[9] Zulkarnaini, Nur A and Ermaliza W 2019 Nitrogen Removal in the Anammox Biofilm Reactor using Palm Fiber as Carrier in Tropical Temperature Operation *J. Ris. Teknol. Pencegah. Pencemaran Ind.* **10** 7–15

[10] Putra R P, Zulkarnaini and Komala P S 2020 Start-Up Anammox Process Using Sludge from Koto Baru Lake as Inoculum *Teknol. Lingkung.* **21** 138–46

[11] Zulkarnaini, Yujie Q, Yamamoto-Ikemoto R and Matsuura N 2018 One-stage nitritation/anammox process using a biofilm reactor with two-inflow *J. Water Environ. Technol.* **16**

[12] Zulkarnaini, Afrianita R and Putra I H 2020 Application of Anammox Process in Nitrogen Removal Using Up-Flow Anaerobic Sludge Blanket Reactor *J. Teknol. Lingkung.* **21** 31–9

[13] Kindaichi T, Awata T, Mugimoto Y, Rathnayake R M L D, Kasahara S and Satoh H 2016 Effects of organic matter in livestock manure digester liquid on microbial community structure and in situ activity of anammox granules *Chemosphere* **159** 300–7

[14] Strous M, Kuenen J G and Jetten M S M 1999 Key Physiology of Anaerobic Ammonium Oxidation Key Physiology of Anaerobic Ammonium Oxidation *Appl. Environ. Microbiol.* **65** 0–3

[15] Schmidt I, Sliekers O, Schmid M, Bock E, Fuerst J, Kuenen J G, Jetten M S M and Strous M 2003 New concepts of microbial treatment processes for the nitrogen removal in wastewater *FEMS Microbiol. Rev.* **27** 481–92

Effect of Occupational Noise Exposure to Work-Fatigue of Indonesian Crumb Rubber Plants

T Ihsan[1], N Afriani[1], T Edwin[1], V Derosya[2], P N Sari[3]

[1]Department of Environmental Engineering, Faculty of Engineering, Universitas Andalas, Kampus Unand Limau Manis, Padang, West Sumatra 25163, Indonesia
[2]Department of Agroindustrial Technology, Faculty of Agriculture, Universitas Andalas
[3]Department of Environmental and Occupational Health, Faculty of Public Health, Universitas Andalas

Corresponding author:taufiqihsan@eng.unand.ac.id

Abstract. The purpose of this research was to analyze the effect of noise on the level of work-fatigue in the production area of three crumb rubber plants at Padang City, Indonesia. It also examines the impact of age, period work, marital status, and education level on work-fatigue. Occupational noise in the working area was measured using a Sound Level Meter, while work-fatigue was measured using the Indonesian fatigue measurement questionnaire (KAPUK2). The respondents of this study were workers in these plants, 213 people in the wet-area, and 135 people in the dry-area. Results showed that the average noise of the production area had exceeded the threshold value, which was 90.98 dB. There were 26.19% experienced mild-fatigue, 71.43% moderate-fatigue, and 2.38% heavy-fatigue. Noise, age, and working experience were a related and significant influence on work-fatigue, where the probability value was respectively 0.046, 0.001, and 0.000 ($p \leq 0.05$). The results of multiple regressions showed that occupational noise was the most influential factor in the occurrence of work-fatigue with an odds ratio of 3.92.

1. Introduction

Noise can be identified as an unwanted voice. This state conveys that noise is voice pollution. In addition, the noise has been ranked second among several environmental triggers for public health impacts [1], including hearing loss. In general, hearing loss is the fourth highest cause of disability, and there are an estimated 466 million people with hearing loss [2]. In developing countries, occupational noise exposure is the second most self-reported occupational disease with functional, social, emotional, and economic impacts on industrial workers [3]. This situation is caused by exposure to sound levels or duration that damage cochlear hair cells, and may have been advanced at that time to cause sufficient disability [4]. It can develop gradually over time with excessive noise exposure resulting from the manufacturing industry [5].

Exposure to excessive work noise can cause hearing loss, which is a partial or complete hearing loss in one or both ears as a result of work [6]. Worldwide, 16% of crippling hearing loss in adults is caused by work noise. Hearing loss caused by noise at work is a common occupational disease. Hearing loss is irreversible and irreversible; however, hearing loss can be prevented by health management in the workplace. In recent years, many studies have emerged in order to avoid people from being affected by noise. Not only hearing loss, constant exposure to noise also causes health problems such as sleep disorders [7]–[10], cardiovascular effects [11], [12], learning disorders [13], [14]and chemical hypertension of heart disease [15]. "Burden of Disease from Environmental Noise," as the most recent document published by the World Health Organization (2018) [2], has provided evidence-based descriptions of the consequences of environmental noise on health. According to the report, noise is considered as one of the most severe general physical factors that contribute to an increased risk of environmental and occupational health hazards in the workplace.

Nowadays, economic development in most countries leads to extraordinary growth in industrial processes and technological advancements, thus impacting living standards. Most industries make less effort to increase the consequences of their environmental performance as a result of various production activities that produce noise [16]. However, people were faced with attacks of noise pollution as one of the critical problems in their work environment and residence. This technological growth leads to health complications, which include: hearing loss and mental balance [17]–[19]. Research has shown that noise from factories has severe consequences for changes in worker performance and productivity [20], [21] as well as poor communication resulting in increased chances of work accidents [22]. Production quality and volume are essential factors in the production process, which can be hampered by poor environmental conditions such as noise, thus affecting the efficiency of a company [23], [24]. Noise is a universal occupational health problem today, especially in developing countries. This situation is worse in these countries because less effort is spent on noise control [25]. Work hazards have indeed declined in developed countries, due to the introduction of protective measures. Data on developing countries are not available, but the evidence shows that average sound levels are well above the recommended employment threshold in many developed countries [26]. Thus, the average noise level in developing countries will probably increase significantly because noise control and protection are not always carried out in conjunction with industrialization. This is one of the common problems associated with manufacturing that can cause large expenditures and severe health problems [27].

In Indonesia, the noise quality standard for the industrial work environment was 85 dB (A) (8 hours/day), which was regulated based on the Regulation of the Indonesian Health Ministry Number 70 of 2016 concerning Standards and Requirements for Industrial Work Environment Health. When comparing standards with the National Institute for Occupational Safety and Health (NIOSH), then this Indonesian standard was quite stringent. The NIOSH standard sets 85 dB for the allowable time of 16 hours per day as the Recommended Exposure Limit (REL). It makes specific recommendations on key elements of an effective hearing conservation program. Besides, the OSHA standard also determines 15 minutes. Time Weighted Average (TWA) exposure should not be exceeded at any time during the working day, even if the overall eight hours of TWA is below the Threshold Value and the Total Weighted value (TLV-TWA). Workers may not be exposed more than four times per day to concentrations between TLV-TWA and Occupation Exposure Limit (OEL). The NIOSH standard was a more protective limit for health. It was estimated that about one in four workers exposed to 90 dB eight hours per day for 40 years of work would suffer a hearing loss that can be compensated for by noise, compared to only about one in twelve workers who exposed to NIOSH REL 85 dB The highest opening time of 85 decibels was eight hours, while the noise level was 110 dB, the maximum opening time should not exceed one minute and 29 seconds [6].

In terms of application, negligence was still ordinary in Indonesia. Including in the Padang city, the capital of the West Sumatra province. In this city, there were three crumb rubber plants with huge production for the export of Indonesian rubber. The production process carried out by these three companies uses tools that produced loud noises and cause noise. Noise is one of the causes of work-fatigue. Work-fatigue was an individual's total response to psychosocial stress experienced in a certain period, and work-fatigue tends to reduce the achievement or motivation of the worker concerned. Work-fatigue was not only related to physical and psychological fatigue but is more related to the feeling of fatigue, decreased motivation, and decreased work productivity [28]. Subjective work-fatigue is a feeling of fatigue that is felt by workers, besides it results in reduced motivation and can reduce mental and physical activity to a certain degree. Subjective work-fatigue measurement can use the fatigue measurement questionnaire, namely, in the Indonesian Language is "Kuesioner Alat Ukur Perasaan Kelelahan Kerja" (KAUPK2). This questionnaire method adopts the IFRC and Fatigue Assessment Scale/FAS. The KAUPK2 questionnaire, with 17 questions, aims to find out all unpleasant feelings in workers who experience work-fatigue. The advantage of the KAUPK2 method is simple, valid, reliable, and in Indonesian (suitable for Indonesian workers) [28].

This study aims to analyze the level of noise and the effect of work environment noise on work-fatigue, which occurs in the production area of the three crumb rubber plants located in the city of Padang. During 2019, there were 11 work accidents in the production area of these plants. The cause of these cases was due to employee negligence due to fatigue.

2. Materials and methods

This research was carried out in production areas in three crumb rubber plants in Padang City, West Sumatra, Indonesia, which are along the Padang City by-Pass Road. Research time was conducted in November 2019 to January 2020 (before COVID-19 pandemic in Padang), including the time of field observations, distribution of questionnaire sheets (KAUPK2), sampling, data processing, and analysis.

The observation was carried out by direct observation in the production area by seeing and observing the work carried out in the production process as well as interviews conducted to get a general picture of the data needed in this study. The interview was explicitly for workers who work in the production area. Respondents were all workers and divided into 213 people in wet areas and 135 people in dry areas.

Noise measurements were carried out using a Sound Level Meter (SLM) tool in the production area for eight hours during working hours (08.00 AM - 04.00 PM). Noise measurement in each factory has five sampling points (where workers carry out work activities close to the source of noise), three points in the wet-area, and two points in the dry-area. The wet-area sampling point consists of point 1 (breaker), point 2 (hammermill), and point 3 (creeper). While the sampling point in the dry-area consists of point 4 (cutter) and point 5 (dryer), each location was carried out four times when data collection, namely:

- L1 was taken at 09.00 AM representing the hours of 08.00 - 10:00 AM;
- L2 was taken at 11.00 AM representing 10:00 AM - 12.00 PM;
- L3 was taken at 1.00 PM representing 12.00 - 02.00 PM;
- L4 was taken at 03.00 PM representing hours 02.00 - 04.00 PM.

Furthermore, the points were measured using SLM for 10 minutes, with readings carried out every five seconds to obtain 120 data points, and the results were recorded. The determination of this measurement point refers to Indonesian Standard No 7231 of 2009[29].

Statistical tests to see the effect between variables with work-fatigue, using the Pearson Product Moment correlation method. Testing multiple regression analysis needs to be done to see which variable most influences work to fatigue in the crumb rubber industries

3. Results and discussion

3.1. Noise Measurement Results

The results of noise measurements in production areas (wet and dry areas) at each plant can be seen in the following table 1.

Table 1. Noise measurement results in production areas.

Plant	Noise Measurement (dB)	Mean ± SD
Wet-Area		
I	88.1	
II	90.1	89.9 ± 1.8
III	91.7	
Dry-Area		
I	94.9	
II	85.1	90.1 ± 4.9
III	90.2	

Based on table 1 shows that the noise at measurement points in both wet and dry areas in each factory has exceeded the Threshold Value (TLV) based on the Regulation of the Indonesian Health Ministry Number 70 of 2016. The allowable TLV was 85 dB. The average noise in the wet-area is 89.9 dB, while the dry-area is 90.1 dB.

3.2. Characteristics of Respondents

Workers who were respondents of this study were 348 people. Table 2 shows the characteristics of respondents.

Table 2. Baseline characteristics of respondents.

Characteristics		Percentage (%)
	Teenager (<25)	10.71
Age (year)	Mature (25-45)	65.48
	Old (>45)	23.81
	< 5	19.05
Working Experience (Year)	5-10	26.19
	> 10	54.76
Marital Status	Unmarried	11.90
	Married	88.10
Education	High School	26.19
	Senior High School	73.81

Based on table 2, it can be seen that the most age categories are adults with a range of 26 - 45 years (65.48%), wherein that age range workers are in productive age at work. The most considerable frequency of tenure was in the> 10-year group by 54.76%. This data shows that during the work period, workers are more experienced in carrying out their work. The percentage of the frequency of married workers was 88.1%. This number was because workers were in the age range that should already have a family. Furthermore, most of the last education level of workers was high school with 73.81%

3.3. Fatigue Measurement Results using the KAUPK2

Categories on this questionnaire were divided into three, namely mild work-fatigue with a score of 17-45, moderate work-fatigue with a score of 46-74, and heavy work-fatigue with a score of 75 - 102 [28]. The frequency distribution of work-fatigue categories can be seen in figure 1. Based on figure 1, it can be seen that the level of work-fatigue was in the moderate category (71.43%). Workers who experience heavy fatigue category were workers who work in wet-area, especially near creeper and hammermill machines.

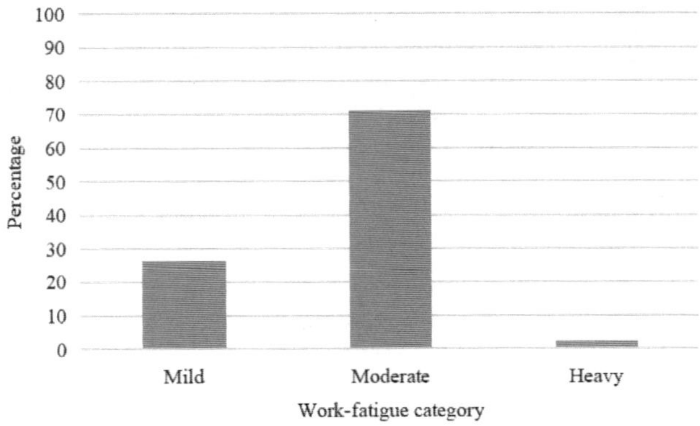

Figure 1. Distribution for work-fatigue.

3.4. Relationships and Effects of Noise on Work-fatigue

Based on a statistical analysis of correlation and the t-test of noise and work-fatigue produces a probability value (p-value) of 0.046 ($p \leq 0.05$) and a t-value of 2.024> t-table of 1.989. This value shows that noise is significantly related and has an influence on work-fatigue in crumb rubber plants of Padang.

In addition to hearing loss, noise exposure can cause non-hearing effects, such as disturbed perception, sleep disturbance, and cardiovascular disorders [30], [31]. Noise exposure can cause an increase in the release of stress hormones, imbalances in organ homeostasis, changes in the cardiovascular system and metabolism, and worsening risk factors for heart disease (high blood pressure, serum lipid disorders, blood viscosity), and increased blood glucose [32], [33]. Noise as an unpleasant sound, so that disturbs humans physically and physiologically [34], [35]. The noise was a general health hazard in the workplace that causes discomfort, to stress to exhaustion. It is natural that in some studies, the noise was the most challenging feature to handle, because of this subjective phenomenon [36], [37].

3.5. Relationship and Effect of Age on Work-fatigue

The results of the calculation of the correlation value and age t-test on work-fatigue produce a p-value of 0.001 ($p \leq 0.05$) and a value of t-value 3.493> t-table 1.989. This value shows that age has a relationship and a significant effect on work-fatigue.

More workers in this age range can potentially increase work productivity. This condition was an advantage for the company. If the worker is in the age range, then work productivity decreases, because the body has undergone tissue changes, so the body's strength has decreased. Age can affect maximum power in one's muscles, where the peak of muscle tightness is in the age range of 25-35 years, and one's muscle strength will decrease by around 15-25% in the age range of 50-60 years [38]–[40].

As humans get older, the feeling of fatigue a person feels will get faster. They are usually caused by a decreased physical state such as sharp eyesight, hearing, the speed of distinguishing things, making decisions, and the ability to remember short-term [41].

3.6. Relationship and Effect of Working Experience on Work-fatigue

Based on a statistical analysis of correlation and t-test work period of work, fatigue produces a p-value of 0,000 ($p \leq 0.05$) and a t-value of 3.916> t-table of 1.989. That value means the years of service have a relationship and have a significant effect on work-fatigue.

The working experience was one of the factors causing work-fatigue. Because the longer a person works will increase a feeling of saturation that is felt because the job done is monotonous and carried out repeatedly. Longer work will cause fatigue and boredom. The longer a person works, the faster the adaptation process. This condition causes the pressure felt by workers to be more constant. This adaptation process provides both positive and negative effects for workers, where the positive impact is that it can improve work performance or increase work activities and can reduce tensions felt by workers. While the adverse effect is the body's resistance will decrease with a growing work period so that the body feels tired quickly. Benefits for industries with experienced workers in completing their work so that obstacles in working can be rapidly overcome. The longer a person works, the more qualified he is in doing his job [42]–[45].

3.7 Relationship and Effect of Marital Status on Work-fatigue

Based on the statistical analysis, the correlation of marital status to work-fatigue yields a p-value of 0.189 ($p > 0.05$). Marital status can affect the quality of work, both positive and negative influences. The positive effect is that married workers will be more enthusiastic about doing their job because of the support and encouragement of the family and the desire to make their family happy. In contrast, the negative effect of married workers on the quality of work will decrease because family problems will affect the worker. In this study, marital status did not have an impact on work-fatigue at the crumb rubber plants of Padang.

3.8. Relationship and Effect of Education on Work-fatigue

Based on a statistical analysis of correlation and t-test, the level of education on work-fatigue produces a p-value of 0.408 ($p > 0.05$) and a t-value of -0.832 <t-table 1.989. The level of education was proportional to the quality of work. The higher one's knowledge, the wider the way of thinking, the power of initiative becomes more elevated, and it is easier for someone to find efficient ways of completing their work so that the quality of their work also increases. In this study, there was no significant relationship between education level and work exhaustion.

3.9. Multiple Regression Analysis

Based on variables that have an influence on work-fatigue, a multiple regression test is performed to see the odd ratio that affects work-fatigue in this study. Table 3 shows the results of these calculations.

Table 3. Variables effect to work-fatigue.

Variables	Odds Ratio
Noise (X_1)	3.92
Age (X_2)	0.25
Work Experience (X_3)	1.79
Another Factor	15.33

Table 3 shows that the noise variable has the most significant influence at 3.92 times compared to other variables in affecting work-fatigue at these three crumb rubber plants. So the higher the noise level, the higher the potential for work-fatigue in workers.

Noise control efforts that can be done are:

1. Control of the noise source by insulating the source and replacing the use of the machine with more noise-friendly technology;
2. Control over the distance between the source and the worker/receiver. The industry can provide insulation and shielding between workers and noise sources. In addition, the industry can use automatic machinery;

Control of workers/recipients. The industry increases the discipline of workers in using PPE that has been given. Besides, the industry can also regulate and reduce work time by rotating the work area until the shift time

4. Conclusions

The average level of noise that occurs in the production area of three crumb rubber plants of Padang was 90.98 dB already exceeding the TLV set by the Regulation of the Indonesian Health Ministry Number 70 of 2016. Percentage of the level of work-fatigue that occurs in the production area was light work-fatigue 26.19%, moderate work-fatigue 71.43% and heavy work-fatigue 2.38%. Noise, age, and length of work were related and significantly influence work-fatigue, where the p-values were 0.046, 0.001, and 0.000, respectively ($p \leq 0.05$). The multiple regression results explain that noise was the most influential factor in the occurrence of work-fatigue, with an odds ratio of 3.92.

Acknowledgments

This research was supported by Research and Community Service Institutions (LPPM), Universitas Andalas, Padang. The authors are very grateful to all the participants who joined in the study.

References

[1] Ezzati M, Lopez A D, Rodgers A Murray C J L2004 *Comparative Quantification of Health Risks: Sexual and reproductive health* Geneva: World Health Organization p1200
[2] World Health Organization 2018 *Addressing the rising prevalence of hearing loss*(Geneva: World Health Organization) p 40
[3] World Health Organization 2009 *Global health risks: mortality and burden of disease attributable to selected major risks*(Geneva: World Health Organization) p 70
[4] Zaw A K Myat A M Thandar M Htun Y M Aung T H Tun K M and Han Z M 2020 *Saf. Health Work* **11**, 2 p. 199–206
[5] Pelegrin A C Canuet L Rodríguez Á A Pilar M and Morales A2020 *Noise Health* **17** 78 343–349
[6] Occupational Safety and Health (OSHA), 2015 *Occupational safety and health act*(Washington: OSHA) p 28
[7] Miedema H M E and Oudshoorn C G M 2001 *Environ. Health Perspect.* **109** 4 409–416.
[8] Muzet Ã A, 2007 *Sleep Med Rev.* **11** 2 135–142.
[9] Licitra G Fredianelli L Petri D and Angela M2015 *Sci. Total Environ.* **568** 1315-1325
[10] de Kluizenaar Y De Janssen S A Lenthe F J Van and Mackenbach J P,2014 *J. Acoust. Soc. Am* **126** 2 626-533
[11] Babisch W Beule B Schust MKersten NIsing H 2015 *Epimiol.* **16** 1 33–40.
[12] Bluhm G and Eriksson C 2011 *Noise Health.*13 212–216
[13] Lercher P Evans G W and Meis M 2003 *Environ. Behavior.* **35** 6 725–735
[14] Chetoni M Ascari E Bianco F Fredianelli L and Licitra G 2016 *Noise Mapp.* **3** 157–171.
[15] Kempen E Van and Babisch W, 2012*J. Hypertens.* **30** 6 1075–1086
[16] Djekic I and Smigic N, 2013 *J. Clean. Prod.* **41** p 263–269
[17] Fyhri A and Marit G, 2010 *Sci. Total Environ.* **408** 21 p 4935–4942
[18]Garcia S L Smith K J and Palmer C, 2018 *J. Mil. Med.* **183** 9–10 e547–e553
[19] Rahman A and Badayai A2012 *Procedia - Soc. Behav. Sci* **42** 486-495
[20] Melamed S Fried Y and Froom P 2004 *J. Occup. Environ. Med.* **46** 10 1023–1032
[21] Nassiri P Monazam M Dehaghi B F Abadi L I G Zakerian S A and Azam K2013 *Int. J. Occup. Environ. Med.* **4** 2 87–95
[22] Sen T Bhattacharjee P K Banerjee D and Sarkar B2010 *Int.J. Environ. Sci. Dev.* **1** 364–367
[23] Meerding W J Ijzelenberg W Koopmanschap M A Severens J L and Burdorf A, 2005*J. Clin. Epidemiol.* **58** 5 517–523
[24] Kahya Ã E, 2007 *Int. J. Ind. Ergon.* **37** 6 515–523
[25] NIOSH, 1998 *Criteria for a recommended standard - Occupational noise exposure, revised criteria*(Cincinnati: NIOSH) p 126
[26] Berger E H, 2003 *The Noise Manual 5th Edition*(USA: American Industrial Hygiene Association) p 796

[27] Ali S A 2011 *Appl. Acoust.* **72**(4) 221–225.
[28] Maurits L S K 2013 *Introduction of Work-fatigue, fourth ed* (Yogyakarta Indonesia: Amara Books)
[29] The National Standardization Agency of Indonesia, 2009 *Indonesian Standard No 7231 of 2009 concerning the Method of Measurement of Noise Exposure in the Workplace* (Jakarta Indonesia: BSN)
[30] Basner M Babisch W, Davis A, Brink M, Clark C, Janssen S, Stansfeld S 2013 *Lancet.* **383** 1325–1332
[31] Moghadam S R Tizabi M N L, Khanjani N, Emkani M, Manesh V T, Mohammadi A A, Delkhosh M B, Najafi H 2015 *J. Occup. Health Epidemiol.***7** 53–64
[32] Ghotbi M R Khanjani N Barkhordari A Moghadam S R Mozaffari A and Gozashti M H, 2013 *Environ. Monit. Assess.* **185** 8809–8814
[33] Gori T Babisch W Basner M and Mu T2014 *Eur. Heart J.* **35** 829–836.
[34] Zamanian Z Rostami R Hasanzadeh J and Hashemi H2013 *J. Environ. Public Health.* **2013** 1-3
[35] Bolaji B O, Olanipekun M U, Adekunle A A, Adeleke A E 2018 *J. Clean. Prod.* **172** 1800–1806
[36] Yogeswara T, Siddiqui N A, Hamsagar R S, Muenster R N 2013 *J. Ind. Pollut. Control.* **29** 2 185–191
[37] Murphy C F B Pagan-neves L O Wertzner H F and Schochat E2015 *PLoS ONE.* **9** 3 e93091
[38] Enoka R M and Duchateau J, 2008 *J. Physiol.* **586** 1 11-23
[39] Fuller J R Lomond K V Fung J and Côté J N 2009 *J. Electromyogr. Kinesiol.* **19**, 6 p. 1043–1052
[40] Katrine A Sjøgaard G Madeleine P Baare H and Søgaard K, 2005 *J. Electromyogr. Kinesiol.* **15 2** 138-148
[41] Tarwaka, 2010 *Industrial Ergonomics Basic Knowledge of Ergonomics and Application at Work* (Surakarta Indonesia: Harapan Press) p 568
[42] Folkard S and Lombardi D A 2006 *Am. J. Ind. Med.* **49** 11 953-963.
[43] Åkerstedt T, Anund A, Axelsson A, Kecklund G 2014 *J. Sleep Res.* **23** 3 240-252
[44] Caruso C C, Histchcock E M, Dick R B, Russo J M, Schmit J M 2004 *Overtime and extended work shifts: recent findings on illnesses, injuries, and health behaviors* (USA: NIOSH Publications Dissemination) p 49
[45] Ihsan T Rachmatiah I Salami S 2020 *J. Kesehatan Masyarakat* **15** 3 316–323.

IOP Conf. Series: Materials Science and Engineering 1041 (2021) 012039 doi:10.1088/1757-899X/1041/1/012039

Quality improvement of pillow product using Quality Function Deployment (QFD) methods at PT. XYZ

R Ginting[1], A Ishak[1], I T Primsa[1] and A F Malik[1]

[1]Department of Industrial Engineering, Faculty of Engineering, Universitas Sumatera Utara, Medan, Indonesia 20155

Corresponding author: rosnani@usu.ac.id

Abstract. PT XYZ is a company engaged in the manufacturing of bedding goods. One of the products from this company that will be the object of research is the pillow. The number of disability pillow products that occurred during 2018 amounted to 21.69%, which could cause a decrease in company productivity and the shift of consumers of these companies to competing companies. The method used to improve the quality of pillow products is Quality Function Deployment. In QFD Phase I, consumer desires are translated into technical characteristics. The results of QFD Phase I, show that the highest priority technical characteristics are machine capacity with a difficulty level of 4, importance of 20% and estimated cost of 19%. Determination of product critical parts using QFD Phase II. The results of QFD Phase II show the highest priority of critical parts based on the level of difficulty, degree of importance and estimated cost is durability of 20%, measurement accuracy of 15% and fabric dimensions of 15%.

1. Introduction

The rapid development of technology has led to the rapid changes that have occurred in the business world and in the sleeping equipment industry, which is a cycle of rapid development and is continuously needed by society. Sleep quality is a person's satisfaction with sleep so that someone does not show feelings of tiredness, irritability and anxiety, lethargy and apathy[1]. Therefore, humans need time and good sleeping equipment so that their sleep is comfortable and their health is not disturbed.

The National Sleep Foundation is one of the organizations in the United States that is very focused on researching and discussing every sleep problem in America because many citizens are sleep deprived, and the severity of which often experiences insomnia due to discomfort during sleep both in terms of time, room atmosphere and sleeping equipment. This organization conducted research to find out "How Americans Think About Key Elements of Their Bedrooms and to Determine the Effect of the Bedroom Environment on Sleep". Pillows are needed during sleep to support the head in order to achieve harmony in the body. Pillows appear as one of the products that humans need to meet their needs during rest, especially to provide comfort and support human posture, resulting in quality sleep [2].

The same thing also happened to PT Hilon Sumatra, which continued to innovate its products. PT Hilon Sumatera is a company engaged in the manufacture of bedding, which is located at Jalan Jamin Ginting Km.11 No. 64A Medan-Tuntung District. This company produces various types of bedding such as pillows, bolsters, carded fiber (dacron crushed), padding and other bedding products. This company operates with a make to stock system, which means that the company produces bedding according to the amount needed by the company. The company also accepts orders based on requests from consumers and large companies such as hotels, hospitals, villas and other lodging facilities.

The average percentage of defective pillow products at PT XYZ in 2018 was 21.69%. This shows the characteristics of pillow products that are not in accordance with consumer desires. This can lead to a decrease in the productivity of the company and can result in a shift of consumers from the company to competing companies. In order for the company to continue to survive and increase sales in the bedding industry market that is increasingly widespread in the market, the company must be able to maintain and improve the quality of the products it produces by identifying every factor that causes defects in bedding products. pillow. production so that improvements can be found that can be made to address the causes of the disability. In addition, the company must also be able to listen to the voice of the customer (voice of the customer) so that the company can work to meet customer satisfaction by translating it into technical characteristics. The appropriate method to overcome this problem is the integration approach of Quality Function Deployment (QFD)[3–8].

Previous research conducted by Ginting[9] was related to the use of the Quality Function Deployment (QFD) method. This study succeeded in designing a rubber tapping tool product in Langkat Regency, North Sumatra with the Quality Function Deployment (QFD) method and the Kano Model. The QFD method can determine customer needs and translate these needs into technical characteristics so that each functional area and level of the organization can understand and make improvements to achieve goals.

2. Materials and methods

The object of the research being researched by the researcher was a pillow product manufactured by PT XYZ. This research was conducted starting from the initial process of making pillow products to the final process of making the product.

The right method to solve this problem is the integration approach of Quality Function Deployment (QFD) Phase I and Phase II. Quality Function Deployment (QFD) is a way to improve the quality of goods or services by understanding consumer needs and then linking them with technical provisions to produce goods or services at each stage of the manufacture of goods or services produced [10,11]. Quality Function Deployment is a planning tool used to help businesses focus on the needs of their customers when developing design and manufacturing specifications.

Data collection was carried out by the following steps[12]:

- Sampling technique. The sampling technique used in determining this sample size is total sampling. Total sampling is a sampling technique with the same number of samples as the population. The reason for using total sampling is because the entire population is used as the research sample.
- Open Questionnaire. An open questionnaire contains questions that give respondents the freedom to answer according to their opinion on a particular topic. This questionnaire is a form of questions posed to 27 respondents about the assessment of workers in the pillow-making production process.
- Closed Questionnaire. A closed questionnaire is a question that has been given a choice of answers by the questionnaire maker. Respondents may not answer outside the options provided. This closed questionnaire was obtained from an open questionnaire mode.
- Validity and Reliability Test. The validity test is calculated using the Pearson coefficient formula. Reliability test by calculating the variance of each, the formula used in this test is the Alpha Cronbach method.

3. Results and discussion

Identification of consumer needs is done through distributing questionnaires to obtain 10 factors of consumer needs for pillow products. The results of the identification of consumer needs can be seen in table 1.

Table 1. Results of the identification of consumer needs

No	Requirement Variable
1	Fiber is the type of foam or foam that best suits the design of pillow products
2	1 kg is the weight of foam or foam that best suits the design of the pillow product
3	3 cm is the thickness of the foam or foam on the pillow that best suits the design of the pillow product
4	Cotton fabric is the type of fabric that best suits the pillow product design
5	White is the color of the fabric that best suits the design of the pillow product
6	55 x 40 cm is the dimension or size of the fabric that best suits the design of the pillow product
7	50 x 70 cm is the dimension or size of the pillow that best suits the design of the pillow product
8	45 minutes is the length of the process of breaking down raw materials in the required pillow product
9	20 minutes is a long process of sewing the fabric on the pillow product needed
10	The thickness of the foam is a factor that determines the durability of the product that best suits the design of the pillow product

The assessment of the level of importance of the attributes was obtained based on the value of the closed questionnaire mode based on the frequency of the most respondents' answers for each attribute. For example, in attribute 1 the highest mode is 5, so the level of importance for attribute 1 is 5. The level of importance of all attributes can be seen in table 2.

Tabel 2. Variable importance of consumer needs

No	Consumer Needs	Questionnaire Results					Questionnaire Results
		Scale of Measurement					
		5	**4**	**3**	**2**	**1**	
1	Fiber is the type of foam or foam that best suits the design of pillow products	12	10	4	0	1	5
2	1 kg is the weight of foam or foam that best suits the design of the pillow product	12	8	7	0	0	5
3	3 cm is the thickness of the foam or foam on the pillow that best suits the design of the pillow product	8	9	9	0	1	4
4	Cotton fabric is the type of fabric that best suits the pillow product design	12	8	7	0	0	5
5	White is the color of the fabric that best suits the design of the pillow product	7	9	8	3	0	4
6	55 x 40 cm is the dimension or size of the fabric that best suits the design of the pillow product	10	10	7	0	0	4
7	50 x 70 cm is the dimension or size of the pillow that best suits the design of the pillow product	10	9	8	0	0	5
8	45 minutes is the length of the process of breaking down raw materials in the required pillow product	10	12	3	2	0	4
9	20 minutes is a long process of sewing the fabric on the pillow product needed	9	11	6	1	0	4
10	The thickness of the foam is a factor that determines the durability of the product that best suits the design of the pillow product	5	3	8	3	8	3

Determination of product characteristics based on interviews and discussions with factory production supervisors. The next step is to determine the degree of relationship between the technical characteristics of the product and consumer desires. The level of the relationship in question starts

IOP Conf. Series: Materials Science and Engineering 1041 (2021) 012039 doi:10.1088/1757-899X/1041/1/012039

from a strong, medium, weak scale, and is not related to the same level. The level of the relationship in question starts from a strong, medium, weak scale, and is not related at all.

The HoQ matrix is based on the data obtained in the previous step. QFD pillow products can be seen in Figure 1.

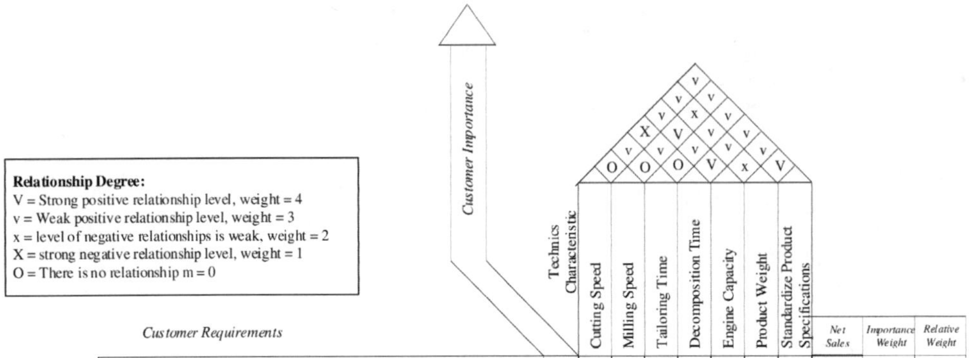

Customer Requirements		Cutting Speed	Milling Speed	Tailoring Time	Decomposition Time	Engine Capacity	Product Weight	Standardize Product Specifications	*Net Sales*	*Importance Weight*	*Relative Weight*
Fiber is the type of foam or foam that best suits the design of pillow products	5	0	9	1	9	9	1	3	1,5	160	12,76
1 kg is the weight of foam or foam that best suits the design of the pillow product	5	0	3	0	9	9	3	9	1,2	165	13,16
3 cm is the thickness of the foam or foam on the pillow that best suits the design of the pillow product	3	0	9	1	9	9	1	9	1,2	114	9,09
Cotton fabric is the type of fabric that best suits the pillow product design	5	9	0	9	1	3	1	1	1,2	120	9,57
White is the color of the fabric that best suits the design of the pillow product	4	1	0	1	0	1	0	1	1	16	1,27
55 x 40 cm is the dimension or size of the fabric that best suits the design of the pillow product	4	9	0	9	1	3	9	9	1	160	12,76
50 x 70 cm is the dimension or size of the pillow that best suits the design of the pillow product	5	9	0	9	1	3	9	9	1	200	15,96
45 minutes is the length of the process of breaking down raw materials in the required pillow product	4	0	9	0	9	9	1	1	1,2	116	9,25
20 minutes is a long process of sewing the fabric on the pillow product needed	4	3	0	9	0	9	9	1	1,2	124	9,89
The thickness of the foam is a factor that determines the durability of the product that best suits the design of the pillow product	3	0	9	1	3	3	9	1	1,2	78	6,22
Units		cm/s	rpm	min	min	kg/hari	cm	-		1253	100
Degree of difficulty		3	2	2	3	4	3	4			
Degrees of Interest		10	13	13	14	20	14	15			
Cost estimation		14	10	10	14	19	14	19			

Figure 1. QFD Phase I Results

Determination of Critical Part of pillow products can be seen in more detail in table 3.

Table 3. Critical Part of Pillow Products

No	Critical Part

1	Dimensions of Fabric
2	Strength of Sewing Thread
3	Foam Material Levels
4	Foam Density
5	Durability
6	Accuracy of Cutting

The Phase II HoQ matrix is made based on data that has been obtained in the previous steps. QFD Phase II pillow products can be seen in figure 2.

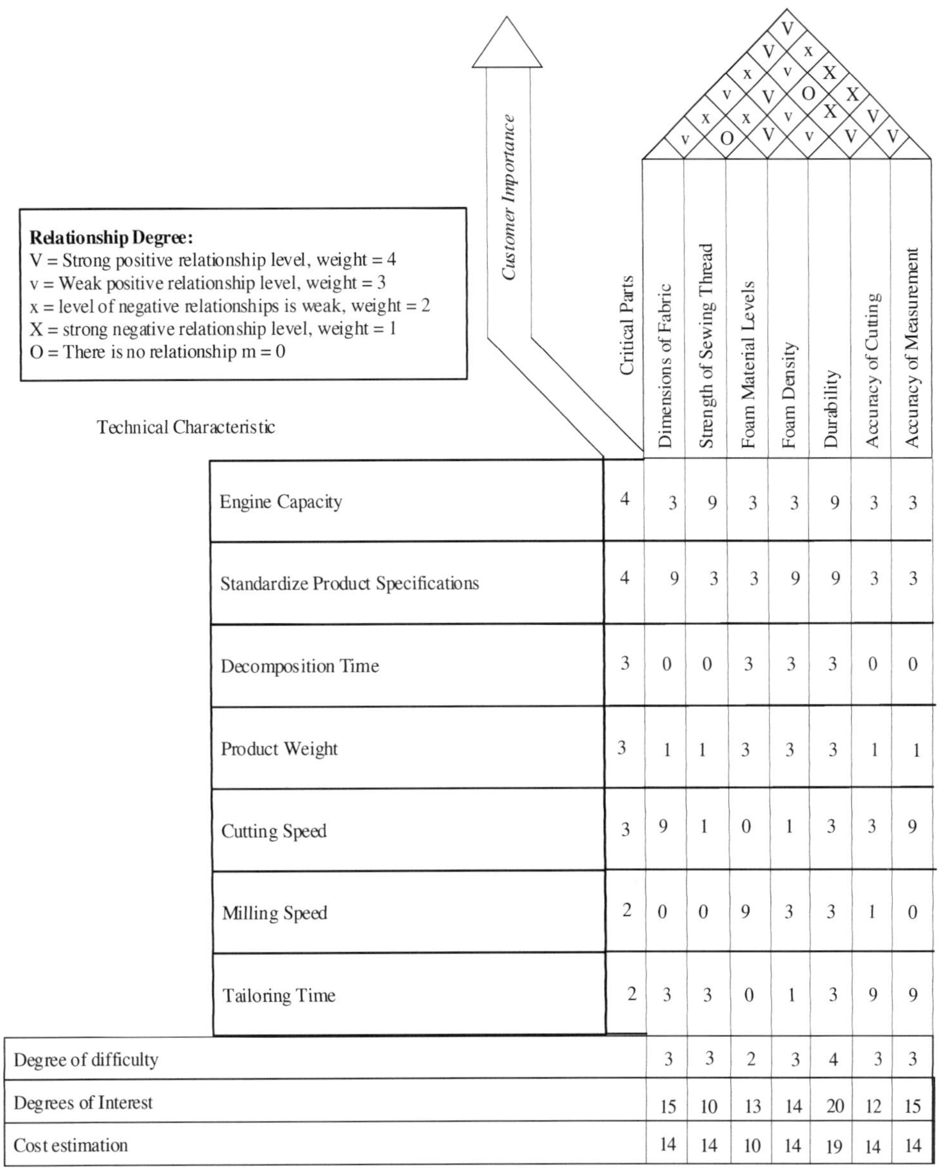

Figure 2. QFD Phase II Results

4. Conclusion

The results of the QFD Phase I pillow products show that the characteristics that have the highest difficulty level are engine capacity 4, an importance level of 20% and an estimated cost of 19%. This shows that machine capacity is an important technical characteristic in the manufacture of pillow products because it can result in the product being rejected if a cutting error occurs. These attributes are included in the Phase II QFD. The results of QFD Phase II show that the characteristics that have the highest difficulty level are durability, measurement accuracy, and fabric dimensions. The resistance difficulty level is 4 and the importance level is 20% and the estimated cost is 19%. This shows that there is a critical part that must be considered in the pillow-making process because it can result in a repetition process and even become a reject product.

Acknowledgments

This study was funded by Research Institutions of University of Sumatera Utara, in accordance with project of Research Funding Scheme of Direktorat Riset dan Pengabdian Masyarakat (DRPM) DIKTI 2020.

References

[1] Alimul A and Aziz A 2006 Kebutuhan Dasar Manusia: Aplikasi Konsep dan Proses Keperawatan *Jakarta Salemba Med.*
[2] National Sleep 2013 *2013 International Bedroom Poll*
[3] Ginting R, Tarigan U and Panjaitan N 2020 Integration of quality function deployment and value engineering: A case study of designing a texon cutting tool *Songklanakarin J. Sci. Technol.* **42** (4)771-779
[4] Ginting R, Ishak A and Widodo 2019 Prioritizing critical parts of crumb rubber product by using quality function deployment (QFD) phase II: Product design matrix *IOP Conf. Ser. Mater. Sci. Eng.***505**
[5] Ginting R, Tarigan U and Panjaitan N 2019 The Application of Quality Function Deployment and Ergonomics: A Case Study for A New Product Design of A Texon Cutting Tool *IOP Conf. Ser. Mater. Sci. Eng.***495** 0–9
[6] Zare Mehrjerdi Y 2010 Quality function deployment and its extensions *Int. J. Qual. Reliab. Manag.***27** 616–40
[7] Rajabi Asadabadi M 2014 A hybrid QFD-based approach in addressing supplier selection problem in product improvement process *Int. J. Ind. Eng. Comput.***5** 543–60
[8] Tarigan U, Ginting R and Siregar I 2018 Determining the need for improvement of infant incubator design with quality function deployment *IOP Conf. Ser. Mater. Sci. Eng.***309** 0–6
[9] Ginting R, Siregar I and Ginting T U H 2015 Perancangan Alat Penyadap Karet Di Kabupaten Langkat Sumatera Utara Dengan Metode Quality Function Deployment (Qfd) Dan Model Kano *J@Ti Undip J. Tek. Ind.***10** 33–40
[10] Cohen L 1995 *Quality function deployment: how to make QFD work for you* (Prentice Hall)
[11] Day R G 1993 *Quality function deployment: Linking a company with its customers* (Asq Press)
[12] Sinulingga S 2014 *Metode Penelitian* (Medan: USU Press)

IOP Conf. Series: Materials Science and Engineering 1041 (2021) 012040 doi:10.1088/1757-899X/1041/1/012040

Defect analysis on PVC pipe using Statistical Quality Control (SQC) approach to reduce defects (Case Study: PT. XYZ)

R Ginting[1], S Supriadi[1]

[1]Industrial Engineering Department, Faculty of Engineering, Universitas Sumatera Utara, Medan, 20155 Indonesia

Corresponding author: rosnani@usu.ac.id, gosupri45@yahoo.com

Abstract. Product quality is an important factor affecting the level of development and progress of a company. Statistic Quality Control (SQC) as a quality control tool for production can help companies whether the products produced are still under control or not from the initial process or the final product. The research was conducted at PT. XYZ to find out the causes of defects that occur in the PVC pipe production process. In the research, it is known that the defects that occur are melt, crack and hollow with the majority of defects that occur are melt and crack. By using a Cause and Effect diagram, it is known that the cause of defects occurs due to human, machine, materials and method factors. Solutions for defects that occur are given in the hope that the company will implement the solutions offered to reduce the number of defects that occur in the production process.

1. Introduction

The quality of products (services) is crucial in assessing the performance of each enterprise or organization. The release of effective and high-quality products allows the company to obtain more additional profits and ensures rapid production development [1]. In the current era of technological development, industrial business is considered to be one of the determining factors in a country to develop, both domestically and internationally. The industrial sector has higher competition in fulfilling consumer demand according to the desired specifications [2]. In order to survive in a competitive market, improving the quality and productivity of a product or process is a must for companies [3]. The quality of the company or business and product productivity are the keys to success for the development of the company's production system because quality is the main factor for consumers in choosing products. The quality of a product produced by a company is determined by size and the size is determined by characteristics [4]. Product quality is an important factor that affects the level of development and progress of a company [5]. Quality control is a strategy that companies undertake in facing global competition with other company products. Quality is a characteristic of a product or service that is desired by consumers and is obtained through a process and continuous improvement [6]. To produce quality products, the company must always check and improve in various stages, one of which is the quality control stage. The quality of a product is not something that can be underestimated. Good quality will be produced through a good process and in accordance with predetermined quality standards based on consumer demand [7].

Based on these conditions, it can be seen the importance of maintaining product quality so that it is possible to use Statistical Quality Control (SQC), which consists of analyzing processes, setting standards, comparing performance, verifying and studying defects, finding and implementing solutions to obtain better improvements. As in the stage of mass productions, statistical process control has a signification contribution in improving quality control [8]. Statistical Quality Control emerged as an alternative to improve and monitor the quality of production, where the control is carried out through an efficient and simple control chart to monitor the average and threshold of the observed characteristics [9]. Several previous studies that have had the SQC method include Nina, et al who conducted research on Statistical Quality Control (SQC) analysis on bread production at Aremania Bakery [10]. Meldayanoor, et al who conducted research on Statistical Quality Control (SQC) analysis as a control and improvement of tortilla product quality at UD. Noor Dina Group [11]. Koyor and Hana who conducted research on quality control of tofu production using the SQC method in UKM H. Musauwimin [12]. Ira who conducted research on quality control analysis using Statistical Quality Control (SQC) at PT. Pratama Abadi Industri (JX) Sukabumi [13].

The research was conducted at PT. XYZ to find out the causes of defects that occur in the PVC pipe production process. According to the results of interviews with the management of PT. XYZ, defects were still found during the production process. This study aims to determine the factors that cause product defects, and later it can become a company reference to reduce the number of defects that occur in the production process.

2. Materials and methods

In this study, there are several steps taken to approach the Statistical Quality Control method so that it can analyse and find the causes of product defects and can improve the quality of PVC pipes.

2.1. Data Collecting

There are two types of data used in this study, namely primary data and secondary data. Primary data is data in a scientific study which is taken directly from the research subject. Primary data were collected based on the results of research on the object of research with work procedures that had been carried out at the time of conducting the research. Data taken in the form of types of defects in products and the number of defects found obtained from direct observations in the field and company reports. Secondary data is data obtained indirectly from sources or research objects. Secondary data are usually obtained from archives, books and previous research. Secondary data were obtained from company-owned reports, books, and scientific journals that support primary data [14].

2.2. Data Processing

Data processing techniques are performed using a Statistical Quality Control approach consisting of histogram, check sheet, scatter diagram, stratification, pareto diagram, cause and effect diagram, and control chart [15].

- Stratification is the process of grouping data into a group that has the same characteristics.
- Check sheet is a form in the form of items to be checked that have been printed in the form with the intention that data can be collected easily and concisely.
- Histogram is a block diagram that is used to show the division of data groups that have been carried out.
- Pareto diagram is a chart containing bar and line charts; Bar charts show data classifications and values, while line charts represent cumulative data totals. The data classification is sorted from left to right according to the order of highest to lowest ranking. The highest ranking is the priority problem or the most important problem to be resolved immediately, while the lowest ranking is a problem that does not have to be resolved immediately.
- Scatter diagram is used to see the correlation of the causes of disability with the types of defects found.
- Control chart is a graph that shows whether the defect found exceeds a predetermined threshold or not. In this study, the control map P is used. The formula used to calculate the threshold is as follows.

- Cause and effect diagram is a diagram that depicts lines and symbols that show the relationship between the causes and effects of a problem, for further corrective action is taken for the problem.

3. Results and Discussion

3.1. Stratification

Stratification is used to group data into groups that have the same characteristics. Based on data obtained from the data collection process, three types of defects were found in PVC pipes, namely melting, cracking, and holes.

3.2. Check sheet

Check sheet is a simple sheet regarding the data collection process carried out which is used to determine the number of defects, types of defects, or other information. Check sheet on PVC pipe can be seen in table 1.

Table 1. PVC pipe check sheet

Period	Total Production	Type of Defect						Number of Defects
		Melt	Percentage	Crack	Percentage	Hollow	Percentage	
January	85.225	1.663	1.95%	1.311	1.54%	573	0,67%	3.547
February	92.884	1.987	2.14%	1.270	1.37%	898	0,97%	4.155
March	86.159	1.239	1.44%	1.381	1.60%	851	0,99%	3.471
April	84.782	1.993	2.35%	1.420	1.67%	957	1,13%	4.370
May	96.956	1.964	2.03%	1.280	1.32%	914	0,94%	4.158
June	97.228	1.785	1.84%	707	0.73%	564	0,58%	3.056
July	92.194	1.538	1.67%	1.372	1.49%	617	0,67%	3.527
August	88.555	1.437	1.62%	609	0.69%	902	1,02%	2.948
September	86.368	1.734	2.01%	1.337	1.55%	557	0,64%	3.628
October	81.622	1.351	1.66%	1.438	1.76%	589	0,72%	3.378
November	84.881	1.944	2.29%	773	0.91%	735	0,87%	3.452
December	91.062	1.420	1.56%	854	0.94%	853	0,94%	3.127
Total	**1.067.916**	**20.055**	**1.88%**	**13.752**	**1.29%**	**9.010**	**0,84%**	**42.817**

Based on the table above, it can be seen that the types of defects starting from the highest number are melted, perforated, and cracked, with a total of 42.817 defects.

3.3. Histogram

The histogram of defects found in PVC pipes can be seen infFigure 1.

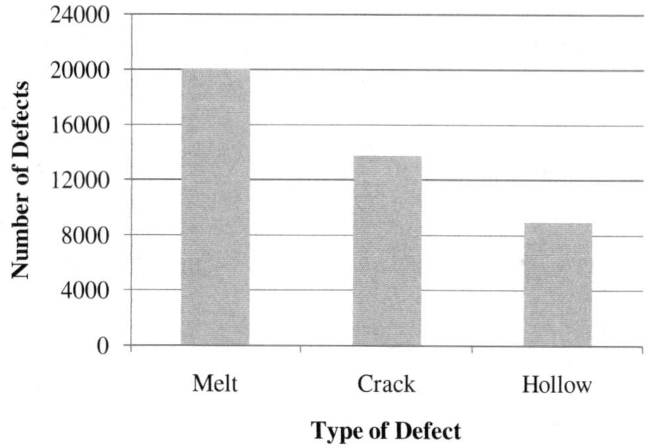

Figure 1. Histogram of PVC pipe

Based on the picture above, it can be seen that the type of melt defect is the most common type of defect, followed by the type of defect with crack and hollow.

3.4. Pareto Diagram

The Pareto chart is a bar graph and a line graph that illustrates how the data types compare to the whole. The function of the Pareto diagram is to identify the main problems for quality improvement from the biggest to the smallest. The results of making a Pareto diagram using Minitab software can be seen in figure 2.

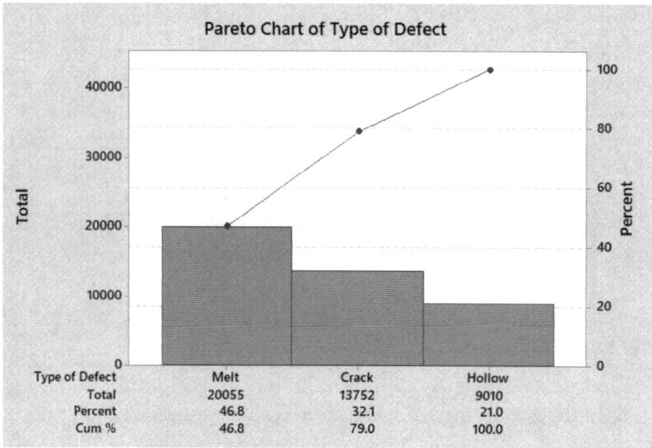

Figure 2. PVC pipe pareto diagram

The Pareto principle as the 80/20 rule means that 80% of defects that occur in PVC pipes are caused by melting (46,8%) and cracks (32,1%) which results in 20% of PVC pipe products produced. melts and cracks have the greatest impact because the cumulative percentages for both are nearly 80%. So that by making repairs at the root of the two defects, it is expected that the defective product level can be reduced.

3.5. Scatter Diagram

The scatter diagram is used to see the correlation of the causes of disability with the types of defects found. The scatter diagram for the types of melt and crack defects is shown in figures 3 and 4.

IOP Conf. Series: Materials Science and Engineering 1041 (2021) 012040 doi:10.1088/1757-899X/1041/1/012040

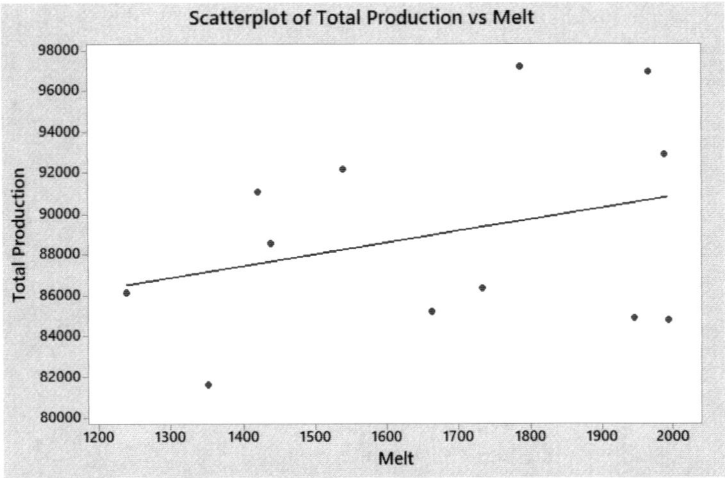

Figure 3. Scatter diagram of production amount with melt defects

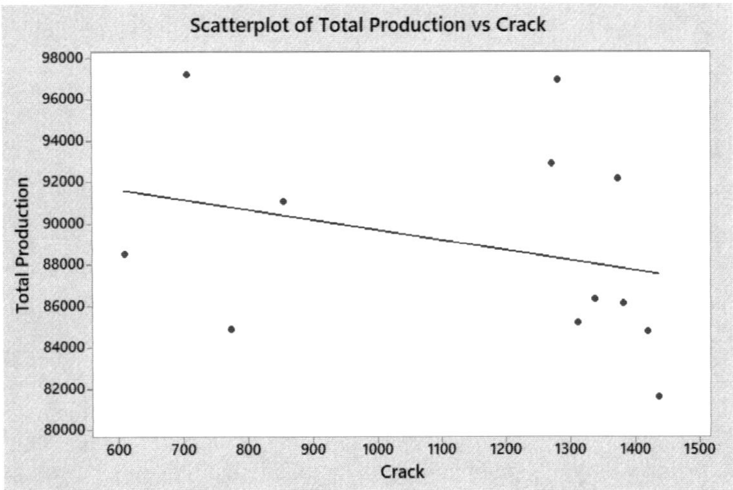

Figure 4. Scatter diagram of production amount with crack defects

Based on the figure above, there is no linear relationship between the amount of production with melting defects and cracking defects. Based on the calculation, it is found that the correlation between the amount of production with melt defects is 0,308647 which means it has a weak unidirectional relationship. Meanwhile, the correlation between the amount of production and crack defects is -0,305186, which means it has a weak opposite relationship.

3.6. Control Chart

The control map is a graph that shows whether the defect found exceeds a predetermined threshold or not. P chart on PVC pipe production can be seen in Figure 5.

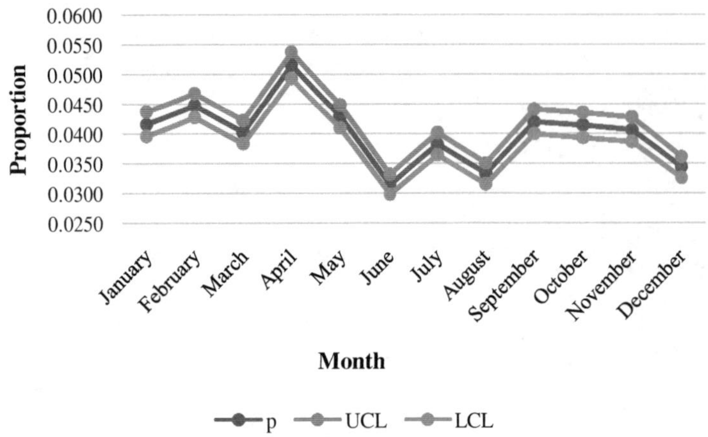

Figure 5. PVCpipe control chart

Based on the picture above, it can be seen that there is no number of defects that exceeds the threshold where the entire period is still within the predetermined limit. then based on the calculation of the capability process, it is known that the Cp value is 95.98%

3.7. Cause and Effect Diagram

Causal diagrams are used to find the biggest contributing factors for defects found in products. The causal diagram for the types of melt and crack defects can be seen in figures 6 and 7.

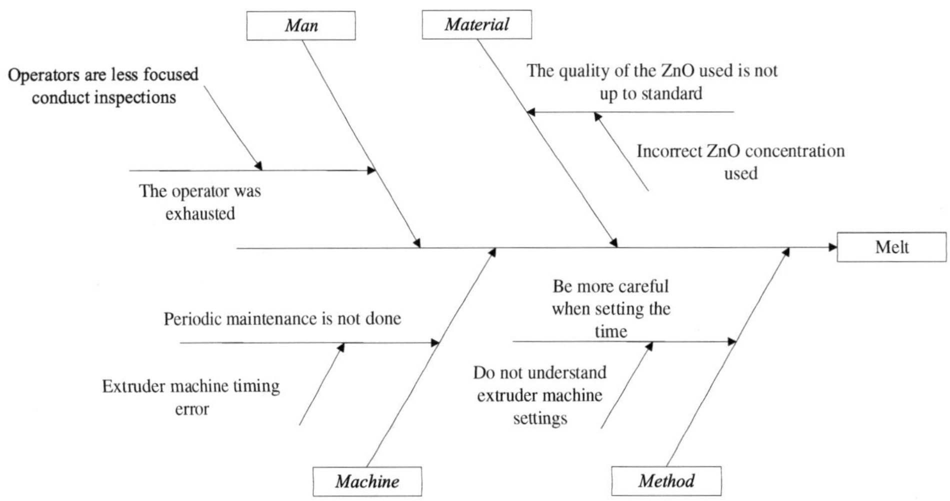

Figure 6. Cause and effect diagram of melting defective PVCpipe

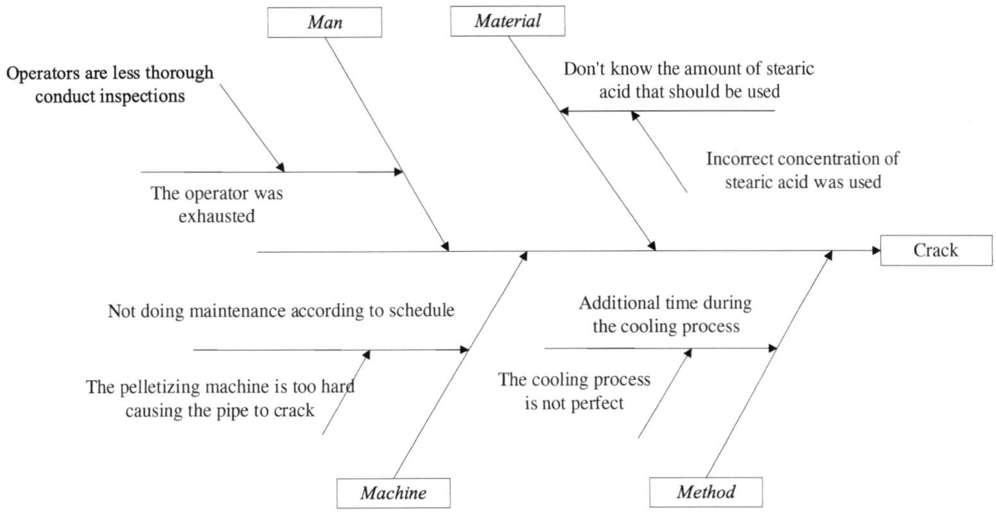

Figure 7. Cause and effect diagram of crack defective PVC pipe

Based on the picture above, it can be seen that the causes of disability can be caused by humans, machines, materials, and the methods used. So that improvements can be made to these four aspects.

4. Conclusion

Every manufacturing company must eliminate the causes that could cause production problems. Defects in products can cause the company to suffer material and immaterial losses. Through the Statistics Quality Control approach, types of defects can be found and ways to deal with these disabilities. In PVC pipe production companies, the defects observed were melting and cracking where nearly 80% of defects found were both defects. The causes of defects originating from humans, machines, materials and methods can be overcome in various ways. Both defects can be overcome by recruiting new employees so as to reduce the effect of fatigue on other employees. In addition, regular maintenance can reduce the occurrence of defects. The raw materials used must comply with standards set by the government and it is recommended to use materials of high quality.

Acknowledgments

The author would like to thank all parties who have helped in providing data and support received. The author also congratulates Mrs. Ir. Rosnani Ginting, MT, Ph. D, IPU, ASEAN Eng. as the head of the Production Systems laboratory, Department of Industrial Engineering, Faculty of Engineering, University of North Sumatra who has guided the author in completing this research. The author also thanks all Production Systems Laboratory assistants who have helped in completing this research so that it can be carried out to completion.

References

[1] Simonenko K V., Bulatova G S, Antropova L B and Varepo L G 2018 Quality Control of the Print with the Application of Statistical Methods *J. Phys. Conf. Ser.* **998**

[2] Kulkarni S, Kulkarni C, Vimal K E K and Jayakrishna K 2019 Statistical Quality Control of Torque Wrenches Used in Automotive Assembly Department *Lecture Notes in Mechanical Engineering* pp 199–208

[3] Runtuwene V E, Massie J D, and Tumewu F 2017 Quality Control Analysis Using Statistical Quality

Control At Pt Massindo Sinar Pratama Manado *J. EMBA J. Ris. Ekon. Manajemen, Bisnis dan Akunt.*

[4] Ramdani A, Satori M and As' ad N R 2020 Perbaikan Kualitas pada Produk Pembuatan Tas Backpack Menggunakan Metode Statistical Quality Control (SQC) dan Failure Mode And Effect Analysis (FMEA)

[5] Rakhmad A 2010 Penerapan Statistical Quality Control (SQC) dalam pengendalian proses produksi batik menggunakan control chart P (Grafik Pengendali P)

[6] Anggraini M and Anggriawan N A 2019 Analisis Pengendalian Kualitas Pengolahan Produk Karet Remah Sir 20 Dengan Pendekatan Statistical Quality Control *J. Rekaya, Teknol. dan Sains*

[7] Andriani D P, Sani M and Qurrota A 2019 Analisis Pengendalian Kualitas Mesh Size Distribution pada Kristal Monosodium Glutamat (MSG) dengan Statistical Quality Control *Proisiding Seniati* 198–204

[8] Saputra T M, Hernadewita H, Yudha A and Saputra P 2019 Quality improvement of molding machine through statistical process control in plastic industry *J. Appl. Res. Ind. Eng.* **6** 87–96

[9] de Andrade M G, Vilas Boas M A, Siqueira J A C, Dieter J, Sato M, Hermes E, Mercante E and Kazue Tokura L 2017 Statistical quality control for the evaluation of the uniformity of microsprinkler irrigation with photovoltaic solar energy *Renew. Sustain. Energy Rev.*

[10] Hairiyah N, Amalia R R and Luliyanti E 2019 Analisis Statistical Quality Control (SQC) pada Produksi Roti di Aremania Bakery Statistical Quality Control (SQC) Analysis of Bread Production at Aremania Bakery *Teknol. dan Manaj. Agroindustri*

[11] Meldayanoor M, Amalia R R and Ramadhani M 2018 Analisis Statistical Quality Control (SQC) Sebagai Pengendalian dan Perbaikan Kualitas Produk Tortilla di UD. Noor Dina Group *J. Teknol. Agro-Industri*

[12] Rujianto K and Wahyuni H C 2019 Pengendalian Kualitas Produk Dengan Menggunakan Metode SQC dan HRA Guna Meningkatkan Hasil Produksi Tahu di IKM H. Musauwimin *PROZIMA (Productivity, Optim. Manuf. Syst. Eng.*

[13] Andespa I 2020 Analisis Pengendalian Mutu Dengan Menggunakan Statistical Quality Control (SQC) Pada Pt.Pratama Abadi Industri (JX) Sukabumi*E-Jurnal Ekon. dan Bisnis Univ. Udayana*

[14] Putri K U, Akil R and Wardhani M K 2019 Strategi PT. Indosat Tbk dalam Mensosialisasikan Nama dan Logo Kepada Customer *Cakrawala - J. Hum.*

[15] Handayani R E W 2020 Statistical Process Control (SPC) Untuk Pengendalian Kualitas Produk Mebel Di Ud . Ihtiar Jaya **6** 50–8

IOP Conf. Series: Materials Science and Engineering 1041 (2021) 012041 doi:10.1088/1757-899X/1041/1/012041

Designing carbon dioxide absorbent and detector products using the Quality Function Deployment (QFD) method

R Ginting[1] and M Ramadhan[1]

[1]Industrial Engineering Department, Faculty of Engineering, Universitas Sumatera Utara, Medan, 20155 Indonesia

Corresponding author: rosnani@usu.ac.id, ramadhanucul@gmail.com

Abstract. Global warming is defined as an increase in the average temperature of the earth's surface caused by an increase in the amount of greenhouse gases in the earth's atmosphere. The largest contributor to greenhouse gases on earth is carbon dioxide (CO_2). It is hoped that this CO_2 detector and absorber tool can solve the problem of harmful gases generated from existing industries. This tool is silver with a size of 30 x 30 cm, and the material used is aluminum and has an additional function in the form of a coil cable. The design of this tool uses the method of brainstorming techniques and sampling techniques by means of open and closed questionnaires to determine the type of product. After that, a market survey is conducted using sampling techniques and determining the validity and reliability of the product with competitor I, competitor II and competitor III. The stages from problem to sub problem are determined to classify the goals that will be made in designing the product. Sub-problem to sub-solution steps are carried out to determine the Quality Function Deployment (QFD) of the product.

1. Introduction

The progress of human civilization causes activities in the industrial sector to be faced with tough challenges, every business actor who wants competition will pay full attention to the quality of the products produced. The quality of the product produced is a very important factor in determining a company. Products are said to be of quality to meet the needs and desires of customers [1]. A high quality product will require high costs and consumers may not necessarily need a product that is too high quality. This view is what makes companies oriented towards consumer needs so that they can provide products according to consumer desires and low prices [2]. Consumers still exist but their purchasing power is still limited, as a result, consumers are becoming more careful in making purchases and determining a product they want [3]. One of the human factors is always trying to create something both tools and other objects to help activities. In the embodiment of these objects, a design or product design is needed [4]. Design is an activity or engineering design that starts from design innovation ideas, or the ability to produce works and creations that can truly describe market demand due to technological research and development [5]. The combination of several functions in one product is often the main focus for managers of product research and development departments to produce product design concepts today [6].

Global warming causes erratic climate change. Global warming is generally defined as an increase in the average temperature of the earth's surface caused by an increase in the amount of greenhouse gases in the earth's atmosphere. The largest contributor to greenhouse gases on earth is carbon dioxide (CO_2) [7]. Increasing the temperature of the earth's surface will result in extreme climate changes on earth. This can result in disruption of forests and other ecosystems, thereby reducing their ability to absorb atmospheric carbon dioxide. Global warming results in melting of icebergs in arctic regions which can cause sea levels to rise. The greenhouse effect will also result in an increase in sea water temperature so that the sea water expands and the sea level rises, which results in an archipelagic country that will have a very big influence [8]. The increase in the concentration of greenhouse gases also causes the height of the infrared radiation emitted by the earth to decrease, so that the amount of infrared emitted into space decreases, while the sun continues to shine. As a result, there is a change in the amount of incoming and outgoing radiation energy in the tropopause layer of the atmosphere, which is expressed in radiative forcing [9]. The most abundant gas produced from each change (oxidation) process is carbon dioxide. Therefore, the CO_2 produced from this transformation process is in the form of a gas, so the earth's atmosphere is a container that will accommodate all this CO_2 waste [10]. Various activities in urban areas such as transportation, households, hotels, industry, and other activities require propulsion and heating energy, most of which are obtained from burning fossil fuels. The process of burning fossil fuels produces carbon dioxide (CO_2) [11]. With the development of industry, more and more harmful gases are produced that can threaten the cleanliness of the air, especially for the health of our bodies. These gases will pollute the air in the environment so that it will unconsciously be inhaled and enter the body, then gradually the body will feel the impact. CO_2 gas is odorless and invisible, therefore humans will not know about the presence of these gases around, humans can only know when they have felt the impact, then a tool is made that can help measure gas in an industrial environment. This tool is expected to solve the solution to the problem of hazardous gases generated from existing industries [12].

The purpose of this research is to design a CO_2 detector and absorber that can measure carbon dioxide levels and will check the carbon dioxide levels which represent the detected CO_2 levels, which are detected to be above predetermined limits, the tool will automatically reduce CO_2. so that it will reduce CO_2 and provide benefits to the existing environmental health and can reduce the increase in the amount of greenhouse gases in the earth's atmosphere.

2. Materials and method

2.1. Data Collecting

This research uses a questionnaire instrument. The questionnaire used is based on the establishment of open and closed questionnaires [13] and market surveys conducted to find out the information needed related to the product design of CO^2 detection and absorber.

2.2. Data Processing

Brainstorming. Brainstorming is a form of discussion in order to gather ideas, opinions, information, knowledge, experiences from all participants to generate ideas and most of the ideas will be discarded to choose the best idea [14]. From the results of brainstorming, 10 attributes were obtained which were then made in the form of questions for making open questionnaires [15].

Nigel Cross Method. The stages in the product design process with Nigel Cross by classifying objectives, determining functions, compiling needs, determining characteristics, determining alternatives, evaluating alternatives and communication.

Sampling was done using probability sampling, namely simple random sampling. Simple random sampling is a sampling method where each element that makes up the population is given the same opportunity to be selected as a sample [16]. This method is done when members of the population are considered homogeneous. This technique can be used when the number of sampling units in a population is not too large [17]. Through this sampling method, the number of samples was determined, then made open questionnaires, closed questionnaires, and AHP questionnaires. The questionnaire was designed to determine what kind of tools are expected in the product design process. Then after the questionnaire was recapitulated, validity and reliability tests were carried out which were useful for determining CO_2 detection and absorber tools.

Questionnaire. A questionnaire is a research or survey tool consisting of a series of written questions, which will be distributed to appropriate respondents, with the aim of obtaining answers or opinions from certain groups of people through personal interviews or in the form of a list of questions [18]. Analytical Hierarchy Process (AHP) is a General Theory of Measurement for finding discrete scale comparison and continuous pair comparison. AHP breaks down complex multi-factor or multi-standard problems into hierarchical arrangements. Hierarchy is defined as an expression of a complex multi-layer structural problem, where the first level is the goal, the second level is the factors, conditions, sub-conditions, etc. Until the last level of alternatives. By using a hierarchical structure, complex problems can be broken down into groups which are then arranged into a hierarchical form so that the problem will appear more structured and systematic [19].

The application of the quality function (QFD). QFD is a way to improve the quality of goods or services, understand what consumers want and relate it to technical characteristics to produce goods or services at each stage of the manufacture of goods or services produced [20]. This QFD provides a strong framework to consider customer expectations and production capabilities quantitatively so that the target specifications for a particular product or service can be in line with what the customer expects. QFD is a quality system that can guarantee satisfaction to customers within the scope of total quality management. In this case, the application of the quality function aims to produce better products [21]. Then, classify the objectives and product determination functions. For sub-problems, the determination of needs, is a way to compare the results of brainstorming and attributes to form a questionnaire recapitulation. Next is to determine product characteristics using QFD based on a questionnaire containing customer needs or an investigation of customer satisfaction with products related to technical characteristics [22].

Then in order to find solutions to any problems arising from the manufacture of the product, steps are determined to generate alternatives, evaluate, and improve detail. From all these steps, it can be determined the characteristics of the manufacture of the product at an appropriate cost.

3. Results and discussion

The result of designing this product is a CO_2 detector and absorber that is efficient, effective, ergonomic and affordable that can be used by workers to reduce CO_2 levels in the room. The design problems contained in the design of this CO_2 detector and absorber include product composition, assembly time, length of assembly, length of soldering, length of measurement and thickness of CO_2 detector and absorber equipment.

3.1. Classification of Purpose, Function, and Determination of Needs

To find a solution to the above problems, there are 3 steps needed so that the problem will be divided into sub-problems, namely the classification of objectives, functions, and determination of needs. The conclusion of the 3 steps in designing a CO_2 detector and absorber product is:

- List of objectives for the design of CO2 detection and absorber tools consisting of Color Detection Tool and CO_2 Absorbent, Size of CO_2 Detectors and Absorbers, Color Indicator for CO_2 Detection and Absorbers, Position of CO_2 Detector and Absorbent Switches, Material for CO_2 Detection and Absorption Equipment, Base material for CO_2 detection and absorber, Inner Material of CO_2 Detector and Absorbent Equipment, Has a cabel winder holder, Material from the cable winder, Color of cable winding material. The Purpose Tree Diagram can be seen in figure 1. below.

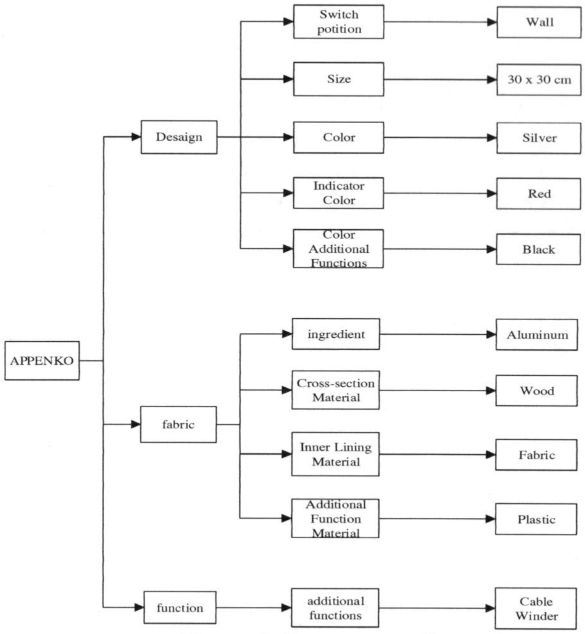

Figure 1. Purpose tree diagram

- The division of functions into essential sub-functions consisting of the sub-function design of the CO2 detector and absorber, MQ135 sensor design sub function, The sub-function of the design of the base section of the CO_2 detector and absorber, Sub function of cable reel design CO_2 Detector and Absorbent Equipment.
- Determine the level of generality to operate which consists of products having a lightweight, comfortable and ergonomic design, The product has a low electrical power (12 volts), The product does not need special care.

3.2. Determination of Characteristics

In this section, the sub-problem of the CO_2 Detector and Absorbent Equipment specification will be found a sub-solution with the steps from the Nigel Cross design, to determine product characteristics. House of Quality CO_2 Detector and Absorbers can be seen in Figure 2 below.

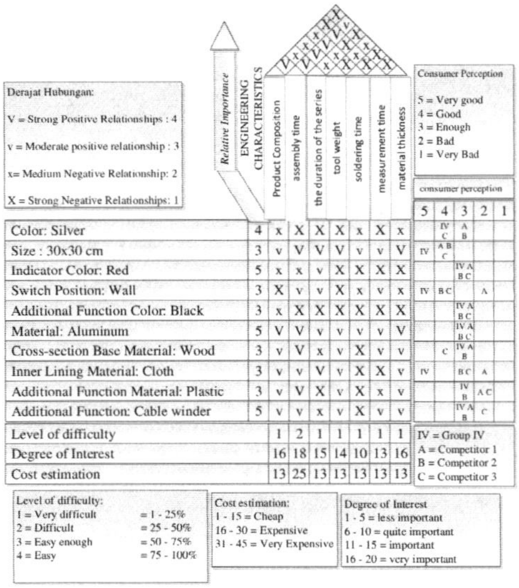

Figure 2. Quality function deployment of CO_2 detection and absorbent equipment

The conclusion from the QFD image of the CO_2 detector and absorber in figure 2, namely:

- The attributes of the carbon dioxide detector and absorption device based on the results of the questionnaire according to consumer desires are as follows : Silver Color, Material of Aluminum, Size 30x30 cm, Indicator Color Red, Cross-sectional Base Material of Wood, Fabric of the Inner Layer, Position of the Switch on the Wall, Additional Functions of Cable Winders, Additional Function Materials made of Plastic, Additional Function Color is BlackDifficulty Level: all technical characteristics are classified as very difficult except for the difficult assembly.
- Degree of Interest: soldering is quite important. The assembly time, equipment weight, soldering time and measurement time are all important, while product composition, assembly time, and material thickness are all very important.
- Estimated Cost: all technical characteristics are cheap except for expensive assembly.

3.3. Sub-Result

The sub-solution is the solution to each problem that occurs, including the selection of attributes for CO_2 Detection and Absorbers which is carried out by using Nigel Cross steps, while maintaining the advantages that are owned and improving the quality of the product.

In this section, there are 3 steps taken so that sub-solutions become solutions, namely generating alternatives, evaluating alternatives, and improving details. Following are the conclusions of the three steps in the process of designing a CO_2 detector and absorber.

3.3.1. Alternative Generation. The alternative generation stage aims to collect as many alternatives as possible that can be used to solve problems in designing CO_2 detection and absorber products, so that the best solution or alternative is then sought. This is done by using the Morphological Charts method with the following steps:

- Make a list of functions or goals that are important to the product.
- Make ways to achieve essential functions.
- Identify a combination of design solutions that can applied.
- Identify the feasibility of a combination of sub-solutions.

The Morphological Chart shows all possible solutions or alternative relationships that can be used in the design of CO_2 detection and absorber tools as shown in the Morphological Chart for CO_2 Detection and Absorbing Equipment products shown in the form of a 10 x 3 matrix, where there are 10 functions that must be achieved and there are 3 possible alternatives. The combination formula used is :

$$3^{10} = 59.049$$

So the total possible combination to achieve these alternatives is 59,049 ways. The Morphological Chart of CO_2 Detection and Absorption Equipment can be seen in table 1 below.

Table1. Morphological chart

No	Characteristic	How To Achieve Function		
		1	2	3
1	Color	Silver	Blue	White
2	material	Aluminum	Iron	Plastic
3	Size	30x30cm	30x20x20cm	15x100x8cm
4	Indicator Color	Red	Red yellow green	Red yellow blue
5	Bahna Cross-section	Wood	Plate	Aluminum
6	Inner Lining Material	Fabric	Plastik	Aluminum
7	Switch Position	Wall	Beside the socket	Beside the product
8	Additional Functions	Cable roll	Fire extinguisher	Alarm
9	Additional Function Material	Plastic	Iron	Wood
10	Color Additional Functions	Black	Red	White

Alternative 1 Alternative 2 Alternative 3

3.3.2. Alternative Evaluation. Alternative evaluation aims to compare the utility values of alternative product designs that are made or made on the basis of performance on the basis of weighted objectives, where the results of the alternative generation step will be evaluated by re-examining the alternatives to be selected so that the best alternative is produced. The method used is the weighted objective method with the AHP scale.

Rating ratings are carried out using Pair Wise Comparison and AHP scale, with data obtained from the importance value for each attribute in QFD. Level I is a Pairwise Appeal Matrix between Primary Attributes, Level II is a Pairwise Appeal Matrix between Design Secondary Attributes, and Level III is a Pairwise Appeal Matrix between Secondary Attributes of Materials and Additional Functions. After the paired appeal matrix is carried out with the AHP scale then weighting is carried out for each level. Weighting for each attribute is needed to determine how the influence of these attributes in product design. Weighting is done by dividing the ranking value of each attribute to the total value of the ranking itself and determining the performance parameters of each attribute. The following performance parameters can be seen in table 2 below.

Table 2. Performance parameters of each attribute

Characteristic	Parameter	Score				
		5	4	3	2	1
Color	Contrast	in stark contrast	Contrast	enough contrast	lack of contrast	not in contrast
Size	Size	30x30 cm	30x20x20 cm	15x100x8 cm	20x100x50 cm	15x100x80 cm
Color Indicator	Contrast	in stark contrast	Contrast	enough contrast	lack of contrast	not in contrast
Switch Position	Conformity	very good	good	pretty good	Kurang baik	not good
Color Additional Functions	Contrast	in stark contrast	Contrast	enough contrast	lack of contrast	not in contrast
Material	Quality	very good	good	pretty good	not good	not good
Cross-section Material	Quality	very good	good	pretty good	not good	not good
Inner Lining Material	Quality	very good	good	pretty good	not good	not good
Additional Function Material	Quality	very good	good	pretty good	not good	not good
Additional Functions	Conformity	very good	good	pretty good	not good	not good

3.3.3. Improve Detail. The final stage of the design process aims to increase product value for consumers and reduce costs incurred by producers. This can be done by using the Value Engineering method. The steps in improving details are as follows :

- List the product components and identify the function of each component as in table 3 below.

Table 3. Data on the components of CO_2 detection and absorption equipment products

Component	Function
Aluminium	As a case / product frame material
Sensor MG811	As a CO_2 sensor
Mini Arduino	As a microcontroller from the AP.PENKO system
fabric	As an isolator
Monitor Running Text	CO_2 level percentage indicator screen
Tombol *on-off*	Switch to turn on and off AP.PENKO
Wood	As AP.PENKO's foundation on the wall
Bolt	The lock between the components and the frame
Mur	The lock between the frame and the base
Cable	As a connection between each component
mini fan	As an air regulator

- Determines the value of the identified function.
  Based on the functions that have been identified, the values are determined based on consumer perceptions. The values of each function assessed based on the suitability of the design against the desires of the upper consumers are as shown in table 4.

Table 4. Value of each function

No	Function	Score	Information
1	Design	good	AP. PENKO has a simple design, with a size that is not too big, light, and easy to move, besides that this tool is also designed ergonomically and practically making it easy to use.
2	ingredient	good	AP. PENKO uses tools that are light and easy to find and assemble such as aluminum, cloth, wood, as well as the main tool in the form of a CO2 sensor and its energy source, namely electrical energy.
3	dimensions	good	In this AP.PENKO, the dimensions of the length of the tool, the length of the wooden base, the length of the cable, and the number of bolts and nuts to be used are determined.
4	Multifunction	good	AP.PENKO is designed with the main function of detecting CO2 in the room, and the additional function is a cable rewinding theme that can make it easier to use.

- Calculate the cost of each component.
  The prices of the main raw materials, additives, and auxiliary materials for the manufacture of the product have been estimated beforehand to determine the selling price of the resulting product. The component prices are assumed as shown in table 5.

Table 5. Cost of each product component for CO_2 detection and absorption equipment per unit of product

Component	Component Prices (IDR)	Number of Components Required	Total price (IDR)
Aluminium	IDR 80.000,00	1 sheet	IDR 80.000,00
Sensor MG811	IDR 300.000,00	1 unit	IDR 300.000,00
Mini Arduino	IDR 100.000,00	1 unit	IDR 100.000,00
fabric	IDR 15.000,00	1 sheet	IDR 15.000,00
Monitor Running Text	IDR 150.000,00	1 unit	IDR 150.000,00
On-offbutton	IDR 10.000,00	1 unit	IDR 10.000,00
Wood	IDR 50.000,00	1 unit	IDR 50.000,00
Bolt	IDR 2.500,00	16 unit	IDR 40.000,00
Nut	IDR 1.000,00	8 unit	IDR 8.000,00
Cable	IDR 3.000,00	2 m	IDR 6.000,00
	Total		**IDR 859.000,00**

3.4. Design Prototype
Prototype is a description or initial model of the product that you want to design on a CO2 detector and absorber. The results of the prototype CO2 detector and absorber device that have been designed can be seen in figure 3 below.

Figure 3. Carbon dioxide detection and absorbing equipment

4. Conclusion
The characteristics of the CO_2 detector and absorber products obtained from the design objectives are silver color with a size of 30 x 30 cm, and the material used is aluminum and has an additional function in the form of a cable coil. For the attributes of this tool product can be divided into several parts. The primary attributes of the CO_2 Detection and Absorbent Equipment product are design, materials and additional functionality. The design attributes are divided into secondary attributes, namely color, size, indicator color, switch position, and additional function color. Material attributes are divided into secondary attributes, namely material, cross-sectional base material, inner layer material and additional function materials. Function attributes are divided into secondary attributes, namely additional functions of cable winders. For QFD, it was found that all the technical characteristics were classified as very difficult except for the relatively difficult assembly time, for the degree of importance that soldering was found to be quite important. The assembly time, equipment weight, soldering time and measurement time are all important, while product composition, assembly time, and material thickness are all very important. and in the estimated product cost, the technical characteristics are classified as cheap except for the expensive assembly. Based on the value engineering step through improving details, it was found that from the existing alternatives there is the best alternative with a total cost of IDR. 895,000.00.

Acknowledgement
The author thanks the production system laboratory assistant for their guidance so we can finish this research well. And my thanks and respect to Mrs. Ir. Rosnani Ginting, MT., Ph.D., IPU., ASEAN Eng. who has guided this research until the research is completed.

References
[1] Rangkuti L A, Rambe A J M and Ginting R 2016 Peningkatan Kualitas Produk Crumb Rubber dengan Menggunakan Metode Quality Function Deployment *J. Tek. Ind. USU* **5** 219661
[2] Anon 2014 Pendekatan Terpadu Menggunakan Metode Qfd-Anp dalam Menentukan Sasaran Peningkatan Rancangan Pada Produk Springbed*J. Tek. Ind. USU*
[3] Soewito Y 2013 Kualitas Produk, Merek dan Desain Pengaruhnya terhadap Keputusan Pembelian Sepeda Motor Yamaha Mio*J. Ris. Ekon. Manajemen, Bisnis dan Akunt.*
[4] Widiasih W and Murnawan H 2016 Penyusunan Konsep untuk Perancangan Produk Pot Portable dengan Pendekatan Quality Function Deployment (QFD) *Seminar Internasional dan Konferensi Nasional IDEC*
[5] Wiraghani S R and Prasnowo M A 2017 Perancangan Dan Pengembangan Produk Alat Potong

Sol Sandal*Tek. Eng. Sains J.*

[6] Hasibuan C F and Sutrisno S 2018 Perancangan Produk Tas Travel Multifungsi Dengan Menggunakan Metode Quality Function Deployment (Qfd) *J. Sist. Tek. Ind.* **19** 40–4

[7] Purba E and Khairunisa A C 2012 Kajian Awal Laju Reaksi Fotosintesis untuk Penyerapan Gas CO2 Menggunakan Mikroalga Tetraselmis Chuii *J. Rekayasa Proses* **6** 7–13

[8] Pratama R 2019 Efek Rumah Kaca Terhadap Bumi *Bul. Utama Tek.*

[9] Pujiastuti D 2010 Analisis Efek Karbon Dioksida (Co2) Terhadap Kenaikan Temperatur Di Bukit Kototabangtahun 2005 – 2009 *J. ILMU Fis. | Univ. ANDALAS*

[10] Nebath E, Pang D and Wuwung J O 2014 Rancang Bangun Alat Pengukur Gas Berbahaya CO Dan CO2 di Lingkungan Industri *J. Tek. Elektro dan Komput.* **3** 65–72

[11] Karim A 2017 Penerapan Metode Brainstorming Pada Mata Pelajaran IPS Untuk Meningkatkan Hasil Belajar Kelas VIII Di SMPN 4 Rumbio Jaya *PEKA***5** 1–12

[12] Marsusanti E, Susilawati S, Nugraha R and Koeswara T S N 2018 Faktor – Faktor Yang Mempengaruhi Pendapatan Driver Gojek *Swabumi*

[13] Muhammad M and Fadilla M 2018 Perancangan Desain Produk Alat Tulis Kantor dengan Menggunakan Metode Quality Function Deployment*J. Sist. Tek. Ind.*

[14] Wibawa G N A, Abapihi B, Laome L and ... 2019 Peningkatan Kompetensi Pendidik Melalui Evaluasi Pembelajaran Berbentuk Kuesioner Online ... *Ilmu Terap.*

[15] Ginting R, Batubara T Y and Widodo W 2018 Desain Ulang Produk Tempat Tissue Multifungsi Dengan Menggunakan Metode Quality Function Deployment*J. Sist. Tek. Ind.*

[16] Bangun A, Ginting R and Tarigan U 2013 Analisis Kualitas Pelayanan Jasa Pendidikan dengan Menggunakan Quality Function Deployment di Jurusan Akuntansi Xyz *J. Tek. Ind. USU***3** 219371

[17] Azza G M and Dores A 2018 Sistem Informasi Manajemen Marketing Tools Serta Penerapan Metode Ahp (Analytical Hierarchy Process) Pada Proses Uji Kualitas Barang (Studi Kasus : PT Edi Indonesia) *J. Cendikia*

[18] Tampubolon H F, MSIE I A R M, Ginting M T and Rosnani I 2013 Optimasi Quality Function Deployment (QFD) berbasis model kano untuk memaksimalkan kepuasan konsumen *J. Tek. Ind. USU***1** 219153

[19] Susana T 1988 Karbon Dioksida *Oceana* **13** 1–11

[20] Ardani F, Ginting R and Ishak A 2014 Perancangan Desain Produk Spring Bed Dengan Menggunakan Metode Quality Function Deployment *J. Tek. Ind. USU* **5** 219632

[21] Deveci M, Öner S C, Canıtez F and Öner M 2019 Evaluation of service quality in public bus transportation using interval-valued intuitionistic fuzzy QFD methodology *Res. Transp. Bus. Manag.*

[22] Sukmawati T 2015 Penyerapan karbon dioksida pada tanaman hutan kota Di Surabaya *LenteraBio Berk. Ilm. Biol.* **4**

IOP Conf. Series: Materials Science and Engineering 1041 (2021) 012042 doi:10.1088/1757-899X/1041/1/012042

Evaluation of e-commerce services quality using Fuzzy AHP and TOPSIS

A Ishak[1], R Ginting[2] and W Wanli[3]

[1]Department of Industrial Engineering, Faculty of Engineering, Universitas Sumatera Utara, Medan, Indonesia.

Corresponding author: aulia.ishak@usu.ac.id

Abstract. Rapid technological developments require people to use information and communication technology to transact goods/services. Based on a survey by the Central Statistics Agency (BPS), of all data collection efforts, only 15.08% are e-commerce entrepreneurs. This shows that business conducted via the internet in Indonesia is still low. The low utilization of e-commerce can be caused by several obstacles, one of which is the customer trust that is built when customer satisfaction is met. This research was conducted to evaluate the quality of e-commerce services using Fuzzy AHP and TOPSIS. The determination of sub-criteria is done using the Delphi method. After that, the calculation of consistency is carried out using the AHP method. The results of calculations using Fuzzy AHP-TOPSIS show that the best e-commerce orders are Tokopedia (0.6206), Bukalapak (0.5208), Elevenia (0.2094) and Shopee (0.1910). Based on the research results, the recommendation for e-commerce with the best service quality is Tokopedia, this is because Tokopedia has the highest weight on the sub-criteria for product return services compared to other e-commerce. The results of this study can be used by other e-commerce companies to improve service quality based on criteria and sub-criteria that are considered important according to expert opinion.

1. Introduction

The rapid development of information and communication technology has an impact on changes in various fields, such as social, economic, political, and cultural, as well as an impact on changes in lifestyle, including consumption patterns and the way people sell and shop. In this era, people make use of information and communication technology to buy and / or sell goods and / or services via the internet. This phenomenon is known as e-commerce.

Based on a survey conducted by the Central Statistics Agency (BPS) with a sample of 3,504 Census Blocks spread across 101 districts/cities in all provinces in Indonesia in 2019, it shows that of all data collection efforts, only 15.08% of business was using e-commerce. In North Sumatra province alone, businesses that use e-commerce are only 12.34%, and those that do not use e-commerce are 87.66% [1].

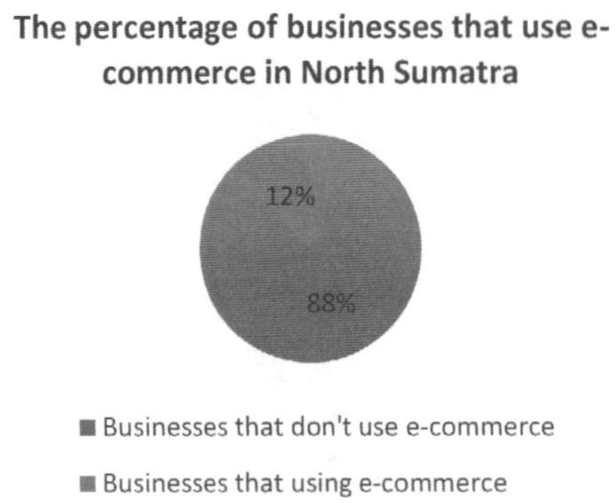

Figure 1. Percentage of businesses using e-commerce in North Sumatra

Based on figure 1. it can be seen that 87.66% businesses that do not conduct e-commerce transactions, the reason most businesses do not do e-commerce is because they are more comfortable selling directly (offline) as much as 79.49%. The recapitulation of reasons for business owners still not doing business online can be seen in table 1.

Table 1. Reasons Why Company Don't Use E-commerce.

Reasons companya don't do e-commerce	Percentage (%)
Not interested in selling online	42,52
It's more convenient to sell in person	70,89
Concerns about security	5,69
Concerns about privacy	6,13
Concerns about technical issues	7,03
Concern about the issue of trust	7,25
Lack of knowledge	21,78

Based on this data, it can be seen that business actors are still comfortable doing business offline, this shows that consumers who transact offline are very large. Several things that make consumers reluctant to make transactions online, such as difficulties in making payments, customer trust, website quality, and so on. Customer trust will be built when customer needs and satisfaction are met through quality standards. So that customer satisfaction becomes an important parameter in building a sustainable business. Quality standards that are described in E-commerce in the form of service quality or website/application facilities that can provide convenience for customers [2].

In fact, based on research by Riris and Yeny, it was found that the growth of e-commerce had a significant effect on the amount of taxes paid. The results of these studies prove that e-commerce has a significant effect on taxpayer compliance where the t statistic is 3.406> 1.96 with a coefficient of 0.273. Where the tax itself has several functions, such as financing state expenditures, opening jobs, controlling the inflation rate, and so on [3].

In 2017, Sutami conducted a research entitled Evaluation of the quality of securities web services, where this study aims to compare the quality of 6 securities websites, namely Danareksa Sekuritas, Valbury Asia securities, Sucomnvest Central Gani, Phintraco Securities, First Asia Capital, and OSO Securities. The criteria used in the assessment of securities websites are user ease, content, accuracy, timeliness of responses, and security. Of the five criteria, 20 sub-criteria are described that are used as an assessment. The results of calculations using the Fuzzy AHP-TOPSIS method show that Valbury Asia Securities is web security with the best service, and this study also describes the advantages and disadvantages of each of these web securities, such as the Danareksa securities web, which has advantages in the overview of the user experiences problems or questions. , while the weakness of this web security is that it is not reliable under the actual conditions [4].

The general objective of this study is to rank e-commerce with the best service quality using the Fuzzy AHP and TOPSIS methods.

2. Materials and methods

2.1. E-commerce Classification

According to Dave Chaffey, the scope of electronic commerce (e-commerce) is narrower than digital business. Often considered simply referring to buying and selling using the Internet; people immediately thought of consumer retail purchases from companies like Amazon. But e-commerce must be considered because all transactions are electronically mediated between an organization and the third parties with whom it deals. By this definition, non-financial transactions such as customer support and requests for further information will also be considered part of e-commerce [5].

Mahir in 2015, classify the E-commerce business in Indonesia [6]. The results of the classification of e-commerce based on business models can be seen in table 2.

Table 2. E-commerce in Indonesia.

No	Business Model	E-commerce
1	Business to Business	Blibli.com
2	Business to Customer	OLX, Zalora.com, Lazada
3	Customer to Customer	Tokopedia, Bukalapak, Elevenia, Shopee
4	Customer to Business	Kitabisa,com, wujudkan.com

2.2. Service Quality Dimension

According to parasuraman, there are 5 dimensions of service quality, which consist of [7]:

- Reliability, In connection with the company's ability to provide accurate services from the first time without making any mistakes and deliver services according to the agreed time.
- Responsiveness, With regard to the willingness and ability of employees to help customers and respond to their requests, as well as informing when services will be provided and then providing services quickly.
- Assurance, With regard to employee behavior, it can foster customer trust in the company and the company can create a sense of security for its customers. Assurance also means that employees are always polite and have the knowledge and skills needed to handle any customer service or problem.
- Empathy, means that the company understands the problems of its customers and acts in the best interests of the customers, while paying personal attention to customers and having comfortable hours of operation.
- Tangible, With regard to the attractiveness of physical facilities, equipment and materials used by the company, as well as the appearance of employees.

2.3. Purposive Sampling

Purposive sampling is a non-probability sampling method that uses specific people (specific target-groups) as a source of data / information. Certain people referred to here are individuals or groups who because of their knowledge, experience, position, etc. make the individual or group a source of information. This individual or group's name is immediately recorded as a respondent without going through a random selection process.

Purposive sampling technique, also called judgment sampling, is a deliberate choice of a participant because of the qualities the participant possesses. This type of sampling is a nonrandom technique that does not require an underlying theory or a number of participants. Simply put, the researcher decides what needs to know and sets out to find people who can and are willing to provide information based on knowledge or experience [8].

2.4. Delphi Method

The Delphi method is a modification of the brainwriting and survey techniques. In this method, panels are used in the movement of communication through several questionnaires contained in writing. The Delphi technique was developed in the early 1950s to elicit expert opinion. The object of this method is to obtain the most reliable consensus from a group of experts [9,10].

2.5. Analytical Hierarchy Process

The Analytical Hierarchy Process is a general theory of measurement. It is used to derive the ratio scale from both separate and continuous comparisons in a tiered hierarchical structure. These comparisons can be drawn from actual measurements or from fundamental scales that reflect the relative strengths of preferences and feelings. AHP is designed to tackle both rational and intuitive to select the best from several alternatives evaluated concerning multiple criteria. In this process, the decision-maker performs a simple pairwise comparison assessment which is then used to develop overall priorities for ranking alternatives.

The simplest form used to construct a decision problem is a hierarchy consisting of three levels: decision objectives at the top level, followed by a second level consisting of criteria by which alternatives, located at the third level, are evaluated. The hierarchical decomposition of complex systems appears to be the basic tool by which the human mind deals with diversity. One organizes the factors that influence decisions in gradual steps from the general, at the top level of the hierarchy, to the special, at the lower level. The purpose of this structure is to make it possible to judge the importance of elements at a certain level concerning some or all elements at adjacent levels above [11].

2.6. Fuzzy AHP

The observed values in real-world problems are often imprecise or vague. Imprecise or vague data may be the result of unquantifiable, incomplete, and non-obtainable information. They are often expressed with bounded intervals, ordinal (rank order) data, or fuzzy numbers.

To effectively handle subjective perceptions and impreciseness, fuzzy numbers are integrated with AHP, allowing the appropriate expression of linguistic evaluation according to Calabrese et al. in 2016. Fuzzy numbers are also used to deal with uncertainties affecting subjective preferences in assessing real-world decision-making problems.

Despite the convenience of AHP in handling both quantitative and qualitative criteria of MCDM problems based on decision makers' judgements, FAHP can reduce or even eliminate the fuzziness; vagueness existing in many decision-making problems may contribute to the imprecise judgements of decision makers in conventional AHP approaches [12–16].

2.7. TOPSIS

Topsis is a multi-criteria decision making method that was first introduced by Yoon and Hwang. TOPSIS uses the principle that the chosen alternative must have the closest distance from the positive ideal solution and the farthest from the negative ideal solution from a geometric point of view by using the Euclidean distance to determine the relative proximity of an alternative to the optimal solution. The ideal positive solution is defined as the sum of all the best values that can be achieved for each attribute, while the negative ideal solution consists of all the worst scores achieved for each attribute [16].

3. Results and discussion

3.1. Determination of sub-criteria

The determination of sub-criteria for evaluating the quality of e-commerce services is carried out using the Delphi method. First of all, the sub-criteria that will be used to assess e-commerce are determined, after which some sub-criteria suggested by experts are added, such as product return services if they are not suitable, e-commerce is quick to respond to customer complaints, free shipping according to conditions, guarantee of customer confidentiality , user experience and ease of payment. In the second round, the experts agreed on the sub-criteria used in the selection of e-commerce, which consisted of 21 sub-criteria and 5 criteria.

Hierarchical structure for evaluating the quality of e-commerce services can be seen on figure 3.

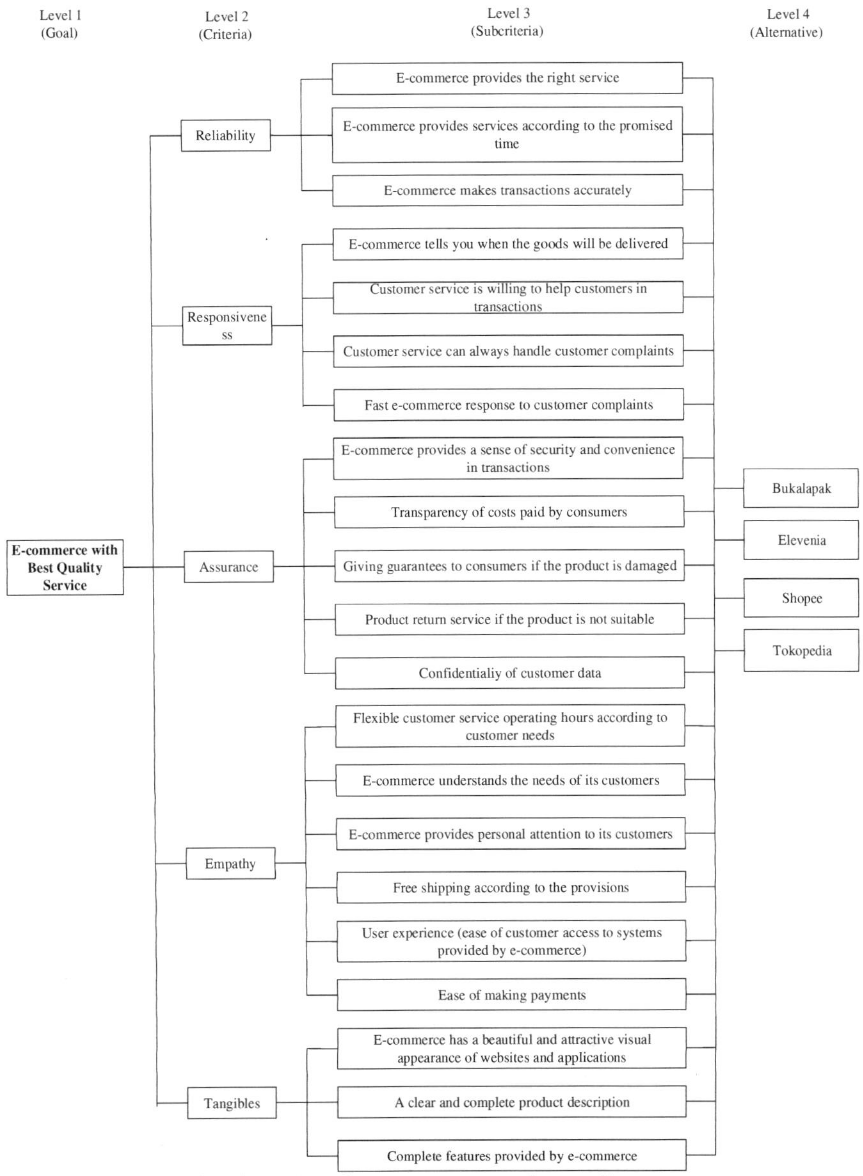

Figure 3. Hierarchy of e-commerce service quality assessment.

3.2. *Fuzzy AHP calculation results*

The results of the weight calculation for each criterion, sub-criterion, and alternative can be seen in table 3.

Table 3. Eigenvector of all level.

Eigenvector					
Level 2		**Level 3**		**Level 4**	
Reliability	0,2290	E-commerce provides the right service	0,5536	Elevenia	0,0953
				Shopee	0,2261
				Tokopedia	0,4721
				Bukalapak	0,2065
		E-commerce provides services according to the promised time	0,1691	Elevenia	0,0938
				Shopee	0,2011
				Tokopedia	0,2579
				Bukalapak	0,4473
Reliability	0,2290	E-commerce provides transaction accuracy	0,2774	Elevenia	0,1733
				Shopee	0,0901
				Tokopedia	0,3751
				Bukalapak	0,3615
Reponsiveness	0,2214	E-commerce tells when the goods will be delivered	0,0744	Elevenia	0,1028
				Shopee	0,1091
				Tokopedia	0,5077
				Bukalapak	0,2804
		Customer service is willing to help customers in transactions	0,2960	Elevenia	0,1448
				Shopee	0,4092
				Tokopedia	0,3204
				Bukalapak	0,1256
		Customer service can always handle customer complaints	0,4803	Elevenia	0,1575
				Shopee	0,1535
				Tokopedia	0,3174
				Bukalapak	0,3717
		Fast e-commerce response to customer complaints	0,1493	Elevenia	0,1289
				Shopee	0,1858
				Tokopedia	0,4587
				Bukalapak	0,2266
Assurance	0,3510	E-commerce provides a sense of security and convenience in transactions	0,0515	Elevenia	0,1474
				Shopee	0,0597
				Tokopedia	0,2691
				Bukalapak	0,5238
		Transparency of costs paid by consumers	0,2511	Elevenia	0,0955
				Shopee	0,1089
				Tokopedia	0,4949
				Bukalapak	0,3007
		Giving guarantees to consumers if the product is damaged	0,2324	Elevenia	0,1250
				Shopee	0,1109
				Tokopedia	0,2578
				Bukalapak	0,5063
		Product return service if the product is not suitable	0,3135	Elevenia	0,0862
				Shopee	0,1442
				Tokopedia	0,4023
				Bukalapak	0,3672

		Confidentiality of customer data	0,1515	Elevenia	0,4983
				Shopee	0,1417
				Tokopedia	0,0798
				Bukalapak	0,2803
		Flexible customer service operating hours according to consumer needs	0,0958	Elevenia	0,0860
				Shopee	0,1446
				Tokopedia	0,3007
				Bukalapak	0,4687
		E-commerce understands the needs of its customers	0,0739	Elevenia	0,0898
				Shopee	0,1486
				Tokopedia	0,4896
				Bukalapak	0,2720
Empathy	0,0727	E-commerce provides personal attention to its customers	0,1974	Elevenia	0,1575
				Shopee	0,0820
				Tokopedia	0,3138
				Bukalapak	0,4467
		Free shipping according to the provisions	0,0492	Elevenia	0,0617
				Shopee	0,1456
				Tokopedia	0,5029
				Bukalapak	0,2898
		User experience (ease of customer access to systems provided by e-commerce)	0,2884	Elevenia	0,0814
				Shopee	0,1563
				Tokopedia	0,2867
				Bukalapak	0,4756
		Ease of making payments	0,2953	Elevenia	0,0956
				Shopee	0,1350
				Tokopedia	0,3300
				Bukalapak	0,4394
Tangible	0,0559	E-commerce has a beautiful and attractive visual appearance of websites and applications	0,4090	Elevenia	0,1672
				Shopee	0,2362
				Tokopedia	0,5129
				Bukalapak	0,0837
		A clear and complete product description	0,4529	Elevenia	0,1053
				Shopee	0,3328
				Tokopedia	0,3801
				Bukalapak	0,1817
		Complete features provided by e-commerce	0,1381	Elevenia	0,1675
				Shopee	0,1005
				Tokopedia	0,5176
				Bukalapak	0,2144

The results of ranking the quality of e-commerce services using fuzzy AHP can be seen in table 4.

Table 4. The rank of e-commerce based on their service quality using fuzzy AHP method.

Alternative	Weight	Percentage	Rank
Elevenia	0,1394	13,94%	IV
Shopee	0,1735	17,35%	III
Tokopedia	0,3680	36,80%	I

Bukalapak	0,3190	31,90%	II
Total	**1,0000**	**100%**	

Based on the table above, it can be seen that Tokopedia is e-commerce with the best service quality. This is because Tokopedia has a high weight on the product return service sub-criteria (0.3135) and correct service delivery (0.5536). Where these sub-criteria are the most important sub-criteria according to experts on each of the criteria, namely Assurance and Reliability. Based on the comparison between criteria, the assurance factor is the most important criterion according to experts among other criteria, followed by reliability, responsiveness, empathy, and tangible.

3.3. Fuzzy AHP-TOPSIS calculation results

The results of calculations using FAHP-TOPSIS can be seen in table 5.

Table 5. Calculation result using fuzzy AHP-TOPSIS.

Alternatives	Preference Value	Rank
Elevenia	0.2094	III
Shopee	0.1910	IV
Tokopedia	0.6206	I
Bukalapak	0.5208	II

The ranking results show that the selected alternative is based on calculations with the TOPSIS method which is e-commerce with the best service quality with a preference value of 0.6206.

Based on the research above, it appears that Tokopedia can be a recommendation for business owners and users to transact online. Where the growth of e-commerce will have a significant effect on the amount of tax paid, and the tax itself is used to finance state expenditures, create jobs, control the inflation rate, and so on.

4. Conclusion

The criteria for selecting e-commerce with the best service quality used in this study consisted of 5 criteria, namely reliability, responsiveness, assurance, empathy and tangible, where the sub-criteria used consisted of: E-commerce provides the right service, E-commerce provides appropriate services the promised time, e-commerce provides transaction accuracy, E-commerce notifies when the goods will be sent, Customer service is willing to help customers in transactions, Customer service can always handle customer complaints, e-commerce responds quickly to customer complaints, e-commerce provides a sense of security and comfort in transactions, transparency of costs paid by consumers, Guarantee to consumers if the product is damaged, product return service if the product is not suitable, confidentiality of customer data, flexible customer service operating hours according to consumer needs, E-commerce understands the needs of its customers, E-commerce delivers personal attention to its customers, free shipping according to the provisions, user experience (easy customer access to systems provided by e-commerce), ease of making payments, e-commerce has a beautiful and attractive visual appearance of websites and applications, product descriptions clear and complete and completeness of the features provided by e-commerce.

Based on the processing carried out by the Fuzzy AHP method, it was found that Tokopedia was chosen as the e-commerce with the best service quality with a weight of 0.3680. Then followed by Bukalapak, Shopee and Elevenia.

Based on the processing carried out by the TOPSIS method, it was found that Tokopedia was also selected as the e-commerce with the best service quality with a weight of 0.6206. Then followed by Bukalapak, Elevenia and Shopee.

References

[1] Badan Pusat Statistik 2019 *Statistik E-Commerce 2019*

[2] Setiawan N, Winati F D, Achmadi R E and Shafira Y P 2016 Analisis Hubungan Antara Kepuasan Pelanggan Berdasarkan Kualitas Layanan E-Commerce Terhadap Loyalitas Pelanggan Dengan Menggunakan Metode Statistik (Studi Kasus Di Website E-Commerce Lazada)

[3] Sitorus R R and Kopong Y 2017 Pengaruh E-Commerce Terhadap Jumlah Pajak Yang Disetor Dengan Kepatuhan Wajib Pajak Sebagai Variabel Intervening *Media Akunt. Perpajak.*

[4] Sutami S, Soedijono B and Henderi H 2017 Evaluasi Kualitas Layanan Web Menggunakan Model Integrasi Fuzzy AHP-TOPSIS *Respati* **12**

[5] Chaffey D, Edmundson-Bird D and Hemphill T 2019 *Digital business and e-commerce management* (Pearson UK)

[6] Pradana M 2016 KLASIFIKASI BISNIS E-COMMERCE DI INDONESIA *MODUS*

[7] Sukaria S 2004 Metode Penelitian. Edisi Ketiga

[8] Etikan I 2016 Comparison of Convenience Sampling and Purposive Sampling *Am. J. Theor. Appl. Stat.*

[9] Ishak A and Ali A Y Bin 2019 Identification criteria and indicators of palm oil industrial liquid waste processing technology *IOP Conference Series: Materials Science and Engineering*

[10] Ishak A and Habibi Rahman Barus W 2020 Application of the delphi method to new renewable energy assessments in power plants in North Sumatra province *IOP Conference Series: Materials Science and Engineering*

[11] Saaty T and Vargas L 2012 *Models, methods, concepts & applications of the analytic hierarchy process*

[12] Emrouznejad A and Ho W 2017 *Fuzzy analytic hierarchy process*

[13] Ishak A and Ali A Y Bin 2017 Decision Support Model for Selection Technologies in Processing of Palm Oil Industrial Liquid Waste *IOP Conference Series: Materials Science and Engineering*

[14] Ishak A, Asfriyati and Akmaliah V 2019 Analytical Hierarchy Process and PROMETHEE as Decision Making Tool: A Review *IOP Conference Series: Materials Science and Engineering*

[15] Ishak A and Akmaliah V 2020 Technology assessment of liquid waste in rubber factory using analytical hierarchy process and promethee methods *AIP Conference Proceedings*

[16] Ishak A and Parinduri S 2019 Supplier Selection Using Analytical Hierarchy Process (AHP) and Technique for Order Preference by Similarity to Ideal Solution (TOPSIS) *2019 1st International Conference on Engineering and Management in Industrial System (ICOEMIS 2019)* (Atlantis Press)

Sustainable plastic waste management strategy: Optimization of plastic manufacturing plant waste (gas and product transition)

F Khair[1], J Matsana[2]

[1]Industrial Engineering Department, BINUS Online Learning, Bina Nusantara University, Jakarta, Indonesia 11480
[2]Industrial Engineering Department, Faculty of Engineering, Bina Nusantara University, Jakarta, Indonesia 11480

Corresponding author: fauzi.khair@binus.ac.id

Abstract. Production waste management has become the key to a sustainable manufacturing operation strategy. This study aims to analyse the waste gas and production waste from the polymerization process, especially the residual exhaust gas from the degassing unit and the final product from the product transition process that does not meet the requirements. Research using qualitative and quantitative methods. The analysis is done by calculating the amount of gas removed from the sweeping or degassing process and calculating the material balance of the production process to analyse the process loss. The results show that the largest contributor to the waste gas is from the degassing process and the pellets from the transition products. In addition, an optimization process is needed to recover exhaust gas with a recovery system and perform a trimming process or a mixing process to reduce the transition product to become prime product.

1. Introduction

Plastics are produced through the polymerization of synthetic or semi-synthetic materials derived from petroleum, natural gas, or coal. Plastics are composed of various elements such as carbon and hydrogen as the main constituent elements, as well as oxygen, nitrogen, chlorine, and sulphur[1]. Plastics are used for food packaging, industrial and health products, building construction, clothing, electronics, automotive and aeronautics[2]. Given the large need for plastics, the plastic industries also face challenges in handling plastic waste [3]. Since 1960, the cumulative total production of plastics has been estimated at around eight billion tons. Due to its non-degradable nature, unless incinerated or recycled, plastic waste will still exist [4, 5].

The biggest problem in plastic manufacturing plants is waste originating from gas flaring and downgrade products, namely the final product that does not meet specification standards [3]. Disposal of waste needs to be handled with further research, better technology, and more qualified resources so that the sustainability of the plastic industry can generate greater opportunities in the future [5]. This disposal activity is carried out through a series of material testing, recycling policies, and an increasing need for technology to separate and recycle plastic waste. This process will be able to produce secondary plastics that are more competitive than by continuing to consume or use pure polymers [4, 5]. Given that plastic waste is non-degradable, the waste will continue to pollute forever. With the recycling technology, plastic waste will be reproduced in good quality without relying on the production of new plastic (virgin material).

Plastic waste has become one of the major environmental issues with its main polymer components being polyethylene and polypropylene [6]. Separation of plastic waste into individual polymers is expensive and complete sorting is often not possible. However, plastic waste (off spec product) can be recycled easily by turning it into polymer blends [7]. However, there are major limitations to PP which has a glass transition below room temperature and a high coefficient of linear thermal expansion. Thus, limiting dimensional stability and high temperature applications. One of the promising methods to reduce this deficiency is to blend or trimming polypropylene with thermoplastic techniques [8].

The polymerization process produces waste in the form of waste gas from the polymerization reaction and exhaust gas from the stripping or degassing process [9]. The exhaust gas is then burned in the flare system and then discharged into the atmosphere through flare networks. The main source of air pollution from industry is exhaust gas from the residual polymerization or gas that is intentionally released to maintain pressure when the polymerization process occurs, and from processes when abnormal situations occur such as plant shutdowns. In addition, the increasing demand for oil and gas production in the world results in gas burning or gas flaring [10].

Flaring or combustion from industry cannot be avoided but can be reduced [10]. Several ways to reduce flaring, for example by implementing procedures on how to adjust the process rate to prevent or minimize flaring [11]. The use of flare gas recovery is something that is very important in the industry. In addition to reducing the greenhouse effect and environmental pollution, it is also very useful in reducing costs or costs of hydrocarbon loss. One of the flare gas recovery techniques that can be applied is to change the gas to liquid with the compression method [11, 12].

Limitations of technology, technology costs, company vision and mission, and human mindset are the main factors outside government regulations that can determine a company's policy in managing assets, optimization and contribution to the implementation of waste management [3, 5, 8]. This study aims to discuss how to recycle wasted gas from the degassing process by separating nitrogen, light gases, and hydrocarbons. So that it can be reused in the degassing process and polymerization raw materials using flare gas recovery technology [11, 12]. This study seeks to design innovations on how to reduce or eliminate transitional products using mixing technology or trimming facilities. Based on the background description that has been stated, a way is needed to seek the benefits obtained from placing the flare recovery unit. Especially for sustainable waste management, an optimal strategy is needed to utilize the trimming facility in reducing or eliminating transitional products.

2. Materials and methods

This research uses the integration of qualitative and quantitative methods in the process of data collection and processing. This research was conducted to analyse the work system and process optimization in the flare recovery unit facility as part of sustainable waste management efforts. This study aims to determine the optimal amount of regression in process waste gas so that it can provide benefits both from a financial and environmental side. As well as finding the best solution in increasing the company's profit by utilizing a mixing facility or trimming facility from non-prime products to prime products. This research was conducted in several plastic production companies in the United Arab Emirates as one of the largest plastics producing countries in the world.

The technical method used is in the form of observation and data collection related to the degassing process. Data analysis was performed on monthly reports on energy balance and total gas wasted from each unit including the degassing system unit. Data collection on the number or total of transitional products resulting from each product changeover, the data is obtained from monthly reports where the number of prime and non-prime products is recorded. Monitoring is carried out on the sweeping process between nitrogen gas and hydrocarbons on the DCS (Distribute Control System) life screen to obtain Life Data. The analysis of the current system under study is focused on the degassing system where the exhaust gas from the degassing is the largest contribution of the wasted gas from the polymerization process. In addition, the running system that is optimized in this study is the production of material transitions when a product change occurs. Data processing is carried out using descriptive graphical models such as pareto charts and fishbone in analysing the main problems and consequences. Data analysis is focused on the degassing process and other activities through graphics and histograms, especially related to reliability, FRU Load, Monthly CO_2 Reduction Chart, Percent HC & N_2 Base on Design Chart, and off grade comparation with or without trimming.

3. Results and discussion

The results of analysis and observation of the degassing system and flare recovery unit obtained data from four polyethylene plants such as the amount of nitrogen sent, the amount of feed gas available, utility consumption such as electricity or power, the cooling system (cooling water, glycol water, propylene refrigerant), plant water, steam, running hours and cost of the production process. Next, an analysis of recover nitrogen and hydrogen is performed and an analysis of the reliability of the system (% online to unplanned shutdown). The scheme and process of the degassing system are shown infigure 1.

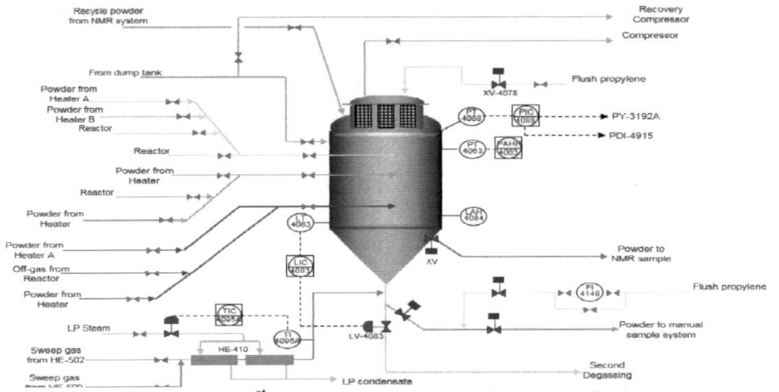

Figure 1. 1st degassing unit - product Receiver

3.1 Technical Analysis - Recover Nitrogen and Hydrocarbons

Figure 2 illustrates how the recovery unit provides the ability to recycle gas or feeds that enter the washing unit. The months of March and May, the chart shows zero point. This is because in that month no feeds were received due to the Plant Shutdown. Recovered hydrocarbons and nitrogen are taken from the analyser and smart flow meter from each membrane stage. Meanwhile, tons of recovered N_2/Hydrocarbons are taken from the calculation of recovered nitrogen / hydrocarbons from the flow meter compared to the total production produced per month.

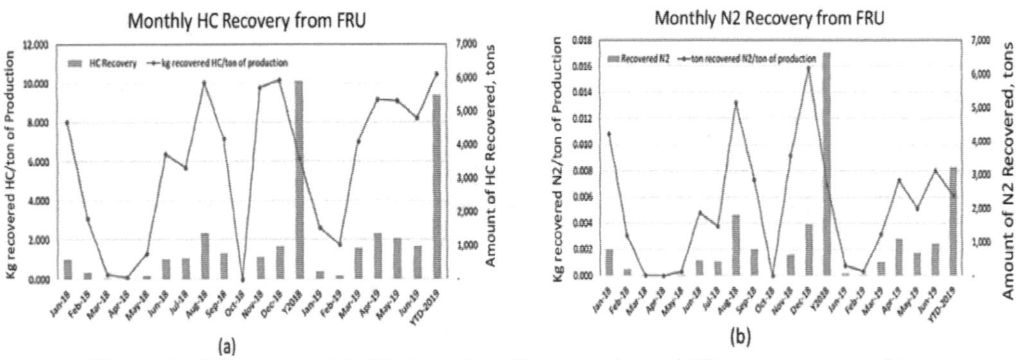

Figure 2. Chart of monthly Hydrocarbon Recovers (a) and Nitrogen recovers (b)

3.2 Technical analysis - reliability of the system (% online to unplanned shutdown)

In Figure 3, there is a very significant difference between 2018 and 2019. In 2018 mechanical failure from compressors contributed the highest to around 76% of the total downtime of 3009 hours from 22 stops, followed by Turn Around about 14.5% and the rest from processes such as clogged filters and gas flow from the washing unit. The pareto analysis result show that the highest bar graph causes the amount of downtime is due to the reasons for both mechanical, electrical and instrument equipment. Damage tó the membrane cartridge, rotor, problematic programable logic control, oil leakage, and heater problems are the most important elements that must be resolved.

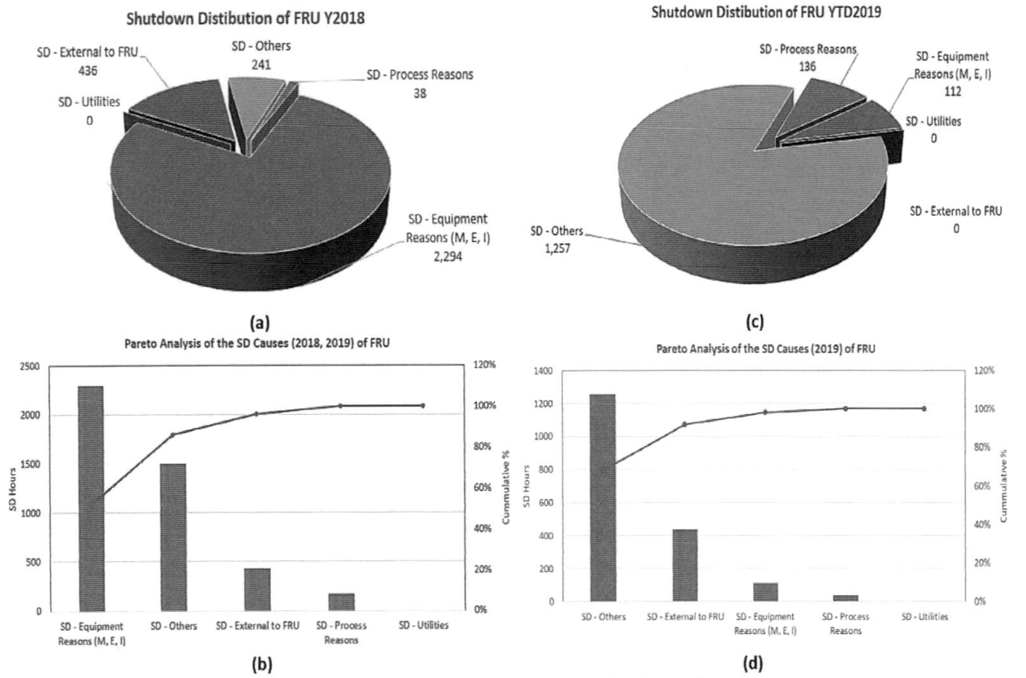

Figure 3. FRU Shutdown Distribution-Pareto Analysis: (a,b) 2018 and (c,d) 2019

The flare recovery unit vendor was brought in to repair the FRU system as a strategy to improve the flare recovery unit's performance. Meanwhile, 2019 shows an improvement. The results of the improvements made at the initial stage of start-up, namely in 2018 on the reliability and operability of the FRU, the total shutdown hour became 1503 hours with 15 stops. The results of the analysis for 2019, the shutdown contribution was caused by others causes such as Plant Turn Around which cannot be avoided, roughly 1257 hours. Plant Turn Around (TA) in this company is scheduled once every 5 years. Plant Turn Around itself is something that must be done by a company, especially a manufacturing company, to restore the direction of the company from a decline in performance or to restore company performance.

The next analysis is how to know that the flare recovery unit will last for the future, so the reliability of the flare recovery unit must be much better than the previous year. Figure 4 shows that the reliability of the flares-recovery unit experienced a very significant and stable improvement, except in May and October where plant shutdowns due to process and equipment were frequent. data is taken from 2018 to June 2019, showing that the average reliability is around 94% for 2019.

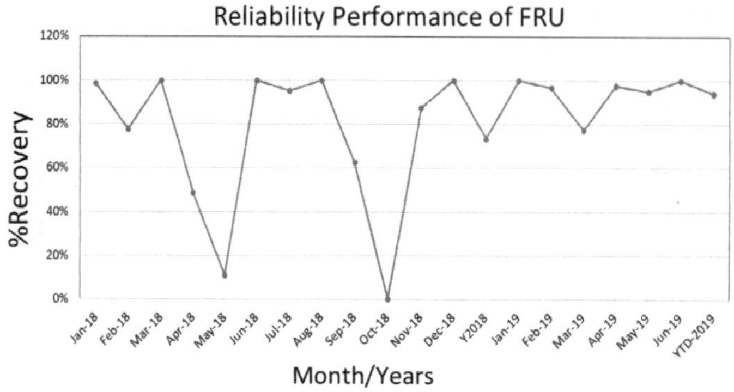

Figure 4. Reliability Performance Chart

3.3 Environment Impact - CO_2 Footprint Analysis

Carbon footprint is the amount of greenhouse gas emissions produced by an organization, event, product or individual. To obtain the carbon footprint of the flare recovery unit, the total hydrocarbons recycled are divided by ethylene as the reference gas most contained in this recycling system:

$$CO_2\ footprint\ (kg) = \left(HC \frac{recover}{C2H4mw}\right) * 2 * CO_2\ wm * 1000 \qquad (1)$$

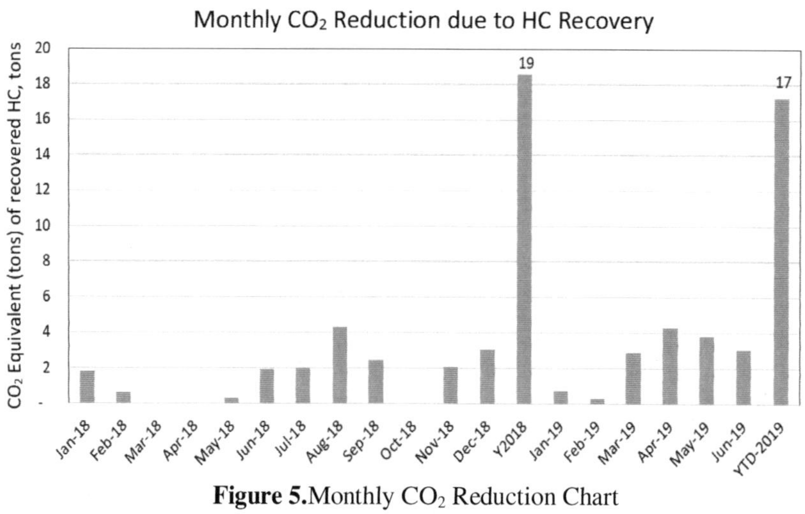

Figure 5. Monthly CO_2 Reduction Chart

Figure 5 show that adding exhaust gas to the flare recovery unit for recycling provides enormous benefits, apart from cost reduction as well as environmental issues. It appears that the greater the ability to recycle hydrocarbons, the more we contribute to reducing emissions, in other words the smaller the amount of CO_2 is wasted if the hydrocarbons are not burned.

3.4 Economic Analysis - Saving of Nitrogen and Hydrocarbon per Ton of Production

Figure 6 shows how the respective percent of nitrogen and hydrocarbon is recycled per month. Percent recovery is obtained from hydrocarbons and nitrogen recovered material against flare recovery recover design value. Hydrocarbon design recover is 2917 ton / month and design recover nitrogen is 5000 kg / month. There are several values on the graph showing a low percent recycled, this is because the factory was in a shutdown that month.

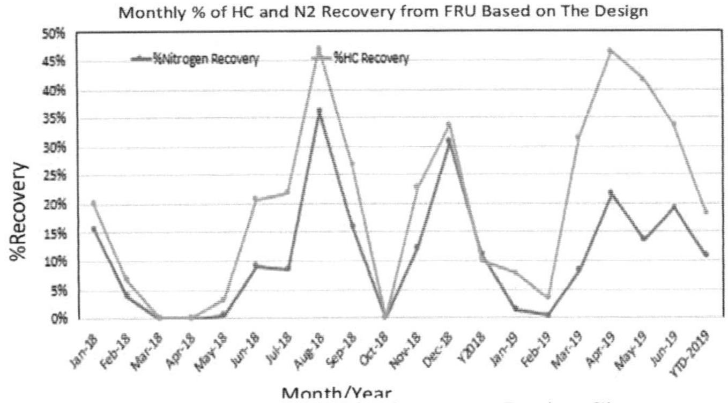

Figure 6. Percent HC and N2 Base on Design Chart

3.5 Economic Analysis - Variable cost Analysis

Design recovery capacity for hydrocarbons that were set was 291 tons and nitrogen at 5000 kg. The optimal operation of the flare recovery unit is the lower the operating costs and the greater the ability to recycle hydrocarbons and nitrogen. When the operating load from recovery is very low the need for utility consumption will remain, so the utility consumption costs will be high. Figure 7 uses the analytical method, from the data points, the regression lines are installed. Both the height R-squared line and the equation can be used to estimate the costs or benefits under different loads.

The break-even point analysis results show the minimum point at which the flare recovery loading or operating load must be or at least 42%. So that the higher the operation or loading of the flare recovery unit, the greater the profit and the lower the production costs. The total benefit line shows equivalent monetary benefits due to hydrocarbon and nitrogen recovery. Whereas the total cost line shows the operating costs of the FRU, excluding labour costs (the same labour is used to manage the FRU) and no additional labour is required.

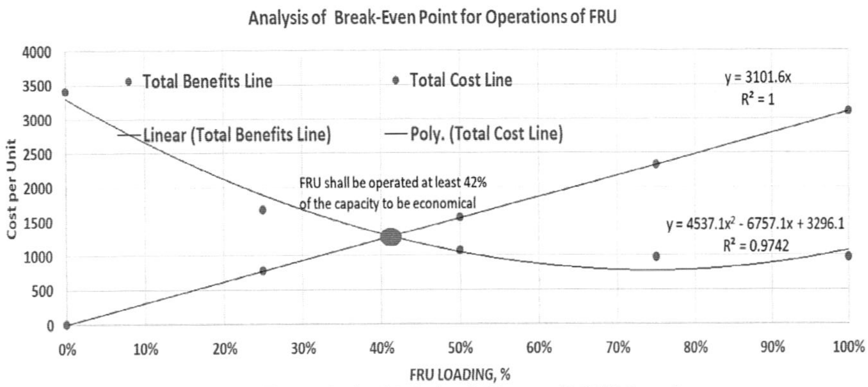

Figure 7. Analytical Method Chart of FRU Load

3.6 Trimming System Data Analysis

Figure 8 shows a monthly comparison of% prime product in plant A and plant B. The trimming facility increases the% prime at both plants for total monthly production showing a significant difference in% prime before and after cutting ($p < 0.05$). % prime increased by an average of 6% to 7% (absolute), which translates to millions of dollars in savings.

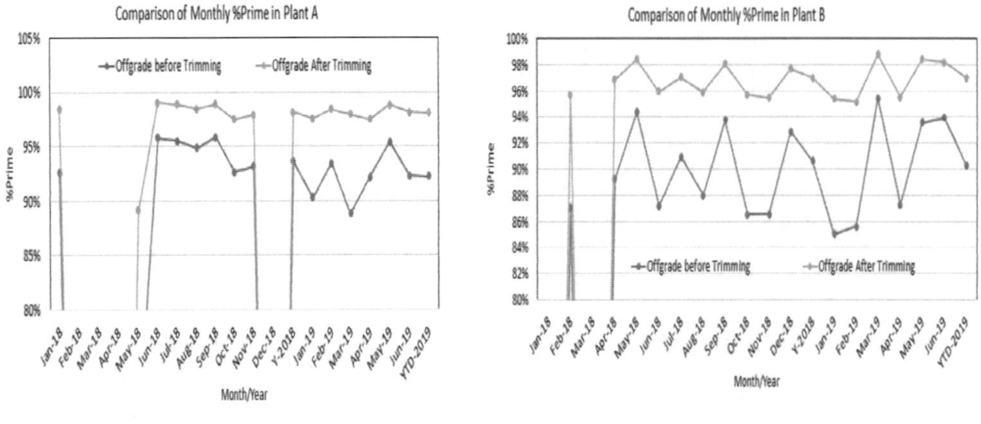

(a) (b)

Figure 8. Off grade Comparation with and without trimming on (a) Plant A and (b) Plant b

Based on several tests that have been carried out in mixing or trimming, it is known that the effect of quality in this case is that the melt flow of the prime product determines the final MFR. The result of methodology regression analysis shows that the quality of the prime MFR has a p-value of 0.04 indicating that there is an effect of the MFR value of the prime product which can influence the final value of the final MFR. From the results, a regression equation can be made as follows:

$$Final\ MFR = 1.49 * \%\ non\ prime * 0.18 * MFR_{nonprime} - 0.29 * \%prime * 0.8 * MFR_{prime} \quad (2)$$

The regression equation results in a prediction of how many final MFRs according to the regression of the final MFR results from offline check or laboratory. Based on available data (production grade mix and the main properties of the grades), a regression equation was developed to approximate the final MFR result based on the volume of non-prime and prime values and MFR of non-prime and prime respectively. Even though, the high p-value is close to 0.05 (indicating insignificance of the factors), the regression equation can be used to generally predict the resulting properties when blending or trimming. The regression calculation results show that the final MFR value is very close to the final MFR value generated from the laboratory results. However, it is recommended, however, to perform a regression analysis based on each prime product and the transition product to be trimmed with a higher amount of data.

3.7 Economic Analysis-saving for upgrade of the transition material

The number of material transitions successfully trimmed varies in number, depending on how many products must be produced for each product. In one month about 800 to 1200 tons of transition material are produced as a product change result. The price difference between transitional products and prime products is around 300 dollars per ton. The percentage (%) of material transitions that can be mixed or successfully trimmed is an input or profit for the company.

The more transitional products that can be trimmed, the more prime products that can be produced, thus the number of company savings or profits will increase. Apart from the product cycle, the length of the campaign of the product as well as the efficiency, reliability, and operability of trimming systems such as feeder performance, silo availability and of course online quality reading (online rheometer) must function properly.

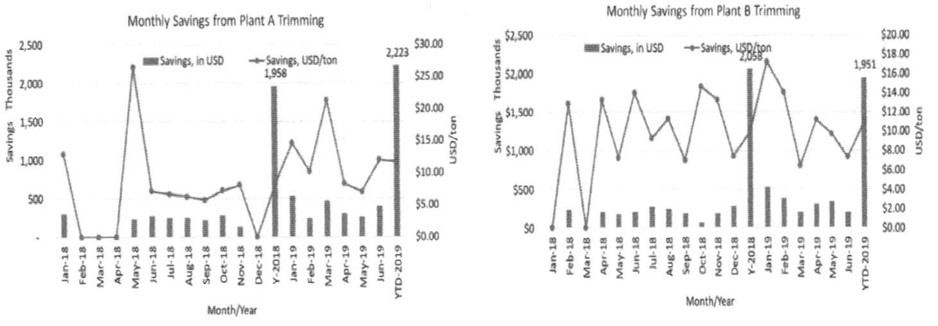

(a) (b)

Figure 9.(a) Plant A and (b) Plant B Saving Due to Trimming

Figure 9 shows the monthly savings after trimming between transitional products or it can be called near prime grade and products that have prime quality. Bar graphs show savings equivalent to million dollars while line charts show savings in units per ton of production. In 2018, the use of trimming was more conservative to confirm or prove if there was an influence on product quality. Trimming shows a significant increase in 2019 after the production and marketing parties are quite well established and successful in managing the trimming system operation while maintaining its quality. Savings in the first six months of 2019 were around 200 thousand USD higher than the savings throughout 2018. Apart from achieving a good working method of the trimming system, the difference in the amount of profit in 2019 was also caused by changes in product demand due to variations in supply and demand.

4. Conclusions

Most of the gas and petrochemical factories that have been running still have not utilized recycling technology for exhaust gas and reducing product offs with the trimming system. The flare recovery unit only accepts recycling from the low flare system, still requires further studies on how to utilize exhaust gases from other sources. Optimization of exhaust gas utilization through the flare recovery unit from continuous flaring resulting from the degassing process in the purge bin has a very significant progress. Only the emergency flaring gas cannot be utilized, while the maximum flare recovery unit intake pressure is below 1 bar. Meanwhilefor intermittent flaring, optimum exhaust gas is recycled in the vent recovery system.

Trimming system is a solution to reduce or even eliminate transition products, making transition materials into prime products. The more transitional products that can be trimmed, the more prime products that can be produced, so that the number of company savings or profits will increase. In this study, it was found that the average final number of prime products increased between 6% to 7% after trimming. Apart from the role of production planning in scheduling product cycles, the reliability and operability of the trimming system are important factors in reducing or eliminating transition materials.

For further research, it is necessary to improve the flare recovery system process and trimming system. it is necessary to develop standard operating procedures, preventive maintenance and corrective maintenance as important elements that must be maintained. For the final product that passes through mixing or trimming, it is advisable to get one additional treatment in the pellet silo, for example, the addition of blending before the product is bagged or packed.

References

[1] Ghayebzadeh M, Taghipour H, Aslani H 2020 Estimation of plastic waste inputs from land into the Persian Gulf and the Gulf of Oman: An environmental disaster, scientific and social concerns *Sci Total Environ.* **733** 138942 Available from:

https://doi.org/10.1016/j.scitotenv.2020.138942
[2] Phanisankar BSS, Vasudeva Rao N, Manikanta JE 2020 Conversion of waste plastic to fuel products. *Mater Today Proc.* Available from: https://doi.org/10.1016/j.matpr.2020.02.880
[3] Awoyera PO, Adesina A 2020 Plastic wastes to construction products: Status, limitations and future perspective *Case Stud Constr Mater.* **12** e00330 Available from: https://doi.org/10.1016/j.cscm.2020.e00330
[4] Jang YC, Lee G, Kwon Y, Lim J hong, Jeong J hyun 2020 Recycling and management practices of plastic packaging waste towards a circular economy in South Korea *Resour Conserv Recycl.* **158** 104798 Available from: https://doi.org/10.1016/j.resconrec.2020.104798
[5] Dilkes-Hoffman LS, Pratt S, Lant PA, Laycock B 2018 The role of biodegradable plastic in solving plastic solid waste accumulation *Plastics to Energy: Fuel, Chemicals, and Sustainability Implications.* Elsevier Inc 469–505 Available from: http://dx.doi.org/10.1016/B978-0-12-813140-4.00019-4
[6] Hamidzadeh Z, Sattari S, Soltanieh M, Vatani A 2020 Development of a multi-objective decision-making model to recover flare gases in a multi flare gases zone *Energy.* **203** 117815 Available from: https://doi.org/10.1016/j.energy.2020.117815
[7] Nezhadfard M, Khalili-Garakani A 2020 Power generation as a useful option for flare gas recovery: Enviro-economic evaluation of different scenarios *Energy* **204** 117940 Available from: https://doi.org/10.1016/j.energy.2020.117940
[8] Karasalihović Sedlar D, Vulin D, Jukić L, Smajla I 2018 Possibility of decreasing CO2 emissions from flaring on a mature oil field *J Pet Sci Eng.* **171** 1302–8 Available from: https://doi.org/10.1016/j.petrol.2018.08.026
[9] Gai H, Wang A, Fang J, Lou HH, Chen D, Li X, et al 2020 Clean combustion and flare minimization to reduce emissions from process industry *Curr Opin Green Sustain Chem.* **23** 38–45 Available from: https://doi.org/10.1016/j.cogsc.2020.04.006
[10] Willyard KA 2020 A license to pollute? Opportunities, incentives, and influences on oil and gas venting and flaring in Texas *Energy Res Soc Sci.* 2020 **62** Available from: https://doi.org/10.1016/j.erss.2019.101381
[11] Hajizadeh A, Mohamadi-Baghmolaei M, Azin R, Osfouri S, Heydari I 2018 Technical and economic evaluation of flare gas recovery in a giant gas refinery *Chem Eng Res Des.* **131** 506–19 Available from: http://dx.doi.org/10.1016/j.cherd.2017.11.026
[12] Comodi G, Renzi M, Rossi M 2016 Energy efficiency improvement in oil refineries through flare gas recovery technique to meet the emission trading targets *Energy.* **109** 1–12

Analysis of Quality Function Deployment (QFD) and Analytical Network Process (ANP) Methods at PT. XYZ

R Ginting[1], A Ishak [1], R Y Tarigan [1] and Alfin F. Malik[1]

[1,]Department of Industrial Engineering, Faculty of Engineering, Universitas Sumatera Utara, Medan, Indonesia 20155

Corresponding author: rosnani@usu.ac.id

Abstract. The product under study is a mattress product. The problem at hand is a mismatch with consumer desires. To minimize this mismatch, an analysis was carried out using the Quality Function Deployment (QFD) and Analytical Network Process (ANP) methods. Where QFD is used to determine customer technical desires and then weighting is carried out using the ANP method. From the calculation results, the highest to the lowest technical characteristics are cutting speed, cutting accuracy, product dimensions, component composition, machine capacity, and durability. The value of the difficulty level of cutting speed is 4 and the importance level is 23% and the estimated cost is 20%. Then from these results, the company must pay more attention to these characteristics because it can have a direct effect on the product. The wrong cutting speed will result in reject products so that it can reduce profits for the company or harm the company.

1. Introduction

PT. XYZ is a company in the bedding industry. The products produced are pillows, bolsters, mattresses, and folding mattresses. The raw materials of this product are pads, fabrics, and threads. One of the company's products that are used as an example of design improvement is a mattress. Mattress products have several types of defects such as broken stitches, skewed stitches, wrong size cuts, dirty fabrics, and uneven foam contents. Disabilities can be influenced by several factors, namely machines/equipment, work methods, people, and materials. Therefore, it is necessary to improve the quality by taking into account several factors.

Previous researchers conducted by Singh, et al. [1] described the application of ANP with QFD. This study discusses the combination of two techniques, namely Quality Function Deployment (QFD) and Analytic Network Process (ANP) which are used in the product development process. Combine QFD and ANP methods to prioritize Technical Characteristics. ANP is a multi-criteria decision-making method used to derive relative priority from individual assessments, which can systematically handle all types of dependency. ANP assists in the QFD matrix which obtains a pairwise comparison matrix and checks the consistency ratio for the level of customer interest, the relationship between Customer Requirements (CR) and Technical Characteristics (EC), and the internal dependence between CR and EC.

The method used in QFD (Quality Function Deployment) and ANP Analytical Network Process). QFD is used to determine consumer desires, while ANP is to look for value and the relationship between technical characteristics and customer needs.

QFD is defined as a structured process or mechanism for determining customer requirements and translating those needs into relevant technical requirements where each functional area at the organizational level can understand and act according to procedures. QFD translates what customers need into the results of an organization. [2-7]

ANP is a qualitative approach method used to find values and relationships between variables horizontally, vertically, and loops. ANP basically has a shape similar to AHP where AHP shows the relationship between variables that form a hierarchy. Not all decision-making problems are based on hierarchy only, but also have dependencies between components in the cluster. In such a problem, Network Process Analytics is used where the network is spread out in all directions and includes rotation between clusters and loops within the same cluster [8].

2. Materials and methods

The research location chosen was PT. XYZ. The types of data used are primary data and secondary data. Primary data sources come from direct interviews with workers or production managers who are people who are directly related to product quality. Primary data is data from open questionnaires, closed questionnaires, technical characteristics questionnaires, technical characteristics relationship questionnaires, and ANP questionnaires, while secondary data sources are data obtained through companies and employees using interview techniques [9].

The data used in this study have data sources that can be primary data and secondary data. Primary data collected by direct observation is data on consumer desires related to the mat production process. The instruments used in collecting data on consumer desires are open questionnaires and closed questionnaires. The data collected are as follows:

- Open Questionnaire
- Closed Questionnaire
- Technical Characteristics Questionnaire
- Relationship Questionnaire Technical Characteristics
- Critical Part Questionnaire
- ANP Questionnaire

Secondary data were obtained based on company documentation data, and obtained by interview with the production department

QFD-ANP steps [10]:

- Step 1: Identify the CR and the importance of each CR
- Step 2: Determine the dependency matrix of parts in CRs with respect to each CR.
- Step 3: Identify technical characteristics
- Step 4: Determine the dependency matrix in the EC associated with each EC
- Step 5: Determine the relationship between ECs associated with each CR by assuming that there is no dependency between ECs
- Step 6: Determine the interdependent priorities of the CR
- Step 7: Determine the interdependent priorities of the EC
- Step 8: determine the overall priority of the EC

3. Results and Discussion

Open and closed questionnaires are used to determine consumer desires. Respondents' answers to the preliminary or open questionnaires obtained several modes that support the question attributes in the second stage questionnaire, namely the closed questionnaire. An open questionnaire provides results in the form of a mode which is then converted into technical characteristics through discussions and interviews with managers. Open and closed questionnaires were distributed to 30 respondents, namely production workers. The ANP questionnaire was given to 1 respondent, namely the Head of Padding Production Manager.

Table 1. Analytical Network Process (ANP) questionnaire recapitulation.

Consumer Voice Cluster	**Foam Padding**			
	Component Composition	Engine Capacity	Durability	
Component Composition	1	5	3	
Engine Capacity	1/5	1	1/4	
Durability	1/3	4	1	
Consumer Voice Cluster	**Knitting Fabric**			
	Component Composition	Durability		
Component Composition	1	3		
Durability	1/3	1		
Consumer Voice Cluster	**Size 100 x 200 cm**			
	Cutting Speed	Accuracy of Cutting	Product Dimensions	
Cutting Speed	1	1/7	1/5	
Accuracy of Cutting	7	1	3	
Product Dimensions	5	1/3	1	
Consumer Voice Cluster	**Foam thickness 7 cm**			
	Component Composition	Engine Capacity	Durability	
Component Composition	1	5	3	
Engine Capacity	1/5	1	1/3	
Durability	1/3	3	1	
Consumer Voice Cluster	**Cream cloth color**			
	Component Composition	Durability		
Component Composition	1	2		
Durability	1/2	1		
Consumer Voice Cluster	**Foam As A Endurance Factor**			
	Component Composition	Engine Capacity	Durability	
Component Composition	1	1/4	1/2	
Engine Capacity	4	1	3	
Durability	2	1/3	1	
Consumer Voice Cluster	**Average 2 Year Product Life**			
	Component Composition	Durability		
Component Composition	1	4		
Durability	1/4	1		
Consumer Voice Cluster	**Foam Quality as a Factor of Technical Specifications**			
	Component Composition	Durability		
Component Composition	1	1/5		
Durability	5	1		
Cluster Technical Characteristic	**Component Composition**			
	Foam Padding	Knitting Fabric	Cream cloth color	Foam Quality
Foam Padding	1	3	5	1/3
Knitting Fabric	1/3	1	5	1/5
Cream cloth color	1/5	1/5	1	1/7
Foam Quality	3	5	7	1
Cluster Technical Characteristic	**Cutting Speed**			
	Knitting Fabric	Size 100 x 200 cm	Foam thickness 7 cm	
Knitting Fabric	1	1/5	1/3	
Size 100 x 200 cm	5	1	5	
Foam thickness 7 cm	3	1/5	1	
Cluster Technical Characteristic	**Accuracy of Cutting**			
	Size 100 x 200 cm	Foam thickness 7 cm		
Size 100 x 200 cm	1	5		
Foam thickness 7 cm	1/5	1		
Cluster Technical Characteristic	**Engine Capacity**			
	Knitting Fabric	Foam thickness 7 cm		
Knitting Fabric	1	3		
Foam thickness 7 cm	1/3	1		
Cluster Technical Characteristic	**Durability**			
	Foam As A Endurance Factor Produk	Average 2 Year Product Life		

IOP Conf. Series: Materials Science and Engineering 1041 (2021) 012044 doi:10.1088/1757-899X/1041/1/012044

Foam As A Endurance Factor Produk	1	4
Average 2 Year Product Life	1/4	1
Cluster Technical Characteristic	**Product Dimensions**	
	Size 100 x 200 cm	Foam thickness 7 cm
Size 100 x 200 cm	1	4
Foam thickness 7 cm	1/4	1

Table 2. Summary of Analytical Network Process (ANP) questionnaire among priorities.

	Degree of importance	Cost Estimation
Degree of importance	1	3
Cost Estimation	1/3	1

The next step after obtaining the mode of each open and closed questionnaire, the next step in building HoQ is to determine the characteristics needed to meet the variable consumer needs for mattress products. Determination of product characteristics based on interviews and discussions with factory production supervisors. The technical characteristics of the product are as follows: (a) Component Composition, (b) Cutting Speed, (c) Cutting Accuracy, (d) Machine Capacity, (f) Durability, and (g) Product Dimensions.

The next stage is to determine the level of relationship between technical characteristics. After determining the relationship between technical characteristics, then determining the level of relationship between the technical characteristics of the product and consumer desires. The relationship level is obtained from the weights which are then calculated using the priority weights which are then calculated using the ANP priority weights for each relationship characteristic with the consumer's voice based on the ANP questionnaire.

The priority weight between elements is arranged into a matrix called a super matrix. Super matrix is used to get the limit of the super matrix using super decision software. Super matrix tables and super matrix constraints can be seen in table 3.

Table 3. Global Weight.

No.	Element	Weight
	Costumer Requirements	
1	Foam Padding	0.053363
2	Knitting Fabric	0.105433
3	Size 100 x 200 cm	0.032457
4	Foam thickness 7 cm	0.025774
5	Cream Cloth Color	0.010079
6	Foam As A Endurance Factor	0.152637
7	Average 2 Year Product Life	0.038159
8	Foam Quality as a Factor of Technical Specifications	0.114555
	Technical Requirements	
1	Component Composition	0.121892
2	Cutting Speed	0.206109
3	Accuracy of Cutting	0.190796
4	Engine Capacity	0.101735
5	Durability	0.103095
6	Product Dimensions	0.174625

Before building a quality house (HOQ) it is necessary to calculate the HoQ performance measurement which consists of three aspects, namely the level of difficulty, the level of importance and the estimated cost. The calculation of these three aspects can be seen in the following description:

- Determination of the level of difficulty. The level of difficulty is determined by the relationship of technical characteristics. The calculation is done by translating all the weights of the relationship value then dividing the weights of each technical characteristic by the number of weights earlier. Furthermore, the level of difficulty is given based on the range of percentages obtained.

- Determination of the degree of importance. Degree of importance is a value indicating the importance of a technical characteristic based on limiting super matrix.
- Cost estimation. The difficulty level factor is used as the basis for cost estimates because the more difficult a technical characteristic is made, the more expensive the allocation of costs will be.

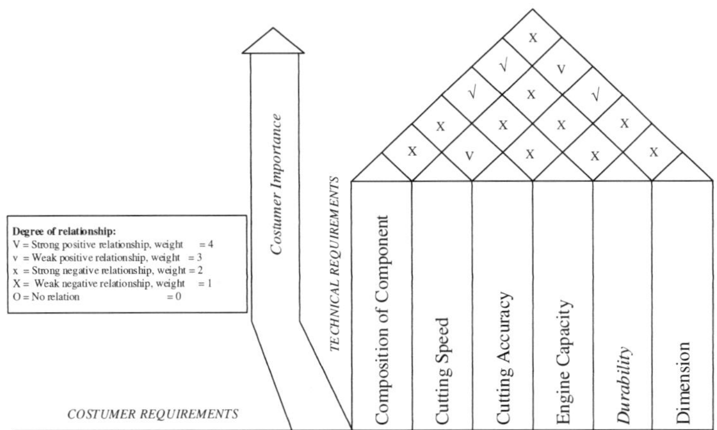

COSTUMER REQUIREMENTS	Costumer Importance	Composition of Component	Cutting Speed	Cutting Accuracy	Engine Capacity	Durability	Dimension
Padding foam is a foam or mattress foam that best suits the mattress product design	5	0,6194	0,0000	0,0000	0,0964	0,2842	0,0000
Knitted fabrics are one of the most suitable mattress types for the mattress product design	4	0,7500	0,0000	0,0000	0,0000	0,2500	0,0000
The mattress size is 100 x 200 cm which best fits the mattress product design	4	0,0000	0,0738	0,6434	0,0000	0,0000	0,2828
7 cm is the thickness of the foam or froth on the mattress that best fits the mattress product design	4	0,6333	0,0000	0,0000	0,1062	0,2605	0,0000
The cream is the color of the mattress that best suits the mattress product design	4	0,6667	0,0000	0,0000	0,0000	0,3333	0,0000
Foam is a factor that determines the durability of a mattress product that best fits the mattress product design	4	0,1373	0,0000	0,0000	0,6232	0,2395	0,0000
2 years is the average length of life of the product that best fits the mattress product design	4	0,8000	0,0000	0,0000	0,0000	0,2000	0,0000
Foam quality is a factor that is considered in terms of technical specifications in mattress product designs	4	0,1667	0,0000	0,0000	0,0000	0,8333	0,0000
Level of difficulty		3	4	4	3	3	4
Degree of importance		14	23	21	11	11	20
Cost estimation		15	20	20	15	15	20

Figure 1. QFD of mattress products

Determination of the level of difficulty, degree of importance and estimated costs can be seen in table 4. HoQ matrix is made based on data that has been obtained in the previous steps. QFD Mattress products can be seen in figure 1.

Table 4. Difficulty levels, interests, and estimated costs.

Difficulty level	3	4	4	3	3	4
Degree of importance	14	23	21	11	11	20
Cost estimation	15	20	20	15	15	20

4. Conclusion

The results of QFD mattress products show that the attributes that have the highest degree of difficulty are the cutting speed, cutting accuracy, product dimensions, component composition, machine capacity and durability. The value of the difficulty level of cutting speed is 4 and the degree of importance is 23% and the estimated cost is 20%. The results of weighting technical characteristics using the Analytical Network Process (ANP) method of the largest weight value are cutting speed, cutting accuracy, product dimensions, component composition, machine capacity and durability.

Acknowledgments

This study was funded by Research Institutions of University of Sumatera Utara, in accordance with project of Research Funding Scheme of Direktorat Riset dan Pengabdian Masyarakat (DRPM) DIKTI 2020.

References

[1] Singh S and Kumar M 2015 *Int. J. Adv. Eng. Res. Appl.* **1**(3) 2454-2377
[2] Ginting R *et al* 2020 *Songklanakarin J. Sci. Technol.* **42**(4) 771-779
[3] Rosnani Ginting *et al* 2019 *IOP Conf. Ser.: Mater. Sci. Eng.* **505** 012090
[4] R. Ginting *et al* 2019 *IOP Conf. Ser.: Mater. Sci. Eng.***495** 012009
[5] Mehrjerdi Y Z 2010 *International Journal of Quality & Reliability Management* **27** (6) pp 616-640
[6] Asadabadi M 2014 *International journal of industrial engineering computations* **5** (4) pp 543-560
[7] U Tarigan et al 2018 *IOP Conf. Ser.: Mater. Sci. Eng.* **309** 012103
[8] Saaty T L and Vargas L G 2006 *Decision making with the analytic network process* (US: Springer Science+ Business Media, LLC)
[9] Sinulingga S 2015 *Metode Penelitian* (Medan: USU Press)
[10] Shih H S and Chen S H 2013 *International Journal of Operations Research* **10**(2) 80-91

Optimizing Inventory Control System of Crumb Rubber Raw Material: a Case Study

E Amrina[1] and A Y Dewi[1]

[1] Department of Industrial Engineering, Faculty of Engineering, Universitas Andalas, Padang, Indonesia

Corresponding author: elita@eng.unand.ac.id

Abstract. Rubber companies often face problems with raw material inventory because obtained from nature. To anticipate shortages of inventory, companies usually buy raw materials in excess quantities. This will result in large inventory costs. This study aims to optimize the inventory control system of crumb rubber raw material in a case rubber company. It started with the demand forecasting of crumb rubber raw material. The level strategy method is then used to determine the production planning of crumb rubber. Finally, the fixed order quantity model is applied to determine the optimal inventory control of crumb rubber raw material. The results show the optimal inventory level of crumb rubber raw material is 988,881 Kg, with a reorder point is 526,570 Kg, and a safety stock is 342,302 Kg. It obtained the total inventory cost is IDR 1,843,689,421 per year. Inventory control is useful for ensuring the smooth production process and preventing excess or shortage of inventory. The results are expected can aid the rubber company in improving inventory control and minimize inventory costs.

1. Introduction

Optimizing the production level of a product is very important for companies using production planning. Production planning is carried out to utilize limited production resources effectively and efficiently, especially to fulfill customer demand and generate profits for the company. In production planning, inventory needs to be highly considered. Inventory is any item or resource that is stored or stocked by an organization [1]. Inventory is one of the most expensive assets in a company, which represents 40% of the invested capital [2]. Companies can reduce inventory costs by optimizing the inventory level (on-hand inventory), but this can cause customer dissatisfaction if the product ordered is not available (out of stock). Therefore, companies must be able to balance inventory investment with customer service levels [3].

Inventory control plays an important role in the cost savings of a company [4]. Without proper inventory handling, it will cause major marketing problems in increasing revenue and retaining customers [5]. Inventory inaccuracy has a significant negative impact on raw material replenishment performance and production control [6]. Raw material inventory planning can be a determinant of manufacturing operations. Incorrect inventory planning can disrupt production activities. Lack of raw materials can stop production or change production schedules, which in turn will increase costs and cause shortages of finished products [3]. The raw material itself is a source of production that supports the continuity of the factory production process. The number of raw materials that exceeds daily needs can accumulate of raw materials in storage areas [7]. This accumulation of raw material costs will result in an increase in inventory costs. However, if the amount of raw materials is less than daily production requirements, it will be difficult to meet the planned production targets. Shortage or excess inventory is a factor that can lead to increased costs. To support inventory control, researchers have developed various inventory control methods in the past that mainly focus on determining optimal production and order quantities, reorder point, and safety stock levels [8].

Rubber companies often face problems with the supply of raw materials because they are obtained from nature. To anticipate shortages of inventory, companies usually buy raw materials in excess quantities. This will result in high inventory costs. PT X is a rubber company in West Sumatra, Indonesia, that produces crumb rubber. Crumb rubber is a semi-finished product from raw rubber with a granular form obtained after the cleaning process and reduction in the size of the natural rubber [9]. In this study, the raw material of crumb rubber is natural rubber from plantations in the form of lumps called *bokar*. PT X applies the make to stock production system. Production planning for crumb rubber is carried out by considering the amount of raw material inventory in the form of natural rubber available in the warehouse and customer needs. Based on the interview with the Quality Assurance Manager, it is found that PT X is still having problems in planning the crumb rubber production and controlling the supply of crumb rubber raw material. The crumb rubber production planning is not in accordance with the actual production. This may be due to fluctuations in customer demand and uncertainty in the supply of raw materials. The problem of raw material inventory arises because there is no determination of the size of the shipping lot from the supplier to the company. Suppliers routinely deliver raw material even though there are sufficient stocks in storage for production.

In order to meet the availability of raw material for crumb rubber, PT X has several suppliers located in West Sumatra and Jambi Provinces. Currently, suppliers deliver natural rubber to companies in different quantities according to the supplier's capacity and company needs. The company has problems in determining the optimal quantity of raw material orders for each supplier, which is related to controlling the availability of raw materials in warehouses. Therefore, it is necessary to optimize the inventory control of crumb rubber raw material.

2. Materials and methods

This research was conducted in PT X, a rubber company in West Sumatra, Indonesia. PT X processes natural rubber into semi-finished rubber, known as crumb rubber. The products are used as raw materials for several well-known tire manufacturing companies such as Cooper, Continental, Goodyear, Dunlop, Bridgestone, and GT-Radial. The company also exports its products to several countries such as the US, Canada, and China. The methodology is divided into three stages, as described below.

2.1. Forecasting of crumb rubber raw material

In this stage, forecasting the demand for crumb rubber raw material is carried out to determine the raw material demand for crumb rubber in the future. This stage begins with the aggregation of the number of raw material shipments from suppliers. The aggregation results are then plotted to identify trends in demand data. Four methods were selected for forecasting crumb rubber raw material. Forecasting error is calculated using the Mean Absolute Percentage Error (MAPE), and the forecasting method with the lowest MAPE value is selected as the best method. Then, the Moving Range method is applied to verify and validate the forecasting results. Furthermore, the forecasting using the best method is performed for the future period. Finally, the disaggregation of forecasting results for crumb rubber raw material is determined to obtain the raw material demand for crumb rubber.

2.2. Production planning of crumb rubber

The production planning of crumb rubber is then performed to determine the total production of crumb rubber in the future period using the level strategy method. In this stage, the forecasting results of crumb rubber raw material are disaggregated to classify the amount of raw material in the form of families into items for each of the suppliers. A total of 20 suppliers routinely send crumb rubber raw material to PT X every month. The suppliers are located in West Sumatra and Jambi. The aggregate percentage of the total shipments for each supplier is then calculated. Next, the demand disaggregation of crumb rubber raw material for each supplier every month is determined based on the percentage of Dry Rubber Content (% DRC) of each supplier.

2.3. *Inventory control of crumb rubber raw material*

The inventory control is carried out to determine the optimal quantity of orders and the optimal ordering frequency of crumb rubber raw material using the Fixed Order Quantity (FOQ) method. The steps of the FOQ method are as follows:

Step 1 Determining the optimal amount of orders (Q*) using equation (1):

$$Q^* = \sqrt{\frac{2.A.D}{h}} \tag{1}$$

where:
Q* = optimal quantity of orders
A = ordering cost
D = demand
h = holding cost

Step 2 Determining the safety stock (ss) using equation (2):

$$ss = z.SD.\sqrt{L} \tag{2}$$

where:
ss = safety stock
z = service level
SD = standard deviation
L = lead time

Step 3 Determining the reorder point using equation (3):

$$ROP = (d.L) + ss \tag{3}$$

where:
ROP = safety stock
d = average demand per month

Step 4 Determining the total inventory costs using equation (4):

$$TC = TC_O + TC_H = \left(\frac{D.A}{Q*}\right) + \left(\left(\frac{1}{2}Q*+ss\right)h\right) \tag{4}$$

where:
TC = total inventory costs
TC_O = total ordering costs
TC_H = total holding costs

This study also uses the Min-Max method to determine the inventory control of crumb rubber raw material in PT X for comparison purposes. The steps of the Min-Max method are as follows:

Step 1 Determining the optimal amount of orders (Q*) using equation (5):

$$Q^* = \text{maximum stock - minimum stock} \tag{5}$$

$$\text{Maximum stock} = 2(d.L) + ss \tag{6}$$

$$\text{Minimum stock} = (d.L) + ss \tag{7}$$

Step 2 Determining the safety stock using equation (8):

$$\text{Safety stock} = (\text{maximum demand} - \text{average demand}) \text{ x lead time} \tag{8}$$

Step 3 Determining the reorder point using equation (9):

$$\text{Reorder point} = (\text{average demand x lead time}) + \text{safety stock} \tag{9}$$

Step 4 Determining the total inventory costs using equation (4).

Based on the results, the quantity and frequency of shipments are determined for each of the crumb rubber raw material suppliers.

3. Results and discussion

3.1. Forecasting of crumb rubber raw material

Forecasting the demand for crumb rubber raw material begins by aggregating raw material shipment data for each supplier every month. Aggregation delivery data is the raw material amount based on the percentage of Dry Rubber Content (% DRC). The % DRC of each supplier varies depending on the quality of the crumb rubber raw material supplied. The aggregation was carried out on all suppliers for 12 periods. The aggregation results for each period are used to determine data patterns. The results of the aggregate demand for crumb rubber raw material in PT X can be seen in figure 1.

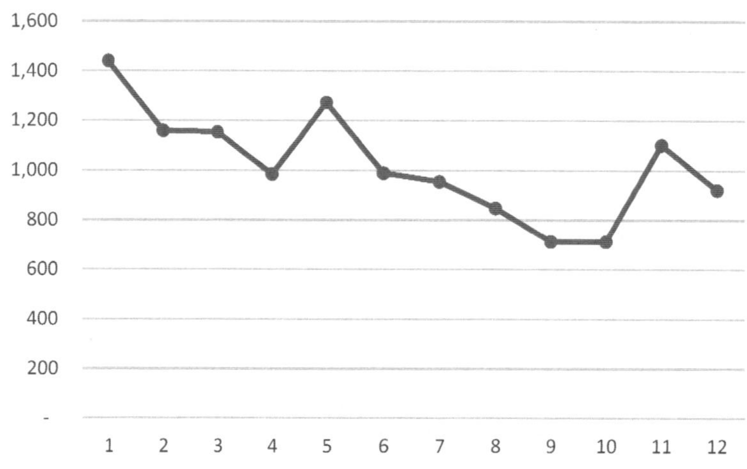

Figure 1. Aggregate demand of crumb rubber raw material.

Based on historical data patterns of raw material demand, forecasting is carried out using four methods, i.e., linear regression, quadratic, exponential, and cyclical trend. The forecasting error is then calculated using the Mean Absolute Percentage Error (MAPE). Based on the MAPE calculation results, the quadratic method with the lowest MAPE value is selected. The results of forecasting errors are presented in table 1.

Table 1. The forecasting error.

Forecasting method	MAPE value
Linear regression	0.1342
Quadratic	0.1024
Exponential	0.4825
Cyclical trend	0.1564

The results of forecasting using the quadratic method are then verified using the Moving Range method. The Moving Range chart is shown in figure 2.

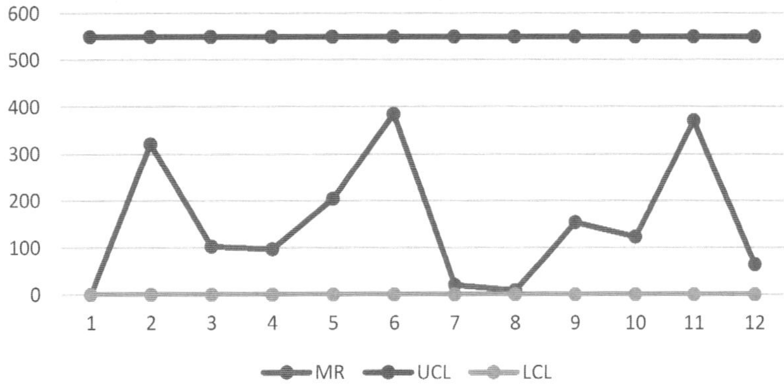

Figure 2. Moving range chart.

It can be seen from figure 2 that the Moving Range value is between the upper control limit (UCL) and lower control limit (LCL). There is no data outside of UCL and LCL. Therefore, the demand forecasting using the quadratic method is then used for inventory planning of crumb rubber raw material. The forecasting results of crumb rubber raw material using the quadratic method are presented in table 2.

Table 2. The forecasting results.

Period	Forecasting (tons)
January	974.470
February	1,037.191
March	1,114.871
April	1,207.511
May	1,315.111
June	1,437.669
July	1,575.188
August	1,727.665
September	1,895.103
October	2,077.500
November	2,274.856
December	2,487.172

3.2. Production planning of crumb rubber

Based on the forecasting results, production planning is performed to determine the total production of crumb rubber in the future period. The level strategy method is used because the company does not recruit or terminate employees and only uses regular working hours without overtime. Based on interviews with production employees of PT X, employees who work every month are permanent operators, and the company applies regular working hours; overtime rarely occurs. The forecasting results of raw material for crumb rubber are disaggregated to classify the amount of raw material in the form of families into items for each supplier. For that purpose, a percentage of the total shipments of each supplier is determined. The aggregate percentage of raw material for crumb rubber can be seen in table 3.

Table 3. The aggregate results of crumb rubber raw material.

Suppliers	Percentage of shipments	Amount of raw material
Supplier-1	27.70%	6,173.908
Supplier-2	12.63%	2,887.543
Supplier-3	5.41%	1,315.703
Supplier-4	18.16%	2,762.253
Supplier-5	0.58%	99.583
Supplier-6	1.91%	374.024
Supplier-7	0.59%	118.671
Supplier-8	1.39%	286.681
Supplier-9	2.24%	328.729
Supplier-10	1.02%	147.731
Supplier-11	2.71%	546.658
Supplier-12	3.26%	747.927
Supplier-13	3.13%	640.282
Supplier-14	2.89%	661.223
Supplier-15	0.72%	139.149
Supplier-16	0.85%	84.136
Supplier-17	0.45%	51.054
Supplier-18	0.85%	108.882
Supplier-19	12.46%	1,514.332
Supplier-20	1.04%	135.839

The current amount of raw material is at the dry rubber level. To obtain the total shipments at the natural rubber level, it is necessary to multiply the % DRC of each supplier. The results of the demand disaggregation of crumb rubber raw material for each month are presented in table 4.

Table 4. The demand disaggregation.

Period	**Demand (tons)**
January	1,414.323
February	1,503.830
March	1,636.894
April	1,742.742
May	1,902.584
June	2,033.237
July	2,214.016
August	2,462.020
September	2,782.624
October	3,050.441
November	3,288.876
December	3,608.663
Total	27,640.248

3.3. Inventory control of crumb rubber raw material

Based on the results of forecasting and production planning, inventory control is carried out to determine the optimal quantity of orders and the optimal ordering frequency of crumb rubber raw material. The inventory control is aimed to reduce the total cost of inventory. The inventory cost is one measure of manufacturing performance [10]. The inventory control of raw material for crumb rubber is conducted using the Fixed Order Quantity method. The Fixed Order Quantity is a method used to determine the specific point (R) where an order is placed in the quantity to be ordered (Q). The calculation of Q uses the formula EOQ (Economic Order Quantity) or also known as Optimal Order Quantity. The steps are as follows:

Step 1 Determining the optimal amount of orders (Q):*

Total demand (D) = 27,640.248 tons/year
Ordering cost (A) = IDR 24,500,000
Holding cost (h) = IDR 1,385,000/tons/year
The optimal amount of orders:

$$Q^* = \sqrt{\frac{2.A.D}{h}} = \sqrt{\frac{2xRp24{,}500{,}000x27{,}640.248}{Rp1{,}385{,}000}} = 988.881$$

Q^* = 988.881 tons = 988,881 kg

Based on the calculations, it obtained the optimal amount of orders for crumb rubber raw material in PT X is 988,881 kg.

Step 2 Determining the safety stock:

Safety stock is an additional supply held to anticipate the possibility of a shortage of crumb rubber raw material. Safety stock is needed in the inventory system to maintain the continuity of the production process. The amount of safety stock depends on the uncertainty of raw material supply and demand. Under normal circumstances, the uncertainty of raw material is represented by the standard deviation of the lead time, which is the time between the company placing orders until the raw materials are received. Conversely, demand uncertainty is usually represented by the standard deviation of demand. The calculation is as follow:

Service level (z) = 1.65
Standard deviation (SD) = 733.467 tons
Lead time (L) = 0.08 month

$$ss = z.SD.\sqrt{L}$$
$$= 1.65 \text{ x } 733.467 \text{ tons x } 0.282 = 342.302 \text{ tons}$$

Based on the calculations, the amount of safety stock for crumb rubber raw material in PT X is 342.302 tons.

Determining the reorder point:
The reorder point indicates when to make reordering crumb rubber raw material. The calculation is as follow:

ROP = (d . L) + ss = (2,303.354 tons/month x 0.08 month) + 342.302 tons = 526.570 tons

The result shows the reordering of crumb rubber raw material in PT X is carried out if the amount of inventory in the warehouse reaches 526,570 tons.

Determine the total inventory costs:
The total inventory costs of crumb rubber raw material using the Fixed Order Quantity method is calculated as follows:

$$TC = TC_O + TC_H = \left(\frac{D.A}{Q^*}\right) + \left(\left(\frac{1}{2}Q^* + ss\right)h\right)$$

$$TC = \left(\frac{27{,}660.248x24{,}500{,}000}{988.881}\right) + \left(\left(\frac{1}{2}x988.881\right)x1{,}850{,}000\right) = \text{Rp } 1{,}843{,}689{,}421$$

It is obtained the total inventory costs of crumb rubber raw material in PT X using the Fixed Order Quantity method is IDR 1,843,689,421. For comparison purposes, this study also uses the Min-Max method to optimize the inventory control of crumb rubber raw material in PT X. The Min-Max method is an inventory control if an inventory in a storage warehouse has passed the minimum limits and is approaching the safety stock limit, thus a reorder must be made. This method is a method applied by a company to control the availability of raw materials in a storage warehouse. The calculations are as follows:

Determining the optimal amount of orders (Q):*
The optimal amount of orders for crumb rubber raw material is calculated as follows:

Minimum stock = (average demand x lead time) + safety stock
= (2,303.354 tons x 0.08 month) + 175.547 Tons = 359.816 tons
Maximum stock = 2 x (average demand x lead time) + safety stock
= 2 x (2,303.354 tons x 0.08 month) + 175.547 tons = 544.084 tons
Optimal amount of orders (Q*) = maximum stock – minimum stock
= 544.084 tons – 359.816 tons = 184.268 tons

Based on the calculation, it is obtained the optimal amount of orders for crumb rubber raw material in PT X is 184.268 tons.

Determining the safety stock:
The safety stock of crumb rubber raw material is calculated as follows:

Safety stock = (maximum demand – average demand) x lead time
= (3,608.663 tons – 2,303.354 tons) x 0.08 month = 175.547 tons

It is obtained the safety stock of crumb rubber raw material in PT X is 175.547 tons.

Determining the reorder point:
The reorder point is calculated as follows:

Reorder point = (average demand x lead time) + safety stock
= (2,303.354 tons x 0.08 month) + 175.547 tons = 359.816 tons

The reorder of crumb rubber raw material will be made if the available quantity in the warehouse is reached 359.816 tons.

Determining the total inventory costs
The total inventory costs of crumb rubber raw material using the Min-Max model is calculated as follows:

$$TC = TC_O + TC_H = \left(\frac{D \times A}{Q^*}\right) + \left(\left(\frac{1}{2}Q^* + ss\right) \times h\right)$$

$$TC = \left(\frac{27{,}660.248x24{,}500{,}000}{184.268}\right) + \left(\left(\frac{1}{2}x184.268 + 175.547\right)x1{,}850{,}000\right) = Rp4{,}045{,}738{,}674$$

Based on the calculation, it is obtained the total inventory costs for crumb rubber raw material in PT X using the Min-Max method is IDR 4,045,738,674. The comparisons between FOQ method and Min-Max method are presented in table 5.

Table 5. The comparisons of FOQ and Min-Max method.

	FOQ method	Min-Max method
Optimal order (tons)	988.881	184.268
Safety stock (tons)	342.302	175.547
Reorder point (tons)	526.571	359.816
Total inventory cost (IDR)	1,843,689,421	4,045,738,674

It can be seen in table 5 that the total inventory cost of the FOQ method is lower than the Min-Max method. Therefore, the FOQ method is chosen as the best method to optimize the inventory control of crumb rubber raw material in PT X. Finally, the disaggregation of demand forecasting for crumb rubber raw material is used to determine the quantity and frequency of shipments for each supplier. The summary of the lot size and frequency of shipments of crumb rubber raw material for each supplier is presented in table 6.

It can be seen from table 6 that supplier-1 has the largest quantity of crumb rubber raw material shipments compared to other suppliers with a value of 73,560 tons and a frequency of 112 times a year. Followed by supplier-19, supplier-2, and supplier-4 with shipping quantities of 57,061 tons, 52,073 tons, and 51,485 tons and a frequency of 60 times, 78 times, and 79 times in 1 year, respectively. The company must pay more attention to all of these suppliers because it highly affects the smoothness of the production process.

Table 6. The quantity and frequency of shipments.

Suppliers	Lot size (tons)	Frequency of shipments/year
Supplier-1	73.560	112
Supplier-2	52.073	79
Supplier-3	35.404	54
Supplier-4	51.485	78
Supplier-5	10.035	15
Supplier-6	19.301	29
Supplier-7	10.711	16
Supplier-8	16.771	26
Supplier-9	20.004	30
Supplier-10	13.536	21
Supplier-11	22.988	35
Supplier-12	26.315	40
Supplier-13	24.521	37
Supplier-14	25.662	39
Supplier-15	11.684	18
Supplier-16	10.215	16
Supplier-17	8.450	13
Supplier-18	12.754	19
Supplier-19	57.061	60
Supplier-20	13.364	27

4. Conclusions

This study has determined the optimal inventory control of crumb rubber raw material in PT X. It is found that the optimal inventory level of crumb rubber raw material is 988,881 Kg. The reorder of crumb rubber raw material will be made if the available quantity in the warehouse is reached 526,570 Kg. The safety stock required for crumb rubber raw material is 342,302 Kg. By applying the optimal inventory control, the total inventory cost for crumb rubber raw material is IDR 1,843,689,421 per year. This study is expected to assist the company in improving the inventory control system for crumb rubber raw material to reduce the total inventory costs and prevent the excess and shortages of inventory. Future research will be analyzing the implementation of optimal inventory control for crumb rubber raw material.

Acknowledgments

The authors would like to thanks Universitas Andalas, Indonesia, and the Directorate General of Higher Education, Ministry of Education and Culture, Republic of Indonesia. The publication of this article is supported by the Faculty of Engineering, Universitas Andalas.

References

[1] Jacobs F R and Chase R B 2018 *Operations and supply chain management* (New York: McGraw-Hill Education).

[2] Tersine R J 1994 *Principles of inventory and materials management* (Englewood Cliffs NJ: Prentice-Hall).

[3] Heizer J, Render B and Munson C 2017 *Operations management: sustainability and supply chain management* (Pearson).

[4] Alizadeh M, Eskandari H and Sajadifar S M 2014 A modified (S-1,S) inventory system for deteriorating items with poisson demand and non-zero lead time *Applied Mathematical Modeling* **38** 699.
[5] Waters D *Inventory control and management (*John Wiley & Sons).
[6] Li M and Wang Z 2017 An integrated replenishment and production control policy under inventory inaccuracy and time-delay *Computers and Operations Research* **88** 137.
[7] Iksan 2018 Perencanaan dan pengendalian produksi dengan menggunakan metode manufacturing resources planning di PT Semen Gresik Tbk. *MATRIK (Manajemen dan Teknik Industri-Produksi)* **7** 47.
[8] Glock C H and Grosse E H 2021 The impact of controllable production rates on the performance of inventory systems: a systematic review of the literature *European Journal of Operational Research* **288** 703.
[9] Tham T C, Hii C L, Onga S P, Chin N L, Abdullah L C and Law C L 2014 Technical review on crumb rubber drying process and the potential of advanced drying technique *Agriculture and Agricultural Science Procedia* **2** 26.
[10] Amrina E and Yusof S M 2010 Manufacturing performance evaluation tool for Malaysian automotive small and medium-sized enterprises *International Journal of Business and Management Science* **3** 195.

IOP Conf. Series: Materials Science and Engineering 1041 (2021) 012046 doi:10.1088/1757-899X/1041/1/012046

Production Scheduling Optimization to Minimize Makespan and the Number of Machines with Mixed Integer Linear Programming

F Farizal[1], D S Gabriel[1], A Rachman[1], I Rinaldi[1]
[1]Department of Industrial Engineering, University of Indonesia, Kampus Baru UI, Depok 16424, Indonesia

Correspondent author: farizal@eng.ui.ac.id

Abstract. In production management field,production efficiency can be improved through minimizing machine downtime, optimizing production scheduling, minimizing inventory of raw materials and finished products. In this research, the production efficiency is improved through optimizing production schedule that has two goals; not only minimizing production makespan but also minimizing the number of machines used. For the purpose, a mixed integer linear programming (MILP) model was developed. The model was used to solve a problem producing 6 products with 4 production lines. In this study, product demand was first forecasted with Autoregressive Integrated Moving Average (ARIMA) method. Optimization results comes up with makespan of 295.4 hours using only one production line. As addition, this research is also dealing with three real life production scenarios.

1. Introduction

Production scheduling belongs to supply chain operation type of decision that sets in weekly or daily basis [1]. Operation decision has a strong impact on overall company profitability and success. The important role of scheduling is even more this day due to the existence of globalization and the increasingly demanding customers. A company may end up into troubles if it fails to fulfill the production deadline or it faces a delay on introducing its new product to the market. On both situations, customers may shift to other brand name product to fulfill their needs. This situation will lead to a significant loss. Scheduling is also important since scheduling is also dealing with using the available limited resources, i.e. machine, man, time, and money in an efficient way[2]. Scheduling is, thus, an optimization problem that has one or more goal to be reached.

Scheduling jobs on machine(s) has many settings and objectives. The simplest scheduling is with single machine. The setting can be, for example, time dependent [3], indivisible orders[4], release date [5]. A more complex scheduling is dealing with parallel machines [2,6,7], multiple different machines, and jobshop[2,8,9]. Scheduling several different jobs to single machine requires the machine to be set up every time changing the job. Since there exist setup decisions in the system, the problem can be formulated as a mixed integerlinear programming(MILP)[10]. MILP will be more efficient if it is solved by B&B algorithm [11,12,13].

[14,15]wrote time series forecasting is an important area of forecasting in which past observations of the same variable are collected and analyzed to develop describing the underlying relationship. One of the most important and widely used time series models is the autoregressive integrated moving average (ARIMA) model.

In practical setting in industry, each machine on production line has its own capacity, age, operating requirement and mode, as well as maintenance program. This variety gives different machine capability and utilization. Production scheduling is quite challenging problem since it has many varieties of settings, big size and different target. The problem is even complicated by for instance engine failure and demand uncertainty.

In this research, production scheduling objective to be achieved is not a single target, but it has two goals, minimizing makespan and minimize the number of machines used.Issues such as replacement of product variations and variety of the number of demands are also addressed. Scheduling problem to minimize makespan and line machine utilization is, as far as the author's knowledge, new and no papers have dealt with. In this study, scheduling is combined with forecasting. The proposed method was used to solve a problem in a fast moving consumer good (FMCG) that produces soaps.

2. Materials and methods

Scheduling problems deal with a number of jobs to be processed at a number of machines[2]. In a production, a job can represent an item of a product. To perform the job, some pieces of data are required, i.e. duration time of the job processed on the machine, the time when the job is ready to be processed, the dead line of the job, and the priority factor. This later data can be ignored if all jobs have the same priority. From the machine environment angle, several scheduling settings might exist. For instance, single machine, identical parallel machine, parallel machine with different speed, flow shop, and job shop. Scheduling jobs in a single machine is the simplest scheduling environment. For this scheduling, the objective might be to minimize completion time, to minimize number of tardy jobs, to minimize makespan, etc. identical parallel machine can be considered as a number of single machines. In a job shop with some machines' environment, each job has its own process to be completed. in a job shop, a job might or might not revisit one same machine. Figure 1 illustrates an example of scheduling 10 products on production line A (machine A is line A in single machine setting).

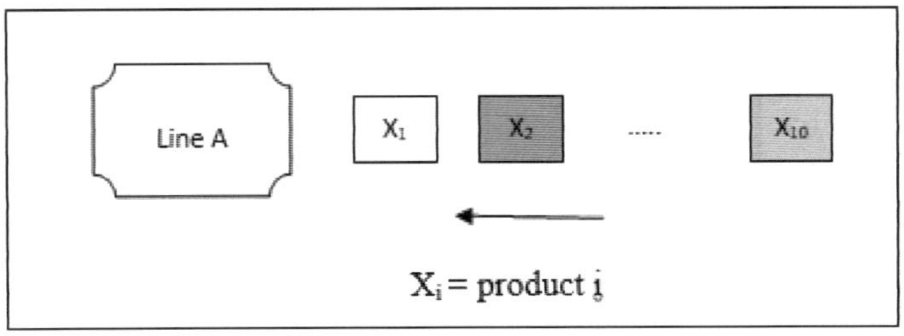

Figure 1. 10 different products are scheduled to be processed on production line A.

2.1. Mathematical model

The following section explain the proposed mathematical model that has two objective functions to reached. Since the model is to minimize, the goal can be achieved by minimizing the number of machines used or minimizing the makespan or minimizing both objectives at the same time.

Indexes:
i= product i = 1, ..., N (N = number of products made)
j = production line j = 1, ..., M (M = the number of machines used)

The decision variables:
M_{ij} = the start time of product i on line j
S_{ij} = time finishedproduct i on line j
d_{ij} = the remainder of the production of I on line j
C = constant
G = constant
Q_i = the number of products i requested

V_j = the speed of the production line j
$\frac{Q_i}{V_j}$= time of production order i
$\mathbf{MS_s}$= *makespan*

$$\text{Min } Z = C\sum_j^M z_j + MS_s \quad \text{................ (1)}$$

St:

$$\sum_i^N x_{ij} \le 1 \qquad \forall j \text{...................... (2)}$$

$$\sum_j^M x_{ij} = 1 \qquad \forall i \text{...................... (3)}$$

$$S_{ij} = M_{ij} + \sum_i^N \frac{Q_i}{V_j} x_{ij} + 0{,}5(x_{ij}) \text{.......... (4)}$$

$$M_i \geq M_{ij} \quad \forall i \quad \forall j \text{ .... (5)}$$

$$S_i \geq S_{ij} \quad \forall i \quad \forall j \text{ .... (6)}$$

$$M_{ij} \le S_{i+1j} \quad \forall i;\ \forall j \text{ .... (7)}$$

$$M_{i+1j} \le S_{ij} \quad \forall i;\ \forall j \text{............................... (8)}$$

$$Gy_{i\,i+1\,j} + M_{i+1\,j} - S_{ij} \ge 0 \forall i; \quad \forall j \text{ .... (9)}$$

$$G(1 - y_{i+1\,1\,j}) + M_{ij} - S_{i+1\,j} \ge 0 \forall i; \quad \forall j \text{ ...(10)}$$

$$M_{i,j} = 0\ \forall i\ ;\ i = 1\ ;\quad \forall j \text{ ....(11)}$$

$$550 - S_{ij} = d_{ij} \forall i\ ;\ \forall j \text{ .... (12)}$$

$$550 \le Sij{*}z_j \forall i\ ;\ \forall j \text{ .... (13)}$$

$$MS_j \ge S_{ij} \forall i\ ;\ \forall j \text{ .... (14)}$$

$$MSS \ge MS_j\ \forall j \text{ .... (15)}$$

$$x_{ij} \in \{0,1\}\ \forall i,j \text{ .... (16)}$$

$$z_j \in \{0,1\}\ \forall j \text{.. (17)}$$

$$y_{ijk} \in \{0,1\}\ \forall i,j,k \text{ .... (18)}$$

Equation (1) is the objective function of the model,i.e minimize the makespan and the number of production lines used.Equations 2 to 18 are the model constraint.Equation (2) isline constraint that says on one-line j at most only one product is done. Equation (3) is product constraint that says a product i can only be worked on one line j. Equation (4) shows the constraints of the time needed to work on the product i, in line j. Equation (5) and equation (6) show the relationship M_{ij} and S_{ij} will be zero if the work was not done on the machine that there is a new variable that is M_i and S_i are positive that the product i worked on one machine j. Equations (7) and (8) are a constraint options, if there are two products i and i + 1 can be done at the same time instance, then the constraints will sequence i after i + 1 or i + 1 after i. Equations (9) and (10) are the linear form of equations (7) and (8). Equation (11) shows a beginning process of time constraint. Equation (12) is the rest of the time constraint. Equation (13) is the line usage constraints. Equation (14) shows *makespan* line j as the longest time of all time completion of work on the line. Equation (15) the *makespan* of the line is the longest time of all existing job completion time. Equations (16, 17, 18) are binary constraints. These constraintswill give an option whether to utilizeor not to utilize the machine.

2.2. Model verification and validation

To be able to use the proposed mathematical model to solve real problems, the model is translated to a computer program with Lingo software version 11.0. To check and to make sure that the computer coding is correct, the computer code is verified and validated. The program is verified if it can be run without error and give an output result. While to validate the program, a small scale problem is prepared and solved manually and using the program. If the results with manual calculation are the same with the results solved by the program computer, the program is validated.

For the purpose of validation, a small scale scheduling problem is shown on tables 1 and 2. Table 1 is listed 4 products with their respective demand. Table 2 tells about 2 production lines with their capacities.

Table 1. Small scale problem demand data.

Product	Demand (box)
i1	8,000
i2	9,000
i3	3,500
i4	8,000

Table 2. Machine capacity data .

Line	Capacity V (box/hour)
j1	100
j2	600

Table 3 shows the processing time of each product if it is produced at the available line. The processing time is calculated by dividing the demand with machine production capacity (Q_i/V_j).

Table 3. The processing time required for product.

Line	Product	Process Time (hour)
j1	i1	80.0
	i2	90.0
	i3	35.0
	i4	80.0
j2	i1	13.3
	i2	15.0
	i3	5.8
	i4	13.3

Table 4 shows that for the lines j1, the smallest processing time required is 35 hours for product i3, while on the line j2 the maximum processing time is 15 hours for product i2. Since the processing time of line j1 is bigger than j2 processing time, j1 is not selected. Table 4 shows the finish time of all products in line j2. The completion time is calculated using Equation (4).

Table 4. Manual calculation result.

Line	Product	Start Time (hour)	Processing Time (hour)	Completion Time (hour)
J2	i1	0.0	13.3	13.8
	i2	13.8	15.0	29.3
	i3	29.3	5.8	35.7
	i4	35.7	13.3	49.5

Validation with Lingo version 11.0 was performed on a computer with Intel (R) Core (TM) i5-3210M CPU @ 2.5 GHz, 4 GB of RAM. Computer operating system is Windows 7. Table 5tabulates the results. Furthermore, if the results on Table 6 is plotted on a Gantt chart, the chart is displayed in Figure 2.

Table 5. Lingo optimization results.

Line	Product	M (hour)	T (hour)	S (hour)
j2	i1	0	13.3	13.8
	i3	13.8	5.8	20.2
	i2	20.2	15	35.7
	i4	35.7	13.3	49.5

Description: M: starting time, T: processing duration time, S: completion time

Since the makespan result of the optimization shown in table4 and the makespan result of the manual calculations shown on Table5 is the same, i.e. 49.5 hours, it is concluded that s the proposed computer algorithm is validated. The algorithm is ready to be used to solve real case problems.

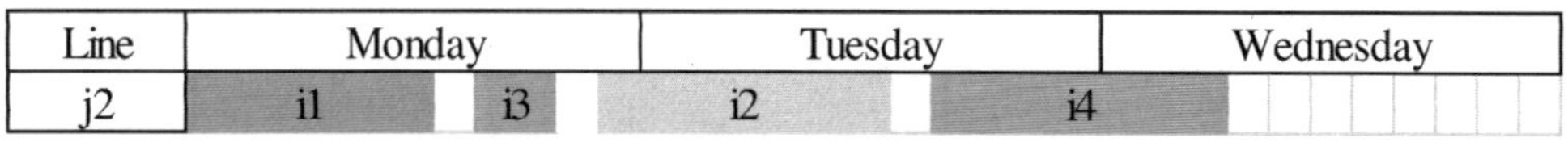

Figure 2. Gantt chart of example problem optimization result.

Gantt chart on figure 2 is the real schedule plotted on the calendar time. The chart tells that one complete production cycle is taking 2.5 working days. It will be 2 changeovers on the first working day and one changeover on the second day.

3. Results and discussion

The model developed was used to determine production schedule of an FMCG company that produces soaps. 6 products will be made on 4 production lines each of which has a single machine. Each line has different speed (different capacity). Table 6 describes the production line capacity of each line.

Table 6. Machine capacity.

Line	box/shift	box/hour
A	1,500	214
B	800	114
C	1,200	171
D	2,500	357

Assumptions used:

- Scheduling is conducted for a product with 6 different colors that have a certain demand quantity (in number of boxes).
- When a machine is producing different color, there is a changeover time of 0.5 hours for the 24-hour product replacement.
- If a given demand meets/exceeds the capacity of the machine, the line can be run all. However, if the demand is granted under a certain line, then it will be a line(s) that can be turned off.
- The production of certain lines in accordance with the capacity of these lines (assuming no product damage in lines).
- Working time for 1 month is 25 days, and 1 day = 22 hours. So that the total working hours within 1 month is 22 x 25 = 550 working hours.

For estimating the number of jobs (the number of products) for each product, a forecasting with ARIMA method is conducted. The forecast uses 27-months historical selling data for the six products. Table 7 shows the forecasting results including their error.

Table 7. Forecasting results with ARIMA.

Product	Arima Model (p,d,q)	MAPE (%)	Demand Forecasting (box)
1	2,0,0	8.534	29,192
2	2,0,0	8.92	23,209
3	2,0,0	10.26	16,182
4	2,0,0	9.369	9,470
5	2,0,0	14.96	14,794
6	2,0,0	12.735	11,530

3.1. Existing scheduling

Currently the production scheduling at the company is developed subjectively by the employees who have experience in making the schedule. The schedule is based on achieving the minimum possible machine changeover. Using the number of products obtained from ARIMA forecasting shown in table 8, the existing company scheduling result is shown in table8.

Table 8. Scheduling results using subjective method.

Line	Product	M (hour)	T (hour)	S (hour)
A	1	0.0	136.4	136.9
B	6	0.0	101.1	101.6
	4	101.6	83.1	185.2
C	3	0.0	94.6	95.1
	5	95.1	86.5	182.1
D	2	0.0	65.0	65.5

Table 8 shows that the makespan of the schedule is 185.2 hours. The table also shows that all the four production lines are utilized. Lines A and D are totally used to make products 1 and 2,

IOP Conf. Series: Materials Science and Engineering 1041 (2021) 012046 doi:10.1088/1757-899X/1041/1/012046

respectively. Line B is used to make products 6 and 4. Products 3 and 5 are processed at line C. Scheduling on Table 9 presented with Gantt chart is shown on figure3.

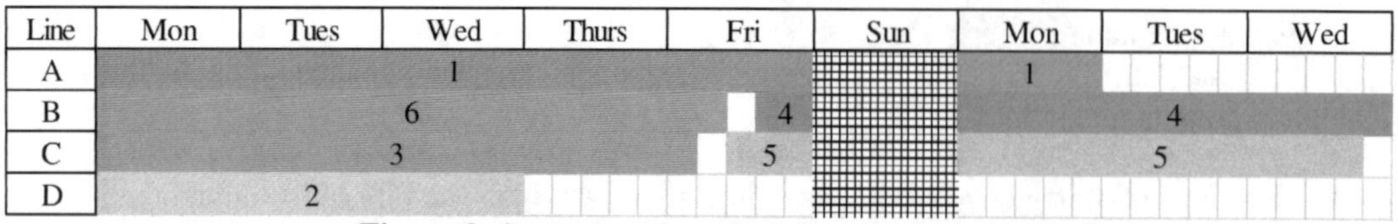

Figure 3. Gantt chart for scheduling product of table 8.

Gantt chart on Figure 3 tells that in order to minimize the number of changeover, the production cycle for one complete production is 8 days; started on Monday and completed on Wednesday. During this cycle, it has 2 times changeover at Lines B and C since those two lines are used to make 2 different products. However, from machine utilization, this schedule is not good. Line D is only utilized for 3 days of 8 days (utilization rate is 37.5%). The machine is idle for 5 days (the idle rate is 62.5%).

3.2. Scheduling with objective function minimizing makespan

When the number of products listed on Table7 is solved with the objective to minimize the makespan, the results are different (see Table9).

Table 9. Scheduling result using optimization to minimize makespan.

Line	Product	M (hour)	T (hour)	S (hour)
A	2	0	108.5	109
B	5	0	129.8	130.3
C	6	0	67.4	67.9
	4	67.9	55.4	123.8
D	1	0	81.8	82.3
	3	82.3	45.3	128.1

Table 10 shows that using optimization approach, the makespanis 130.3 hours. This value is almost 35 hours less than the makespan of subjective scheduling method. This shorter makespan is achieved through utilizing all the four lines like the results of subjective method but with different arrangement. Lines A and B are used dedicatedly to make products 2 and 5, respectively. While lines C and D are used to make products 6 and 4, and 1 and 3, respectively. Scheduling results on table 10 presented with Gantt chart is displayed on figure 4.

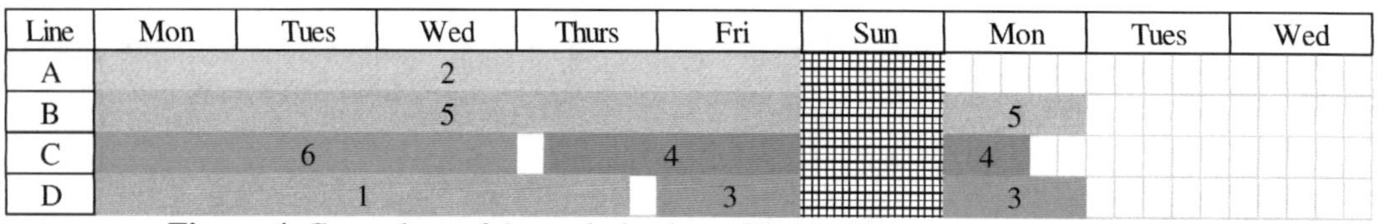

Figure 4. Gantt chart of the optimization of Table 10

Gantt chart on figure 4 informs that the demand on table 8 can be made within 6 working days (Monday to Monday). This production cycle is 2 days shorter than subjective scheduling. Scheduling with objective to minimize makespan, fortunately has also 2 times changeover; one changeover at line C and the other is at line D. Another benefit of this approach is a high machine utilization. Lines B and D have 100% utilization. The lowest machine utilization is Line A which is 83.3%.

3.3. Scheduling with objective function minimizing makespan and total line used

Table 10 shows the optimization results with objective function to minimize makespan and the number of lines being used.

Table 10. Scheduling result using optimization with 2 objective functions.

Line	Product	M (hour)	T (hour)	S (hour)
D	1	0	81.8	82.3
	3	82.3	45.3	128.1
	6	128.1	32.3	160.9
	2	160.9	65	226.4
	5	226.4	41.4	268.3
	4	268.3	26.5	295.4

With the objective function minimizing the number of lines as addition to minimizing makespan, the result on Table 11 shows that the schedule is only using one line, i.e. line D. the schedule is possible since line D has the highest capacity (see Table 6). For this schedule, the makespan is 295.4 hours. When production is only using one machine from some m available parallel machines, the problem become scheduling with single machine problem. Figure 5 shows the Gantt chart for scheduling result of Table11. The figure shows that the production cycle is 3 weeks. During that period, the number of changeovers is 5.

Line	Mon-Thurs	Fri	Sun	Mon	Tues-Wed	Thur-Fri	Sun	Mon	Tues-Wed	Thurs-Fri
D	1	3		3	6	2		2	5	4

Figure 5. Gantt chart optimization results of Table 11

3.4. Cost Perspective Analysis

When dealing with 2 different parameters that giving 2 different results, it is not easy to determine which condition is better. For example is comparing results of optimization methods with one and two objective functions. In this situation, comparison in term of costs can be used. Table 11 showsdata used for conducting cost analysis.

Table 11. Power and labor needed to run the line.

Line	Line Capacity (Box/hour)	Electricity Power (kW)	Man Power (man)
A	214	1,000	12
B	114	723	10
C	171	1,234	12
D	357	1,364	20

Assuming that the 0.5 hour changeover is comparable to produce 2,500 pcs of product, and if a product price is 1,000 IDR, the changeover's worth is equal to 2,500,000 IDR. Based on the minimum government wage regulation, wage per hour for the worker is calculated as 6,090.9 IDR. Electricity

IOP Conf. Series: Materials Science and Engineering 1041 (2021) 012046 doi:10.1088/1757-899X/1041/1/012046

rate is 1,150 IDR per kWh. In this case, the total cost is the operating cost which is the summation of electrical cost to run the lines, wage to pay the workers running the machine and changeover equivalent. Furthermore, the electrical cost is a multiplication of electricity rate (in IDR/kWh), electricity power to run the machine (in kW) and the duration to run the line (in hours). While the wage is a multiplication of the number of workers to operate the line (man), wage rate per hour (IDR/man) and the working duration (hour). Using working duration listed on tables 9, 10, and 11, table 12 displays the summary of the calculation.

Table 12. Summary of the total cost.

Scheduling Model	Cost (IDR)				% saving
	Electrical	Wage	Changeover	Total	
Existing (subjective)	672,580,950	42,575,454	5,000,000	720,156,405	-
Single objective	610,310,175	40,556,927	5,000,000	655,867,102	8.9
Two objectives	463,364,440	55,505,236	15,000,000	533,869,676	25.87

Table 12 clearly shows that the scheduling model with two objective functions (minimize makespan&the number of lines used) will give the least total cost for producing the same product quantity compare to the subjective and single objective scheduling model (minimize makespan only). The model can reduce 25.87% of cost compare to the manual existing one even though the makespan of this model is longer than manual scheduling and single objective scheduling model.

3.5. Sensitivity Analysis

Three scenarios are provided to check the proposed model applicability to provide optimal results. The first is demand increases triple or more scenario, changing production time scenario, and unused production line scenario. The third scenario is for example a situation due to predicted line maintenance activity that should be done without disrupting production scheduling engine.

3.5.1. Scenario 1: Surging demand

This scenario is a situation where additional demand occurs threefold or more than the forecasting value. This situation is not an impossible situation. In real life, demand can increase triple or more when special events occur. For example, during religious holiday or at the end of year. To cope such a situation, companies usually carry safety inventory for a given period[1]. However, if the sudden demand is beyond the safety inventory, company can implement overtime strategy or sub-contracting strategy. In this scenario, say the demand goes up like shown at Table 13. The demand goes up 4 times than the forecast. For this kind of demand, will the existing production capacity be able to handle? If the answer is yes, what is the production arrangement and the schedule?

Table 13. Scenario 1 surging demand.

Product	Demand (Box)	
	Forecasting	Scenario 1
1	29,192	89,564
2	23,209	73,826
3	16,182	71,975
4	9,470	51,554
5	14,794	58,290
6	11,530	68,740

Product	Demand (Box)	
	Forecasting	Scenario 1
Total	104,377	413,949

Table 14 is the optimization result for the scenario displayed on Table 13. The result show that the current production lines are able to handle a demand spike of 4 folds if the machine is scheduled properly. Table 14 clearly shows that to handle the high jump of the demand, all the four lanes are utilized with the makespan 540.1 hours. This makespan makes the production lines almost in 100% utilization. The makespan is approaching the total machine available time, i.e. 550 hours. For this scenario, Lines A, B, and C are dedicated to produce products 2, 5, and 1, respectively. While the rest of products, i.e. products 3, 4, and 6 are scheduled at Line D with the production sequence as product 6, product 3, and product 4.

Table 14. Scheduling result for Scenario 1.

Line	Product	M (hour)	T (hour)	S (hour)
A	2	0	345	345.5
B	5	0	511.3	511.8
C	1	0	523,8	524.3
	6	0	192.5	193
D	3	193	201.6	395.2
	4	395.2	144.4	540.1

3.5.2. Scenario 2: Changing production time

This scenario happened when the total available time for the lines to be operated is only 295.4 hours instead of 505 hours. This scenario happens for instance if national holidays are observed at one particular month. The holiday will reduce the regular production time. The scenario may frequently happen in Indonesia so it should be anticipated.

Table 16 shows the optimization scheduling results for this scenario. The results indicate to use the two largest capacity machines, i.e. lines D and A. Line D is used for producing products 3 and 2, while line A is for products 1, 5, 6, and 4. The makespan for this scheduling scheme is 185.1 hours. The two lines almost utilized at the same utilization rate.

Table 15. Scheduling result for Scenario 2.

Line	Product	M (hour)	T (hour)	S (hour)
D	3	0	75.6	76.1
	2	76.1	108.5	185.1
A	1	0	81.8	82.3
	5	82.3	41.4	124.2
	6	124.2	32.3	157
	4	157	26.5	184

3.5.3. Scenario 3: Production line is not being used

Scenario 3 is a situation where the number of products needed is normal but a line with the largest capacity is broken or turned off for maintenance. If this scenario happens, what is the optimum scheduling to be used? In this research, the scenario is only checked for the largest capacity machine since this scenario will make sure the production will be fulfilled if a lesser capacity machine was broken.

Table 16 shows the optimization result for scenario 3. The production can be scheduled with the second largest capacity machine, Line A. The makespan for this scenario is 490.7 hours and the production sequence is product 1, 3, 2, 5, 4, and 6.

Table 16. Scheduling result for Scenario 3.

Line	Product	M (hour)	T (hour)	S (hour)
A	1	0	136.4	136.9
	3	136.9	75.6	213
	2	213	108.5	322
	5	322	69.1	391.6
	4	391.6	44.3	436.4
	6	436.4	53.9	490.7

4. Conclusion

The proposed scheduling mathematical model with an objective function to minimize makespan and the number of lines can produce a better makespan and better machine utilization compare to a single optimization model and subjective approach. From cost point of view, the model provides a smaller fee than a model with just to minimize makespan. The model can be used in various settings with a single machine scheduling and/or parallel machine scheduling.

In the case study, the model with two objective functions may obtain makespan of 295.4 hours with a single line machine. It can reduce 25.87% of cost compare to the manual existing schedule.

References

[1] Chopra S (Kellog S of M and Meindl P (Kepos C 2016 *Supply Chain Management: Strategy, Planning, and Operation* (Essex: Pearson)

[2] Pinedo M L (Stern S of B 2008 *Scheduling: Theory, Algorithms, and Systems* (New York: Springer)

[3] Kuo W H and Yang D L 2006 Single-machine group scheduling with a time-dependent learning effect *Comput. Oper. Res.***33** 2099–112

[4] Yang D L, Hou Y T and Kuo W H 2017 A note on a single-machine lot scheduling problem with indivisible orders *Comput. Oper. Res.***79** 34–8

[5] Oyetunji E O 2012 Minimizing the Sum of Makespan and Maximum Tardiness on Single Machine with Release Dates **2** 1623–32

[6] Jiang X, Lee K and Pinedo M L 2020 Ideal schedules in parallel machine settings *Eur. J. Oper. Res.* 1–13

[7] Bektur G and Saraç T 2019 A mathematical model and heuristic algorithms for an unrelated parallel machine scheduling problem with sequence-dependent setup times, machine eligibility restrictions and a common server *Comput. Oper. Res.***103** 46–63

[8] Penn M and Raviv T 2009 An algorithm for the maximum revenue jobshop problem *Eur. J. Oper. Res.***193** 437–50

[9] Defersha F M and Bayat Movahed S 2018 Linear programming assisted (not embedded) genetic algorithm for flexible jobshop scheduling with lot streaming *Comput. Ind. Eng.***117** 319–35

[10] Ramezanian R, Rahmani D and Barzinpour F 2012 An aggregate production planning model for two phase production systems: Solving with genetic algorithm and tabu search *Expert Syst. Appl.* **39** 1256–63

[11] Hillier F (Stanford U and Lieberman G (Late of S U 2001 *Introduction to Operations Reserach* (New York: Mc.Graw Hill)

[12] Benttaleb M, Hnaien F and Yalaoui F 2016 Two-machine job shop problem for makespan minimization under availability constraint *IFAC-PapersOnLine* **49** 132–7

[13] Yin Y, Wu C C, Wu W H, Hsu C J and Wu W H 2013 A branch-and-bound procedure for a single-machine earliness scheduling problem with two agents *Appl. Soft Comput. J.* **13** 1042–54

[14] Zhang P G 2003 Time series forecasting using a hybrid ARIMA and neural network model *Neurocomputing* **50** 159–75

[15] Montgomery D (Arizona S U, Jennings C (Bank of A and Kulahci M (Technical U of D 2008 *Introduction to Time Series Analysis and Forecasting* (Hoboken, NJ: John Wiley & Sons)

IOP Conf. Series: Materials Science and Engineering 1041 (2021) 012047 doi:10.1088/1757-899X/1041/1/012047

Applying system dynamic for predicting the strengths, weaknesses, opportunities, and treats of Patchouli Oil Agroindustry in West Sumatra

D Rahmayanti[1], R A Hadiguna[2], Santosa[3], N Nazir[4]

[1]Department of Industrial Engineering, Faculty of Engineering, Universitas Andalas, Padang, 25163, Indonesia
[2]Department of Industrial Engineering, Faculty of Engineering, Universitas Andalas, Padang, 25163, Indonesia
[3]Department of Agricultural Engineering, Faculty of Agricultural Technology, Universitas Andalas, Padang, 25163, Indonesia
[4]Department of Agricultural Product Technology, Faculty of Agricultural Technology, Universitas Andalas, Padang, 25163, Indonesia

Corresponding author: dina@eng.unand.ac.id

Abstract. The patchouli oil agroindustry is one industry that has the potential to be developed in West Sumatera. This study aims to analyze the future condition of the patchouli oil agroindustry in West Sumatera. As an initial step of the formulation strategy for developing patchouli oil agroindustry in West Sumatra. This study conducted a SWOT analysis using a dynamic system approach. The system dynamic is very suitable for the patchouli oil agroindustry's design development model because of the complex system. Other studies that have been conducted with SWOT analysis uses qualitative data obtained from corporate informants. In this study, SWOT data is collected based on the output derived from the dynamic system model. The system dynamics model run by using Powersim software with graphical output. Graphics can present internal and external conditions and the patchouli oil industry. Graphic trends that tend to increase will be categorized as strengths, while declining trends are classified as weaknesses. Based on the simulation results, there are six strengths, six weaknesses, two opportunities, and three threats of patchouli oil agroindustry forces in West Sumatra.

1. Introduction

Agroindustry is an industry of processing agricultural products (plants and animals), including physical processes and chemical transformation, storage, packaging, and distribution [1]. Vikash (2018) argues that agroindustry is an industry that produces products whose main ingredients come from plants and animals [2]. Agroindustry can be classified into four, which include; Agroindustry that processing of agricultural products, producing equipment and machinery, agricultural inputs like fertilizer, pesticides, herbicides, and agricultural sector services (supporting services) [3].

Patchouli oil agroindustry is one industry that has the potential to be developed in West Pasaman because it has a comparative advantage in procuring raw materials. This agroindustry can absorb a lot of labor, and its processing technology is quite simple so that it is quickly developed [4]. Raw materials are available sufficiently because the hilly geographical West Pasaman is very suitable for patchouli plants. In terms of processing, technology is relatively simple, processing traditional using drums heated with wood fuel. In terms of high employment and a transparent market, Indonesia is the world's largest patchouli exporter. Statistics show that the West Pasaman Regency has been the central patchouli oil-producing region in the West Sumatra for the past five years, followed by the Mentawai and Pasaman districts [5].

Patchouli oil is a type of essential oil with a high fixation ability, so it is widely used in making perfumes, detergents, and hair conditioners [6]. Fixation materials are y necessary for fragrance products so that fragrance lasts longer. Essential oil is a type of oil that distilled from various kinds of plants such as lemongrass, fragrant roots, cloves, cinnamon, patchouli, roses, and other types of plants that give aroma [7]. Especially for essential oils obtained from the sources, stems, and leaves of plants extracted first [8][9].

Patchouli oil is a type of export commodity with the highest percentage compared to other essential oils. Indonesia supplies around 90% of the world's needs or 1600 tons per year [10]. The United States is the largest patchouli oil importer in the world. Patchouli oil needs in the United States come from several countries, especially countries in Asia and Europe. Besides being the largest patchouli oil importer globally, the United States is also the most significant oil market for Indonesia [11].

Based on previous research, patchouli oil agroindustry in Indonesia has not developed well because it fails to meet market opportunities, such as research conducted by Winarti (2005) and Purnomo (2007). Patchouli oil market opportunities are very wide open, while the patchouli oil agroindustry has not met the market demand optimally [14]. Based on research conducted by Rahmayanti (2018) and Rahmayanti (2019), the amount of patchouli oil production in West Sumatra last year has continued to decrease [29][30]. Reducing patchouli oil production is also supported by the National Statistics Agency data 2019 [31]. For this reason, it is necessary to start increasing the amount of patchouli oil production in West Sumatra. The developed agroindustry strategy must be by the characteristics of each agroindustry. The management strategy stages are strategy formulation, strategy implementation, resource implementation, and strategy evaluation [32]. Strategy formulation: includes activities to develop the organization's vision and mission, discuss opportunities and challenges of external organizations, determine the strengths and weaknesses of internal organizations, develop long-term organizational goals, make alternative strategies for the organization, and choose the specific approach at the initial stages of structuring the plan, internal and external analysis of the company done.

SWOT analysis organizes strengths, weaknesses, opportunities, and threats into a ranked list. Strengths and Weaknesses are internal come from internal agroindustry. Things that the owner can control and can change. Opportunities (opportunities) and Threats (threats) are external things that affect business or things outside agroindustry in a broader market. The owner can take advantage of opportunities and protect from dangers, but they cannot change them.

Various studies have been done in the development of agroindustry using SWOT analysis. Gu Wan-rong (2013) conducted a SWOT analysis to obtain a strategy for developing the corn industry in China in Heilongjiang Province [22]. Evalia (2015) has reviewed the plan of developing palm sugar agroindustry in Lareh Sago Halaban District with the SWOT method [33]. In the SWOT analysis, it is using qualitative data from company informants. In contrast to research using SWOT analysis, this study conducted a SWOT analysis with a quantitative approach. SWOT is interpreted based on the dynamic system output that has been designed for the development of patchouli oil agroindustry in West Sumatra.

This study aims to analyze the conditions of patchouli oil agroindustry in West Sumatra. The analysis was carried out by determining the strengths, weaknesses, opportunities, and threats of patchouli oil agroindustry in West Sumatra. It is an initial step in developing the strategy for developing patchouli oil agroindustry in West Sumatra. The analysis will do using system dynamics. System dynamics is a method used to describe, model, and simulate a dynamic system (from time to time continually changing). In the system dynamic, it also thinks systemically. It means that it is not just one central part of solving a problem, but its effects on all related to the problem [15]. Systems dynamics have advantages compared to other methods. Its model can provide more reliable estimates than the statistical model, provides a way to understand the causes of mechanical behavior, detect early changes in industrial structure and determine the factors that predict behavior significantly and sensitive [16]. Dynamic system models allow the determination of scenarios that make sense as input for the company's.

2. Materials and methods

2.1. Materials

Agroindustry is a broad subsector that covers upstream industries from the agricultural sector to the downstream sector. The upstream industry is an industry that produces farm equipment and machinery and industrial production facilities used in the process of agricultural cultivation. In contrast, the downstream sector is an industry that processes agricultural products into raw materials or goods that are ready for consumption or is a postharvest industry and processing of farm products. The three main areas of agroindustry activity, according to Austin 1992, are:

a. Marketing, marketing related to consumer preferences, market segmentation, demand forecasting, product pricing, distribution channels, and competitiveness are addressed.
b. Procurement, procurement related to the relationship between the stages of production and processing, the method of managing the flow of raw materials from the land to the factory.
c. Process, factors related to processing and technology selection issues, plant location, inventory management, processing inputs, and operational considerations.

Many previous studies have been carried out related to the development of agroindustry. Some other research on agroindustry development can be seen in table 1.

2.2. Methods

The stages carried out in this study are:

1. Surveys, interviews related to patchouli oil production stages starting from the supply of raw materials to exporters
2. Designing a patchouli agroindustry system based on surveys and interviews conducted
3. Designing a stock-flow diagram (SFD) as one step in creating a system dynamics. SFD is designed in more detail than Causal Loop Diagrams (CLD) because it contains mathematical logic. In this study, five SFDs were intended, namely the patchouli plantation subsystem, the farmer's subsystem, the intermediary trader subsystem, the collector trader subsystem, and the exporter's subsystem.
4. The next step is running the program. The program is created using Powersim. Before the program run, they are ensuring that the application has been validated and verified. Dynamic system verification can be done with the Powersim software itself. Powersim Studio 2005 software has fulfilled the needs of automatic and real-time verification. The red question mark on the SFD identifies an error in the program. Errors can occur due to syntax or logic between incorrect variables. Validation in this study was done by checking the factors that influence agroindustry development that has been included in each Powersim output. This process is done by asking the opinions of experts. In this study, seven experts were selected, consisting of practitioners and academics.
5. Output analysis. Powersim output will be categorized into strengths, weaknesses, opportunities, and threats (SWOT). Graphic trends that tend to increase will be classed as strengths while declining trends are classified as weaknesses.

The data consists of primary and secondary data, primary data obtained through interviews and surveys with farmers, suppliers, and traders, while the primary data collected from the Central Statistics Agency.

IOP Conf. Series: Materials Science and Engineering 1041 (2021) 012047 doi:10.1088/1757-899X/1041/1/012047

Table1. Agroindustry Development Research.

Researcher	Topic	Objective	Object	Tools	Development Aspects
Suryaningrat (2016)	Procurement of raw materials in managing the agroindustry supply chain of fruit processing in Indonesia	Evaluate current conditions of supply chain management procurement systems in Indonesia Fruit processing industry	Fruit processing agroindustry	Correlation analysis	Supply
PisaniadanScrocco (2016)	New business opportunities for farmers in Peru	Comparing new business opportunities such as the cultivation of asparagus, avocados, oranges with current agriculture namely cotton	Farmers in Peru	Making scenarios and analyzing sensitivity	Supplier
	Influencing political situation Policy, stability	Determine the effect of situation instability on import base	*Import Base Industry*		Policy
Machfoedz (2015)	On the development of agro-industry in Indonesia.	Industry and agro-based industry	*Agro Base Industry*	Literature review	
Ginting (2015)	Improve agro-industry performance through empowerment entrepreneurs and resources	Design a conceptual model to improve agro-industry competitiveness in Indonesia	Agro-industry sector	Literature review	Marketing, product innovation Technical
Gu Wan-rong, et al (2013)	Developing a strategy for corn agroindustry in China	Stimulate the development of the corn industry in Heilongjiang Province, increase the capacity of wheat production, as well as economic growth and increase the income of local farmers.	Wheat Agroindustry	SWOT analysis	
Pongchanun (2017)	Paper industrydevelopment strategy	Determining strategies to deal with global climate change and environmental challenges taking place in the paper processing industry, much traditional pulp and paper industries are required to create new business opportunities.	The pulp and paper industry	*Dynamic capabilities approach*	Organization Management
Teresa, et al (2017)	Improve food security governance in Mexico	Identify, describe, and analyze elements of food security governance in Mexico.	Food industry	Literature review	Supply, Technical
Augustine (2010)	Challenging for the corn seed industry in South Africa	Increased corn seed production	Corn seed industry	Literature review	Supply
Ferreira, et al (2016)	It is increasing the competitiveness of citrus agribusiness	Integrated production planning to increase competitiveness	Citrus agribusiness	System dynamic	Technical
Supriatna, et al (2004)	Development of clove leaf oil agroindustry	Analyze the initial planning system of an essential oil refining agroindustry from clove leaves in the North Sulawesi region, and analyze its feasibility from the technical aspects, management, and aspects financially	Clove leaf oil agroindustry	Plant layout design and econometric methods	Technical, management, financial
Indrawanto and Mauludi (2004)	Patchouli Industry Development Strategy Indonesia	Determine the development strategy for the Indonesian patchouli oil industry so that Indonesia can more enjoy the added value of the patchouli oil industry as a significant producer	Patchouli oil industry	Financial analysis	Financial

IOP Conf. Series: Materials Science and Engineering 1041 (2021) 012047 doi:10.1088/1757-899X/1041/1/012047

3. Results and discussion

The system will divide into five subsystems based on actors involved in the patchouli oil supply. The subsystems are patchouli plantation subsystems, patchouli farmers subsystems, intermediary trader subsystems, collector trader subsystems, and exporters. The addition of the patchouli planting area depends on the demand for patchouli in West Pasaman. Unfortunately, there are no specific data about the need for patchouli in West Pasaman. In this research, patchouli in West Pasaman is predicted based on patchouli oil production in the last few years, coupled with a commitment to increase patchouli oil production from the government in a certain percentage. The addition of new planting areas will undoubtedly impact the decline in agricultural land and plantations in West Pasaman. Patchouli plants that are newly planted will increase the size of patchouli plants that have not been produced even though the percentage does not reach one hundred percent. There are several percent of patchouli affected by pests. Increased pest infestation also results in a decrease in patchouli planting areas. Patchouli plants that produce after a few years will grow the patchouli field that does not provide, so rejuvenation needs to be done. In addition to the area of the plant that does not produce, rejuvenation is also influenced by pest attacks.

Furthermore, patchouli plants that have undergone rejuvenation will increase patchouli planting areas that have not been produced. And so on so that it forms a cycle. Patchouli planting area will increase the amount of harvest in the form of wet patchouli. However, the amount of patchouli harvest is also influenced by price. Farmers tend to harvest patchouli plants when the purchase price of patchouli oil is high at the farm level. If prices are low, farmers are more likely to leave them alone.

The amount of wet patchouli harvest, acquisition of patchouli, weather, patchouli seed types, and patchouli distillation equipment influence Patchouli oil production. Patchouli oil can be directly sold by farmers or stored as inventory. This condition usually occurs because of market prices. Farmers tend to hold patchouli oil for a while and will sell it if the purchase price of patchouli at the farm level is high. It increased acquisition and increase in patchouli oil inventory. Will result in an increase in the maximum amount of supply from farmers. The maximum supply amount and the amount of patchouli oil demand influence the amount of patchouli oil delivery. Increased demand will cause the number of shipments to increase and decrease in inventory. Besides the price, the increase in demand for patchouli oil also affects the wet patchouli harvest.

Slightly different from patchouli farmers, temporary traders in patchouli oil agro-industry are variable and tend to change. Most temporary traders only make this activity a side business. If the purchase price of patchouli oil rises, the number of intermediary traders will also increase. Increasing the purchase of patchouli oil to farmers will increase the stock and patchouli oil shipments. The longer the storage time for patchouli oil, the amount of patchouli oil stocks in intermediary traders will also increase.

Collecting traders will get patchouli oil supplies from farmers and intermediary traders. Collecting traders have the maximum quantity to send to exporters because they have stock. The amount of stock and the number of demand influence the amount of patchouli oil shipment to exporters.

In this study, foreign demand assumes the amount of patchouli oil exports. Increasing it will cause an increase in the amount of patchouli oil shipments. Increasing the amount of patchouli oil shipments has a positive relationship with the amount of patchouli oil order to the collector level. Increased shipping will also reduce the amount of storage and storage time. However, an increase in patchouli oil orders to intermediary traders over time causes an increase in inventory. This condition happens if the purchase price of patchouli oil in the world has decreased. Figure 1 shows the SFD of Patchouli Oil Agroindustry in West Sumatera.

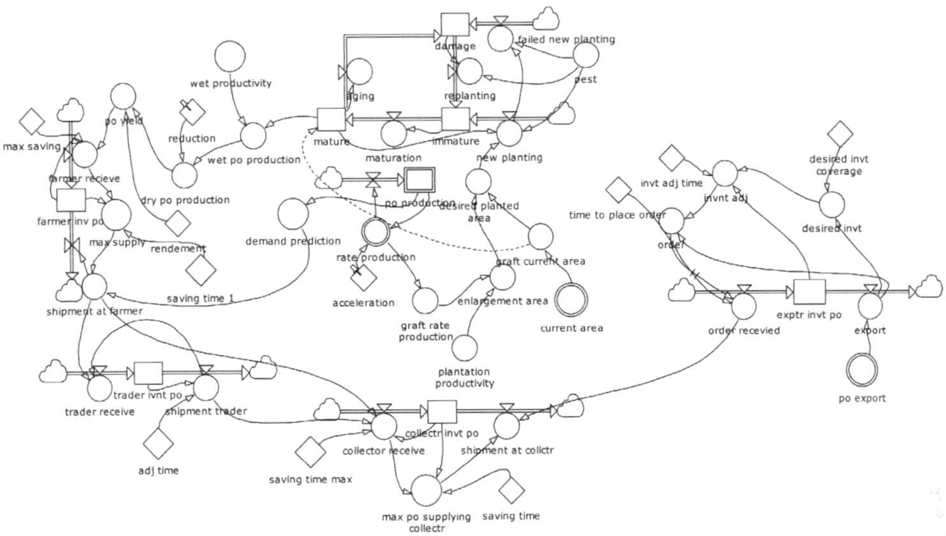

Figure 1. SFD of Patchouli Oil Agroindustry.

The next stage is to carry out the validation and verification process. In this study, the SFD was free from mistakes because no question marks were found; therefore, it concluded that the program had verified. Validation does through expertise. They will check whether all factors affect patchouli oil agroindustry in the design of the system dynamics model. From the results of testing seven experts, the model is valid because it contains all the factors that influence patchouli oil agroindustry development in West Sumatra. The program is running for the next six years, from 2020 to 2026. The results of the dynamic system output can be seen in figure 2.

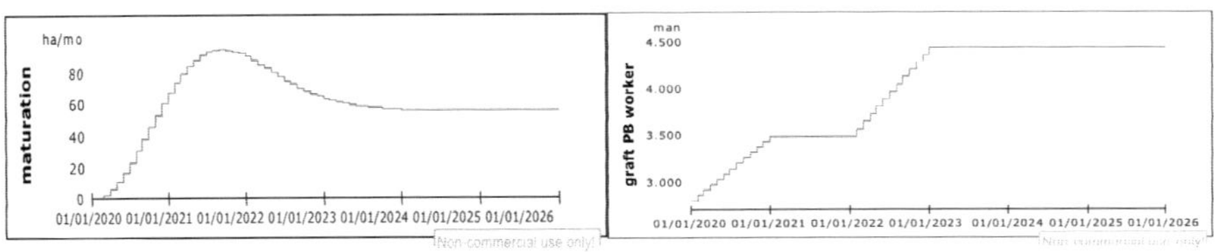

(a) Simulationo of Patchouli Productive Area Number **(b)** Simulation of Worker Number

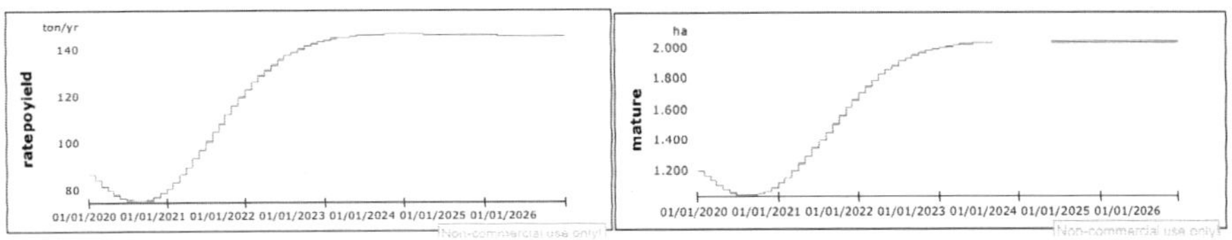

(c) Simulation of Patchouli Oil ProductionNumber **(d)** Simulation of Planting Area Number

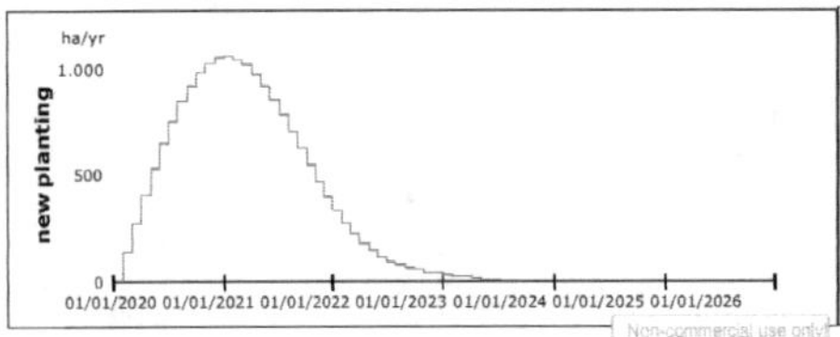

(e) Simulasi of New Planting Area

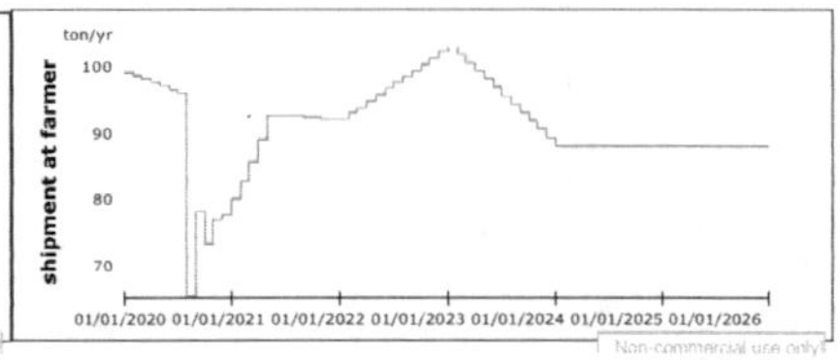

(f) Simulation of Selling Product at Farmer Level

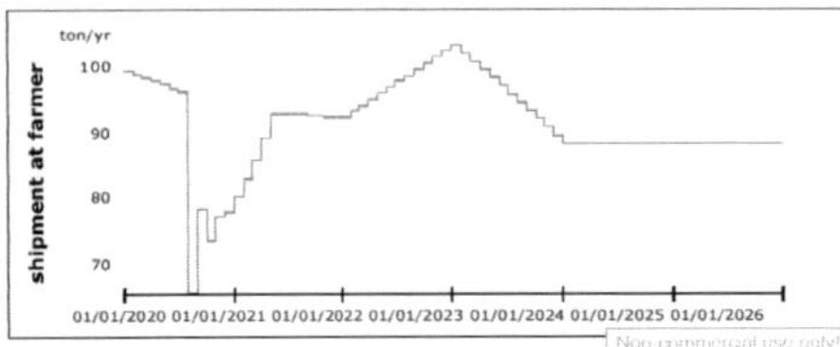

(g) Simulation of Selling Product at Intermediate Level

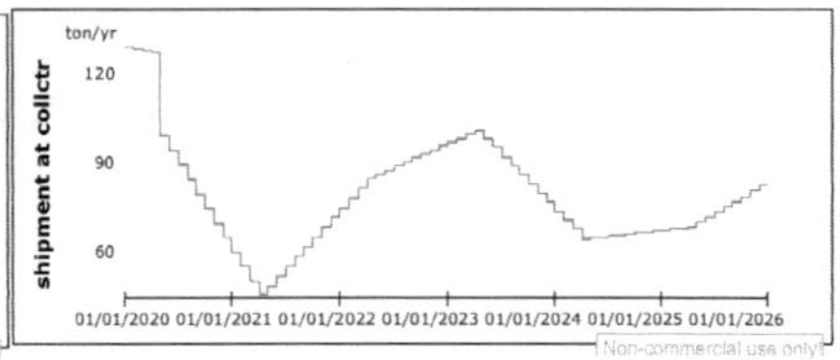

(h) Simulation of Selling Product at Collector Level

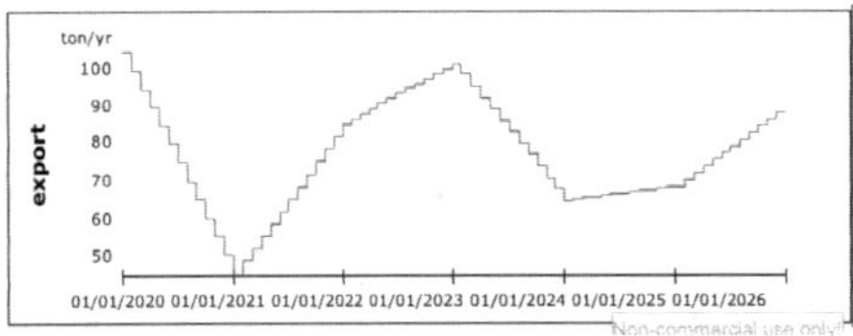

(i) Simulation of Ptachouli Oil Export Number

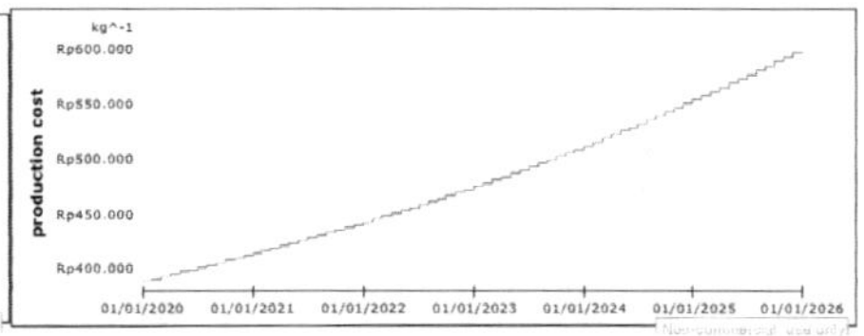

(j) Simulation of Patchouli Production Cost at Farmer Level

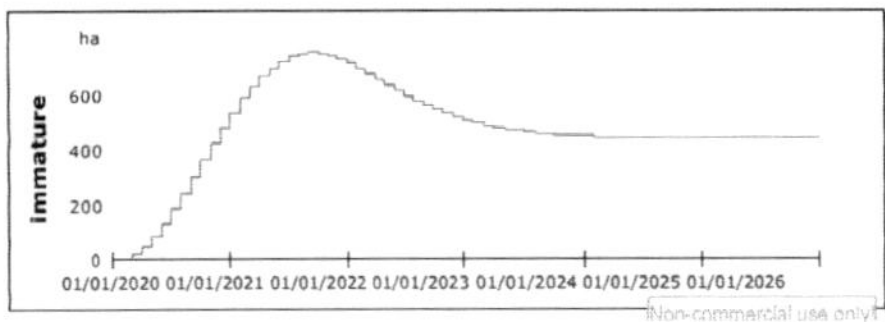

(k) Simulation of Non Productive Area

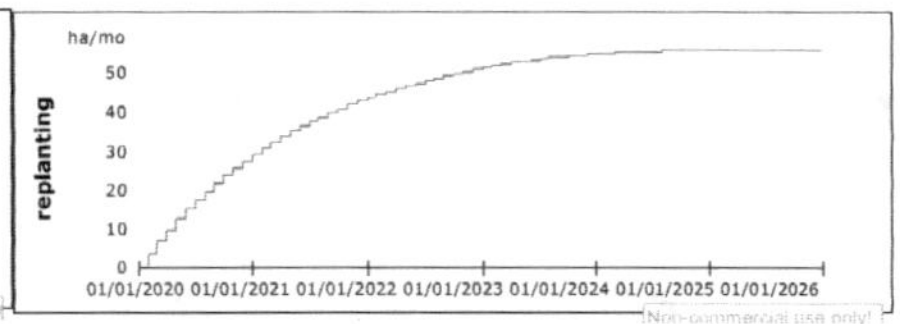

(l) Simulation of Replanting Area

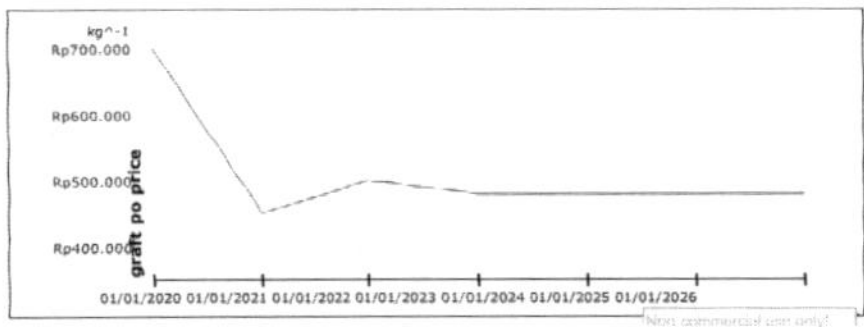

(m) Simulation of Selling Price at Farmer Level

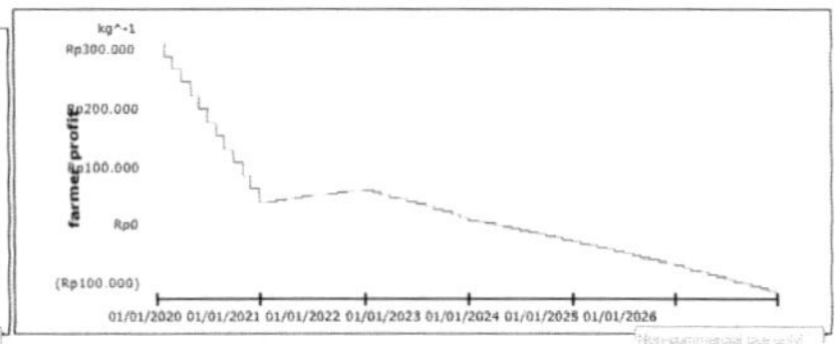

(n) Simulation of Profit at Farmer Level

Figure 2. Output System Dinamics from Powersim Software.

Based on Figure 3(a), the number of patchouli plants ready to be harvested has increased from 2020 to 2022, for the following year, the number of patchouli productive areas has decreased slightly. In contrast, for the next few years, it tends to be stable. Patchouli's prolific area trend for the next few years is relatively stable. The number of patchouli plants ready to be harvested is one of the strengths for agroindustry development. The guaranteed supply of raw materials in wet patchouli is available in increasing quantities and is relatively stable.

Figure 3(b) shows that the number of workers continues to increase from 2020 to 2023, while the following years are relatively stable. The existence of adequate human resources is one of the factors driving the development of patchouli oil agroindustry. In terms of education level, based on data from the Central Statistics Agency 2018, the number of residents of West Pasaman who have taken tertiary education has increased every year. The increasing number of college graduates can be an indicator of the workforce's ability in West Pasaman to increase. The existence of resources is one factor in the industry [34][35].

The amount of patchouli productive area is directly proportional to the amount of patchouli oil produced. Thus, the amount of patchouli oil production in the West Pasaman region for the next few years tends to increase and is relatively stable. Increasing of patchouli production means that this is one factor that is a force for the development of patchouli oil agroindustry. Raw materials are the most critical factor for the sustainability of industry [36][37].

The total patchouli planting area is the total number of patchouli planting areas currently available in West Pasaman, both productive and non-productive. Figure 3(d) presents the total patchouli planting area for the next few years. The total patchouli planting area has increased in the first three years, while the following years are relatively stable from the simulation results. Increasing the number of patchouli planting areas is one of the strengths of patchouli oil agroindustry in West Pasaman. Patchouli's total planting area is one indicator that can guarantee the availability of patchouli oil raw material stock in West Pasaman.

Figure 3(e) shows that the patchouli planting area at the beginning of the first year has increased. Furthermore, it has decreased. In 2023, the number of new planting areas will be zero. This condition means that there is no opening of a new planting area for patchouli. Patchouli production for 2023 and the following year only relies on the rejuvenation of existing land. The patchouli planting area is one of the weaknesses for the patchouli oil agroindustry in West Pasaman.

From the simulation results seen between 2020 and 2021, the number of shipments has decreased steeply. This condition is possible because the amount of patchouli inventory at the collector level is still sufficient to meet exporters' demand. At the same time, exporters also have supplies to reach the foreign market. Another factor that triggered the absence of buying and selling transactions was the amount of patchouli oil exports, which declined because Indonesia's position, especially West Pasaman, as the world's largest patchouli oil exporter, has been replaced. Patchouli oil sales patterns at the farm level tend to be unstable and decline. The decline can be a weakness and inhibit the development of the patchouli oil agroindustry in West Pasaman.

The amount of patchouli oil sales at the collectors level tends to fluctuate. Patchouli oil sales decreased from 2020 to 2021. Then it increased the next two years and so on to form an up and downtrend. The fluctuation of patchouli oil sales at the level of intermediary traders was affected by the demand from exporters. This indefinite amount of order has become a weakness in the development of the patchouli oil agroindustry. Patchouli oil exports tend to fluctuate in the next few periods. The same is the case with the number of exports a few years before. Unstable and fluctuating export conditions are one of the threats to the development of patchouli oil agroindustry.

Production costs consist of fixed and variable costs. Based on historical data from various components of production costs, Production costs histories an increase every time. Figure 39(j) presents an increase in production price, which is linear over time. Based on the calculation of the current production cost, the cost of production is still under IDR 400,000/kg. This condition means that farmers can still benefit from seeing the current patchouli selling price. But if the situation of rising costs always occurs while the purchase price of patchouli at the farm level does not experience an increase, then in the future, it is certain that farmers will not benefit. The rise in production prices is undoubtedly a weakness as well as a barrier to patchouli oil agroindustry.

The purchase price at the farm level is predicted based on patchouli oil costs in recent years. As shown in Figure 3(m), patchouli oil prices only experienced an increase in the early years. In the following years, it tends to decrease and be stable. From interviews with farmers, it was found that patchouli oil prices often ranged between IDR 400.000 to IDR. 500,000, only in specific periods of a sudden increase and then decrease again.

There is no domestic oil demand in West Pasaman because collectors only sell patchouli oil to exporters located in Padang or Medan. In this study, the number of domestic patchouli oil demand for the next few years is assumed to be non-existent because there is no patchouli oil derivative industry in West Sumatra. The absence of a domestic market for patchouli in West Pasaman could be one of the opportunities to establish the patchouli oil derivative industry.

The profit of patchouli oil sales depends on the purchase price of patchouli oil and patchouli oil production costs. The rising production costs and the relatively low and stable patchouli oil rate have resulted in a downward trend in profits. The simulation results in Figure 3(n) predict that in 2023 farmers will not benefit, and even the following year, farmers will suffer losses. The decline in farmers' profits is a weakness in the development of patchouli oil agroindustry in the future. For this reason, the government must ensure participation in controlling patchouli purchase prices at the farm level. If the increase in the number of goods for production is assumed to follow historical price increases and the selling price of patchouli oil also follows historical data, from the simulation results, it appears that farmers will not benefit for the next few years. This simulation results can be a picture for the government to determine the strategy that should be done in the matter of the benefits of patchouli farmers.

4. Conclusion

This study conducted a SWOT analysis using a dynamic system approach. The system dynamic is very suitable for the patchouli oil agroindustry's design development model because of the complex system. Other studies that have been conducted with SWOT analysis uses qualitative data obtained from corporate informants. In this study, SWOT data is collected based on the output derived from the dynamic system model. The system dynamics model run by using Powersim software with graphical output. Graphics can present internal and external conditions and the patchouli oil industry. Graphic trends that tend to increase will be categorized as strengths, while declining trends are classified as weaknesses.

This research can analyze the internal and external conditions of the patchouli oil agroindustry in West Sumatra. This research has investigated the strengths, weaknesses, and threats of patchouli oil agroindustry development in West Sumatra. Based on the simulation results, there are six strengths, six weaknesses, two opportunities, and three threats of patchouli oil agroindustry forces in West Sumatra. These strengths are the increasing patchouli productive area, increasing patchouli oil production, the number of labor has tended to grow, increasing the total planting area, increasing the unproductive planting, and increasing the patchouli replanting area. The weakness of patchouli oil agroindustry in West Sumatra is the decreasing patchouli planting area and decreasing patchouli oil sales at the farm level. Other factors for deficiency are reducing the number of patchouli oil shipments at the level of intermediary traders, reduce the amount of patchouli oil shipments at the level of collector traders, production costs tend to fall increase, farmers' profit tends to decrease. Opportunities for developing patchouli oil agroindustry in West Sumatra are domestic sales none and he market available well. The threat of patchouli oil agroindustry development in West Sumatra is the world's unstable patchouli oil price, the number of patchouli exports tends to fluctuate, and patchouli purchase price oil in farmers does not increase.

Reference

[1] Austin, J, E. 1981. "Agroindustrial Project Analysis. The economic development institute of the world bank," *The Johns Hopkins University Press.*

[2] Vikash, P., et al. 2018. Chapter 3: Implication of nanoscience in the food processing and agricultural industries impact of nanoscience in the food industry, pp: 57-85.

[3] Udayana, G.B. 2011. Peran agroindustri dalam pembangunan pertanian. Singhadwala, Edisi 44. [In Indonesian]

[4] Junaedi, A., dan Hidayat, A. 2010. Uji asal sumber bibit nilam (*pogostemon cablin benth*) di Pasaman Barat Sumatera Barat. *Peneliti di Balai Penelitian Hutan Penghasil Serat, Kuok.* [In Indonesian]

[5] National Statistics Agency. 2018.

[6] Swamy and Sinniah. 2016. "Patchouli (pogostemon cablin benth.): botany, agro technology and biotechnological aspects", *Industrial Cropsa and Products*, vol. 87, pp. 161–176.

[7] Rahmayanti, D. et al. 2018. "Determining the profit margin of "patchouli oil" supply chain: case study in Indonesia," *International Journal on Advanced Science Engineering Information and Technology*, Vol. 8, no. 2, pp. 483-488.

[8] Kusuma, H.S. and Mahfud, M. 2016. "Microwave hydro distillation for extraction of essential oil from pogostemon cablin benth: analysis and modelling of extraction kinetics," *Journal of Applied Research on Medicinal and Aromatic Plants,* vol. 4, pp. 46–54.

[9] Bey,Z. H. et.al. 2016. "Essential oils composition, antibacterial and antioxidant activities of hydro distillated extract of eucalyptus globulus fruits," *Industrial Crops and Products*,

vol 89, pp.167–175.
[10] http://www.kemenperin.go.id/artikel/1921/pemasok-90-bahan-baku-dunia,-Tapi-RI.
[11] Sari, P, N dan Hartono, S. 2010. "Analisis eksport minyak nilam Indonesia ke Amerika Serikat," *Agro Ekonomi*, vol 17, No 1, Juni 2010, hal 19-28. [In Indonesian]
[12] Winarti, C., Laksmanahardja, M.P. dan Sumangat, D. 2005. "Kajian status pengembangan agroindustri minyak nilam terhadap tingkat kepuasan petani di Majalengka," *Jurnal Pascapanen*, vol 2. No 2, pp: 36-44. [In Indonesian]
[13] Purnomo, D. 2005. Analisis kebijakan pengembangan agroindustri minyak atsiri di Jawa Barat (Studi Kasus Komoditas Minyak Nilam). Master Theses Industrial Engineering and Management ITB. [In Indonesian]
[14] Hendrastuti., Eryatno, Rusli, M.S., dan Soedarsono, J.W. 2012. "Optimisasi Penentuan Kesepakatan Harga Nilam pada Rantai Pasok Minyak Nilam di Kabupaten Kuningan," *Jurnal Agrotek*, vol 6, no 1. [In Indonesian]
[15] D. Wijaya. Perencanaan produksi menggunakan teknik simulasi dinamis (Studi Kasus PT.Agro Palindo Sakti Sumatera Selatan). TugasAkhir. Fakultas Teknik Program Studi Teknik Industri. Universitas Bina Darma Palembang. [In Indonesian]
[16] Lyneis, James M. 2000. System dynamics for market forecasting and structural analysis.
[17] Oxa Axella dan Erma Suryani. 2012. "Aplikasi model sistem dinamik untuk menganalisis permintaandan ketersediaan listrik sektor industri (Studi Kasus :JawaTimur)," *Jurnal Teknik ITS*, vol. 1, (Sept, 2012) ISSN: 2301-9271. [In Indonesian]
[18] Suryaningrat, I. B. 2016. "Raw material procurement on agroindustrial supply chain managementA case survey of fruit processing industries in Indonesi," *Original Research Article. Agriculture and Agricultural Science Procedia*, vol. 9, pp. 253-257.
[19] Pisania, E, dan Scrocco, S. 2016. "Building new income opportunities for small-farmers in Peru: the case of native and naturally colored cotton," *Agriculture and Agricultural Science Procedia*, vol. 8, 2016, pp. 426-432.
[20] Machfoedz, M. M. 2015. "Stabilizing and decentralizing the growth through agro-industrial development," *Agriculture and Agricultural Science Procedia*, vol. 3, 2015, pp. 20-25.
[21] Ginting, G. 2015. Open innovation model: empowering entrepreneurial orientation and utilizing network resources as determinant for internationalization performance of small medium agroindustry," *Agriculture and Agricultural Science Procedia*, vol. 3, 2015, pp 56-61.
[22] Gu Wan-rong, Yi, J., Yao, M., Jian-guo, W., Xian-long, Z., Jin, L and Shi, W. 2013. "SWOT analysis and development strategies of maize industry in Heilongjiang Province," *Journal of Northeast Agricultural University*, vol. 20, no. 1, March 2013, pp.76-84.
[23] Pongchanun, L. 2017. Opening up new strategic options in the pulp and paper industry: Case biorefineries. *Computers and Electronics in Agriculture*, vol.135, pp.11–22.
[24] Teresa Shamah-Levya, Verónica Mundo-Rosas, María Margarita Flores-De la Vega, and CassioLuiselli-Fernández. 2017. "Food security governance in Mexico: How can it be improved?," *Global Food Security*,vol. 14, September 2017, pp 73-78.
[25] Augustine, S, L. 2010. "Challenges of the maize seed industry in eastern and southern Africa: A compelling case for private–public intervention to promote growth," *Food Policy*, vol. 35, pp. 323–331.
[26] Ferreira, J. O., dan Batalha, M. O., D, J. C. 2016. "Integrated planning model for citrus agribusiness system using systems," *Computers and Electronics in Agriculture*, vol. 126, pp. 1–11.
[27] Supriatna, A.S., Rambitan, U.K., Sumangat, D., Nurdjannah, N. 2004. "Analisis sistem perencanaan model pengembangan agroindustri minyak daun cengkeh: Studi Kasus Di Sulawesi Utara," *Buletin Balai Besar Penelitian dan Pengembangan Pascapanen Pertanian*, vol. 15, no. 1.
[28] Indrawanto, C., dan Mauludi, L. 2004. "Strategi pengembangan industri nilam indonesia," *Balai Penelitian Tanaman Rempah dan Obat. Perkembangan Teknologi*, vol.16, no. 2. [In Indonesian]
[29] Rahmayanti, D. et. al. 2017. "Model Konseptual Pengembangan Agroindustri Minyak Nilamdi Pasaman Barat Menggunakan Sistem Dinamik," *Jurnal Teknologi dan Manajemen Agroindustri*, vol. 6, no. 3, pp. 126-132. [In Indonesian]
[30] Rahmayanti, D. et. al. 2019. "Model kebijakan produksi perkebunan nilam dengan menggunakan sistem dinamik," *Jurnal Optimasi Sistem Industri*, Vol. 18, no. 1, pp. 65-74. [In Indonesian]
[31] National Statistics Agenc. 2019.
[32] David, F. R. 2009. Manajemen Strategis Konsep. (Ed.12). Salemba Empat, Jakarta. [In

Indonesian]

[33] Evalia, A, N. 2015. "Strategi pengembangan agroindustri gula semut aren,"*Jurnal Manajemen dan Agribisnis*, vol. 12, no 1. [In Indonesian]

[34] Tariq, M., and Johannes, E. 2019. "Augmented reality in support of industry 4.0-implementation challenges and success factors," *Robotics and Computer-Integrated Manufacturing*, vol 58, August 2019, pp 181-195.

[35] Luis Maurício Martins de Resende., et al. 2018. "Critical success factors in coopetition: Evidence on a business network," *Industrial Marketing Management,* vol. 68, January 2018, pp 177-187.

[36] Jiali Song, et al. 2019. "Material flow analysis on critical raw materials of lithium-ion batteries in China," *Journal of Cleaner Production*, vol. 2151 April 2019, pp. 570-581.

[37] Ferro, P. 2019. "Materials selection in a critical raw materials perspective Materials & Design", *International Journal of Physics Research and Applications*, vol 3, no.1, pp 017-019.

The role of surfactant to enhance photocatalyst performance on phenol degradation in TiO_2-CNT composite-modified CNT (Cetyltrimethylammonium bromide)

D Heltina[1], U Avisa[1] and M I Fermi[1]

[1]Riau University, Department of Chemical Engineering, Pekanbaru, 28293, Riau, Indonesia

Corresponding author:desi.heltina@lecturer.unri.ac.id

Abstract. The TiO_2-CNT composite – modified CNT (Cetyltrimethylammonium bromide) photocatalyst material was successfully prepared. The composites material was characterized by Scanning Electron Miscroscopy (SEM), X-ray Diffraction (XRD), Fourier Transform Infrared absorption spectroscopy (FTIR), and Breneur Emmet Teller (BET). The TiO_2-CNT composite- modification CNT (Cetyltrimethylammonium bromide) formation obtained are discussed in detail. The highest performance effectiveness of photocatalyst material composites for degrading phenol was obtained on phenol concentration 10 mg/L and degradation temperature at 50°C was 98,74 %

1. Introduction

Phenol is a dangerous type of industrial waste. Phenol can be found in wastes from various industries such as polymer, plastics, pulp and paper, and textile industries [1,2]. The past few years photocatalysis process has become the choice in degrading waste compounds such as phenol because it is a process that can degrade phenol to CO_2 and H_2O [3,4].

The photocatalyst material which is widely used for the photocatalysis process is titanium dioxide (TiO_2). Titanium dioxide (TiO_2) has a band gap energy greater than 3 eV (anatase = 3.2 eV is equivalent to 380 nm and rutile = 3.1 eV is equivalent to 410 nm) so it makes TiO_2 is active when exposed to UV light [5]. In the photocatalyst material oxidation and reduction processes occur which results in the hole reacting with H_2O contained in solution to form hydroxyl radicals (•OH). This hydroxyl radical is a strong oxidizer that can degrade organic compounds to CO_2, H_2O and halide ions. Electrons on the surface of TiO_2 can react to form superoxide radicals (•O_2-) which can also degrade organic compounds in solution [6]. But the effectiveness of the performance of TiO_2 is hampered by the rate recombination between electron pairs and holes [7].

Photocatalytic activity is increased when the use of TiO_2-CNT composites in the process of photocatalysis [8]. When TiO_2 binds CNT to form a composite, CNT encourages the transfer of electrons from the TiO_2 conduction band to the surface of the CNT. So the CNT receives and stores photogeneration electrons and inhibits the recombination of electron and holes. These electrons can be transferred to other electron acceptors such as oxygen molecules to form reactive oxygen species (O_2-,H_2O_2, and •OH) that can degrade organic pollutants [9].

Studies show that there are weaknesses in TiO_2-CNT composites such as low CNT solubility levels. Cetyltrimethylammonium bromide (CTAB) is a type of cationic surfactant that is widely used to increase the stability of CNT in solution [10]. The surface of the CNT is hydrophobic. Addition of surfactant is a way to increase stable dispersion between CNT and TiO_2. Surfactants have a molecular structure consisting of hydrophilic and hydrophobic groups. Hydrophilic groups can be well dispersed in water and organic solvents, while hydrophobic groups will form bonds with CNT [11].

The purposes of this research are to synthesize and obtain the characterization of TiO_2-CNT composites – modified CNT(Cetyltrimethylammonium bromide) and obtain the performance effectiveness in degrading phenol.

2. Material and methods

The synthesized of TiO_2-CNT composite-modified CNT (CTAB) were carried out with two main processes, which were surface modification of CNT in the CTAB surfactant and then compiled it with TiO_2. The performance of TiO_2-CNT composite-modified CNT (CTAB) was tested for phenol degradation.

2.1. Materials

TiO_2 (P25), CNT for the research was multi walled carbon nanotube (MWCNT) type purchased from Dong Yang (HK) Int'l Group Limited of diameter 10-30 nm, length of 5-15 μm, and purity of 99%, Cetyltrimethylammonium bromide (CTAB), 0.1 M HNO_3 and aquades.

2.2. CNT Modification with CTAB (Cetyltrimethylammonium bromide) Surfactant

1 gram of carbon nanotube (CNT) was added with CTAB surfactant of 0.5 gram in 100 mL aquades. Solution was then sonicated for 1 hour. Then solution was dried.

2.3. Synthesis of TiO_2-CNT Composite -Modified CNT (Cetyltrimethylammonium bromide)

0.015 gram of CNT(CTAB) was dissolved in 100 ml aquades. The mixture was sonicated for 30 minutes. Then re-sonication for 30 minutes with the addition of 1 gram of TiO_2. The pH of mixture was adjusted until pH is 3 by added 0.1 M HNO_3 and after that the mixture stirred by using magnetic stirrer 300 rpm for 3 hours. The mixture then dried in temperature of 100-110°C for 2 hours. Sample calcined at a high temperature 400°C for 2 hours in a nitrogen atmosphere.

2.4. Performance test of TiO_2-CNT composite-modified CNT (CTAB) for phenol degradation.

TiO_2-CNT composite-modified CNT (CTAB) each of 0.3 gram was put into 300 mL of phenol liquid of 10 ppm, 20 ppm and 30 ppm. The experiment was conducted for 270 minutes (4,5 hour), in which first 30 minutes was conducted without light (light off) and next 240 minutes the photodegradation using a 60 watt UV lamp as photon source (light on) was conducted and degradation temperatures of 30°C, 40°C, and 50°C. Photocatalyst used for the process was 0.3 gram powder and continuously stirred during experiment in order to make photocatalyst particles and phenol distributed evenly in sample solution. Quantitative result of degrade phenol was acquired by using UV-Vis Spectrophotometer in accordance with procedure on SNI 06-6989.21-2004.

2.5. Charateristic

Sample was characterized by using SEM, BET, FT-IR , XRD, and UV-vis spectroscopy.

3. Results and discussion

Figure 1 shows SEM images of the TiO_2-CNT composite- modified CNT (CTAB). These look TiO_2 particles are well distribute among the CNT and there are some pasticles of TiO_2 gathered together to form small groups. The spread of CNT is not clearly visible because of the small amount of CNT than TiO_2. Even distribution of particles showed that the synthesis of TiO_2-CNT composite- modified CNT (CTAB) was successful.

Figure 1. SEM images of TiO_2-CNT composite–modified CNT (CTAB).

The BET surface area of the TiO_2, CNT, TiO_2-CNT, TiO_2-CNT composite –modified CNT (Cetyltrimethylammonium bromide) composite is shown in table 1. Based on table 1 it can be seen that an increase in the surface area of TiO_2 photocatalysts after the addition of CNT. This can be seen from research by Miandoab and Fatemi [12] where an increase in the surface area of TiO_2-CNT is 62.3 m^2/g. Increased surface area shows that CNT can affect the surface area of photocatalysts. This statement is in accordance with the results of research Huang et al.[8] that CNT can provide greater specific surface area in photocatalysts. In this research, the surface area of photocatalyst is 84.75 m^2/g which is bigger than the research conducted by Miandoab and Fatemi [12]. Surface modification of CNT with CTAB surfactant resulted in an increase in surface area on TiO_2-CNT composites. According to Manilo et al. [10] CTAB surfactant is one type of cationic surfactant that is widely used to improve the stability of CNT in solution, so that it can help CNT dispersion in solution. so that this can be proven by increasing the surface area of the TiO_2-CNT composite (Cetyltrimethylammonium bromide) obtained.

Table 1. The surface area for photocatalyst materials.

Sample	Surface Area (m^2/g)
[a]TiO_2 P25	53.6
[b]CNT	200
[c]TiO_2-CNT	62.3
TiO_2-CNT–modified CNT (CTAB)	84.75

[a]Slamet et al. [13]
[b]Heltina et al. [14]
[c]Miandoab and Fatemi [12]

The FTIR spectra of TiO_2-CNT composite-modified CNT (CTAB) is given in figure 2. Peaks at 3188.47 cm^{-1} and 3853.94 cm^{-1} indicates the presence of the –OH groups. The -OH group indicates that water molecules are adsorbed on the surface of the photocatalyst [15]. According to Linsebigler et al.[16] -OH groups in water molecules can act as electron donors from holes formed by the induction of light and will produce hydroxyl radicals (•OH) which are very helpful in increasing photocatalytic activity. At wavelengths of 2953.14 cm^{-1} it is a C-H bond from the surface of the CNT [17]. At a wavelength of 2361.94 cm^{-1} it is a C-C bond [18]. The bonds C = O and C = C are present at wavelengths of 1549.87 cm^{-1} and 1647.28 cm^{-1}[19]. Ti-O-Ti and Ti-O-C bonds are found at wavelengths below 798 cm^{-1} [20]. According to Lee et al.[21] the formation of Ti-O-C bonds between

TiO_2 and CNT causes close contact between TiO_2 and CNT thus offering an efficient route for electron transfer from the TiO_2 conduction band to the surface of the CNT.

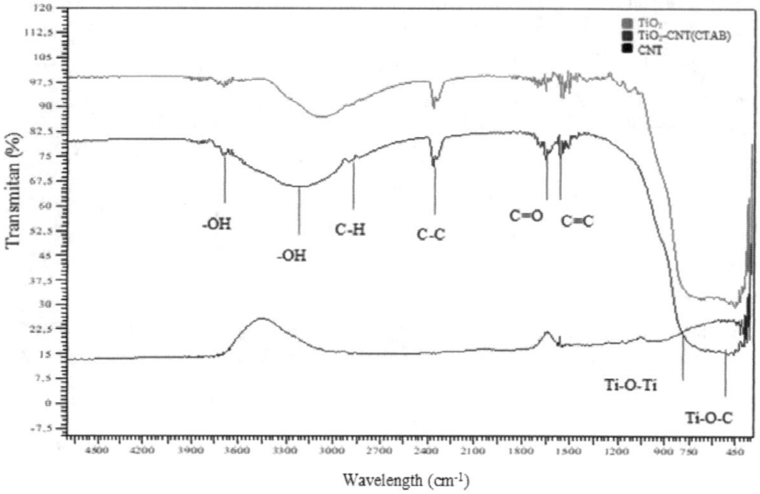

Figure 2. FTIR images of TiO_2-CNT composite–modified CNT (CTAB).

The crystal structure of the TiO_2-CNT composite-modified CNT (CTAB) can be investigated using the XRD pattern as shown in figure 3. The diffraction peaks at 25.29°, 48.03°, 37.83°, 53.94°, 55.05°, 62.71°, 68.88°, 70.22°, and 75.08° correspond to the (101), (200), (004), (105), (211), (204), (116), (220) and (215) reflections of anatase, respectively. The peaks at 2θ values of 27.42° is related to rutile (110). But besides these peaks there are also peaks centered on 2θ; 26° which indicates the presence of carbon structures (002) that overlaps with the anatase (101) crystal peaks [8].

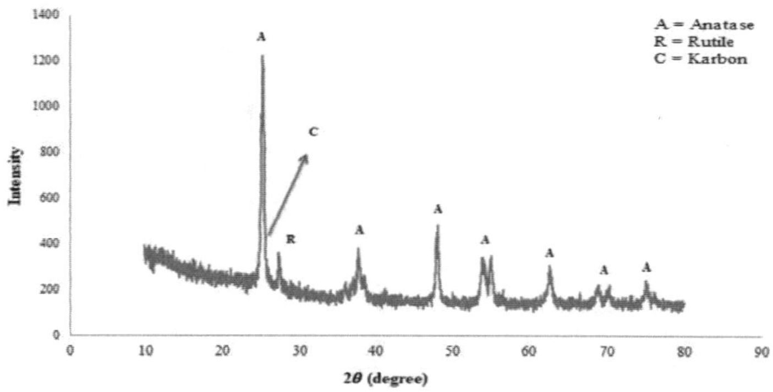

Figure 3. XRD patterns of TiO_2-CNT composite-modified CNT (CTAB).

From the XRD analysis the crystal size of the TiO_2-CNT composite-modified CNT (CTAB) was obtained using the Scherer Equation as inequation (1).

$$D = \frac{0.9\lambda}{\beta \cos\theta} \quad (1)$$

In order to obtain the crystal size and crystal fraction of the TiO_2-carbon nanotube (Cetyltrimethylammonium bromide) composites shown in table2. And it is known that the mixture of anatase and rutile crystals has a synergistic effect and increased photocatalytic activity compared to pure phase [22].

Table 2. Size and crystalline fraction of TiO_2-CNT composite–modified CNT (CTAB).

	Anatase	Rutile
Crystal size (nm)	17.73	23.72
Crystal fraction (%)	14.22	85.78

The effect of phenol concentration on the performance of photocatalyst material to degrade phenol was studied. The research variables were carried out with different initial phenol concentrations of 10, 20 and 30 mg/L, respectively.

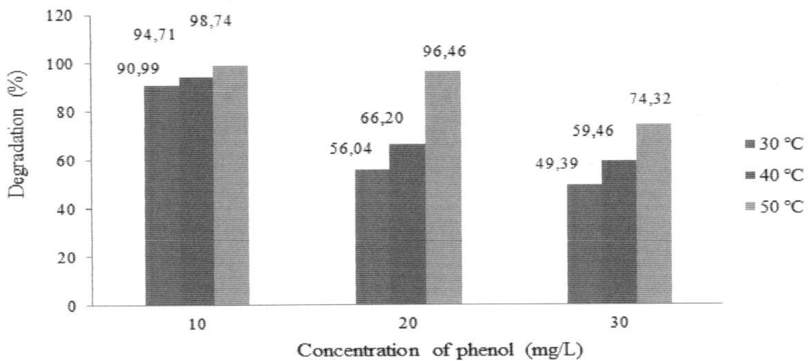

Figure 4. Effectof phenol concentration on the performance of composites to degrade phenols.

Figure 4 shows that the percentage degradation of phenol has decreased with increase in initial concentration of phenol. It was found that 98.74% degradation of phenol occurred within 240 minutes with degradation temperature 30°C. At lower concentrations it can be observed that high rates of degradation. the use of a fixed amount of catalyst with increasing phenol concentration makes the hydroxyl radical (•OH) produced on the surface of the catalyst also constant while the amount of phenol that needs to be degraded increases so that the amount of hydroxyl radical (•OH) that is available is unable to completely degrade phenol. In addition, this opinion is also supported by the results of research Chong et al.[23] that high phenol concentrations can reduce the absorption of photons by photocatalyst particles, because increasing phenol molecules cover the surface of photocatalysts thereby reducing the activity of photocatalysts which causes a decrease in the performance of photocatalysts.

The effect of temperature degradation on the performance of photocatalyst material to decrease phenol was also studied. One of the factors that also affects photocalytic degradation is the temperature of degradation. The experiments carried out with variations in the degradation temperature of 30, 40 and 50°C, respectively. Based on figure 5, the percentage increase in degradation occurs with increasing temperature degradation used. It can be seen that at the initial concentration of 10 ppm phenol there was an increase in the percentage of degradation from temperatures 30°C, 40°C and 50°C.

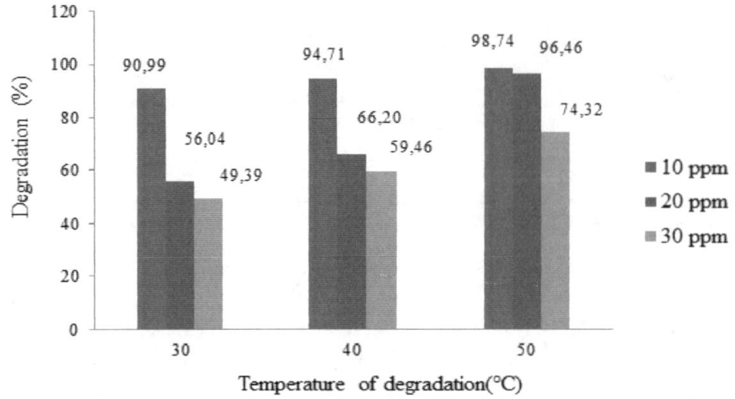

Figure 5. Effect of temperature on the performance of composites to degrade phenols.

According to Kumar and Pandey [24], rising temperatures result in increased interactions that occur between phenol molecules and the surface of the photocatalyst. This opinion is also supported by the research of Mamba et al.[25] that the best temperature range for the photocatalysis process is 20°C-80°C. An increase in reaction temperature generally increases photocatalysis activity but when the degradation temperature is above 80°C it encourages recombination of the photocatalyst material and a further decrease in temperature to 0°C will increase activation energy. Therefore the degradation temperature of 20°C-80°C is the best range for the degradation process to occur [24].

Figure 6 below shows the performance of photocatalysts in degrading phenols from TiO_2, TiO_2-CNT composite-modified CNT (CTAB) and CNT with phenol degradation percentage of 89%, 98.74% and 29% respectively. Performance TiO_2-CNT composite-modified CNT (CTAB) has a higher photocatalyst performance than other materials. This degradation result is also supported by the results of SEM characterization on composites characterized by the presence of well distributed TiO_2 particles between CNT and also the presence of Ti-O-C bonds in the FTIR characterization results indicating close contact between TiO_2 and CNT thus offering an efficient route for electron transfer from the TiO_2 conduction band to the surface of the CNT [21].

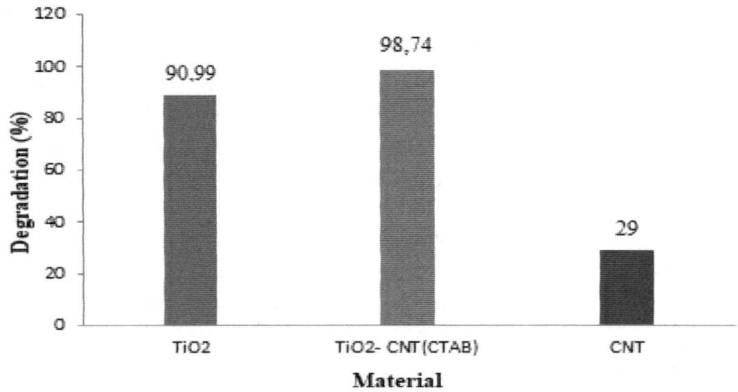

Figure 6. Effectifity of performance of TiO_2-CNT composite – modified CNT (CTAB) to phenol degradation.

4. Conclusion

The TiO_2-CNT composite–modified CNT (Cetyltrimethylammonium bromide) photocatalyst material has been successfully synthesized. Effectivity of performance of photocatalyst material composite was influenced by cristanility, morphology and function group. The highest performance effectiveness of photocatalyst material composites for degrading phenol was obtained on phenol concentration 10 mg/L and degredation temperature at 50°C was 98.74 %.

Acknowledgement

The authors would like to thank DRPM Universitas Riau and Directorate General of Higher Education (DGHE) Indonesian Ministry of National Education for the financial support of this research (Hibah) Penelitian Dasar, no. 395/UN.19.5.1.3/PT.01.03/2020.

References

[1] Dargahi A, Mohammadi M, Amirian F, Karami A and Almasi A 2017 Phenol removal from oil refinery wastewater using anaerobic stabilization pond modeling and process optimization using response surface methodology (RSM) *Desalin. Water Treat.***87** 199–208

[2] Pradeep N V., Anupama S, Navya K, Shalini H N, Idris M and Hampannavar U S 2015 Biological removal of phenol from wastewaters: a mini review *Appl. Water Sci.***5** 105–12

[3] Liwei D 2007 City environment art and sustainable development *Chinese J. Popul. Resour. Environ.***5** 63–6

[4] Tao Y, Cheng Z L, Ting K E and Yin X J 2013 Photocatalytic Degradation of Phenol Using a Nanocatalyst: The Mechanism and Kinetics *J. Catal.***2013** 1–6

[5] Hermawan P, Pranowo H D and Kartini I 2011 Physical Characterization Of Ni(Ii) Doped Tio_2 Nanocrystal By Sol-Gel Process *Indones. J. Chem.***11** 135–9

[6] Hoffmann M R, Martin S T, Choi W and Bahnemann D W 1995 Environmental Applications of Semiconductor Photocatalysis *Chem. Rev.***95** 69–96

[7] Tan T L, Lai C W, Hong S L and Rashid S A 2018 New insights into the photocatalytic endocrine disruptors dimethyl phathalate esters degradation by UV/MWCNTs-TiO2 nanocomposites *J. Photochem. Photobiol. A Chem.***364** 177–89

[8] Huang L, Chan Q, Wu X, Wang H and Liu Y 2012 The simultaneous photocatalytic degradation of phenol and reduction of Cr(VI) by TiO 2/CNTs *J. Ind. Eng. Chem.***18** 574–80

[9] Liu X, Wang M, Zhang S and Pan B 2013 Application potential of carbon nanotubes in water treatment: A review *J. Environ. Sci. (China)***25** 1263–80

[10] Manilo M V., Lebovka N and Barany S 2017 Combined effect of cetyltrimethylammonium bromide and laponite platelets on colloidal stability of carbon nanotubes in aqueous suspensions *J. Mol. Liq.***235** 104–10

[11] Heltina D 2017 Modification of Carbon Nanotube for Synthesis of Titania Nanotube (Tint)-Carbon Nanotube (Cnt) Composite *J. Energy, Mech. Mater. Manuf. Eng.***1** 29–35

[12] Soroodan Miandoab E and Fatemi S 2015 Upgrading TiO2 Photoactivity under Visible Light by Synthesis of MWCNT/TiO2 Nanocomposite *Int. J. Nanosci. Nanotechnol.***11** 1–12

[13] Slamet A R and W 2005 Pengolahan Limbah Cr(Vi) Dan Fenol Dengan Fotokatalis Serbuk Tio 2 Dan Cuo/Tio_2*Prosiding SNTPK Jakarta*

[14] Heltina D, Wulan P P D K and Slamet 2015 Synthesis and characterization of titania nanotube-carbon nanotube composite for degradation of phenol *Int. J. Technol.***6** 1137–45

[15] Bao N, Li Y, Wei Z, Yin G and Niu J 2011 Adsorption of dyes on hierarchical mesoporous TiO2 fibers and its enhanced photocatalytic properties *J. Phys. Chem. C***115** 5708–19

[16] Linsebigler A L, Lu G and Yates J T 1995 Photocatalysis on TiO2 Surfaces: Principles, Mechanisms, and Selected Results *Chem. Rev.***95** 735–58

[17] Askari M B, Tavakoli Banizi Z, Seifi M, Bagheri Dehaghi S and Veisi P 2017 Synthesis of TiO2 nanoparticles and decorated multi-wall carbon nanotube (MWCNT) with anatase TiO2 nanoparticles and study of optical properties and structural characterization of

TiO2/MWCNT nanocomposite *Optik (Stuttg).* **149** 447–54

[18] Marques Neto J O, Bellato C R, De Souza C H F, Da Silva R C and Rocha P A 2017 Synthesis, characterization and enhanced photocatalytic activity of iron oxide/carbon nanotube/Ag-doped TiO2 nanocomposites *J. Braz. Chem. Soc.* **28** 2301–12

[19] Saleh T A, Siddiqui M N and Al-Arfaj A A 2014 Synthesis of multiwalled carbon nanotubes-titania nanomaterial for desulfurization of model fuel *J. Nanomater.* **2014**

[20] X. Pan, Y. Zhao, S. Liu, C. L. Korzeniewski S W and Z F 2012 Supporting Information Supporting Information *ACS Appl. Mater. Interfaces* **2** 1–5

[21] Lee D H, Park J G, Kyoung J C, Choi H J and Kim D W 2008 Preparation of brookite-type TiO2/carbon nanocomposite electrodes for application to Li ion batteries *Eur. J. Inorg. Chem.* 878–82

[22] Scanlon D O, Dunnill C W, Buckeridge J, Shevlin S A, Logsdail A J, Woodley S M, Catlow C R A, Powell M J, Palgrave R G, Parkin I P, Watson G W, Keal T W, Sherwood P, Walsh A and Sokol A A 2013 Band alignment of rutile and anatase TiO2 *Nat. Mater.* **12** 798–801

[23] Chong M N, Jin B, Chow C W K and Saint C 2010 Recent developments in photocatalytic water treatment technology: A review *Water Res.* **44** 2997–3027

[24] Kumar A 2017 A Review on the Factors Affecting the Photocatalytic Degradation of Hazardous Materials *Mater. Sci. Eng. Int. J.* **1**

[25] Mamba G, Mamo M A, Mbianda X Y and Mishra A K 2014 Nd,N,S-TiO2 decorated on reduced graphene oxide for a visible light active photocatalyst for dye degradation: Comparison to Its MWCNT/Nd,N,S-TiO2 Analogue *Ind. Eng. Chem. Res.* **53** 14329–38

Effect of rock sugar as a pore-forming material on the physical and mechanical properties of hydroxyapatite scaffold

A Indra[1,2], F Hadi[2], I H Mulyadi[2], J Affi[2], Gunawarman[2*]

[1] Department of Mechanical Engineering, Institut Teknologi Padang, KP Olo Padang, Indonesia.
[2] Department of Mechanical Engineering, Universitas Andalas, Limau Manis Padang, Indonesia

Corresponding author: gunawarman@eng.unand.ac.id

Abstract. Hydroxyapatite (HA) ceramic scaffold had been made using the pressureless sintering method for bone scaffold applications. Micron- and nano-sized HA as the primary material with the addition of polyvinyl alcohol (PVA) as a binder and rock sugar (RS) sized 298-420 μm as pore-forming material had been prepared. RS was applied to produce interconnecting pores in the HA scaffold. The green body was made by uniaxial pressing method at a pressure of 200 MPa. Preheating was carried out at a temperature of 700°C with a holding time of 1 h, aimed at eliminating PVA and RS, and then the temperature was increased to 1200°C with a holding time of 2 h. During the sintering process, the heating rate and cooling rate were maintained at 3°C/min. Physical and mechanical properties were tested respectively, including porosity and compressive strength testing. The results showed that the 23.0% porosity obtained a compressive strength value of 15.6 MPa at a ratio of HA:RS of 65:35 wt%. The highest porosity of 48.4% was obtained with a compressive strength of 4.9 MPa at a ratio of 55:45 wt%. In this study, the maximum addition of RS was only at 45 wt%, while the addition of RS reaching 50 wt% or more could affect the tangential bonding between HA powders during the sintering process.

1. Introduction

Ceramic-based products, such as bioceramics as implant materials, have been widely applied. Hydroxyapatite (HA) is a bioceramic material that has the potential to be used as an implant material for its biocompatible properties and similarity to human bone elements [1–9]. One of the implant applications being developed is the bone scaffold, which functions to replace cancellous bone.

Ideally, a scaffold has a level of strength which is comparable to the mechanical strength of bone. A bone's biomechanical system consists of cortical bone, which has a solid and hard surface with a compressive strength of 100-200 MPa, and cancellous bone with pores inside having a compressive strength of 2-20 MPa. The sizes of scaffold range from 100 to 400 μm to facilitate adherence of cells for bone growth [10,11] and approximately 70% porosity [12]. For this reason, a scaffold design that is accurate and has properties that match the original bone is required.

Some conflicting properties are still problematic in many studies conducted to make bone scaffold a cancellous bone replacement implant. Bone scaffold requires high porosity, which is approximately 70%, and relatively large pore size, namely 100-400 μm. At the same time, it also requires a relatively high compressive strength of 2-20 MPa. The results of the study mapping showed that when the physical properties (pore size and percentage of porosity) meet the requirements, the mechanical properties (compressive strength) do not; and vice versa.

Many ways have been developed in the manufacture of bone scaffolds, such as the method of soaking polymer foam made in the form of slurry by mixing it in a PVA solution [13,14], the polymer sponge moulding method [15,16], the 3D printing method [17–19], as well as uniaxial pressing and isostatic pressing methods [20,21]. Bone scaffold manufacturing needs a particular material to make pores, to produce an appropriate level of porosity. Sodium chloride (NaCl) [20], polymethyl methacrylic (PMMA)[21–23], and other pore-forming materials have been developed. The main requirement of the pore-making materials is that they can melt during the sintering process, resulting in pores in the scaffold produced. Rock sugar (RS) has the potential as a pore-forming material. It can be made accordingly to the needs of both shape and size. Besides, its hardness helps it withstand the pressure when making a green body.

In this study, the primary material for making bone scaffold was micron- and nano-sized HA powder. PVA was also used as a binder. RS grains were added variously to obtain porosity on the bone scaffold. The powder metallurgy method using uniaxial pressing at a pressure of 200 MPa was applied for the manufacture of the green body. The sintering process was carried out in an air environment to obtain bone scaffold targeted.

2. Materials and method

2.1. Materials

The bioceramics used were analytical commercial HA sized 2.5 μm and 200 nm and obtained from Sigma-Aldrich Co, USA. The binder material was technical PVA, with alcohol as a diluent, obtained from LQ Lab, Indonesia. All materials were used as they were in their initial conditions without any preliminary treatment. Pure water (distilled water) used throughout the research process was obtained from Brataco, Indonesia.

2.2. Method

2.2.1. Sample preparation. HA powder was prepared in sizes of 2.5 μm and 200 nm at a ratio of 80:20 wt% [3]. Then, 10 wt% PVA [24] and alcohol twice the weight of HA were added. The materials were mixed for 2 h using a rotary mixer with an alumina ball mill in it. After the mixing process, HA was dried up for 48 h to remove the alcohol content. The HA powder grains were refined in a rotary mixer for 2 h with an alumina ball mill inside. To get the appropriate porosity for the targeted bone scaffold, RS was added variously to the HA powder with ratios of 65:35, 60:40, 55:45, and 50:50 wt%, respectively. Previously, the RS was prepared in sizes of 297-420 μm by sieving at 40 and 50 mesh. HA and RS at the mentioned ratios were mixed in a rotary mixer for 2 h to obtain a homogeneous mixture. The green body was made in the form of pellets with a diameter of 8mm and a thickness of 4mm with a uniaxial pressing process at a pressure of 200 MPa. In the sintering process, preheating was carried out at a temperature of 700°C with a holding time of 1 h to eliminate PVA and RS. The sintering was continued to a temperature of 1200°C with a holding time of 2 h. During the sintering process, the heating rate and cooling rate were maintained at 3°C/min.

2.2.2. Characterisation of sample. Linear shrinkage test was carried out in two ways, namely diameter shrinkage and weight shrinkage tests, by comparing the green body with the sintered body. The resulting porosity levels were tested for all variations of the addition of RS. Furthermore, the distribution of porosity was observed using scanning electron microscopy (SEM-Hitachi Horiba S-3400 N, Japan) and digital microscopy (DM-CMOS Controller 2MP AMCAP, China). Compressive strength testing on the bone scaffold was carried out using a compressive test tool (adinitp-UTM20kN, Indonesia), referring to ASTM-C1424 Standard for Monotonic Compressive Strength of Advanced Ceramics.

3. Results and discussion

3.1. Linear shrinkage

Figure 1 shows the linear shrinkage that occurred, where the test was carried out in two ways including diameter shrinkage and weight shrinkage tests. Linear shrinkage could only be observed in HA:RS variations with ratios of 65:35, 60:40, and 55:45 wt%, respectively. Meanwhile, the ratio of 50:50 wt% bone scaffold sample was not successful to observe (the sample broke after the sintering process) due to the high amount of the RS added. After testing separately to melt down RS in the furnace, RS showed the behaviour to produce bubbles before melting at temperatures above 200°C. This caused the tangential release of HA powder during the sintering process at the addition of 50 wt% RS. Figure 1 shows that the shrinkage diameter for the three ratios of the addition of RS ranged from 15-16%. No difference occurred because the diameter shrinkage was more dominantly influenced by the sintering temperature. Similar values of diameter shrinkage have also been reported in previous

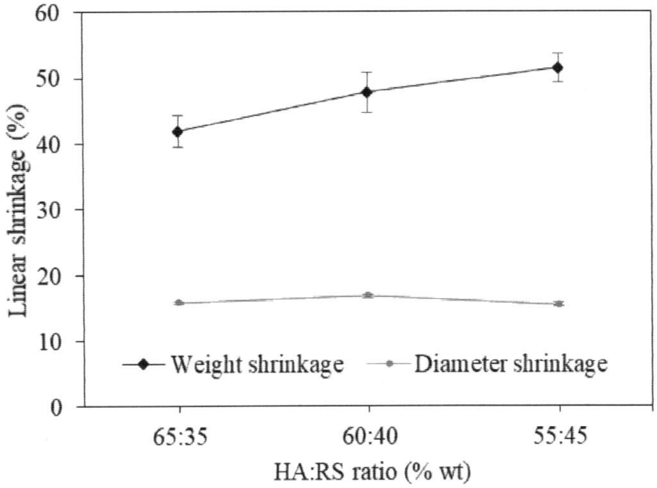

Figure 1. Effect of adding RS on linear shrinkage of HA scaffold

studies [25,26]. Weight shrinkage increased from 41.9% to 51.5% in the ratios of 65:35 and 55:45 wt%, respectively. The high weight shrinkage was due to RS and PVA, which completely melt at a temperature of 700°C, resulting in pores on the HA scaffold after sintering.

3.2. Porosity

Figure 2 shows the effect of the addition of RS as a pore-forming material on the porosity of the HA scaffold. The increased porosity was directly proportional to the addition of RS up to the highest ratio, namely 55:45 wt%. At a HA:RS ratio of 65:35 wt%, the porosity was 23% and continued to increase to 36.9% and 48.4% at ratios of 60:40 wt% and 55:45 wt%, respectively. Based on figure 2, increasing the amount of RS as a pore-making material was thought to increase the percentage of porosity, in accordance with the porosity standard of a bone scaffold product, i.e. 70% [27,28]. However, the addition of RS as a pore material can only be done at a maximum of 45 wt%. This is due to the behaviour of RS which produces bubbles when it melts, thus damaging the tangential bonds between the HA powders. At RS liquid temperature, around 200°C, HA has not yet undergone grain bonds, so it has the potential to release the tangential bonds that were generated during the making of the green body. The resulting pores already have interconnections, as shown in figure 3.

IOP Conf. Series: Materials Science and Engineering 1041 (2021) 012049 doi:10.1088/1757-899X/1041/1/012049

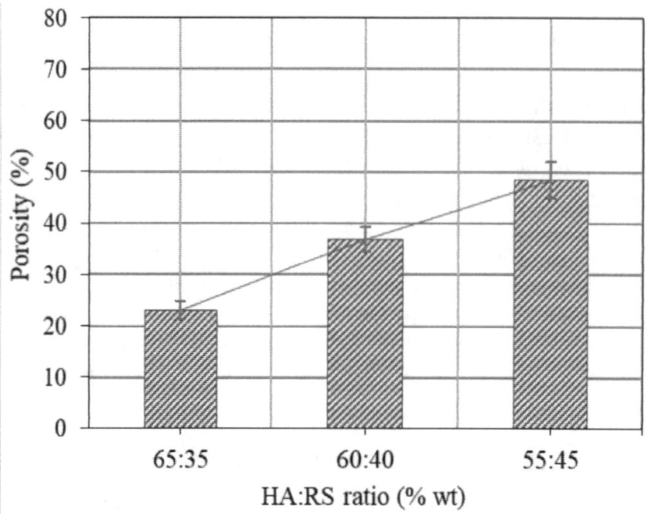

Figure 2. Effect of adding RS on porosity of HA scaffold

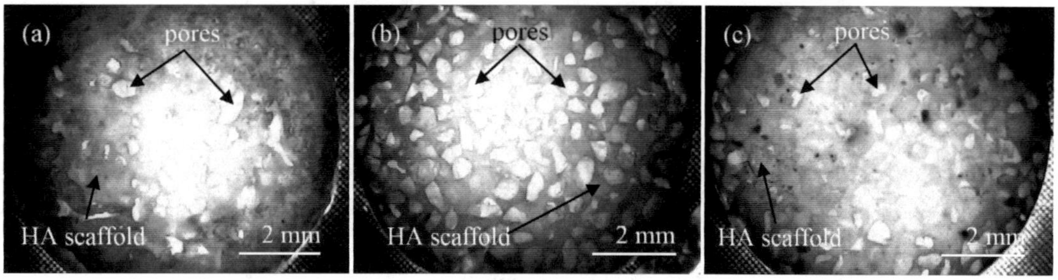

Figure 3. DM image of HA scaffold with variations of HA:RS (a) 65:35 wt%, (b) 60:40 wt%, (c) 55:45 wt%.

3.3. Macrostructural characteristics

Figure 4 shows the macrostructure and shape of the pores produced from the variations of the amounts of RS added as a pore-making material. Figure 4 ($a_{1,2}$) indicates the HA scaffold with the addition of 35 wt% RS. SEM image shows the shape of the pores with total porosity of 23%. The figure also indicates the presence of interconnecting pores but in a small number. Figure 4 ($b_{1,2}$) shows the HA scaffold with the addition of 40 wt% RS. The number of pores increased with the porosity of 36.9%. The interconnection between the pores also increased. Figure 4 ($c_{1,2}$) shows the HA scaffold with the addition of 45 wt% RS. The porosity reached 48.4%. The interconnections between the pores could be seen more than others. Overall, the shape of the pores was not uniform, and some even had sharp corners. This form possibly had high-stress concentration when given a compressive load. A pore-making agent is necessary to prepare with a shape that has no angle or with a shape close to round. The interconnection between pores is also shown in Figure 3. The DM image was carried out by providing light on the bottom of the HA scaffold. The light seemed to penetrate the HA scaffold, which attested to the interconnection between the pores.

3.4. Compressive strength

One of the main mechanical properties of HA scaffold testing is compressive strength, where the minimum requirement is 2 MPa. Figure 5 shows the compressive strength value of the HA scaffold with the various amounts of the RS added. The compressive strength of the HA scaffold is inversely proportional to the amount of the RS added. At the addition of 35 wt% RS, the compressive strength

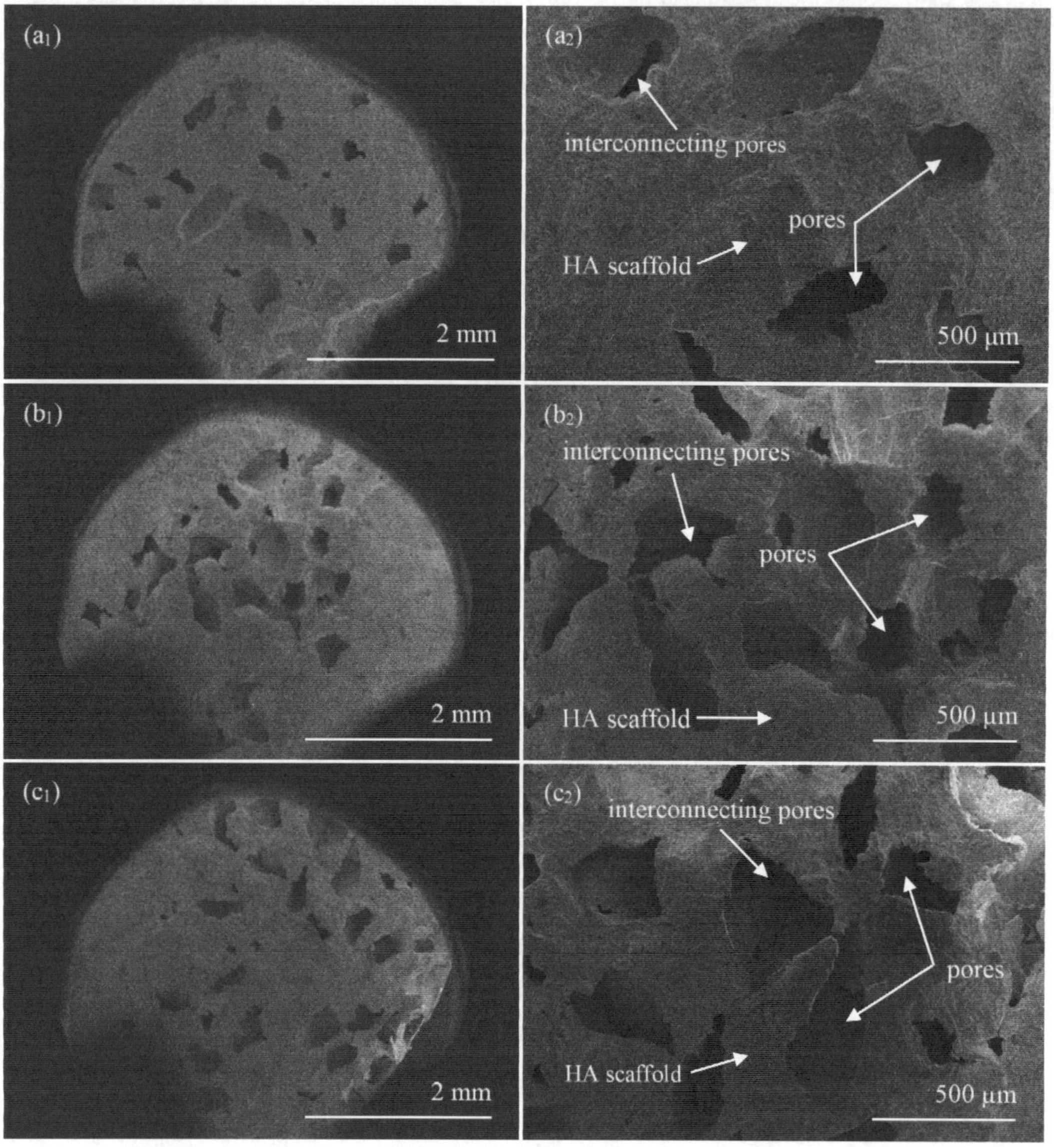

Figure 4. SEM image of HA scaffold with variations of HA:RS ($a_{1,2}$) 65:35 wt%, ($b_{1,2}$) 60:40 wt%, ($c_{1,2}$) 55:45 wt%.

value was 15.6 MPa and continued to decrease significantly to 4.9 MPa with the addition of 45 wt% RS. Theoretically, the lowest compressive strength in this study met the minimum standard of a scaffold, i.e. 2 MPa [29]. However, according to the data in figure 2, the resulting porosity only reached 48.4%, still below the porosity requirement of 70% [28]. Figure 6 shows the relationship between compressive strength and porosity. Compressive strength decreased as RS decreased, and, vice versa, porosity increased as RS increased. From the overall results of this study, it is necessary to consider the selection of other materials as pore-making materials. It is required that pore-forming materials, when melting, do not affect the tangential bonds between the HA powders. As a comparison, another study using a spherical pore material obtained a compressive strength value in nano-sized HA scaffold of about 31.4 MPa at 50% porosity [21] and 1.1 MPa at a porosity above 80% [30]. A summary of the physical and mechanical properties of the HA scaffold can be seen in table 1.

Table 1. Summary of physical and mechanical properties of HA scaffold

Ratio of HA:RS (wt%)	Linear shrinkage (%)		Porosity (%)	Compressive strength (MPa)
	Diameter	Weigh		
65:35	15.81 ±0.14	41.9 ±2.42	23.0 ±1.82	15.6 ±1.3
60:40	16.75 ±0.19	47.8 ±3.08	36.9 ±2.40	11.5 ±0.7
55:45	15.50 ±0.20	51.5 ±2.19	48.4 ±3.59	4.9 ±0.2

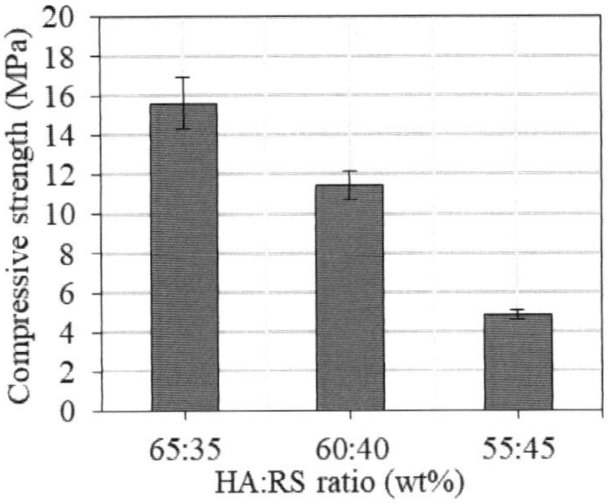

Figure 5. Effect of adding RS on compressive strength of HA scaffold

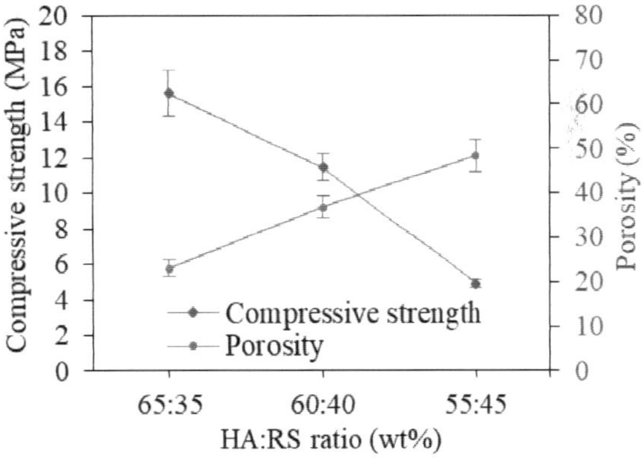

Figure 6. Effect of adding RS on compressive strength and porosity of HA scaffold

4. Conclusion

The addition of RS as a pore-forming material to manufacture HA scaffold has succeeded in obtaining pores with total porosity of 48.4% with a compressive strength value of 4.9 MPa. The maximum amount of the RS added should be 45 wt%. The addition of RS as much as 50 wt% or more can affect the tangential bond between HA powders during the sintering process.

Acknowledgment

This research was funded by *Hibah Penelitian Disertasi Doktor* of the National Research and Innovation Agency, the Ministry of Research and Technology, Indonesia with Contract No. 163/SP2H/AMD/LT/DRPM/2020.

References

[1] Sivaperumal V R, Mani R, Nachiappan M S and Arumugam K 2017 Direct hydrothermal synthesis of hydroxyapatite/alumina nanocomposite *Mater. Charact.* **134** 416–21
[2] Khoshsima S, Yilmaz B, Tezcaner A and Evis Z 2016 Structural, mechanical and biological properties of hydroxyapatite-zirconia-lanthanum oxide composites *Ceram. Int.* **42** 15773–9
[3] Indra A, Firdaus R, Mulyadi I H, Affi J and Gunawarman 2020 Enhancing the physical and mechanical properties of pellet-shaped hydroxyapatite by controlling micron- and nano-sized powder ratios *Ceram. Int.* **46** 15882–8

[4] Catalgol Z 2019 Sintering effect on borosilicate glass-bovine hydroxyapatite composites *J. Aust. Ceram. Soc.***55** 1075–9

[5] Nuswantoro N F, Gunawarman, Saputra M R, Nanda I P, Idris M H and Arafat A 2018 Microstructure analysis of hydroxyapatite coating on stainless steel 316L using investment casting technique for implant application *Int. J. Adv. Sci. Eng. Inf. Technol.***8** 2168–74

[6] Szcześ A, Hołysz L and Chibowski E 2017 Synthesis of hydroxyapatite for biomedical applications *Adv. Colloid Interface Sci.***249** 321–30

[7] Galler K M, D'Souza R N, Hartgerink J D and Schmalz G 2011 Scaffolds for dental pulp tissue engineering. *Adv. Dent. Res.***23** 333–9

[8] Kolmas J, Krukowski S, Laskus A and Jurkitewicz M 2016 Synthetic hydroxyapatite in pharmaceutical applications *Ceram. Int.***42** 2472–87

[9] Esmaeili S, Akbari Aghdam H, Motififard M, Saber-Samandari S, Montazeran A H, Bigonah M, Sheikhbahaei E and Khandan A 2020 A porous polymeric–hydroxyapatite scaffold used for femur fractures treatment: fabrication, analysis, and simulation *Eur. J. Orthop. Surg. Traumatol.***30** 123–31

[10] Whang K, Healy K E, Elenz D R, Nam E K, Tsai D C, Thomas C H, Nuber G W, Glorieux F H, Travers R and Sprague S M 1999 Engineering Bone Regeneration with Bioabsorbable Scaffolds with Novel Microarchitecture *Tissue Eng.***5** 35–51

[11] Li Z, Ramay H R, Hauch K D, Xiao D and Zhang M 2005 Chitosan-alginate hybrid scaffolds for bone tissue engineering *Biomaterials***26** 3919–28

[12] Wake M C, Patrick C W and Mikos A G 1994 Pore morphology effects on the fibrovascular tissue growth in porous polymer substrates *Cell Transplant.***3** 339–43

[13] Swain S K 2009 Processing of Porous Hydroxyapatite Scaffold Processing of Porous Hydroxyapatite Scaffold *Rourkela*

[14] Miao X, Tan D M, Li J, Xiao Y and Crawford R 2008 Mechanical and biological properties of hydroxyapatite/tricalcium phosphate scaffolds coated with poly(lactic-co-glycolic acid) *Acta Biomater.***4** 638–45

[15] Jo, I.-H., K.-H. Shin, Y.-M. Soon, Y.-H. Koh, J.-H. Lee and H-E K 2009 "Highly Porous Hydroxyapatite Scaffolds with Elongated Pores using Stretched Polymeric Sponges as Novel Template." *Mater. Lett.***63**, no. **20**

[16] Guda, T., S. Oh, M.R. Appleford and J L O 2012 "Bilayer Hydroxyapatite Scaffolds for Maxillofacial Bone Tissue Engineering." *Int. J. oral Maxillofac. Implant. 27***no. 2**, p.

[17] Shao H, He J, Lin T, Zhang Z, Zhang Y and Liu S 2019 3D gel-printing of hydroxyapatite scaffold for bone tissue engineering *Ceram. Int.***45** 1163–70

[18] Wang H, Wu G, Zhang J, Zhou K, Yin B, Su X, Qiu G, Yang G, Zhang X, Zhou G and Wu Z 2016 Osteogenic effect of controlled released rhBMP-2 in 3D printed porous hydroxyapatite scaffold *Colloids Surfaces B Biointerfaces***141** 491–8

[19] Trachtenberg J E, Placone J K, Smith B T, Fisher J P and Mikos A G 2017 Extrusion-based 3D printing of poly(propylene fumarate) scaffolds with hydroxyapatite gradients *J. Biomater. Sci. Polym. Ed.***28** 532–54

[20] Eilbagi M, Emadi R, Raeissi K, Kharaziha M and Valiani A 2016 Mechanical and cytotoxicity evaluation of nanostructured hydroxyapatite-bredigite scaffolds for bone regeneration *Mater. Sci. Eng. C***68** 603–12

[21] Gain A K, Zhang L and Liu W 2015 Microstructure and material properties of porous hydroxyapatite-zirconia nanocomposites using polymethyl methacrylate powders *Mater. Des.***67** 136–44

[22] Radha G, Balakumar S, Venkatesan B and Vellaichamy E 2017 A novel nano-hydroxyapatite — PMMA hybrid scaffolds adopted by conjugated thermal induced phase separation (TIPS) and wet-chemical approach: Analysis of its mechanical and biological properties *Mater. Sci. Eng. C***75** 221–8

[23] Mahammod B P, Barua E, Deb P, Deoghare A B and Pandey K M 2020 Investigation of

Physico-mechanical Behavior, Permeability and Wall Shear Stress of Porous HA/PMMA Composite Bone Scaffold *Arab. J. Sci. Eng.* **45** 5505–15

[24] Indra A, Setiawan R B, Mulyadi I H, Affi J and Gunawarman 2019 The Effect of PVA Addition as Binders on the Properties of Hydroxyapatite Sintered Body *IOP Conference Series: Materials Science and Engineering, Submitted for publication.* (IOP Conference Series: Materials Science and Engineering, Submitted for publication.)

[25] Pei X, Ma L, Zhang B, Sun J, Sun Y, Fan Y, Gou Z, Zhou C and Zhang X 2017 Creating hierarchical porosity hydroxyapatite scaffolds with osteoinduction by three-dimensional printing and microwave sintering *Biofabrication* **9** 0–24

[26] Zeng Y, Yan Y, Yan H, Liu C, Li P, Dong P, Zhao Y and Chen J 2018 3D printing of hydroxyapatite scaffolds with good mechanical and biocompatible properties by digital light processing *J. Mater. Sci.* **53** 6291–301

[27] Keaveny T M 2004 *'Standard Handbook of Biomedical Engineering and Design, McGraw Hill.*

[28] Fierz, F.C., F. Beckmann, M. Huser, S.H. Irsen, B. Leukers, F. Witte, Ö. Degistirici, A. Andronache, M. Thie and B M 2008 "The Morphology of Anisotropic 3D-printed Hydroxyapatite Scaffolds." *Biomaterials* **29**, no. **28**

[29] Nawawi N A, Alqap A S F and Sopyan I 2011 Recent Progress on Hydroxyapatite-Based Dense Biomaterials for Load Bearing Bone Substitutes 63–80

[30] Wang Q, Li J, Xu T, Lu X, Zhi W and Weng J 2017 Porous hydroxyapatite scaffolds containing dual microspheres based on poly(lactide-co-glycolide) and chitosan for bone regeneration *Mater. Lett.* **188** 387–91

Comparison of Natural Fiber Types as Reinforcement Material on Composite Mechanical Properties via Carbon Nanotubes Addition

Y Yolanda[1] and Praswasti P D K Wulan[1]

[1]Sustainable Energy Research Group, Department of Chemical Engineering, Faculty of Engineering, Universitas Indonesia, Jl. Margonda Raya, PondokCina, KecamatanBeji, Kota Depok, Jawa Barat 16424, Indonesia

Corresponding author:yogyoland@gmail.com

Abstract. Composites with natural fiber reinforcement have been widely used in various field. Natural fibers have the advantages of easy recycling, environmentally-friendly characters, renewable, non-toxic, inexpensive, high toughness, good insulation against heat and noise, and good thermal properties. The aim of this paper is compare natural fibers used as reinforcing material on composite with an epoxy matrix and the addition of carbon nanotubes (CNT). The natural fibers used are oil palm empty fruit bunches (OPEFB), bagasse, and hemp fiber. Alkalization treatment with NaOH was carried out on natural fibers to remove impurities on the fiber surface and activate hydroxyl groups. CNTs need to be functionalized to achieve the desired compatibility. Functionalization is carried out by the mild acid oxidation method using nitric acid and hydrogen peroxide. Silane coupling agent treatment is carried out on natural fibers and CNTs to improve compatibility with the matrix. The result of the bending strength test of OPEFB, bagasse, and hemp fiber was 509.94 MPa, 36.22 MPa, and 18.12, respectively, in addition to CNT 0.5% mass. The bending strength of OPEFB fiberismore significant than bagasse and hemp fiber, so it has the opportunity to be developed in the automotive industry.

1. Introduction

The development of material technology has grown rapidly. The polymer has replaced many conventional materials such as metals in various applications due to advantages such as ease of processing, lightweight, higher productivity, and reduced costs. For many of these applications, polymer properties are modified using fillers and fibers to conform to high strength modulus requirements [1].

Composites with natural fiber reinforcement have been widely used in various fields such as automobiles, furniture, heating, and construction industries. Natural fibers have the advantages of low density, easy recycling, environmentally-friendly characters, renewable, non-toxic, inexpensive, high toughness, good insulation against heat and noise, good thermal properties, easy separation, and low abrasiveness[2]. The high content of chemical compounds in cellulose in natural fibers makes natural fibers suitable for modern applications [3]. Some natural fibers used as composite materials reinforcement are oil palm empty fruit bunch (OPEFB) [4], bagasse [5], and hemp fiber [6].

Engineering materials on the nanometer scale can produce various materials with various properties. One of the materials is carbon nanotubes (CNT) due to their excellent mechanical strength and electronic properties [7]. The CNT's specific strength is 10 to 100 times stronger than the strongest steel [8].

The purpose of this paper is to compare natural fibers used as reinforcing materials to increase the mechanical strength of composites with the addition of carbon nanotubes and epoxy resin as a matrix. The natural fibers used are oil palm empty fruit bunches, bagasse, and hemp fiber. Natural fibers are first subjected to an alkaline treatment process then followed by surface treatment of the fibers with a silane coupling agent, after that the fibers are formed into chopped strands. CNTs need to be functionalized to improve compatibility with the polymer matrix [9]. After that, the CNT is mixed with liquid epoxy resin and then the epoxy/CNT-natural fiber composite is formed. FTIR characterization is carried out on natural fiber and CNT to analyze the functional group that exist in the substance. A bending test is carried out to see the performance of composites.

2. Materials and methods

2.1. *Materials*

2.1.1. Oil Palm Empty Fruit Bunch. The most massive waste produced from the oil palm industry is empty oil palm bunches and palm fronds [10]. The bunch is where the oil palm fruit sticks. After the oil palm fruit is separated for processing, it produces waste in the form of EFB. EFB waste can be used as a raw material for making composites because palm oil bunch fibers contain a lot of wood chemical components such as cellulose (45.8%), hemicellulose (31.6%), and lignin (22.6%) [11]. Cellulose can be used as a source of natural fibers in the formation of composite materials.

2.1.2. Bagasse. The main sugarcane product is sugar, but the by-product in the form of bagasse is 35-40% of each processed sugarcane [12]. Bagasse is a fibrous material that is obtained as a residue from sugarcane after being crushed for extraction. Bagasse contains chemical components such as cellulose 40-50%, hemicellulose 25-35%, lignin 17-20%, and several other components such as 0.8% wax and 2.3% ash [13], [14]. The use of bagasse as reinforcement in composite production is a good for increasing the bagasse's economic value [15].

2.1.3. Hemp Fiber. Hemp is one of the long-lived plants, grows well in areas with warm and humid weather with rainfall that is almost evenly distributed throughout the year [16]. Hemp fiber is the bastfiber of the tree obtained from the inner bark of the hemp stem. Hemp has been used as a textile fiber in China and South Asia for centuries and has been known in Ancient Egypt [17]. Hemp fiber has good strength, durability, and absorption. One of the uses of hemp fiber is as an alternative material for cotton in using textiles to reduce cotton imports. Hemp fiber has chemical components such as cellulose (75.6%), hemicellulose (10.7%), and lignin (6.6%) [18].

2.2. *Methods*

2.2.1. Alkalization Treatment. Cellulose and hemicellulose in the plant body are generally physically encased by lignin and pectin [19]. Lignin is insoluble in water and stable in nature and acts as adhesive connecting cellulose and hemicellulose [20]. Alkalization is the most widely used method to remove the lignin and oil content covering the fiber's outer surface [21] and increase the number of free hydroxyl groups, thereby increasing the fiber and resin's interface bond [22]. The alkalization process is carried out by immersing the fibers in an alkaline solution (NaOH) with different treatments at time and temperature [23].The alkalization process will raise cellulose and remove impurities on the surface of natural fibers such as lignin and pectin to improve the fiber's mechanical properties [24]. Plant fiber can be obtained more easily from the plant body after the adhesive components in the form of lignin and pectin dissolve in the cooking solution. Cooking fiber with a 10% weight NaOH solution at 100^0C can dissolve most of the lignin found in plant components so that the fibers can be easily separated

[11]. The high-temperature of NaOH can damage the fiber's hydrogen bond structure to increase the surface area of the fiber[25].

2.2.2. Natural Fiber Surface Treatment. The effectiveness of cellulose fiber strengthening in polymers is very dependent on the compatibility between the two materials. Chemical treatment such as acid etching and a coupling agent can be applied to increase the surface interaction of the composite's constituent materials. The hydroxide group produces the hydrophilic nature of natural fibers and the hydrophobic nature of the matrix makes the two combinations incompatible. In this case, silane surface treatment can improve the interface adhesion between natural fibers and the matrix. Silane molecules have bifunctional groups, where one of them reacts with the fiber and the other with the polymer. Therefore, the silane surface treatment reduces the number of hydroxyl groups between the natural fiber interface and the matrix [26]. One part of the silane will react with cellulose on the fiber's surface, and one part of the other will bind to the matrix [27].The tensile and flexural strength of the composite with modified fibers is increased when the silane coupling agent is used [28]. The fiber treatment and functionalization CNT using a silane coupling agent increase the matrix's compatibility. Good interfacial bonding between the reinforcement and the matrix provides improved mechanical properties [29]. Alkali treatment of fibers and CNT functionalization with the mild acid oxidation method is first carried out to activate the fiber and CNT's hydroxyl groups, which will then bind to the silane coupling agent [30]. Surface treatment using a silane coupling agent increase mechanical strength due to good dispersion and adhesion to the epoxy matrix [31].

2.2.3. Carbon Nanotube Treatment. CNTs can act as reinforcing agents in polymers. The conductivity, strength, elasticity, robustness, and resistance of the composites formed can be increased substantially with the CNT's [32]. The surface energy of CNTs is significantly different from matrices such as organic solvents or polymers, and CNTs have no chemical affinity for the organic matrix [33]. Pure CNT is a hydrophobic and have low slubility in solvent. Therefore, CNT needs to be functionalized to make CNT dispersible in solution [34].The CNT functionalization method can be carried out in 2 ways; the first is mild acid oxidation using HNO_3 and H_2O_2 to oxidize pure CNT, which will form a hydroxyl functional group on the CNT surface which functions as a reaction site [35]. Oxidation with acid can perform CNT purification, disperse, and surface activation simultaneously. Oxidation with 3M HNO_3 and 30% v/v H_2O_2 and ultrasonic vibrations have been shown to functionally function CNT surfaces without damaging the structure of the nanotubes [36]. The second method used is CNT silanization using a silane coupling agent by dissolving the silane coupling agent solution in ethanol so that it forms a silanol group due to hydrolysis, then forms a covalent bond with the OH group on the CNT surface to form a Si-O-C bond [37]. CNT surface treatment with a silane coupling agent has also been shown to increase the flexural strength and crack strength of epoxy matrix-based composites [38] to increase composite's tensile strength with an epoxy matrix [39].

3. Results and discussion

FTIR characterization was carried out on natural fibers and CNT, in its pure and modified form to analyze functional group in those subtanceA bending test is done to analyze mechanical strength of composite.

3.1. Surface Modification

Transmittance graphs of natural fiber obtained from FTIR spectroscopy, both pure and modified with GPTMS.

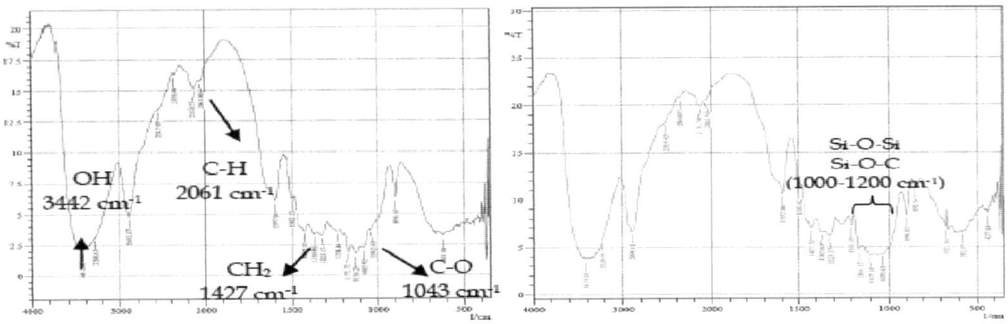

(a) (b)

Figure 1. Transmittance graph of OPEFB (a) pure, (b) modified [40].

Figure 1 shows several peaks that prove the functional group characteristics, -OH at 3442 cm^{-1}, C-H at 2061 cm^{-1}, CH_2 1427 cm^{-1}, C-O 1043 cm^{-1}. FTIR spectra of modified OPEFB not only showed the same peak as pure, but also inceasing intensity of peak at 1000-1200 cm^{-1}wavelenghth[40] which means there is indicating the presence of Si-OH bonds on natural fiber [31].

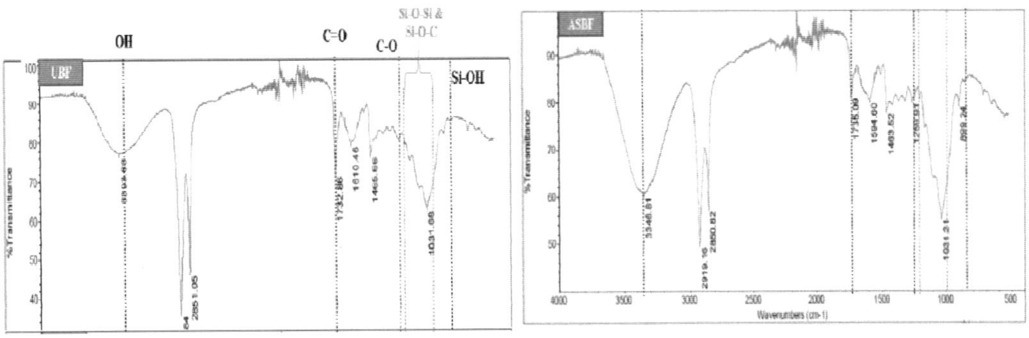

(a) (b)

Figure 2. Transmittance graph of bagasse (a) pure, (b) modified [5].

Figure 2 shows C-O wavelengths at 1250 cm^{-1}, C=O at 1735 cm^{-1}, and -OH at 3300 cm^{-1}. Surface treatment of sugarcane pulp fiber showed the presence of Si-O-Si and Si-O-C groups in waves 1000-1200 cm^{-1}[5]. Si-O-Si bonds are formed due to cross-links between coupling agents, while Si-O-C bonds are formed due to chemical reactions between coupling agents and fibers [41].

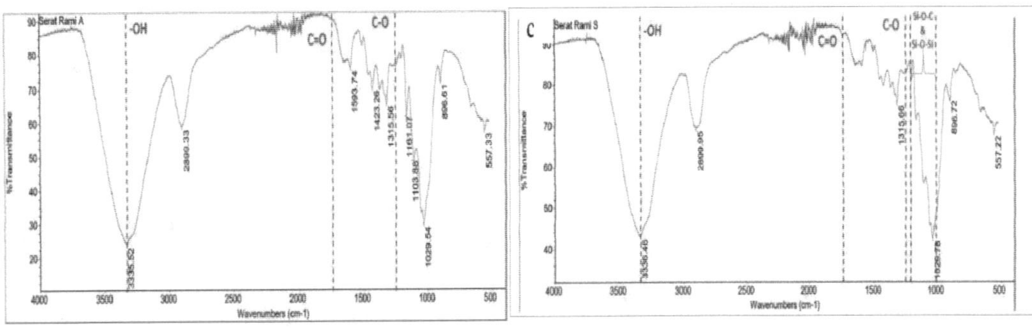

(a) (b)

Figure 3. Transmittance graph of hemp fiber (a) pure, (b) modified [6].

Figure 3 shows the functional group's wavelength -OH at 3331 cm^{-1}, C-O at 1242 cm^{-1}, C=O at 1733 cm^{-1}. Si-O-Si and Si-O-C bonds at 1000-1200 cm^{-1}[6]. There is a reaction between hydrolyzed silane and hemp fiber to produce a Si-O-C strain in the hemp fiber [42].

The Si-O-C covalent bond will react with the polymer matrix thereby increasing the fiber-matrix interface [43]. Treatment of the fiber surface with a silane coupling agent forms a covalent bond between the silane and the fiber surface, thus forming a composite that has better mechanical strength because it has good stress transfer between the fiber and the epoxy matrix [29].

3.2. Functionalization CNT

FTIR analysis was carried out on the CNT before and after the functionalization process. The CNT transmission graph in figure 4 shows the wavelength of -OH at 3455 cm^{-1}, C=O at 1654 cm^{-1}, and C-O at 1171 cm^{-1} so it can be confirmed that functionalization CNT by acid has OH group [40]. Silane coupling agent GPTMS will react rapidly with water forming silanol group. This group react rapidly with OH group at the surface of oxidized CNT to form polymeric bond Si-O-Si [41]. The presence of Si-O-Si bonding was confirmed by FTIR analysis where there are peaks at 1000-1200 cm^{-1} wavelength [40].

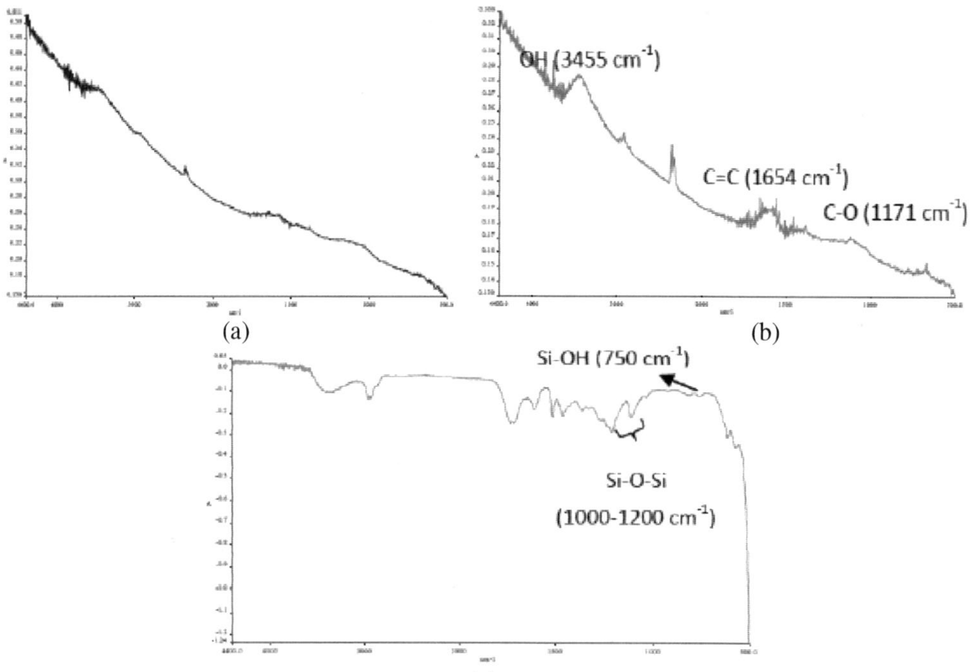

Figure 4. Transmittance graph of funtionalization CNT (a) pure, (b) mild acid oxidation treatment, (c) silanecouling agent treatment [40].

3.3. Composite Molding and Bending Test

CNT addition 0.5% and fiber form chopped strand are done to make effective composition for composite. The standard used for the bending test is ASTM D790 using 3 point loading. Specimens were prepared to be 80 mm x 25.4 mm x 3.2 mm for OPEFB [4] and 125 mm x 12.7 mm x 3.2 mm for bagasse and hemp fiber[5],[6]. The specimen is supported by two cantilevers in a determined support span. The pressure is applied in the middle part of the specimen; by a metal bar until the specimen gets fatigued and cannot hold any more pressure load.

Table1. Bending Test Result.

Natural Fiber	Bending Strength (MPa)		Reference
	Epoxy + Natural Fiber	Epoxy + 0.5% CNT + Natural Fiber	
OPEFB	464.14	509.94	Syahrudin, U. P, 2016
Bagasse	14.45	36.22	Daffa, M, 2020
Hemp fiber	35.55	18.12	Raihan, S. D, 2020

Table 1 shows the bending strength of each composite. The addition of CNT can increase the bending strength of the OPEFB fibers and bagasse [4],[5]. The bending strength of bagasse was 150.65%, more significant than OPEFB, which increased by 9.86%. However, the increase in bending strength in bagasse does not exceed the bending strength of the OPEFB fibers. The addition of CNT to hemp fiber decreased mechanical strength by 48.96% [6]. The addition of a smaller number of CNTs harms the interface bond between the CNT and the matrix and the Van der Waals bond between the fiber and matrix. Likewise, adding too much CNT can show negative results [45]. So we need an optimal point for adding CNT to the hemp fiber composite. The bending test results show that the formed composite can be applied to the automotive industry, especially as a car bumper.

4. Conclusion

Surface treatment of natural fibers and CNTs with a silane coupling agent successfully modifies the surface and helps strengthen the bonds between composite components. The addition of 0.5% CNT to OPEFB increases the bending strength by 9.86% and bagasse by 150.65%. Meanwhile, the addition of 0.5% CNT to hemp fiber reduces the bending strength by 48.96%. The bending strength of CNT composites with OPEFB fibers was 509.94 MPa, bagasse 36.22 MPa, and hemp fibers 18.12 MPa. CNT-OPEFB composites have greater bending strength than bagasse and hemp fibers and can be developed in the automotive industry.

Acknowledgments

We thank the Directorate of Research and Public Service, Universitas Indonesia, for funding this research in the scheme PublikasiTerindeksInternasional (PUTI) SaintekesTahunAnggaran 2020 Nomor: NKB-2463/UN2.RST/HKP.05.00/2020.

References

[1] Yashas Gowda T G, Sanjay M R, Subrahmanya Bhat K, Madhu P, Senthamaraikannan P, and Yogesha B 2018 Polymer matrix-natural fiber composites: An overview *Cogent Engineering* doi: 10.1080/23311916.2018.1446667

[2] Sapuan S M 2014 *Tropical Natural Fibre Composites: Properties, Manufacture and Applications*

[3] Nair A B and Joseph R 2014 Eco-friendly bio-composites using natural rubber (NR) matrices and natural fiber reinforcements *Chemistry, Manufacture and Applications of Natural Rubber*

[4] Syahrudin U P 2016 *Pengaruh Variasi Adisi Carbon Nanotube dan Bentuk Serat Tandan Kosong Kelapa Sawit Terhadap Peningkatan Kekuatan Mekanis Komposit Epoksi* Depok Jawa Barat Departemen Teknik Kimia Universitas Indonesia

[5] Daffa M 2020 Pengaruh Adisi Carbon Nanotube Terhadap Kekuatan Mekanis Komposit Serat Ampas Tebu (Bagasse) dengan Matriks Epoksi Depok Jawa Barat Departemen Teknik Kimia Universitas Indonesia

[6] Raihan S D 2020 Pengaruh Adisi Carbon Nanotube Terhadap Kekuatan Tekuk Komposit Epoksi Serat Rami Dalam Pemanfaatan Socket Prosthesis Depok Jawa Barat Departemen Teknik Kimia Universitas Indonesia

[7] Kim H and Lee S 2018 Characterization of carbon nanofiber (CNF)/polymer composite coated on cotton fabrics prepared with various circuit patterns *Fash. Text.* **5** 1 p 7 doi: 10.1186/s40691-017-0120-2
[8] Chen L, Ozisik R and Schadler L S 2010 The influence of carbon nanotube aspect ratio on the foam morphology of MWNT/PMMA nanocomposite foams *Polymer (Guildf).* **51** 11 pp 2368–2375 doi: https://doi.org/10.1016/j.polymer.2010.03.042
[9] Hill D E, Lin Y, Rao A M, Allard L F and Sun Y P 2002 Functionalization of carbon nanotubes with polystyrene *Macromolecules* doi: 10.1021/ma020855r
[10] Nadia A, Fauziah A and Mayori E 2017 Potential of Lignocellulosic Oil Palm Waste in South Kalimantan for Bioethanol and Xylitol Production *J. Inov. Pendidik. Sains*
[11] Anggraini D and Roliadi H 2011 PEMBUATAN PULP DARI TANDAN KOSONG KELAPA SAWIT UNTUK KARTON PADA SKALA USAHA KECIL *J. Penelit. Has. Hutan* doi: 10.20886/jphh.2011.29.3.211-225
[12] Misran E 2005 Industri Tebu Menuju Zero Waste Industry *J. Teknol. Proses*
[13] de Barros R da R O, Paredes R de S, Endo T, Da Silva Bon E P and Lee S H 2013 Association of wet disk milling and ozonolysis as pretreatment for enzymatic saccharification of sugarcane bagasse and straw *Bioresour. Technol.* doi: 10.1016/j.biortech.2013.03.009
[14] Motaung T E and Anandjiwala R D 2015 Effect of alkali and acid treatment on thermal degradation kinetics of sugar cane bagasse *Ind. Crops Prod.* doi: 10.1016/j.indcrop.2015.05.062
[15] Xiong W 2018 Bagasse composites: A review of material preparation, attributes, and affecting factors *Journal of Thermoplastic Composite Materials*. doi: 10.1177/0892705717734596
[16] Ilham M M and Istiqlaliyah H 2019 Pemanfaatan Serat Rami (Boehmeria Nivea) Sebagai Bahan Komposit Bermatrik Polimer *J. Mesin Nusant.*
[17] Jose S, Rajna S and Ghosh P 2016 Ramie Fibre Processing and Value Addition *Asian J. Text.* doi: 10.3923/ajt.2017.1.9
[18] Wang B, Sain M and Oksman K 2007 Study of structural morphology of hemp fiber from the micro to the nanoscale *Appl. Compos. Mater.* doi: 10.1007/s10443-006-9032-9
[19] Holtzapple M T 2003 HEMICELLULOSES *Encyclopedia of Food Sciences and Nutrition*
[20] Akin D E 2008 Plant cell wall aromatics: Influence on degradation of biomass *Biofuels, Bioproducts and Biorefining* doi: 10.1002/bbb.76
[21] Pradana M A, Ardhyananta H and Farid M 2017 Pemisahan Selulosa dari Lignin Serat Tandan Kosong Kelapa Sawit dengan Proses Alkalisasi untuk Penguat Bahan Komposit Penyerap Suara *J. Tek. ITS* doi: 10.12962/j23373539.v6i2.24559
[22] Owen M 2014 THE EFFECTS OF ALKALI TREATMENT ON THE MECHANICAL PROPERTIES OF JUTE FABRIC REINFORCED EPOXY COMPOSITES **4**
[23] Li X, Tabil L G and Panigrahi S 2007 Chemical treatments of natural fiber for use in natural fiber-reinforced composites: A review *Journal of Polymers and the Environment* doi: 10.1007/s10924-006-0042-3
[24] Fuqua M A, Huo S and Ulven C A 2012 Natural fiber reinforced composites *Polymer Reviews* doi: 10.1080/15583724.2012.705409
[25] Wang X, Chang L, Shi X and Wang L 2019 Effect of hot-alkali treatment on the structure composition of jute fabrics and mechanical properties of laminated composites *Materials (Basel).* doi: 10.3390/ma12091386
[26] Akil H M, Omar M F, Mazuki A A M, Safiee S, Ishak Z A M and Abu Bakar A 2011 Kenaf fiber reinforced composites: A review *Materials and Design* doi: 10.1016/j.matdes.2011.04.008
[27] Hermanson G T 2008 *Bioconjugate Techniques*
[28] Arrakhiz F Z *et al.* 2012 Mechanical properties of high density polyethylene reinforced with chemically modified coir fibers: Impact of chemical treatments *Mater. Des.* doi: 10.1016/j.matdes.2012.01.020

[29] Dilfi A K F, Balan A, Bin H, Xian G and Thomas S 2018 Effect of surface modification of jute fiber on the mechanical properties and durability of jute fiber-reinforced epoxy composites *Polym. Compos.* doi: 10.1002/pc.24817

[30] Ouarhim W, Zari N, Bouhfid R and Qaiss A E K 2018 Mechanical performance of natural fibers-based thermosetting composites *Mechanical and Physical Testing of Biocomposites, Fibre-Reinforced Composites and Hybrid Composites*

[31] Lu J, Askeland P and Drzal L T 2008 Surface modification of microfibrillated cellulose for epoxy composite applications *Polymer (Guildf).* doi: 10.1016/j.polymer.2008.01.028

[32] Bal S and Samal S S 2007 Carbon nanotube reinforced polymer composites - A state of the art *Bull. Mater. Sci.* doi: 10.1007/s12034-007-0061-2

[33] Jeon I Y, Chang D, Nanjundan A K and Baek J B 2011 Functionalization of Carbon Nanotubes

[34] Sears K *et al.* 2010 Recent developments in carbon nanotube membranes for water purification and gas separation *Materials (Basel).* doi: 10.3390/ma3010127

[35] Avilés F, Cauich-Rodríguez J V, Rodríguez-González J A and May-Pat A 2011 Oxidation and silanization of MWCNTs for MWCNT/vinyl ester composites *Express Polym. Lett.* doi: 10.3144/expresspolymlett.2011.75

[36] Avilés F, Cauich-Rodríguez J V, Moo-Tah L, May-Pat A and Vargas-Coronado R 2009 Evaluation of mild acid oxidation treatments for MWCNT functionalization *Carbon N. Y.* doi: 10.1016/j.carbon.2009.06.044

[37] Bai Y *et al.* 2018 Carbon nanotube bundles with tensile strength over 80 GPa *Nat. Nanotechnol.* doi: 10.1038/s41565-018-0141-z

[38] Kim M T, Rhee K Y, Park S J and Hui D 2012 Effects of silane-modified carbon nanotubes on flexural and fracture behaviors of carbon nanotube-modified epoxy/basalt composites *Compos. Part B Eng.* doi: 10.1016/j.compositesb.2011.12.007

[39] Lee J H, Rhee K Y and Park S J 2011 Silane modification of carbon nanotubes and its effects on the material properties of carbon/CNT/epoxy three-phase composites *Compos. Part A Appl. Sci. Manuf.* doi: 10.1016/j.compositesa.2011.01.004

[40] Handaya Saputra A *et al.* 2014 Mechanical Reinforcements Of Composites Made From Fiber Of Fruit Bunch Palm Oil By Adding Carbon Nanotube *Artic. Int. J. Sci. Technol. Res.*

[41] Hong H, Xiao R, Guo Q, Liu H and Zhang H 2019 Quantitively characterizing the chemical composition of tailored bagasse fiber and its effect on the thermal and mechanical properties of polylactic acid-based composites *Polymers (Basel).* doi: 10.3390/polym11101567

[42] Wladyka-Przybylak M *et al.* 2015 Synergistic effect of modified natural fibres with halogen-free fire retardants in reducing flammability of composites *J. Biobased Mater. Bioenergy* doi: 10.1166/jbmb.2015.1506

[43] Oushabi A, Oudrhiri Hassani F, Abboud Y, Sair S, Tanane O and El Bouari A 2018 Improvement of the interface bonding between date palm fibers and polymeric matrices using alkali-silane treatments *Int. J. Ind. Chem.* doi: 10.1007/s40090-018-0162-3

[44] Velasco-Santos C, Martinez-Hernandez A L and Castano V M 2011 Silanization of Carbon Nanotubes: Surface Modification and Polymer Nanocomposites *Carbon Nanotubes - Polymer Nanocomposites*

[45] Randjbaran E, Zahari R, Majid D and Mazlan N 2017 Reasons of adding carbon nanotubes into composite systems - review paper *Mech. Mech. Eng.*

IOP Conf. Series: Materials Science and Engineering 1041 (2021) 012051 doi:10.1088/1757-899X/1041/1/012051

Improvement in the accuracy of the dynamic behaviour prediction of a bolted structure using a simplified finite element model and model updating

R Omar[1,2], M N A Rani[1*], M A Yunus[1], M Rusli[3], M A S A Shah[1] and W I Izhan W I Mirza[1]

[1] Structural Dynamics Analysis & Validation, Faculty of Mechanical Engineering, Universiti Teknologi MARA (UiTM), 40000 Shah Alam, Selangor, Malaysia
[2] Machinery Technology Centre, SIRIM Industrial Research, 44200 Hulu Selangor, Selangor, Malaysia
[3] Mechanical Engineering Department, Faculty of Engineering, Universitas Andalas, Kampus Limau Manis, Padang 25163, Indonesia

Corresponding author: mnarani@uitm.edu.my

Abstract. Bolted joints in assembled structures contribute to the dynamic behaviour of the structures but also increase uncertainties in modelling and prediction of the dynamic behaviour. This paper presents a modelling scheme able to reduce uncertainties and increase the accuracy of the prediction. A finite element (FE) model of a bolted joint structure was developed using CBUSH, CBEAM and 3D elements representing the bolts, contacting interfaces, and structural components, respectively. Normal modes analysis was performed on the FE model to calculate the natural frequencies and mode shapes. Sensitivity analysis was formulated and used in identifying the highly sensitive design parameters of the FE model. Model updating was adopted in correlating the FE model with experimental modal analysis (EMA) and producing an updated FE model. It was found that the updated FE model is capable of predicting the dynamic behaviour of the bolted joint structure with 95 per cent accuracy. This proposed scheme demonstrated potential for applicability to the dynamic behaviour analysis of complex bolted structures.

1. Introduction

Bolted joints remain as one of the attractive mechanical fasteners in engineering applications due togreat load-carrying capability, economical pricing, handling and maintenance flexibility. A bolted joint is typically built-up from related components (head, stud and nut), all of which are made up for assembling two or more structural components to form an assembled structure. Their presence in automotive and aerospace body structures have a very remarkable effect on the dynamic behaviour of the structures [1,2]. Failure to accurately evaluate the effect could result indirectly impact the success of final products [3,4].

With regard to the modelling of bolted joints, various finite element based modelling schemes ranging from simplified to detailed have been proposed and investigated [5]. However, a simple and practical model, which is based on 1D elements, has been a preferable model among researchers[6–8].

IOP Conf. Series: Materials Science and Engineering 1041 (2021) 012051 doi:10.1088/1757-899X/1041/1/012051

The scenario is generally driven by the fact that a detailed model, which is developed using 3D elements, tends to be computationally expensive for analysing the dynamic behaviour of large structure[5,9].

Modelling of bolted joints using FE 1D elements such as CBAR [10,11], CBEAM [6,12,13], CELAS [10,11] and CBUSH [14–16] has been extensively utilised to reduce computational time without dramatically undermining the accuracy of prediction. Recently, the authors proposed CBEAM and CBUSH elements based FE representation of bolted joints for the prediction of the dynamic behaviour of a bolted joint structure [17]. However, the proposed schemes have been found to have a few notable limitations as a result of the difficulty in creating reliable FE models of the bolt joints itself. Therefore, a deeper understanding of other potential parameters affecting the modelling accuracy of bolted joints and the dynamic behaviour of bolted structures is crucial and demands more efficient and accurate methods.

The objective of this study is to propose a modelling scheme of bolted joints that predicts the dynamic behaviour of a bolted structure better than the models proposed in the literature. The accuracy of the proposed scheme is evaluated by comparing the predicted results with experimental modal analysis (EMA) counterparts. FE model updating is used to update the initial FE results in view of EMA data to enhance the accuracy and reliability of the FE model of the assembled structure with bolted joints.

2. Materials and methods

Figure 1shows the proposed FE modelling scheme for the bolted structure that is identical to the one used in the previous study [17,18]. The structure consisted of two identical plates, namely Plate A and Plate B, respectively, and the plates were connected by two M10 stainless steel bolts and nuts. Both plates were made of Aluminium 6061 and had the dimensions of 380 mm in length, 45 mm in width and 6 mm thick. Two embossed areas with a diameter of 18 mm and a thickness of 2 mm were introduced to each plate, to minimise the contact interfaces between the two plates [18].

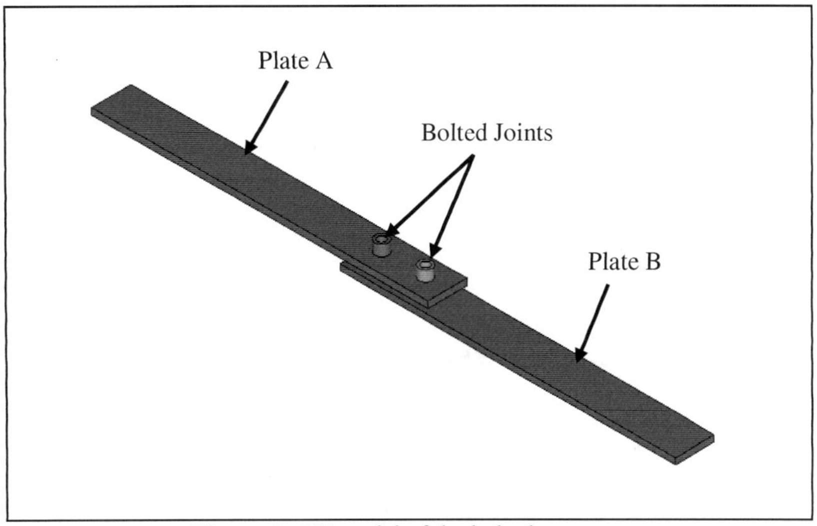

Figure 1. FE model of the bolted structure

HEX-6 3D elements were used in modelling of Plates A and B. The plates were discretised into 65660 nodes and 52560 elements. The number of nodes and elements were selected based on the mesh independent study [19,20]. The nominal values of the material properties used in the modelling are as tabulated in Table 1.

Table 1. Material properties for Plates A and B

Property	Nominal Value	Unit
Young's Modulus	69000	MPa
Poisson's Ratio	0.29	Unitless
Density	2700	kg/m^3

The proposed bolted joint modelling used in this study was developed based on CBEAM, RBE2 and CBUSH elements. CBEAM elements were used to represent the bolt shanks and RBE2 elements to represent the bolt heads and nut, respectively. Meanwhile, the contact interfaces between the embossed areas were modelled using CBUSH spring elements. Table 2 shows the material and geometrical properties of the 1D bolt model.

Table 2. Material and geometrical properties of 1D bolt model

Property	Parameter	Nominal value	Unit
Material of Bolts	Young's Modulus, E	190000	MPa
(PBEAM)	Poisson's Ratio, v	0.33	Unitless
Contact Interface 1	Axial Stiffness - Z, K_1	1676	N/mm
(PBUSH)	Shear Stiffness - X, K_2	4334	N/mm
	Shear Stiffness - Y, K_3	4334	N/mm
Contact Interface 2	Axial Stiffness - Z, K_1	1676	N/mm
(PBUSH)	Shear Stiffness - X, K_2	4334	N/mm
	Shear Stiffness - Y, K_3	4334	N/mm

Normal modes analysis was performed to calculate the natural frequencies and mode shapes of the FE model of the bolted structure. The calculated natural frequencies and mode shapes were compared and validated with the measured counterparts prior to performing model updating. The measured natural frequencies and mode shapes of the bolted structure used in this study are obtained from previously published papers [18,21]. The equipment packages used in the measurement were the LMS SCADAS system, LMS Test.Lab software, impact hammer & accelerometers [22]. The structure was suspended from the test rig by using rubber bands to simulate free-free boundary conditions [23,24]. The illustration of the experimental set-up is shown in Figure 2.

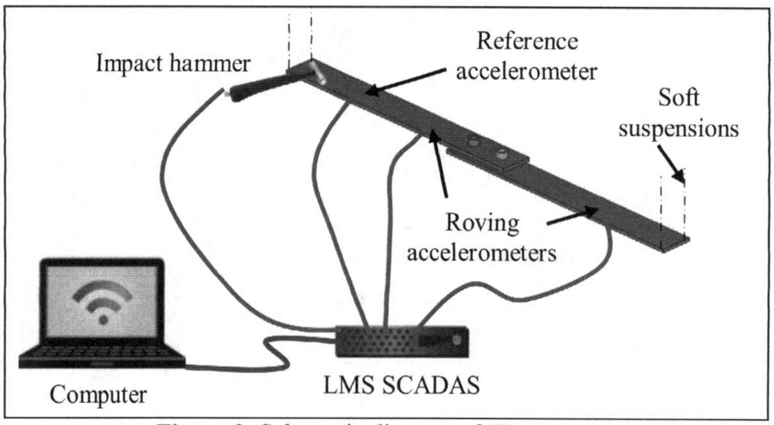

Figure 2. Schematic diagram of EMA set-up

Attempts to represent bolted joints using simplified FE modelling approaches have been reported in the literatures[5,13,19]. None of the attempts had produced a reasonable predicted result. However, one common suggestion given to improve the accuracy of the predicted results is using model updating and sensitivity analysis. In addition, sensitivity analysis should be performed based on engineering judgement and critical observation, which are the essential element of selecting the potential updating parameters [9].

In this study, the FE model updating was carried out by using MSC Nastran solver with SOL 200 script. The updating was aimed to improve the correlation results between the experimental and predicted data of the bolted structure [25]. In this study, the discrepancies between the experimental and predicted eigenvalues were used as the objective function of the updating that may be mathematically expressed as [26,27]:

$$min \sum_{j=1}^{m} \left(\frac{\omega_i^{FE}}{\omega_i^{EXP}} - 1 \right)^2 \tag{1}$$

where ω_i^{FE} is the ith predicted natural frequency and ω_i^{EXP} represents the ith experimental natural frequency.

3. Results and discussion

Developing accurate and efficient modelling of bolted joints for investigating the dynamic behaviour of bolted structures has been found to be very challenging. Predicted dynamic behaviour obtained in previous studies [5,28] using the finite element method revealed that 3D modelling had a reasonable predictive accuracy capability, but was very computationally expensive for large complex structures, which are prevalent in the automotive and aerospace industry [9]. Therefore, 1D modelling is an approach that many engineers would describe as being an attractive alternative to 3D modelling in terms of efficiency and accuracy. As reported by Tanlak et al. [6], 1D modelling offered greater efficiency for and the possibility of obtaining improved accuracy for predicting the dynamic behaviour of the bolted structure.

In this study, the dynamic behaviour of the bolted structure, which is natural frequencies and mode shapes, was obtained using experimental modal analysis (EMA) and the proposed 1D modelling scheme. The proposed modelling scheme was adopted in developing the initial FE model from which the dynamic behaviour of the bolted structure was computed. Model updating was then employed to improve to the initial FE model. Table 3 shows the results of 10 modes obtained using EMA, initial FE and updated FE models.

The recorded total error of 18.6% showed that the initial FE model was incapable of reproducing the experimental dynamic behaviour. In addition, all predicted natural frequencies of the initial FE were lower than the experimental counterparts. The results of the comparison indicate that the input properties used in the initial FE model, particularly that of the bolt shanks and contact interfaces may be underestimated.

In the effort of reducing the total error, model updating was used to correlate the initial FE model of the bolted structure with the EMA model, as is the case for researchers [29,30] adopting model updating to match the initial FE models with the experimental counterparts. Sensitivity analysis was performed on the initial FE model to identify updating parameters that have high sensitivity to changes in the objective response, which is natural frequencies [9]. Three parameters, which are the material properties of bolt shank, contact interface 1 and contact interface 2 were identified and used as the updating parameters in correlating the initial FE model. Table 4 shows the nominal and updated values of the updating parameters.

Table 3. Results of a comparison between EMA, initial FE and updated FE natural frequencies of the bolted structure

Mode	I. EMA (Hz)	II. Initial FE (Hz)	III. Error between I and II (%)	IV. Updated FE (Hz)	V. Error between I and IV (%)
1	69.37	67.86	2.18	68.96	0.59
2	189.40	189.56	0.08	189.88	0.25
3	374.10	365.96	2.18	372.00	0.56
4	440.70	417.87	5.18	435.03	1.29
5	580.40	574.51	1.01	579.71	0.12
6	643.20	623.86	3.01	636.23	1.08
7	963.10	946.20	1.75	962.32	0.08
8	973.20	966.25	0.71	973.18	0.00
9	1168.00	1141.19	2.30	1164.34	0.31
10	1381.00	1378.34	0.19	1379.91	0.08
	Total Error		18.60		4.37

Table 4. Updated parameters for the bolted structure

Property	Parameter	Nominal value	Updated value	Unit
Material of Bolts (PBEAM)	Young's Modulus, E	190000	200393	MPa
	Poisson' Ratio, v	0.33	0.30	Unitless
Contact Interface 1 (PBUSH)	Axial Stiffness - Z, K_1	1676	14966.68	N/mm
	Shear Stiffness - X, K_2	4334	37662.46	N/mm
	Shear Stiffness - Y, K_3	4334	37922.50	N/mm
Contact Interface 2 (PBUSH)	Axial Stiffness - Z, K_1	1676	14899.64	N/mm
	Shear Stiffness - X, K_2	4334	37185.72	N/mm
	Shear Stiffness - Y, K_3	4334	36665.64	N/mm

A remarkable improvement was recorded in the results of the comparison between EMA and updated FE model, in which the total error had been successfully minimised from 18.6% to 4.37%. The achievement showed that the proposed modelling scheme coupled with model updating was capable of accurately and efficiently predicting the dynamic behaviour of the bolted structure. Table 5 and Table 6 show the comparison of mode shapes obtained from the EMA model and updated FE model of the bolted structure. The calculated MAC values indicate that all predicted mode shapes are in a good correlation with EMA counterparts.

IOP Conf. Series: Materials Science and Engineering 1041 (2021) 012051 doi:10.1088/1757-899X/1041/1/012051

Table 5. Results of a comparison between EMA and updated FE of 1st to 5th mode shapes of the bolted structure

Mode	EMA	Updated FE	MAC
1	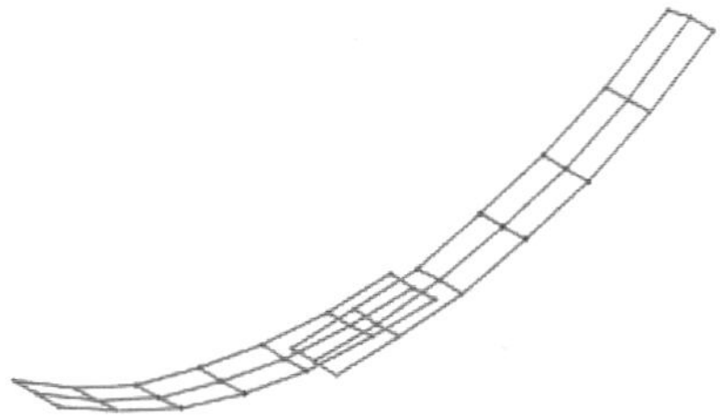		98.1
2	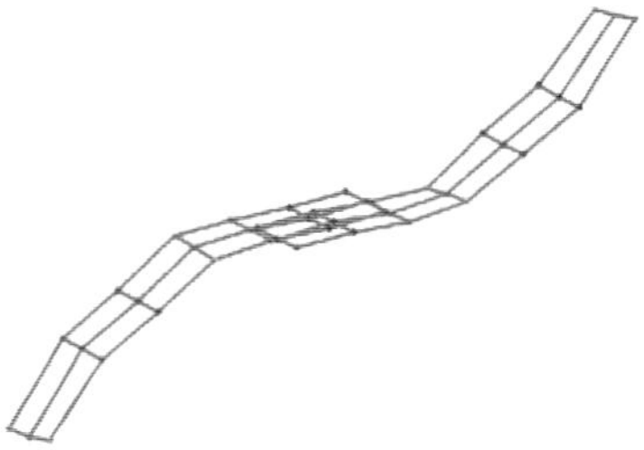		99.6
3	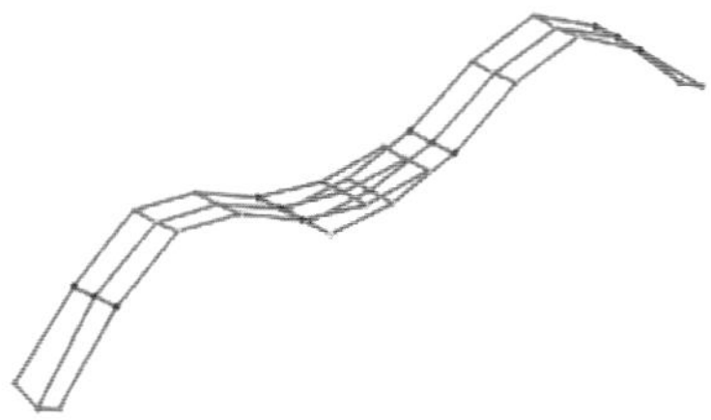		98.1
4	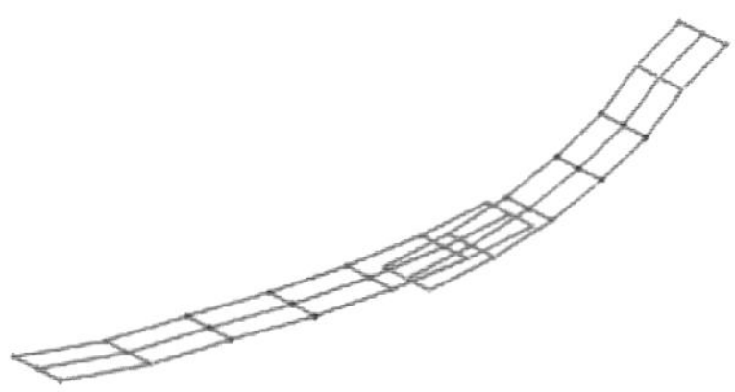	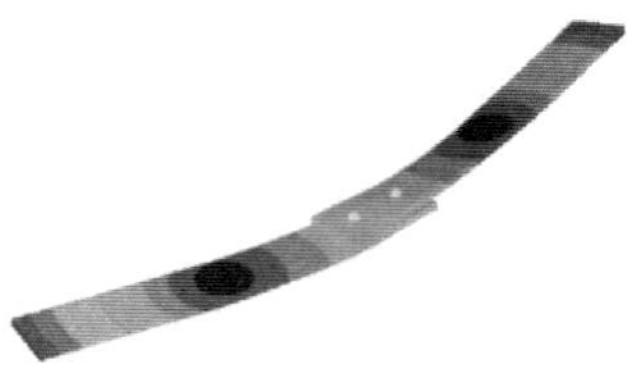	96.2
5	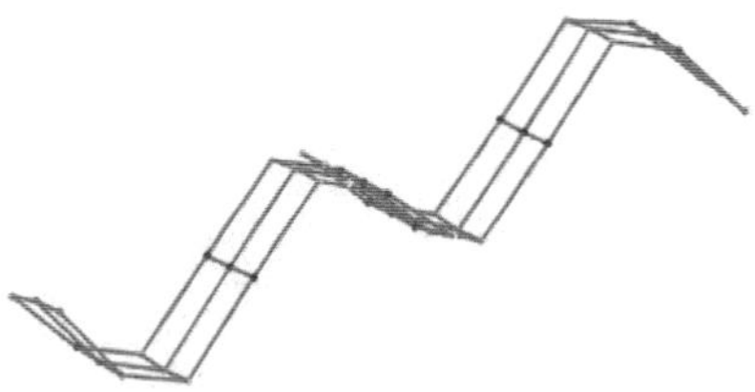		95.5

IOP Conf. Series: Materials Science and Engineering 1041 (2021) 012051 doi:10.1088/1757-899X/1041/1/012051

Table 6. Results of a comparison between EMA and updated FE of 6th to 10th mode shapes of the bolted structure

Mode	EMA	FE Analysis	MAC
6	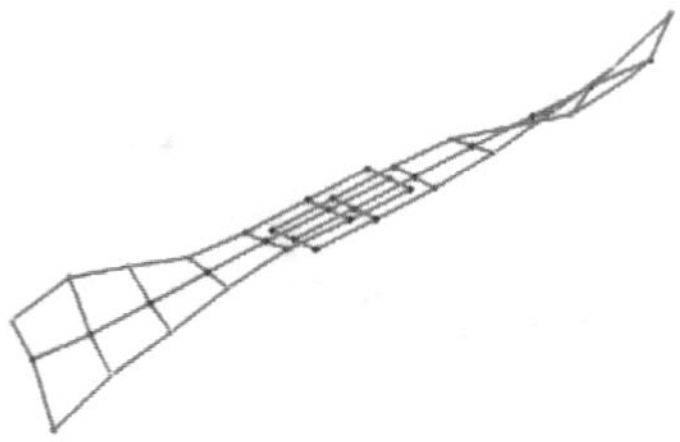		95.3
7	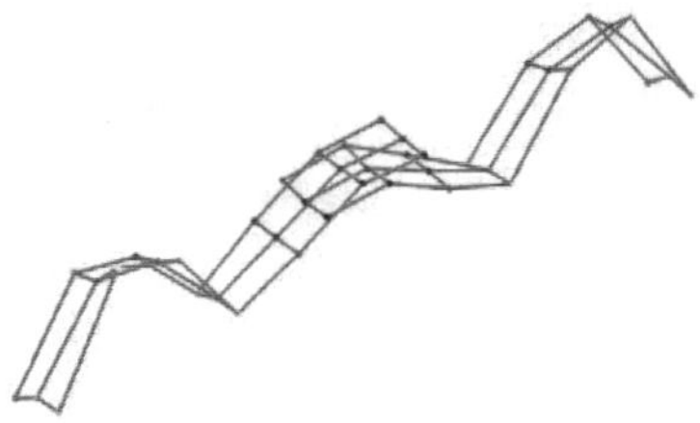		97.9
8	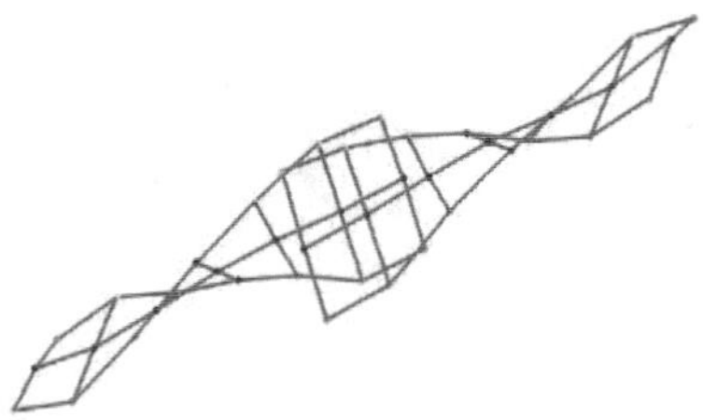		97.0
9	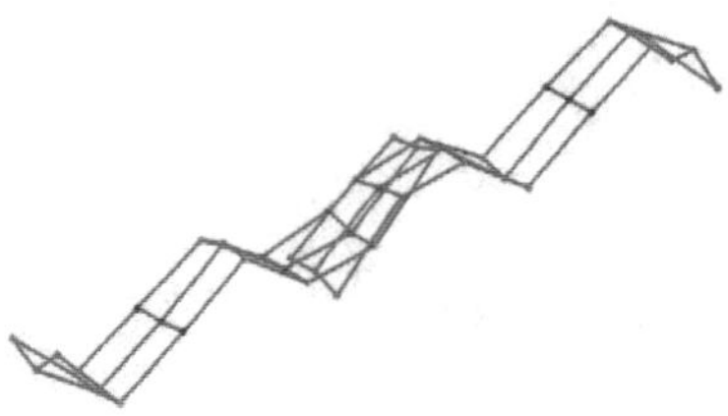		95.7
10	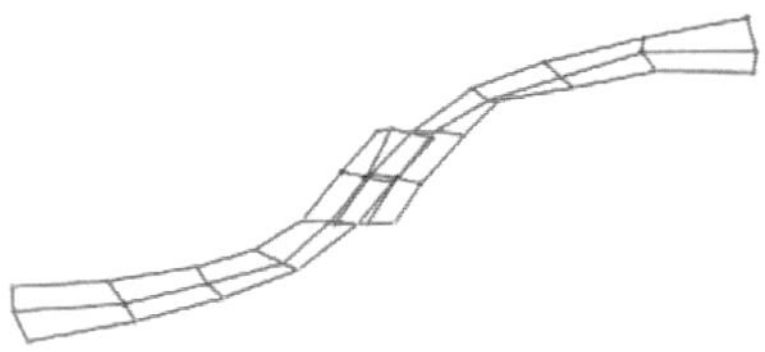		98.6

4. Conclusions

In this study, a simplified modelling scheme for bolted joints to predict the dynamic behaviour of a bolted structure in finite element analysis is presented. The proposed scheme reduces the number of DOFs used in modelling of the bolted structure and increases the accuracy of the prediction.

Percentage accuracy level achieved using the proposed scheme was about 95 per cent of the predicted natural frequencies and was above 96 per cent on average for the predicted mode shapes. The most significant advantage of the proposed scheme over other existing modelling techniques lies in the use of different types of 1D elements to represent the complexity and contact relations of bolted joints.

Acknowledgements

The authors would like to acknowledge the Ministry of Education Malaysia and Universiti Teknologi MARA for providing financial support for this study through the Fundamental Research Grant Scheme (FRGS600-IRMI/FRGS 5/3 (335/2019)). They would like to express their gratitude for technical support given by the members of Structural Dynamics Analysis & Validation (SDAV) and Machinery Technology Centre (MTC).

References

[1] Ahmadian H, Mottershead J E, James S, Friswell M I and Reece C A 2006 Modelling and updating of large surface-to-surface joints in the AWE-MACE structure *Mech. Syst. Signal Process.* **20** 868–80
[2] Mirza W I I W I, Rani M N A, Yunus M A, Ayub M A, Sani M S M and Mohd Zin M S 2019 Frequency based substructuring for structure with double bolted joints: A case study *Int. J. Automot. Mech. Eng.* **16** 6188–99
[3] Cara F J, Juan J and Alarcón E 2014 Estimating the modal parameters from multiple measurement setups using a joint state space model *Mech. Syst. Signal Process.* **43** 171–91
[4] Daouk S, Louf F, Cluzel C, Dorival O, Champaney L and Audebert S 2017 Study of the dynamic behavior of a bolted joint under heavy loadings *J. Sound Vib.* **392** 307–24
[5] Omar R, Abdul Rani M N, Yunus M A, Wan Iskandar Mirza W I I, Mohd Zin M S, Rani M N A, Yunus M A, Mirza W I I W I and Zin M S M 2018 Efficient finite element modelling for the investigation of the dynamic behaviour of a structure with bolted joints *AIP Conf. Proc.* **1952** 20082
[6] Tanlak N, Sonmez F O and Talay E 2011 Detailed and simplified models of bolted joints under impact loading *J. Strain Anal. Eng. Des.* **46** 213–25
[7] Chandregowda S and Ganga Reddy C R 2018 Evaluation of fastener stiffness modelling methods for aircraft structural joints *AIP Conf. Proc.* **1943** 020001
[8] Liu Y and Glass G 2013 Effects of mesh density on finite element analysis *SAE Tech. Pap. 2013-01-13*
[9] Mottershead J E, Link M and Friswell M I 2011 The sensitivity method in finite element model updating: A tutorial *Mech. Syst. Signal Process.* **25** 2275–96
[10] Izham M H N and Sani M S M 2019 Model updating of frame structure with bolted joints *J. Phys. Conf. Ser.* **1262** 012023
[11] Rutman A, Boshers C, Pearce L and Parady J 2007 Fastener modeling for joining parts modeled by shell and solid elements *Americas Virtual Product Development Conference* (Detroit, MI) pp 1–26
[12] Grzejda R 2017 Modelling bolted joints using a simplified bolt model *J. Mech. Transp. Eng.* **69** 29–37
[13] Omar R, Rani M N A, Yunus M A, Isa A A M and Rahman Z A 2019 Updating of the dynamic behaviour of an assembled structure with bolted joints in light of EMA results *J. Phys. Conf. Ser.* **1262** 012016
[14] Uysal I, Suer M and Ozkol I 2011 Finite element modelling and analysis of simple lap joints

Proceedings of 5th International Conference on Recent Advances in Space Technologies - RAST2011 pp 68–72
[15] Omar R, Abdul Rani M N, Yunus M A, Wan Iskandar Mirza W I I and Mohd Zin M S 2019 Investigation into joint stiffness for finite element modelling of the dynamic behaviour of a structure with bolted joints *IOP Conference Series: Materials Science and Engineering* vol 506 p 012009
[16] Omar R, Abdul Rani M N and Yunus M A 2019 Finite element modelling and updating of a bolted structure using elements representing the stiffness members, bolts and affected areas of the joints *J. Phys. Conf. Ser.* **1262** 012005
[17] Omar R, Abdul Rani M N and Yunus M A 2020 Representation of bolted joints in a structure using finite element modelling and model updating *J. Mech. Eng. Sci.* **14** 7141–51
[18] Omar R, Abdul Rani M N, Wan Iskandar Mirza W I I, Yunus M A and Othman M H 2017 Finite element modelling and updating for bolted lap joints *J. Mech. Eng.* **SI 4** 202–22
[19] Omar R, Abdul Rani M N, Yunus M A, Mat Isa A A, Wan Iskandar Mirza W I I, Mohd Zin M S and Roslan L 2018 Investigation of mesh size effect on dynamic behaviour of an assembled structure with bolted joints using finite element method *Int. J. Automot. Mech. Eng.* **15** 5695–708
[20] Sazzad M M, Azad M S, Islam M T and Rahman F I 2017 Effect of mesh size of floor slab against lateral loads while using Etabs program *Int. J. Adv. Struct. Geotech. Eng.* **6** 40–4
[21] Mirza W I I W I, Rani M N A, Ayub M A, Yunus M A, Omar R and Mohd Zin M S 2018 Innovative FRF measurement technique for frequency based substructuring method *AIP Conference Proceedings* vol 1952 p 020010
[22] Aziz Shah M A S, Yunus M A, Saman A M, Abdul Rani M N, Bahari A R and Peter C 2019 Finite Element Model Updating of Laser Stitch Welded Structure *IOP Conf. Ser. Mater. Sci. Eng.* **506** 012010
[23] Lepoittevin G and Kress G 2011 Finite element model updating of vibrating structures under free-free boundary conditions for modal damping prediction *Mech. Syst. Signal Process.* **25** 2203–18
[24] Aziz Shah M A S, Yunus M A, Abdul Rani M N, Tormodi N A, Mat Isa A A and Wan Iskandar Mirza W I I 2018 Closed mode separation of the symmetrical structure *J. Mech. Eng.* **SI 5** 88–97
[25] Mottershead J E and Friswell M I 1993 Model updating in structural dynamics: A survey *J. Sound Vib.* **167** 347–75
[26] Abdul Rani M N, Kasolang S, Othman M H, Yunus M A, Wan Iskandar Mirza W I I and Ouyang H 2016 Finite element modelling and modal based updating for the dynamic behaviour of a laser spot welded structure *23rd International Congress on Sound and Vibration* (Athens, Greece: International Institute of Acoustics and Vibrations) pp 315–22
[27] Abdul Rani M N, Ouyang H, Yunus M A and Aminudin B A 2013 Model updating for a thin steel sheet welded structure *20th International Congress on Sound and Vibration 2013* (International Institute of Acoustics and Vibrations) pp 1–8
[28] Kim J, Yoon J-C and Kang B-S 2007 Finite element analysis and modeling of structure with bolted joints *Appl. Math. Model.* **31** 895–911
[29] Saeed S, Ahmad N and Farhad A 2014 Finite element model updating of interfaces in single-lap bolted joints using modal test data *21st International Congress on Sound and Vibration*
[30] Omar R, Abdul Rani M N, Yunus M A, Wan Iskandar Mirza W I I and Mohd Zin M S 2018 Finite element modelling and updating of a non-identical built-up structure with bolted joints *25th International Congress on Sound and Vibration (ICSV 25)* (Hiroshima: International Institute of Acoustics & Vibration) pp 4624–31

IOP Conf. Series: Materials Science and Engineering 1041 (2021) 012052 doi:10.1088/1757-899X/1041/1/012052

Assessment of Rotational Accelerometer Mountings for Rotational Frequency Response Function Measurement

W I I W I M[1], M N A Rani[1,3], M I M Rosli[1], M A Yunus[1,3], R Febrina[2] and M A S A Shah[1]

[1] Structural Dynamics Analysis & Validation (SDAV), Faculty of Mechanical Engineering, Universiti Teknologi MARA (UiTM), 40000 Shah Alam, Selangor, Malaysia

[2] Department of Civil Engineering, Malahayati University, Lampung, Indonesia

[3]Institute for Infrastructure Engineering and Sustainable Management (IIESM) Universiti Teknologi MARA (UiTM), 40450 Shah Alam, Selangor

Corresponding author: mnarani@uitm.edu.my

Abstract. Kistler Kshear 8840 piezoelectric rotational accelerometer is designed with an M5 bolt type of mounting and a 5mm diameter of through or pre-tapped hole is required in a structure under test. However, the mounting is not applicable for some cases, and other types of mounting need to be used. This work investigates the applicability and accuracy of five different configurations of mounting for Kistler Ksher 8840 piezoelectric rotational accelerometer for the measurement of the rotational FRFs of a steel plate. The mounting configurations consist of 3D printed PLA and ABS mounting, bee's wax, steel, and bolt mounting. The rotational FRF of the steel plate was measured by using impact testing method. The applicability and accuracy of the mountings were evaluated by comparing the measured FRFs with the finite element (FE) FRFs counterparts. The comparisons of the results revealed that using the bee's wax mounting had produced the most accurate and least noisy FRF. It was also found that using the 3D printed PLA mounting had contributed to the highest contamination of noise and several spurious peaks in the measured FRFs. These findings provide useful information for selecting the appropriate types and designs of mounting for the Kistler 8840 rotational accelerometer.

1. Introduction

Rotational FRFs have important contributions in certain structural dynamics applications such as structural modification, substructuring and model validation. Although the rotational FRFs represents 75% of all the existing FRF matrix, they are often omitted because they are difficult to measure [1]. Several rotational FRF measurement techniques have been proposed and used by researchers [2,3]to estimate the rotational accelerations using two spatially separated and sensitivity-matched accelerometers. For example, [4,5] showed that the rotational FRFs were indirectly obtained using T-block mass attachment The system was then decoupled to obtain the rotational FRF. This approach is also known as transmission simulator [6]. Another s similar striking method employed in estimating rotational FRFs is the use of an X-block attachment with several translational accelerometers demonstrated in [7]. The X-block attachment was used for the investigation of a structural modification of a helicopter tail. Although there are several techniques to directly measure the rotational data, the proposed techniques usually require special equipment such as laser vibrometer rotational accelerometers, which may not be readily available in most laboratories [8].

Recently, the use of piezoelectric direct rotational accelerometer has gained much attention among the structural dynamicists. A Kistler Kshear 8840 piezoelectric rotational accelerometer, as shown in figure 1, provides researchers with a great opportunity to perform the measurement of the rotational FRFs of a test structure [9]. The sensor is based on a very stable quartz crystal and can be powered by any commercially available 20–30 VDC power supply. The technical specifications of the rotational accelerometer is presented in table 1 and the details can be found in [9,10].

Figure 1. Kistler Kshear 8840 rotational accelerometer.

Table 1. Technical specification for Kistler Kshear 8840.

Specification	Value	Unit
Acceleration Range	±150	$krad/s^2$
Maximum Limit	±200	$krad/s^2$
Sensitivity	35	µV/rad /s
Frequency response, ±10%	0 - 2000	Hz
Source Voltage	20-30	V
Source Current	4	mA

The comparison of Kistler 8840 rotational accelerometer with the indirect T-block technique was shown in [10]. It has been reported that the piezoelectric rotational accelerometer is far more accurate and reliable but with slightly higher levels of noise in signal compared to the indirect method. The resulting higher level of noise is observed and found to be linked to the types of mounting configurations. Therefore, the installation and mounting configurations of rotational accelerometers on a structure under test are crucial in ensuring the accuracy of the measurement. This is because mounting configurations may differ, with some transferring information more efficiently than others. Previous studies [11–13] have showed that stud mounting is perceived to be the best for translational accelerometers, but the conclusion cannot be directly applied for Kistler 8840 rotational accelerometers. This is because the accelerometer comes with an M5 mounting bolt and definitely a 5mm diameter of through or pre-tapped hole is required in a structure under test for the accelerometer attachment. However, in some cases, especially when drilling the structure is not an option, therefore, other types of mounting need to be used for the measurement of rotational FRFs.

IOP Conf. Series: Materials Science and Engineering 1041 (2021) 012052 doi:10.1088/1757-899X/1041/1/012052

This work investigates the applicability and accuracy of five different configurations of mounting for a Kistler 8840 piezoelectric rotational accelerometer used in the measurement of the rotational FRFs of a steel plate. The mounting configurations consist of 3D printed PLA and ABS mounting, bee's wax, steel, and screw mounting.

2. Materials and methods

Five types of mounting configurations were investigated in this work. The mounting configurations are presented in figure 2 and are as follow;

1. 3D printed PLA mounting
2. 3D printed ABS mounting
3. Bee's wax mounting
4. Steel mounting
5. Bolt mounting

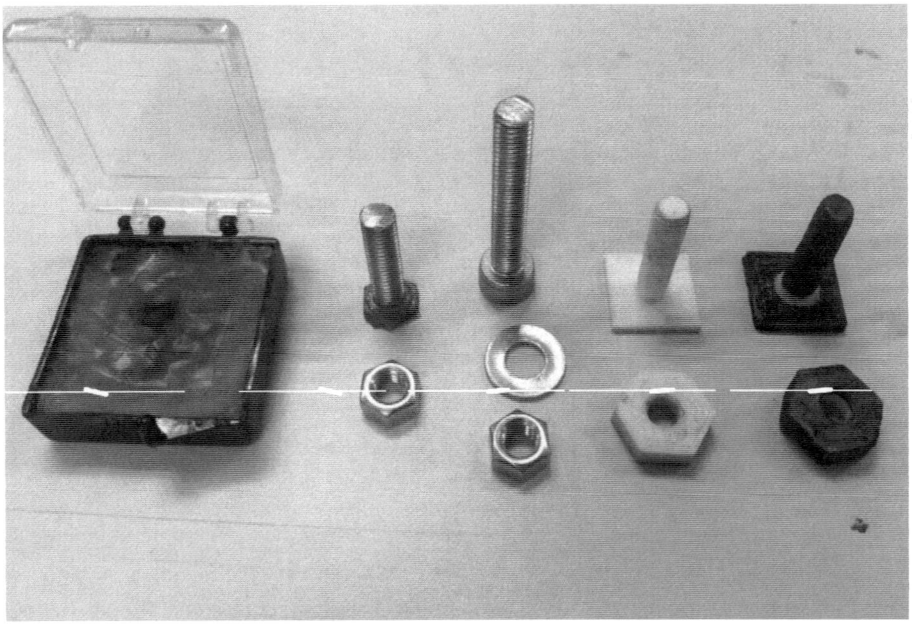

Figure 2. Rotational accelerometer mountings.

The PLA and ABS mounting were developed by using CAD and printed using a 3D printer. Both PLA and ABS mounting were 5mm in diameter and 15mm in length. Meanwhile, the steel mounting was an M5 steel bolt with the bolt head being removed. The PLA, ABS and steel mounting types were attached to the test structure using adhesive. The mounting configurations are presented in figure 3.

(a) (b) (c) (d) (e)

IOP Conf. Series: Materials Science and Engineering 1041 (2021) 012052 doi:10.1088/1757-899X/1041/1/012052

Figure 3. Mounted rotational accelerometer on using respective (a) 3D printed PLA (b) 3D printed ABS (c) steel s(d) bolt (e) bee's wax mounting.

2.1 EMA for the test structure

The applicability and accuracy of the mounting configurations for a rotational accelerometer were evaluated by observing the FRFs obtained from experimental modal analysis (EMA). In this work, the impact testing method was performed to obtain rotational FRFs. The test configuration is illustrated in figure 4. The experimental work was performed by hanging the steel plate with soft suspensions to mimic a free-free configuration [14,15]. The experimental configuration was designed as to the finite element model [16]. A Kistler 8840 rotational accelerometer was used to measure the dynamic data, and a 21.65mV/N modal hammer was used to excite the structure. Leuven Measurement System (LMS) data acquisition was used to acquire the data. The frequency bandwidth was set between 0 Hz to 2000 Hz with 1 Hz frequency resolution. In this work, only rotational x and y-axis were measured. The measured rotational FRFs from impact testing were then validated with the FE counterparts.

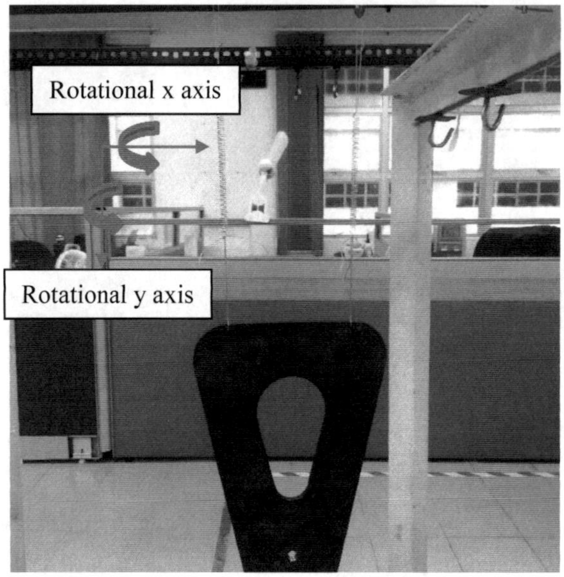

Figure 4. test configuration of the steel plate.

2.2 FE modelling and FRF synthesis method.

MSC Patran was used in the development of the FE model of the steel plate. The FE model consisted of 54660 elements and 70115 nodes, as shown in Figure . The material properties were defined as follows; density = 7850 kg/m3, Young's modulus = 70GPa and Poisson's ratio = 0.33. The dynamic behaviour of the FE model was then calculated by using normal modes solution.

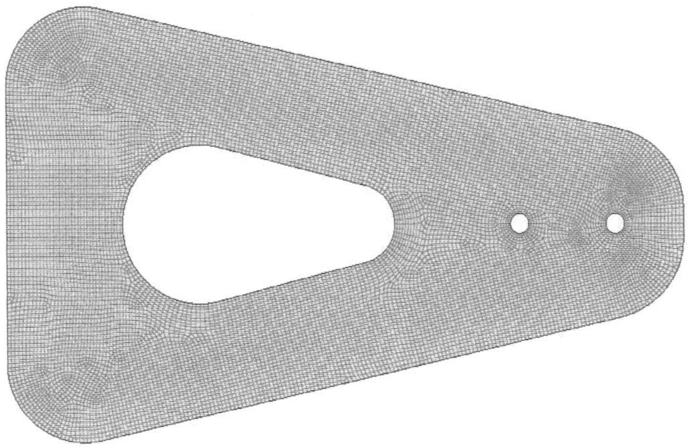

Figure 5. Finite element model of the steel plate.

The FE FRF was derived using FRF synthesis method. For this method, the synthesized FRF matrix $H_{syn}(\omega_k)$ and mode shapes are expressed by:

$$H_{syn}(\omega_k) = \sum_{i=1}^{N} \frac{\{\emptyset\}_i\{\emptyset\}_i^T}{(\omega_{n_i}^2 - \omega_k^2) + j2\xi_i\omega_k\omega_{n_i}} \quad (1)$$

where N is the number of calculated modes, $\{\emptyset\}_i$ is the *i*th mass normalised mode shapes, ω_{n_i} is *i*th natural frequency and ξ_i is the *i*th modal damping ratio. The FRF derived from the FE model was compared with the measured counterparts for validation purposes.From the derived FE FRF, four resonance peaks at the rotational x FRF and seven resonance peaks at the rotational y FRF were identified.

3. Results and discussion

Figures 6,7,8,9 and 10 present the rotational FRFs measured in the x and y axis using 5 types of mounting configurations. From the figures, all the resonance peaks in the rotational x-axis were successfully measured using all type of the mounting configurations. However, using 3D printed ABS mounting had led to the presence of two spurious peaks in the frequency range between 1000 to 1500 Hz as shown in figure 6. The presence of the spurious peaks may be because resonances occurred in the mounting. Meanwhile, all the resonance peaks in the rotational y-axis were successfully measured using the steel, bolt, and bee's wax mounting types, as shown in figures 8, 9 and 10. However, a resonance peak for the 6th mode within the frequency range between 1800-2000 Hz was not visible in the measured rotational y FRFs using 3D printed ABS and PLA mounting as shown in figures 6 and 7.

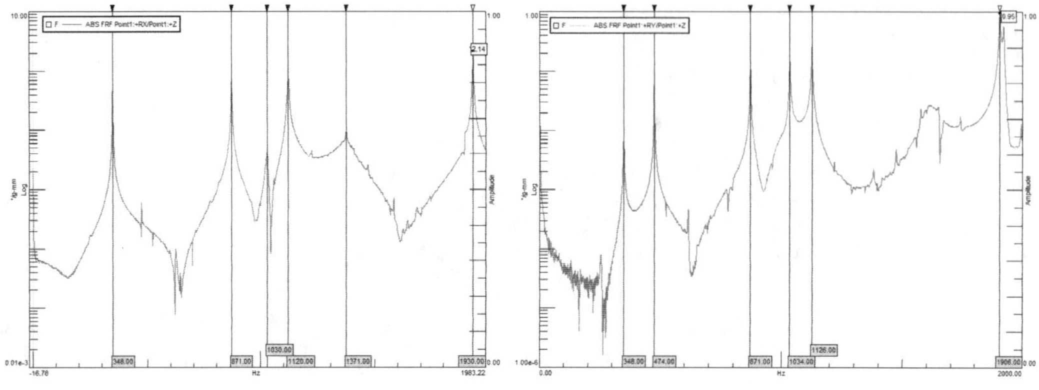

(a) (b)

Figure 6. Measured (a) rotational x (b) rotational y FRFs using the 3D printed ABS mounting configuration.

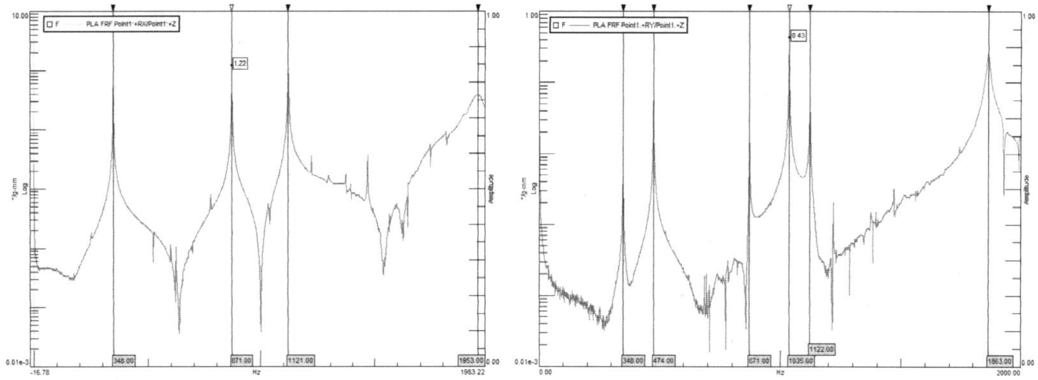

(a) (b)

Figure 7. Measured (a) rotational x (b) rotational y FRFs using the 3D printed PLA mounting configuration.

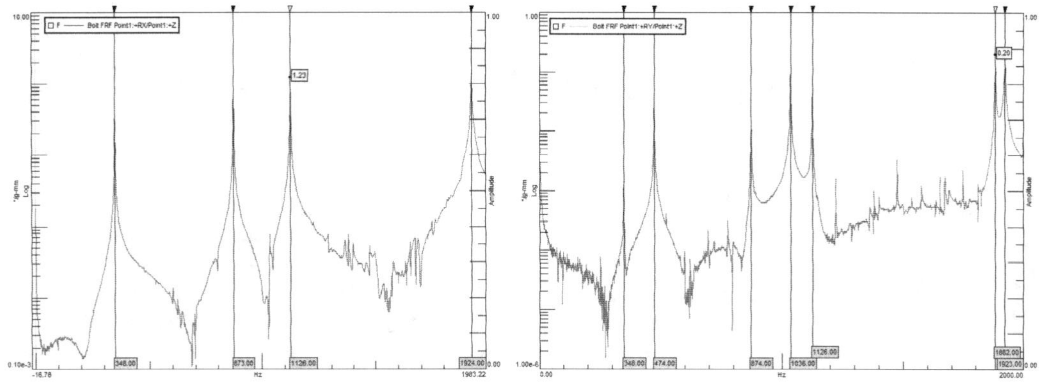

(a) (b)

Figure 8. Measured (a) rotational x (b) rotational y FRFs using the bolt mounting configuration.

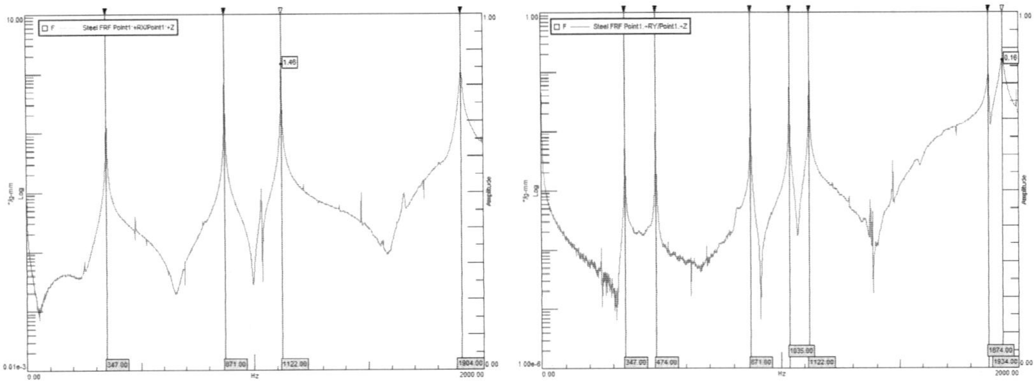

(a) (b)

Figure 9. Measured (a) rotational x (b) rotational y FRFs using the steel mounting configuration.

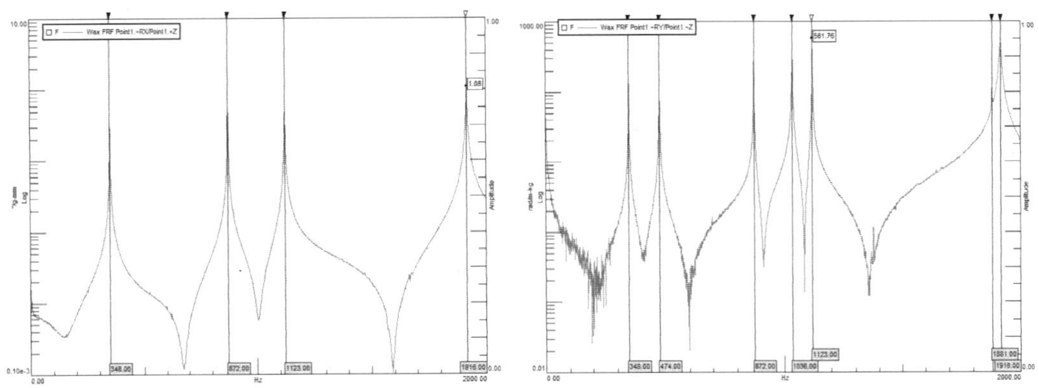

(a) (b)

Figure 10. Measured (a) rotational x (b) rotational y FRFs using the bee's wax mounting configuration.

All the resonance peaks for the rotational FRFs in the x and y-axes were successfully measured using the steel, bolt stud and bee's wax mounting types as presented in figures 8 to 10. However, it was shown that using bee's wax appeared to be the least noisy and produced the highest quality of FRF compared to the other mounting types. It was also found that the pattern of resonance and anti-resonance calculated was also almost similar to the FE counterparts, as shown in figures 11 and 12. Therefore, bee's wax is the most suitable mounting configuration for the rotational accelerometer and impact testing purposes.

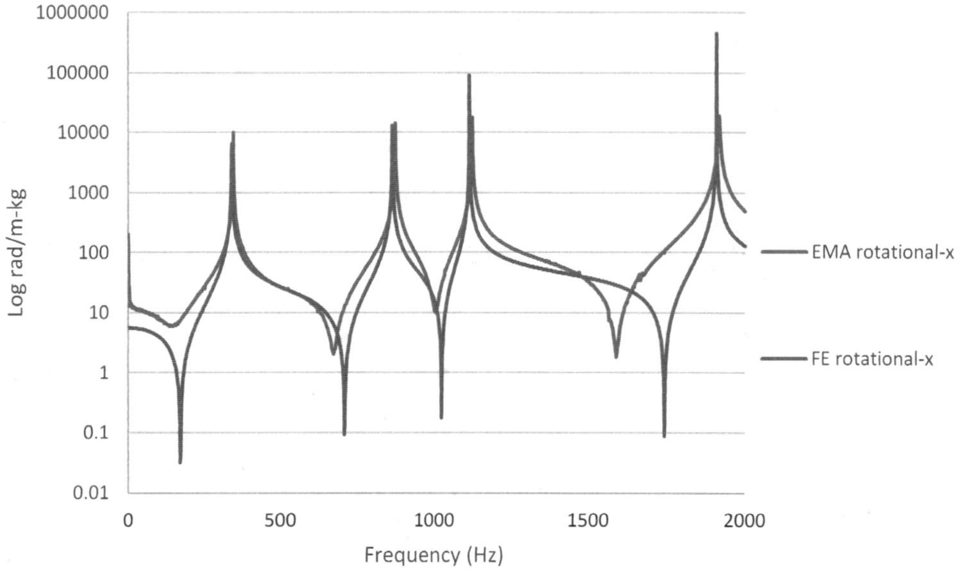

Figure 11. Comparisons of rotational x FE and measured FRF using bee's wax mounting configuration.

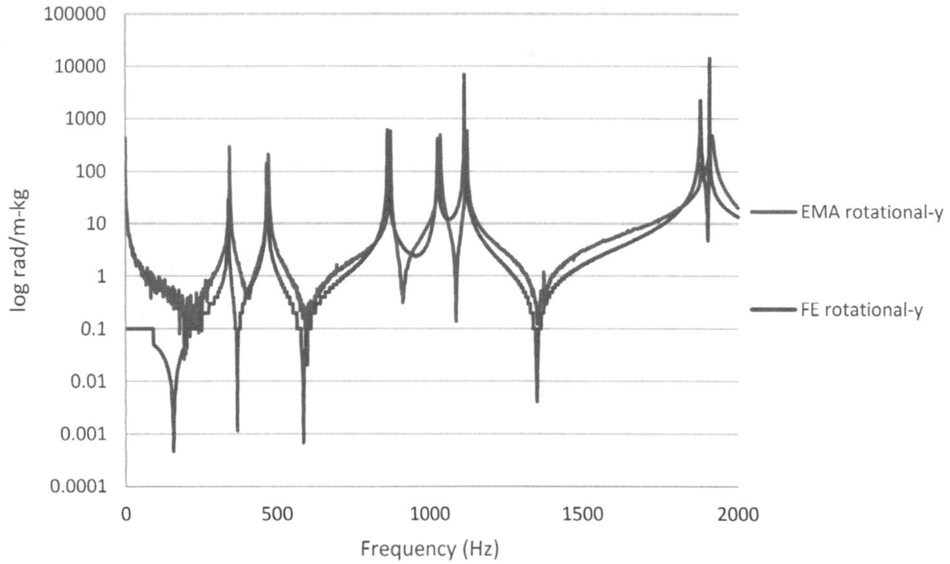

Figure 12. Comparisons of rotational y FE and measured FRF using bee's wax mounting configuration.

4. Conclusions

The applicability and accuracy of five different configurations of mounting for Kistler 8840 piezoelectric rotational accelerometer in the measurement of the rotational FRFs of a steel plate were investigated. The mounting configurations consisted of 3D printed PLA and ABS mounting, bee's wax, steel, and bolt mounting. The measured rotational FRFs using the mounting configurations was observed and compared with the FE FRF counterparts. The comparisons of the FRFs revealed that all the five mounting configurations were capable of measuring all the resonance peaks in the rotational x-axis. It was found that bee's wax as the mounting had produced the most accurate and least noisy FRF. On the other hand, using both 3D printed mounting had produced several spurious resonance peaks and poor quality of FRF. The outcome of this study suggests that using bee's wax as the mounting configuration for the rotational accelerometer is the most suitable mounting for impact testing purposes. These findings are thought to provide useful information for selecting the appropriate types mounting for the Kistler 8840 rotational accelerometer.

Acknowledgements
The authors are gratefully indebted to the Malaysian Ministry of Higher Education for providing financial support for this study through the Fundamental Research Grant Scheme (FRGS600-IRMI/FRGS 5/3 (335/2019).

References
[1] Wan Iskandar Mirza W I I, Abdul Rani M N, Yunus M A, Sani M S M and Mohd Zin M S 2019 The Prediction of the Dynamic Behaviour of a Structure Using Model Updating and Frequency Based Substructuring *IOP Conference Series: Materials Science and Engineering***506**
[2] Avitabile P and O'Callahan J 2003 Frequency response function expansion for unmeasured translation and rotation DOFS for impedance modelling applications *Mechanical Systems and Signal Processing***17** 723–45
[3] Montalv D 2004 Estimation of the rotational terms of the dynamic response matrix *Shock and Vibration***11** 333–50
[4] Tsai S H, Ouyang H and Chang J Y 2019 Identification of torsional receptances *Mechanical Systems and Signal Processing***126** 116–36
[5] Lv B, Ouyang H, Li W, Shuai Z and Wang G 2016 An indirect torsional vibration receptance measurement method for shaft structures *Journal of Sound and Vibration***372** 11–30
[6] Konjerla K C 2016 FRF Based Experimental – Analytical Dynamic Substructuring Using Transmission Simulator
[7] Mottershead J E, Ghandchi Tehrani M, Stancioiu D, James S and Shahverdi H 2006 Structural modification of a helicopter tailcone *Journal of Sound and Vibration***298** 366–84
[8] Gibbons T J, Öztürk E and Sims N D 2018 Rotational degree-of-freedom synthesis: An optimised finite difference method for non-exact data *Journal of Sound and Vibration***412** 207–21
[9] Dumont M and Kinsley N 2016 Rotational Accelerometers and Their Usage in Investigating Shaker Head Rotations *Sensor and Instrumentation***5** 85–92
[10] Drozg A, Rogelj J, Čepon G and Boltežar M 2018 On the performance of direct piezoelectric rotational accelerometers in experimental structural dynamics *Measurement: Journal of the International Measurement Confederation***127** 292–8
[11] Yanqiu H, Juan C and Peng Y L 2010 Experimental analysis of accelerometer installation error *2010 International Conference on Computer, Mechatronics, Control and Electronic Engineering (CMCE)* 40–2
[12] Dumont M, Cook A and Kinsley N 2016 Acceleration measurement optimization: Mounting considerations and sensor mass effect *Conference Proceedings of the Society for Experimental Mechanics Series***10** 61–71
[13] Shokrollahi S, Adel F and Ahmadian H 2017 An investigation into the accelerometer mounting effects on signal transmissibility in modal measurements *Scientia Iranica***24** 2436–44
[14] Wan Iskandar Mirza W I ., Abdul Rani M N, Yunus M A, Ayub M A, Sani M S M and Mohd Zin M S 2019 Frequency Based Substructuringfor Structure with Double Bolted Joints: A

Case Study *International Journal of Automotive and Mechanical Engineering* **16** 6188–99

[15] Wan Iskandar Mirza W I ., Abdul Rani M N, Yunus M A, Omar R and Mohd Zin M S 1394 Alternative Scheme for Frequency Response Function Measurement of Experimental-Analytical Dynamic Substructuring *Journal of Mechanical Engineering and Science* **13** 4946–57

[16] Omar R, Abdul Rani M N, Yunus M A, Mat Isa A A, Wan Iskandar Mirza W I ., Mohd Zin M S and Roslan L 2018 Investigation of Mesh Size Effect on Dynamic Behaviour of an Assembled Structure with Bolted Jointsusing Finite Element Method *International Journal of Automotive and Mechanical Engineering* **15** 5695–708

IOP Conf. Series: Materials Science and Engineering 1041 (2021) 012053 doi:10.1088/1757-899X/1041/1/012053

Effect of changes in the regeneration temperature and the regeneration airflow velocity on the system performance of the solid desiccant air conditioning

D A Saputra[1,2], Y Osaka[3], T Tsujiguchi[3], A Kodama[3]

[1]Graduate School of Natural Science and Technology, Kanazawa University, Kakumamachi, Kanazawa, Ishikawa 920-1192, Japan
[2]Department of Mechanical Engineering, Faculty of Engineering, Andalas University, Padang, West Sumatera 25163, Indonesia
[3]Faculty of Mechanical Engineering, Institute of Science and Engineering, Kanazawa University, Kakumamachi, Kanazawa, Ishikawa 920-1192, Japan

Corresponding author: akodama@se.kanazawa-u.ac.jp

Abstract. This paper presents the effect of changes in operating parameters on the solid desiccant air conditioning system performance. It has consisted of the regeneration temperature and the regeneration airflow velocity parameters. These parameters were examined in two operation modes, the former ranging from 2 m/s–0.83 m/s at constant regeneration temperature, and the latter ranging from 75°C–46.4°C at constant regeneration airflow velocity. The results indicate a smaller decline in dehumidification performance for the constant regeneration temperature operating mode than the constant airflow velocity operating mode. On the contrary, the system performance and the recovery energy ratio at the latter operating mode gives a higher value than the former operating mode. These results suggest that when the dehumidification performance becomes the main function in the system, the regeneration airflow velocity should not be too low to avoid a significant reduction in system performance.

1. Introduction

Since the 1930s, solid desiccant technology has been used in humidity control for industrial applications in the United States [1]. Unfortunately, the use of this technology for heating, ventilation, and air conditioning (HVAC) applications is not feasible because performance and the additional cost do not scale. In the early 1960s, a new method for improving the performance of solid desiccant systems was invented by N.A. Pennington [2] and Carl. G. Munters [3]. Introduced by Pennington, a rotary heat exchanger matrix in the form of a wheel was impregnated with a desiccant material to make an adiabatic regenerative dehumidifier when coupled to a heat source and a double evaporative cooler. This operating cycle is widely known as the ventilation cycle. Up to now, several studies on the solid desiccant air conditioning (SDAC) have been carried out focusing on system performance [4], desiccant materials [5], and strategy to implement.

One of the highlights of this technology is its ability to reduce energy consumption by utilizing a low-grade heat source. Besides, the advantage of SDAC is to reduce the initial equipment cost because this system can separate the handling of latent heat loads and sensible heat loads. Even the desiccant material able to absorb the undesirable particle, and thus increase indoor air quality. Therefore, the

SDAC system is simpler, safer, more durable, and easier to implement for air conditioning applications [6,7]. SDAC comprises a rotor that impregnated with desiccant material, a heater, and a sensible heat exchanger. The rotor needs to be regenerated by flowing the hot air with a specific temperature range. This process defines as the regeneration process, which is intended to break the bond between moisture and desiccant material in the rotor. With the presence of this process, the desiccant wheel can operate continuously and keep its performance.

In our previous work, we have investigated the effect of changes in the regeneration heat input on the performance of the desiccant wheel. The regeneration heat input is considered to rise and drop based on changing the load factor. The changing load factors are represented by a variation of the regeneration temperature (T_{reg}) and the regeneration airflow velocity (V_{reg}) that examined in two operation modes [8]. The result showed that the changes in heat input are influencing the decreases in the desiccant wheel performance. Comparison among two operating modes, the constant regeneration air temperature operation had a smaller drop in dehumidifying performance when the amount of heat input to the regeneration air decreased. It was also found that the optimal rotor rotation speed is proportional to the regeneration airflow rate that can be proposed as a solution for recovering the decline of the dehumidification performance.

In this paper, we present the further investigation of the performance of SDAC with a combination of a sensible heat exchanger (SHE). We want to clarify the effect of changes in the regeneration temperature (T_{reg}) and the regeneration airflow velocity (V_{reg}) on the system performance of SDAC. We expected that the existence of SHE is not only to reduce the outlet temperature of the process air but also preheats the regeneration airflow that can reduce the external energy input for the regeneration process and improve the system performance. Two operating modes, a constant T_{reg} mode with changing of V_{reg} from 2 m/s to 0.83 m/s, and a constant V_{reg} mode at various T_{reg} from 75°C–46.4°C, were used as the operating parameters to examine the effectiveness of SHE on the system performance. Finally, this investigation can use as a reference for implementing the SDAC in real applications.

2. System description

The experimental apparatus (figure 1) consists of a desiccant rotor, a sensible heat exchanger, a set of control devices, and a measurement instrument. The system was designed with two flow paths: the regeneration side and the adsorption side. Each flow path has a heater, an air control box, a blower, and a regeneration hot air. The desiccant rotor has a diameter of 0.32 m and a width of 0.2 m with an effective cross-sectional area of 0.066 m^2. It is formed from a fiberglass-based honeycomb matrix impregnated with silica gel of about 50 wt%. The crossflow heat exchanger type plate was adopted as the sensible heat equipment to further remove more sensible load of the outlet process air.

The desiccant wheel was divided into two chambers. In both chambers, six thermocouples (Type-T, accuracy = ±0.5°C) were installed for air temperature measurements at the inlet and outlet of the desiccant rotor and distributed uniformly at angular intervals of 30°. The thermocouples were also installed at the inlet and outlet of the hot and fresh air that flows in the sensible heat exchanger. A chilled mirror dew-point hygrometer (Optica, GE Sensing & Inspection Technologies and GE Optimization and Control Co., Ltd., Boston, United States., accuracy = ±0.2°C in dew point) was used to measure the absolute humidity at the inlet and outlet of the wheel. The airflow rate of the system was monitored using a pitot tube-type flow meter (New Aero Eye, Wetmaster Co., Ltd., Tokyo, Japan., accuracy = ±3%RD). From the measured airflow, the surface air velocity was calculated from the cross-sectional area of the adsorption (or regeneration) zone of the desiccant wheel.

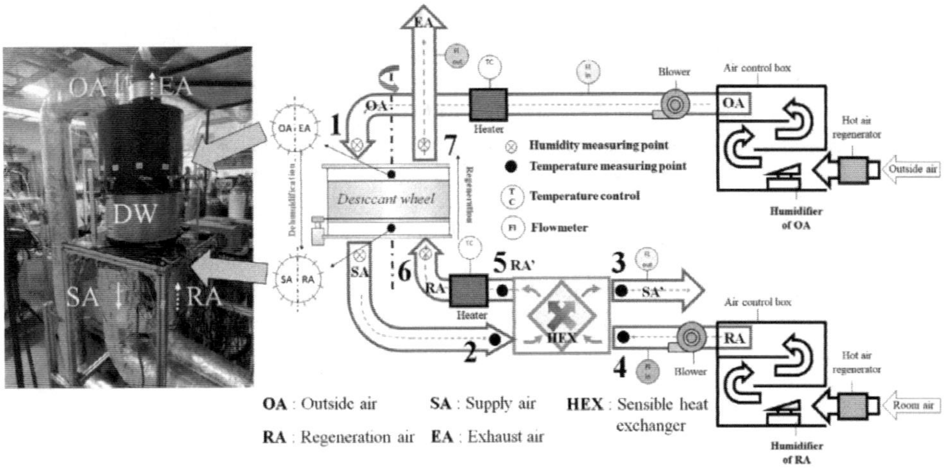

Figure 1. A schematic diagram of the experimental apparatus.

2.1 Experimental conditions

The inlet condition of the outside air (OA) and the regeneration air (RA) were represented by air states 1 and 4 respectively, as shown in figure 2. The inlet OA condition was fixed at 30°C and 15 g/kgDA, while the RA condition was set at 26°C and 10.5 g/kgDA. To reach the temperature and humidity targets on both conditions, the air control box that equipped with the humidifier and heater was controlled separately in each air handling unit. The adsorption process can be seen on the airflow from the air state 1 to the air state 3, and the regeneration process is shown at the airflow state conditions of 4-7. The outside air at state point 1 enters the desiccant wheel (DW) with a constant process airflow rate (V_{pro}) of 2 m/s, in which it is dehumidified and heated by the regeneration air at state point 6. At state point 2 to 3, the hot dry air is then cooled in the heat exchanger, using a crossflow type sensible heat exchanger.

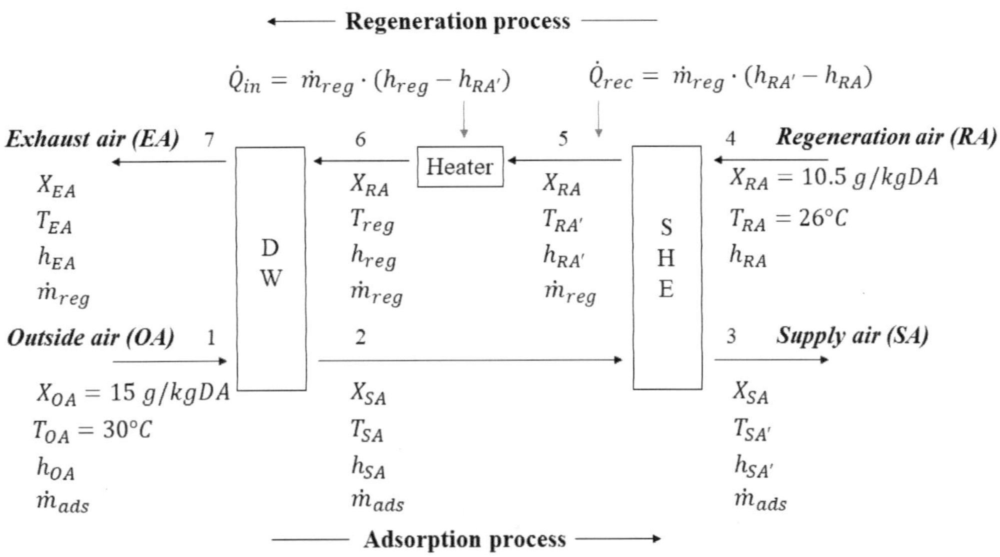

Figure 2. The inlet conditions for the adsorption and regeneration air.

Meanwhile, the regeneration air at state point 4 is preheated in the sensible heat exchanger, using the waste heat air at state point 2. Further, the preheated regeneration air at state point 5 is heated until the predetermined regeneration temperature through a heater. Then, the air state at point 6 is passed through the desiccant wheel at which desiccant material dries up moisture from the adsorption process air and transferred to the hot regenerated air. Finally, the leaving air at the desiccant wheel (the air state at point 7) is exhausted to the ambient air. The regeneration air temperature (T_{reg}) and the regeneration airflow velocity (V_{reg}) fitted to a specified of values that depend on the operating modes. In the constant regeneration air temperature operating mode, V_{reg} has a varied value from 2 m/s to 0.83 m/s, while T_{reg} was kept constant at 75°C. For the constant regeneration airflow velocity operating mode, T_{reg} was varied from 75°C to 46.4°C with maintaining V_{reg} at 2 m/s. In addition, the rotor rotation speed was fixed at 35 rph.

2.2 Performance indexes

In order to investigate the effect of changes in T_{reg} and V_{reg} on the system performance of SDAC, the following indexes were analyzed.

(1) Dehumidification performance (ΔX)

The dehumidification capacity represents the ability of the desiccant wheel to remove moisture from processed air; here, ΔX is defined as the absolute humidity difference between the inlet (X_{OA}) and outlet (X_{SA}) of the processed air. Specifically, we obtain

$$\Delta X = X_{OA} - X_{SA} \text{ [g/kgDA]}, \quad (1)$$

(2) Efficiency of the sensible heat exchanger (ε_{SHE})

The heat transfer efficiency of the sensible heat exchanger refers to the air heating through the wheel on the adsorption process and the thermal energy supplied for the regeneration process [9] that can be calculated as follows:

$$\varepsilon_{SHE} = (T_{SA} - T_{SA'})/(T_{SA} - T_{RA}) \quad (2)$$

where T_{SA}, $T_{SA'}$, and T_{RA} are the outlet temperature of the process air at state point 2 (°C), the outlet temperature of the process air at state point 3 or the air state after the SHE (°C), and the inlet temperature of the regeneration air at state point 4 (°C), respectively.

(3) Recovery energy ratio (*RER*)

The recovery energy ratio defined as the ratio between the heat recovery (Q_{rec}, (kW)) and the total energy (Q_{tot}, (kW)) that used in the regeneration process. It was estimated as follows:

$$RER = \frac{Q_{rec}}{(Q_{rec} + Q_{in})} \quad (3)$$

$$Q_{rec} = \dot{m}_{reg}(h_{RA'} - h_{RA}) \quad (4)$$

$$Q_{in} = \dot{m}_{reg}(h_{reg} - h_{RA'}) \quad (5)$$

where, h_{reg}, $h_{RA'}$, h_{RA} are the enthalpy of the regeneration air at state point 6 or the air state after the heater (kJ/kg), and the enthalpy of the preheated regeneration air at state point 5 or the air state before entering heater (kJ/kg), and the enthalpy of the inlet regeneration air or the air state at point 4 (kJ/kg), respectively.

(4) Coefficient of performance (*COP*)

COP is the ratio between the cooling capacity (Q_a) and the total heat input (Q_{in}). *COP* represents the overall performance of the desiccant system that calculated from the following equation.

$$COP = Q_a/Q_{in} \quad (6)$$

$$Q_a = \dot{m}_{ads}(h_{OA} - h_{SA'}) \quad (7)$$

where h_{OA} and $h_{SA'}$ are the enthalpy of the outside air (OA) at state point 1 (kJ/kg) and the supply air (SA) at state point 3 (kJ/kg), respectively.

This experiment was substantiated by verifying the mass conservation (ΔM) between the adsorption and regeneration sides using the estimates of airflow rates and the difference in the absolute humidity between inlet and outlet. The mass conservation equation reads [6]

$$\Delta M = \frac{\dot{m}_{ads}(x_{OA}-x_{SA})-\dot{m}_{re}(x_{EA}-x_{RA})}{\dot{m}_{ads}(x_{OA}-x_{SA})}, \quad (8)$$

where $\dot{m}_{ads}$ and $\dot{m}_{re}$ are the mass flow rates [kg-DA/s] in the adsorption and the regeneration sides, x_{OA}, x_{SA}, x_{EA}, and x_{RA} are the absolute humidity of the outside air, the supply air, the exhaust air, and the regeneration air, respectively. In all experiments, it was confirmed that ΔM was less than 5%. Taking into account the experimental uncertainty for the measuring instrument, the maximum amounts of relative uncertainty obtained for ΔX, ε_{SHE}, RER, and *COP* are 2.7%, 1.0%, 3.1% and 2.2%, respectively.

3. Result & Discussions

3.1 Effect of the constant regeneration temperature operating mode

In the first experiment, the regeneration temperature (T_{reg}) was set to 75°C while the regeneration airflow velocity (V_{reg}) was varied from 2 m/s to 0.83 m/s (figure 3). The rotor rotation speed was fixed at 35 rph. As shown in figure 3, the dehumidification performance (ΔX) of the desiccant declined along with a reduction in the mass flow rate of the regeneration stream. This condition diminishes the capacity of the desiccant wheel to take more moisture from processed air. The decline in dehumidification performance is not so high as long as the airflow velocity is between 2 m/s and 1.25 m/s or about 1.8 g/kgDA in absolute humidity. However, when the regeneration airflow velocity is 0.83 m/s, the air state of the exhaust after regeneration reaches a limit value, and the dehumidification performance is constrained to maintain the material balance. This phenomenon causes a significant decline in dehumidification performance of 3.7 g/kgDA. This result suggests that the dehumidification performance declines sharply when the regeneration airflow rate is too low. A similar trend is also seen in COP, which increases as the V_{reg} increase. This means that the rise in V_{reg} causes the ratio between the regeneration heat input (Q_{in}) and the cooling capacity (Q_a) to increase.

Furthermore, the ratio between the heat recovery and the total energy input, the recovery energy ratio (RER) of the system, is shown in figure 4 at the left vertical axis. The result shows that the changes in V_{reg} from 2 m/s to 0.83 m/s affect the RER drops from 53% to 42% or a drop of 21%. It can be explained that the reduction in the amount of regeneration air that less than the amount of adsorption air (process air) leads to an imbalance in performance between the adsorption process and the regeneration process. This means that the rotor on the adsorption side could not extract more moisture from the processed air because of the rotor starting to saturate. Consequently, the amount of regeneration heat is more used to heat the rotor rather than heating the regeneration air that causes the supply air temperature at point 2 to decrease. This condition directly declines the effectiveness of SHE, as shown in Figure 4 (on the right vertical axis). Thus, we can conclude that the increment in V_{reg} affects a rise in the energy recovery ratio (RER) and the effectiveness of SHE, and conversely.

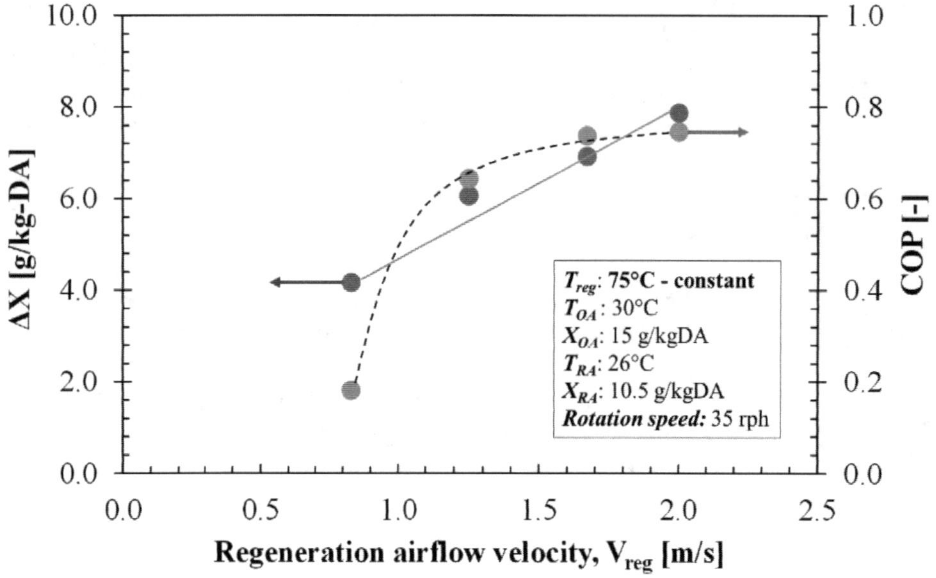

Figure 3. Effect of changes in V_{reg} on the dehumidification performance (ΔX) and COP at the constant T_{reg} operating mode.

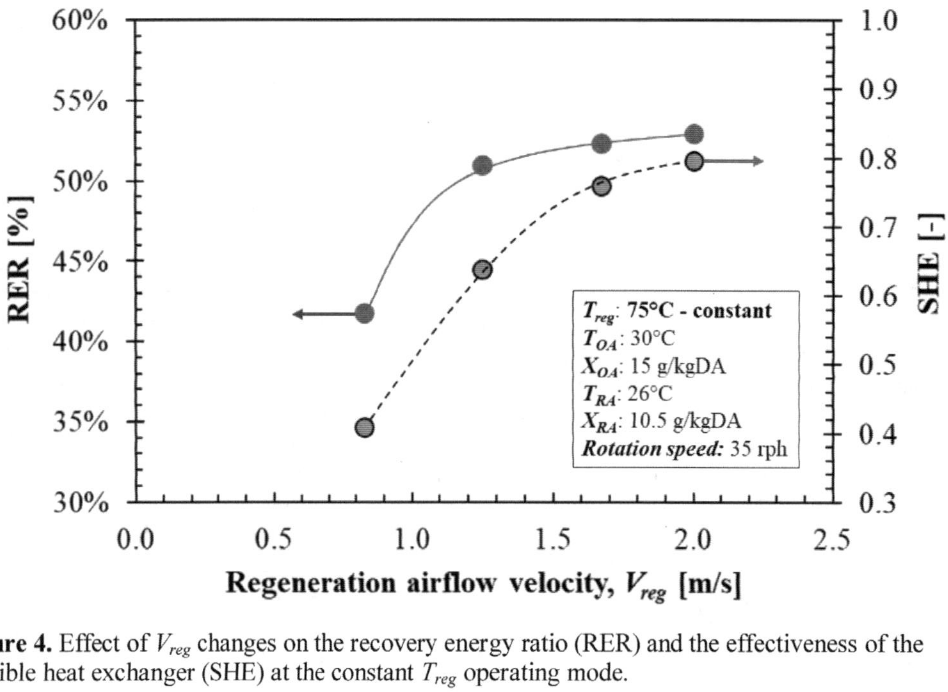

Figure 4. Effect of V_{reg} changes on the recovery energy ratio (RER) and the effectiveness of the sensible heat exchanger (SHE) at the constant T_{reg} operating mode.

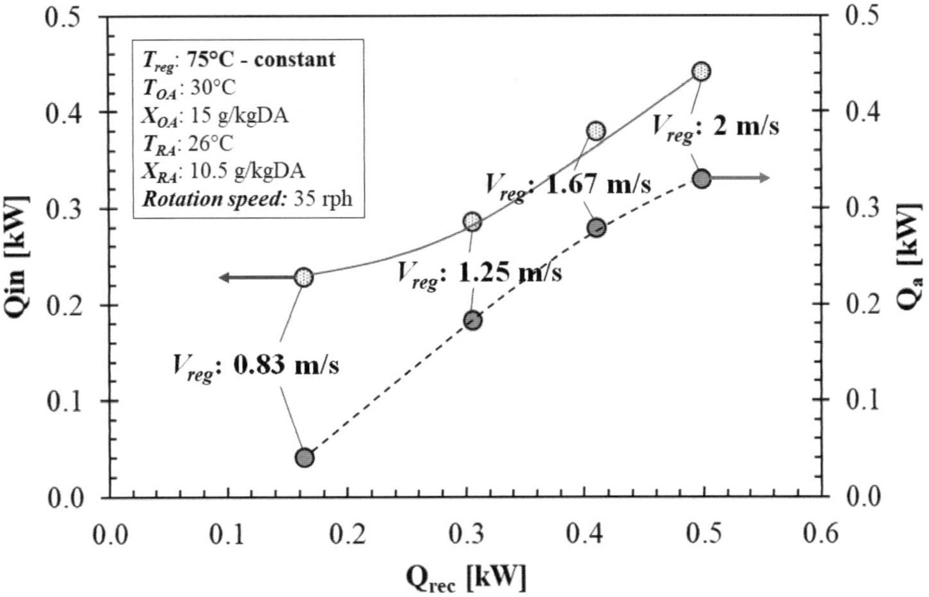

Figure 5. Effect of changes in V_{reg} on the regeneration heat input (Q_{in}), the heat recovery (Q_{rec}), and the cooling capacity (Q_a).

figure 5 shows that a rise in V_{reg} from 0.83 m/s to 2 m/s leads to an increase in recovering waste heat from the rotor from 0.16 kW to 0.5 kW. The presence of the sensible heat exchanger is not only to recover the waste heat but also to increase the cooling capacity. This condition can be seen in figure 5 (in the right vertical axis) that presents the cooling capacity (Q_a) enhancement along with the rise in V_{reg}. As regards the regeneration heat input (Q_{in}), the rise in V_{reg} causes an increase in the enthalpy of the regeneration air that affects the regeneration heat input to increase. Thus, a rise in V_{reg} at the constant T_{reg} causes an increase in Q_{rec}, Q_{in}, and Q_a.

3.2 Effect of the constant regeneration airflow velocity operating mode

figure 6 shows the dehumidification performance (ΔX) of the SDAC system when the regeneration air temperature (T_{reg}) varies between 75°C to 46.4°C. This operating mode was examined at a constant regeneration airflow velocity of 2 m/s. The absolute humidity of the regeneration air was fixed at 10.5 g/kgDA, while the regeneration temperature was varied. The result shows that the amount of dehumidification tends to drop of 4 g/kgDA as the temperature of regeneration is decreased from 75°C to 46.4°C. It can be explained that a decrease in regeneration temperature is followed by a rise in the dehumidification limit. This means that there is a decrease in the adsorption thrust force that causes the dehumidification performance to decline. However, there is a decrease in system performance (COP) as an increased regeneration temperature. This is due to an increase in the amount of heat to achieve a higher T_{reg} that leads to an increase in the heat input of the system.

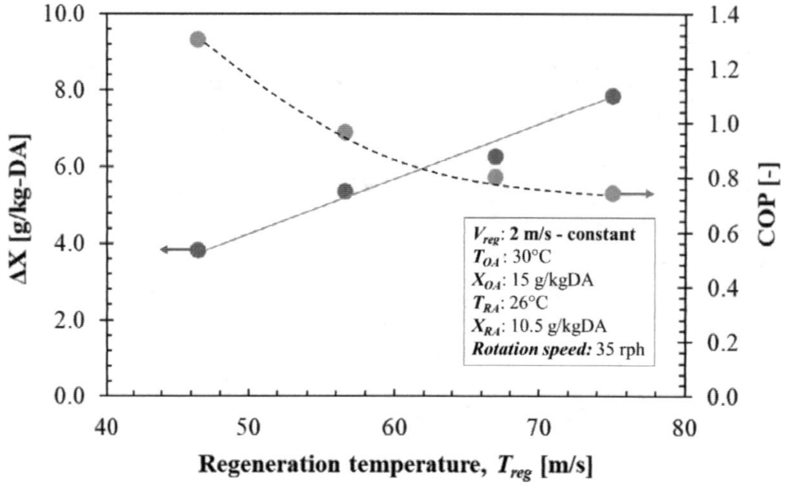

Figure 6. Effect of changes in T_{reg} on the dehumidification performance (ΔX) and COP at the constant V_{reg} operating mode

In figure 7, the effectiveness of SHE is almost not affected by changes in T_{reg}. This result indicates that the effectiveness of SHE is more influenced by V_{reg} than T_{reg}. However, the recovery energy ratio (RER) increases when the regeneration temperature decreases. The reduction in T_{reg} from 75°C to 46.4°C causes an increase in RER from 53% to 61% or about 16%. Furthermore, figure 8 shows that the increase in T_{reg} affects the heat recovery (Q_{rec}), the regeneration heat input (Q_{in}), and the cooling capacity (Q_a) to increase. However, it was found that at the lowest T_{reg} of 46.4°C, the cooling capacity (Q_a) is higher than the regeneration heat input (Q_{in}). As a result, there is a significant increase in system performance (COP) and recovery energy ratio (RER) when the lowest of T_{reg} is used. However, it should be noted that the dehumidification performance gives the most deficient performance. Therefore, it is necessary to find out the optimal operating mode by considering the changes in V_{reg} and T_{reg} at various rotor rotation speed.

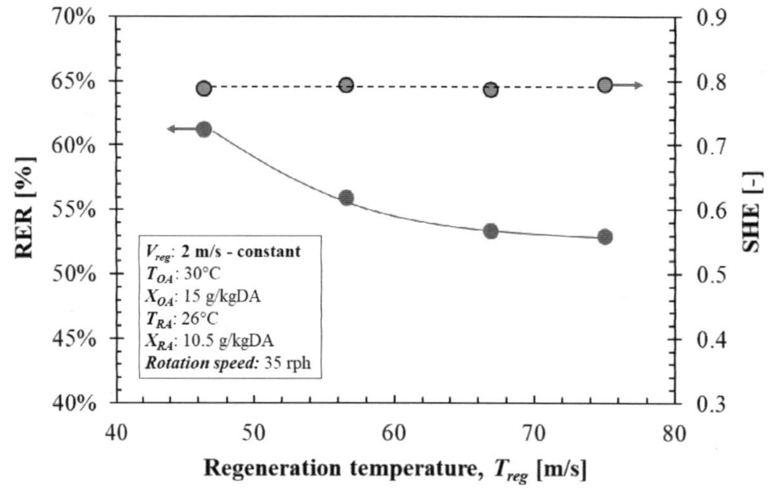

Figure 7. Effect of T_{reg} changes on the recovery energy ratio (RER) and the effectiveness of the sensible heat exchanger (SHE) at the constant V_{reg} operating mode

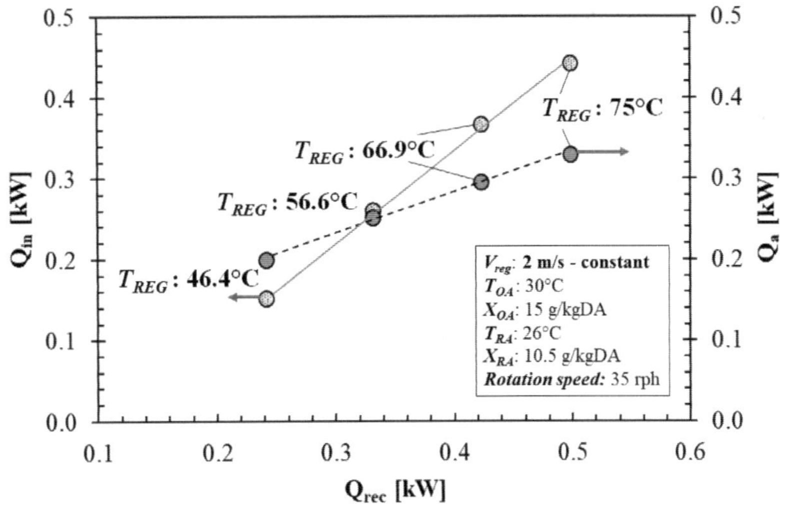

Figure 8. Effect of changes in T_{reg} on the regeneration heat input (Q_{in}), the heat recovery (Q_{rec}), and the cooling capacity (Q_a).

4. Conclusions

In this present work, the effect of changes in the regeneration airflow velocity (V_{reg}) and the regeneration temperature (T_{reg}) on the system performance of the solid desiccant air conditioning has been obtained in two operating modes, i.e., the constant regeneration temperature mode and the constant regeneration airflow velocity mode. Regarding the dehumidification performance, the constant regeneration temperature (T_{reg}) operating mode offers better performance than that for the constant airflow velocity (V_{reg}) operating mode. On the contrary, the system performance (COP) and the recovery energy ratio (RER) at the constant V_{reg} operating mode gives a higher value than the constant T_{reg} operating mode. These results suggest that when the dehumidification performance becomes the main function of the system, the reduction in V_{reg} should be considered to avoid a significant reduction in the system performance. Further investigations are needed to find out the optimal operating mode by considering the changes in V_{reg} and T_{reg} at various rotor rotation speed. By understanding its behavior, we can determine the effective configuration that provides the optimal performance of the solid desiccant air conditioning system.

Nomenclature

COP	coefficient of performance [-]	*Abbreviations*	
h	enthalpy [kJ/kg]	DW	desiccant wheel
ΔM	mass conservation [%]	EA	exhaust air
$\dot{m}$	regeneration mass flow rate [kg/s]	OA	outside air
$\dot{m}_{ads}$	adsorption mass flow rate [kg-DA/s]	RA	regeneration air
$\dot{m}_{reg}$	regeneration mass flow rate [kg-DA/s]	RA’	regeneration air after pre-heat
Q_{in}	regeneration heat input [kW]		(the air state at point 5)
Q_{rec}	heat recovery [kW]	rph	rotations per hour
Q_{tot}	total regeneration heat input [kW]	SA	supply air
RER	recovery energy ratio [%]	SA’	final supply air (the air state at
T	temperature [°C]		point 3)

T_{reg}	regeneration temperature [°C]
V_{pro}	adsorption inlet air velocity [m/s]
V_{reg}	regeneration inlet air velocity [m/s]
X	absolute humidity [g/kg-DA]
ΔX	dehumidification capacity [g/kg-DA]

References

[1] Mei V C, Chen F C, Lavan Z, Collier, R. K J and Meckler G 1992 An assessment of desiccant cooling and dehumidification technology 127
[2] Fennington N A 1955 Humidity Changer For Air Conditioning 9
[3] Munters C G 1968 Method for the manufacture of gas conditioning packing *US Pat. Off.*
[4] Comino F, Táboas F, Peci F and Adana M R De 2020 Detailed experimental analysis of the energy performance of a desiccant wheel activated at low temperature *Appl. Therm. Eng.* **178** 115580
[5] Wu X N, Ge T S, Dai Y J and Wang R Z 2019 Investigation on novel desiccant wheel using wood pulp fiber paper with high coating ratio as matrix *Energy* **176** 493–504
[6] Tsujiguchi T, Osaka Y and Kodama A 2017 Feasibility study of simultaneous heating and dehumidification using an adsorbent desiccant wheel with humidity swing *Appl. Therm. Eng.* **117** 437–42
[7] Vivekh P, Kumja M, Bui D T and Chua K J 2018 Recent developments in solid desiccant coated heat exchangers – A review *Appl. Energy* **229** 778–803
[8] Saputra D A, Osaka Y, Tsujiguchi T, Haruki M, Kumita M and Kodama A 2020 Experimental investigation of desiccant wheel dehumidification control method for changes in regeneration heat input *Energy* **205** 118109
[9] Sheng Y, Zhang Y, Sun Y, Fang L, Nie J and Ma L 2014 Experimental analysis and regression prediction of desiccant wheel behavior in high temperature heat pump and desiccant wheel air-conditioning system *Energy Build.* **80** 358–65

Effect of Particle Size on Adhesion Strength of Bovine Hydroxyapatite Layer on Ti-12Cr Coated by using Electrophoretic Deposition (EPD) Method

Gunawarman[1], I H Mulyadi[1], Z Arif[1], N F Nuswantoro[2], J Affi[1], M Niinomi[3, 4, 5, 6, 7]

[1]Mechanical Engineering Department, Engineering Faculty, UniversitasAndalas, Padang, Indonesia
[2]Biomedical Science Department, Faculty of Medicine, UniversitasAndalas, Padang, Indonesia
[3]Institute for Materials Research, Tohoku University, Sendai, Japan
[4]Department of Materials and Manufacturing Science, Graduate School of Engineering, Osaka University, Osaka, Japan
[5]Department of Material Science and Engineering, Graduate School of Science and Technology, Meijo University, Nagoya, Japan
[6]Institute of Materials and Systems for Sustainability, Nagoya University, Nagoya, Japan
[7]Faculty of Chemistry, Materials and Bioengineering, Kansai University, Osaka, Japan

*Corresponding Author: gunawarman@eng.unand.ac.id

Abstract. Hydroxyapatite (HA) extracted from bovine bones (called as natural HA) was used to coat a relatively new developed titanium alloy, Ti-12Cr, by using the electrophoretic deposition (EPD) method. This is to improve biocompatibility and bioactivity properties of the material to achieve optimal osseointegration in orthopaedic implant applications. There are three particle sizes of the natural HA used in this study (25 μm, 63 μm, and 125 μm) which aims to determine their effect on morphology, structure, and the strength of the resulting coating adhesion. The coating process was carried out at a voltage of 5 Volt for 5 minutes. The resulting layer morphologies and surface coverage were observed using an optical microscope. The increase in sample mass was measured using digital scales to determine the amount of the particles deposition. The coating thickness was measured using coating thickness gauges, and adhesion strength the coating layer was measured by using the cross-cut tape test method. The results of this study indicate that the HA particle size influences significantly on the quality of the coating produced after the EPD process. It is found that the coated Ti-12Cr with a small particle size has better surface properties as compared to the coarse one. Therefore, small size natural HA particles seemmore suitablefor implant applications.

1. Introduction

The use of hydroxyapatite (HA) coating onthe surface of metalsespecially titanium alloys has been attracted some researchers in order to improve the biocompatibility and bioactivity properties of the material for orthopedic implants [1]–[3]. This is due to the inert nature of titaniumthat is unable to produce adequate osseointegrationduring implantation, and it is not optimal corrosion resistance in human body environment [4], [5]. One of favor method for coating titanium is Electrophoretic Deposition (EPD) due to its simplicity, low cost and able to coat any shape.In the coating process using the EPD method, many factors will affect the quality of the coating layer. These factors include the value of the voltage used, the duration of coating, the concentration of the suspension, the pH of the suspension, the temperature of the suspension, the crystallinity of HA, the presence of the dispersant, and the particle size of the HA used [6]–[8]. The HA particle size used is believed to affect the quality of the resulting coating because it is related to the ability of deposition and the level of adhesion between the particles and the metal substrate. Furthermore, HA particle size will also influence the immune response generated by the body after implantation [9]. So, it is necessary to investigate the effect of particle sizes on the surface properties of coating layer. There are three sizes of HA particles used in this study which aims to see their effect on the morphology, structure and the adhesion strength of the coating layer.

HA particles used in this study extracted from bovine bones. The use of bovine bone HA aims to find an alternative material that is cheaper and easier to obtain as compared to the commercial one that is relatively more expensive and it is difficult to have in developing countries. The important things such kind natural HA has biocompatibility and bioactivity levels that are not inferior to the commercial one [10]–[12]. The use of the bovine boneHA is also expected to provide a solution in the utilization of waste bonesat the slaughtering livestock.

EPD is a technique that utilizes an electrophoretic mechanism to move charged particles suspended in a solution under the influence of an electric field to be deposited on the substrate to form a thin or thick layer. This method has potential to be used in both small and large scale industries [13]. EPD has two anodes that is anode and cathode to connect electric current. Cathode acts as working electrode where the sampleis located to connect electric current. The charged HA particles are scattered in the ethanol liquid and move towards the working electrode due to electrophoresis, then deposits into sample surface due to coagulation of the HA particles [13]. A new type of titanium alloy that is claimed as low cost titanium alloy, Ti-12Cr, is used as the sample. This alloy has been relatively new developed for spinal implants due to its ability forchanging modulus of elasticity.

2. Materials and methods

2.1. *Sample Preparation*

The sample used was a plate-shaped Ti12Cr titanium alloy with a size of 10 mm x 10 mm. The sample was then grindedslightly with sandpapers to remove any dirties that might have stuck to it during the storage period. After that, the sample is solution treated to remove residual stresses and restore the β phase of the material. After solution treatment, the sample is grinded again to remove the crusts formed during heat treatment by using #100, #600, #1000, and #1500 emery paper, until the surface of the sample is smooth and shiny.

2.2. *Preparation of HA suspension*

The HA powders used for coating material is derived from previously prepared bovine bones. There are three group of particle sizes of HA used in this study, that is 25 μm, 63 μm, and 125 μm which aims to determine their effect on morphology, structure, and the strength of the resulting coating adhesion. HA suspension was thenprepared by mixing 4 g of HA powders into 100 mL 96% ethanol which is then stirred using a magnetic stirrer for 60 minutes for homogenization. The acidity level of the suspension was kept at pH 4 by adding 25% HNO_3.

2.3. *HA coating using EPD*

The coating process was preceded by pre-treatment of the Ti12Cr sample. Pre-treatment is done by immersing the sample using 96% ethanol and ethanol in the ultrasonic cleaner sequentially for 15 minutes. After that, it was continued by immersing the sample in 25% HNO3 for 30 minutes followed by soaking in NaOH for 60 minutes. The sample was then dried on a hot plate magnetic stirrer at 50°C for 5 minutes. The weight of sample was conducted before coating process.

The coating process is carried out using the EPD apparatus by attaching the sample to the cathode, in this apparatus the anode used is a carbon rod. The cathode and anode are connected to the power supply which is then set at a voltage of 5 volts, the sample and carbon rod are then immersed in HA suspension and spaced about 10 mm. The coating time used was 5 minutes. The sample was then sintered with a temperature of 800°C and a holding time of 10 minutes in a vacuum furnace. The mass of the sample was then weighed after sintering to determine the difference between mass before and after coating process. The thickness of the HA layer was measured by using the coating thickness gauges. The surface coverage was calculated after taking photos using an optical microscope. The micrograph was then processed using *ImageJ* software. The adhesion strength of the HA layer was measured using the cross-cut tape test method.

2.4. Data Analysis

The qualitative data were determined from the images of the HA coating layer on Ti12Cr surface. The obtained data were the increase in sample mass, HA layer thickness, HA layer surface coverage, and the percentage of damage to the HA layer after the cross-cut tape test. ANOVA test was conducted to determine the difference in the average value of each quantitative data in each HA particle size group. Statistical tests were performed using R statistic software.

3. Results and discussion

3.1. Morphology of the HA layer on the Ti12Cr surface

Figure1 shows the morphology of the HA layer on the Ti12Cr surface after the coating process. These images show that the difference in the size of the particles used to coat Ti12Cr resulted in different layer morphologies. HA with an average particle size of 25 μm showed a much better coating ability as compared to HA powder with larger particle size. HA powder with a particle size of 63 μm lookslike quite difficult to coat on the Ti12Cr surface because there were still many empty gaps that had not been filled. Meanwhile, the powder with a particle size of 125 μm seems unable to coat properly on the surface of the material. There were only a few particles attached.

Coating using the EPD method utilizes electric current as energy to deposit coating particles on the surface of the substrate to be coated. The most influential factors in this process are the voltage and coating time [14]. Meanwhile, in this study, the voltage and time used were constant at 5 volts and 5 minutes and the coating powder used had different particle sizes. The results of this study indicate that the smaller particles look easier to coat the surface of the Ti12Cr substrate by utilizing the available energy. The results of this study also confirmed that the trend for the coating process that continues to use coating particles that are smaller and even up to the nanoscale is because it will produce a much better layer morphology [15].

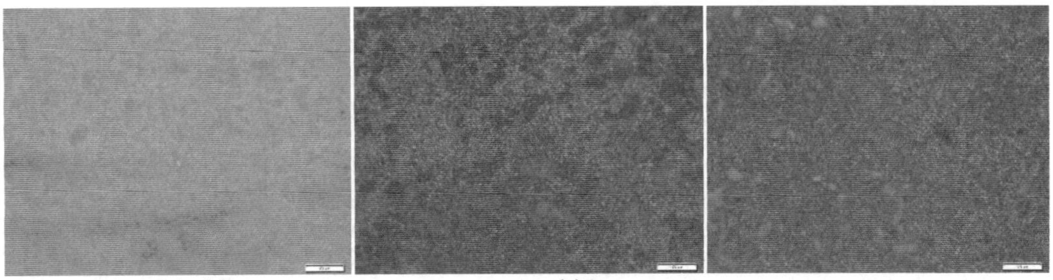

(a) (b) (c)

Figure 1. Morphology of the HA layer on the Ti12Cr surface, (a) 25 μm particles, (b) 63 μm particles, and (c) 125 μm particles.

From figure 1, it can be seen that HA powder with fine particle size produces an even layer morphology and there are no visible gaps or particle buildups. Meanwhile, the powder with a particle size of 63 μm was not able to coat evenly because there were many empty gaps and there was a lot of uneven particle buildup on the surface of the Ti12Cr substrate. Furthermore, there is another assumption which states that it is difficult for large particles to coat the surface of the substrate due to the small tangential contact that occurs between each HA particle which results in weak bonds formed so that the HA particles are easy to release and do not completely coat the substrate surface [16].

3.2. Mass Growth of Samples

Figure 2 shows the effect of particle sizes on the mass growth of the sample (or additional mass) after coating with the natural HA powders. It can be seen that the highest layer mass is found in samples with HA particles measuring 63 μm, while samples with HA particles measuring 25 μm and 125 μm did not appear to have significant layer mass and tends to be very low. This difference is probably due to differences in the ability of energy from electric currents to deposit HA particles. Logically, particles with smaller sizes will be easier to be positioned on the surface of the substrate compared to larger particle sizes because the energy required is also smaller. However, in this study, there was an oddity because samples coated with HA with a particle size of 63 μm had a much higher mass as compared to samples coated with HA with a particle size of 25 μm. While samples coated with HA with a particle size of 125 μm tended to have mass relatively low.This is probably because HA particles with the size 25 μm do have a much lower mass than particles with larger sizes so that when measured using a balance, the resulting mass is much lower even though the HA particles have covered the entire surface of Ti12Cr (figure 1). (a)). Meanwhile, the HA particles measuring 63 μm had a higher mass when measured even though the entire surface of the Ti12Cr sample was not completely coated (figure 1 (b)). Based on these results, it can be ascertained that the increase in sample mass cannot be used as a single reference to assess whether the coating process has given the desired results because the high mass gain does not necessarily indicate good or perfect coverage even though the mass increase can be used as a reference for the occurrence of HA particle deposition processes on the substrate to be coated.

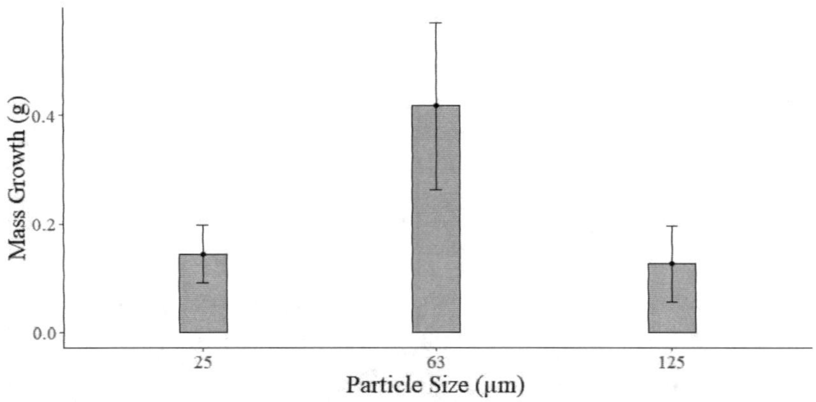

Figure 2. Relationship between HA particle size and the increase in sample mass.

The statistical tests performed using the ANOVA test, shows that each sample group with different HA particle sizes do not have a significant difference in mass gain (F-value = 2.54; p-value = 0.15). Thus, it could be concluded that HA particle size do not give a significant effect on the mass increase of the sample. A study shows that voltage greatly affects the deposition of HA particles onto the surface of titanium, coating carried out at constant voltage will deposit large particles and the formed layer will be porous, while dynamic voltage will produce a layer that has a gradient from large particles to large particles. small size [6], [7]. Other studies have also stated that the particle size greatly affects the quality of the HA layer produced by the EPD method because it greatly affects the mobility of charged particles in the coating process [2].

3.3. HA coating thickness

Figure3 shows the relationship between particle size and the HA coating thickness for the indicated particle size. It can be seen that there is no significant difference in layer thickness between each HA particle size group. The HA particle size used as a coating does not have a significant effect on the coating thickness. In this study, the voltage and coating time given are constant at 5 volts and 5 minutes so that the amount of energy and time given to each sample is the same as the expectation that the particle size used will make a difference, but the particle size of the coating layer does not give a significant effect on the thickness of the layer.

However, other assumptions can be used to explain this phenomenon. In the coating process using the EPD method, the layer formed consists of small particles that are arranged homogeneously and evenly layer by layer. The thickness of the HA layer formed will certainly be the same as the diameter of the coating particles or their accumulation if the particles accumulate to form layers. Thus, the layer formed by particles with a size of 25 μm requires more than four layers to reach a thickness of more than 100 μm while HA particles with a size of 125 μm only need one layer to achieve the same thickness. Thus, although the deposition of HA particles with a size of 125 μm tends to be lower, they can have the same thickness due to the particle size itself.

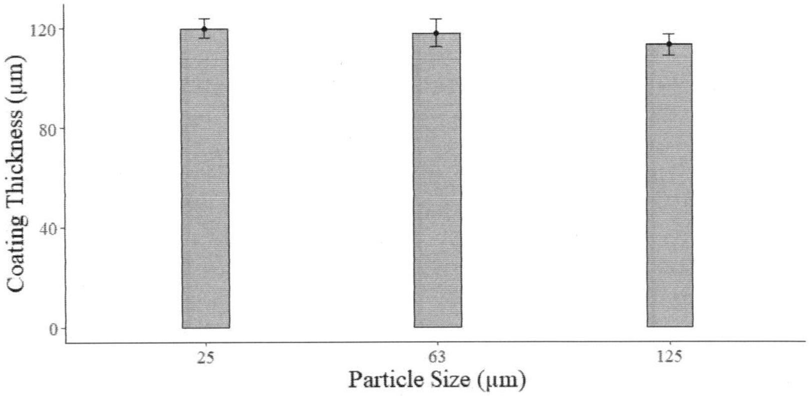

Figure 3. Relationship between HA particle size and the coating thickness.

The results of statistical analysis using the ANOVA test also shows the same conclusion (F-value = 0.56, p-value = 0.59), namely that there is no significant difference in layer thickness between the groups of HA particle size used as a coating material. The layer thickness is very much influenced by the dynamics of the applied voltage [8]. The higher the applied voltage, the thicker the resulting layer will be, and vice versa, the lower voltage will produce a thin layer [17]–[19]. The thickness of the resulting coating is not much difference between each treatment group since the voltage is constant at 5V.

3.4. Surface Coverage of HA coating

Figure4 shows a graph of the relationship between the HA particle size used in the coating process and the surface coverage that produced. It can be seen that there is a significant difference between the 25 μm particle size group and the other groups. The results of statistical tests using the ANOVA test also shows that there is a significant difference in the mean surface coverage values generated between the particle size groups (F-value = 4.34; p-value = 0.06). These results indicate that the smaller particle size is easier to form severe layer than the larger particle size.

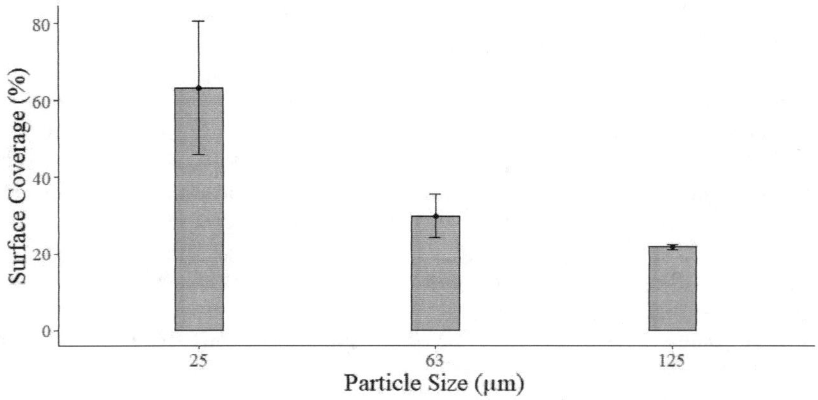

Figure 4. Relationship between HA particle size and surface coverage.

The availability of energy produced by a voltage of 5 volt appears to be able for distributing HA particles with a size of 25 μm across the surface of the Ti12Cr substrate as evidenced by the surface coverage value that can exceed 80%. However, for HA particles with a larger size, it looks very difficult to be positioned using the available energy. It is necessary to know proportional particle size so that the resulting layer is better. In connection with the results of this study, it can be assumed that 25 μm is a proportional particle size for use. So far it has been known that the voltage and coating time are important factors affecting the surface coverage ability of HA particles on the surface of titanium implants [14], [15], [17]–[20]. Besides, the HA particle size seem also giving an equally important effect.

3.5. Damage Percentage of HA layer after Cross-Cut Tape Test

Figure 5 is a graph the relationship between the HA particle size used as a coating material to the percentage of damage produced after the Cross-Cut Tape Test. It can be seen that there is a significant difference in the percentage of damage between each group of particle size. The highest damage was found in the group of particles with a size of 125 μm and the lowest one is in the group of particle sizes of 25 μm. The results of statistical tests using the ANOVA test also shows the same conclusion, there is a significant difference in the percentage of damage between each treatment group HA particle size (F-value = 354.44; p-value = 5.91×10^{-7}). This result indicates that the smaller the particle size will increase the adhesion strength of the HA particles to the surface of the Ti12Cr substrate. One of the assumptions that can be used to explain this result is that the smaller particles will have a greater tangential contact between the particles so that the bonds formed are also stronger when compared to larger particles. Logically, HA particles with a large size will have a bond with the surrounding particles but also have gaps that are large enough on the non-bonded side, whereas in smaller particles the gap will also get smaller so that the contact that occurs is much more area and the bonds formed are getting stronger. One study stated that the adhesion strength of HA particles on the surface of titanium is influenced by the uniform distribution of porosity [21]. Samples coated with HA particles of 125 μm and 63 μm have a low surface coverage value that means they also have a low porosity distribution, thereby reduces adhesion strength. Residual stress also has a major effect on the weakening of the adhesion of HA particles on the titanium surface. The easy release layer shows that the residual stress on the surface of the material is also quite high [22], [23].

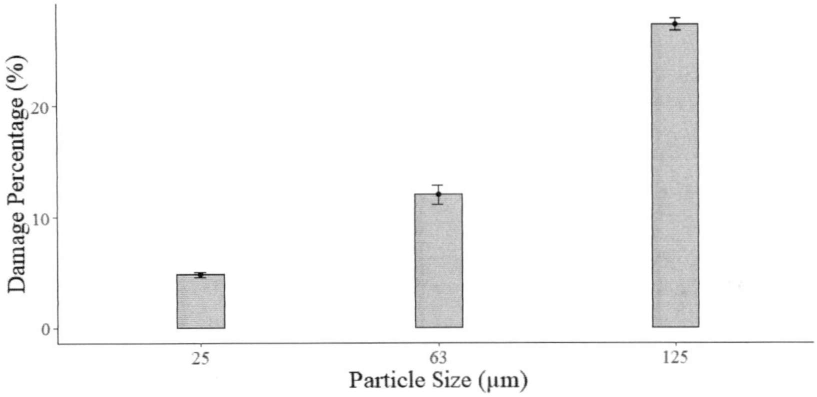

Figure5. Relationship between HA particle size and the percentage of damage to the HA layer produced after the Cros Cut Tape Test.

4. Conclusions

In orderto improve surface properties of new-developed titanium alloys, Ti-12Cr, a bovine bone HA has been coated to the metal surface by using EPD method. There are three group of particle sizes this kind natural HA used in this study, that is 25 µm, 63 µm, and 125 µm. The coating process was carried out at a voltage of 5 Volt for 5 minutes. The physical and mechanical properties of the coating layer were then examined and tested, respectively, using adequate equipment. The following conclusions are obtained:

1. Particle size has a significant effect on the surface performances includingmass growth and surface coverage, but almost no effect on the thickness of coating layer.
2. Adhesion bonding of the coating layer resulting from a small particle size is much better than that of the coarse one.
3. Natural HA coating layer with the size of 25 µm or less is recommended to use in order to improve significantly biocompatibility and bioactivity of Ti-12Cr.

5. Acknowledgement

First author thanks to Directorate General Higher Education, Ministry of Education and Culture, Republik of Indonesia for financial support under Applied Research Grant Contract No. T/35/UN.16.17/PT.01.03/AMD/PTUPTMaterialMaju/2020.

References

[1] Bose S, Tarafder S, and Bandyopadhyay A 2015 Hydroxyapatite coatings for metallic implants in Hydroxyapatite (HAP) for Biomedical Applicationsvol7(Elsevier Ltd.) pp 143–157
[2] Mohseni E, Zalnezhad E, and Bushroa A R2013 Comparative Investigation on the Adhesion of Hydroxyapatite coating on Ti-6Al-4V Implant: A Review Paper*Int. J. Adhes* Adhes
[3] Ji H, Ponton C B, and Marquis P M 1992Microstructural characterization of hydroxyapatite coating on titanium *J. Mater. Sci. Mater. Med.***3** pp 283–287
[4] Chen F et al. 2007Biocompatibility of Electrophoretical Deposition of Nanostructured Hydroxyapatite Coating on Roughen Titanium Surface: In Vitro Evaluation Using Mesenchymal Stem Cells *J. Biomed. Mater. Res. Part B Appl. Biomater.***82B** pp 183–191
[5] Aksakal B, Gavgali M, and Dikici B 2010The effect of Coating Thickness on Corrosion Resistance of Hydroxyapatite Coated Ti6Al4V and 316L SS Implants *J. Mater. Eng. Perform.***19** 6 pp 894–899

[6] Meng X, Kwon T, Yang Y, Ong J L, and Kim K 2006Effects of Applied Voltages on Hydroxyapatite Coating of Titanium by Electrophoretic Deposition*J. Biomed. Mater. Res. Part B Appl. Biomater.***78B** pp 373–377

[7] Meng X, Kwon T, and Kim K 2008Hydroxyapatite coating by electrophoretic deposition at dynamic voltage *Dent. Mater. J.***27** 5 pp 666–671

[8] Rad T A, Solati-Hashjin M, Osman N A A, and Faghihi S 2014Improved bio-physical performance of hydroxyapatite coatings obtained by electrophoretic deposition at dynamic voltage*Ceram. Int.* pp 1–11

[9] Lebre F, Sridharan R, Sawkins M J, Kelly D J, Brien F J O, and Lavelle E C 2017The shape and size of hydroxyapatite particles dictate inflammatory responses following implantation *Nat. Sci. Reports***7** 2922 pp 1–13

[10] Kusrini E and Sontang M 2012Characterization of x-ray diffraction and electron spin resonance : Effects of sintering time and temperature on bovine hydroxyapatite *Radiat. Phys. Chem.***81** pp 118–125

[11] Werber K, Brauer R B, Weiß W, and Becker K 2000Osseous Integration of Bovine Hydroxyapatite Ceramic in Metaphyseal Bone Defects of the Distal Radius *J. Hand Surg. Am.***25A** 5 pp 833–841

[12] Gao Y et al. 2006Characterization and osteoblast-like cell compatibility of porous scaffolds : bovine hydroxyapatite and novel hydroxyapatite artificial bone *J. Mater. Sci. Mater. Med.***17** pp 815–823

[13] Boccaccini A R, Keim S, Ma R, Li Y, and Zhitomirsky I 2010Electrophoretic deposition of biomaterials *J. R. Soc. Interface***7** pp S581–S613

[14] Gunawarman et al.2019Hydroxyapatite Coatings on Titanium Alloy TNTZ using Electrophoretic Deposition *IOP Conference Series: Materials Science and Engineering***602**pp 1–11

[15] Nuswantoro N F, Budiman I, Septiawarman A, Djong H T, Manjas M, and Gunawarman 2019Effect of Applied Voltage and Coating Time on Nano Hydroxyapatite Coating on Titanium Alloy Ti6Al4V Using Electrophoretic Deposition for Orthopaedic Implant Application *IOP Conference Series: Materials Science and Engineering***547** pp 1–11

[16] Indra A, Firdaus R, Mulyadi I H, Affi J, and Gunawarman 2020Enhancing the physical and mechanical properties of pellet-shaped hydroxyapatite by controlling micron- and nano-sized powder ratios *Ceram. Int.*

[17] Fajri H et al. 2020Electrophoretic Deposition (EPD) of Natural Hydroxyapatite Coatings on Titanium Ti-29Nb-13Ta-4.6Zr Substrates for Implant Material *Mater. Sci. Forum***1000** pp 123–131

[18] Juliadmi D et al. 2020The Coating of Bovine-Source Hydroxyapatite on Titanium Alloy (Ti-6Al-4V ELI) Using Electrophoretic Deposition for Biomedical Application *Mater. Sci. Forum***1000** pp 97–106

[19] Nuswantoro N F et al. 2020Electrophoretic Deposition Performance of Hydroxyapatite Coating on Titanium Alloys for Orthopedic Implant Application *Mater. Sci. Forum***1000** pp 69–81

[20] Nuswantoro N F, Maulana I, Djong H T, Manjas M, and Gunawarman 2018Hydroxyapatite Coating on New Type Titanium, TNTZ, Using Electrophoretic Deposition *J. Ocean. Mech. Aerosp.***56** 1 pp 1–4

[21] Yang Y C and E Chang 2003The bonding of plasma-sprayed hydroxyapatite coatings to titanium: effect of processing, porosity and residual stress *Thin Solid Film***444** pp 260–275

[22] Yang Y C and Chang E 2001Influence of residual stress on bonding strength and fracture of plasma-sprayed hydroxyapatite coatings on Ti-6Al-4V substrate *Biomaterials***22** pp 1827–1836

[23] Tsui Y C, Doyle C, and Clyne T W 1998Plasma sprayed hydroxyapatite coatings on titanium substrates Part 1: Mechanical properties and residual stress levels *Biomaterials***19** pp 2015–2029

IOP Conf. Series: Materials Science and Engineering 1041 (2021) 012055 doi:10.1088/1757-899X/1041/1/012055

Mechanical Properties and Microstructure of Mg-1.6Gd Alloys Reduction 80% Hot Rolling as Implant Materials

O Susanti[1], A Rahmadino[1]

[1]Metallurgy Laboratory, Mechanical Engineering Department, Universitas Andalas Padang, Indonesia

Corresponding author: oknovia.s@gmail.com

Abstract. Mg-Gd alloys is of great interest for many potential application especially in implant material. Higher mechanical properties is always the goal pursued by reserachers for the materials which generally have relatively still low strength,low ductile and loe hardness. From this aspect , the present paper reviews the recent report of a kind of Mg-Gd rolled which have been able to achieve strength compared with Mg-Gd casting alloys. The aim of this paper is to investigate the mechanical properties and microstructure forMg-Gd alloys that are rolled at high tempetaure with severe reduction. Mg-Gd rolled with 80% reduction will be applied by unidirectional and cross direction with purpose to find varian characteristic. The Temperature range are 400,450,500 and 550°C. The results of this experiment show that the highest hardness number obtained by single rolling specimen as 64.86 VHN at 550 °C. The highest ultimate tensile strength numbers obtained by cross rolling with 163.22 MPa at 550 °C. For the grain size, single rolling specimen has the smallest number with 46.33μm at 500 °C. Overall, temperature 500-550 °C give a big impact to improve Mg-1.6Gd alloys.

1. Introduction

Performance of Mg alloys is very significant to develop as biomaterial application. Moreover, it well known that the rare earth (RE) are known to offer excellent much properties. Magnesium-rare earth allys have been a great developed that the alloys could significantly improve the strength and corrosion ressitance. Recently, it has been reported that addition Gd is effective for improving mechanical properties by solid solution strengthening and precipitate strengthening. Mg5(Gd, Y) alloys have observed at temperature of 480 °C for 4 hours shows the only some Mg3(Gd 1.26 Y 0.74) particles [1].Cizek *et al*also reported that formation of the metastable β' phase of Mg-15 alloys causes a strong hardening and the peak hardening occur at remarkable lower temperature. This research also obtained that the number of precipitation on the grain caused by various thickness reduction [2]. Janacek *et al* have investigated on the Mg-22 alloys that the precipitation of stable Mg5Gd phase and the phase have higher hardness than magnesium matrix. The Mg-22Gd also result the grain refinement and lot of twinning after the alloys was processed by higher pressure torsion (HPT) a room temperature and the pressure of 2 Gpa [3]. Mg-20Gd sample by casting tehnique resulted as-cast ingot mainly contain α-Mg solid solution and Mg5Gd phase as represented by Peng *at al*. Tensile strength of Mg-20Gd at room temperature 254 Mpa and elongation at less than 2 %. Meanwhile, the fracture syrface of the as-cast alloys is typical cleavage fracture [4, 5]. Kubasek *et al*explained the structures of the as-cast Mg-5Gd alloys were mostly by fine α-Mg dendrites and the formation of Mg5Gd phase. The Mg-Gd alloys was significantly harmfull effect on corrosion resistance [6]. It well known that Gd has a wide range of alloy composition and heat treatment by its large solubility Of 23.49 wt% at eutectic temperature for changing of mechanical properties. Following that , Hort et al observed that the different addition of Gd and heat treatment resulted also various mechanical properties and corrosion behaviour which can be used for many applications. Beside the addition of Gd in magnesium, futhermore, thermo-mechanical process can also enhance mechanical properties and corrosion resistance. They reported that the addition of Gd above 10 wt% can actually improve strength but another side, the much more Gd elements in magnesium can decrease the elongation. Therefore, some researchers have recommended by the small addition of gadolonium on magnesium by thermo-mehanical process which it exhibited excellent mechanical properties. Susanti *et al* reported that the hot extrusion of Mg-1.6Gd alloys revealed fine grain size after hot extrusion on temperature recrystallization. The increase of mechanical propeties on the Mg-1.6Gd alloys compared as-cast condition. Although, the strength and hardness of the alloys decreased as the extrusion temperature increased [7]. Another thermo-mechanical process also carried out on Mg-1.6Gd alloys by hot rolling unidirectiional rolling (UR) and cross rolling (CR) with 95 % reduction. The mechanical properties greatly exhibited by refinement grain size [8, 9].

It is the purpose of this paper, another severe hot rolling processing in term of UR and CR with a total reduction of 80 % is connected to Kg-1.6Gd (wt %) alloys.The mechanical properties of the different direction of the alloys will be investigated and discussed.

2. Materials and methods

Samples of hot rolling were prepared on Mg-1.6Gd alloys ingot with two methods : Unidirectional (UR) and Cross rolling (CR) with dimension of 50 mm x 25 mmx 10 mm as reported by paper before [8]. Before experiment, the alloys ingots were homogenized for 5 hours at a temperature of 560 °C. The temperature of the hot rolled sample was measured and monitored with a thermocouple. Hot rolling was carried out in a temperature range from 400 to 550 °C and held for 15 min before pass rolling. A total reduction of 80 % at 40% per pass (for two times)with speed of 10 mm min−1 which resulted in final rolled sheet with at least 3 mm in thickness. The hardness of rolling specimens were prepared with 10 mm x10 mm size and determined from 10 individual measurements, each using a Vickers Hardness testers with 300 g load for a dwellingtime of 15 s. Flat tensile specimens were machined from the hot rolled material in accordance with the ASTM E8-04 standard. The tensile specimens having a gauge length of 24 mm n the plate were cut from the sheet with tensile axes lying parallel to the direction. . Tensile tests were carried out with an uniaxial tensile testing machine (Gotech AI-7000 LA 10) at room temperature. Microstructure test will be examined by standard metallography procedure. An etching solution for microstructure test was 4 mL HNO_3 + 100 mL ethanol. Furthermore, the fracture surface morphology of the rolling samples wereobserved by scanning electron microscopy (SEM).

3. Results and discussion

The grain size number showed in figure 1. Generally, the grain size of unidirectional rolling has smaller grain size than cross rolling specimen. The smallest grain size of UR specimen is about 46.33 μm at 500 °C while for cross rolling is about 54.27 μm at 500 °C. Meanwhile, the largest grain size for UR specimen is about 65.63 μm at 400 °C and 65.63 μm for CR specimen at 450 °C. As shown in Figure 1, the grain size decline significantly at 450 °C for UR specimen and 500 °C for CR specimen. At those temperature, the hot rolling process give a big impact for grain size as refinement microstructure. The grains size increase at temperature 550 °C which the UR specimen has 56.31 μm and the CR specimen has 52.47 μm.

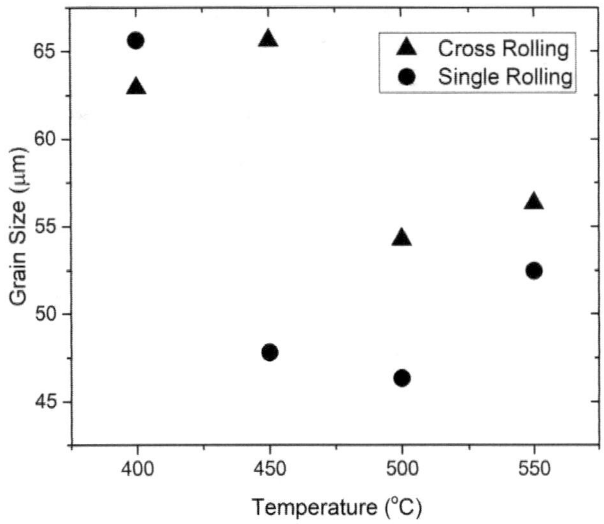

Figure 1. The grain of the hot rolled Mg-1.6Gd alloys for Ur and CR.

Figure 2 and figure 3 show the microstructure of the hot rolled Mg-1.6Gd alloys for UR and CR specimen. The UR specimen has the shape of long grain while CR specimen has an equiaxed grain. Precipitates occuron the hot rolled specimen. However, it were only found on CR specimen. The there are precipitates in grain where it was also explained by Hort et al. (2010) as Mg5Gd. Both of grain size and shape are very important which can result the good mechanical properties for Mg-1.6Gd alloys.

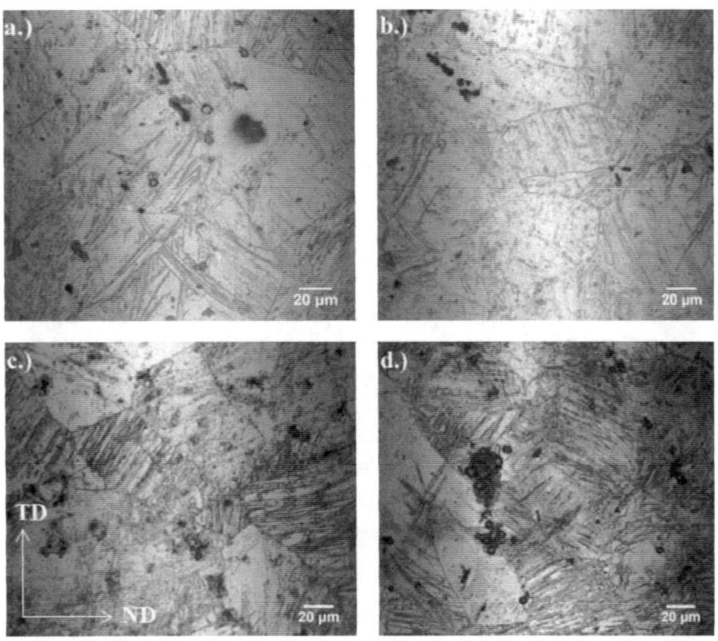

Figure 2. Microstructure of the hot rolled Mg-1.6Gd Alloys at various temperature with reduction of 80 % : a) 400°C, b). 450°C, c). 500°C, d). 550°C of Cross Rolling (CR)

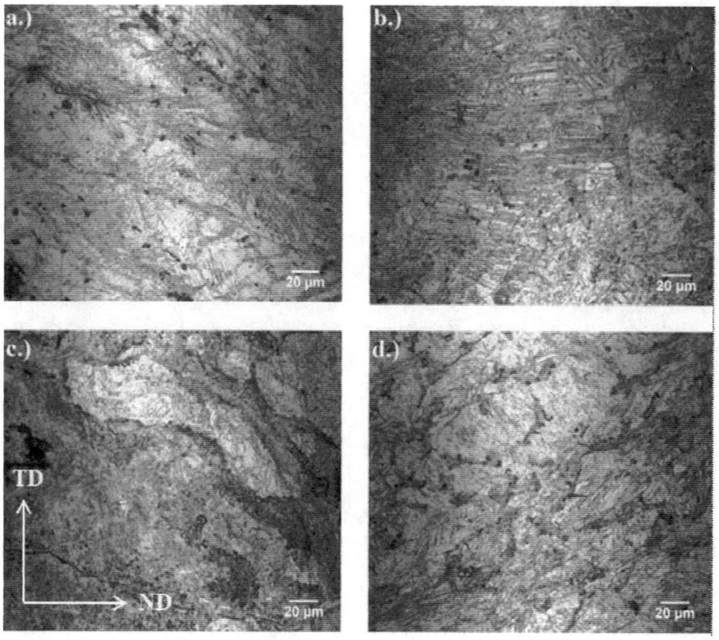

Figure 3. Microstructure of hot rolled Mg-1.6Gd Alloys at various temperature with reduction of 80 % : a) 400°C, b). 450°C, c). 500°C, d). 550°C of Unidirectional Rolling (UR)

Figure 4(a) showed the hardness number for both of UR and CR specimen. The hardness increases as temperature increases where the highest hardness of UR is about of 58.02 VHN and 64.86 VHN for CRat temperature 550 ºC. UR has bigger number than CR because UR specimen has smaller grain size compared to CR specimen. It well known that the grains size are contributedto mechanical properties of material. Both of the UR and CR specimen are better than pure magnesium that it has the hardness of about 30 VHN. Figure 4(b) shows the ultimate strength for both of UR and CR specimen. The highest strength for UR is 159MPa at temperature of 550 ºC and the lowest is 121 at temperature of 400 ºC. In the other hand, CR has the highest strength aboutof 163 MPa and the lowest is 126 MPa at 550 ºC and 400 ºC, respectively. The ultimate strength increases as temperature increases of the hot rolled Mg-1.6Gd alloys. CR is influenced by precipitates and the shape of grain, so that the strength valueis bigger than UR. Precipitates in the grain can hold the movement of atomic andthen the dislocation is so hard to move. Figure 4(c) shows the yield strength for both of UR and CR specimen. As shown in Figure , the yield strength increases as rolling temperature increases. The yield strength of CR is bigger compared to the UR. The largest yield strength for CR is 138 MPa while the UR is 135MPa. This result is related to previous discussion that CR has bigger ultimate strength than UR. It caused by load that can be received by CR specimen. The yield strength value also indicated the properties of material. In this case, CR specimen is more elastic than UR. Figure 4(d) shows the elongation for both of UR and CR specimen. Generally, it shows that the CR elongation is bigger than UR CR has elongation of 9.8% and 5.4% as the lowest elongation value. Meanwhile, UR has the higest elongation of 6.7% and 2.9 % for the lowest one. This result affected by the grain size of material where CR specimen has the smallest grain size compared to UR. It proves that the grain size of material can affect the mechanical properties.

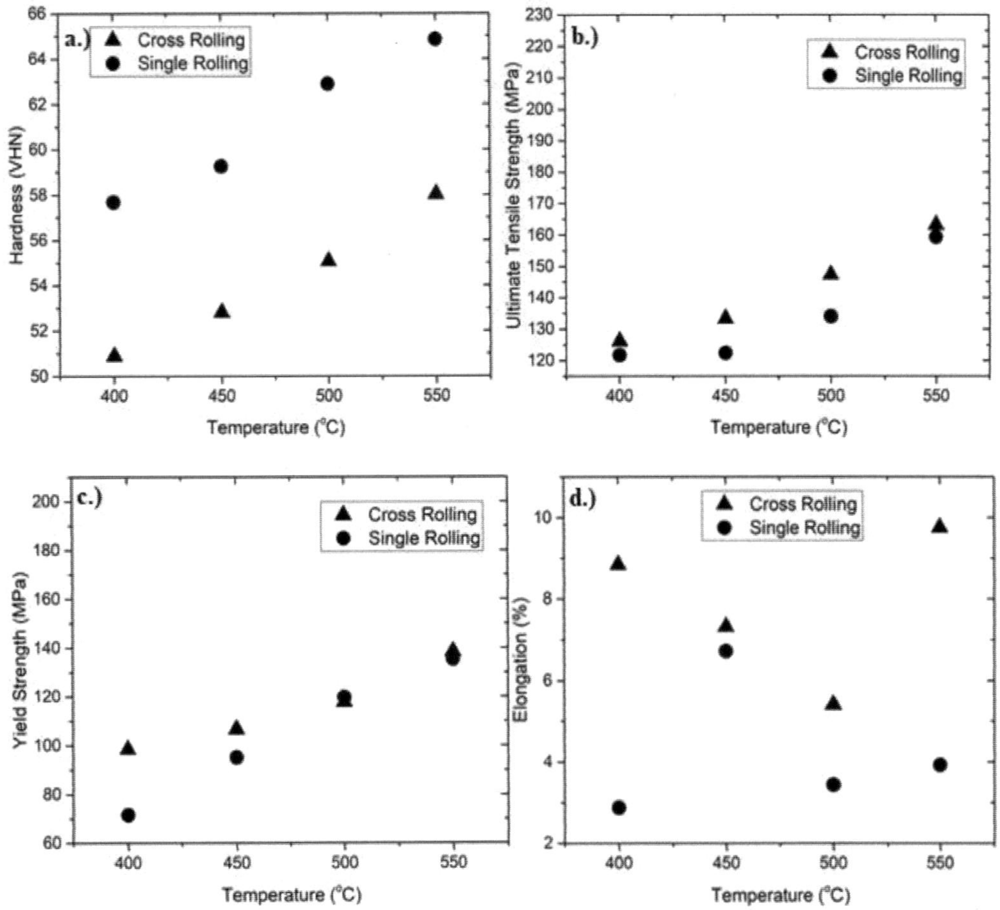

Figure 4. Various mechanical properties of Mg-1.6Gd alloys: a.) Hardness, b.) Ultimate Tensile Strength (UTS), c.) Yield Strength (YS) and d.) Elongation.

4. Conclusion

The mechanical properties of Mg-1.6Gd alloys can be exhibited by hot rolling at temperatures in the range of there crystallization temperature. The finement microstructure is obtained at low temperature and the grain size increase at higher temperature. Th fomation of precipitates occur at cross rolling which are related with mechanical properties. The strength and yield strengh of the CR has higher than unidirectional rolling of about 163 MPa and 138 MPa, respectively. Meanwhile, the hardness of UR specimens are generally higher than CR due to by the small grain size. Overall, the addition of Gd on magnesium alloys could give the improvement of mechanical properties as material implant

Acknowledgments

This study is supported by the Directorate of Research and Community Service, Universitas Andalas and through, Hibah Riset Fakultas Teknik 2020.

References

[1] Zhang L et al. 2011 Microstructure and mechanical properties of as-cast and heat treated Mg-15Gd-3Y alloy *Journal of Rare Earths* **29** 1 p 77-82.

[2] Čížek J et al. 2007 Influence of deformation on precipitation process in Mg–15 wt.% Gd alloy *Journal of alloys and compounds* **430** 1-2 p 92-96

[3] Janeček M et al. 2017 Mechanical properties and microstructure of magnesium alloy Mg22Gd processed by severe plastic deformation
[4] Peng Q et al. 2009 Microstructure and mechanical properties of high performance Mg–Gd based alloys *Materials & Design* **30** 2 p 292-296.
[5] Peng Q et al. *Microstructures and properties of melt-spun and as-cast Mg-20Gd binary alloy.* Journal of Rare Earths, 2006. **24** 4 p 466-470.
[6] Kubásek J and Vojtěch D 2013 Structural and corrosion characterization of biodegradable Mg–RE (RE= Gd, Y, Nd) alloys *Transactions of Nonferrous Metals Society of China* **23** 5 p 1215-1225
[7] Susanti O., Harjanto S and Mochtar M A 2015 Microstructure and Mechanical Properties of Extruded Mg-1.6 Gd as Prospective Degradable Implant Materials *Advanced Materials Research* Trans Tech Publ
[8] Susanti O, Mochtar M A and Harjanto S 2017 Effect of severe reduction on mechanical properties during hot rolling of Mg–1.6 Gd alloys *Materials Research Express* **4** 3 p 034007
[9] Susanti O, Mochtar M A and Harjanto S 2017 Microstructure Transformation of Mg-1.6 Gd during Hot Rolled at High Deformation Ratio *IOP Conference Series: Materials Science and Engineering* IOP Publishing

IOP Conf. Series: Materials Science and Engineering 1041 (2021) 012056 doi:10.1088/1757-899X/1041/1/012056

Lift Study of a Flapping MAV Wing Design Based on Bat Wings

S Suhaimi[1], S Shuib[1*], A H Kadarman[2], H Yusoff[3]

[1]Faculty of Mechanical Engineering, UniversitiTeknologi MARA Shah Alam, 40450, Shah Alam, Selangor, Malaysia.
[2]School of Aerospace Engineering, UniversitiSains Malaysia, Engineering Campus, 14300, NibongTebal, Pulau Pinang, Malaysia.
[3]Faculty of Mechanical Engineering, UniversitiTeknologi MARA, CawanganPulau Pinang, 13500, PermatangPauh, Pulau Pinang, MalaysiaType the author addresses here

Corresponding Author's Email: *solehuddin2455@uitm.edu.my

Abstract. Bio-inspiration is a design ideation method that uses observations found in nature to solve a mechanical problem. In this study, a simulation study of a micro air vehicle (MAV) wing design that is based on bat wings was done and validated experimentally. The wing model was generated using an in vivo observation where the margin wing shape was traced and then used to generate a flat wing with a margin shape of a batwing. The lift of the wing was studied at 4m/s wind speed with a flapping angle of 55°, and at the angle of attack from 0° to 35°. The simulation was done using ANSYS V1.6 software and the experimental study was done using an open circuit subsonic wind tunnel. It was observed that the wing has a stall angle of 25° and a CLmax of 3.2. The result also shows that the difference between the simulation and the observation result to be small with 0.2% at the smallest point and 5% at the largest point. This shows that the simulation model was valid and can be used for future wing design.

1. Introduction

Bio-inspiration or bio-mimicking is a design ideation process where knowledge and observations from nature are used to solve problems from different fields, in this case, to solve mechanical problems. While similar, bio-inspiration and biomimicry serve different purposes. Bio-inspiration refers to producing a design that is inspired by functions and forms found in nature. While biomimicry refers to copying as close as possible to nature. Usually, Biomimicry is done to further the understanding of nature or for scientific purposes while Bio-inspiration is used to produce a novel design or for practical and engineering purposes [1].

In this study, Bio-inspiration was used to design a wing for Micro Air Vehicles (MAV). MAVs are defined as unmanned aerial vehicles that have a wingspan that is less than 15mm [2]. One of the main difficulties involved with MAV design is balancing between efficiency and aerodynamic performance [3]. This study uses Bat wings as a source of inspiration since bats are known for its ability to fly in small and tight spaces while still able to forage in far areas. This shows that bats are manoeuvrable while still efficient [4].

Using bat wings for the source of bioinspiration and biomimicry are far from novel. In a study done by S. Bahlman[5] and by D. Yin [6], a series of actuators and truss was used to create a robot bat wing. The skin that was used as a membrane wing with a recreated bat wing skeleton. In a different study done by S. Wang [7], a 3D wing shape was used recreated from a 3D scan of an in-vivo bat wing and then was used in a simulation study of a batwing. These two approaches were done as a scientific work to understand the mechanics of bat flight and not to produce a flying device. Most of the previous works that involve in designing a flying MAV uses a conventional wing shape like an elliptical wing [8] or a flat straight wing[9]. The closest work that was done that uses a bat shape for the development of a flying MAV was done by H. Yusoff[10] where a flat wing with batwing margin was used. However, the wing margin shape that was used was an estimated wing shape and not based on an in-vivo wing. There is still a gap in the previous work where there is yet a wing that is developed for a flying MAV that is closely mimicking the forms of a batwing.

The problems of biomimicry are due to the complexity of nature where a realistic recreation is impractical. Therefore, a type of simplification is unavoidable. The purpose of this study is to investigate the lift generation of a flapping flat wing with a batwing shape margin. The results of the study will then serve as a foundation for the design of a flying MAV wing design.

2. Materials and methods

For this study to be a basis for a future flying MAV design, a simulation study was done that was validated by experimental work. A wing design was first generated that is based on the forms of a batwing. An FSI CFD simulation was done to study the lift generation of the wing. The simulation result was then validated by an experimental work using a subsonic wind tunnel.

2.1. Wing geometry

As mentioned before a one to one mimicry of an in-vivo batwing is difficult because the wing shape of a bat wing is extraordinarily complex. It is also difficult since there are no standard dimensions to a natural bat wing. The problem of complexity can be overcome through simplification where the wing that was used in this study was considered as a flat wing. The only shape that is copied from a batwing was its margin shape.

Since there are no standard dimensions for a batwing can be solved by taking an average observation of a batwing. In this study, an observation of batwing done by Hubel, T.Y. et al., 2009 [11] was used as a source, as seen in figure 1.

Figure 1. The average wing used as the base for wing geometry generation [11].

The image was uploaded into the Computational Aided Design (CAD) software, CATIA. The wing shape was then sectioned into a 50 sections grid and vertices were then generated at the margin or edge of the wing (figure 2(a)). The generated was then used to draw a wing model (figure 2(b)).

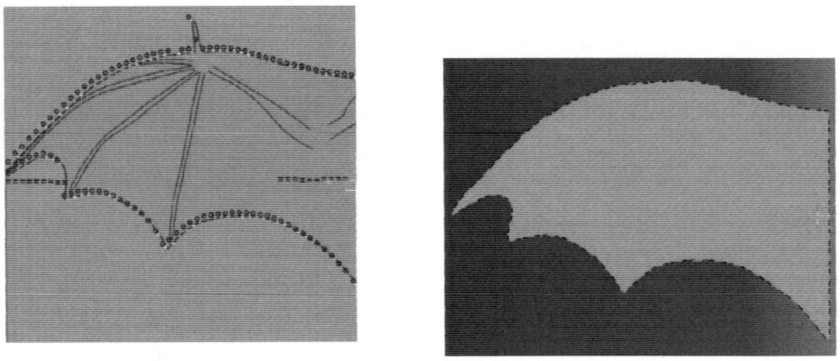

(a) (b)

Figure 2. Wing generation process where; (a) is the margin tracing process and (b) is the generated wing geometry.

The final wing geometry will then be used in a Computational Fluid Dynamics (CFD) simulation where the lift generation will be calculated. The wing geometry will also be then 3D printed and the lift generation will be measured in an experimental study. Table 1 shows the specifications of the final wing geometry.

Table 1. Wing geometry specifications.

Dimensions	Values
Wing surface area	13cm^2
Chord length	0.34cm
Wing thickness	1mm
Root to tip length	6cm

2.2. Operating Conditions
Since the study was meant to closely mimic the flight pattern of a bat wing, the tested flight conditions were based on previous observation. According to E.K.V. Kalko [4], it was observed that at flight have several modes of flight which are: search flight, approach flight, capture flight, and retrieval of prey. The flight mode that this study is interested in is search flight where it was observed the flight speed to be at 3 m/s to 4 m/s. Therefore, the chosen flight speed is 4m/s, the flight frequency at 8.5Hz and finally the flapping angle 55°. The angle of attack of the study that was studied is 0° to 35°.

2.3. Simulation Method
The CFD simulation software that was used for this study was ANSYS V.16 utilising the Transient Structure, Fluent, and Systems Coupling modules. This study also used a simulation method called 2-way Fluid-Structure Interaction (FSI) method where the motion of the wing was calculated by the Finite Element Analysis (FEA) module (Transient Structure module), which is fed into the CFD module (Fluent) that calculates the fluid motion and the resultant force is then fed back into the FEA module. This back and forth calculation were done by the System Coupling module until convergence is achieved. Figure 3 shows the module layout for the simulation.

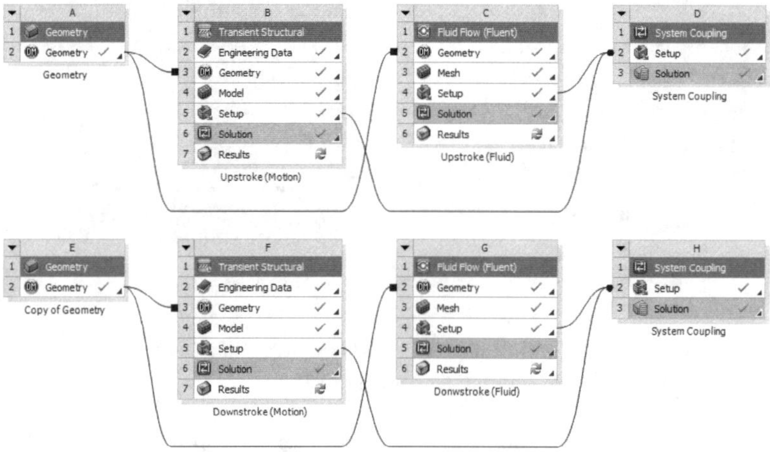

Figure 3. The modules layout for the FSI simulation

A common problem that is involved with FSI simulation is the problem of generated Negative Mesh which will cause an error in the simulation to occur. To avoid error this occurring, coarse, unstructured mesh was generated where71490 tetrahedral elements were generated (figure 4).

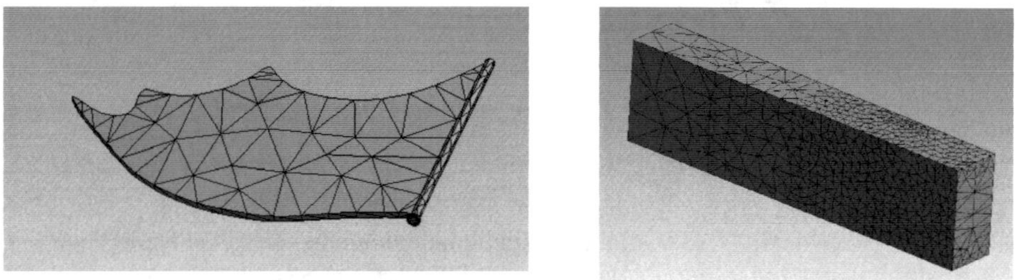

(a) (b)

Figure 4. The generated mesh where; (a) is the wing mesh and, (b) flow field mesh.

Due to the limitations involved with ANSYS software, the simulation can't calculate the motion to be calculated in a single simulation run. The upstroke and the down-stroke wing motion must be simulated separately. This can be overcome by simulating the down-stroke motion first and the result is transferred to the upstroke simulation as an initial condition.

Another precaution step taken to overcome a Negative Mesh error to occur is by using two flow field in the simulation. The internal fluid flow has a dimension of 0.1mX0.2mX0.2m and the external fluid flow have a dimension 0.6mX0.4mX2.5m. The flow field can be seen in figure 5.

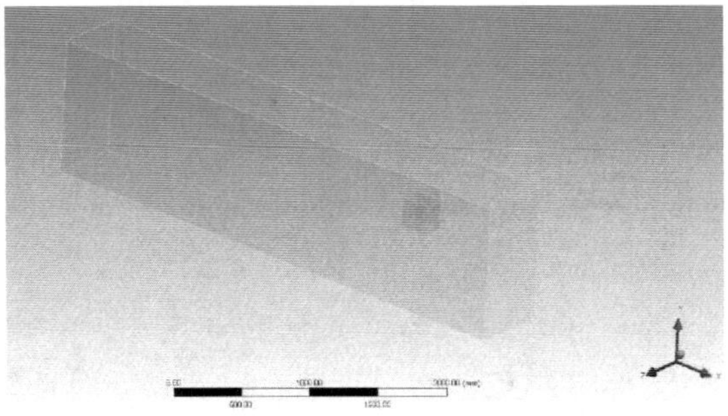

Figure 5. Flow field used for the simulation.

The chosen flapping condition was 8.5Hz this is achieved by setting the time step to be at 0.05882s for both the down-stroke motion and the upstroke motion. The sub-steps were set at 0.00098033s and the simulation uses 60 sub-steps.

The ANSYS software uses Navier-Stokes equation as its base equation for the calculation process which can be expressed as:

$$\frac{\partial u}{\partial x}+\frac{\partial v}{\partial y}+\frac{\partial w}{\partial z} = 0 \quad (1)$$

$$\frac{\partial u}{\partial \tau}+u\frac{\partial u}{\partial x}+v\frac{\partial u}{\partial y}+w\frac{\partial u}{\partial z} = -\frac{\partial p}{\partial x}+\frac{1}{Re}\nabla^2 u \quad (2)$$

$$\frac{\partial v}{\partial \tau}+u\frac{\partial v}{\partial x}+v\frac{\partial v}{\partial y}+w\frac{\partial v}{\partial z} = -\frac{\partial p}{\partial y}+\frac{1}{Re}\nabla^2 v \quad (3)$$

$$\frac{\partial w}{\partial \tau}+u\frac{\partial w}{\partial x}+v\frac{\partial w}{\partial y}+w\frac{\partial w}{\partial z} = -\frac{\partial p}{\partial z}+\frac{1}{Re}\nabla^2 w \quad (4)$$

Where;
u, v, and w are non-dimensional velocity components
p is the non-dimensional pressure
Re is the Reynolds number

The turbulence model that was used was the enhanced near-wall treatment k-epsilon model. Enhanced near-wall treatment was chosen to overcome the poor mesh quality that was generated. The simulation will then measure the lift generated by the flapping wing where the simulation defined as the force generated in the y-direction.

2.4. Experimental Method

For the experimental study, the wing model was fabricated using 3D printing method with PLA as its material. The flapping motion was achieved by using a 3D printed flapping mechanism that uses a series of gears and truss and powered a brushless DC motor. The flapping mechanism can be seen in figure 6.

Figure 6. Flapping mechanism

A sub-sonic open-circuit wind tunnel was used to generate the wind speed. The generated lift was measured using a series of strain gauge which was attached to a Kyowa PCD-30A Data Acquisition System (DAQ) that processes the signal and transfer them to a PC. Data processing process was done using the Kyowa Software where the 10,000 data points of measured lift in Newton (N) was recorded for 10 seconds (figure 7).

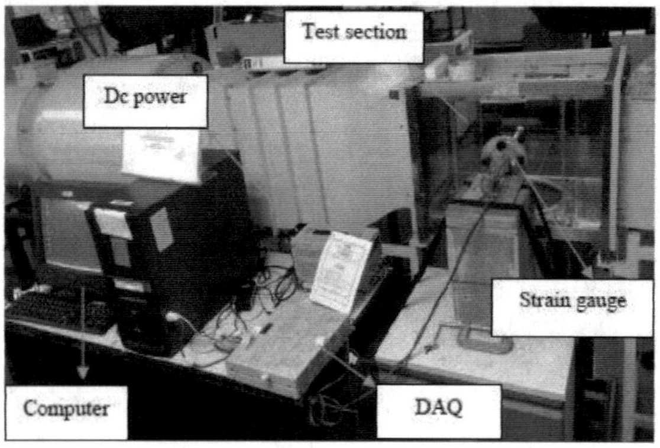

Figure 7. The experimental setup.

The recorded data was then processed using DAQ software to filter out the mechanical noise generated by the flapping mechanism and then transferred as a numerical value in excel. An average value of the measured data was then calculated, and the value was then converted to the lift coefficient by using the following equation:

$$C_L = \frac{L}{\rho V^2 A/_2} \quad (5)$$

Where;
C_L is the coefficient of lift
L is the generated lift
ρ is the air density
V is the flight speed
A is the wing surface area

3. Results and discussion

Figure 8 shows the observed result for both the simulation and the experimental study. Both results show a similar trend where the wing is observed to exhibit a normal lift curve where the generated lift increases with increasing AoA before the lift decays out and decreases at the stall angle, 25°. The lift then did not increase as much after that point. The wing also has a maximum lift coefficient C_{Lmax} of 3.2.

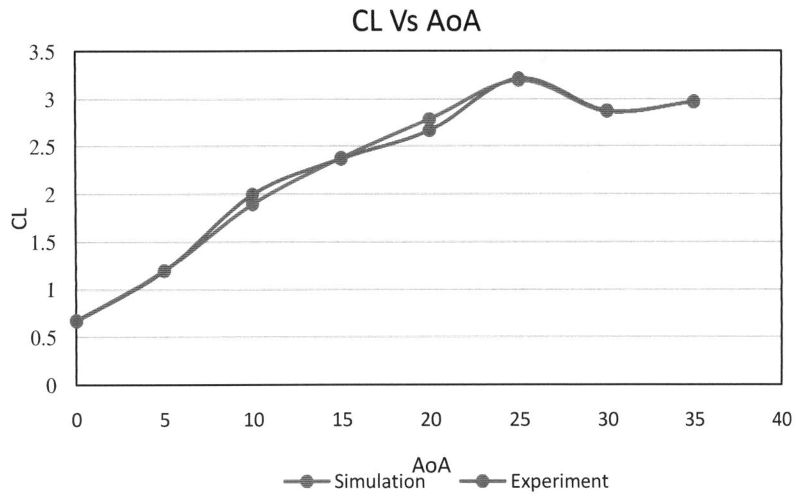

Figure 8. Lift generation over the different angle of attack for both simulation and experimental study.

In terms of the difference between the simulation study and the experimental measurement, it is shown that there are small differences between the two. The smallest difference is at 0.2% while the largest difference between is found to be at 5%. Since the difference between the simulation and the experimental study shows that the simulation calculation is validated and can be used as a foundation for future design and study.

4. Conclusions

The main purpose of the study was to serve as a basis for the future design of a flying flapping wing MAV inspired by natural bat wings. The results showed that the wing geometry was able to generate a significant lift generation with a 25° stall angle which shows the potential of the wing design. More importantly, the CFD model results show a close agreement with the experimental result. This means that the CFD model is valid to be used for future studies of new wing designs and flight conditions.

Acknowledgements

The authors would like to express their gratitude to Research Management Institute (RMI) of UiTM and Ministry of Education, Malaysia for financial support through Research GIP GRANT awards 600-IRMI 5/3/GIP(056/2019).

IOP Conf. Series: Materials Science and Engineering 1041 (2021) 012056 doi:10.1088/1757-899X/1041/1/012056

References

[1] Whitesides G M 2015 Bioinspiration : something for everyone

[2] Menezes A A and Kabamba P T 2016 Efficient and resilient micro air vehicle flapping wing gait evolution for hover and trajectory control *Eng. Appl. Artif. Intell.* **54** 1–16

[3] Galinski C and Zbikowski R 2007 Some problems of micro air vehicles development *Bull. Polish Acad. Sci. Sci.* **55** 91–8

[4] Kalko E K V. 1995 Insect pursuit, prey capture and echolocation in pipistrelle bats (Microchirptera) 861–80

[5] Bahlman J W, Swartz S M and Breuer K S 2013 Design and characterization of a multi-articulated robotic bat wing *Bioinspir. Biomim.* **8** 16009

[6] Yin D and Zhang Z 2016 The inertial power and inertial force of robotic and natural bat wing *Comptes Rendus - Mec.*

[7] Wang J, Zhang C, Wu Z, Wharton J and Ren L 2017 Numerical study on reduction of aerodynamic noise around an airfoil with biomimetic structures

[8] Wu J, Qiu J and Zhang Y 2017 Automated Kinematics Measurement and Aerodynamics of a Bioinspired Flapping Rotary Wing *J. Bionic Eng.* **14** 726–37

[9] Guo S, Li H, Zhou C, Zhang Y L, He Y and Wu J H 2018 Analysis and experiment of a bio-inspired flyable micro flapping wing rotor *Aerosp. Sci. Technol.* **79** 506–17

[10] Yusoff H, Iswadi N, Zulkifly A H, Firdaus S M, Abdullah M Z, Abdullah M K and Suhaimid S 2015 Lift performance of a cambered wing for aerodynamic performance enhancement of the flapping wing *J. Teknol.* **75** 43–7

[11] Hubel T Y, Hristov N I, Swartz S M and Breuer K S 2010 Time-resolved wake structure and kinematics of bat flight *Anim. Locomot.* 371–81

IOP Conf. Series: Materials Science and Engineering 1041 (2021) 012057 doi:10.1088/1757-899X/1041/1/012057

Crack Analysis on Ti-6Al-4V ELI Coated with Commercial Hydroxyapatite for Implant Material: Literature Review

S Ardhy[1,2], Gunawarman[2], J Affi[2], Y Yetri[3]
[1]Mechanical Engineering Department, Dharma Andalas University, Padang, Indonesia
[2]Mechanical Engineering Department, Universitas Andalas, Padang, Indonesia
[3]Mechanical Engineering Department, Politeknik Negeri Padang, Indonesia

Corresponding author: gunawarman@eng.unand.ac.id

Abstract. Ti-6Al-4V Extra Low Intertitial (ELI), a type of titanium alloy that is commonly used for implant material. However, use of Ti-6Al-4V ELI still has some problems. The main one is oxide layer on the surface can be detached during application due to corrosion process. Such corrosion debris is a toxic substance to the human body. It is, therefore, necessary to modify the surface of the implant to avoid corrosion process by using coating method. A good coating material for orthopedic implants comes from ceramics, namely hydroxyapatite (HA). However, there are also an obstacle, HA coating on titanium, sometimes not evenly across the surface, causing cracks on the coating layer surface. There are two methods that are often used in coating materials, namely Electrophoretic Deposition (EPD) and Dip Coating. This study analyzed number of literatures using a searching engine of international reputable journals Science Direct, Pubmed on biomaterials and biomedicine fields using the keywords of Ti-6Al-4V ELI, hydroxyapatite, cracks, EPD method and Dip Coating. This literature review may useful for the readers and researchers, in order to choose the best method for crack reduction in the coated implants. Dip Coating method is better than EPD method in point of view simplicity and cost. Moreover, the coating layer is evenly distributed, no build-up of HA in several surface areas, so that cracks on surface are minimized.

1. Introduction

Ti-6Al-4V Extra Low Intertitial (ELI), an $\alpha+\beta$ type of titanium alloy, is usually used for biomedical applications. This is due to this alloy has better physical and mechanical properties than stainless steel and pure titanium; namely lighter, more ductile, stronger, lower modulus of elasticity, and better biocompatibility. Currently, Ti-6Al-4V ELI is common for dental implants as well as for orthopedic implants such as bone fixation (plates, screws, pins), spine fixation and artificial joints [1]. However, the use of Ti-6Al-4V ELI still causes problems. Kwok [2] exposed that the use of Ti-6Al-4V ELI has a risk of corrosive attack by body fluids, and the release of metal titanium oxide (TiO_2) particles. Such particles are dangerous because has adverse effects on the surrounding tissue. The oxide layer on surface can be released due to corrosion process so that it is toxic to the body. This condition can also result in reducing growth of new bone tissue. Another problem is the metal surface is not bioactive, so it is unable to form osseointegration (the link between the bone and the implant material) in the body. It is necessary, therefore, to modify the surface of material by using surface coating method. This coating aims to avoid direct material contact with body tissues, provides protection against the release of unsafe ions from material, and accelerates the occurrence of osseointegration, stimulates the growth of new bone tissue to repair damaged bone tissue. If there is no osseointegration, the implant will have looseness, and then causing of implant failures. A good coating material for orthopedic implants comes from ceramics, namely hydroxyapatite (HA). HA has bioactive, inert and non-toxic properties. The HA coating aims to prevent such bone loosening. In addition, it also aims to protect the surface of implant from corrosion and direct contact to the body tissues [3].

HA is one of the inorganic compounds or calcium phosphate compounds, which make up the hard tissue of the human body such as bones and teeth. HA is biocompatible because it has a structure that resembles the matrix structure of the human bone element $Ca_{10}(PO_4)_6(OH)_2$. In

IOP Conf. Series: Materials Science and Engineering 1041 (2021) 012057 doi:10.1088/1757-899X/1041/1/012057

addition, HA also has bioactive properties, and it can accelerate the regeneration of bone tissue during implantation. According to Cheng et al. [4], Ti-6Al-4V ELI coated with HA has a higher corrosion resistance level than non-coated one. With this good level of corrosion resistance, it is able to provide a positive biological response to the body. However, there is still a barrier. HA coating layer on titanium, sometimes are not evenly across the surface. Juliadmi [5,6] found that coating Ti-6Al-4V with commercial HA (particle size 10 μm) by using Electrophoretic Deposition (EPD) method, produced in an unequal layer over the whole surface of the substrate and there is a buildup of HA and cracks in several surface areas. A larger the particle size of HA produces higher cracks on the layer [7,8].

HA coating of micro particles on Ti-6Al-4V ELI with the EPD method, the layer is easily brittle and many cracks on the surface. Another method that is widely used for hydroxyapatite coating is the dip coating method. This method has a number of advantages, including low cost of coating, easy coating method, can control the desirable coating thickness and can be carried out at room temperature, so it does not change the properties of HA.

From a number of literature studies [8-12], it is found that Dip Coating method can produce a better evenness of the surface layer than the EPD method. In addition to the HA layer that is evenly distributed and without cracks, the implant material must also contain a good porous (8). This cavity is generated from the heating process (sintering). This cavity serves as the retention of bone cells during implantation. The immersion speed of dip coating method has significant effect on the characteristics of Ti-6Al-4V hydroxyapatite-gelatin coating layer, where the increase of immersion speed, an even surface layer of the implant is obtained, so that the attachment of the layer to the substrate is stronger [10]. The same trend is also obtained for HA coated SS316L where no cracks is obtained at the coating layer [11]. It is also found that the layer thickness increases with increasing drawing speed of the sample from the bath [12]. For instance, the withdrawal speed of 5 cm/minute (0.083 mm / s), resulting in a denser HA structure and severe layer on the surface of Ti-6Al-4V. However, it is necessary to prove that the hydroxyapatite coating layer is not easily to cracke under pressure of fluid during application.

This study analyzed number of literatures using a searching engine of international reputable journals Science Direct, Pubmed on biomaterials and biomedicine fields using the keywords of Ti-6Al-4V ELI, hydroxyapatite, cracks, EPD method and Dip Coating. This study to identify relevant studies regarding to choose the best method for crack reduction in the coated implants. The inclusion criteria used for this review involved the coating layer studies written in English language up to May 2018. Three of the authors independently evaluated the titles and abstracts of potentially relevant previous studies. Selected studies were individually read and analyzed concerning the focus of the present review, which was to discuss and summarize important key publications regarding crack analysis on Ti-6Al-4V ELI coated with commercial hydroxyapatite for implant material. Author names, journal, publication year, objectives, methods, and main outcomes were retrieved from the selected relevant articles.

2. Results and discussion

Dip Coating method consists of three steps as shown in figure 1. First, the immersion step. In this case, the implant material is dipped in HA solution. Second, the implant material that has been dyed is pulled upwards. The pulling velocity factor will determine the thickness and evenness of the surface layer of implant material obtained. In this method, it is necessary to pay attention to controlling the withdrawal speed, in order to obtain the desired coating results. Third, the evaporation process, in which the solvent evaporates from the liquid, forming a thin layer.

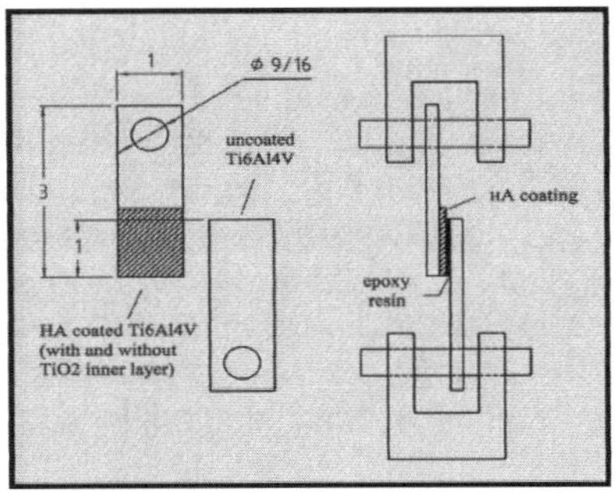

Figure 1. Schematic of dip coating method [13]

Coating Ti-6Al-4V ELI with micro-HA (particle size 10 μm) by using Electrophoretic Deposition (EPD) method, produced in an unequal layer over the whole surface of Ti-6Al-4V ELI and there is a buildup of HA and cracks in several surface areas [5,6. Not only that, the HA layer also disclosed some cracks [7]. Nevertheless, the coating of Ti-6Al-4V ELI with HA 125 μm, the HA layer is also not evenly distributed and this will simultaneously cause cracks to appear on the surface of Ti-6Al-4V ELI.

A larger the particle size of HA produces higher cracks on the layer [7]. The larger of HA particle size has the greater cavity (porosity). The 125 μm size of HA particles is quite large, the ethanol used as the HA solvent, trapped in the cavity formed during the HA coating process, is also quite large. This will cause cracks on the surface of layer during the sintering process, because the trapped ethanol evaporates quickly.

Similar results were obtained from the study of Gnanavela et al [8]. HA coating of micro particles on Ti-6Al-4V ELI using EPD method, obtained the best morphological results at a voltage of 8 volts and a time of 8 minutes, but the layer is easily brittle and many cracks on surface. The cracks in the HA layer were also caused by the nature of the ceramics not having good crack toughness during the sintering process.

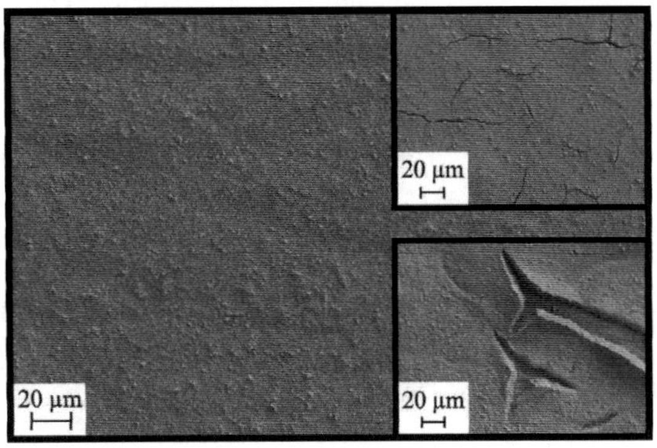

Figure 2. Cracked on Ti-6Al-4V ELI surface [8].

This cavity is generated from the heating process (sintering). This cavity serves as the retention of bone cells during implantation. The immersion speed of dip coating method has significant effect on the characteristics of Ti-6Al-4V hydroxyapatite-gelatin coating layer, where the increase of immersion speed, an even surface layer of the implant is obtained, so that the attachment of the layer to the substrate is stronger [10]. The variations in the withdrawal speed used were 0.089 mm/s, 0.120 mm/s, 0.178 mm/s and 0.198 mm/s. The higher the withdrawal speed, an even surface layer of the implant is obtained, so that the attachment of sample to the substrate is stronger. At the variation of withdrawal speed of 0.198 mm/s, the thickness and shear strength values were 125.28 μm and 13.96 MPa. In this study, commercial HA mixed with 30 percent gelatin was used.

The same trend is also obtained for HA coated SS316L where no cracks is obtained at the coating layer [11]. It is also found that the layer thickness increases with increasing drawing speed of the sample from the bath [12]. For instance, the withdrawal speed of 5 cm/minute (0.083 mm/s), resulting in a denser HA structure and severe layer on the surface of Ti-6Al-4V. However, it is necessary to prove that the hydroxyapatite coating layer is not easily to cracke under pressure of fluid during application. Acquired the HA layer evenly on the surface of Stainless Steel (SS) 316L as an implant material with dip coating method, there were no cracks. The HA used is commercial HA plus additive solvent P_2O_5, Na_2CO_3 and KH_2PO_4, and cracks were not found when the sintering temperature was 600^0 C.

Figure 3. Ti-6Al-4V ELI has a crack free surface [13].

Sintering affects the density of bilayers hydroxyapatite coatings. Although the high temperature treatment at 900°C improves densification and reduces porosity, the coatings crack followed by peeling of the certain layers. High porosity, one of the EPD process flaws, encourages corrosion and releases titanium material ions induced by body environment. Porosity reduction by high temperature promote cracks might be caused by the difference of material thermal coefficient [14]. It will effect expansion and shrinkage difference between coating and material at heating and cooling of sintering process thus resulting thermal stress [15,16]. The Ti-6Al-4V materials have thermal coefficient ~10.3µm/mK lower than hydroxyapatites (~14 µK/mK) that effects resulting of residual tensile stresses promoted the cracking [15]. It proved that crack propagation to used sintering rather than EPD process [17].

Nanoparticle and micro particle hydroxyapatite well bonds to Ti-6Al-4V ELI material surface when sintering was given at 800°C temperature. The non-cracks of nanoparticle hydroxyapatite role as the protecting toward possibility of material corrosion and acting in biology activity at implantation. The loosening micro particle hydroxyapatite may be dissolute acting directly for bone remodeling through the forming of interfacial coatings. Using nanoparticle hydroxyapatite as intermediate between micro particle hydroxyapatite and material surfaces have suitability of thermal coefficient with titanium material cause of number atoms at grain boundaries increasing particle mobility [18]. It caused sintering with 800°C temperature for 10 minutes have not cracks propagation. The previous study reports that the 800°C temperature for 1 hour is optimum condition for hydroxyapatite sintering in vacuum condition because not causes structural hydroxyapatite changes and cracks [19,20] with adhesion strength 47 ± 0.5 MPa [21].

4. Conclusions

1. Dip Coating method is better than EPD method in point of view simplicity and cost. EPD coating method produces an uneven layer over whole of Ti-6Al-4V ELI surface and there is

a buildup of HA in some surface areas. The HA layer also showed some cracks. The larger the particle size of the coating, the more cracks will occur in layer.
2. Dip Coating method can control thickness of the HA layer, namely by varying the speed of implant material withdrawal during immersion. The coating layer is evenly distributed, no build-up of HA in several surface areas, so that cracks on surface are minimized.

References

[1] Boyer R and Welsch G 1994 *Materials Properties Handbook: Titanium Alloys* (USA: ASM International)
[2] Kwok W, Swain M, Ruys A, Milthorpeand B and Sorrell C 2009 Characterization and Corrosion Behavior of Hydroxyapatite Coatings on Ti6al4v Fabricated by Electrophoretic Deposition *China. Department of Electromechanical Engineering, University of Macau.* **45** 11-19
[3] Zhang S, Chen F, Lam W, Lin C, Qiu G, Wu Z, Luk K and Lu W 2013 Hydroxyapatite Coatings for Biomedical Applications *Advances in Materials Science and Engineering.* 169-73
[4] Cheng C, Wang X, Poh C, Tan H, Soe M, Zhang S and Wang W 2009 Characterization and Corrosion Behavior Of Hydroxyapatite Coatings on Ti6Al4V Fabricated by Electrophoretic Deposition *China. Department of Electromechanical Engineering, University of Macau.* **8** 1099-1108
[5] Fernanda S, María E and Galárraga 2019 Zirconia surface modifications for implant dentistry *Materials Science & Engineering.* **98** 1294–1305
[6] Xia W, Fu L and Engqvist H 2017 Critical cracking thickness of calcium phosphates biomimetic coating Verification via a Singh-Tirumkudulu model *Ceram Int.* **43** 15729–34
[7] Gnanavela S, Ponnusamy S and Mohanc L 2018 Biocompatible response of hydroxyapatite coated on near-β titanium alloys *Biocatalysis and Agricultural Biotechnology.* **15** 364–69
[8] Li JP, Habibovic P, Wilson C, Wijn J and Blitterswijk C 2007 Bone ingrowth in porous titanium implants produced by 3D fiber deposition *Biomaterials.* **28** 2810-20
[9] Rachmawati D, Fadli A and Amri A 2019 *Efek Variasi Kecepatan Penarikan Terhadap Karakteristik Ti-6Al-4V Terlapis Hidroksiapatit-Gelatin Menggunakan Metode Dip Coating Sebagai Implan Tulang* (Surabaya: Universitas Airlangga Press) 238-57
[10] Mulya N, Fadli A and Amri A 2018 *Pengaruh Penambahan Hidroksiapatit dan Waktu Pencelupan Terhadap Pelapisan Stainless Steel 316L Dengan Metode Dip Coating* (Pekanbaru: Universitas Riau) 659-66
[11] Jafari S, Mahdi M and Idris J 2012 *Thick Hydroxyapatite Coating on Ti-6Al-4V through Dip Coating Method* (Malaysia: Universiti Teknologi Malaysia) 238-57
[12] Habibovic P, Barrère F, Blitterswijk G and Layrolle P 2002 Biomimetic hydroxyapatite coating on metal implants. *Journal of Biomedical Materials Research Part B: Applied Biomaterials.* **83** 17-22
[13] Abbasi S, Golestani F, Rezaie H and Mirhosseini S 2012 MAO-derived hydroxyapatite/TiO_2 nanostructured multi-layer coatings on titanium substrate *Applied Surface Science.* **26** 37-42
[14] Wei M, Swain M, Ruys B and Milthorpeand S 1999 Solution ripening of hydroxyapatite nanoparticles: Effects on electrophoretic deposition *Biomedical Materials Research.* **45** 11-19
[15] Yamashita K, Yonehara X, Ding M and Umegaki M 1998 Electrophoretic coating of multilayered apatite composite on alumina ceramics *Biomedical Materials Research.* **43** 46–53
[16] Kollath V, Chen R, Closset J, Luyten K, Traina S, Mullens A, Boccaccini and Cloots R 2013 AC vs DC Electrophoretic deposition of hydroxyapatite on titanium *European Ceramic Society.* **33** 2715-21
[17] Yamashita K, Yonehara X, Ding M, Nagai T and Matsuda M 1998 Electrophoretic coating of multilayered apatite composite on alumina ceramics *Biomedical Materials Research.* **43** 46–53
[18] Montenero A, Ganapi F, Ferrari M, Cesari E, Salvioli L, Mattogno S and Fini M 2000 Sol-gel derived hydroxyapatite coatings on titanium substrate *Materials Science.* **35** 2791-97
[19] Zhang Q, Chen J, Feng J, Cao Y, Deng C and Zhang X 2003 Dissolution and mineralization behaviors of HA coatings *Biomaterials.* **24** 4741–48
[20] Chen F, Lam WM, Lin CJ, Qiu GX, Wu ZH, Luk KD and Lu WW 2007 Biocompatibility of Electrophoretical Deposition of Nanostructured Hydroxyapatite Coating on Roughen Titanium

IOP Conf. Series: Materials Science and Engineering 1041 (2021) 012057 doi:10.1088/1757-899X/1041/1/012057

Surface: In Vitro Evaluation Using Mesenchymal StemCells *Biomedical Materials Research Part B: Applied Biomaterials*. 183-91

[21] Chen F, Lam WM, Lin CJ, Qiu GX, Wu ZH, Luk KD and Lu WW 2007 Biocompatibility of Electrophoretical Deposition of Nanostructured Hydroxyapatite Coating on Roughen Titanium Surface: In Vitro Evaluation Using Mesenchymal StemCells (*Biomedical Materials Research Part B: Applied Biomaterials*) pp 183-91

Andalas Medical Robot Assistance (AMIRA) serves Covid-19 isolation patients

F Ridwan[1], S Syamsuddin[1], A Fathan[1], A A Ananta[1], G A Bintang[1], A Jordan[1], Eka M Ningsih[1], M Fadel[1]

[1]Department of Mechanical, Engineering Faculty, Andalas University, Padang, Indonesia

Corresponding author: firmanridwan@ft.unand.ac.id

Abstract. The world is currently alert to the spread of a virus known as the coronavirus. Corona Virus or Severe acute respiratory syndrome coronavirus 2 (SARS-CoV-2) is a collection of viruses that can infect the human respiratory system until death. The spread of the virus is very fast and easy just by coming in contact with sufferers until it has contaminated almost all countries. For places prone to being exposed to viruses, especially in isolation rooms for Covid sufferers, the medical staff is terrified so that they are always on their toes; moreover, dozens of medical workers have died in handling patients.For this reason, innovation is needed to reduce contact between medical personnel and Covid-19 patients by replacing the duty of humans with robots so that the risk of infected medical workers decreases and avoids the level of work fatigue on the medical worker. The team from Andalas University created and designed a robot that could solve these problems. The robot was designed to perform activities carried out by nurses in serving patient's needs. This robot has the necessary features: food delivery, the patient room's smart position, protecting certain areas with disinfectants, reading the patient's body temperature, and a reminder of the patient's medication schedule anti-collision and providing other smart features which needed by the hospital. This robot is called Andalas Medical Robot Assistant or abbreviated as AMIRA.

1. Introduction

COVID 19 (Corona Virus Disease 2019) is a pandemic disease caused by nonsegmented, single-stranded, positive-sense RNA genomes coronaviruses. COVID 19 emerged in Wuhan, China, at the end of 2019. In the early 2020 disease spread rapidly in China and then across the globe. The liquid spreads coronavirus in the respiratory system, such as mucus. Corona infection is transferred from contaminated air or exposed to a virus that is patched to the object. Coronavirus has many symptoms such as fever, hard to breathe, cough, etc. [1]. An infected person takes 2 – 14 days to realized its symptoms after infection.

In Indonesia, coronaviruses appear in early February 2020. Indonesia's government takes a serious step by enacting or large-scale social restriction (PSBB) in March 2020. By 31 March 2020, COVID 19 cases reached 1.528 cases with 136 death toll. Indonesia's mortality rate also more prominent than the world mortality rate by 4,68% and 3.79%, respectively per 3 august 2020 [2]. As of 2 September 2020, Indonesia has 180.646 confirmed cases, with 7.616 deaths and 129.971 recovered across 34 provinces [3]. One of the most significant outbreak causes this blown up number is hospital cluster.

Hospital cluster is one of the biggest spread COVID cases. More than 24.000 infected from hospital. From all cases, 63% cases came from a hospital outbreak. One of the causes of big number of

COVID 19 cases in hospital is contact between the patient and the medical team. Daily checking task is the source of communication between patient and medical team such as food delivery, clothes delivery, temperature check, etc. This is a serious risk that can be minimized by using a robot as a deliver assistant.

The hospital cluster is one of the biggest causes of the spread of COVID. More than 24,000 were infected from the hospital outbreak [4]. Of all cases, 63% of cases were from hospital outbreaks. One of the reasons for a large number of COVID 19 cases in hospitals is patient contact with the medical team. Daily check-up tasks are a communication source between the patient and the medical team, such as food delivery, clothing delivery, temperature checks, etc. This is a serious risk that can be minimized by using a robot as a delivery assistant.

2. Materials and methods

AGV (Automated Guided Vehicle) has played a primary role in increasing its efficiency, especially in transportation. AGV plays a role in smart logistics, such as material distribution. The user only needs to assign a delivery task to AGV, AGV delivers goods, product, or material to a designated location. Most AGV system use the priority-based single AGV order task method. AGV received the task with the highest prioritization. When the task is done, the next arranged task can be accepted [5]. By using this method, AGV can do tasks done by medical staff such as food delivery, clothes delivery, room monitoring, etc.

Andalas Medical Robot Assistant (AMIRA) is an AGV that is built to do that stuff. AMIRA will be equipped with an adequate sensor, controller, and actuator in AGV to function correctly.

2.1. Local Positioning System

The navigation of AGV AMIRA is accomplished by using a joystick located in the nurse station. Raspberry Pi 4 is connected to Logitech Extreme 3D Pro by using a WIFI connection. Raspberry Pi 4 sends data to Arduino Uno to control the navigation of AGV.

Infrared Radiation (IR) was installed to detect the surround of AGV. IR is used to track a person or object through the receiver and emitters. When there is an object with a range 1 m from AGV, the IR sensor will send a signal to Arduino that is transferred to Raspberry Pi 4.

Another technology installed to produce higher precision of the local positioning system is using WIFI technology. WIFI module receiver installed in AGV and WIFI module transmitter installed in the wall. By using the distance between the receiver and every transmitter, AGV location can be mapped effectively [6]. Navigation system of AMIRA can be seen in figures 1 and 2.

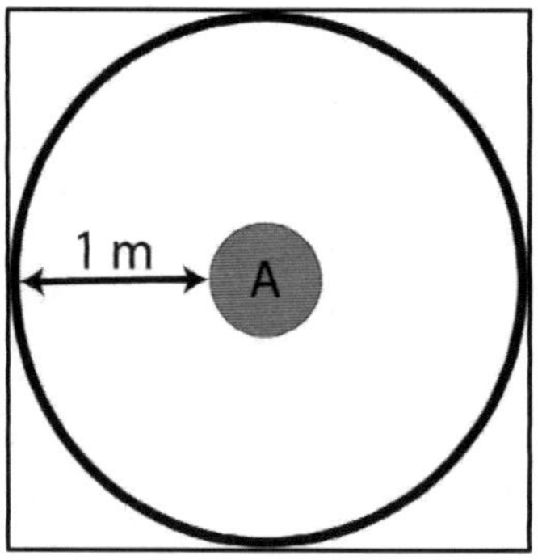

Figure 1. IR sensor Range

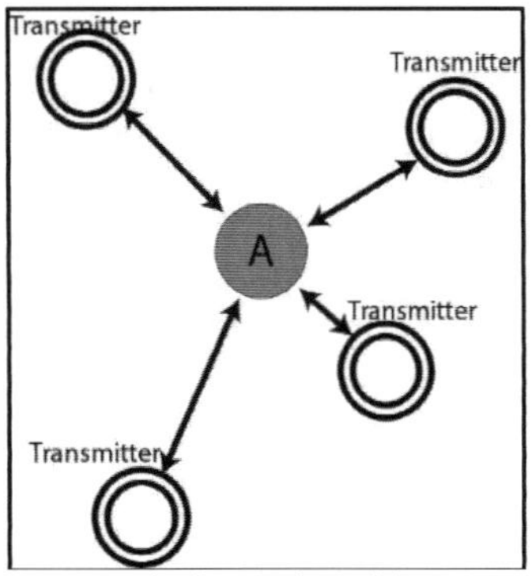

Figure 2 Wi Fi Local Position System

2.2. Temperature Measurement

One of the symptoms of COVID 19 is fever, so the patient's temperature needs to be regularly checked by the medical team. The medical team was tasked to check the temperature of patients vulnerable to expose COVID 19 viruses. AMIRA is installed temperature sensor to measure the body temperature of the patient. The temperature sensor will help medical team check the temperature of patient remotely. AMIRA will lead to patient location, patient pointed forehead 10 – 20 cm to the sensor. This way, the patient temperature was measured and displayed both in AMIRA LCD and nurse station LCD. The scanning process of temperature measurement by AMIRA can be seen in the figure 3.

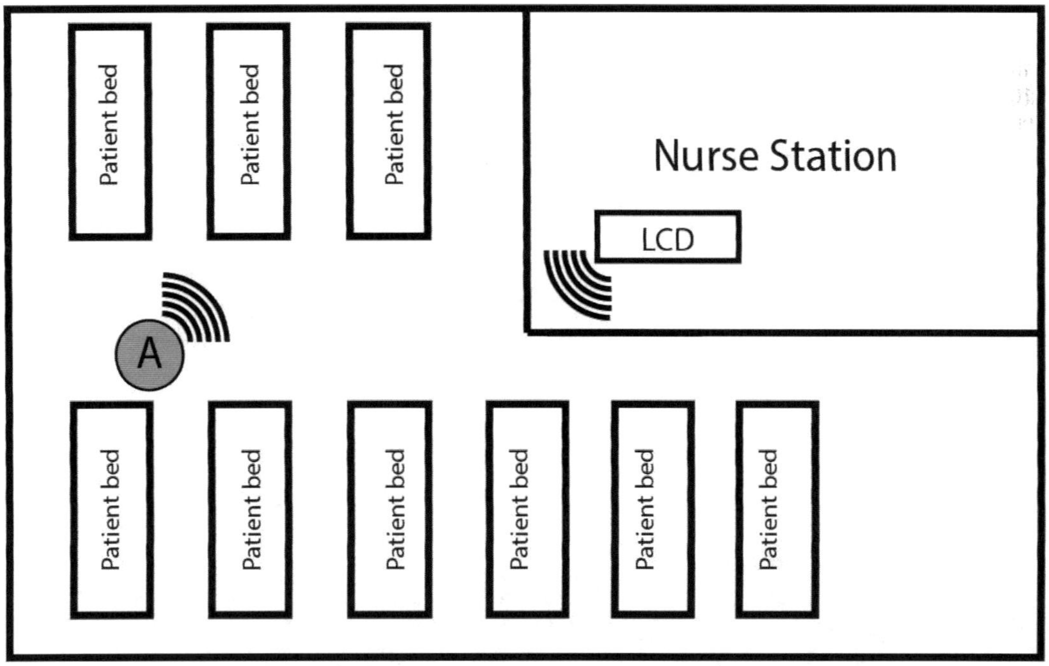

Figure 3. Temperature Measurement Remotely

2.3. *Anti-collision system*

AMIRA is installed with an anti-collision system by using an IR sensor and vibration sensor. IR sensor is based on reflected infrared from the object in front of the sensor. IR sensor is placed around AMIRA so that whenever an object in a range of 60 cm of AMIRA, AMIRA will immediately stop. The vibration sensor is placed around the bottom part of AMIRA, behind the collision pad. This collision pad is made from soft rubber, so it will absorb the vibration when the collision occurs. Damped by the pad, goods in the upper part of AMIRA will stay at its place when the crash occurs. The vibration sensor will warn the operator via WIFI connection that a collision has occurred. The anti-collision system can be seen in figure 4.

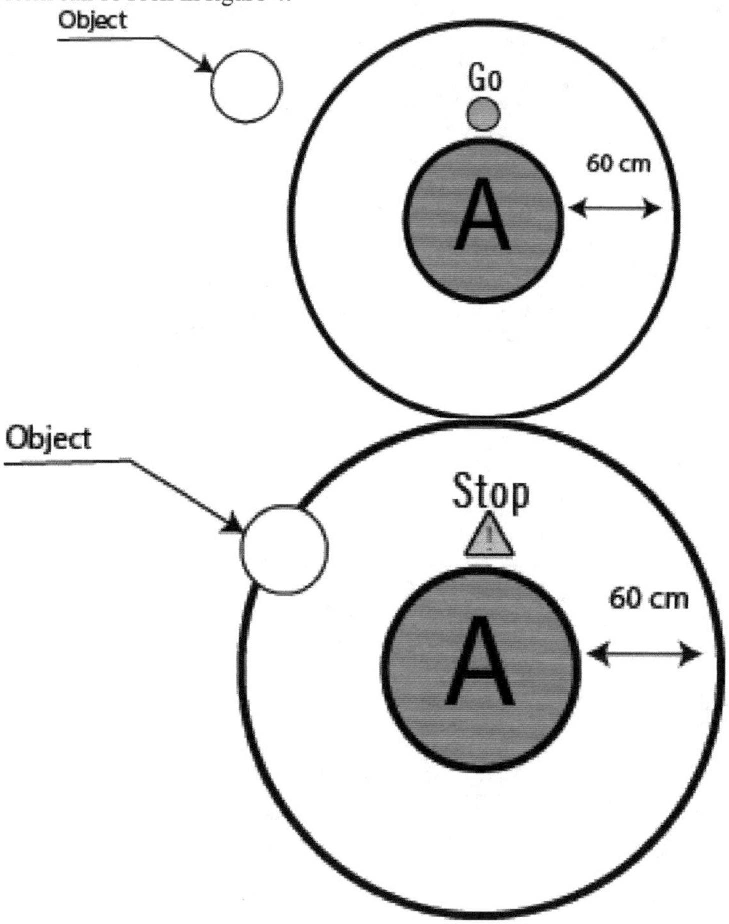

Figure 4. Anti-collision system

2.4. *Hands-free Disinfectant*

Disinfectant is chemical agents that are designated to inactivate or destroy microorganisms on an inert surface. Coronavirus spread mostly by respiratory droplets and next to contaminating surface by droplets. Disinfectant helped to reduce the deployment of coronaviruses by reducing coronavirus contaminations on the [7]. AMIRA is installed with a hands-free disinfectant located in the upper part, left side, and right side to prevent coronavirus contaminations. The disinfectant will shoot when the disinfectant channel's sensor detects an object. Apart from being a hands-free disinfectant, Amira is also equipped with room disinfectant spraying. Hans-free disinfectant and room sprayer can be seen in figure 5.

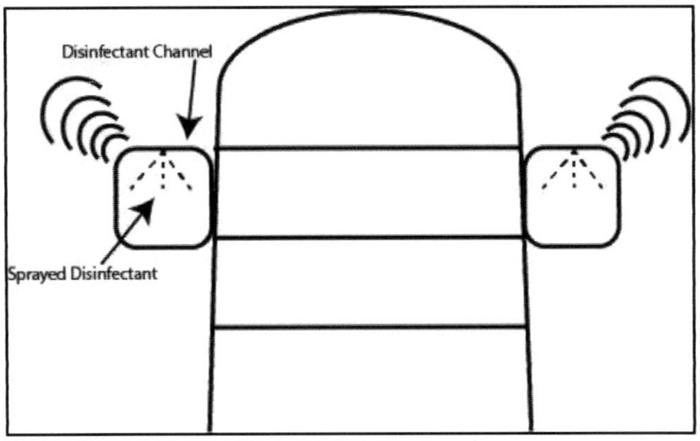

Figure 5. Hans-free disinfectant and room sprayer

2.5. Emergency System

Sometimes, AMIRA is needed urgently, like emergencies situation of the patient. When this happens, AMIRA needs to navigate unhindered. To satisfy these urgencies, AMIRA is installed with an emergency siren. This siren equipped with rotating light and resonate loud sound to alert people. The operator activates this siren. The siren is placed on the top, as seen the figure 6.

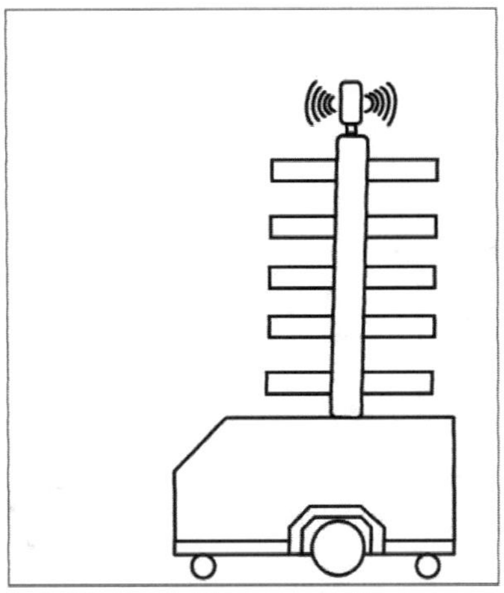

Figure 6. Emergency System

IOP Conf. Series: Materials Science and Engineering 1041 (2021) 012058 doi:10.1088/1757-899X/1041/1/012058

2.6. Auto Pilot
AMIRA can be switched between manual and automatic pilot mode. Manual mode is controlled using a Logitech Extreme 3D joystick from the operator's station. The joystick connects to AMIRA via a WIFI connection. Manual mode works in areas outside the range of the WIFI router. Another feature installed in AMIRA is Auto Pilot mode. This mode works by saving frequent commands to the Raspberry Pi's memory. The operator activates the autopilot mode. In this auto mode, a map of the location in a specific area is stored in the controller. The position of the transmitter assists it at a predetermined location. The transmitter sends a signal to the controller via the receiver. To overcome the collision on the underside, eight proximity sensors are applied. Anti-collision is placed on the front bumper.

2.7. Scheduled Delivery
The medical team has a schedule of visits, such as the patient's maternity needs, daily tasks, etc. The more frequent direct contact between the medical team and the patient will increase the exposure to respiratory droplets from coronavirus patients. AMIRA took over the task by programming the robot according to its commands and duties. AMIRA is equipped with a scheduled dispatch program that allows AMIRA to navigate and move to the operator's designated positions. AMIRA will send beverages to coronavirus patients so that the medical team's direct contact with patients will be reduced. Cameras attached to AMIRA can scan QR Codes. This QR code is placed in the designated scheduled delivery area. AMIRA navigates to the patient room area, and then the robot scans the QR Code that has been prepared attached to the door. In this situation, the robot has already known the patient's identity in the room. The robot alerts the patient via a loudspeaker. Then the patient will pick the item up from the AMIRA shelf. Patients must press one of the buttons on the monitor screen, stating that the goods have been received. by pressing the button, one task is completed and ready to perform the next task. Using QR Code technology, this task can be accomplished. The layout of the delivery map can be seen in figure 7.

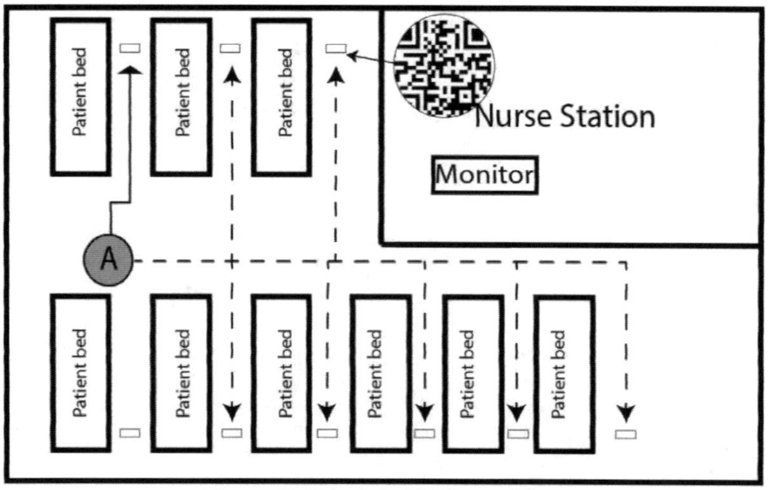

Figure 7. Scheduled Delivery Map

3. Results and discussion
This robot has two control systems, namely manual and automatic. The manual method is an application on the operator's computer to control the robot's motion via a joystick. The application is

built using the ECLIPSE IDE application in the Java programming language. The robot has a mini-computer that can communicate with the operator's computer. The communication system application is built using Python language.

Meanwhile, the Atmega328 microcontroller controls sensors and motors in the robot using the C/C++ programming language. The automatic system allows the robot to deliver goods on a scheduled basis without operator intervention as the robot controller. A local positioning system assists robot navigation. The system is built using several WIFI stations that are placed at predetermined points. The robot will calculate its position with these points by utilizing a large signal strength or RSSI (Receive Signal Strength Indicator). RSSI values are processed by the SVM (Support Vector Machine) algorithm. When the robot is in front of the isolation room door, the camera on the robot will read the QR code and call the patient according to the data in the database built with PHP and MySQL. AMIRA's system development process is ongoing and is planned to be completed by the end of the year. A rough prototype of Amira can be seen in figure 8.

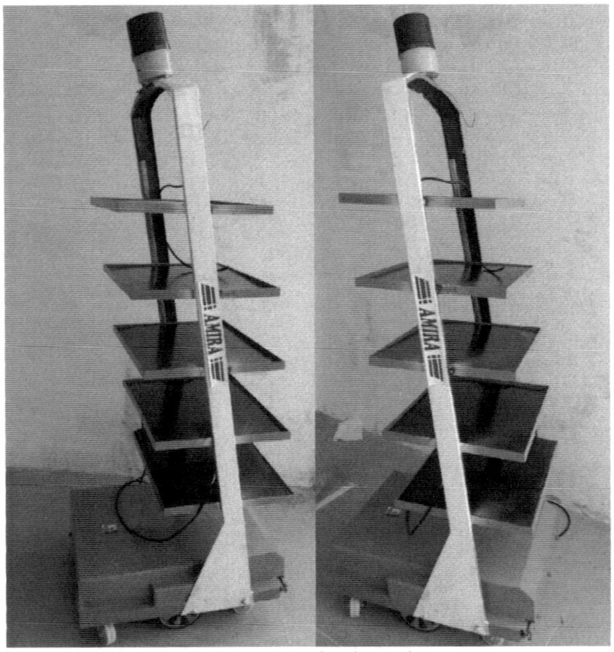

Figure 8. Amira in action

4. Conclusions

AMIRA's features can at least reduce direct contact between the medical team and COVID patients. AMIRA can play an essential role in reducing the spread of the virus. It is crucial to develop robots towards artificial intelligence to provide services to patients by offering hospital features, such as robots that can communicate with patients, entertain patients, etc. Therefore, the involvement of various fields of science is required to achieve this.

Acknowledgment

Publication of this article is supported by Research and Community Service Institutions of Andalas University grant with contract no.: T/3/UN.6.7/PP.RPP/LPPM/2020 the fiscal year 2020

References

[1] Unhale S S, Ansar Q B, Sanap S, Thakhre S and Wadatkar S 2020 a Review on Corona Virus (Covid-19) *World J. Pharm. Life Sci.* **6** 109–15
[2] Nugraheny D E 2020 Satgas: Angka Kematian akibat Covid-19 di Indonesia Masih Lampaui Persentase Global
[3] WHO Indonesia 2020 Coronavirus Disease Situation Report World Health Organization *World Heal. Organ.* **19** 1–17
[4] Chandra Gian Asmara 24.000 Tertular, Rumah Sakit Jadi Klaster Covid-19 Terbesar *22 Sept. 2020 1656*
[5] Wu G and Sun X 2020 AGV task distribution study *J. Phys. Conf. Ser.* **1486**
[6] Hasan H S, Hussein M, Saad S M and Dzahir M A M 2018 An overview of local positioning system: Technologies, techniques and applications *Int. J. Eng. Technol.* **7** 1–5
[7] Fathizadeh H, Maroufi P, Momen-Heravi M, Dao S, Köse Ş, Ganbarov K, Pagliano P, Espsoito S and Kafil H S 2020 Protection and disinfection policies against sars-cov-2 (Covid-19) *Infez. Med.* **28** 185–91

Preparation of avocado leaf simplicia macro/nano particles by using high energy ball mill

F Ridwan[1], M Fadel[1]

[1]Mechanical Engineering Department, Faculty of Engineering, Universitas Andalas, Kampus Limau Manis, Pauh, Padang, West Sumatera 25163, Indonesia

Corresponding author: firmanridwan@ft.unand.ac.id

Abstract. Taguchi method was used to examine high energy ball mill characteristics when producing micro/nano-particles of Simplicia powder using Orthogonal Array (OA) L25. Two factors and five levels were selected for the smallest diameter. The average fineness of the powder produced by high energy ball mill was calculated using the ASTM E11 standard Retsch vibrating sieve machine. The smoothest's fineness value was obtained at a working time of 30 minutes and a shaft rotation speed of 578 rpm with a powder fineness of 181μm. The coarsest fineness value was obtained at 10 minutes of working time and 488 rpm rotating speed with 336μm powder fineness. Obtained the smallest particle diameter has the potential for better absorption of nutrients into the body.

1. Introduction

Nowadays, many researchers compete to create material according to what they want with a particular function. Powder-shaped materials are in great demand. We have encountered much use of powdered materials in medicine, food production, electronics, health, cosmetics, and many more.

The body very well digests powdered herbs because the drugs' absorption into the body is outstanding. One of the most popular medicinal plants and properties is an avocado leaf. Avocado leaf extract can significantly reduce blood pressure in people with hypertension and reduce urea and keratin levels in the kidneys. Avocado leaves contribute to lowering blood pressure through its vasorelaxant effect[1].

Avocado leaf powder, which is widely consumed in the community, is still rough in size. The body does not effectively digest the raw powder because its absorption into the body is not optimal. Finer particle is good for the body's digestion because of better absorption of the drug into the body.

High energy Ball Mill can reduce the powder's size in the grinding bowl by grinding balls by utilizing the impact energy on the grinding bowl to produce friction and impact effects[2]. The Ball mill machine works to destroy the material through a collision between grinding media (crushing balls). The rough material is put into the vial (tube), filled with the ball media, and the balls will collide with the material to be refined[3].

2. Materials and methods

2.1 Size Reduction

Size reduction reduces larger or coarser particles to smaller or finer particles [4]. There are two methods of reducing particle size, namely the Top-Down method and the Bottom-Up method. The Top-Down approach is a mechanical method for reducing the size of the material. This method uses mechanical and kinetic friction from the grinding media, which is transferred to the material to be reduced (reduction). This method is widely used by industry. The bottom-up method uses a chemical reaction, in which the material is made into a gas, after which the material is synthesized onto the substrate by stacking atoms with each other to form crystals, these crystals accumulate with each other, resulting in a very small synthesis of the material [5].

2.2 Size Reduction Methods

There are three kinds of forces used to reduce the size of the particle. The three types of forces are compressive, impact, and shear force [6]. The crushing method, which is often used to reduce particle size, is shown in table 1.

Table 1. Method to Shrink Material Size.

Metode	Machine	Approximate particle size (μm)
Cutting	• *Scissors shears*	100 – 80.000
compression	• *Roller mill* • *Prestel-Mortar*	50 – 10.000
Impact	• *Hammermill*	50 – 8000
Attrition	• *Collodial mill*	1 – 5000
Impact and attrition fluid energy mill	• *Ball mill*	1 - 2000

2.3 Particle Size Observation

To filter particle size produced by the unbalance mass ball mill machine, the Retsch vibratory sieve machine is used to filter the smallest size based on ASTM E11 standards [8]. For this purpose, sieve numbers 35, 60, 120, and 230 are selected. The parameters used in this process are the sifting time for 10 minutes and vibration amplitude of 80. The Retsch vibratory sieve machine can be seen in Figure 1.

Figure 1. *Retsch Vibrating Sieve Machine.*

The equation for determining the particle distribution can be seen in the equation [9].

IOP Conf. Series: Materials Science and Engineering 1041 (2021) 012059 doi:10.1088/1757-899X/1041/1/012059

$$\text{No. Powder Fineness} = \left(\frac{\Sigma(\text{weight x multiplier })}{\Sigma\ \text{weight}}\right)$$

The average powder fineness measure was sought by interpolating the powder fineness value in the standard sieve analysis net by Standard sieve analysis mesh in table 2 [10]and Multiplier Standard Sieve Analysis Mesh in table 3[9].

Table 2 Standard *Sieve Analysis Mesh.*

*mesh*size	Aperture (μm)
18	1000
20	850
25	710
30	600
35	500
40	425
45	300
50	255
60	250
70	212
80	180
100	150
120	125
140	106
150	100
170	90
200	75
230	63
270	53
325	45
400	38
450	32
500	25
600	20

Table 3. Multiplier Standard *Sieve Analysis Mesh.*

Sieve size	Multiplier
6	3
12	5
20	10

30	20
40	30
50	40
70	50
100	70
140	100
200	140
270	200

2.4 Taguchi Method

In this study, the Taguchi method is used for the design of the experiment. The Taguchi method is chosen because this method can find the minimum number of experiments to be carried out from the selected factors and levels to produce more efficient, accurate, and straightforward investigations[11]. In this study, two factors and five levels are selected to be tested on a high energy ball mill. The factors are working time and shaft rotation speed. Each factor has five levels: 10 minutes 15 minutes, 20 minutes 25 minutes, 30 minutes, and 488 rpm, 511 rpm, 533 rpm, 555 rpm, and 578 rpm, respectively. Based on the orthogonal array for two factors and five levels, L25 (5^2) is chosen with 25 trials. In the experiments, the grinding media's material is 304 stainless steel with 5 mm in diameter. For validation purposes, the test is repeated four times. Analysis of Variance (ANOVA) is used to see the most significant factors and to conclude the rejection or acceptance of the factors [12]. Table 4 shows the orthogonal array of various levels on 2 factors [13].

Table 4. Orthogonal Array of Various Levels on 2 Factors.

P = → L = ↓	2				3					4						5						
	Taguchi, P=2, L=5				Taguchi, P=3, L=5					Taguchi, P=4, L=5						Taguchi, P=5, L=5						
	Run#	a	b	X	Run#	a	b	c	X	Run#	a	b	c	d	X	Run#	a	b	c	d	e	X
	1	1	1	X_1	1	1	1	1	X_1	1	1	1	1	1	X_1	1	1	1	1	1	1	X_1
	2	1	2	X_2	2	1	2	2	X_2	2	1	2	2	2	X_2	2	1	2	2	2	2	X_2
	3	1	3	X_3	3	1	3	3	X_3	3	1	3	3	3	X_3	3	1	3	3	3	3	X_3
	4	1	4	X_4	4	1	4	4	X_4	4	1	4	4	4	X_4	4	1	4	4	4	4	X_4
	5	1	5	X_5	5	1	5	5	X_5	5	1	5	5	5	X_5	5	1	5	5	5	5	X_5
	6	2	1	X_6	6	2	1	2	X_6	6	2	1	2	3	X_6	6	2	1	2	3	4	X_6
5	7	2	2	X_7	7	2	2	3	X_7	7	2	2	3	4	X_7	7	2	2	3	4	5	X_7
	8	2	3	X_8	8	2	3	4	X_8	8	2	3	4	5	X_8	8	2	3	4	5	1	X_8
	9	2	4	X_9	9	2	4	5	X_9	9	2	4	5	1	X_9	9	2	4	5	1	2	X_9
	10	2	5	X_{10}	10	2	5	1	X_{10}	10	2	5	1	2	X_{10}	10	2	5	1	2	3	X_{10}
	11	3	1	X_{11}	11	3	1	3	X_{11}	11	3	1	3	5	X_{11}	11	3	1	3	5	2	X_{11}
	12	3	2	X_{12}	12	3	2	4	X_{12}	12	3	2	4	1	X_{12}	12	3	2	4	1	3	X_{12}
	13	3	3	X_{13}	13	3	3	5	X_{13}	13	3	3	5	2	X_{13}	13	3	3	5	2	4	X_{13}
	14	3	4	X_{14}	14	3	4	1	X_{14}	14	3	4	1	3	X_{14}	14	3	4	1	3	5	X_{14}

15	3	5	X_{15}	15	3	5	2	X_{15}	15	3	5	2	4	X_{15}	15	3	5	2	4	1	X_{15}
16	4	1	X_{16}	16	4	1	4	X_{16}	16	4	1	4	2	X_{16}	16	4	1	4	2	5	X_{16}
17	4	2	X_{17}	17	4	2	5	X_{17}	17	4	2	5	3	X_{17}	17	4	2	5	3	1	X_{17}
18	4	3	X_{18}	18	4	3	1	X_{18}	18	4	3	1	4	X_{18}	18	4	3	1	4	2	X_{18}
19	4	4	X_{19}	19	4	4	2	X_{19}	19	4	4	2	5	X_{19}	19	4	4	2	5	3	X_{19}
20	4	5	X_{20}	20	4	5	3	X_{20}	20	4	5	3	1	X_{20}	20	4	5	3	1	4	X_{20}
21	5	1	X_{21}	21	5	1	5	X_{21}	21	5	1	5	4	X_{21}	21	5	1	5	4	3	X_{21}
22	5	2	X_{22}	22	5	2	1	X_{22}	22	5	2	1	5	X_{22}	22	5	2	1	5	4	X_{22}
23	5	3	X_{23}	23	5	3	2	X_{23}	23	5	3	2	1	X_{23}	23	5	3	2	1	5	X_{23}
24	5	4	X_{24}	24	5	4	3	X_{24}	24	5	4	3	2	X_{24}	24	5	4	3	2	1	X_{24}
25	5	5	X_{25}	25	5	5	4	X_{25}	25	5	5	4	3	X_{25}	25	5	5	4	3	2	X_{25}

3. Results and discussion

This experiment uses the Orthogonal Array L25 (5^2). ANOVA is used to determine which factors have the most influence on particle size. Since the desired tendency for the particles to be the smallest, the type S/N ratio of "the smaller, the better" is chosen. These factors were retested to confirm whether the experiment was still within the ANOVA tolerance limit of 5%. The experimental data can be shown in table 5.

Table 5. Research Data on Powder Fineness.

No	Research variable		POWDER FINE
	Working time	Shaft Rotation Speed	
1	10 minutes	488 rpm	336 μm
2	10 minutes	511 rpm	328 μm
3	10 minutes	533 rpm	327 μm
4	10 minutes	555 rpm	329 μm
5	10 minutes	578 rpm	320 μm
6	15 minutes	488 rpm	295 μm
7	15 minutes	511 rpm	281 μm
8	15 minutes	533 rpm	280 μm
9	15 minutes	555 rpm	275 μm
10	15 minutes	578 rpm	272 μm
11	20 minutes	488 rpm	250 μm
12	20 minutes	511 rpm	257 μm
13	20 minutes	533 rpm	253 μm
14	20 minutes	555 rpm	250 μm
15	20 minutes	578 rpm	247 μm
16	25 minutes	488 rpm	231 μm
17	25 minutes	511 rpm	225 μm

18	25 minutes	533 rpm	217 μm
19	25 minutes	555 rpm	228μm
20	25 minutes	578 rpm	204 μm
21	30 minutes	488 rpm	202 μm
22	30 minutes	511 rpm	190 μm
23	30 minutes	533 rpm	200 μm
24	30 minutes	555 rpm	199 μm
25	30 minutes	578 rpm	181 μm

The relationship between the fineness of the powder obtained on the variation in working time and the shaft's rotational speed can be seen in figure 2 and figure 3.

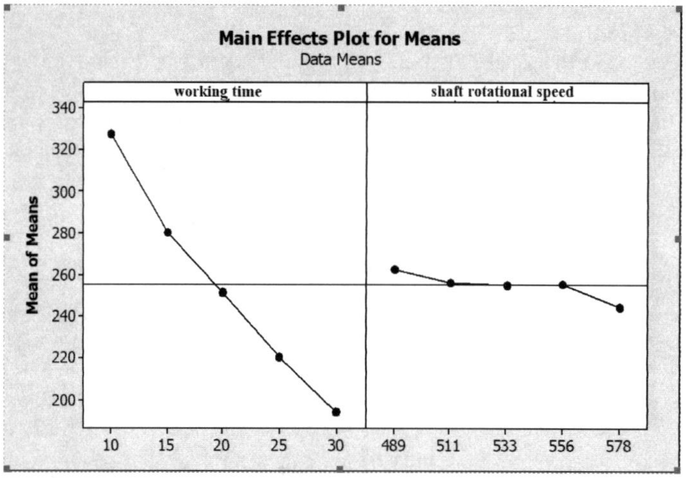

Figure 2. Value of Powder Fineness Against Machine Working Time and Unbalance Shaft Rotation Speed.

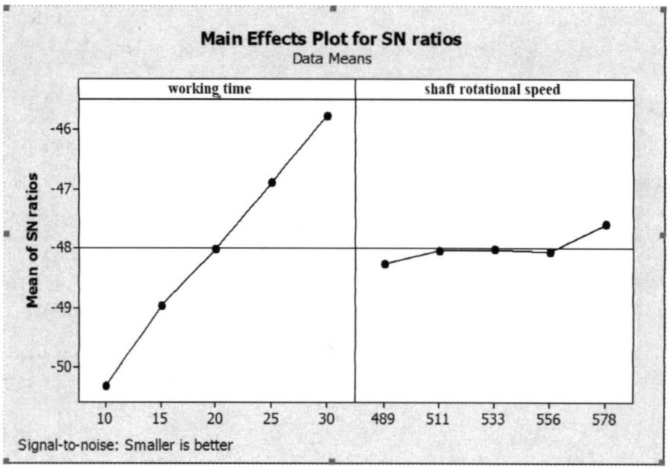

Figure 3. S/N Ratio (Smaller is Better) Powder Fineness Against Machine Uptime and Unbalance Shaft Rotation speed

In the graph shown in figure 2, it can be observed that the working time variable results in a more significant variation in particle. While variable of shaft rotational speed is not too substantial. The rough avocado leaf simplicia powder was obtained at a shaft rotational speed of 489 rpm and a working time of 10 minutes. The smallest avocado leaf simplicia powder was obtained at a shaft rotational speed of 578 rpm and a working time of 30 minutes.

The characteristics of the experimental data can be seen using Signal Noise to Ratio (SN Ratio). In this study, the type of SN Ratio quality characteristics chosen is the smaller is a better approach. This characteristic has a quality with a limit value of zero so that a value close to zero is the desired value [12].

Based on the graph shown in figure 3, the working time for 10 minutes and the shaft rotation speed of 489 rpm has the worst SN Ratio value because the SN Ratio value is farthest from the zero value, while the 30 minutes working time and 578 rpm shaft rotation speed have SN values. The ratio is close to zero. To determine whether the research variables were accepted or rejected, and to determine the most dominant variable, the Analysis of Variance (ANOVA) was used. ANOVA analysis of the refinement of avocado leaf simplicia powder can be seen in table 6.

Table 6. ANOVA for Response to Size of Powder Fineness (Smaller is Better)

source	DF	Seq SS	Adj MS	F-Value	P-Value	% Contribution
Working time	4	62,6369	15,6592	346,06	0,000	97,05 %
Shaft rotational speed	4	1,1329	0,2832	6,26	0,003	2,95 %
Residual Error	16	0,7240	0,0452			
Total	24	64,4937				

Analysis of Variance is used to analyze the parameters that affect the experiment. ANOVA produces a p-value that needs to be compared with the hypothesis, which states, "H0 is residual data with a normal distribution (variables do not affect particle size) and H1 is residual data not normally distributed (variables affect particle size)". From the ANOVA response analysis for the refinement of avocado leaf Simplicia powder, it can be seen that the P-value on the working time variable has a value of 0,000. In contrast, the P-value on the shaft rotation speed variable has a value of 0.033. The hypothesis taken from this study is to reject H0 and accept H1 so that it can be said that the working time variable and the shaft rotation speed affect the fineness of the powder. Multiple regression analysis of the research variables can be written, as shown in Equation 1. The results of observations with a stereo microscope can be seen in figure 4.

Powder fineness = 472 - (6.54 working time) - (0.162 shaft rotation speed) Eq. 1

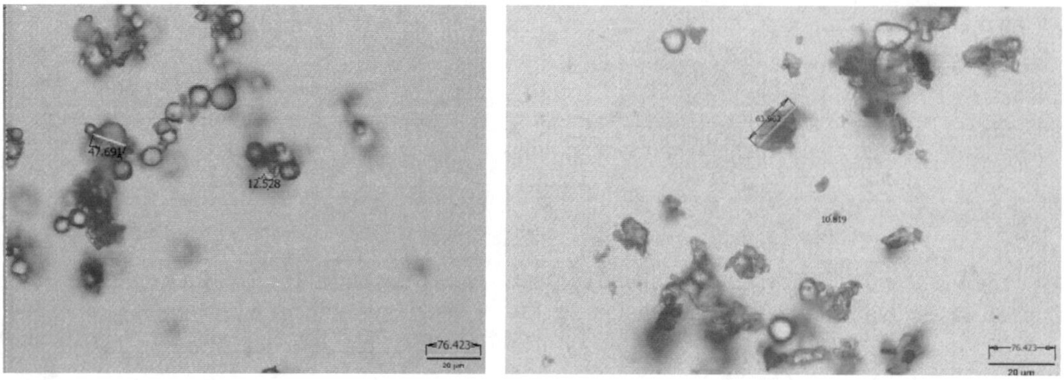

Figure 4. the results of observations with a stereo microscope

4. Conclusion

Taguchi method was used to examine high energy ball mill characteristics when producing micro/nano-particle of Simplicia powder. Two factors and five levels are selected for the smallest diameter. The smoothest avocado leaf Simplicia powder was obtained at 181 μm with a shaft rotation speed of 578 rpm and a working time of 30 minutes from the data obtained. The coarsest avocado leaf Simplicia powder was obtained at 336 μm with a shaft rotation speed of 489 rpm and a working time of 10 minutes. The avocado leaf Simplicia powder's fineness was obtained at the longest working time and the fastest shaft rotation. Working time is very influential in the fineness of Simplicia powder. The contribution value of machine working time has a value of 97.18%. The shaft's rotation speed has a contribution value of 2.82%—smaller avocado leaf Simplicia powder results in faster drug absorption to the body.

References

[1] Owolabi M A 2010 Bioactivity of The Phytoconstituents of The Leaves Of Persea Americana *J. Med. Plants Res.* **4**

[2] Fritsch G 2012 Operating Manual, Translation of the original, pulverisette 6 classic line *Manuf. Lab. Instrum. Ind.*

[3] Chusnul Azhari 2017 *The Effect of Machine Rotation on the Result of Clay Powder in a Ceramic Grinding Machine* (Sekolah Tinggi Teknologi Mandala Bandung)

[4] Rakesh P P 2008 An Overview Of Size Reduction Technologies In The Field Of Pharmaceutical Manufacturing *Asian J. Pharmatic*

[5] M.Arole S . M 2014 Fabrication Of Nanomaterial By Top-Down and Bottom-Up Approaches - An Overview *mMaterial Sci.*

[6] Aftin A 2013 *Engineering for Processing and Preservation of Agricultural Products* (Purkokerto)

[7] German M R 1984 *Powder Metallurgy Science* (USA: Metal Powder Industries Federation)

[8] Gerry D P 2016 *Preparation and Characteristics of Nano-Bioceramic Powder from Eggshells Done by Ball Mill Process and Intermediate Heating* (Universitas Andalas)

[9] Dwiyanto 2010 *The Effect of Differences in Casting Modulus of Castings on Hardness and Microstructure of the Results of the Casting Process of Aluminum Alloy Sand Casting* (Universitas Andalas)

[10] Feng H Test Sieve - Stainless steel or Base Frame

[11] Niebel 2017 Taguchi's Robust Design Methods *Manuf. Process Integer Prod. Process* 1–9

[12] Wulandari A A 2016 Application of the Taguchi Method for Multi Response Cases Using the Gray Relational Analysis Approach and Principal Component Analysis *J. Gussian* **5**

[13] Cimbala J 2014 *Taguchi Orthogonal Arrays* (Penn State University)

This Page Intentionally Left Blank.

Structure modeling and analysis for improving the strength of horizontal pump system

I Haryanto[1], E Yohana[1], I Mahdi[1], I H Sihotang[1], K-H Choi[2]

[1]Department of Mechanical Engineering, Faculty of Engineering, Diponegoro University Jl. Prof. Sudharto, SH. Kampus UNDIP Tembalang Semarang 50239

[2]College of Engineering, Pukyong National University, 365 Sinseon-ro, Nam-gu, Busan 608-739, Korea

Corresponding author: ignatiushasudungan@gmail.com

Abstract. Generally, pumps are used as equipment to move liquids from one place to another, such as in power plants, petroleum companies, etc. One of the kinds of the pump is the Horizontal Pumping System which consists of a motor, trust room, cradle, pump, and base. This study is motivated by the problem appearing of load increase at the suction and discharge nozzle so that it exceeds the allowable load. That load surges cause the pump structure to receive large vibrations, which results in damage to some Horizontal Pump System structures. Therefore, structural analysis on the Horizontal Pumping System to determine the displacement and durability of each cradle are necessary. The purpose of this study is to determine the strength of the pump structure and design additional tools to strengthen the pump structure. In this study, SOLIDWORKS is used for 3D modeling of the Horizontal Pump System structure. Meanwhile, the strength analysis of the structure is conducted using ANSYS Static Structural. Based on the analysis results, there is a quite large displacement in the pump section associated with the suction nozzle. On the other hand, the cradle resistance is still in the elastic deformation area with a maximum value of 24881.97 psi ($\sigma y = 3.01x104$ psi). Therefore, development is needed by adding a cradle in the pump section associated with a suction nozzle. By adding this cradle there is a reduction in displacement experienced by the pump. Pump displacement changes to 0.036 - 0.097 inches in x-axis, 0.009 - 0.021 inches in y-axis, and 0.001 - 0.005 inches in z-axis.

1. Introduction

A pump is a machine that is widely used in industry, especially in the petroleum industry. The pump functions to move the fluid from one place to another through the piping medium by adding energy to the fluid being moved and takes place continuously. Based on its working principle, pumps can be grouped into two, namely positive working pumps (positive displacement pumps) and dynamic working pumps (non positive displacement pumps) [1]. One type of dynamic pump is the Horizontal Pumping System. Horizontal Pumping System, which consists of a motor, trust chamber, cradle, pump, and base. The motor functions to provide power to the shaft to drive the pump. The trust chamber serves to hold the shaft in the axial direction to prevent damage to several parts that are connected. Cradle and base serve as support and restraints for pump movement.

In this study, the suction and discharge nozzles experience an increase in load so that the structure of the pump experiences stress and displacement, which causes the Horizontal Pumping System structure to experience damage in several parts. In the European Patent Application (EPA) publication, the Horizontal Pumping System was created to solve the pump problem which was found to be a failure in removing viscous sand service oil [2]. Failure to move the oil is caused by the thickness and density of oil that is greater than air. During the fluid transfer process, there will be a large vibration, so the firm structure must also have large rigidity. The current horizontal pumping system structural design still needs proper development to produce a pump with the appropriate structure. Therefore, it is very important to analyze and develop the stiffness of the pump reporting the Horizontal Pumping System structure.

Displacement and stress in the Horizontal Pumping System can be estimated by performing statistical analysis. For this purpose, the ANSYS software is applied for modeling and analysis of the FEM Horizontal Pumping System. The analysis process is carried out under the existing geometry, boundary conditions, and load conditions. According to Merad Boudia et al [3], the connection in a structure is very influential in maintaining the rating of the load received. In the Horizontal Pumping System, the joints used are bolts and nuts. Bolts and nuts are designed to work together when the internally threaded fastener (nut) is screwed to the externally threaded fastener (bolt). So that the bolts are forced to accept the tension and the bolt and nut heads are maintained in the joint to maintain the connection with a certain tension. This maintained stress is called bolt tension [4]. Therefore, bolt tightening is also taken into account in the analysis process.

Research by Seong Hyeon Lee et al in 2020 [5], conducted a pump strength analysis using structural analysis. Based on the results of the structural analysis, the number of displacement, and stress on the pump parts, Seong Hyeon Lee suggested developing a pump part that experienced a large displacement and stress exceeding the yield strength of the pump. If the damage is found in one of the pump holders (cradle and base) Horizontal Pumping System, it will cause damage to other pump parts so that a shutdown can occur. Structure analysis itself is a study that studies deformed behavior against a given force or load. Several branch of the implementation of structural analysis such as analysis of building bridges, machines, tools, etc. [6]. The structural analysis combines the discipline between the fields of mechanics, materials engineering, and engineering mathematics to calculate internal strength, stress, structural deformation, bearing reactions, and stability.

The purpose of this study is to analyze the structural strength and structure development of the Horizontal Pumping System. The SOLIDWORKS application is used to draw 3D models and ANSYS Static Structural as computation in analyzing the Horizontal Pumping System structure. The results of the research are to find out how much is the displacement and the stress on the pump. If a large displacement and/or the stress exceeds the yield strength of the material occurs, then it is necessary to develop the Horizontal Pumping System structure.

2. Materials and methods

2.1. Geometry and Materials

The geometry and material of the Horizontal Pumping System are obtained based on the results of measurements. Horizontal Pumping System geometry includes the motor, trust chamber, cradle, pump, and base. The Horizontal Pumping System geometry is modeled using SOLIDWORKS, the modeling results can be seen in figure 1. The materials used are shown in table 1.

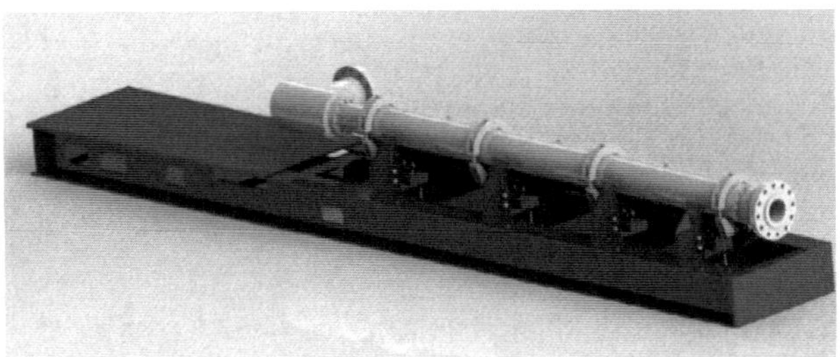

Figure 1. Assembly Horizontal Pumping System

Table 1. Horizontal Material Pumping System

Material	Part
GS A525	Cradle and Base
SS 304	Suction Nozzle, Discharge Nozzle, andpump

SS 316	Bolts and nuts

In this study, the cradle is divided into 5 parts, namely cradle A, cradle B, cradle C, cradle D, and cradle V. The position of each cradle can be seen in figure 2.

Figure 2. Position 1) Cradle A, 2) Cradle B, 3) Cradle C, 4) Cradle D, and 5) Cradle V on a Horizontal Pumping System

2.2. *Numeric Equations*

The structural analysis method uses equilibrium, stress, strain equation, and Young's modulus. The final step is to express the relationship between the external forces acting on the discrete point and the inner forces. In general, this requires that the equations of force and moment of equilibrium be satisfied along the three independent axes shown, as shown by Equations 1 and 2.

$$\sum \boldsymbol{F_x} = \mathbf{0}\,;\sum \boldsymbol{F_y} = \mathbf{0}\,;\sum \boldsymbol{F_z} = \mathbf{0} \quad (1)$$

$$\sum \boldsymbol{M_x} = \mathbf{0}\,;\sum \boldsymbol{M_y} = \mathbf{0}\,;\sum \boldsymbol{M_z} = \mathbf{0} \quad (2)$$

For loads that are still in the elastic region, the relation between stress and strain is stated by Hooke's law as formulated in Equation 3 below:

$$\sigma = \varepsilon\, E \quad (3)$$

Where E indicates the modulus of elasticity while ε is the strain which is defined as the ratio of the increase in length due to the force applied to the initial length of the object before the force is applied, namely:

$$\varepsilon = \Delta L / L \quad (4)$$

The degree of structural change or strain depends on the amount of stress that is applied [7].

2.3. *Analysis Procedure*

The analysis was performed using ANSYS Static Structural software. In this study, a structural analysis was carried out and the results showed how much displacement and stress on the pump was. Load and moment data on the suction and discharge nozzles are entered as input parameters of the Horizontal Pumping System boundary conditions. Load and moment data can be seen in table 2.

Table 2. Loads and Moments on the Suction Nozzle and Discharge Nozzle

Nozzle	Fx (lb)	Fy (lb)	Fz (lb)	Mx (ft.lb)	My (ft.lb)	Mz (ft.lb)
Suction	-2406	344	-1069	2178,3	-803,6	-6197,6
Discharge	2775	2775	0	-507,5	507,5	0

The bolt pretensioning is entered as the input parameter to maintain the pump position. The Bolt Pretensioning is assumed to be 75% of the ideal, following Table 3 is how big the bolt pretensioning is used in the Horizontal Pumping System.

Tabel 3. Bolt Pretensioning on a Horizontal Pumping System [8]

Bolt Pretensioning			
Diameter (in)	Torsion (ft.lbs)	Tensioning (lbf)	Pretensioning (lbf)
¾	129	955.5	716,625
½	45	500	375

Besides, Fix Support is assumed to be at the base of the pump as load support and a barrier to the pump movement so that it does not experience a displacement.

2.4. Meshing

Meshing is a method for dividing an object to be simulated into small elements. One of the most important things in a mesh is the size of the mesh elements applied to the model. The smaller the mesh size, the more accurate the results will be, but it requires more computing power and time than the larger mesh sizes. Therefore, it is necessary to determine the correct size of the mesh elements.

The mesh generation is carried out using ANSYS Static Structural as shown in figure 3. The mesh generation uses fine meshing to obtain mesh convergence. Triangular meshing is considered because it matches the geometry and material characteristics of the pump. So the Horizontal Pumping System is distributed to 275,033 elements which contain 528,450 nodes. The average skewness of the mesh has a value of 0.34926.

Figure 3. Geometry meshing

2.5. Validation

In this study as validation, the results of research by Seong Hyeon Lee et al, conducted in 2020 [5] were used. One of the methods used in this research is static load analysis, which investigates the stress intensity of the primary heat transport system (PHTS) pump using ANSYS structural analysis. The results of Seong Hyeon Lee's research were in the form of displacement and stress on the primary heat transport system(PHTS) pump parts and looking for the parts that experienced the greatest stress. The results of the validation using the program can be seen in table 4.

Table 4. Displacement and Stress Validation

Analysis		Lee et. al.	Present	Error
Displacement (mm)	Max	$0.201 x 10^{-3}$	$0.213 x 10^{-3}$	6%
	Min	0	0	0%
Stress (Pa)	Max	$0.943 x 10^{-4}$	$0.952 x 10^{-4}$	1%
	Min	$31. x 10^{6}$	$31.7 x 10^{6}$	1%

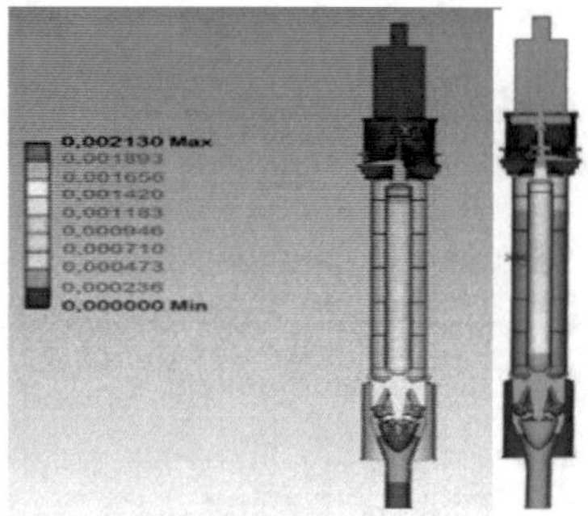

(i) (ii)

Figure 4. Analysis Results Displacement (i)Present dan (ii)Lee etal.

From table 4, it appears that the differences that occur in the displacement and stress analysis between the results obtained in this study and Lee et. al. small enough so that the results of the analysis in this study can be accepted. Furthermore, based on figure 4, it can be seen how much displacement is in each part of the pump.

3. Results and discussion

3.1. Displacement in the Horizontal Pumping System Structure

In this research, the forces and moments come from the increase in fluid load in the suction nozzle and discharge nozzle in the Horizontal Pumping System. The results of the structural analysis show the magnitude of the displacement-based on equilibriumequation, stress, strain, and Young's modulus. The results of the analysis of the Horizontal Pumping System structure can be seen in figure 5.

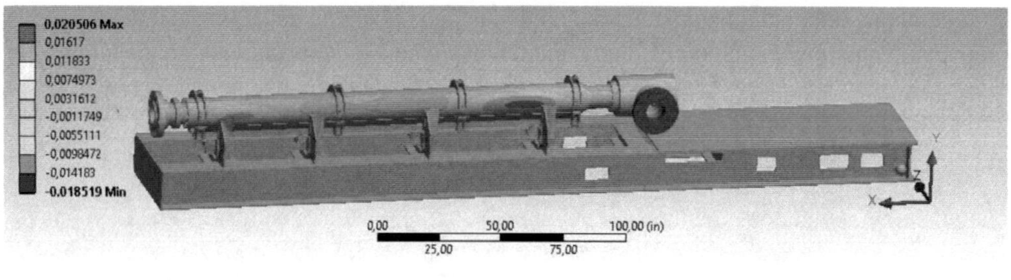

Figure 5.Displacement on the z-axis.

The increase in the load of the Horizontal Pumping System causes stress-strain on the pump. So that there is displacement in several pump parts. Based on Figure 5, it is found that the largest displacement occurs in the Horizontal Pumping System located on the side of the pump which is connected to the suction nozzle.These results are the same as the research conducted by Pramod H Kadam et al [9]. The largest displacement occurs at the suction nozzle. Very large displacement occurs because the load increase exceeds the allowable load on the Horizontal Pumping System structure. The amount of displacement at each nozzle can be seen in table 5.

Table 5. Displacement on the Suction Nozzle and Discharge Nozzle

Nozzle	Displacement (in)		
	X	Y	Z
Suction	0,084 - 0,115	0,116 - 0,226	0,007 - 0,011
Discharge	(-0,006) - (-0,009)	-	0,003 - 0,007

Based on the results of the structural analysis regarding displacement, it was found that the suction nozzle had a very large displacement, so it was necessary to make improvements to the side of the pump that was close to the suction nozzle.

3.2. Stress on the Horizontal Pumping System Structure

Stress can be defined as the result of the ratio between the force/load received by an object to the area of the object being deformed. The results of the structural analysis show the amount of stress on the Horizontal Pumping System as shown in Figure 6.

Figure 6. Stress on the Horizontal Pumping System.

Based on the results of the structural analysis, it was found that the greatest stress was found on the middle side of the Horizontal Pumping System. Stress is caused by an increase in the load on the suction nozzle and discharge nozzle of the Horizontal Pumping System. The part of the pump that plays a very important role in maintaining the displacement due to an increase in load is the cradle. Research by M. Z. Norhirni et al in 2011 [10], conducted in an axial piston pump design, the swash plate plays an important role in controlling the displacement of the pump. Likewise, the cradle-mounted design. Therefore, it is necessary to see how much stress each cradle is experiencing.

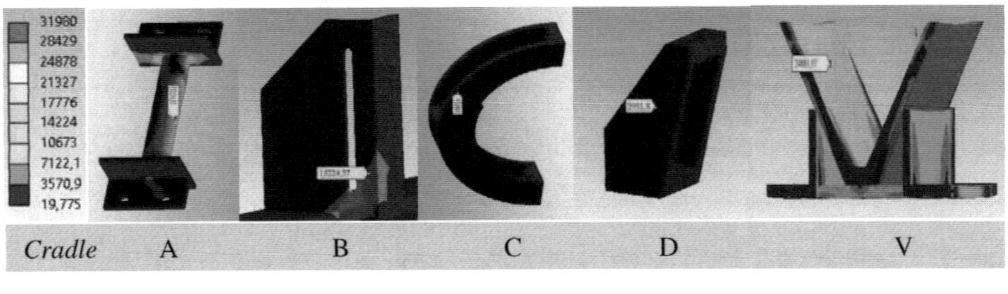

Figure 7. The stress of each Cradle on the Horizontal Pumping System.

The stress of each cradle in the Horizontal Pumping System can be seen in Figure 7. The increase in load on the suction nozzle and discharge nozzle causes the cradle to stretch and tighten. The greatest stress is experienced at cradle V because cradle V is the center of support for all cradles, the amount of stress from each cradle can be seen in Figure 7. After knowing the amount of stress in each cradle, it is necessary to analyze the strength of the cradle by comparing stress received by the yield strength of the cradle material. The material used in the cradle is galvanized steel with a yield strength of 3.04×10^4 Psi. Based on the type of material as described on the Azo Materials website [11], it can be concluded that the cradle is still in the elastic zone so it is still safe to use and does not require improvement.

3.3. Additional Cradle Creation

An additional cradle is made to maintain the position of the pump to prevent displacement. The additional cradle position will be placed in the part of the pump that is connected to the suction nozzle. The material used in this cradle is a solid rod fabricated in a circular shape with galvanized steel material as seen in Figure 8.

Figure 8. Additional Cradle on the pump.

3.4. Displacement on the Horizontal Pumping System After Improvement

After the improvement of the Horizontal Pumping System by adding a cradle on the side of the pump associated with the suction nozzle, the results are shown in figure 9. The results of the comparison of changes in the amount of displacement can be seen in table 6. Displacement changes occur because of the center of displacement force that is held by the cradle, so that displacement on the pump can be minimized.

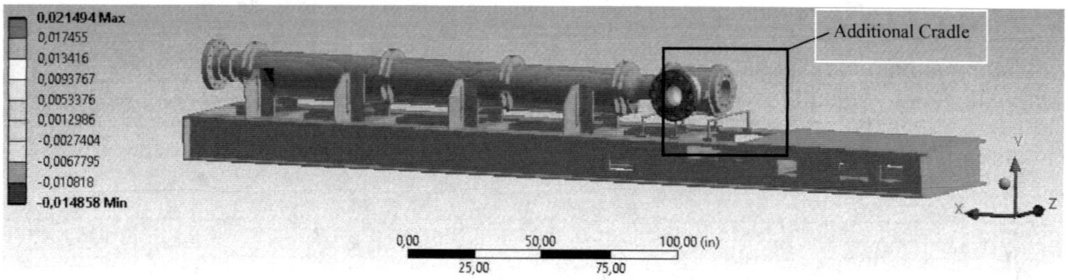

Figure 9. Z-axis Displacement After Improvement.

Table 6. Comparison of Displacement After and Before Improvement

Nozzle		Displacement (in)		
		X	Y	Z
Suction	Before	0,084 - 0,115	0,116 - 0,226	0,007 - 0,011
After		0,036 – 0,097	0,009 - 0,021	0,005 - 0,009
Discharge	Before	(-0,006) - (-0,009)	-	0,003 - 0,007
After		(-0,006) - (-0,009)	-	0,001 - 0,005

4. Conclusion

This study uses structural analysis to determine how much displacement in several parts of the pump and stress on the pump cradle. Based on the structural analysis, it is found that the largest displacement is located on the pump side associated with the suction nozzle, which is 0.084 - 0.115 on the x-axis, 0.116 - 0.226 on the y-axis, and 0.007 - 0.011 on the z-axis. So that improvement is needed by adding a cradle to the pump to hold the pump so that a large displacement does not occur. After the improvement, there was a significant change in the pump displacement to 0.036 - 0.097 on the x-axis, 0.009 - 0.021 on the y-axis, and 0.001 - 0.005 on the z-axis. So it can be concluded that the additional cradle can reduce the amount of displacement at the pump.

Reference

[1] Hariady S 2014 Analisa Kerusakan Pompa Sentrifugal 53-101C WTU Sungai Gerong Pt. Pertamina RU III Plaju *J. Desiminasi Teknol.* **2** 29–42

[2] Krowski P J 1998 Horizontal Pumping Systems *Eur. Pantent Appl.* **1** 1–18

[3] Merad Boudia S B, Boumechra N, Bouchair A and Missoum A 2020 Modeling of bolted endplate beam-to-column joints with various stiffeners *J. Constr. Steel Res.* **167**

[4] Childs P R . 2004 *Mechanical design* (200 Wheeler Road. Burlington: Elsevier Ltd)

[5] Lee S H, Lee J H, Kim S K, Kim J B and Kim T W 2020 Improvement of aseismic performance of a PGSFR PHTS pump *Nucl. Eng. Technol.* **52** 1847–61

[6] Hibbeler R 2015 *Structural Analysis* (New Jersey)

[7] Callister W D 1991 Materials science and engineering: An introduction (2nd edition) *Mater. Des.* **12** 59

[8] Engineering Edge 2018 Torque Values Stainless Steel Bolt Table *Eng. Edge*

[9] Pramod H Kadam and Sanjay S Belkar 2016 Baseplate Structural Stiffness Analysis of API BB5 Type Centrifugal Pump for FPSO Application *Int. J. Eng. Res.* **V5**

[10] Norhirni M Z, Hamdi M, Nurmaya Musa S, Saw L H, Mardi N A and Hilman N 2011 Load and stress analysis for the swash plate of an axial piston pump/motor *J. Dyn. Syst. Meas. Control. Trans. ASME*

[11] Azo Materials 2002 The Effect of Galvanizing on Steel Strength

Comparison of Physical and Mechanical Properties of Antifouling PVDF Membranes by Titanium Dioxide and PES Membrane by Titanium Dioxide

A Mataram[1], Nukman[1], J D Nasution[1], M I Riadi[1], M Pataras[2], B B Aditya[2], G D Septano[1], N Anisya[1], N A Nadiyah[1]
[1]Department of Mechanical Engineering, Faculty of Engineering, Universitas Sriwijaya, Palembang, Indonesia.
[2]Department of Civil Engineering, Faculty of Engineering, Universitas Sriwijaya, Palembang, Indonesia.

Corresponding author: amataram@unsri.ac.id

Abstract. The main factor to support daily life is water, but the fact is there are approximately 2.1 billion people currently threatened with a shortage of clean water. Polyethersulfone (PES) and Polyvinylidene Fluoride (PVDF) are polymers that are commonly used for membrane fabrication because they have extraordinary properties in the screening process. The addition of addictive substances into the membrane formation process is no less important as supporting the ability to be even better in membranes Titanium Dioxide (TiO_2) as an additive with properties relevant to the two polymers which can cover the shortcomings of the two polymers. The printing process of PVDF and PES membranes with the addition of TiO_2 uses a copper plate which is flowed with an electric field of 15000V DC. The results of the surface morphology of Scanning Electron Microscopy (SEM) both showed extraordinary changes compared without modification to the printing process as in previous studies. The tensile strength test results showed uniformity with the membrane morphology and an increase in the maximum stress value of the membrane to 3.86 MPa.Surface roughness values can help identify membrane impurity properties which are also relevant to the results of water treatment performance tests and contact angles. Finally, the process of forming membranes with the additional modification of the printing process can open new avenues to maximize water treatment performance and service life on the membrane.

1. Introduction

In the past ten years, researchers have been intensively researching membrane fabrication. The research is a form of addressing water shortages as part of humans' primary needs. Polymer poly ether sulfone (PES) and poly vinyl dene fluoride (PVDF) are materials commonly used. As has been done by previous studies [1,2] , PVDF and PES are membrane-making materials that have extraordinary filtration capabilities and are very beneficial in terms of function and cost.

However, behind the advantages possessed by the two polymer materials, PES and PVDF have the same disadvantages, namely their impurity properties. These impurity properties affect the filtration process up to the life of the membrane. The use of addictive substances to the polymer will also make the fabrication of better polymers. Among those often used as addictive substances is titanium dioxide (TiO_2). The titanium dioxide (TiO_2) role as an enhancer of impurity resistance that has been confirmed by [7] becomes an auxiliary electric field to overcome impurities in the polymer. Not only acts as an enhancer of impurity resistance, but titanium dioxide (TiO_2) also acts as an increase in hydrophilicity, antibacterial, and self-cleaning of membranes [8]. In the study [9], [10], mixing titanium dioxide as an intermediary increased the hydrophilicity of the membrane. Also, the nature of titanium dioxide (TiO_2) has high electrical conductivity and the presence of hydroxy groups on the surface so that it has a high electrostatic interaction will play a role with a combination of modification of the addition of electric field currents in the membrane molding process [11], [12]. Despite the many advantages of titanium dioxide (TiO_2), it also has its drawbacks, [13] confirmed that mixing titanium dioxide (TiO_2) with a weight of 1wt% creates a lump called agglomeration and is one of the factors that can make the membrane worse. So in this study, the ratio of titanium dioxide (TiO_2) is locked with a composition of 1wt%.

2. Material and methods

PES and PVDF polymers, N, N-Dimethylformamide (DMF) solvents, and titanium dioxide (TiO_2) addictive substances obtained from Aldrich chemical (Indonesia) with no further purification processes. The compositions used are PES (Pure 25wt%, Mixed 20wt% & 30wt%), PVDF (Mixed 10wt%, 15wt%, 20wt%), TiO_2 1wt%. Membrane formation begins with the process of mixing the PVDF polymers into the DMF solvent with Titanium dioxide and PES polymers into the DMF solvent with Titanium dioxide. Dissolution of membrane-making materials ranging from polymer, additives, and DMF using a magnetic stirrer with stirring for 8 hours. Modification of the membrane with the ready solution will continue with the flat sheet printing process. The membrane was modified using a copper plate energized by an electric field of 15000 V dc

3. Results and discussion

In the current work, the test result shown the mechanical and physical properties of the PVDF with titanium dioxide membrane and PES with titanium dioxide membrane each specimen using a membrane tensile testing and for the physical properties using a scanning electron microscope to see the morphology of the membrane. Morphology for PVDF with titanium dioxide and PES with titanium dioxide shown in figure 1 and figure 2.

Based on figure 1 and figure 2 (a) provides information on the concentration of PES@TiO_2 20wt% inhibited pore fabrication, this can be due to agglomeration of the membrane. Whereas the Pure PES 25wt% and PES@TiO_2 30wt% significantly showed clearer pore formation. Figure 3.1 (b) the fine and tight pore size is observed. However, there are significant differences in the PVDF@TiO_2 membrane concentration of 10wt%, pore bonds have a large size compared to other concentrations so that it will affect the mechanical strength of the membrane. The pore size can be further caused by the addition of a 15000V DC electric field to help in pore leveling to prevent clumping of the membrane.

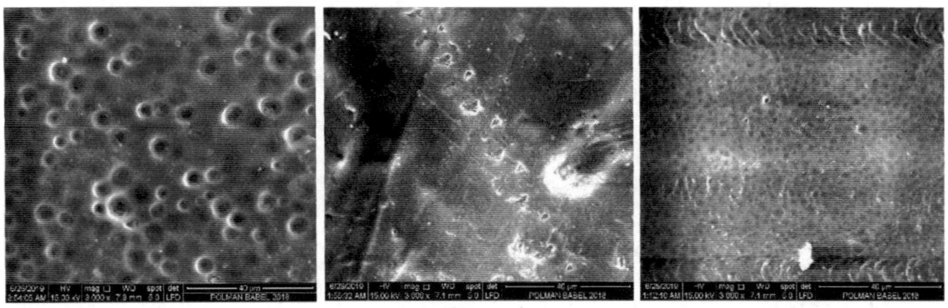

(a) (b) (c)

Figure 1. SEM Polymer Observation PES with 3000 times Magnification (a) Pure PES 25wt% (b) PES@TiO_2 20wt% (c) PES@TiO_2 30wt%.

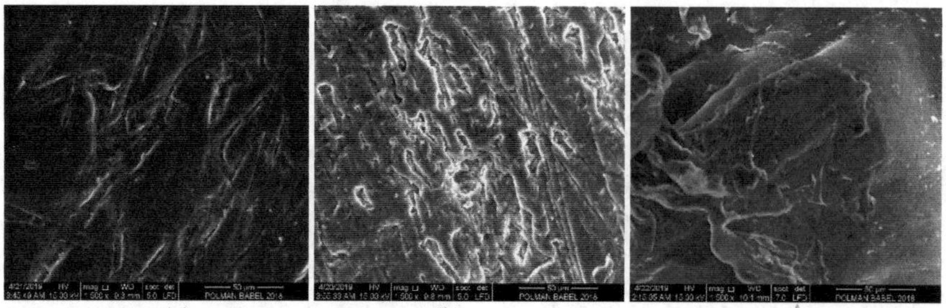

(a) (b) (c)

Figure 2. SEM Polymers PVDF Observation with 1500 times Magnification (a) PVDF@TiO_2 10wt% (b) PVDF@TiO_2 15wt% (c) PVDF@TiO_2 20wt%.

For the tensile stregnth of the membranes are using adhession tearing stress to see the tenacity value of the membranes. The result of the tensile strength of the membranes shown in figure 3.

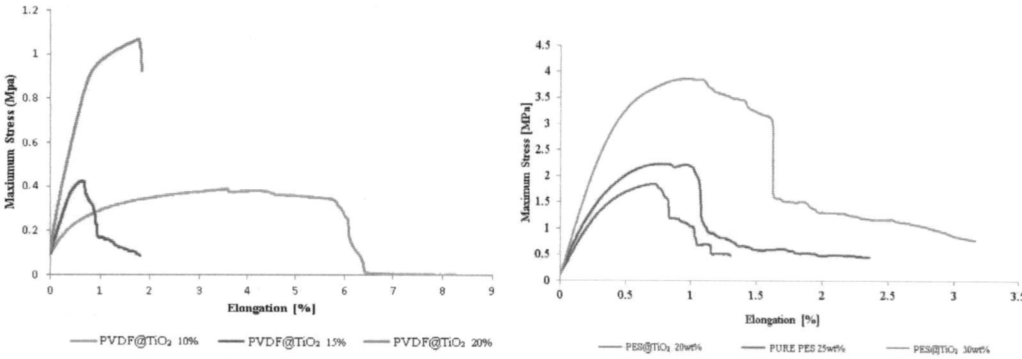

(a) (b)

Figure 3. Tensile Strength result of (a) PVDF with TiO and (b) PES with TiO.

Based on figure 3 (a) The highest tenacity value of the three samples above is at a 30wt% PES@TiO_2 concentration of 0.92%. Then followed by Pure PES 25wt% of 0.78% and PES@TiO_2 20wt%. Whereas in figure 3.2 (b) PVDF@TiO_2 shows an increase in value from 0.453 MPa to 0.5 MPa when the specimens with PVDF fiber composition. As a result, the effect of adding TiO_2 on PES and PVDF significantly increases the mechanical strength of the membrane, which also influences membrane life [15].

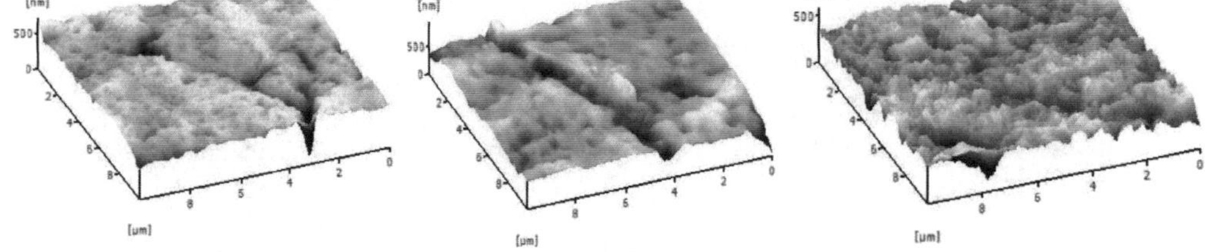

(a) (b) (c)

Figure 4. Surface Roughness test result (a)PVDF@TiO_2 10wt% (b) PVDF @$TiO_2$15wt (c) PVDF@TiO_2 20wt%

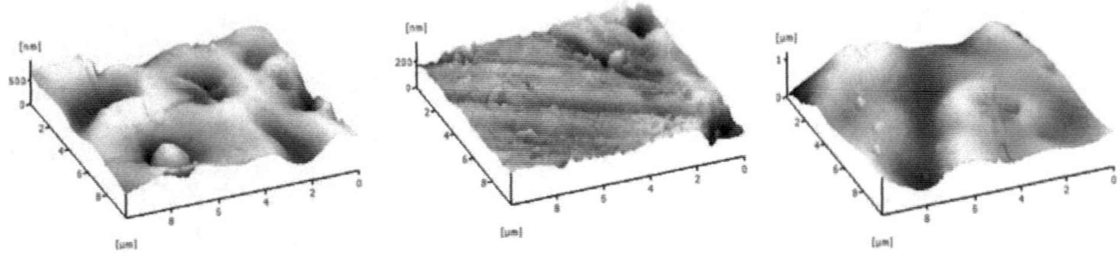

(a) (b) (c)

Figure 5. Surface Roughness test result (a) Pure PES 25wt% (b)PES@TiO220wt% (c) PES@TiO2 30wt%.

Surface roughness of each composition of PVDF and PES also shown the different. For the surface roughness structure of PVDF and PES we can see in figure 4 and figure 5. and the measurement of every surface roughness in table 1.

Table 1. PVDF and PES Surface Roughness measurement.

Membranes	Surface Rougness Measurement		
	R_a(nm)	R_q(nm)	R_t(nm)
PVDF with TiO, 10wt%	11.45	13.76	65.31
PVDF with TiO, 15wt%	19.05	24.73	122.1
PVDF with TiO, 20wt%	30.83	38.36	155.3
Pure PES, 25wt%	21.39	27.45	126.6
PES With TiO, 20wt%	17.22	21.88	88.37
PES With TiO, 30wt%	10.08	13.28	63.92

In table 1 PES@TiO_2 20wt% shows the best surface roughness value among others. The PES@TiO_2 20wt% mixed membrane identifies that with the modification of the addition of an electric field it can make equal distribution of processes. Whereas in Table 4.2 PVDF@TiO_2 10% can reduce the roughness value of the membrane surface compared with other mixed concentrations, this is related to the flux value that has been obtained and the anti-fouling properties of the membrane [16], [17]. The membranes are also working on water treatment performance to see the durability of membranes by fluks measurement of each membrane. The result of performance of water treatmen test can seen in figure 6.

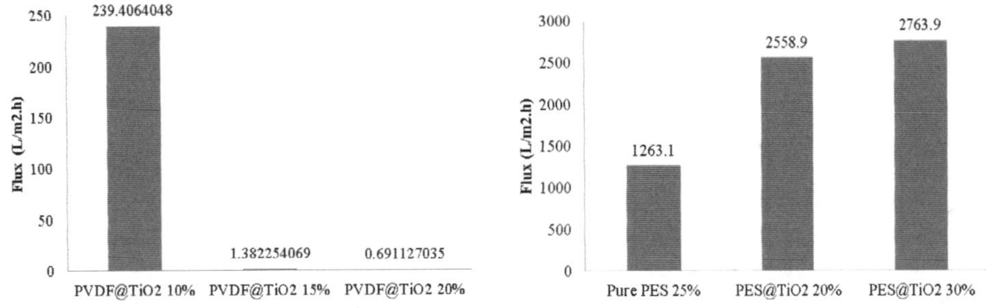

(a) (b)

Figure 6. Water Treatment Performance Test Results (a) PVDF (b) PES.

In research [18]–[21] states that the results of water permeability or flux from the microfiltration membrane are more than 1000 L/m^2.h.bar. Where the use of microfiltration membranes is the process of separating suspended solids or bacteria of water. Also completing this is the beginning of the wate filtration process before proceeding on the Ultrafiltration membrane process and Reverse Osmosis. The results obtained by the PES membrane are also directly proportional to the above statement with the best results at a concentration of PES@TiO_2, 20wt%, 2558.9 L/m^2.h.bar. and for PVDF@TiO_2 membranes at a concentration of 10wt% with the highest flux value of 239.41 L/m^2.h.bar.
For the contact angle of membranes also testing to see the degree of membranes can filtering the water the contact angle of PVDF and PES membranes can be seen ini figure 7.

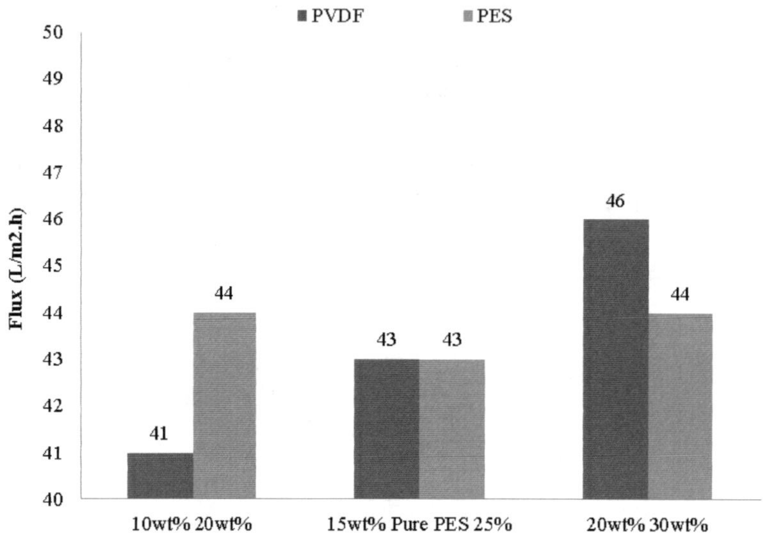

Figure 7. Contact Angle Test Results of PVDF and PES.

The highest hydrophilicity was found in PVDF@TiO_2 10wt% which reached 41° then 43° for mixing Pure PES 25wt%. This proves that a mixture of PES and PVDF membranes with the addition of TiO2 using the electric field method can reach a contact angle that is similar or even lower compared to previous studies [22]–[25].

4. Conclusion

A mixture of PES and PVDF membranes with the addition of TiO_2 was successfully made using the 15000V DC electric field method and showed remarkable results. The results of observations using SEM shows that the morphological structure of the membrane has an equal and fine pore size from low to high concentrations, this is directly proportional to the tensile strength of PES and PVDF membranes, the higher the concentration, the greater the maximum stress obtained. While surface roughness has values related to flux and membrane hydrophilicity. The higher the pure water flux, the lower the hydrophilicity. The value of pure water flux results from the study shows that the addition of TiO_2 and the use of a 15000V DC Electric Field are some of the effective methods to improve membrane capability both in terms of water filtration performance and membrane lifetime.

Acknowledgments

We are grateful for financial support of research funds from the budget of Universitas Sriwijaya.

References

[1] A. L. Ahmad, J. Sugumaran, and N. F. Shoparwe, 2018 "Antifouling Properties of PES Membranes by Blending with ZnO Nanoparticles and NMP–Acetone Mixture as Solvent," *Membranes (Basel).*, vol. 8, no. 4, p. 131,

[2] S. Gee, B. Johnson, and A. L. Smith, 2018 "Optimizing electrospinning parameters for piezoelectric PVDF nanofiber membranes," *J. Memb. Sci.*, vol. 563, no. February, pp. 804–812.

[3] H. T. Bhatti *et al.*, 2018 "Graphene Oxide-PES-Based Mixed Matrix Membranes for Controllable Antibacterial Activity against Salmonella typhi and Water Treatment," *Int. J. Polym. Sci.*, vol.

[4] J. P. Méricq, J. Mendret, S. Brosillon, and C. Faur, 2015 "High performance PVDF-TiO2 membranes for water treatment," *Chem. Eng. Sci.*, vol. 123, pp. 283–291, 2015.

[5] C. Zhao, J. Xue, F. Ran, and S. Sun, "Modification of polyethersulfone membranes - A review of methods," *Prog. Mater. Sci.*, vol. 58, no. 1, pp. 76–150.

[6] N. Nasrollahi, V. Vatanpour, S. Aber, and N. M. Mahmoodi, 2018 Preparation and characterization of a novel polyethersulfone (PES) ultrafiltration membrane modified with a CuO/ZnO nanocomposite to improve permeability and antifouling properties. Elsevier B.V.

[7] T. Bae, I. Kim, and T. Tak, 2006 "Preparation and characterization of fouling-resistant TiO 2 self-assembled nanocomposite membranes," vol. 275, pp. 1–5.

[8] J. Wang, Y. Wang, J. Zhu, Y. Zhang, J. Liu, and B. Van der Bruggen 2017 "Construction of TiO2@graphene oxide incorporated antifouling nanofiltration membrane with elevated filtration performance," *J. Memb. Sci.*, vol. 533, no. March, pp. 279–288.

[9] M. Safarpour, V. Vatanpour, and A. Khataee 2016 "Preparation and characterization of graphene oxide/TiO2 blended PES nanofiltration membrane with improved antifouling and separation performance," *Desalination.*

[10] S. J. Oh, N. Kim, and Y. T. Lee 2009 "Preparation and characterization of PVDF/TiO2 organic-inorganic composite membranes for fouling resistance improvement," *J. Memb. Sci.*, vol. 345, no. 1–2, pp. 13–20.

[11] S. M. Seyed Shahabadi, H. Rabiee, S. M. Seyedi, A. Mokhtare, and J. A. Brant 2017 "Superhydrophobic dual layer functionalized titanium dioxide/polyvinylidene fluoride-co-hexafluoropropylene (TiO2/PH) nanofibrous membrane for high flux membrane distillation," *J. Memb. Sci.*, vol. 537, no. December 2016, pp. 140–150.

[12] J. Chang, H. Chen, S. Ju, H. Chen, and C. Hwang 2010 "Role of Hydroxyl Groups in the NH x (x = 1 - 3) Adsorption on the TiO 2 Anatase (101) Surface Determined by a First-Principles Study," vol. 26, no. 35, pp. 4813–4821.

[13] Z. Pan, S. Cao, J. Li, Z. Du, and F. Cheng, 2019 "Anti-fouling TiO 2 nanowires membrane for oil/water separation: Synergetic effects of wettability and pore size," *J. Memb. Sci.*, pp. 596–606.

[14] J. William D. Callister 2007 "Callister - Materials Science and Engineering -mechanicalengboy." p. viii.

[15] C. S. Ong, W. J. Lau, P. S. Goh, B. C. Ng, and A. F. Ismail, 2015 "Preparation and characterization of PVDF–PVP–TiO2composite hollow fiber membranes for oily wastewater treatment using submerged membrane system," *Desalin. Water Treat.*, vol. 53, no. 5, pp. 1213–1223.

[16] C. S. Ong, W. J. Lau, P. S. Goh, B. C. Ng, T. Matsuura, and A. F. Ismail 2014 "Effect of PVP Molecular Weights on the Properties of PVDF-TiO 2 Composite Membrane for Oily Wastewater Treatment Process," *Sep. Sci. Technol.*, vol. 49, no. 15, pp. 2303–2314.

[17] E. Yuliwati and A. F. Ismail 2011 "Effect of additives concentration on the surface properties and performance of PVDF ultrafiltration membranes for refinery produced wastewater treatment," *Desalination*, vol. 273, no. 1, pp. 226–234.

[18] A. G. Fane, R. Wang, and M. X. Hu 2015 "Synthetic Membranes for Water Purification : Status and Future Angewandte," no. 150, pp. 2–21.

[19] J. Hong, W. Lee, J. Ha, J. Kim, I. Park, and D. Har, "Smart water grid : desalination water

management platform," *Desalin. Water Treat.*, vol. 57, pp. 1–10, 2016.

[20] N. A. Winata 2016 "Teknologi Membran untuk Purifikasi Air Teknologi Membran untuk Purifikasi Air," no. May.

[21] J. Zhang *et al.*, 2013 "w .," vol. 448, pp. 81–92.

[22] H. Rawindran *et al.*, 2019 "Simultaneous separation and degradation of surfactants laden in produced water using PVDF/TiO 2 photocatalytic membrane," *J. Clean. Prod.*, vol. 221, pp. 490–501.

[23] S. Mallick, Z. Ahmad, F. Touati, and R. A. Shakoor 2019 "Improvement of humidity sensing properties of PVDF-TiO 2 nanocomposite films using acetone etching," *Sensors Actuators, B Chem.*, pp. 408–413.

[24] M. L. Luo, J. Q. Zhao, W. Tang, and C. S. Pu 2005 "Hydrophilic modification of poly(ether sulfone) ultrafiltration membrane surface by self-assembly of TiO 2 nanoparticles," *Appl. Surf. Sci.*, vol. 249, no. 1–4, pp. 76–84.

[25] V. Vatanpour, S. S. Madaeni, A. R. Khataee, E. Salehi, S. Zinadini, and H. A. Monfared 2012 "TiO 2 embedded mixed matrix PES nanocomposite membranes: Influence of different sizes and types of nanoparticles on antifouling and performance," *Desalination*, vol. 292, pp. 19–29.

Circular Economy Practices by the Informal Sector: an Implementation of Green Manufacturing in Urban Area

A Sutanto[1], **B Yuliandra**[1]

[1]Mechanical Engineering Department, Faculty Engineering, Andalas University Padang

Corresponding author: agussutanto@eng.unand.ac.id

Abstract. A large amount of household e-waste and their recovery activities by the informal sector positively impact the economic, social, and environmental in the urban area because they strongly support sustainable and green manufacturing principles. This research aims to observe the household e-waste collecting model in Padang city, Indonesia. Furthermore, it also wants to find several retain values of the Circular Economy practices by the informal sector for household e-waste when recycled, reused, or remanufactured. The e-waste from the refrigerator, air conditioner, and washing machine is taken as sample cases in this research. The result shows at least two collecting models: the reverse e-waste flow with and without the repair or service shop's involvement. These models include some actors such as (1) end-users, (2) household collectors, (3) repair shops, (4) intermediary collectors, (5) final collectors, and (6) recycling plants. Most of the collectors and repair shops involved are in the informal sector. The result also shows that remanufacturing through rebuilding its functional status as a new performance by reusing e-waste obtains the highest recovery value. Such remanufactured products in the secondary market can reach up to 40% from their first price. Recycling of household e-waste products has the lowest recovery value. However, copper (Cu) from e-waste products has the highest market value, and it follows aluminum and ferrous metals, among other metals, when recycled or reused.

1. Introduction

Green manufacturing is a paradigm in manufacturing industries that want to transform to be more environmentally friendly or have a minimal impact on the environment [1, 2]. For this reason, it takes effort and even additional costs by the manufacturing industry to minimize this environmental impact at each stage of the product life cycle from the development, design, manufacture, usage, and the end of product life [3]. Indicators of green manufacturing implementation at each step of the product life cycle can be seen from producers' and consumers' points of view. They can reduce the environmental burden and streamline the limited resources like materials and energy. The last aspect is viewed as something more profitable for the product maker because it lowers production costs [4]. While efforts to reduce pollutants due to manufacturing activities still do not significantly impact the industrial world, especially in developing countries.

In the product life cycle, the implementation of green manufacturing is the responsibility of the manufacturing industry (as a producer), but when used, it is exchanged to the user's side. The final stages of the product life cycle have become an environmental burden [5]. It further contains an economic value if some items can still be reused, remanufactured, or existing parts as a new material recovered for another product. In developed countries, the role is often carried out by the government to overcome product wastes [6, 7, 8]. Some products that their components are relatively expensive and suitable, even though they have been used up or damaged, can still be rebuilt to the original product's specifications. They use a combination of repaired, reused, and new components into a new remanufactured product by Original Equipment Manufacturer. Some examples are the remanufacturing of medical imaging devices by the Centre for Remanufacturing & Reuse [9] and remanufacturing at the Xerox photocopier [10].

Circular Economy practices in this context are actions, technologies designed, and actors involved in recovering unused products or products that have reached the end of their life. In this study, e-waste comes from household appliances such as refrigerators, washing machines, etc. Activities carried out by several actors related to household electronic waste recovery can be reused, remanufacturing, or recycled [11, 12].Recovery of household e-waste directly impacts for reducing the environmental burden and pollution on the one hand and minimizing the use of material resources by utilizing materials derived from electronic waste, on the other side. The household appliances e-waste, because of their short economic life, will continually increase every year. Some of these electronic wastes still contain components that are eligible to use, or devices still have sensory functions. The material such as Fe, Cu, Zn, Mn, etc. can also be obtained from e-waste. It can be higher than the corresponding amount of material if mined naturally. On the other hand, electronic waste also contains dangerous waste such as Pb, Mg, PCB, arsenic, etc. and can damage the environment [13, 14]. The Circular Economy (CE), through the recovery of electronic waste from urban areas, has the potential to produce mining activities that are more ecologically appropriate and support the concept of green manufacturing.

Uniquely in Indonesia, the actors of Circular Economy practices are informal sectors, and formally they do not get permanent employment. Urban mining activities in Indonesia are like reverse logistics chains carried out by several actors. Some serve as seekers of the used or defective electronic product; some also play a role as sellers of repair services. Those that function to disassemble the product into smaller assemblies. This last activity ends with the recycling processes, which converting wastes into new materials. Urban mining is a circular economy chain [15], making a manufacturing activity model "circular" or forms a closed product life cycle.

Recovery and utilization of e-waste in Indonesia are made in several ways. Handling e-waste by directly reusing some components for the same benefits in other products, known as reuse, is the greenest and most profitable solution from an environmental perspective. Another option is remanufactured that utilizes parts or several e-waste by adding manufacturing operations to restore the performance level of products that have been used or damaged to function again or be equivalent to new products [16,17]. The advantage of this remanufacture process is to increase the efficiency of material use and energy consumption, thus contributing directly to energy saving and waste reduction efforts. Environmental and economic advantages make remanufacturing an essential strategy in various industries [18]. Another e-waste recovery is recycling, which offers opportunities to reduce the environmental impact of e-waste. If appropriately managed, the recycling process can be a source of new material to meet the needs in the industry or reducing raw materials extracted from nature.

In this study, several questions about the actual urban mining carried out by major cities in Indonesia, especially household electronic wastes, need to be explored further. Some of the research questions about the types of household e-wastes targeted for urban mining and the technical background. Who are the actors involved in this activity, and what kind of household e-waste collecting model happened in urban areas? What is the reuse, remanufacture, recycle (3R) policy carried out by urban mining actors, and how much is the retained value by different product recovery processes of the household e-wastes?

2. Materials and methods

2.1. Research Object

This research explores the practice of household e-waste product recovery in the city of Padang, which focuses on three household e-wastes: (1) Air Conditioner, (2) Refrigerator, (3) Washing machine. The product recovery here means regaining the possession of the lost value of a product by recycling, reusing, or remanufacturing. According to the initial survey, there are several actors involved in the urban mining process and Circular Economy practices:

a) End users,
b) Household collectors,
c) Repair shops,
d) Intermediary collectors,
e) Final collectors,
f) Recycling plants.

A household e-waste collecting model is based on the interaction between those six actors. The circular economy activities discussed in this study are limited to product recovery activities, such as remanufacturing, reuse, and recycling. Remanufacturing is an option in which the household e-waste will enter the reverse flow channel at a particular manufacturing stage. At this stage, household e-waste products would be disassembled, reassessed, replaced, remanufactured, and reassembled to bring the obsolete product to specific quality and reliability levels. Reuse is the activity to retrieve one or more valuable parts from the household e-waste and use it to repair other used products. Recycling is the activity to recover the material of e-waste by recycling plant into a new material that is lower in quality than the original one.

2.2. Method

The research was conducted with the following steps:

a) Survey to identify the household e-waste, which the urban miner often uses in the city of Padang, Indonesia, and some actors involved in this activity. The survey was conducted on eight household collectors using motorized rickshaws or pick-up cars to inquire about the availability of e-waste, eight informal repair shops, eight intermediate and final collectors that collect metals and plastics from e-waste for recycling. The research also identifies the household e-waste collecting patterns, which usually occurred in Padang.
b) Survey to know the product recovery values obtained by the respective urban mining actors by different product recovery scenarios. The amount of money still retained by some household e-waste when recycled, reused, or remanufactured. The survey is conducted by asking direct questions to the actors involved. For instance, the question about the average price of used or obsolete goods (AC, washing machine dan refrigerator) from the end-user. How to quickly evaluate the e-waste based on their conditions and determine the retained value? What is the e-waste selling price in the repair shop? Which items are still valuable to reuse from sample e-waste? What is the selling price of remanufactured e-waste (case: AC, washing machine, and refrigerator) in the secondary market? What is the market value of Cu, Al, Ferrous metal, or plastic for recycling?
c) Sampling and disassembling the e-waste in the laboratory and separating them based on material types. It then identifies the material composition of the samples and obtains the retained value by recycling. For this purpose, two samples are used for each e-waste (AC, refrigerator, and wash machine). Each e-waste should be disassembled into parts, and then they must be weighed and determined the material type (e.g., ferrous, non-ferrous, plastic, Al, Cu) and their retained value by recycling.

3. Results and discussion

3.1. Household E-WasteCollecting Model

A household e-waste collecting model as an implementation of Circular Economy (CE) practices in the urban area in Padang is illustrated in figure 1. This collection model is similar for three different household e-wastes samples (air conditioner, washing machines, refrigerators). At least, there are two collecting models of household e-waste in Padang city, Indonesia: (1) the e-waste flows with the involvement of the repair shop, and (2) without the participation of the repair shop. If the reverse flow includes the repair shop, most e-waste (AC, washing machines, and refrigerators) still function or have slightly damaged. Most of the components can be reused or repaired or can be reassembled using a combination of reused, repaired, and new elements into a remanufacturing product. The second model without including the repair shop is characterized by heavy damaged or obsolete products. The e-waste is generally disassembled, and the components are separated and processed based on the material types. The parts are then utilized by recycling.

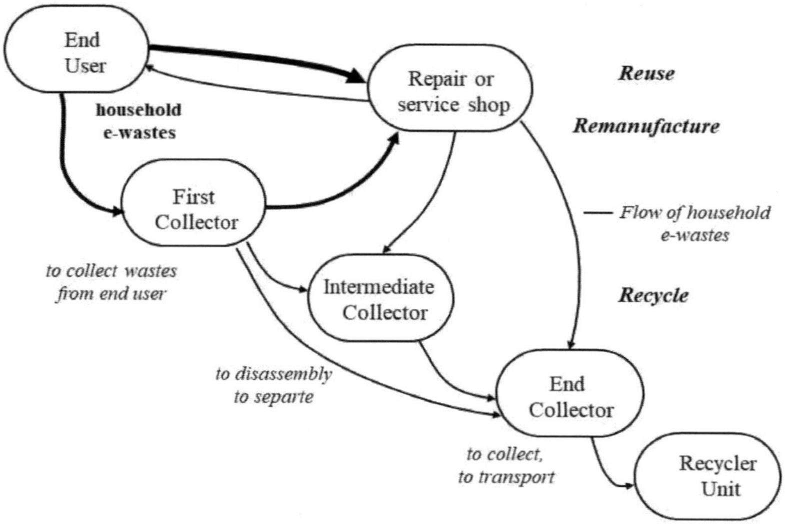

Figure 1. Household e-waste collecting model as a CE practice

There are six main actors involves in this model: (1) end-users, (2) household collectors, (3) repair shops, (4) intermediary collectors, (5) final collectors, and (6) Recycling units. End-users are household electronic product consumers where e-waste originates. Household or First collectors are the agents that collect e-waste from door to door. Usually, these informal collectors ride motorized rickshaws or pick-up cars, as shown in figure 2. Repair shops are those who receive e-waste from end-users or household collectors. Intermediary collectors are those who collect e-waste to be recycled. Final collectors act as material collectors and transport materials to recycling units.

The CE practices by the informal sector start with end users as the source of household e-waste products. Two scenarios can happen: either the end-user sold the e-waste to household collectors or directly to informal repair or service shops. Usually, the repair shops then conduct an assessment to determine whether the e-waste will be reused, remanufactured, or recycled. The choice to remanufacture or reuse is selected when e-waste (or at least their parts) are still in good condition. E-waste parts cannot be used as input to remanufacture or reuse process then sold to intermediary collectors. They disassemble and separate several obsolete parts based on their material type to be recycled. Those parts next sold to final collectors who subsequently transport them to recycling plants.

Figure 2. First collector transportation mode

3.2. Material Composition of E-Waste Products

The material composition of e-waste samples is identified by disassembling them in the laboratory. The type of material from each component is identified and weighed. The material composition in percentage is shown in figure 4. Based on observation, the air conditioner consists of approximately 63% ferrous metals, 20% plastic, 8% copper, 6% aluminum, and 3% of other materials. A similar case can also be seen in the refrigerator, where the most common material is also ferrous metals (58%), followed by plastics (26%), copper, and aluminum (roughly 2%), although the percentage is slightly different. A different result can be seen in the washing machine, where the highest rate is owned by plastics (69%), followed by ferrous metals (22%), copper (6%), and other non-ferrous metals (3%).

There is the market potential for copper, aluminum, and ferrous metals, and each of them has a different market value by CE actors. Copper has the most significant market value. It can be sold at roughly IDR 60-70K per kg by e-waste collectors. Aluminum is the second best (can be sold at roughly IDR 16K per kg), followed by ferrous metals (can only be sold at roughly IDR 2.5K per kg). Material compositions, along with their prices, are used to calculate product recovery value by recycling.

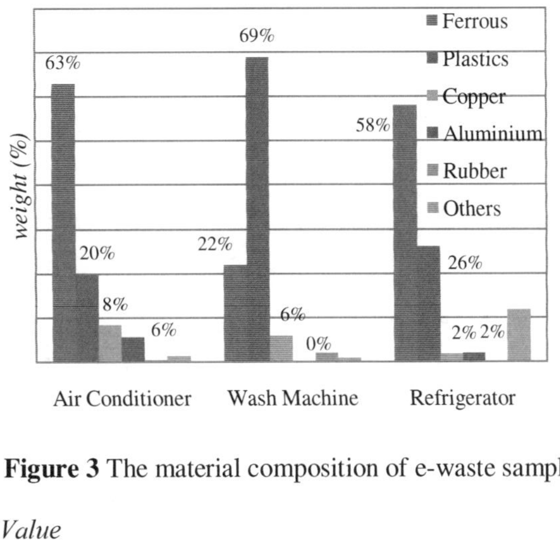

Figure 3 The material composition of e-waste samples

3.3. Product Recovery Value

Product recovery value is the price that still exists or retained at their end of life stage on the household e-waste products. Product recovery is related to the action of obtaining possession of the lost value again after product usage. The product recovery options included reuse, recycle, or remanufacture. A higher value can be expected if most of the used product parts are still in good condition, which means they can be used again through reuse or remanufacturing. Used products that have been damaged beyond repair can only be recycled; thus, it has less value. This product recovery value is highly dependent on the used product functionality and condition.

Table 1 shows some used components that are normally reused as spare parts for remanufacturing in CE practices in Padang. Most of these items obtained from urban mining activities and used to rebuild the second-hand product, either by using them to replace the broken part(s) in used product or taking the valuable parts (parts that still in good condition and can function properly) and remanufactured them with other useful parts from another used products and or new parts. An essential item for the remanufacturing process is the product or unit body. This item serves as the base of assembly operations in remanufacturing. Repair shop conducted a remanufacturing process only if the product or unit body is available.

Table 1. Items are frequently reused or remanufactured in CE practices by informal sectors

Part or Assembly	e-Wastes
Thermostat, evaporator, compressor, overload relay	Refrigerator
Motor unit, compressor, copper pipe, run capacitor, fan unit, thermistor, evaporator unit	AC
Wash motor, spin motor, timer control	Washing Machine
Body	Most of e-wastes *)

*) main assembly for remanufacturing

As an example, the illustration of remanufacturing by the informal sector is explained as follows. Used or damaged refrigerators are obtained from household collectors or directly from the end-user. The used refrigerator was dismantled and cleared. Based on the identification of damage and a comprehensive valuation, the repair shop party can make repairs and or replacements with new components or other components from other refrigerators. In this process, several manufacturing operations are also carried out, such as cutting, forming, or connecting parts or materials so that the refrigerator can operate functionally again. Then do the assembly again and put forward the performance of the "new refrigerator" performance. This remanufactured product is sold with certain guarantees.

The main requirement for this remanufacturing is the main body of the product that is still in good condition. Some components or units in a used refrigerator are often reused by the repair shops, such as thermostats, evaporators, compressors, or overload relays. The retained value of all reuse components can reach 18% of new refrigerators' prices with the same type and brand. The selling price of remanufactured refrigerators in the secondary market can get 40% of the new refrigerators' purchase price. This remanufacturing process also leaves less unused material to be recycled.

The product recovery values from the remanufacturing of used refrigerators and washing machines are obtained from these products' selling prices in the secondary market. They can reach up to 40% to the initial price. Product recovery value from direct reuse of some valuable parts that can function appropriately from the waste of the refrigerators and washing machines can reach up to 18%. While recycling the ferrous metal, plastic, or copper from that e-waste still has its retained value up to 5% to the product's initial price. Used air conditioners (AC) produced lower recovery value than the refrigerator and washing machine if they are remanufactured and recycled as well by the informal sector. From the survey, the remanufactured AC reassembled by AC service/repair shop is sold up to 30% of their initial price, while recycling only 3%. The interesting thing here is although the air conditioner has a lower recovery value by remanufacturing and recycling, it has the highest recovery value from the reuse process. This may be related to the lower middle class's sparse use of air conditioners because the largest market for remanufactured and reuse products is the lower middle class. Table 2 shows product recovery values obtained from each reuse, remanufacture, and recycling process of CE practices by informal sector. The product recovery value obtained from the remanufacturing process is equivalent to the corresponding second-hand product selling price.

Table 2. Product recovery values of CE practices by informal sectors

e-waste	Recycle	Reuse	Remanufacture
Air conditioner	≤ 3%	≤ 22%	≤30%
Refrigerator	≤ 5%	≤ 18%	≤40%
Washing machine	2.5 - 5%	13 - 18%	≤40%

These results indicate two main points: First, there is considerable economic potential in household e-waste products. This potential can be utilized to improve social welfare. Second, remanufacture is a process that can obtain the highest product recovery value, followed by reuse and recycle. This is because the remanufacturing methods involve more value-added production activities than the reuse and recycle process. The implication is that greater economic benefits are obtained by actors who carry out remanufacturing processes or, in this case, the informal repair shop.

4. Conclusion

The Circular Economy (CE) practices for recovery of household electronic waste in urban areas in Padang is unique because many actors are involved in this activity. These activities provide economic benefits, as well as employing the low-income informal sector in Indonesia. For green manufacturing in the final stages of the product life cycle, it will have an excellent effect on the environment and a very sustainable circular economy activity carried out by the community and the informal sector. The practice and the potential of this recovery of e-waste are clearly shown in this article. This work aims to discover the informal remanufacturing potential carried out by the informal repair shops or other urban mining agents in Padang. Besides, extending product life is also outlook research that will be carried out to find environmentally and friendly Circular Economy practices in the urban areas.

Acknowledgments

The authors thank to the Department of Mechanical Engineering, Faculty of Engineering Andalas University for partially financial support on this work.

References

[1] Minhaj A A and ShrivastavaR L 2013 *World Review of Science, Technology and Sustainable Development* **10**17
[2] Paul I D, Bhole G P and Chaudhari J R 2014 *Proc. Materials Science***16** 1644
[3] Deif A M 2011 *J. of Cleaner Production***19** 1553
[4] Mittala V K and Sangwan K S 2014 *Procedia CIRP***17** 559
[5] Hidayat R and Mua'lim 2019 *Int. J. of Geomate***16** 145
[6] Babu R, Parande A K, and Basha A C 2007 *J. of Waste Management***25** 307
[7] Cucchiella F, D'Adamo I, Lenny Koh, S C and Rosa P 2015 *Renewable and Sustainable Energy Reviews* 51 263
[8] Iqbal M, Breivik K, Syed J Hand MalikRN 2015*Environmental Pollution***207** 308
[9] Centre for Remanufacturing & Reuse 2006 *The Notified Budy Bulletin MHRA***6**
[10] King A, Barker S, and Cosgrove A 2007 *Proc. of 16th Int. Conf. on Engineering Design* 743
[11] Gutberlet J 2015 *Waste Management***45** 22
[12] Zeng X, Mathews J A and Li J 2018 *Environmental Science Technology***52** 4835
[13] Kwonpongsagoon S, Jareemit S and Kanchanapiya P 2017 *Int. J. of Geomate***12** 8
[14] Kaya M 2006 *Waste Management***57** 64
[15] Tong X Tao D 2016 *Conservation and Recycling***107** 10

[16] Matsumoto M, Yang S, Martisen K and Kainuma Y 2016 *Int. J. of Precision Engineering and Manufacturing-Green Technology* **3** 129
[17] King A, Burgess S C, Ijomah W L and Mcmahon C 2005 Sustainable *Development* **14** 257
[18] Charter M, and Gray C 2008 *Int. J. of Product Development* **6** 375

IOP Conf. Series: Materials Science and Engineering 1041 (2021) 012063 doi:10.1088/1757-899X/1041/1/012063

Preparation Smallest PVA/Binahong Leaf Nanofiber for Wound Dressing Applications by Varying Electrospinning Variable using Taguchi Method

F Ridwan[1] and A A Ananta[1]

[1]Mechanical Engineering Department, Faculty of Engineering, Andalas University, Kampus Limau Manis, Pauh, Padang, West Sumatera 25163, Indonesia

Corresponding author: firmanridwan@ft.unand.ac.id

Abstract. Binahong leaf is one of the herbal plants known to have interesting properties such as accelerating wound healing, anti-bacterial, etc. Converting Binahong leaves into fiber will increase their effectiveness because they are easier to absorb. The electrospinning method is used to convert Binahong polymers into Binahong fibers. The Taguchi method is used for experimental design with three factors and three levels. The resulting fiber properties vary with each variation factor. The smallest fiber obtained was 459,993 nm with tip and collector distance, voltage, and weight% PVA, respectively, 8 cm, 20 kV, and 8% (98.6922 P). Wound dressing by using Binahong nanofiber is possible.

1. Introduction

A wound is a condition where the skin tissue is exposed to air. This condition can lead to infection and if not treated promptly, can lead to other complications. One treatment is to dress the wound as a barrier to bacteria. Isolated injuries from the air are expected to avoid bacteria, grow new tissue, and accelerate wound healing. Traditional people believe that Binahong leaves accelerate wound healing, and research has proven this[1].

Nanofibers are objects with a very high ratio of length to diameter. The diameter of the nanofibers is tiny. Nanofibers present new opportunities in many aspects of human life. Tissue scaffolding, drug delivery, filtration, protective clothing, wound dressings, sensor, and nanocomposite applications are some of the nanofibers' applications [2].

1.1. Binahong

Andira cardifolia (Ten.) Steenis or so-called Binahong is a long-lasting vine that can reach 5 meters in length. Binahong has other names such as Boussingaultia cordifolia (Ten), Boussingaultia gracilis Miers, madeira vine (English), dheng san chi (China), gondola (Indonesia) [3]. Binahong is a traditional herbal that originated from the United States. Binahong is a traditional herbal medicine from the United States. Binahong is used in conventional medicine, such as wound treatment, wound surgery, etc. The Binahong plant is shown in figure 1.

Figure 1. Binahong Plants.

Binahong has four parts of plant morphology, namely leaves, stems, roots, and flowers. The Binahong plant characteristics are single leaf, stacked, very short (subsessile) stem, cordata, 5-10 cm long, 3-7 cm wide, pointed tip, emerginatus, flat side, and edible. Binahong is a soft, cylindrical stem, vines around, and is red. Binahong roots are rhizome and soft. Binahong flower is a compound flower with a rhizome, long stem, grows under the leaves, and smells nice [3].

Binahong is used as a medicine because it has an alluring effect. Based on its chemical components, Binahong has Saponins, Flavonoids, Terpenoids, Steroids, Alkaloids. Saponins act as cleansers and antiseptics, which prevent the growth of microorganisms, and Saponins also play a role in the formation of Collagen I, which accelerates wound healing [4]. Flavonoids have lipophilic protein that helps destroy the microbe possible to infection causes on wound [5]. Triterpenoids help to regenerate infected cells properly. Alkaloids have anti-bacterial properties. This component does not exist in every Binahong morphology. The complete features are shown in table 1 [5].

Table1. Chemical Contents of Morphology Binahong [3].

Part of Binahong	Chemical Contents				
	Saponin	Flavonoid	Terpenoid	Steroid	Alkaloid
Leaf	+	+	+	+	+
Stem	+	+	+	+	+
Root	+	+	+	-	+
Flower	+	+	+	-	-

1.1. Electrospinning

Electrospinning is a method used to continuously make long fibers with a submicron diameter to a nanometer level using a high-potential (high voltage) electric field. The electrospinning technique is easy to apply at the laboratory level and can be scaled to the industrial level. The fibers produced by Electrospinning have many potential applications. The resulting fiber has many advantages over ordinary fibers, such as a higher surface area. Nonwoven mat with Electrospinning has a small pore size, high porosity, and high surface area. Another advantage of using the electrospinning technique is that the nanofiber composition can be controlled to achieve more surface functionality flexibility. With these advantages, many fiber applications are generated by Electrospinning, such as filtration, scaffold construction, tissue engineering, wound dressings, energy conversion and storage, protective clothing, sensors, drug delivery, and many more. [6].

Electrospinning's working principle is that when a voltage is applied to the syringe, the voltage will push the polymer solution inside to the ground. Electrospinning has three main components, sprayer, collector, and high voltage [7]. A simple electrospinning system is shown in figure 2. Pushed positive-charged polymer solution produces jet polymer when penetrates the collector. This jet is conical and named Taylor Cone. Taylor cone can be seen in figure 3.

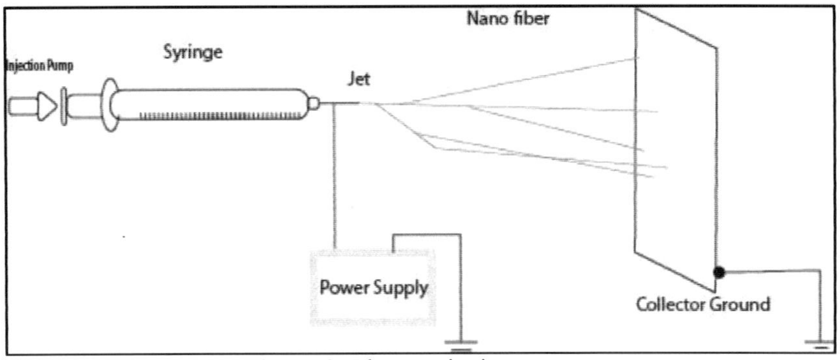

Figure 2. Electrospinning system.

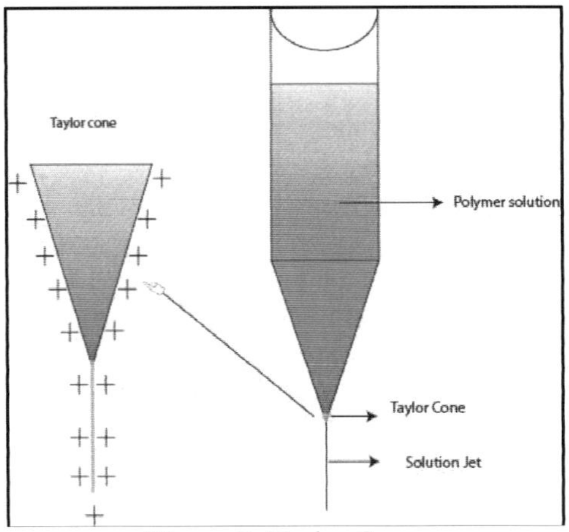

Figure 3. Taylor Cone.

1.2. Parameter of Electrospinning

The manufacture of fibers by Electrospinning has advantages, including controllable surface characteristics. To control these characteristics, it is necessary to know the parameters involved in the electrospinning process. The combination of different electrospinning parameters will result in other fiber characteristics, such as pores, larger fibers, different construction shapes, and so on.

1.2.1. Solutions viscosity.

The polymer concentration will affect the resulting fibers, such as the formation of beads. Higher polymer concentrations result in fewer beads. Contrary to the solution's viscosity, the resulting fiber is less likely to bead if the viscosity increases. But higher density results in larger fiber diameter. The solution's viscosity needs to be high enough for the beads to dissipate but not too large for a larger fiber diameter[8].

1.2.2. Solution molecular weigh

The molecular weight of the polymer is the number of polymer chains attached to the solution. The molecular weight will affect the electrical properties of the polymer, such as surface tension and conductivity. The low molecular weight results in granules' formation, but the higher the molecular weight results in a larger fiber diameter[8].

1.2.3. Voltage
The applied voltage plays an essential role in building the Taylor Cone. Taylor cone will be formed if the applied voltage exceeds the surface tension at the needle tip. When the applied voltage is higher, the resulting fiber will likely be smaller in diameter but coarser. Conversely, if the applied voltage is low, the fiber will be bigger

1.2.4. Tip to collector distance
Fiber morphology is also influenced by the distance between the tip and the collector. It takes sufficient time to allow the fibers to dry before they reach the collector, but if the distance between the tip and collector is too far or too close, beading is produced. The optimal distance between the tip and the collector is required to produce fine fibers.

1.2.5. Flow rate
The flow rate of polymer from the syringe affects jet velocity and material transfer rate. Increased flow rate produces a bigger diameter of pore and beads [9]. A low flow rate makes finer fiber, but it takes a long time since the material transfer rate is also low.

1.2.6. Ambient parameter
The ambient parameter is a parameter created outside the electrospinning system. The ambient parameters are humidity, temperature, etc. It appears that when the temperature is higher, and the humidity is higher, it results in circular pores on the fiber surface

To summarize the parameter, table 2 shows the parameters on Electrospinning.

Table 2. Parameter on electrospinning[10].

Parameter	**Effect on fiber morphology**
Solution concentration/viscosity	↓ – bead formation, ↑ – larger fibers
Conductivity/solution charge density	↑ – uniform bead-free fibers, smaller fibers
Surface tension	No conclusive link
Polymer molecular weight/viscosity	↑ – reduced the number of beads
Flow rate	↓ – smaller fibers, ↑ – larger fibers
Field strength/voltage	↓ – larger fibers, ↑ – smaller fibers
Distance between needle tip and collector	↓ – smaller fibers, ↑ – larger fibers
Ambient parameters (temperature and humidity)	↑ temperature – ↓ viscosity, ↑ humidity – the appearance of circular pores on the fibers

2. Materials and methods
To produce Binahong Leaf fiber, equipment preparation is required. Two aspects need to be prepared. Binahong polymer and electrospinning machines.

2.1. Binahong polymer preparations.
Binahong polymers need to be prepared before using the electrospinning machine because Electrospinning only works on viscous fluids. The steps to obtain Binahong polymer are first that the Binahong leaves need to be extracted. This is done by grinding the Binahong leaves and soaking them

in 95% methanol for two days. Binahong polymer was obtained by stirring and slowly adding polyvinyl alcohol until the polymer was formed. Next, calculate the viscosity of the polymer using Stoke's Law. Table 3 shows the Binahong polymer characteristic.

Table 3. Binahong polymer characteristic

Sample (%wt)	Graduated Cylinder Glass Volume (mL)	Distance ball travel (m)	Time (s)	Viscosity (P)
PVA 8%	20	0.2	21.07	98.6922
PVA 10%	20	0.2	60.28	282.365
PVA 12%	20	0.2	72.00	337.249

2.2. Electrospinning setup
The electrospinning machine also needs to be set up. This setup consists of preparing the syringe pump mechanism, the collector mechanism, the high voltage power supply attached to the syringe and the collector. Figure 4 shows the electrospinning machine ready to operate.

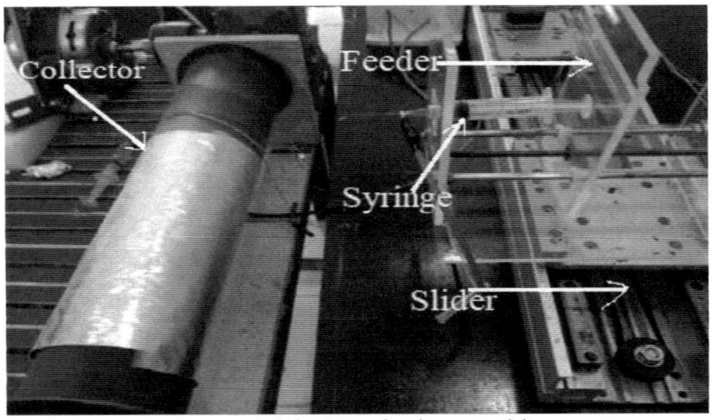

Figure 4. Electrospinning machine

2.3. Observation
Once the Binahong leaf fiber was produced, the surface and morphology of the Binahong fiber are observed. This observation was carried out using an Olympus stereo optical microscope located at the Metallurgical Laboratory of the Mechanical Engineering Department, Andalas University.

3. Results and discussion
The Taguchi method is used to design a more robust experiment[11]. By using the Taguchi Orthogonal Array (OA) L9 (3^3). ANOVA was established to determine which parameters influence the investigation. After the optimum level parameter is obtained, a confirmatory test is carried out to confirm whether the experiment is within the confidence limits (5%).

3.1. Taguchi method of experiment.
The Taguchi method is applied to make experiments more efficient, simple, and robust. The orthogonal array used is L9 with three factors and three levels. Table 4 shows the data experiment, and figure 5 shows the results.

Table 4. Data experiment

No	Distance (cm)	Voltage (kV)	%wt PVA	Average Fiber Diameter (μm)
1	8	15	8%	0,51948
2	8	20	10%	0,60092
3	8	30	12%	0,86287
4	12	15	10%	0,73960
5	12	20	12%	0,92450
6	12	30	8%	0,59309
7	16	15	12%	1,23780
8	16	20	8%	0,69337
9	16	30	10%	0,77042

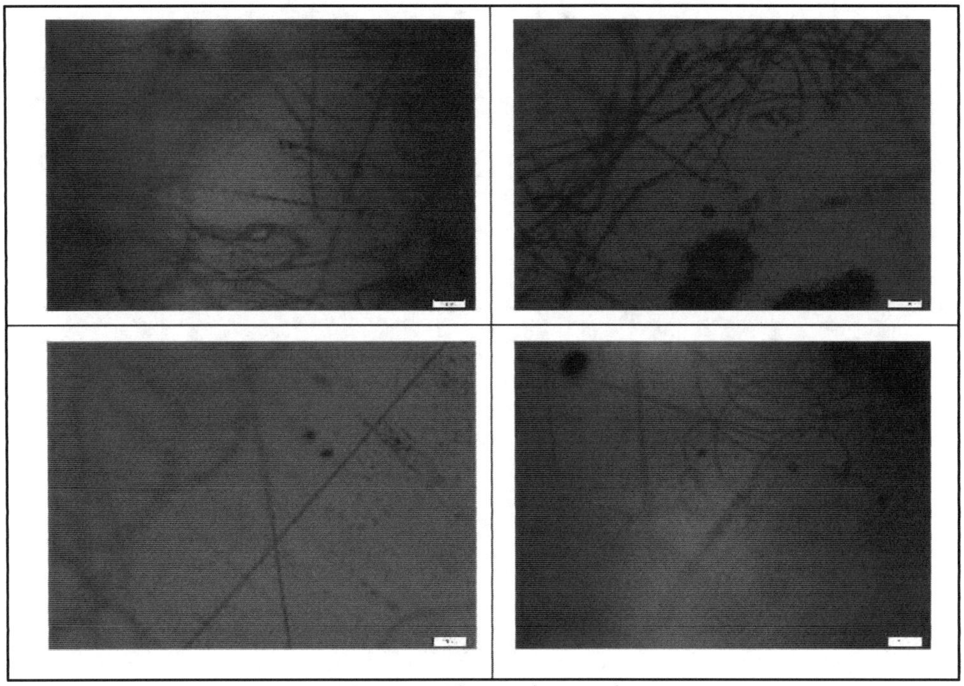

Figure 5. Observation results

The Taguchi method also generates a mean effect and SN ratio plot. In this study, the "smaller is better" S/N ratio was chosen. Means effect shows the best level for each factor, where the S/N ratio shows the best signal variability and noise. Figure 6 and figure 7 show the effect plots for the mean and S/N ratio, respectively.

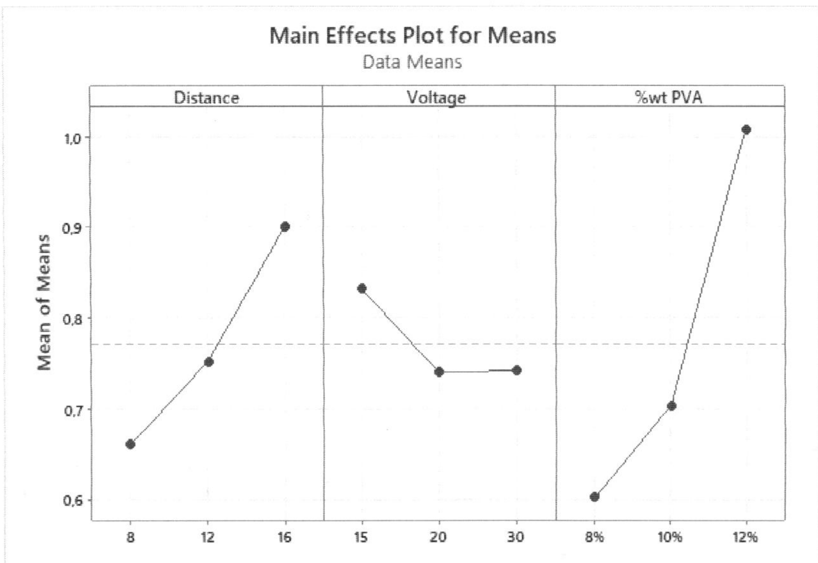

Figure 6. Main Effects Means.

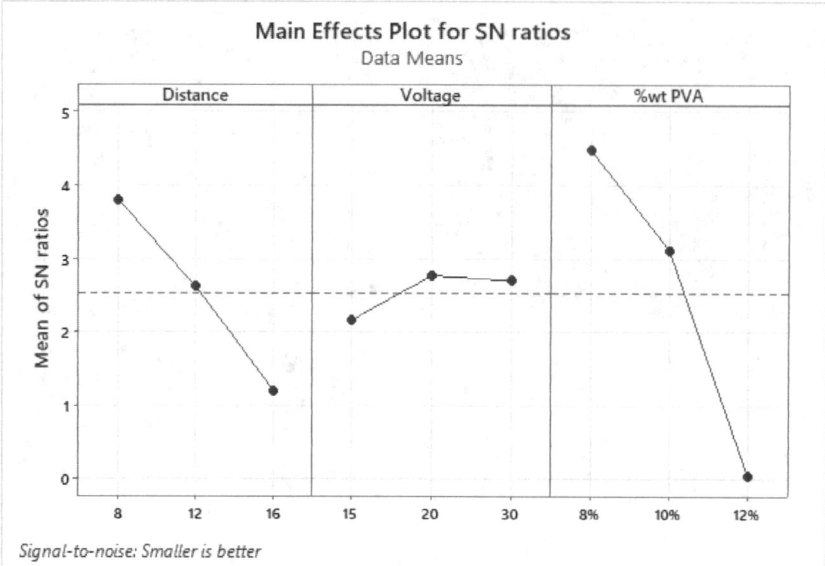

Figure 7. Main Effects SN ratio.

Based on figure 6,%wt PVA is the most dominant factor for this experiment, followed by the distance between the tip and the collector and the voltage. Besides using the Taguchi prediction, the smallest fiber is 0.459993 μm when the length, voltage, and % wt PVA are 8 cm, 20 kV, and 8%, respectively.

Figure 7 shows at a distance of 16 cm, 15 kV voltage, and 12% wt PVA that the lowest SN ratio is close to zero so that the least significant experiment occurs at this level. On the other hand, at a distance of 8 cm, the voltage of 20 kV and% wt PVA of 8% indicates that the highest SN ratio is further away from zero so that the most significant test occurs at that level.

Because the L9 parameter (8cm, 20kv, 8%) was not carried out in the experiment, it is necessary to do a confirmation test to confirm whether the resulting fiber is the smallest.

3.2. ANOVA

Analysis of Variance is used to analyze the parameters that affect the experiment. ANOVA produces a p-value that needs to be compared with the hypothesis, which states, "H0 is residual data with a normal distribution (variables do not affect fiber diameter) and H1 is residual data not normally distributed (variables affect fiber diameter)". Table 5 shows the ANOVA data.

Table 5. ANOVA data

Source	DF	Seq SS	Adj SS	Adj MS	F	P	% Contribution
Distance (cm)	2	12,099	12,0987	6,0494	21,7	0,049	23,2595293
Voltage (kV)	2	0,8253	0,8253	0,4126	1,48	0,440	1,5866241
%wt PVA	2	38,535	38,5346	19,2673	69,11	0,017	74,0820631
Residual Error	2	0,5576	0,5576	0,2788			1,0719758
Total	8	52,016					100

In this study, the research criteria were H1 was accepted, and H0 was rejected, if the P-value was less than α (α <0.05). With this criterion, voltage will not be accepted as a parameter that affects this study. The distance and% wt PVA were accepted.

3.3. Confirmation test.

A confirmatory test was conducted to find out whether the experiment was confident enough. Confirmation test was carried out by Binahong polymer electrospinning with the distance, voltage, and% wt PVA of 8 cm, 20 kV, and 8%, respectively. The predictive value is 0.459993 μm. Figure 8 shows the confirmation test results.

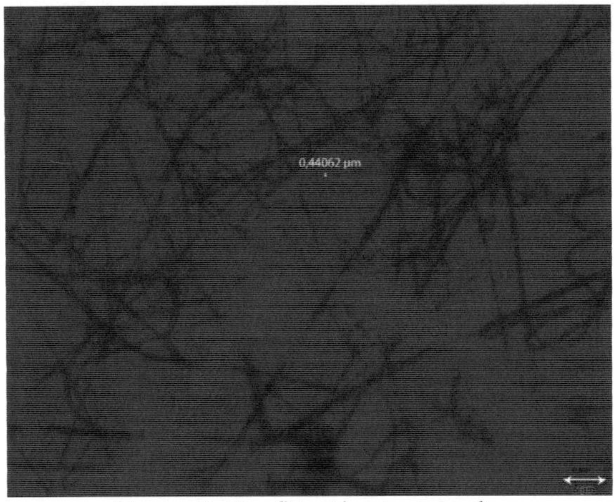

Figure 8. Confirmation test result.

The confirmation test results were 0.44062 μm, and the predicted value was 0.459993 μm. The confirmation test is still within the confidence limit (5%). According to the Taguchi method, the results are acceptable. From this confirmatory test, this study is confident.

4. Conclusions

The Taguchi method was used to test the electrospinning behavior when producing Binahong polymer fibers. Three factors and three levels are selected for the smallest diameter. From the data obtained, it can be concluded that the smallest fiber diameter was reached at 0.44062 μm, where the variation of distance, voltage, and viscosity of PVA is 8 cm, 20 kV, and 8% (98.6922 P), respectively. The largest fiber diameter was achieved at 1.2031330 μm, where the variation of distance, voltage, and viscosity of PVA were 16 cm, 15 kV, and 12% (337,249 P), respectively. The closer the distance between the spinner's end and the collector, the smaller the fiber diameter. Less viscous polymers result in smaller fiber diameter. From ANOVA, tip and collector distance contributed to fiber diameter by 23.2595293%, and viscosity contributed to fiber diameter by 74.0820631%. The smallest fiber established faster absorption on the body, therefore faster-wound healing. Further researches needed to confirm the characteristic of Binahong nanofiber wound dressings and other material possibilities.

References

[1] G P Y A br, Basyuni M and Lesmana I 2004 Phytochemical Screening of Secondary Metabolite in Binahong Leaves (Anredera cordifolia (Ten.) Steenis) to Test In Vitro Inhibitation Growth of Aeromonas hydrophila 1–2
[2] Park J S 2010 Electrospinning and its applications *Adv. Nat. Sci. Nanosci. Nanotechnol.* **1**
[3] Bruce 2013 Kajian Teori Binahong *J. Chem. Inf. Model.* **53** 1689–99

[4] Hasiib E, Riyanti R and Hartono M 2015 Pengaruh Pemberian Ekstrak Daun Binahong (Anredera Cordifolia (Ten.) Steenis) Dalam Air Minum Terhadap Performa Broiler *J. Ilm. Peternak. Terpadu* **3** 233359

[5] Purwasih R and Safitri F A 2018 The potency of Binahong leaves (Anredera cordifolia (Ten.) Steenis) to recovery process of wound in the livestock **5** 211–5

[6] Karakas H 2014 Electrospinning of Nanofibers and Their Applications *MDT "Electrospinning"* **3** 1–35

[7] Al-Hazeem N Z A 2012 Nanofibers and Electrospinning Method *Intech* **i** 13

[8] Kurečič M and Sfiligoj Smole M 2013 Electrospinning: Nanofibre Production Method *Tekstilec* **56** 4–12

[9] Subbiah T, Bhat G S, Tock R W, Parameswaran S and Ramkumar S S 2005 Electrospinning of nanofibers *J. Appl. Polym. Sci.* **96** 557–69

[10] R. M, K. T, J. J and Jin S 2011 Electrospun Nanofibers in Tissue Engineering *Nanofibers - Prod. Prop. Funct. Appl.*

[11] Wysk, R. A., Niebel, B. W., Cohen, P. H., and Simpson T W 2000 Taguchi's Robust Design Method *Manuf. Process. Integr. Prod. Process Des.* 1–9

IOP Conf. Series: Materials Science and Engineering 1041 (2021) 012064 doi:10.1088/1757-899X/1041/1/012064

An Improved Method for Dynamic Behaviour Prediction of Carbon Fibre Reinforced Epoxy (CFRE) using Finite Element Model Updating

M A S A Shah[1], M A Yunus[1], M N A Rani[1], W I I W I Mirza[1] and M. S. M. Sani[2]

[1] Structural Dynamics Analysis & Validation, Faculty of Mechanical Engineering, Universiti Teknologi MARA (UiTM), 40000 Shah Alam, Selangor, Malaysia
[2] Faculty of Mechanical Engineering, Universiti Malaysia Pahang, 26600 Pekan, Pahang, Malaysia.

Corresponding author: mayunus@uitm.edu.my

Abstract. The dynamic behaviour prediction scheme of carbon fibre composite structures are extremely challenging to be performed due to the geometrical ply-by-ply properties and the ply orientation issues. In this paper, a laminated carbon fibre reinforced epoxy plate with elastic and linear layers was used with aim to investigate the dynamic behaviour of the carbon fibre reinforced epoxy using practical modelling scheme via finite element modelling and updating method. A simplified FE model of the CFRE plate was developed using shell element properties known as PSHELL to represent geometrical laminated properties. Subsequently, the dynamic behaviour of the FE model was calculated using Nastran SOL103. Potential updating parameters of the FE model was analysed and identified using the sensitivity analysis. The model updating method was then used to update the initial FE model of the CFRE plate based on the experimental modal analysis (EMA) result. The comparison of the results were used in verifying the accuracy of the updated FE model of the CFRE plate. The result suggested that the PSHELL properties can be used efficiently to represent the geometrical laminated properties of the CFRE plate without considering the geometrical effect of ply-by-ply properties.

1. Introduction

Carbon fibre composite materials are increasingly used in replacing the conventional materials such as steel and aluminium in many engineering supplications such as in the automotive industries for boosting the fuel economy while maintaining the safety and performance. Carbon fibre structures are usually made up from two or more constituent materials and laminated together to produce anisotropic properties and allows the manufactures to produce more complex shapes of with a higher strength-to-weight ratios, corrosion resistance, and workability [1]. However, the anisotropic material behaviour of the laminated carbon fibre complicates the structural dynamic behaviour prediction in which the information of the direction-dependent mechanical properties of the composite materials are often limited [2]. Furthermore, the end-product of the carbon fibre structure is always found not precise as the mould shape because the manufacturing process of the structure induced distortion to the structure and causing the geometrical variations. Therefore, to analyse the dynamics behaviour of the laminated composites are complicated and computationally expensive due to the material behaviour and geometrical variations such as a thermal expansion mismatch between the fibres and resin, chemical shrinkage of the resin during polymerisation, and moisture absorption [3].

There are several approaches has been suggested by researchers in predicting the dynamic behaviour of carbon fibre structure [4–6]. Different type of FE models to represent carbon fibre structure have been studied with various modelling schemes from detailed to simplified models. Previously, detailed modelling using solid (3D) elements has found to be the most schemed to be used by research as presented by [7,8]. There are models in which thin layer elements are used with considering the ply thickness, orientations and individual ply material properties [9]. Most of the researchers modelled composite structures using PCOMP method and PCOMPS method in which the layers properties and detailed meshing predominant [10]. However, a simple and practical modelling has been preferred in which most the previous studies developed a computationally expensive and time consuming modelling schemes for predicting the dynamic behaviour of composite structure.

In this study, the dynamics behaviour of carbon fibre reinforced epoxy is investigated numerically and experimentally. The finite element method is used to calculate the natural frequencies and mode shapes of the initial model of carbon fibre reinforced epoxy and experimental modal analysis (EMA) is used to obtain the experimental result. The low stiffness suspension spring was used to perform the EMA to separate between rigid body and deformation modes of the composite structure because the stiffness and damping of the suspension cord may have significantly altered the modal parameters of the elastic modes of the test structure, if the separation between rigid body and deformation modes was not achieved [11]. In modelling the carbon composite structure, material properties of fibre and matrix of the carbon fibre composite structure are simplified as an orthotropic material and they are modelled as PSHELL element. The validity of the finite element modelling using the PSHELL element was compared with the experimental model. The finite element model updating was used to improve the accuracy of the initial finite element model of the carbon composite structure as close as the experimental model.

2. Materials and methods

2.1. Finite element modelling of the carbon fibre reinforced epoxy (CFRE) plate

The natural frequencies and mode shapes of the composite plate were calculated by using the finite element (FE) method. In the FE modelling, the eight layers of the composite were modelled as a single layer 2D element as shown in figure 1. The composite plate was discretised into 1658 2D shell elements with 2mm uniform thickness profile. The 2D shell elements were defined as orthotropic material properties. The material properties in table 1, which were obtained in [12], were used as the material properties of the FE model. It is worth noting that there are no boundary conditions applied to the FE model.

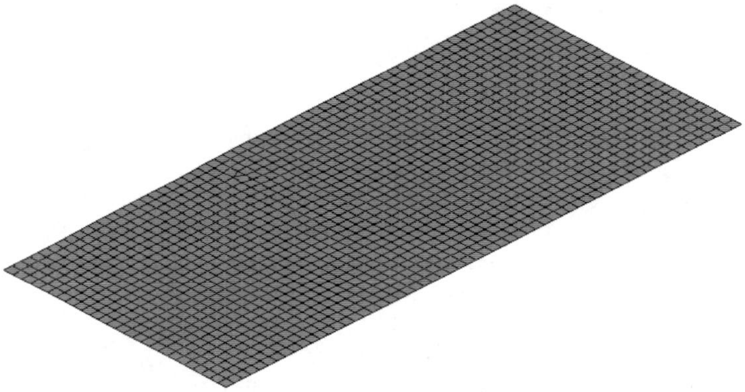

Figure 1. FE model of the CFRE plate.

Table 1. Material and geometrical properties of the CFRE plate [12].

Property	Value	Unit
Young's Modulus in longitudinal direction, E_{11}	40776.5	MPa
Young's Modulus in longitudinal direction, E_{22}	40776.5	MPa
In-plane shear modulus, G_{12}	14197.9	MPa
Poisson's ratio, v1	0.436	Dimensionless
Density	1421.9	kg/m^3
Thickness	2	mm

MSC NASTRAN SOL103 normal modes solution was used in calculating the natural frequencies and mode shapes of the carbon fibre reinforced epoxy. The calculated results were then compared and validated with the corresponding experimental results prior to performing the model updating method.

2.2. Experimental modal analysis (EMA) of the carbon fibre reinforced epoxy (CFRE) plate

In this work, the natural frequencies and mode shapes of the composite plate were measured for validation and model updating purposes. The measured data was obtained by performing EMA. An impact hammer, accelerometer, and Leuven Measurement System (LMS) data acquisition were used in the EMA. The sensitivity of the impact hammer and accelerometer were 11.76mV/N and 10mV/g respectively. The frequency range for the experimental work was between 0 and 1000 Hz. Roving hammer was selected as the excitation method in order to minimise the mass loading issues in the physical structure of the composite plate as the weight of accelerometers might significantly influenced the measured data as reported in [13] and [14]. The test structure was suspended with two soft rubber bands to simulate free-free boundary conditions. Figure 2 presents EMA set-up for the CFRE plate.

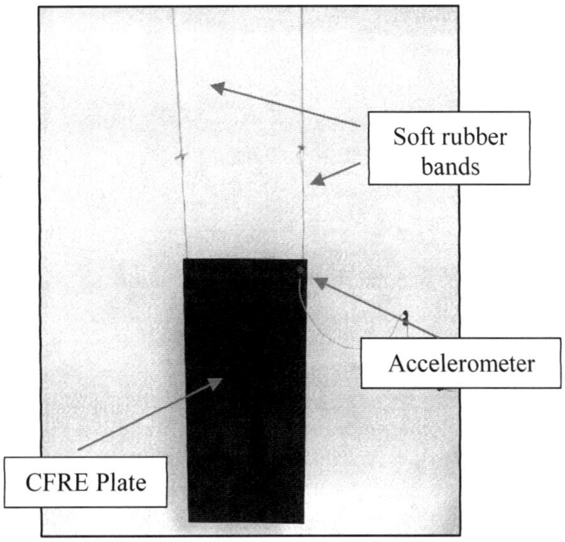

Figure 2. Test set-up of the composite plate.

2.3. Finite element model updating of the carbon fibre reinforced epoxy (CFRE) plate

The main goal of optimization scheme using model updating method is to improve the correlation results between experimental data and predicted data [15–17]. The discrepancies between the experimental and predicted eigenvalue data is used as a penalty function in the updating procedure. This method also allows a wide choice of the updating parameters to be corrected using the sensitivity analysis method.

In the model updating method, mode pairing algorithm is mandatory to be conducted to ensure that the predicted and experimental eigenvalues correspond to the same physical mode. However, it is difficult to pairing the mode shapes and natural frequencies correctly especially in the presence of close modes or repetitive roots in the structure under study [18]. Moreover, inaccurate FE modelling with highly uncertain parameters may lead to the inaccurate predicted mode shapes and thus contribute to the high complexity in the mode pairing process. Failure to follow the procedure of pairing the modes will lead to questionable updated results and the worst case is that the updated results do not reflect any physical meaning. One way to avoid the issue is to use the modal assurance criterion (MAC) which is one of the best methods of verifying the mode shapes are paired correctly [19]. The MAC is normally presented in matrix form is used by calculating experimental ($\boldsymbol{\phi}_{\mathrm{m}}$) and finite element ($\boldsymbol{\phi}_{\mathrm{a}}$) mode shapes using Eq. (1).

$$\mathrm{MAC} = (\boldsymbol{\phi}_{\mathrm{m}}\boldsymbol{\phi}_{\mathrm{a}}) = \frac{\left|\boldsymbol{\phi}_{\mathrm{m}}^{\mathrm{T}}\boldsymbol{\phi}_{\mathrm{a}}\right|^2}{(\boldsymbol{\phi}_{\mathrm{a}}^{\mathrm{T}}\boldsymbol{\phi}_{\mathrm{a}})(\boldsymbol{\phi}_{\mathrm{m}}^{\mathrm{T}}\boldsymbol{\phi}_{\mathrm{m}})} \tag{1}$$

In this study, the updating work for the CFRE plate were conducted by studying two different input properties namely, geometrical and material properties. In the geometrical properties, thickness of the plate was used as potential updating parameters meanwhile, for the material properties, Young's modulus in the longitudinal direction, Young's modulus in the transverse direction, the in-plane shear modulus, Poisson's ratio, and density were used as potential updating parameters.

In the sensitivity analysis, the sensitivity of the mentioned potential updating parameters to the structural responses which are the natural frequencies of the CFRE plate were study and identified. Hence, the updating procedure was used to correct the structural properties with high sensitivity values by using the Nastran SOL 200 script which can be expressed as follow:

$$\mathbf{S}_{\mathrm{i}} = \frac{\partial\lambda_{\mathrm{i}}}{\partial\boldsymbol{\theta}} = \boldsymbol{\phi}_{\mathrm{i}}^{\mathrm{T}}\left[\frac{\partial\mathbf{K}}{\partial\theta_{\mathrm{j}}} - \lambda_{\mathrm{i}}\frac{\partial\mathbf{M}}{\partial\theta_{\mathrm{j}}}\right]\boldsymbol{\phi}_{\mathbf{i}} \tag{2}$$

where $\mathbf{S}_{\mathrm{i}}$ is the m x n sensitivity matrix at the ith iteration which represent as rates of change of the structural eigenvalues $\boldsymbol{\lambda}_{\mathrm{i}}$ with respect to the changes in the parameters $\partial\boldsymbol{\theta}$. Meanwhile the objective function can be expressed as

$$J = min\sum_{j=1}^{m}\left(\frac{\lambda_i^{FE}}{\lambda_i^{EXP}} - 1\right)^2 \tag{3}$$

where λ_i^{FE} is the ith predicted natural frequency and λ_i^{EXP} represents the ith experimental natural frequency.

3. Results and discussion

In this study, the dynamic behaviour of composite structure using carbon fibre reinforced epoxy (CFRE) were obtained experimentally using experimental modal analysis (EMA) and numerically by using finite element modelling (FEM). These works were conducted as an attempt to predict the natural frequencies and mode shapes of the CFRE and reconcile the predicted data in the light of the EMA data. After that, finite element model updating method was used to improve the accuracy of the predicted model.

The result tabulated in table 1 comprises the comparison of first 10 natural frequencies of initial finite element model for CFRE with EMA data. It was found that the total error of the first frequencies modes recorded is 42.61 percent (Column III.). Accordingly, the largest errors recorded is on 3th mode and 9th mode with error of 7.30 percent and 7.54 percent. In Table 2, the comparisons of mode shapes between initial FE with experimental counterpart in form of MAC matrix have indicate that, there are 2 swap modes in initial FE which are between mode 3 and mode 4, mode 9 and mode 10. The swap modes have contributed to the large errors in the predicted frequencies, specifically in mode 3 and mode 9. Moreover, the result in table 2 shows that the predicted mode shapes of initial FE were not in good agreement with EMA data. It was found that the MAC values recorded in mode 4, mode 7, mode 9 and mode 10 were below 70 percent. Therefore, it is essential to improve the accuracy of the initial FE in light of the EMA data using model updating prior to be used for subsequence analysis [20].

In model updating method, it is important to identify updated parameters that are only sensitive to the structural response [21]. In this work, several potential updated parameters were used in sensitivity analysis such as plate thickness *T*, Young's modulus in longitudinal direction E_{11}, Young's Modulus in transverse direction E_{22}, Shear Modulus in-plane, G_{12}, Poisson' Ratio, v_{12} and density, *P*. MSC Nastran SOL 200 was used to calculate the sensitivity of the response toward potential updated parameters. Table 1 (Column VI.) shows the result of the updated FE in comparison with EMA data. A reasonable achievement is seen in the attempt with the total error managed to reduce from 42.61 percent to 25.05 percent with no swapping mode. The updated values of the updating parameters are listed in table 4. However, The MAC matrix in table 3 indicated that, predicted mode 4, mode 9 and mode 10 still failed to be in good agreement with EMA data.

Table 2. Comparison of experimental and FE analysis of CFRE.

	I.	II.	III.	V.	VI.
Mode	Experimental (Hz)	Initial FE (Hz)	Error (I-II)/I (%)	Updated FE (Hz)	Error (I-IV)/I (%)
1	129.71	125.29	3.41	126.46	2.50
2	161.99	166.18	2.58	163.82	1.13
3	339.92	***364.74***	***7.30***	350.86	3.22
4	376.19	***351.15***	***6.66***	***358.86***	***4.61***
5	597.02	628.11	5.21	617.24	3.39
6	674.91	647.10	4.12	663.67	1.67
7	732.65	***750.11***	***2.38***	723.18	1.29
8	753.05	763.85	1.43	766.30	1.76
9	933.48	***1003.91***	***7.54***	***965.96***	***3.48***
10	964.26	***983.38***	***1.98***	***983.62***	***2.01***
	Total Error		42.61		25.05

Table 3. MAC Matrix for initial FE model of CFRE.

Mode	**1**	**2**	**3**	**4**	**5**	**6**	**7**	**8**	**9**	**10**
1	***91***	3.8	0.4	0	0.4	0.3	1.4	0.3	6.8	9.5
2	7.7	***94.6***	1	0	1.5	0.2	1	0.1	0	0.3
3	0	0	14.7	***64.9***	0.3	0	0.5	1	0.9	0.7
4	0	0	***81.2***	34.5	1.6	0	0.4	0.3	0.4	0.9
5	0.3	2.2	0.1	0	***88.7***	0.7	8.8	0.2	1.2	0.7
6	0	0	0	0	0.3	***91.8***	0.8	0.3	1.9	4.1
7	3.5	0	0	0.1	4.6	25.4	***59.4***	7.1	0.5	1.2
8	0.1	0	0.2	0.1	0.5	0.1	19.3	***88.5***	3.2	0.2
9	0	0	3.5	2.6	0.2	0	0.2	0.1	31.6	***50.4***
10	24.3	0.6	0	0	2.2	2.7	0.3	0	***54.1***	45.1

The results of updated natural frequencies in table 1 (Column IV.) and updated mode shapes in table 3 strongly suggest that the updated parameters used in model updating method as listed in table 4 are still not good enough to improving the accuracy of the predicted FE of the CFRE. These results suggest that there is structural uncertainty particularly in the experimental work that might need to be consider in the modelling process.

Table 4. MAC Matrix for Updated FE model of CFRE.

Mode	1	2	3	4	5	6	7	8	9	10
1	***91***	1.3	0.4	0	0.6	0.3	1.1	0.4	7	9.3
2	3.6	***93.1***	0.7	0	1.5	0.2	1.4	0.1	0	0.7
3	0.2	0.3	***81.6***	4.6	0.9	0	0.7	1	1.3	1.6
4	0.2	0.1	30.7	***65.3***	0.3	0	0	0	0	0.1
5	0.2	0.7	0.4	0.1	***93.1***	3.1	1.3	0.4	0.4	0
6	0	0.1	0.1	0	2.1	***93.5***	0.1	0	2.5	4.7
7	1.1	0.7	0	0.4	0	7.1	***87.6***	18.4	0.3	0.5
8	1.7	0.1	0	0	0.8	10.4	0.3	***89.6***	0.8	0.8
9	7.7	1.1	0.9	2.9	0.1	0.1	0.4	0.6	***68.7***	0.5
10	14.4	0.3	1.5	0.9	0.4	1.2	0	0.2	5.8	***67.2***

Table 5. Updated parameters of CFRE model.

Property	Parameter	Nominal value	Updated value	Unit
PSHELL	Thickness, T	2.00	2.03	mm
MAT8 (Material)	Young's Modulus in transverse direction, E_{22}	40776.6	40601.26	MPa
	Shear Modulus in-plane, G_{12}	14198	13326.24	MPa
	Poisson' Ratio, v_{12}	0.436	0.291	Unitless

4. Conclusions

A practical modelling scheme for investigating the dynamic behaviour of the carbon fibre reinforced epoxy (CFRE) plate using finite element modelling and updating method is presented in this study. The presented study suggest that the proposed scheme improve the accuracy of the dynamic behaviour prediction for the CFRE plate in light with experimental (EMA) data. The biggest achievement in the proposed scheme is the scheme abilities in improving the mode swapping issue and reducing the natural frequencies total error up to 40 percent. The advantage of the proposed scheme over other published methods are the abilities in minimising the complexity in finite element modelling of composite structure by implemented the PSHELL properties method coupled with model updating method.

Acknowledgements

The authors wish to acknowledge the Malaysia Ministry of Higher Education (MOHE), Research Management Institute (RMI) and Institute of Graduate Studies of Universiti Teknologi MARA (UiTM) for providing financial support for this research through the fundamental research grant scheme (FRGS) 600-IRMI/FRGS 5/3 (172/2019). They would also like to extend their sincere gratitude for the support and help given by the all SDAV members.

References

[1] Vanaerschot A, Lomov S, Moens D and Vandepitte D 2015 Variability in Composite Materials Properties *Appl. Mech. Mater.* **807** 23–33

[2] Argyris J . 1955 Aircraft Engineering and Aerospace Technology *Aircr. Eng. Aerosp. Technol.* 347–56

[3] Dey S, Mukhopadhyay T, Sahu S K, Li G, Rabitz H and Adhikari S 2015 Thermal uncertainty quantification in frequency responses of laminated composite plates *Compos. Part B Eng.* **80** 186–97

[4] Mishra I and Sahu S K 2015 Modal analysis of woven fiber composite plates with different boundary conditions *Int. J. Struct. Stab. Dyn.* **15**

[5] Stochino F and Cazzani A 2013 Modal analysis of laminated composite plates by a new hybrid assumed strain finite element 1–10

[6] Khdeir A A and Reddy J N 1999 Free vibrations of laminated composite plates using second-order shear deformation theory *Comput. Struct.* **71** 617–26

[7] Cuadrado M, Pernas-Sánchez J, Artero-Guerrero J A and Varas D 2020 Model updating of uncertain parameters of carbon/epoxy composite plates using digital image correlation for full-field vibration measurement *Meas. J. Int. Meas. Confed.* **159** 107783

[8] Cuadrado M, Artero-Guerrero J A, Pernas-Sánchez J and Varas D 2019 Model updating of uncertain parameters of carbon/epoxy composite plates from experimental modal data *J. Sound Vib.* **455** 380–401

[9] Dey S, Mukhopadhyay T, Haddad Khodaparast H, Kerfriden P and Adhikari S 2015 Rotational and ply-level uncertainty in response of composite shallow conical shells *Compos. Struct.* **131** 594–605

[10] Hutchings A and Palodichuk M 2014 *A Comparison of Composite Modeling Techniques Siemens PLM Connection 2014*

[11] Bahari A R, Yunus M A, Abdul Rani M N, Nalisa A and Shah M A S A 2019 Investigation on the effects of suspension stiffness using experimental modal analysis and finite element model updating *IOP Conf. Ser. Mater. Sci. Eng.* **506**

[12] Ahmad Basri A B, Chae D W and Lee H 2020 Investigation of the dynamic characteristics of a carbon-fiber-reinforced epoxy with adhesive-jointed structure *Compos. Struct.* **247** 112499

[13] Shah M A S A, Yunus M A, Saman A M, Rani M N A, Bahari A R and Peter C 2019 Finite Element Model Updating of Laser Stitch Welded Structure *IOP Conf. Ser. Mater. Sci. Eng.* **506** 012010

[14] Shah M A S A, Yunus M A, Rani M N A, Ayub M A, Sani M S M and Saman A M 2019 Structural modification and evaluation using finite element reconciliation method *AIP Conf. Proc.* **2059**

[15] Yunus M A, Rani M N A, Sani M S M and Shah M A S A 2018 Finite element model updating of riveted joints of simplified model aircraft structure *AIP Conf. Proc.* **1952** 20013

[16] Yunus M A, Ouyang H, Rani M N A and Isa A A M 2013 Finite element modelling and updating of bolted joints in a thin sheet metal structure *20th International Congress on Sound and Vibration 2013, ICSV 2013* vol 2 pp 1146–52

[17] Yunus M A, Aziz Shah M A S, Abdul Rani M N, Wan Iskandar Mirza W I I and Kushairi S 2017 Experimental and Finite Element Analysis of Laser Stitch Welded Structure *J. Mech. Eng.* **4** 155–62

[18] Mottershead J E, Link M and Friswell M I 2011 The sensitivity method in finite element model updating: A tutorial *Mech. Syst. Signal Process.* **25** 2275–96

[19] R. Omar, M. N. Abdul Rani, M. A. Yunus , A. A. Mat Isa, W. I. I. Wan Iskandar Mirza M S M Z and L R 2018 Investigation of Mesh Size Effect on Dynamic Behaviour of an Assembled Structure with Bolted Joints using Finite Element Method *Int. J. Automot. Mech. Eng.* **15** 5695–708

[20] Simoen E, De Roeck G and Lombaert G 2015 Dealing with uncertainty in model updating for

damage assessment: A review *Mech. Syst. Signal Process.* **56** 123–49
[21] Petersen Ø W and Øiseth O 2017 Sensitivity-based finite element model updating of a pontoon bridge *Eng. Struct.* **150** 573–84

IOP Conf. Series: Materials Science and Engineering 1041 (2021) 012065 doi:10.1088/1757-899X/1041/1/012065

Experimental Studies of Application Passive Momentum Exchange Impact Damper *(PMEID)* on UAV'S Landing Gear

Darmawan[1,2], Lovely Son[1], Mulyadi Bur[1], Nurmansyah[1]
[1]Department of Mechanical Engineering,Andalas University, Padang, Indonesia
[2]Department of Electrical Engineering, Andalas University,Padang, Indonesia

Corresponding author: wawan@eng.unand.ac.id

Abstract. Unmanned aircraft or UAV (Unmanned Aerial Vehicle) is a future technology that continues to be developing. This research examines the structural vibration acceleration response of the aircraft's nose landing gear. The landing gear is the part of the aircraft that receives the most significant shock load during landing. This is due to the magnitude of the maximum acceleration of the landing gear vibration when experiencing shock loads. This phenomenon causes the landing gear damaged due to excessive vibration. Therefore, uncontrolled shock loads can damage the landing gear structure and even pierce its frame structure. A momentum transfer method was applied to the landing gear design to reduce the maximum acceleration amplitude of vibration due to experiencing shock loads during landing. The research experiment was carried out using the principle of passive momentum transfer; the test was carried out by dropping the landing gear at a certain height. When the landing gear experiences a shock load, kinetic energy and momentum will be transferred to the PMEID system so that the landing gear remains in a stable condition. The experiment gives an output in the form of the vibration acceleration amplitude of the landing gear measured by the accelerometer sensor on the landing gear's main mass. The vibration response will then be sent to the voltage amplifier and forwarded to the digital signal processing and displayed on a computer monitor. The test results prove the application of the momentum transfer method in the landing gear design can reduce the maximum amplitude of vibration acceleration when landing.

1. Introduction

UAV (Unmanned Aerial Vehicle) is a future technology that is continuously has been developing. The use of drones is currently widely used in various fields, such as agriculture, military, industry, the arts, and other areas. The deficiencies in unmanned aircraft design continue to develop, not a few studies and studies have been carrying out with different studies. In this study, an analysis of shock vibration control in unmanned aircraft during landing was conduct. When landing, the aircraft receives a shock load due to the influence of the aircraft; the shock load is magnified by the acceleration of the aircraft when it hits the ground. When landing, the landing gear supports the aircraft's entire weight; thus, the aircraft hits the ground hard and receives a huge shock load. This large shock load produces uncontrollable shocks, causing damage to the landing gear itself and the unmanned aircraft structure, in conditions large enough to affect the safety of the environment around the vehicle [1]. Control of shock vibrations in the landing gear is carrying out to reduce vibrations so that when the plane lands, the vibrations that propagate the airframe will not cause damage due to excessive vibrations [2]. In addition to causing damage to the fuselage, excessive shock vibrations can cause loss of sensitivity to aircraft electronic components. The aircraft's electronic components that lose the sensitivity make the aircraft difficult to control and lose orientation in carrying out the mission. The thing that is feared by the shock vibration is the high amplitude of the maximum acceleration of the vibrations that occur when receiving the shock load. The amplitude is the maximum displacement measured from the equilibrium point of the wave. The magnitude of the shock force determines the magnitude of the maximum acceleration amplitude of the vibration. Various methods of controlling shock vibrations on geological landing have been implementing to reduce the maximum acceleration amplitude. A new method used to reduce the vibration amplitude due to shock loads on the landing gear is to use the passive momentum transfer method. The use of PMEID (Passive Momentum Exchange Impact Damper) in vibration control has been carrying out because the system receives a huge shock vibration. This huge shock vibration causes vibration control to be damaged. This is due to the high maximum amplitude of acceleration that occurs before the vibration is damped. PMEID is a method used to reduce shock loads by reducing the amplitude of the maximum acceleration that occurs in the system. This method uses the principle of momentum transfer on the third pendulum swing; in this case, PMEID is analogous to the third ball, which absorbs energy and momentum from the second ball after being hit by the first ball [3]. In previous studies, the application of PMEID to floor dampers has been showing to reduce the shock vibration response by 25% [1]. In this final project, the method applied to the experimental model of unmanned aircraft gearing. The energy and momentum received by the landing gear will be transferred to the PMEID system so that it can reduce the maximum acceleration amplitude due to the shock force that occurs in the landing gear system.

PMEID (Passive Momentum Exchange Impact Damper) is a shock vibration control method using passive momentum transfer. The basic concept of PMEID comes from MEID (Momentum Exchange Impact Damper), where MEID uses the three pendulums swing [3]. The following shows the basic working principle of MEID momentum transfer in figure 1.

(a) Before Collision (b) After collision

Figure 1 Analogy of the 3 Pendulum Swing System.

Figure 1 shows that when the collision occurs between the first pendulum and the second pendulum, the first pendulum's momentum and kinetic energy have transferred to the second pendulum. The same condition applies to the collision that occurs between the second pendulum and the third pendulum. It can conclude that the momentum and kinetic energy in the first pendulum has transferred to the third pendulum; this energy and momentum transfer mechanism causes the second pendulum to be maintained in a stable condition. In designing a landing gear system using PMEID, the first pendulum is analogous to the shock load's excitation force. The second pendulum is analogous to the main system in the form of a U-profile spring, and the third pendulum is analogous to the mass of the PMEID system.

In designing shock absorbers based on landing gear momentum changes, the PMEID system's mass and spring characteristics affect the vibrations to has moved. Following is the working principle of PMEID on landing gear can be seen in figure 2.

Figure 2 shows the PMEID system landing gear mechanism, which is considered to have linear rigidity. When a landing made and a collision, momentum and kinetic energy on the wheel were transferred to a U-shaped leaf spring. The same thing happens to the U profile and the second mass because the U profile and the second mass are in contact with each other when the collision occurs momentum and the profile's kinetic energy. U is transferred to the second mass. In contrast to the 3 pendulums swing principle, when a collision occurs due to the shock load, the U profile experiences a deflection because the leaf springs have stiffness. The transfer of energy and momentum in the landing gear can be maximized by designing a U spring with linear stiffness and low deflection. In the second mass, a spring is attached, which gives the second mass a tendency to move because personal frequencies influence it due to the stiffness and the second mass's influence quickly. The contact spring stiffness value determines the contact time between the U profile and the second mass. The amount of energy and momentum transfer was determined by the value of the second mass used [3].

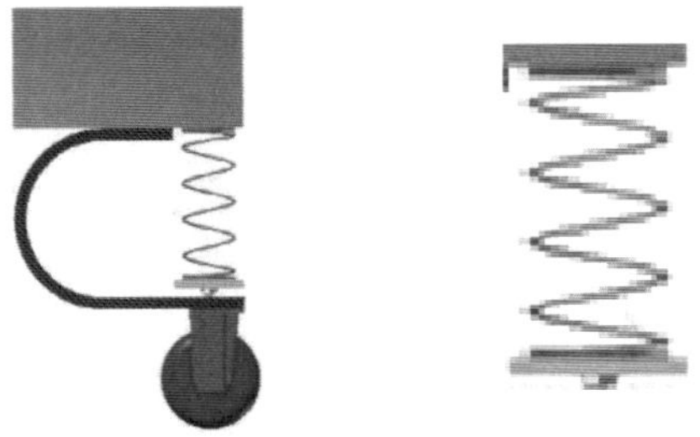

(a)Nose Landing Gear (b) PMEID System

Figure 2 PMEID Nose Landing Gear Mechanism

2. Materials and methods

In this research, the experimental was conductedto prove the effect of kinetic energy transfer and momentum on reducing shock vibrations in unmanned aircraft's landing gear. The initial step is designing the landing gear main spring (U profile). The U profile is designed with high stiffness to produce an elastic impact on the system. The vibration would maximally transfer to the PMEID system before being absorbed by the nose landing gear.

The next step is designing a PMEID shock absorber system. In this stage, all aspects that can affect the damper's performance were considered, such as contact mass and spring stiffness. In addition, at this stage, the PMEID system's position against the excitation force due to the collision is calculated.

For the next stage, a test instrument is made based on the design that has been making. After the test equipment was complete, a test was carried out to see the PMEID system's performance on the landing gear. Tests are carried out for several parameter configurations of the mass of the contact springs and the excitation force's magnitude, which is determined by the height of H = 100 mm. The test results are then analysed and compared with the designed contact mass configuration.

3. Results and discussion

3.1.Personal Frequency Test Results

3.1.1 Personal Frequency Testing with Simulations

The landing gear response dynamics in personal frequency and vibration mode were simulated with the finite element method using Autodesk Inventor software. The use of a finite element simulation method and modal analysis showing in figure 3.

The calculation of the personal frequency in modal coordinates determined by applying the four lowest nose landing gear modes was obtained from the simulation. See the lowest mode simulation results on the nose landing gear in the form of a U profile, where the lowest mode occurs at the first personal frequency of the system at 47 Hz. The value of the first private frequency in the simulation then used as a parameter for testing the personal frequency experimentally.

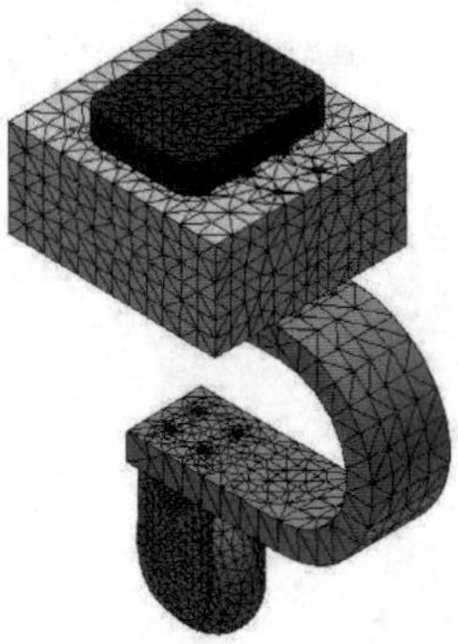

Figure3. Finite Element Model of Nose Landing Gear.

3.1.2 Personal Frequency Testing with Experiments

A personal frequency calculation performs in the experiment obtained from the FRF (frequency response function) graph. Figure 4.2 is the FRF graph from the comparison of the measured landing gear acceleration response to the magnitude of the impulse force exerted by the impact hummer.

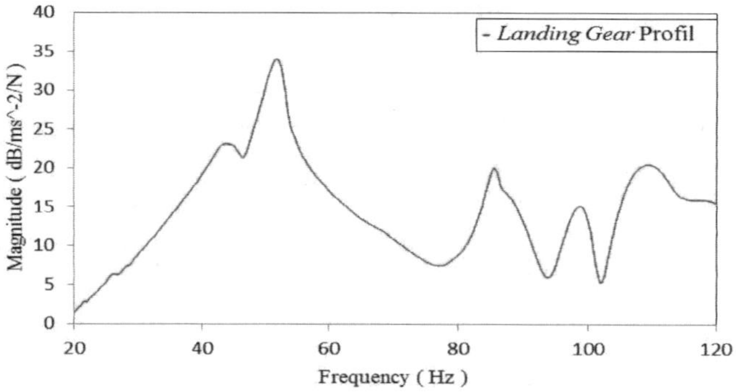

Figure 4. FRF Nose Landing Gear Graph.

Figure 4, it is seen that the results of experimental testing that have been carried out provide the lowest four-mode values for the nose landing gear system, namely 51.50 Hz in the first mode, 85.51 Hz in second, 98.51 Hz in the third mode, and 109.50 Hz in the fourth mode.

The differences in simulated and experimental personal frequency testing results can be analysed on the first personal frequency. The experiment result is 4 Hz higher. It can be due to the higher system stiffness in the experiment because it is influenced by the bolt stiffness and the test equipment's slider stiffness.

3.2 The Result of U Profile Spring Stiffness Testing

3.2.1 Stress Testing with Static Analysis

Static simulation of nose landing gear aims to find the PMEID strut nose landing gear component's stiffness value. A stress analysis simulation performed using Autodesk Inventor software to determine the nose landing gear's stiffness value. The simulation carried out using static analysis of U-profile springs by looking at the deflection that occurs when a force applied to the U-profile spring. The deflection and force parameters are used as input values for the strength calculation in the excel. This stiffness value is needed as a basic parameter for PMEID landing gear design. The material in the strut nose landing gear is assumed to be isotropic. In this simulation, an E-fibre glass composite was used with a Young modulus of 25 GPa [12]. The U-profile landing gear strut nose element model when applied loading is shown in figure 5.

.

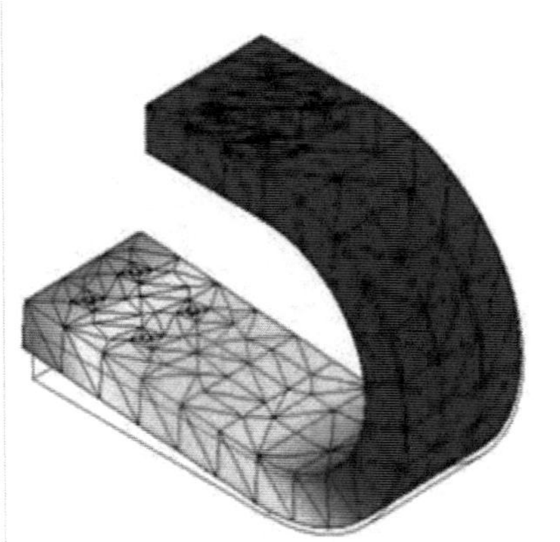

Figure 5.U Profile Spring Finite Element Model.

Based on the simulation results, it obtained that the U profile spring stiffness is 194,924 N/m, the U profile spring stiffness value used as a comparison with the stiffness of the experimental results.

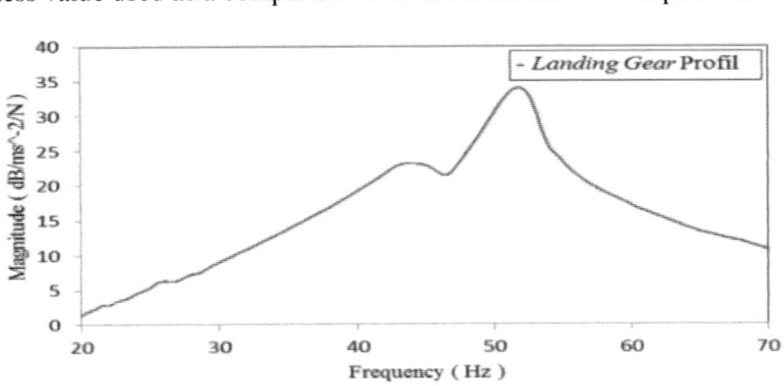

Figure 6. First Vibration Mode FRF Graph.

3.2.2 Stress Testing with Dynamic Analysis

Figure 6 shows the experimental results of the FRF (frequency response function) graph. As a parameter in determining the first personal frequency of the landing gear system.The first personal frequency value from the experiment is 51.50 Hz. The personal frequency value used to find the U profile spring stiffness and the U profile spring stiffness obtained of 209,201 N/m. The U profile spring stiffness value is the main parameter in the nose landing gear shock design absorbers on changes of momentum. Based on the results of stiffness calculations by simulation and experiment, the stiffness value of the static analysis simulation results is 194,924 N/m, while the value from the experimental test results is 209,201 N/m, with the measurement accuracy (error) of the simulation and experiment of 7.3%

3.3 Test Results Without PMEID System

The experiment shows the results of the measurement of the maximum acceleration amplitude on the landing gear without using the PMEID system. Measurements were carried out in five tests with a system drop height of 100 mm. The acceleration value of the nose landing gear (U profile) can be seen in figure 7.

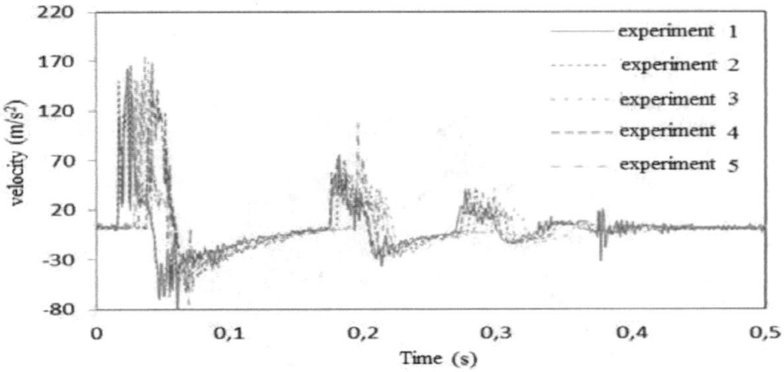

Figure 7. Test Results Without PMEID

In figure 7, it can be analysed that the maximum acceleration amplitude value of the nose landing gear in the form of a U profile spring is around 170 m/s^2. Significant acceleration occurs when the landing gear hits the ground at a very short contact time. A negative acceleration value indicates that the landing gear lifted upwards, the landing gear is lifted three times for 0.4 seconds before the condition is stable. The average landing gear maximum acceleration amplitude without PMEID is 168.22 m/s^2.

3.4 Test Results with PMEID

PMEID testing was carried out by varying the contact mass of PMEID, namely 125 g, 165 g, and 185 g. This test was conducted to see the decrease in the maximum acceleration amplitude response of the nose landing gear on the influence of the PMEID system's personal frequency due to variations in contact mass. The following result is a landing gear test with PMEID mass variations to the maximum acceleration amplitude landing gear.

3.4.1 PMEID Mass Contact 125 g

In tests with PMEID, a contact mass of 125 g was used. The experiment was carried out five times, with the landing gear falling height of 100 mm. So that it can be seen the landing gear maximum acceleration amplitude response with the PMEID system in figure 8.

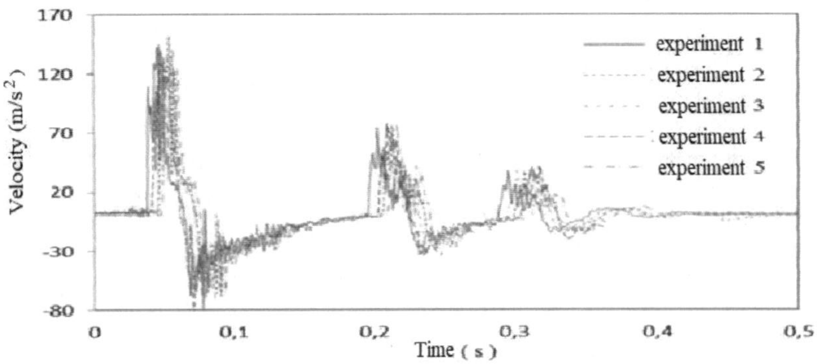

Figure 8. PMEID Test Results for Contact Mass 125 g

In Figure 8, it can be analysed that the maximum acceleration amplitude of the nose landing gear for the PMEID system with a mass of 125 g is around 130 m/s^2. A negative acceleration value indicates that the landing gear lifted upwards, the landing gear is lifted three times for 0.4 seconds before the condition is stable. The mean of the maximum acceleration amplitude of 125 g of PMEID is 144.90 m/s^2.

3.4.2 PMEID Mass Contact 165 g

In the experiment, it can be seen that the landing gear acceleration response after testing five times the variation of the contact mass of 165 g. The effect of landing gear acceleration after being dropped 100 mm on the nose landing gear's acceleration amplitude can be seen in figure 9.

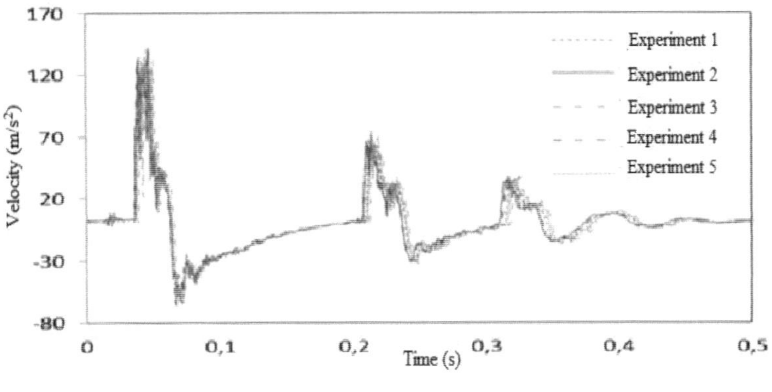

Figure 9. PMEID Test Results for Contact Mass 165 g.

In Figure 9, it can be analysed that the value of the maximum acceleration amplitude of the nose landing gear with the PMEID system contact mass of 165 g tends to decrease. The acceleration's negative value indicates that the landing gear lifted upwards, the landing gear is lifted three times for 0.4 seconds before the condition is stable. -The mean of the maximum acceleration amplitude of the contact mass of the PMEID system 165 g is 130.51 m/s^2. It can be analysed that the variation of the PMEID contact mass of 165 g can reduce the maximum acceleration amplitude by 22.41% of the landing gear acceleration without the PMEID system.

3.4.3 PMEID Mass Contact 185 g

In the test, the total addition of contact mass increased to 85 g. This mass is the last condition to determine the effect of mass variation on the decrease in landing gear acceleration amplitude. The effect of landing gear acceleration after a contact mass variation of 185 g is given to the nose landing gear's acceleration amplitude when receiving a shock load can be seen in figure 10.

In figure 10 can be analysedthat the value of the maximum acceleration amplitude of the nose landing gear with PMEID contact masses around 130 m/s^2. A negative acceleration value indicates that the landing gear lifted upwards, the landing gear is lifted three times for 0.4 seconds before the condition is stable. The average maximum acceleration amplitude without PMEID is 128.50 m/s^2. It can be analysed that the addition of the PMEID contact mass of 185 g can reduce the maximum acceleration amplitude by 23.06% of the landing gear acceleration without the PMEID system.

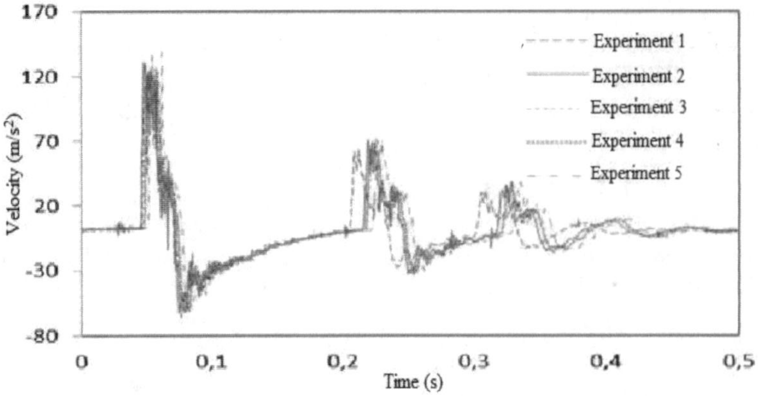

Figure 10. PMEID Test Results for Contact Mass 185 g.

3.5 Comparison Result of Landing Gear Shock Response without PMEID (U Profile) and PMEID Landing Gear

Figure 11 shows the performance of each test condition. The test result compares the amplitude value of the maximum landing gear acceleration without PMEID with the mass variation of PMEID landing gear.

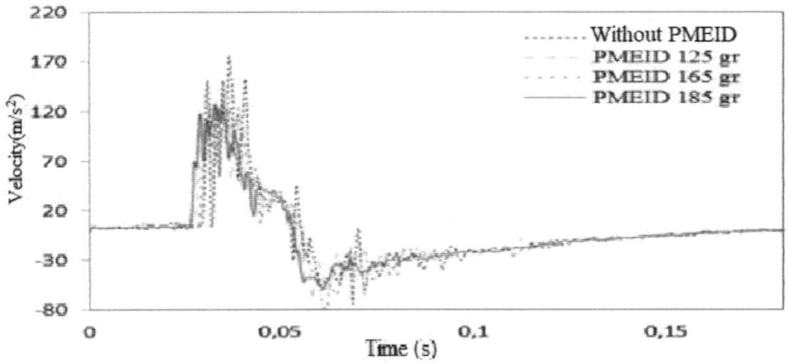

Figure 11. Comparison of PMEID Landing Gear Acceleration Amplitude

In Figure 4.9, it can be analysed that the maximum acceleration amplitude of the landing gear using the PMEID system has decreased with each additional contact mass. The decrease in the response of the maximum acceleration amplitude of the vibration at the mass variation of 125 g is 13.86%. The mass variation of 165 g resulted in a decrease of 22.41%, the mass of 185 g, and a decrease of 23.06%. This result explains that the PMEID method effectively reduces the shock load on the nose landing gear.

4. Conclusion

Based on the results of experiments that have carried out, several conclusions obtained, namely:

a. It has obtained an experimental model of shock vibration control with a passive momentum change method in an unmanned aircraft's landing gear, which can reduce the shock vibration response by up to 23.06%.

b. It found that the comparison of the shock vibration response in a landing gear structure without a damper that has a higher maximum acceleration amplitude response than a landing gear structure with a passive momentum change damper.

IOP Conf. Series: Materials Science and Engineering 1041 (2021) 012065 doi:10.1088/1757-899X/1041/1/012065

References

[1] Lovely Son, Makoto Kawachi, Hiroshi Matsuhisa. 2007. "Reducing Floor Impact Vibration And Sound Using A Momentum Exchange Impact Damper". Journal of system design and dynamics. Vol 1, no.1.

[2] RikiArdiansyah, Donny Hidayat, Dan AfidNugroho. "Design and Analysis of Main Landing Gear Frame of LSU-05By Finite Element Method". LembagaPenerbangandanAntariksa Nasional (LAPAN).

[3] Lovely Son, Mulyadi Bur, MeifalRusli. 2016. "A New Concept For UAV Landing Gear Shock Vibration Control Using Pre-Straining Spring Momentum exchange impact damper" Journal of vibration and control.

[4] WillyantoAnggono, Ian HardiantoSiahaan, dkk. "AnalisaDistribusiTeganganPadaModifikasi Axle Sleeve Landing Gear PesawatTerbangDenganMetodeElemenHingga". Surabaya : Mechanical Engineering Universitas Petra Christian.

[5] Rajesh,A., Abhay,BT., 2015. "Design and Analysis Aircraft Nose and Nose Landing Gear". Journal of Aeronautics & Aerospace Engineering Vol.4, Issue 2.

[6] LembagaPenerbangandanAntariksa Nasional."LapanPamerkanPesawat LSU-02 NGLD". Lapan.go.id (diaksespada 22/10/2019)

[7] Thomson, William T., &Marrie Dillon Dahleh. 1998."Theory Vibration With Applications5th Edition". Prentice Hall. United States of America.

[8] Ferdinand P. Beer, E. Russel Johnston, Jr. "Mechanics for Engineers, Dynamics". Singapore : Mc-Graw Hill.

[9] Mulyadi Bur.1999."Diktat PenuntunKuliahGetaranMekanik". Padang :UniversitasAndalas.

[10] Mohammad Sadraey, Daniel Webster College. 2013. "Air craft Design A systems Engineering Approach". New Hampshire, USA : Wiley.

[11] GusmanArifPandy "AnalisisStruktur Landing Gear DenganPatran- Nastran PadaPesawatNirawak LSU-05". Padang :TeknikMesinUniversitasAndalas.

[12] Muhammad AdiKusnanda, Dony Hidayat.2018. "Simulasi Drop Test Nose Landing Gear Lsu-02 MenggunakanPendekatan Multi-Body Simulation (Mbs) Rigid Models". PustekbangLapan, Rumpin : Bogor.

AUTHOR INDEX